THEORY AND METHODS OF STRUCTURAL ANALYSIS

THEORY AND METHODS OF STRUCTURAL ANALYSIS

Ziad M. Elias

Professor of Civil Engineering
University of Washington, Seattle

A Wiley-Interscience Publication

JOHN WILEY & SONS

New York · Chichester · Brisbane · Toronto · Singapore

Library of Congress Cataloging in Publication Data:

Elias, Ziad M.
Theory and methods of structural analysis.

"A Wiley-Interscience publication."
Bibliography: p.
Includes index.
1. Structures, Theory of. I. Title.
TA645.E38 1985 624.1'71 85-13765
ISBN 0-471-89768-X

Printed in the United States of America

10 9 8 7 6 5 4 3 2 1

To my Father

PREFACE

Structural analysis is a part of applied mechanics and has a particular feature that sets it apart from classical analytical methods. In classical analytical methods, the formulation of governing equations is only a starting point followed by the main subject of developing particular solution techniques suitable for various classes of problems and of studying the behavior of some important solutions. By contrast, in modern structural analysis, the formulation of governing equations is the main problem whose solution reduces usually to that of a set of linear or linearized simultaneous equations. This feature has become dominant through the development of matrix methods and the finite element approach and has brought into the field of structural analysis such previously separate fields of study as plates, shells, and three-dimensional problems. It is also making it a routine practice for the stress analyst to describe the data of a problem and to obtain a numerical solution through a computer program.

The more automation expands, however, the more important it becomes to the student of structural analysis and to the users of computer programs to be well grounded in the underlying theory.

The matrix formulation of structural analysis is now recognized as the proper vehicle for a unified treatment to the wide field of structural elements and structural types. If the general nonlinear theory is to be presented, however, the unified treatment becomes necessarily more abstract and less accessible and still leaves much development to be made for particular structural types in order to attain a practical understanding of the subject.

The purpose of this text is to concentrate first on one structural type whose general nonlinear theory could be developed in detail but whose formulation embodies not only the general concepts governing the field but also the structural form of the governing equations. This is the plane frame made of slender elements undergoing a general nonlinear state of plane deformation.

Space structures are treated by generalizing, to the extent possible, methods and equations developed for planar structures and by new developments when necessary. In particular, a general geometrically nonlinear theory for spatial behavior is developed.

The text was written partially for classroom instruction but also as an attempt to present the subject matter as it follows from basic concepts and laws of mechanics applied to an assemblage of beam elements. A particular aspect of the development is that deformation theory is formulated directly for a finite beam element. This approach has turned out to be rich in consequences especially in formulating nonlinear theories in both planar and spatial behaviors.

The text contains material at different levels but is organized in a way that separates linear from nonlinear theories. Guidance for its use is offered in Chapter 1.

Apart from an introductory chapter, the contents may be divided logically into three parts. The first (Chapters 2 to 7) deals with the structural member in plane deformation, the second (Chapters 8 to 13) deals with the plane behavior of the planar assemblage of members, and the third (Chapters 14 to 16) deals with space structures.

Following the general static-kinematic theory of the member, the simplified second-order theory and the linear theory are developed. The description of member flexibility and stiffness properties is treated first for linear and then for nonlinear behavior with the effect of shear deformation included. Methods include solution of governing differential equations and the application of principles of virtual displacements and of virtual forces developed earlier in the general member theory. Both of these principles are applied in particular to treat the problem of member instability. Member analysis ends with a chapter devoted to the topic of transformations and of development of effective member matrices that allow us to adopt a unified approach to the treatment, in the second part of the text, of an assemblage of members, including all particular cases of member-to-joint connections.

The second part of the text starts with a chapter covering the formulation of governing equations of structural analysis irrespective of particular methods of solution such as the displacement and force methods. This is followed by a chapter on the static-kinematic properties of structures that are independent of material properties, with a treatment of kinematic instability and the analytical determination of mechanism modes and then the treatment of static indeterminacy and the analytical determination of self-equilibriating modes.

The following chapters are devoted about equally to linear and nonlinear analysis by both the displacement and the force methods. The presentation is made systematic so as to be suitable for programming without going into programming techniques or aspects of efficient data processing. A systematic approach is also presented for efficient "hand-matrix" methods suitable for small-size problems. The inextensional behavior of frames and the behavior of trussed frames are treated in the linear and the geometrically nonlinear theories from the points of view of the techniques of analysis, the justification of the approximations involved, and structural behavior. Geometrically nonlinear behavior is treated in detail, including identification of various degrees of nonlinearity, the treatment of members with variable axial force, and the instability problem.

The third part of the text deals with space structures. Chapter 14 contains the linear theory and limited aspects of geometric nonlinearity. In Chapter 15 a general geometrically nonlinear theory of the structural member is developed, and the foundations for a discrete formulation suitable for the direct stiffness method are laid. Member stability problems are formulated by differential equations and by variational methods. A particular section deals with the method of virtual forces in stability analysis. Chapter 16 is devoted to the direct stiffness method for general geometrically nonlinear analysis with applications to linearized stability problems.

In attempting to develop this text from basic principles, I have not devoted the time required to find out and give credit to all those who first discovered certain governing laws or formulated certain methods. I am deeply indebted to them and to many others whose writings, teaching, or personal contact have shaped my thinking and my approach to the subject matter.

Special thanks go to my wife Ralda for her continued support and forbearance and for typing the manuscript.

ZIAD M. ELIAS

Seattle, Washington
January 1986

CONTENTS

NOTATION

A general guideline to notation is given in Chapter 1, Sections 6 to 9. The list that follows contains the main symbols used in the text with reference to the sections or equations where they are defined. Definitions of symbols used only locally are given as they occur and are not listed. Further reference to the text is needed for symbols with subscripts or superscripts.

The tilde $\sim$ is used in Chapter 15 to represent vector cross-product algebra and is defined in Section 15/2.

a	Member end, Section 2/2.1
$\bar{\mathbf{a}}, \mathbf{a}, \mathbf{a}_v$	Connectivity matrices, Sections 8/2.2, 2.3, 2.4
A, A_s	Cross-sectional area, shear area, Section 4/6
b	Member end, Section 2/2.1
$\mathbf{b}$	Member static matrix, Sections 2/3.2, 3/2.2, 3/3.2 Structure static matrix, Section 9/2
B	Bimoment, Section 14/3.3, 15/10
$\mathbf{B}, \bar{\mathbf{B}}, \mathbf{B}_\varphi$	Member static-kinematic matrix, Sections 2/3, 3/2.2, 3/3, 6/2.6, 7/2.2, 15/7
$\mathbf{c}_1, \mathbf{c}_2, \mathbf{c}_3$	Member geometric matrices, Section 6/2.6, 7/6.4
$\mathbf{c}$	Strain-displacement matrix, Section 5/4 Structure static-kinematic matrix, Section 9/2
$\vec{\mathbf{e}}, \vec{\mathbf{e}}_0$	Unit base vectors, Sections 14/2, 15/2
E	Young's modulus
$f_{ij}, \mathbf{f}$	Flexibility coefficients, flexibility matrix, Sections 5/2, 6/2, 14/3.4
$\mathbf{F}$	Structure flexibility matrix, Sections 11/2.3, 3, 7 Force components, Section 15/6
$\mathbf{g}, \mathbf{g}_s$	Geometric matrix, Section 6/2.1, 3.1
G	Shear modulus

$\mathbf{G}, \mathbf{G}_{\varphi}$	Moment components, Section 15/6, 7
I, I_2, I_3	Second moments of area
J	St-Venant's torsional constant
$k_{ij}, \mathbf{k}, \mathbf{k}_{\varphi}$	Stiffness coefficients, reduced stiffness matrix, Sections 5/3, 6/2, 6/3, 14/3.4, 15/8
$\mathbf{k}_c, \bar{\mathbf{k}}_c, \mathbf{k}_{c\varphi}$	Complete stiffness matrix, Sections 5/3, 7/2.3, 15/8
$\mathbf{k}_G, \mathbf{k}_{G\varphi}$	Geometric stiffness matrix, Sections 8/5.3, 7/6.4, 15/8
$\mathbf{k}_{gp}, \mathbf{k}_{Gp}$	Geometric stiffness matrix associated with axial span load, Section 6/3.1
$\mathbf{k}_s$	Member bending stiffness matrix in beam-column theory, Section 6/3.1
$\mathbf{k}_{sG}$	Geometric stiffness matrix, Section 6/3.1
$\mathbf{k}_{GF}$	Geometric stiffness matrix, Section 6/3.3
K	Effective length factor, Section 6/5.1 Warping constant, Section 14/3.3
$\mathbf{K}$	Structure stiffness matrix, Sections 8/2.2, 10/1
$\mathbf{K}_G, \mathbf{K}_{GF}$	Structure geometric stiffness matrices, Section 8/5.3
l	Member length in undeformed state
$m, \mathbf{m}$	Moment load intensity, Sections 2/3.1, 3/2.2, 14/3.1, 15/6
$M, \mathbf{M}$	Bending moment, Section 2/3.2; internal moments, Sections 14/3.1, 15/6
n	Axis normal to cross sections, Sections 2/2.2, 2.3
$\mathbf{N}$	Stress resultants, Sections 2/3.2, 2/5.3, 14/3.1, 15/6
$\mathbf{N}_p$	Particular equilibrium solution for **N**, Sections 2/3.2, 3/2.2, 3/3.2
$\mathbf{p}$	Load intensity, Sections 2/3.1, 14/3.1, 15/6
$\mathbf{P}$	Point load, Sections 3/2.2
$\mathbf{P}^a$	Resultant at point *a*, Section 2/3.1
$q, \mathbf{q}$	Shear flow, Section 4/5 Generalized displacements, Sections 5/4, 6/3, 7/1, 15/13 Force load intensity, Section 14/3.1
$Q, \mathbf{Q}$	First moment of area, Sections 4/3, 5 Generalized forces, Sections 5/4, 6/3.1, 7/1
$\mathbf{r}$	Joint displacements, Section 8/2.1, 14/4
$\mathbf{R}$	Joint external forces, Section 8/2.2
$\mathbf{R}_F, \mathbf{R}_p$	Equivalent joint forces, Sections 8/2.2, 2.4
s	Axis in plane of cross section, Section 2/2.2, 2.3
$\mathbf{s}, \bar{\mathbf{s}}$	Member end displacements, Sections 2/2.1, 7/2.1, 14/3.2, 15/7
$\mathbf{S}, \bar{\mathbf{S}}, \mathbf{S}_{\varphi}$	Member end forces, Sections 2/3.1, 7/2.1, 14/3.1, 15/7
$\bar{\mathbf{S}}_N, \mathbf{S}_{1N}$	Nonlinear term in stiffness equations, Sections 7/4, 8/5.2
$\mathbf{S}_F$	Fixed end forces, Sections 5/3, 6/2.6

$\mathbf{S}_p$	Particular solution of member equilibrium equations, Sections 2/3.1, 3/2.2
$\mathbf{T}$	General symbol for transformation matrix
$\mathbf{T}_b^a$	Static transformation matrix from point b to point a, Section 2/3.1
$u, \mathbf{u}$	Displacements, Sections 2/2.1, 14/3.2, 15/5.1
U	Internal work density, strain energy density, Sections 2/5.2, 5.3
U^*	Internal complementary work density, complementary strain energy density, Sections 2/5.2, 5.3
$\bar{U}$	Strain energy, Section 5/5
$\bar{U}_{\text{ext}}$	External potential, Section 6/5.3
$\bar{U}^*$	Complementary strain energy, Section 5/5
v	Transverse displacement, Section 2/2.1
$v_i, \mathbf{v}$	Member deformations, Sections 2/2.2, 14/3.2, 15/7 Incremental displacements, Section 15/15
$\mathbf{v}_I, \mathbf{v}_p, \mathbf{v}_0$	Initial deformations, Section 5/2
v_{1N}, v'_{1N}	Nonlinear part of axial deformation, Sections 8/5.3, 12/8
$\bar{V}$	External potential, Section 15/13
$V_i, \mathbf{V}, \mathbf{V}_\varphi$	Member statical redundants or reduced forces, Sections 2/3.1, 8/5.3, 14/3.1, 15/6, 7
$\mathbf{V}_F, \mathbf{V}'_F$	Fixed end forces, Sections 5/3, 6/3.3
W, W^*	Work, complementary work, Section 2/5.2
x, y, z	Fixed local axes, Section 2/2.1
X, Y, Z	Global axes, Section 2/2.1
$\mathbf{X}$	Statical redundants, Section 9/2
α	Parameter for shear deformation, Section 6/2.1 Coefficient of thermal expansion, Section 4/4
$\boldsymbol{\alpha}, \bar{\boldsymbol{\alpha}}, \boldsymbol{\alpha}_v$	Static-kinematic matrix, Sections 2/3.2, 4.3 Generalized connectivity matrix, Section 8/4
β	Nondimensional constant for shear deformation, Section 5/2 Angle defining orientation of local axes, Section 14/2
β_2, β_3	Geometric constants, Eqs. (15/10–8h, i)
$\boldsymbol{\beta}$	Vector transformation matrix, Sections 2/3.2, 4.3
γ	Shear strain angle, Fig. 2/2.3–1, Section 2/5.3 Load parameter, Section 6/5.4
δ	Symbol for first variation, differential or infinitesimal quantity, Section 1/9
$\delta\vec{\omega}$	Infinitesimal rotation vector, Sections 14/5.1, 15/3
δ^2	Symbol for second variation, Section 1/9
Δ	Symbol for finite variation, Section 1/9
$\boldsymbol{\Delta}$	Deformations in primary structure, Section 9/3

$\boldsymbol{\varepsilon}$	Strain matrix, Section 2/2.3, 14/3.2, 15/5.3
$\varepsilon_n, \varepsilon_s$	Normal and shear strains, Section 2/2.3
ε_1	Axial strain, Sections 2/2.3, 15/10; normal strain, Section 15/5.6
$\varepsilon_2, \varepsilon_3$	Shear strains, Sections 14/3.2, 15/5.6
η	Axis in the plane of a cross section, Section 4/5
θ	Rotation of tangent to centerline, Fig. 2/2.3–1 Member orientation in initial state, Fig. 7/2.1–1
θ_1	Incremental rotation, Section 15/15
θ_2, θ_3	Displacement derivatives, Section 16/2
$\mathscr{x}$	Rigidity matrix, Section 2/5.3 Tangent rigidity matrix, Section 4/9
λ	Stability parameter, Sections 6/2.1, 15/10, 15/12
$\boldsymbol{\lambda}$	Coordinate transformation matrix, Sections 7/2.1, 14/2, 15/2
$\vec{\mu}, \vec{\boldsymbol{\mu}}, \boldsymbol{\mu}$	Section 15/2
ξ	Coordinate on member x axis
Π	Total potential energy, Section 6/5.3, 15/14
σ	Axial stress, Sections 4/1, 15/10
σ_y	Yield stress, Section 4/8
ρ, ρ_0	Member chord rotation, Section 2/2.2 Rotation of bound reference, Section 6/3.3 Rank of matrix $\mathbf{a}_v$, Section 9/2 Geometric constant, Eq. (15/10-8g)
τ	Shear stress, Section 4/5
$\varphi, \vec{\varphi}, \boldsymbol{\varphi}$	Rotation of cross section, Sections 2/2.1, 14/3.2, 15/2, 15/5.1
$\boldsymbol{\Phi}$	Compliance matrix $= \mathscr{x}^{-1}$ Normal deformations modes, Section 10/11
χ	Curvature strain, Section 2/2.3
$\chi_i, \boldsymbol{\chi}$	Curvature strains and twist, Sections 14/3.2, 15/5.3
ψ	Finite rotation, Section 15/2
$\boldsymbol{\psi}$	Shape functions, Section 5/4

THEORY AND METHODS OF STRUCTURAL ANALYSIS

1

INTRODUCTION

1. OBJECT AND ROLE OF STRUCTURAL ANALYSIS

The object of a problem in structural analysis is to determine stresses, strains, and displacements of interest in a given structure, caused by a given loading condition. Traditionally structural analysis dealt with frame structures made of beam elements such as trusses and rigid frames, but it has now a wider scope in which a structure may be any constructed facility made of solid materials. This widening of the field of structural analysis is due to the finite-element approach which formulates governing equations for a structure by treating it as an assemblage of parts or elements.

As part of structural engineering, structural analysis is a component in the process of structural design. It allows to check whether a proposed design meets the requirements of adequate resistance to loading conditions and, if necessary, to revise a proposed design until such requirements are met.

As a theory structural analysis is represented by a system of governing equations developed from the basic concepts and physical laws of solid mechanics. Governing equations and methods of analysis form the basis for describing structural behavior as a whole as well as for obtaining numerical solutions to particular problems.

2. FORMULATION OF STRUCTURAL ANALYSIS

Structural analysis is part of applied mechanics and is based on the concepts and physical laws of this general field. Accordingly its theory will be formulated in the three classical parts of mechanics represented, respectively, by deformation-displacement relations, equilibrium equations, and material constitutive relations between forces and deformations. A fundamental property to be observed is the applicability of the principle of virtual work to the geometric and statical equations.

A basic aspect of modern structural analysis is an approach that treats a structure as an assemblage of finite elements. The formulation of governing equations is thereby separated into two parts pertaining, respectively, to the general element and to the assemblage, and each part is formed in turn of geometric, statical, and constitutive equations. Further the element force-deformation relations form the constitutive part of the assemblage equations.

The mathematical modeling of a frame into an assemblage of elements corresponds directly to its physical construction from structural members. However, a member could also be subdivided into several beam elements. The final object of the analysis is to be able to analyze each element independently after determining a sufficient number of displacement or force unknowns at its ends, thus providing boundary conditions to element governing equations. These unknowns are obtained from the solution of the assemblage equations which are reduced to a set of linear or linearized equations. Methods of forming assemblage governing equations and their specialization to particular types of structures and solutions are considered to be the main problem of structural analysis, as distinct from the subsequent independent analysis of the elements. The formulation of assemblage equations is also presented as a systematic procedure applicable to any particular structure in a given class. It may thus be regarded as the solution of a general problem.

3. SCOPE OF THE TEXT

The text deals specifically with frame structures in static equilibrium. Its basic concepts and methods are, however, applicable within the wider field of finite-element methods. The matrix form of governing equations and general methods of analysis such as the stiffness and the flexibility methods are applicable to general structures.

The development of the subject is mostly self-contained, starting with the development of structural member theory before addressing the problem of an assemblage of members.

The assumed mathematical background consists mainly of elementary matrix and linear algebra, with elements of differential and integral calculus and ordinary differential equations. In mechanics a fundamental background in statics, kinematics, and mechanics of deformable bodies is assumed. Although this background is usually achieved in the first two years of engineering study, the text is intended for students who have also a background in structural analysis and primarily for graduate study.

The scope of structural behavior covers linearly elastic and elements of inelastic behavior, planar and spatial structures, and geometrically linear and nonlinear behaviors including the particular problem of stability of equilibrium.

Methods of analysis, as distinct from general governing equations, include the stiffness and the flexibility methods and are developed each for linear, nonlinear, and stability analysis.

4. ORGANIZATION AND USE OF THE TEXT

Apart from this introductory chapter, the text may be separated into three parts:

1. Chapters 2 through 7. The structural member in plane deformation, linear and nonlinear theories.
2. Chapters 8 through 13. Plane structures, linear and nonlinear theories.
3. Chapters 14 through 16. Space structures, linear and nonlinear theories.

A list of contents appears at the start of each chapter.

Portions of the text may be selected for courses on linear structural analysis and on geometrically nonlinear and stability analysis. A course of study may be identified by the contents associated with the member, the assemblage of members, and the solution methods, and a corresponding selection may then be made from the division of chapters mentioned here.

Except for Chapters 2 and 3, linear theories and methods precede the nonlinear treatment and thus allow a progression from the simpler to the more advanced material. This order is reversed in Chapters 2 and 3 which contain the geometric and statical parts of element theory in plane deformation, the principles of virtual work, and a short section devoted to the general form of constitutive equations. For a course of study limited to linear structural analysis, the study of Chapter 2 may be limited to definitions, notations, and general concepts and properties, without entering into the detail of the nonlinear equations. The basis of the linear theory and corresponding equations may then be seen in Chapter 3. A similar approach may be used if the nonlinear theory is to be limited to the second-order or beam-column theory.

Since the text is not intended for a first course in structural analysis, some material should be already known to the student and could be reviewed quickly. The manner in which such material fits into the general subject could still be of interest, however.

For a course of study primarily devoted to geometrically nonlinear behavior, linear theory and behavior form an essential background. Back references in the text should help the student to review such material if necessary.

5. MATRIX FORMULATION

Matrix notation and matrix algebra have become the mathematical language of structural analysis for several reasons. Perhaps the most obvious one is the suitability for programming. Mathematically, however, matrix algebra is the natural language for linear relationships, which are fundamental to structural analysis. The ability of matrix formulation to convey the abstract concept and generality of a relationship or a transformation, as well as to represent a specific computational procedure, serves both the theoretical and applied aspects of learning the subject. This is the point of view of matrix formulation taken in this text. It should be pointed out, however, that in both manual and programming applications, the matrix representation of a relationship does not always provide the most efficient computational procedure. Modified procedures are devised in programming for efficient use of storage space and execution of arithmetic operations. Similarly other computational procedures may be devised which are more efficient for manual solutions. Such particular procedures may be referred to as manual matrix methods.

The text uses matrix formulation throughout and detailed equations when appropriate. It requires from the student not only knowledge of the basic arithmetic operations of matrix algebra and an ability for mathematical abstraction but also the ability to relate a partitioning of a matrix to a physical partitioning of the structure or of its variables.

A problem in studying structural analysis by matrix methods is how to solve problems and exercises when the operations involved are too long and tedious to do by hand. Resort to a prewritten computer program is useful for that part of the subject that deals with processing the results of the analysis and for gaining practical experience through exposure to different structures and behaviors under load. It does not help, however, in learning the theory and formulation of the subject matter. Resort to direct programming using a general programming language such as FORTRAN may be impractical or distractive from the main purpose of the study. Use of a higher-level programming language specifically designed for matrix operations would be an appropriate tool. Ideally all three approaches—the manual, the programming, and the preprogrammed—would be used. There are numerous programs for structural analysis but only a few higher-order programming languages for matrix operations. One such language, SMIS or Symbolic Matrix Interpretive System, was developed at the University of California at Berkeley and was further developed at the Department of Civil Engineering at the University of Washington, Seattle. A similar but more capable program that has evolved from SMIS is CAL 80.

A sample of SMIS operations is presented in what follows in order to illustrate the type of language and its capabilities.

Input-Output Operations

LOAD $A \quad n_1 \quad n_2$

Matrix named A of n_1 rows and n_2 columns is defined from input of $n_1 \times n_2$ data values.

PRINT $A \quad n_1$

Matrix A is printed preceded by n_1 title lines supplied as data for this operation.

REMARK n_1

n_1 lines are supplied and are printed.

Logical Operations

ZERO $A \quad n_1 \quad n_2$

A null matrix A of n_1 rows and n_2 columns is defined.

DELETE A

Matrix A is deleted.

DUPL $A \quad B$

B is defined equal to A.

TRANS $A \quad B$

B is defined equal to the transpose of A.

STOSM $A \quad B \quad n_1 \quad n_2$

B is stored in A placing the first element of B in position (n_1, n_2) in A.

ADDSM $A \quad B \quad n_1 \quad n_2$

B is added into A with element (1, 1) of B being added to element (n_1, n_2) of A.

RMVSM $A \quad B \quad n_1 \quad n_2 \quad n_3 \quad n_4$

Submatrix named B of size $n_3 \times n_4$ is removed from A starting at element (n_1, n_2).

DELRC $A \quad B \quad n_1 \quad n_2$

Matrix B is formed by deleting n_1 rows and n_2 columns from A and compacting the result. The row and column numbers to be deleted are supplied as data to this operation.

MERGE $A \quad B \quad n_1$

Matrix B is merged into A. The merging operation involves storing or adding elements of B into locations in A according to merging parameters supplied as data. The merging parameters specify the row numbers in A into which the successive rows of B are stored or added, and the column numbers in A into which the successive elements of a row of B are stored or added. The value of n_1 determines the type of merging that takes place.

STODG $A \quad B$

Row matrix B is stored on the main diagonal of A.

RMVDG $A \quad B$

A row matrix B is formed from the main diagonal of A.

Arithmetic Operations

ADD $A \quad B$

$A + B$ replaces A.

SUB A B
$A - B$ replaces A.

MULT A B C
Matrix C is formed equal to AB.

TRMPY A B C
$C = A^T B$

INVERT A B
A^{-1} replaces A and $A^{-1}B$ replaces B. If B is left blank, only inversion takes place.

SCALE A n s
A is multiplied by the scalar $s \times 10^n$

DGADD A B
Row matrix B is added to the main diagonal of square matrix A.

DGSUB A B
Row matrix B is substracted from the main diagonal of square matrix A.

DGPRE A B
BA replaces A, where B is a diagonal matrix stored as a row matrix.

DGPOST A B
AB replaces A, where B is a diagonal matrix stored as a row matrix.

DGMPY A B
AB replaces A, where A and B are diagonal matrices stored as row matrices.

Structural Member Commands

FORMK A n_1 n_2
Stiffness matrices are formed for n_2 members and are stored as submatrices on the main diagonal of A. n_1 specifies the type of the stiffness matrices. Member properties are supplied as data.

6. MATRIX NOTATION

Matrices are represented by boldface letters. Except when emphasis is desired, there is no special indication for column, row, square, or rectangular matrices. These characteristics depend on the nature of the matrices involved. Definition matrices, or matrices of state variables such as displacements, strains, and forces, are defined as column matrices. Relationship matrices are generally rectangular or square, and their size is determined from the row sizes of the column matrices being related. A row matrix occurs in representing a scalar as a linear combination of several column-ordered variables. Relationship matrices are sometimes called transformation matrices. An invertible transformation requires a square nonsingular matrix.

When the notation describes a specific ordering of elements into a matrix, the symbol $\{\quad\}$ is reserved for column ordering, $\lfloor\quad\rfloor$ is reserved for row ordering, and $\lceil\quad\rfloor$ is reserved for diagonal ordering.

Matrices may have subscripts and superscripts. Numerical subscripts or superscripts may be separated by commas for clarity. The superscript T is reserved for the transpose operation and occurs last if other superscripts are present. However, in conjunction with the inverse operation both $\mathbf{A}^{-1T}$ and $\mathbf{A}^{T-1}$ may be used. The symbol $\mathbf{I}$ is reserved for identity matrices and $\mathbf{I}_n$ indicates size $n \times n$. An exception occurs for $\mathbf{I}_1 = \{1 \ \ 0 \ \ 0\}$. Usually, a letter superscript does not mean raising to a power. Exceptions such as occurs in polynomial expressions are recognized by the context. Numerical superscripts usually indicate raising to a power except in specific examples where the number is the value taken by a letter superscript.

Elements of matrices are generally refered to by the (nonboldface) matrix symbol with subscripts. For a rectangular matrix the first subscript refers to the row number and the second to the column number. These subscripts precede any subscript used for the matrix as a whole. This latter subscript may be deleted in order to alleviate the notation.

Superscripts are generally used for partitioning of member matrices into parts associated with member ends and for partitioning of structure matrices by members, member ends, or joints.

7. IDENTIFICATION OF EQUATIONS, FIGURES, SECTIONS, AND SUBSECTIONS

An equation is identified by section or subsection number followed by a dash (–), then by a sequential number for that section or subsection. For example, Eq. (8–6) would refer to the equation labeled (8–6) in Section 8 of the current chapter, that is, the chapter in which the reference occurs. Equation (2.3–6) would refer to subsection 2.3 of Section 2. For reference to an equation in another chapter, the equation number is preceded by the chapter number followed by a slash (/). Thus the reference Eq. (3/2.3–6) would refer to Eq. (2.3–6) of Chapter 3. A similar system is used for identifying and referencing figures, sections, and subsections.

8. NOTATION FOR DERIVATIVES

A comma followed by x or t indicates derivative with respect to x or t. These are the only cases where a comma indicates a derivative. If $\mathbf{u}$ is function of x, $\mathbf{u}_{,x}$ is obtained by taking the derivative of each element of $\mathbf{u}$. A prime and d/dx may also be used for derivatives, but the prime is identified for that purpose in order to avoid confusion.

To outline the notation for partial derivatives, let $\mathbf{s} = \{s_1 \; s_2 \; \ldots\}$ be a set of independent variables. $\partial/\partial\mathbf{s}$ is defined as the column matrix operator

$$\frac{\partial}{\partial \mathbf{s}} = \left\{ \frac{\partial}{\partial s_1} \quad \frac{\partial}{\partial s_2} \cdots \right\} \tag{8–1a}$$

The $\partial/\partial\mathbf{s}$ may be applied to a scalar function $f(\mathbf{s})$, and in general to a row matrix function of $\mathbf{s}$, following the rule of matrix multiplication. Thus

$$\frac{\partial f}{\partial \mathbf{s}} = \left\{ \frac{\partial f}{\partial s_1} \quad \frac{\partial f}{\partial s_2} \quad \cdots \right\} \tag{8–1b}$$

If $\mathbf{f} = \{f_1 \; f_2 \; f_3\}$ then $\mathbf{f}^T$ is a row matrix, and

$$\frac{\partial \mathbf{f}^T}{\partial \mathbf{s}} = \begin{bmatrix} \dfrac{\partial f_1}{\partial s_1} & \dfrac{\partial f_2}{\partial s_1} & \dfrac{\partial f_3}{\partial s_1} \\ \dfrac{\partial f_1}{\partial s_2} & \dfrac{\partial f_2}{\partial s_2} & \dfrac{\partial f_3}{\partial s_2} \\ \vdots & \vdots & \vdots \end{bmatrix} \tag{8–1c}$$

The notation for the transpose of the preceding operations is

$$\frac{\partial f}{\partial \mathbf{s}^T} = \left(\frac{\partial f}{\partial \mathbf{s}} \right)^T = \llcorner \frac{\partial f}{\partial s_1} \quad \frac{\partial f}{\partial s_2} \quad \cdots \lrcorner \tag{8–2a}$$

$$\frac{\partial \mathbf{f}}{\partial \mathbf{s}^T} = \left(\frac{\partial \mathbf{f}^T}{\partial \mathbf{s}} \right)^T = \begin{bmatrix} \dfrac{\partial f_1}{\partial s_1} & \dfrac{\partial f_1}{\partial s_2} & \cdots \\ \dfrac{\partial f_2}{\partial s_1} & \dfrac{\partial f_2}{\partial s_2} & \cdots \\ \dfrac{\partial f_3}{\partial s_1} & \dfrac{\partial f_3}{\partial s_2} & \cdots \end{bmatrix} \tag{8–2b}$$

The $\partial/\partial\mathbf{s}^T$ is thus an operator that may be applied to a column matrix and that produces from each of its elements a row of partial derivatives. The notation is extended to second derivatives by writing

$$\frac{\partial}{\partial \mathbf{s}}\left(\frac{\partial f}{\partial \mathbf{s}^T} \right) = \frac{\partial}{\partial \mathbf{s}^T}\left(\frac{\partial f}{\partial \mathbf{s}} \right) = \frac{\partial^2 f}{\partial \mathbf{s} \partial \mathbf{s}^T} = \frac{\partial^2 f}{\partial \mathbf{s}^T \partial \mathbf{s}} \tag{8–3}$$

Equation (8–3) defines a symmetric matrix whose (i, j) element is $\partial^2 f/\partial s_i \partial s_j$. Note that the notation in Eq. (8–3) may be used only for a scalar function.

The preceding is now applied to the quadratic form

$$f = \tfrac{1}{2}\mathbf{s}^T\mathbf{a}\mathbf{s} = \sum_i \sum_j \tfrac{1}{2} a_{ij} s_i s_j \tag{8–4}$$

where $\mathbf{s}$ is a set of n independent variables, and $\mathbf{a}$ is an $n \times n$ matrix of constants a_{ij}. Being a scalar f is equal to its transpose $\tfrac{1}{2}\mathbf{s}^T\mathbf{a}^T\mathbf{s}$. Thus $\mathbf{s}^T(\mathbf{a}-\mathbf{a}^T)\mathbf{s}$ vanishes for any $\mathbf{s}$, and only a symmetric matrix $\mathbf{a}$ need be considered. With $\mathbf{a} = \mathbf{a}^T$, application of Eqs. (8–1b) and (8–3) yields

$$\frac{\partial f}{\partial \mathbf{s}} = \mathbf{a}\mathbf{s} \tag{8–5}$$

$$\frac{\partial^2 f}{\partial \mathbf{s}\partial \mathbf{s}^T} = \mathbf{a} \tag{8–6a}$$

This last equation is equivalent to

$$a_{ij} = \frac{\partial^2 f}{\partial s_i \partial s_j} \tag{8–6b}$$

Consider next a bilinear form

$$f = \mathbf{s}^T\mathbf{b}\mathbf{t} = \mathbf{t}^T\mathbf{b}^T\mathbf{s} \tag{8–7}$$

where $\mathbf{s}$ and $\mathbf{t}$ are sets of m and n independent variables, respectively, and b is an $m \times n$ matrix of constants b_{ij}. Application of Eq. (8–1b) yields

$$\frac{\partial f}{\partial \mathbf{s}} = \mathbf{b}\mathbf{t} \tag{8–8a}$$

$$\frac{\partial f}{\partial \mathbf{t}} = \mathbf{b}^T\mathbf{s} \tag{8–8b}$$

Further application of derivative operations yields

$$\frac{\partial}{\partial \mathbf{s}}\left(\frac{\partial f}{\partial \mathbf{t}^T}\right) = \frac{\partial}{\partial \mathbf{t}^T}\left(\frac{\partial f}{\partial \mathbf{s}}\right) = \mathbf{b} \tag{8–9a}$$

$$\frac{\partial}{\partial \mathbf{t}}\left(\frac{\partial f}{\partial \mathbf{s}^T}\right) = \frac{\partial}{\partial \mathbf{s}^T}\left(\frac{\partial f}{\partial \mathbf{t}}\right) = \mathbf{b}^T \tag{8–9b}$$

and

$$b_{ij} = \frac{\partial^2 f}{\partial s_i \partial t_j} \tag{8–9c}$$

9. VARIATIONAL OPERATIONS AND TERMINOLOGY

The Taylor expansion of a scalar function f of independent variables $\mathbf{s}$ has the form

$$f(\mathbf{s}+\delta\mathbf{s}) = f + \delta f + \tfrac{1}{2}\delta^2 f + \dots. \tag{9–1}$$

where $\delta\mathbf{s}$ is another set of independent variables, and

$$\delta f = \frac{\partial f}{\partial \mathbf{s}^T}\delta\mathbf{s} = \delta\mathbf{s}^T\frac{\partial f}{\partial \mathbf{s}} \tag{9–2}$$

$$\delta^2 f = \delta\mathbf{s}^T\frac{\partial^2 f}{\partial\mathbf{s}\partial\mathbf{s}^T}\delta\mathbf{s} \tag{9–3}$$

Equation (9–1) may be written for a matrix $\mathbf{f}$ and applies then to each element of that matrix. However, the first expression of δf in Eq. (9–2) may be generalized only to a column matrix, and the second expression only to a row matrix. Equation (9–3) applies only to the individual scalar elements of a matrix.

If $\delta\mathbf{s}$ is considered an arbitrary variation of $\mathbf{s}$, then $\Delta f = f(\mathbf{s}+\delta\mathbf{s}) - f(\mathbf{s})$ is called the total variation of f. δf and $\delta^2 f$ are called first and second variations, respectively. δ may be considered a variational operator that produces the first variation as $\mathbf{s}$ is varied by $\delta\mathbf{s}$. If Eq. (9–1) is applied to $\mathbf{s}$, we obtain $\delta^2\mathbf{s} = 0$, and since $\delta\mathbf{s}$ is independent of $\mathbf{s}$, we can write $\delta(\delta\mathbf{s}) = \delta^2\mathbf{s} = 0$. δ has the property $\delta(fg) = f\delta g + g\delta f$. If δ is applied to δf in Eq. (9–2), it is seen that

$$\delta^2 f = \delta(\delta f) \tag{9–4}$$

If elements of $\mathbf{s}$ are now considered as functions of new independent variables $\mathbf{s}'$, δf and $\delta^2 f$ are by definition linear and of second order in $\delta\mathbf{s}'$, respectively, in the Taylor expansion of $f(\mathbf{s}'+\delta\mathbf{s}')$. Equation (9–2) remains valid, but

$$\delta^2 f = \delta\mathbf{s}^T\frac{\partial^2 f}{\partial\mathbf{s}\partial\mathbf{s}^T}\delta\mathbf{s} + \frac{\partial f}{\partial\mathbf{s}^T}\delta^2\mathbf{s} \tag{9–5}$$

and $\delta\mathbf{s}$ and $\delta^2\mathbf{s}$ are the first and second variations of $\mathbf{s}$ with respect to the independent variables $\mathbf{s}'$.

In the case of a quadratic form $f = \frac{1}{2}\mathbf{s}^T\mathbf{a}\mathbf{s}$ with $\mathbf{a} = \mathbf{a}^T$, we obtain

$$\delta f = \delta\mathbf{s}^T\mathbf{a}\mathbf{s} = \mathbf{s}^T\mathbf{a}\delta\mathbf{s} \tag{9–6a}$$

$$\delta^2 f = \delta(\delta f) = \delta\mathbf{s}^T\mathbf{a}\delta\mathbf{s} \tag{9–6b}$$

In the case of the bilinear form of Eq. (8–7), we obtain

$$\delta f = \delta\mathbf{s}^T\mathbf{b}\mathbf{t} + \mathbf{s}^T\mathbf{b}\delta\mathbf{t} \tag{9–7a}$$

$$\delta^2 f = \delta(\delta f) = 2\delta\mathbf{s}^T\mathbf{b}\delta\mathbf{t} \tag{9–7b}$$

The variation $\delta\mathbf{s}$ is identical with the differential $d\mathbf{s}$, and the first variation δf is identical with the total differential df. The variational operator δ is used, however, in a more general context than d in expressing the variation of a functional.

Let $U(\mathbf{u}, \mathbf{u}_{,x}, \mathbf{u}_{,xx})$ be an expression depending on a function $\mathbf{u}$ and its derivatives with respect to an independent variable $0 \leqslant x \leqslant l$, and let

$$\bar{U} = \int_0^l U(\mathbf{u}, \mathbf{u}_{,x}, \mathbf{u}_{,xx})dx$$

$\bar{U}$ is a scalar that depends on the function $\mathbf{u}$. It is called a functional of $\mathbf{u}$. Let $\delta\mathbf{u}$ be an arbitrary function of x considered as a variation of $\mathbf{u}$. The total variation $\Delta\mathbf{u}$ of U, as $\mathbf{u}$ varies by $\delta\mathbf{u}$, is evaluated at any x in the same way as for a function of three independent variables $\mathbf{u}$, $\mathbf{u}_{,x}$, and $\mathbf{u}_{,xx}$. U has thus a Taylor expansion of the form

$$\Delta U = \delta U + \tfrac{1}{2}\delta^2 U + \ldots \tag{9–8}$$

where δU is linear in $(\mathbf{u}, \mathbf{u}_{,x}, \mathbf{u}_{,xx})$, $\delta^2 U$ is of second order, and so on. The total variation $\Delta\bar{U}$ is obtained by integration of ΔU as

$$\Delta\bar{U} = \delta\bar{U} + \tfrac{1}{2}\delta^2\bar{U} + \ldots \tag{9–9}$$

where the successive terms are linear, of second order, and so on, in terms of $\delta\mathbf{u}$ and its derivatives. As seen in the case of a function $f, \delta(\delta\mathbf{u}) = \delta^2\mathbf{u} = 0$ and $\delta^2\bar{U} = \delta(\delta\bar{U})$. However, if elements of $\delta\mathbf{u}$ cease to be arbitrary because elements of $\mathbf{u}$ are made dependent on new functions $\mathbf{u}'$, then $\delta(\delta\mathbf{u}) = \delta^2\mathbf{u}$ is not necessarily zero. First, second, and subsequent variations are now defined in terms of $\delta\mathbf{u}'$ and its derivatives.

If $\mathbf{u}$ is expressed linearly in terms of independent parameters $\mathbf{s}$ as, for example, by a truncated series in functions of x whose coefficients are $\mathbf{s}$, $\bar{U}$ turns into a function of independent variables $\mathbf{s}$, and its variations reduce to those defined earlier for such a function.

δ is used mostly as a variational operator as outlined earlier. In some instances, however, it is also used to indicate an infinitesimal quantity of first order which is not necessarily a first variation. This is the case, for example, if δW is a linear expression in $\delta\mathbf{s}$ which is not necessarily an exact differential of some function of $\mathbf{s}$. The symbol δW is then a single symbol—that is, there may be no quantity called W.

2

EXACT THEORY OF PLANE DEFORMATION

1. INTRODUCTION

The theory to be developed is formed of the three basic parts of a mechanical theory; the geometric analysis of deformation, the statical analysis of forces, and the constitutive relations representing material properties. The first two parts are independent of the material and are referred to as static-geometric analysis. They are developed in detail in Sections 1 through 4. Constitutive relations are discussed in general terms in Section 5, and Section 6 gives a summary and purpose to the total formulation.

The structural member to be considered is a straight beam as defined in introductory texts on mechanics of solids [1]. Its geometry in the natural undeformed state is defined by an axis of length l and by the shape and dimensions of cross sections orthogonal to that axis. For a prismatic or uniform member the centroidal line of the cross sections is conveniently chosen as the axis. The static-geometric analysis to be developed is valid for a general member of variable cross section for which only the orientation of the axis is assumed given. Any material line parallel to that orientation could be chosen as member axis.

The engineering theory of beams is based on the Euler-Bernoulli assumption by which a cross section displaces as a rigid body. In the usual assumption, flexural shear deformation is neglected by assuming that the cross sections remain orthogonal to the deformed axis. By allowing the cross sections not to remain orthogonal to the beam axis, a more general theory that includes shear deformation may be developed.

An exact static-geometric theory for plane deformations will be developed in this chapter based on the generalized Euler-Bernoulli assumption. The theory is labeled exact because the kinematics of rigid body motion and statics in the deformed configuration are treated without approximations.

The approach to the development of beam theory to be followed differs from what is usually found in publications on that subject. In particular, the starting concept will be that of deformations for a finite member. Strains will then appear from a particular application of that concept to a beam element of infinitesimal length. In statical analysis the notion of work will be used to ensure a correspondence between static and geometric quantities. This is done first for a finite member and is then specialized to an infinitesimal element. The principle of virtual displacements, which is part of any well-formulated static-geometric theory, will be a natural outcome of this approach and will serve as a basis for developing consistent simplified theories of the structural member.

The present chapter is limited to a study of plane deformation. The general exact theory is left to Chapter 15.

2. DEFORMATION ANALYSIS

2.1. Displacements

An assemblage of members forming a structure is referred as a whole to a fixed right-handed global reference frame (X, Y, Z). Each member is also referred to a fixed right-handed local reference frame (x, y, z) with the x axis coinciding with the member axis in the undeformed state. For a plane structure, all local (x, y) axes lie in one plane which is taken as the (X, Y) plane, Fig. 2.1–1. In plane deformation, the rigid body displacement of a typical cross section consists of the superposition of a translation $\vec{u}$ in the (X, Y) plane and of a rotation φ about an axis perpendicular to that plane. The reference point at which $\vec{u}$ is defined may be chosen arbitrarily. It will be chosen at the intersection of the cross section with the

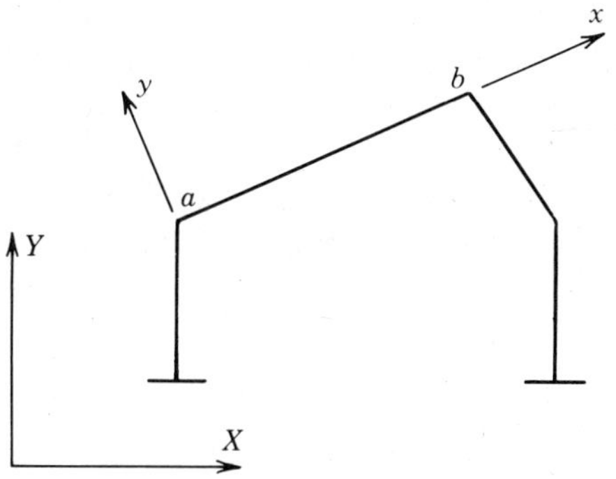

FIG. 2.1–1. Global and local reference frames.

member axis. Letting u and v be the components of $\vec{u}$ on the local axes, the displacement matrix of a cross section is defined as $\mathbf{u} = \{u \ v \ \varphi\}$. The sign convention for φ is according to the right-hand rule. $\mathbf{u}$ is a function $\mathbf{u}(x)$ of the coordinate x locating a cross section in the undeformed geometry.

The ends of a general member of length l are denoted a and b, respectively, and the x axis is oriented from a to b with $x = 0$ at a and $x = l$ at b. The notation to be adopted for the member end displacements is

$$\mathbf{u}(0) = \mathbf{s}^a = \{s_1^a \ s_2^a \ s_3^a\} = \{s_1 \ s_2 \ s_3\}$$
$$\mathbf{u}(l) = \mathbf{s}^b = \{s_1^b \ s_2^b \ s_3^b\} = \{s_4 \ s_5 \ s_6\}$$

and

$$\mathbf{s} = \{\mathbf{s}^a \ \mathbf{s}^b\}$$

For graphical representation a member is reduced to its axis, and a cross section is shown as the intersection of its plane with the plane of deformation.

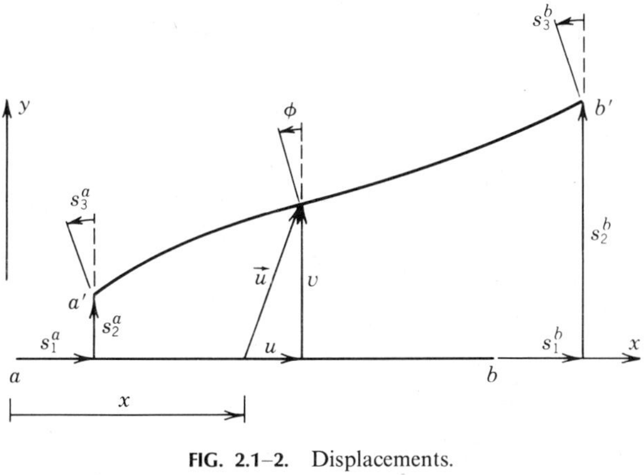

FIG. 2.1–2. Displacements.

Figure 2.1–2 shows a member *ab* in its undeformed and deformed states. Cross sections at *a*, *b*, and *c* are shown with their corresponding displacements.

In representing rotations graphically, it is sometimes convenient to show them as angles between the *x* axis and normals to the displaced cross sections. In a theory including shear deformation a cross section does not remain orthogonal to the deformed axis. Its normal should not then be confused with the tangent to the deformed axis.

2.2. Deformations

Deformations are defined as displacements relative to a "member-bound reference frame." If the member undergoes as a whole a rigid body displacement, it carries with it the bound reference frame, and no deformation occurs. Two such reference frames will be considered and will be referred to, respectively, as the cantilever type and the alternate type.

Deformations of Cantilever Type

The member-bound reference is rigidly bound to the cross section at end *a*. Its axes (n, s, z) coincide in the undeformed state with the local axes (x, y, z). In the deformed state axis *n* remains normal to the cross section at *a*. Relative to this reference, the member is a cantilever fixed at *a*. Member deformations $\mathbf{v} = \{v_1\ v_2\ v_3\}$ are by definition the relative displacements at end *b*. v_1 and v_2 are the components of a translational deformation vector $\vec{v}$. In Fig. 2.2–1 the deformed state of the member may be considered as resulting from the motion of

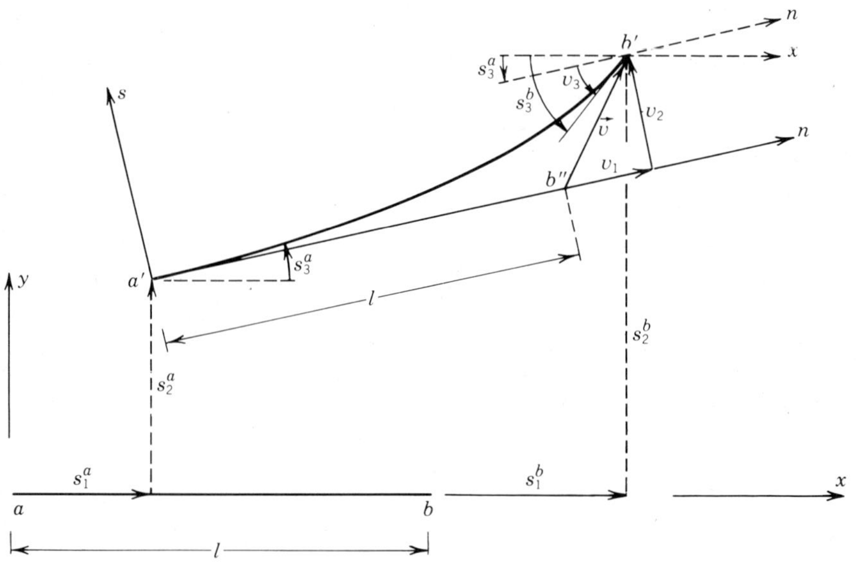

FIG. 2.2–1. Deformations, cantilever type.

the member-bound reference that carries the member into $a'b''$ followed by displacements relative to the member-bound reference. v_1 and v_2 are expressed in terms of member end displacements as components on n and s of the vector $\overrightarrow{b''b'} = \overrightarrow{b''a'} + \overrightarrow{a'a} + \overrightarrow{ab} + \overrightarrow{bb'}$, and v_3 is equal to $s_3^b - s_3^a$. There comes

$$\begin{aligned} v_1 &= -l(1 - \cos s_3^a) + (s_1^b - s_1^a)\cos s_3^a + (s_2^b - s_2^a)\sin s_3^a \\ v_2 &= -l\sin s_3^a - (s_1^b - s_1^a)\sin s_3^a + (s_2^b - s_2^a)\cos s_3^a \\ v_3 &= s_3^b - s_3^a \end{aligned} \qquad (2.2\text{–}1)$$

Deformations of Alternate Type

The member-bound reference of the alternate type has axes (x_1, x_2, z) centered at the reference point in cross section a, with axis x_1 remaining along the chord joining the member ends. Relative to this reference, the member is simply supported with a pin at end a and roller at end b. Member deformations consist of relative displacement v_1 at end b along the chord and of the end rotations v_2 and v_3 relative to the chord. Letting ρ be the chord rotation and l' be the length of the chord, we can write, using Fig. 2.2–2,

$$\begin{aligned} \cos\rho &= \frac{(l + s_1^b - s_1^a)}{l'} \\ \sin\rho &= \frac{(s_2^b - s_2^a)}{l'} \\ l' &= [(s_2^b - s_2^a)^2 + (l + s_1^b - s_1^a)^2]^{\frac{1}{2}} \end{aligned} \qquad (2.2\text{–}2)$$

and

$$\begin{aligned} v_1 &= l' - l \\ v_2 &= s_3^a - \rho \\ v_3 &= s_3^b - \rho \end{aligned} \qquad (2.2\text{–}3)$$

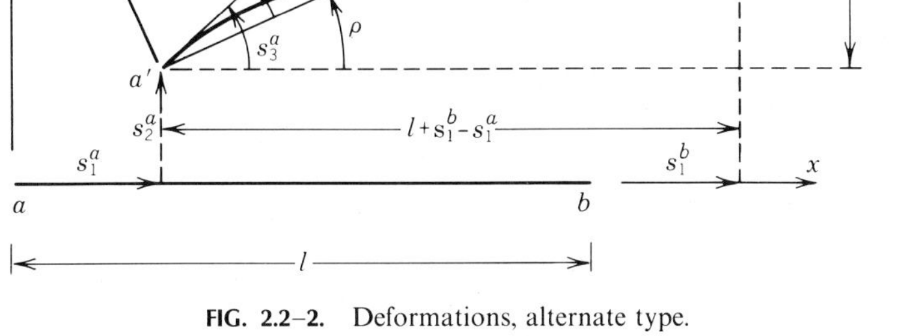

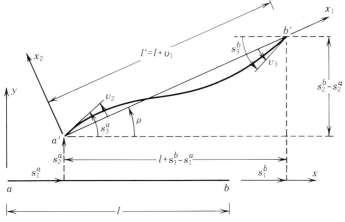

FIG. 2.2–2. Deformations, alternate type.

2.3. Strains

Strains in beam theory define the deformation of a beam slice contained between two cross sections at x and $x + dx$. Such a beam slice is considered as a member of infinitesimal length dx with end a at x and end b at $x + dx$. Deformations of the cantilever type for the beam slice are obtained from Eqs. (2.2–1) by replacing l with dx, $\mathbf{s}^a$ with $\mathbf{u}$, and $\mathbf{s}^b$ with $\mathbf{u} + d\mathbf{u}$. v_1, v_2, and v_3 become infinitesimals in dx whose principal parts are denoted, respectively, $\varepsilon_n dx$, $\varepsilon_s dx$, and χdx. We thus obtain the strain-displacement relations

$$\begin{aligned} \varepsilon_n &= (1 + u_{,x}) \cos \varphi + v_{,x} \sin \varphi - 1 \\ \varepsilon_s &= -(1 + u_{,x}) \sin \varphi + v_{,x} \cos \varphi \\ \chi &= \varphi_{,x} \end{aligned} \tag{2.3–1}$$

ε_n and ε_s are translational strains called, respectively, normal and shear strain, and χ is a curvature or bending strain. The geometric representation of $\varepsilon_n dx$ and $\varepsilon_s dx$ is shown in Fig. 2.3–1. χdx is the rotation increment $d\varphi$ and is not shown.

Of interest in beam theory is the change in length of axial material fibers. Fiber ab along the beam axis deforms into $a'b'$ of length $(1 + \varepsilon_1)dx$. In Fig. 2.3–1 vector $\overrightarrow{a'b'}$ has the orthogonal components $(1 + \varepsilon_n)dx$ and $\varepsilon_s dx$. It also has on the x and y

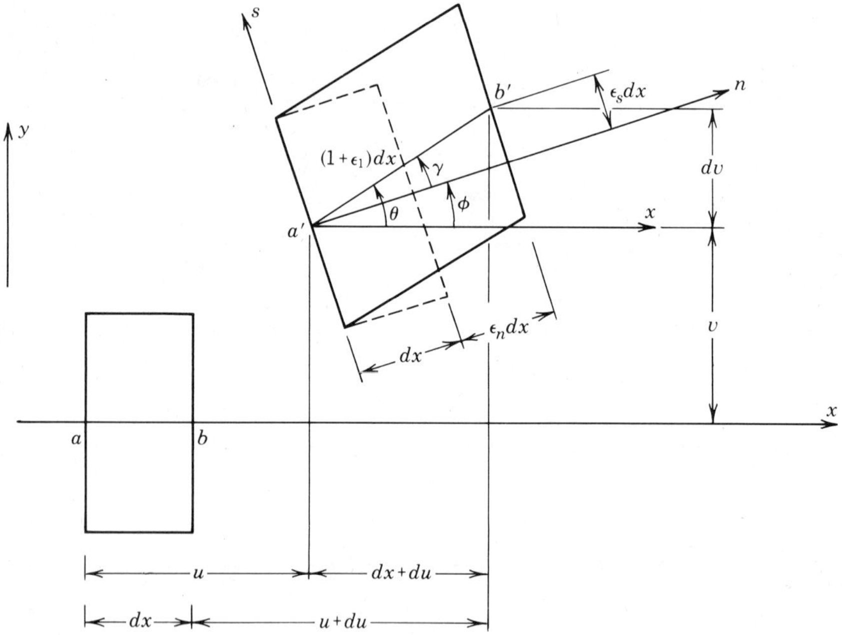

FIG. 2.3–1. Strains.

axes the components $(dx + du)$ and dv. We thus can write

$$(1 + \varepsilon_1)^2 = (1 + \varepsilon_n)^2 + \varepsilon_s^2 \tag{2.3–2a}$$

and

$$(1 + \varepsilon_1)^2 = (1 + u_{,x})^2 + (v_{,x})^2 \tag{2.3–2b}$$

To determine the strain distribution over the cross section at x, we consider the displacements over the cross section at $(x + dx)$ relative to the (n, s) axes attached to the cross section at x. This relative rigid body displacement has the translations $(\varepsilon_n dx, \varepsilon_s dx)$ and the rotation χdx. Since the rotation is infinitesimal, the translations at an arbitrary point (y, z), per unit dx, are

$$\varepsilon_n' = \varepsilon_n - y\chi \tag{2.3–3a}$$

$$\varepsilon_s' = \varepsilon_s \tag{2.3–3b}$$

It is thus found that ε_s and χ are independent of the reference point and that ε_n' varies linearly with the initial coordinate y. The axial strain ε_1' at a general point (y, z) is related to ε_n' and ε_s' by an equation similar to Eq. (2.3–2a) whence

$$(1 + \varepsilon_1')^2 = (1 + \varepsilon_n - y\chi)^2 + \varepsilon_s^2 \tag{2.3–4}$$

By contrast with ε_n', ε_1' is nonlinear in y unless ε_s is zero, in which case axial and normal strains coincide.

3. STATICAL ANALYSIS

3.1. Equilibrium of External Forces

The deformed member considered in Section 2 is now considered to be in equilibrium under the application of some external loading. The external forces acting on a beam slice of initial length dx have a resultant $\vec{p}\,dx$ and a moment at

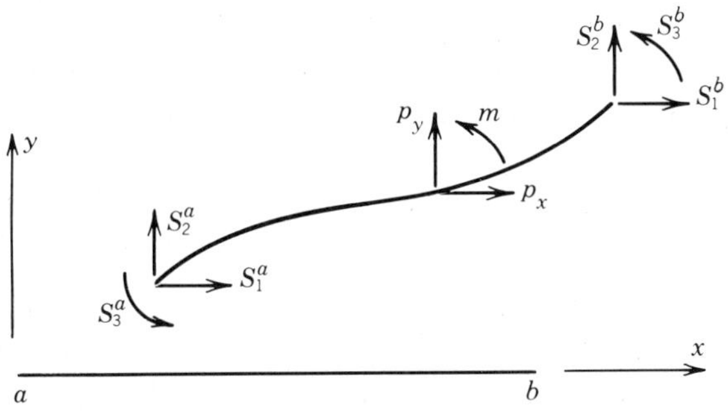

FIG. 3.1–1. Member end forces and span load.

the beam axis $\vec{m}dx$. The load components of interest in plane deformation, defined on the local axes, are $\mathbf{p} = \{p_x \ p_y \ m_z\}$. The member, isolated from the structure, is also acted upon at ends a and b by stresses having resultants $\mathbf{S}^a = \{S_1^a \ S_2^a \ S_3^a\}$ and $\mathbf{S}^b = \{S_1^b \ S_2^b \ S_3^b\}$, respectively, Fig. 3.1–1. Elements of $\mathbf{S} = \{\mathbf{S}^a \ \mathbf{S}^b\}$ are called member end forces, and elements of $\mathbf{p}$ are called span load or line load intensities. There is a correspondance as vector components between elements of $\mathbf{S}$ and $\mathbf{s}$ and elements of $\mathbf{p}$ and $\mathbf{u}$.

The planar force system $(\mathbf{S}, \mathbf{p})$ is governed by two force equilibrium equations and one moment equilibrium equation. Taking the moment equation about point a, the equilibrium equations in the local axes have the form

$$\mathbf{S}^a + \mathbf{T}_b^a \mathbf{S}^b + \mathbf{P}^a = 0 \qquad (3.1\text{–}1a)$$

where

$$\mathbf{T}_b^a = \begin{bmatrix} 1 & 0 & 0 \\ 0 & 1 & 0 \\ (s_2^a - s_2^b) & (l + s_1^b - s_1^a) & 1 \end{bmatrix} \qquad (3.1\text{–}1b)$$

$$\mathbf{P}^a = \int_0^l \mathbf{T}_x^a \mathbf{p} \, dx \qquad (3.1\text{–}1c)$$

$\mathbf{T}_b^a$ is the static transformation matrix from b to a. It transforms $\mathbf{S}^b$ into its statical equivalents $\mathbf{T}_b^a \mathbf{S}^b$ at point a. $\mathbf{T}_x^a$ in Eq. (3.1–*1c*) is defined similarly as the static transformation from point x to point a.

Typically a member is connected at both ends to the rest of the structure so that all elements of $\mathbf{S}$ are unknown. Assuming $\mathbf{p}$ and $\mathbf{u}$ given, Eqs. (3.1–1a) form a linear system of three equations in six unknowns $\mathbf{S}$. The system is thus statically indeterminate to the third degree. It may be solved in terms of three statically arbitrary quantities called member statical redundants or reduced forces and denoted $\mathbf{V} = \{V_1 \ V_2 \ V_3\}$. The general form of this solution is

$$\mathbf{S} = \mathbf{S}_p + \mathbf{B}\mathbf{V} \qquad (3.1\text{–}2)$$

where $\mathbf{S}_p$ is a particular solution of Eqs. (3.1–1a) depending on $\mathbf{P}^a$, and $\mathbf{BV}$ is the general solution of the homogeneous system obtained by setting $\mathbf{P}^a = 0$.

There are various ways of choosing the three elements of $\mathbf{V}$. The choice to be made establishes a correspondance as vector components between elements of $\mathbf{V}$ and the deformations, elements of $\mathbf{v}$.

Relative to a member-bound reference, the member is supported so that three end displacements are zero and the remaining three are the deformations. Elements of $\mathbf{V} = \{V_1 \ V_2 \ V_3\}$ are chosen as end force components applied in the directions of the deformations $\mathbf{v} = \{v_1 \ v_2 \ v_3\}$, respectively. We thus define member statical redundants of the cantilever type, Fig. 3.1–2, and of the alternate type, Fig. 3.1–3. The remaining end forces are reactions at the supports in the member-bound reference, in equilibrium with $\mathbf{V}$ and $\mathbf{p}$. In Eq. (3.1–2) $\mathbf{S}_p$ is the set

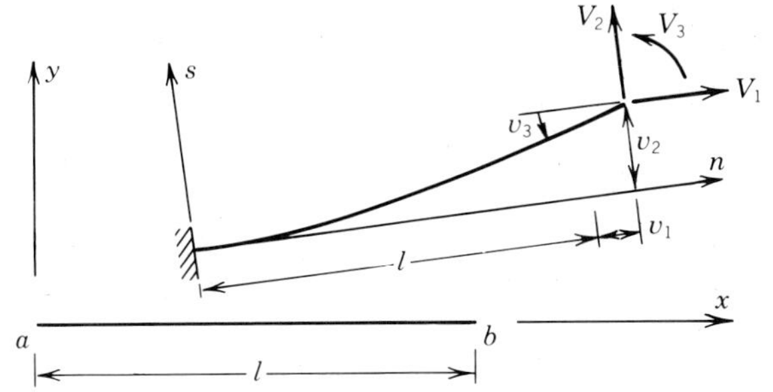

FIG. 3.1–2. Member statical redundants, cantilever type.

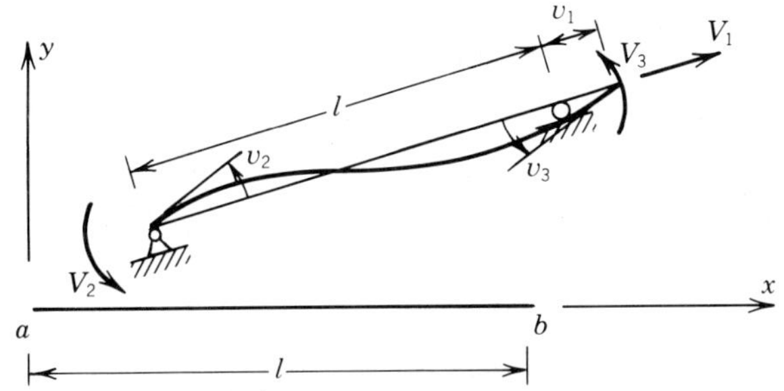

FIG. 3.1–3. Member statical redundants, alternate type.

of reactions at the supports in equilibrium with $\mathbf{p}$ and $\mathbf{V} = 0$, resolved on the local axes. $\mathbf{BV}$ is formed by transforming $\mathbf{V}$ and the reactions due to $\mathbf{V}$ to the local axes. It is a self-equilibriating force system. For the cantilever type, $\mathbf{S}_p^a = -\mathbf{P}^a$ and $\mathbf{S}_p^b = 0$, so that

$$\mathbf{S}_p = (\mathbf{S}_p)_c = \{-\mathbf{P}^a \quad 0\}$$

$\mathbf{BV}$ is partitioned by member ends into $\mathbf{S}^a = \mathbf{B}^a\mathbf{V}$ and $\mathbf{S}^b = \mathbf{B}^b\mathbf{V}$. For the cantilever type, $\mathbf{S}^b$ is obtained by a vector component transformation of the elements of $\mathbf{V}$ to the local axes. Thus

$$\mathbf{B}^b = (\mathbf{B}^b)_c = \begin{bmatrix} \cos s_3^a & -\sin s_3^a & 0 \\ \sin s_3^a & \cos s_3^a & 0 \\ 0 & 0 & 1 \end{bmatrix} \tag{3.1–3a}$$

Since $\mathbf{B}^a\mathbf{V}$ and $\mathbf{B}^b\mathbf{V}$ satisfy Eq. (3.1–1a) with $\mathbf{P}^a = 0$ and $\mathbf{V}$ arbitrary, there comes $\mathbf{B}^a = -\mathbf{T}_b^a\mathbf{B}^b$, and the result is

$$\mathbf{B}^a = (\mathbf{B}^a)_c = \begin{bmatrix} -\cos s_3^a & \sin s_3^a & 0 \\ -\sin s_3^a & -\cos s_3^a & 0 \\ v_2 & -(l + v_1) & -1 \end{bmatrix} \tag{3.1–3b}$$

The determination of $\mathbf{S}_p$, $\mathbf{B}^a$, and $\mathbf{B}^b$ of the alternate type is left as an exercise. **B** will be referred to as the member static-kinematic matrix. In Section 4.1 a kinematic role will be discovered for **B** which will allow its derivation from the deformation-displacement relations.

3.2. Determination of Internal Forces

Consider a member cut in two parts by a cross section (A) at x. The interaction stress between these two parts has a resultant force $\vec{N}$ and a resultant moment $\vec{M}$ at the reference point. $\vec{N}$ and $\vec{M}$ are defined as acting on the positive face of the cut, namely that which belongs to the part of the member between end a and (A). By the principle of action and reaction the actions over (A) on the part between (A) and end b are $-\vec{N}$ and $-\vec{M}$. In plane deformation the stress resultants of interest are the force components in the (x, y) plane and the moment component in the z direction. The components of $\vec{N}$ and $\vec{M}$ on the local axes are denoted $\mathbf{S}^x = \{N_x \ N_y \ M\}$, Fig. 3.2–1. At end b, $\mathbf{S}^x = \mathbf{S}^b$, and at end a, $\mathbf{S}^x = -\mathbf{S}^a$. The stress resultants that correspond as vector components to the strains $\{\varepsilon_n \ \varepsilon_s \ \chi\}$ are defined as components on the (n, s, z) axes attached to the cross section (A) and denoted $\mathbf{N} = \{N_n \ N_s \ M\}$, Fig. 3.2–1. **N** and $\mathbf{S}^x$ are related through the vector component transformation relation

$$\mathbf{S}^x = \boldsymbol{\beta}\mathbf{N} = \begin{bmatrix} \cos\varphi & -\sin\varphi & 0 \\ \sin\varphi & \cos\varphi & 0 \\ 0 & 0 & 1 \end{bmatrix} \begin{Bmatrix} N_n \\ N_s \\ M \end{Bmatrix} \tag{3.2–1}$$

The objective of member statical analysis is to determine **N** at any cross section

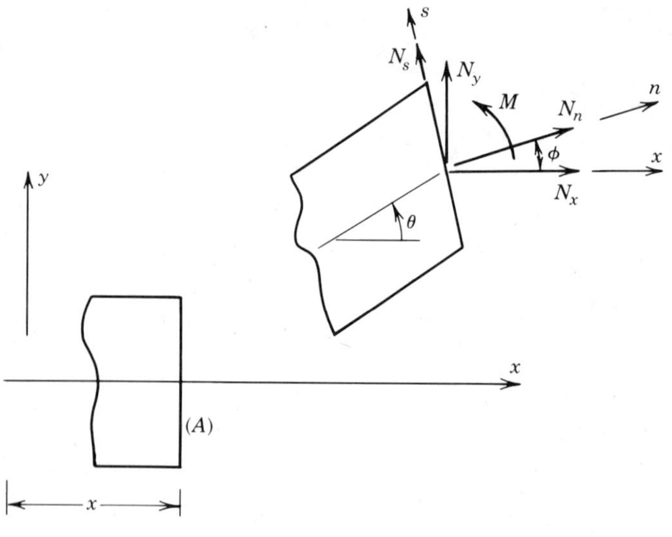

FIG. 3.2–1. Stress resultants.

(A). Equilibrium of the part of the member between (A) and end b makes $\mathbf{N}$ statically equivalent to the external forces acting between (A) and end b. Static transformation of these forces to point x yields

$$\mathbf{S}^x = \mathbf{T}_b^x \mathbf{S}^b + \int_x^l \mathbf{T}_\xi^x \mathbf{p}\, d\xi \tag{3.2–2}$$

With $\mathbf{N} = \boldsymbol{\beta}^{-1}\mathbf{S}^x$ and $\mathbf{S}^b = \mathbf{S}_p^b + \mathbf{B}^b\mathbf{V}$, $\mathbf{N}$ is expressed in terms of $\mathbf{V}$ in the form

$$\mathbf{N} = \mathbf{N}_p + \mathbf{b}\mathbf{V} \tag{3.2–3a}$$

where

$$\mathbf{N}_p = \boldsymbol{\beta}^{-1}[\mathbf{T}_b^x \mathbf{S}^b + \int_x^l \mathbf{T}_\xi^x \mathbf{p}\, d\xi] \tag{3.2–3b}$$

$$\mathbf{b} = \boldsymbol{\beta}^{-1}\mathbf{T}_b^x \mathbf{B}^b \tag{3.2–3c}$$

Since $\mathbf{N}$ and $\mathbf{V}$ are force components with respect to axes bound to the member, Eqs. (3.2–3) depend effectively only on displacements relative to the member-bound reference and may be expressed accordingly by redefining all quantities on their right-hand sides relative to that reference.

In the present exact statical analysis the determination of $\mathbf{N}$ through Eq. (3.2–3) must await that of the deformed shape as well as that of the member statical redundants.

For later purposes we now derive differential equations of equilibrium governing $\mathbf{N}$. The beam slice in Fig. 2.3–1 is acted upon by $-\mathbf{S}^x$ at point a', by $\mathbf{S}^x + d\mathbf{S}^x$ at point b', and by the span load $\mathbf{p}\,dx$. The components of the lever arm $a'b'$ are indicated in the figure on both the local axes (x, y) and the bound axes (n, s). The moment of these forces about point a' may be evaluated directly in terms of the components N_n and N_s using the lever arm components $(1 + \varepsilon_n)dx$ and $\varepsilon_s dx$. Replacing $\mathbf{S}^x$ with $\boldsymbol{\beta}\mathbf{N}$, the equilibrium equations on the local axes, divided by dx, take the form

$$(\boldsymbol{\beta}\mathbf{N})_{,x} - \boldsymbol{\alpha}\mathbf{N} + \mathbf{p} = 0 \tag{3.2–4}$$

where

$$\boldsymbol{\alpha} = \begin{bmatrix} 0 & 0 & 0 \\ 0 & 0 & 0 \\ \varepsilon_s & -(1+\varepsilon_n) & 0 \end{bmatrix} \tag{3.2–5}$$

Premultiplication of Eq. (3.2–4) by $\boldsymbol{\beta}^{-1}$ yields component equations on the (n, s, z) axes. We obtain

$$\mathbf{N}_{,x} + \boldsymbol{\beta}^{-1}(\boldsymbol{\beta}_{,x} - \boldsymbol{\alpha})\mathbf{N} + \boldsymbol{\beta}^{-1}\mathbf{p} = 0 \tag{3.2–6}$$

or, in detail,

$$\begin{Bmatrix} N_{n,x} \\ N_{s,x} \\ M_{,x} \end{Bmatrix} + \begin{bmatrix} 0 & -\chi & 0 \\ \chi & 0 & 0 \\ -\varepsilon_s & 1+\varepsilon_n & 0 \end{bmatrix} \begin{Bmatrix} N_n \\ N_s \\ M \end{Bmatrix} + \begin{Bmatrix} p_n \\ p_s \\ m \end{Bmatrix} = 0 \tag{3.2–7}$$

4. PRINCIPLES OF VIRTUAL WORK

4.1. Preliminaries

The concept of work plays a fundamental role in a mechanical theory and occurs in two aspects associated, respectively, with virtual work and actual work. Virtual work is considered first after a brief review of its definition in the general context of mechanics.

The virtual work of a force $\vec{F}$ applied at a material point of position vector $\vec{r}$ is the scalar product $\vec{F}\cdot\delta\vec{r}$ where $\delta\vec{r}$ is an arbitrary infinitesimal displacement taking place from position $\vec{r}$. $\delta\vec{r}$ may be considered as an infinitesimal variation of $\vec{r}$. The work is termed virtual because $\delta\vec{r}$ is not necessarily caused by anything but represents only a geometrically possible displacement.

Consider now a body subjected to a system of forces and undergoing an infinitesimal rigid body displacement. Such a displacement is defined in terms of the displacement $\delta\vec{r}$ of some arbitrarily chosen reference point O and a free rotation vector $\delta\vec{\omega}$. The displacement of an arbitrary point P of position vector $\vec{r}'$ is

$$\delta\vec{r}' = \delta\vec{r} + \overrightarrow{\delta\omega} \times \overrightarrow{OP} \tag{4.1–1}$$

A force $\vec{F}$ acting at P does the virtual work $\vec{F}\cdot\delta\vec{r}' = \vec{F}\cdot\delta\vec{r} + \vec{F}\cdot(\delta\vec{\omega} \times \overrightarrow{OP})$ or, by the permutability of the mixed vector product, $\vec{F}\cdot\delta\vec{r} + (\overrightarrow{OP} \times \vec{F})\cdot\delta\vec{\omega}$. The virtual work of a system of forces acting on the rigid body is the sum of the virtual work of the individual forces and takes the form

$$\delta W = \vec{R}\cdot\delta\vec{r} + \vec{M}\cdot\delta\vec{\omega} \tag{4.1–2}$$

where $\vec{R} = \Sigma\vec{F}$ is the resultant force and $\vec{M} = \Sigma\overrightarrow{OP} \times \vec{F}$ is the resultant moment at the reference point O.

The first application of the concept of virtual work pertains to the equilibrium equations of a free body which are

$$\begin{aligned} \vec{R} &= 0 \\ \vec{M} &= 0 \end{aligned} \tag{4.1–3}$$

It is seen through Eq. (4.1–2) that if the force system is in equilibrium—that is, if $\vec{R} = 0$ and $\vec{M} = 0$—then $\delta W = 0$. Conversely, if $\delta W = 0$ for arbitrary $\delta\vec{r}$ and $\delta\vec{\omega}$, then $\vec{R} = 0$ and $\vec{M} = 0$ because, if $\vec{R} \neq 0$ or if $\vec{M} \neq 0$, a choice of $\delta\vec{r}$ and $\delta\vec{\omega}$ in the directions of $\vec{R}$ and $\vec{M}$, respectively, would yield $\delta W > 0$.

The preceding is a presentation of the principle of virtual displacements for a rigid body, namely that $\delta W = 0$ is a necessary and sufficient condition for a force system to be in equilibrium. Note that the body need not be rigid but that only rigid body virtual displacements are considered. Accordingly Eqs. (4.1–3) are called rigid body equilibrium equations.

The concept of virtual work will be applied in subsequent sections to the deformable member and will lead to two principles that represent integral formulations of equilibrium and of geometric compatibility conditions. Some preliminary results will first be established in what follows.

A virtual displacement of a member taking place from a deformed state $\mathbf{u} = \{u \ v \ \varphi\}$ is defined by means of an infinitesimal and arbitrary $\delta\mathbf{u} = \{\delta u \ \delta v \ \delta\varphi\}$ function of x. $\delta\mathbf{u}$ may be considered as an infinitesimal variation of $\mathbf{u}$, and $\delta\varphi$, seen as a vector component in the z direction, coincides with the infinitesimal rotation vector $\overrightarrow{\delta\omega}$ of rigid body motion. (This latter property is particular to planar displacements and cannot be generalized to spatial rigid body motion.) The virtual work of the external load acting on a beam slice is the scalar product of vectors having as orthogonal components $\mathbf{p}dx$ and $\delta\mathbf{u}$ and is thus $\mathbf{p}^T\delta\mathbf{u}dx$ or, equivalently, $\delta\mathbf{u}^T\mathbf{p}dx$. At end a the virtual work of $\mathbf{S}^a$ is $\mathbf{S}^{aT}\delta\mathbf{s}^a$, and similarly at end b. The virtual work of $\mathbf{S} = \{\mathbf{S}^a \ \mathbf{S}^b\}$ is $\mathbf{S}^{aT}\delta\mathbf{s}^a + \mathbf{S}^{bT}\delta\mathbf{s}^b$ or $\mathbf{S}^T\delta\mathbf{s}$. The virtual work of all external forces acting on the member is called the external virtual work and is thus

$$\delta W_{\text{ext}} = \mathbf{S}^T\delta\mathbf{s} + \int_0^l \mathbf{p}^T\delta\mathbf{u}dx \tag{4.1-4}$$

δW_{ext} may also be expressed in terms of displacements relative to a member-bound reference. A result of kinematics is that the infinitesimal translational displacement $\overrightarrow{\delta u}$ is the vector sum of the displacement $\overrightarrow{\delta u_r}$ relative to the member-bound reference and of the carrying displacement of the reference. The carrying displacement is a rigid body displacement for the member as a whole as it is carried by the member-bound reference. Similarly the rotation $\delta\varphi$ is the sum of the relative rotation $\delta\varphi_r$ and of the carrying rotation. Since the external force system is in equilibrium, its virtual work through the carrying displacement vanishes so that δW_{ext} is equal to the virtual work through the relative displacements. The member is now viewed supported in its member-bound reference and subjected to the member redundants $\mathbf{V}$ and to the span load. The virtual work through the relative displacements is

$$\delta W_{\text{ext}} = \mathbf{V}^T\delta\mathbf{v} + \int_0^l \mathbf{p}_r^T\delta\mathbf{u}_r dx \tag{4.1-5}$$

where $\delta\mathbf{v}$ are virtual deformations, and $\mathbf{p}_r$ and $\delta\mathbf{u}_r$ are components on orthogonal axes of the span load and of the relative displacements, respectively.

The virtual deformations $\delta\mathbf{v}$ are infinitesimal variations of $\mathbf{v}$ induced by $\delta\mathbf{s}$. $\delta\mathbf{v}$ is thus the total differential, or first variation, of $\mathbf{v}$. If the equilibrium equation $\mathbf{S} = \mathbf{S}_p + \mathbf{BV}$ is substituted into Eq. (4.1–4), the term in $\mathbf{V}$ is found to be $\mathbf{V}^T\mathbf{B}^T\delta\mathbf{s}$, and this must also be $\mathbf{V}^T\delta\mathbf{v}$. Thus

$$\delta\mathbf{v} = \mathbf{B}^T\delta\mathbf{s} \tag{4.1-6a}$$

and, $\delta\mathbf{v}$ being the differential of $\mathbf{v}$,

$$\mathbf{B}^T = \frac{\partial \mathbf{v}}{\partial \mathbf{s}^T} \tag{4.1–6b}$$

It is a matter of verification that **B**, as determined by statics, coincides with that obtained according to Eq. (4.1–6b) through partial differentiation of the deformation-displacement relations. **B** may now be called the member static-kinematic matrix.

4.2. Equation of Virtual Work

The equation of virtual work to be established shortly is a characteristic property of an arbitrary system of external and internal forces in equilibrium, and an arbitrary system of geometrically compatible incremental displacements and strains. It plays a fundamental role in the theoretical formulation of constitutive equations and in the derivation of member stiffness and flexibility relations.

Let $\delta\mathbf{u}$ be a continuous virtual displacement, and let the member be separated into any number of segments or slices of length dx. The internal forces, acting at the common cross section of two contiguous slices, are opposite vectors whose virtual work through the continuous virtual displacement of the cross section vanishes. It follows that the sum of the external virtual work for the slices is equal to the external virtual work for the member as a whole. This sum is now evaluated in the limit as $dx \to 0$, that is, as an integral, and is called the internal virtual work δW_{int}.

The external virtual work for a beam slice is obtained by specializing Eq. (4.1–5) to a member of infinitesimal length dx and keeping only infinitesimals of first order in dx. For a beam slice dx, referred to the bound axes (n, s, z) at cross section x, $\mathbf{V}$ reduces to $\mathbf{N}$ and $\mathbf{v}$ reduces to $\boldsymbol{\varepsilon}dx$, so that $\mathbf{V}^T\delta\mathbf{v}$ reduces to $\mathbf{N}^T\delta\boldsymbol{\varepsilon}dx$. The virtual work of the span load $\mathbf{p}dx$ is an infinitesimal of second order in dx because $\delta\mathbf{u}_r$ is of the order of $\delta\mathbf{v}$, or $\delta\boldsymbol{\varepsilon}dx$. The internal virtual work is thus found to be

$$\delta W_{\text{int}} = \int_0^l \mathbf{N}^T \delta\boldsymbol{\varepsilon}\, dx \tag{4.2–1}$$

and is equal, as seen before, to δW_{ext}. The equation of virtual work is thus

$$\int_0^l \mathbf{N}^T \delta\boldsymbol{\varepsilon}\, dx = \mathbf{V}^T\delta\mathbf{v} + \int_0^l \mathbf{p}_r^T \delta\mathbf{u}_r\, dx \tag{4.2–2a}$$

or equivalently,

$$\int_0^l \mathbf{N}^T \delta\boldsymbol{\varepsilon}\, dx = \mathbf{S}^T\delta\mathbf{s} + \int_0^l \mathbf{p}^T \delta\mathbf{u}\, dx \tag{4.2–2b}$$

The equation of virtual work may be used to derive a geometric compatibility equation between incremental strains $\delta\boldsymbol{\varepsilon}$ and incremental deformations $\delta\mathbf{v}$ taking place from some deformed state. For this purpose a virtual force system is devised in which $\mathbf{p} = 0$ and $\mathbf{N}$ is expressed in the form of Eq. (3.2–3a), or $\mathbf{N} = \mathbf{bV}$. Eq. (4.2–2a) reduces to

$$\mathbf{V}^T \delta\mathbf{v} = \mathbf{V}^T \int_0^l \mathbf{b}^T \delta\boldsymbol{\varepsilon}\, dx$$

$\mathbf{V}$ being arbitrary, there results

$$\delta\mathbf{v} = \int_0^l \mathbf{b}^T \delta\boldsymbol{\varepsilon}\, dx \tag{4.2–3}$$

Eq. (4.2–3) will provide a method for evaluating actual incremental deformations caused by a load increment, following the establishment of material constitutive properties which allow $\delta\boldsymbol{\varepsilon}$ to be expressed in terms of internal force increments $\delta\mathbf{N}$.

In a similar manner a geometric relation giving incremental relative displacements $\delta\mathbf{u}_r$ at cross section x, in terms of $\delta\boldsymbol{\varepsilon}$ may be derived. A virtual load consisting of concentrated actions $\mathbf{P}_r$ applied at x, with $\mathbf{V} = 0$, is considered. The internal forces in equilibrium with $\mathbf{P}_r$ have the form $\mathbf{N}_\xi = \mathbf{b}_x^\xi \mathbf{P}_r$ where ξ replaces x as a position coordinate. $\mathbf{P}_r$ being arbitrary, Eq. (4.2–2a) yields

$$\delta\mathbf{u}_r = \int_0^l \mathbf{b}_x^{\xi T} \delta\boldsymbol{\varepsilon}\, d\xi \tag{4.2–4}$$

4.3. Principles of Virtual Work

The approach followed in the previous section concentrates on showing how the equation of virtual work results from subdividing the member into parts, without entering into more mathematical developments than necessary. Such developments are considered in what follows and are accompanied by a presentation of two principles that formulate in integral form conditions of static equilibrium and incremental geometric compatibility.

The expression of $\delta\boldsymbol{\varepsilon}$ in terms of $\delta\mathbf{u}$ in Eq. (4.2–2b) may be obtained by incrementing Eqs. (2.3–1) and keeping the first-order terms in $\delta\mathbf{u}$. It is called the first variation of $\boldsymbol{\varepsilon}$. The coefficient matrices of this expression have a static-kinematic character which is better demonstrated by deriving $\delta\boldsymbol{\varepsilon}$ from the property that for a beam slice dx, $\mathbf{v} = \boldsymbol{\varepsilon}\, dx$ and $\delta\mathbf{v} = \delta\boldsymbol{\varepsilon}\, dx$. Applying the relation $\delta\mathbf{v} = \mathbf{B}^{aT} \delta\mathbf{s}^a + \mathbf{B}^{bT} \delta\mathbf{s}^b$ to a beam slice, we let $\delta\mathbf{s}^a = \delta\mathbf{u}$ and $\delta\mathbf{s}^b = \delta(\mathbf{u} + d\mathbf{u})$, or

$\delta\mathbf{s}^b = \delta\mathbf{u} + (\delta\mathbf{u}_{,x})dx$. Then

$$\delta\boldsymbol{\varepsilon} = \lim \frac{\delta\mathbf{v}}{dx} = \lim \frac{(\mathbf{B}^a + \mathbf{B}^b)^T}{dx}\delta\mathbf{u} + \lim \mathbf{B}^{bT}\delta\mathbf{u}_{,x}$$

The coefficient of $\delta\mathbf{u}$ tends to $\boldsymbol{\alpha}^T$, where $\boldsymbol{\alpha}$ is the matrix appearing in the differential equations of equilibrium. $\mathbf{B}^b$, which for the cantilever type is a coordinate transformation matrix, tends at point x to $\boldsymbol{\beta}$, as defined in Eq. (3.2–1). The result is

$$\delta\boldsymbol{\varepsilon} = \boldsymbol{\beta}^T\delta\mathbf{u}_{,x} + \boldsymbol{\alpha}^T\delta\mathbf{u}$$

The fact that $\boldsymbol{\beta}$ and $\boldsymbol{\alpha}$ occur also in the differential equation of equilibrium could have been found by deriving this equation by application of the equilibrium equation $\mathbf{S} = \mathbf{BV} + \mathbf{S}_p$ to a beam slice.

Continuity requirements of $\mathbf{N}, \delta\boldsymbol{\varepsilon}$, and $\delta\mathbf{u}$ are now discussed. In establishing that the external virtual work for a beam slice is $\mathbf{N}^T\delta\boldsymbol{\varepsilon}dx$, it was implicitly assumed that $\mathbf{N}$ is a continuous function of x. If there is a concentrated action at point x, then $\mathbf{N}$ has a jump discontinuity at that point. For the beam slice containing point x, the jump discontinuity in $\mathbf{N}$ equilibriates the concentrated action so that the external virtual work for the slice remains infinitesimal in dx and does not affect the expression of δW_{int}. In the expression of δW_{ext} a concentrated action may be represented as a singularity function [1], or its virtual work may be written separately from the integral.

In deriving the equation of virtual work, it had to be assumed that $\delta\mathbf{u}$ is a continuous function of x. This minimum continuity requirement corresponds to at least a piecewise continuous $\delta\boldsymbol{\varepsilon}$.

In summary, $\mathbf{p}$ may contain singularity functions representing concentrated loads and $\mathbf{N}$ and $\delta\boldsymbol{\varepsilon}$ may be piecewise continuous.

Some terminology is now adopted in order to state and prove succinctly the principles of virtual work.

A force system $(\mathbf{N}, \mathbf{p}, \mathbf{S})$ is said to be statically admissible in a certain deformed state if it satisfies the differential equation of equilibrium and the statical end conditions, or

$$(\boldsymbol{\beta}\mathbf{N})_{,x} - \boldsymbol{\alpha}\mathbf{N} + \mathbf{p} = 0 \tag{4.3–1a}$$

$$\{(\boldsymbol{\beta}\mathbf{N})_{x=0} + \mathbf{S}^a \quad (\boldsymbol{\beta}\mathbf{N})_{x=l} - \mathbf{S}^b\} = 0 \tag{4.3–1b}$$

An incremental kinematic system $(\delta\boldsymbol{\varepsilon}, \delta\mathbf{u}, \delta\mathbf{s})$ is said to be geometrically admissible if

$$\delta\boldsymbol{\varepsilon} = \boldsymbol{\beta}^T\delta\mathbf{u}_{,x} + \boldsymbol{\alpha}^T\delta\mathbf{u} \tag{4.3–2a}$$

and

$$\{(\delta\mathbf{u})_{x=0} - \delta\mathbf{s}^a \quad (\delta\mathbf{u})_{x=l} - \delta\mathbf{s}^b\} = 0 \tag{4.3–2b}$$

Principle of Virtual Displacements

The principle of virtual displacements may be stated as follows:

The equation of virtual work, in which the kinematic system is arbitrary but geometrically admissible, is a necessary and sufficient condition for a force system to be statically admissible.

The proof of the principle relies on integration by parts of $\mathbf{N}^T\delta\boldsymbol{\varepsilon}$, which gives

$$\int_0^l \mathbf{N}^T(\boldsymbol{\beta}^T\delta\mathbf{u}_{,x} + \boldsymbol{\alpha}^T\delta\mathbf{u})dx = (\boldsymbol{\beta}\mathbf{N})^T\delta\mathbf{u}\Big|_0^l - \int_0^l [(\boldsymbol{\beta}\mathbf{N})_{,x} - \boldsymbol{\alpha}\mathbf{N}]^T\delta\mathbf{u}dx \qquad (4.3\text{–}3a)$$

By putting all terms in Eq. (4.2–2b) on the right-hand side, there comes

$$\int_0^l [(\boldsymbol{\beta}\mathbf{N})_{,x} - \boldsymbol{\alpha}\mathbf{N} + \mathbf{p}]^T\delta\mathbf{u}dx + [\mathbf{S}^a + (\boldsymbol{\beta}\mathbf{N})_{x=0}]^T\delta\mathbf{s}^a + [\mathbf{S}^b - (\boldsymbol{\beta}\mathbf{N})_{x=l}]^T\delta\mathbf{s}^b = 0 \qquad (4.3\text{–}3b)$$

If the force system is statically admissible, the coefficients of $\delta\mathbf{u}$, $\delta\mathbf{s}^a$, and $\delta\mathbf{s}^b$ vanish. Conversely, if Eq. (4.3–3b) is satisfied for any $\delta\mathbf{u}$, $\delta\mathbf{s}^a$, and $\delta\mathbf{s}^b$, the coefficients of these quantities must vanish because otherwise it is possible to choose each of the quantities $\delta\mathbf{u}, \delta\mathbf{s}^a$, and $\delta\mathbf{s}^b$ to be proportional to its coefficient, and this would make the left-hand side of Eq. (4.3–3b) nonzero.

Principle of Virtual Forces

The principle of virtual forces may be stated as follows:

The equation of virtual work, in which the force system is arbitrary but statically admissible, is a necessary and sufficient condition for an incremental kinematic system to be geometrically admissible.

An arbitrary force system is also referred to as a system of virtual forces.

The necessary part of the principle does not differ from that of the principle of virtual displacements since in both cases the kinematic and static systems are admissible.

In the sufficient part, $\mathbf{p}$ satisfies the differential equation of equilibrium. Thus $\mathbf{p}^T\delta\mathbf{u}dx$ is equal to the last term in the integration-by-parts relation (4.3–3a). Also $(\boldsymbol{\beta}\mathbf{N})$ satisfies the end static conditions (4.3–1b). Putting all terms in Eq. (4.2–2b) on the left-hand side, there comes

$$\int_0^l \mathbf{N}^T(\delta\boldsymbol{\varepsilon} - \boldsymbol{\beta}^T\delta\mathbf{u}_{,x} - \boldsymbol{\alpha}^T\delta\mathbf{u})dx - \mathbf{S}^{aT}(\delta\mathbf{s}^a - \delta\mathbf{u}_{x=0}) - \mathbf{S}^{bT}(\delta\mathbf{s}^b - \delta\mathbf{u}_{x=l}) = 0$$

For the foregoing equation to hold for arbitrary $\mathbf{N}$, $\mathbf{S}^a$, and $\mathbf{S}^b$, the coefficients of these quantities must vanish, and the result is the geometric admissibility conditions.

5. CONSTITUTIVE EQUATIONS

5.1. Introduction

Constitutive equations are relations between $\mathbf{N}$ and $\boldsymbol{\varepsilon}$ representing material properties. They form with static and kinematic equations a system of governing equations for the structural member.

Constitutive equations may be classified in several ways. They could be finite or incremental, elastic or inelastic, linear or nonlinear and time dependent or independent. Only time-independent relations are considered in this text.

Following general notions on work in Section 5.2, the rest of Section 5 is devoted to what characterizes beam constitutive properties for an elastic or conservative material, with application to the linearly elastic case.

5.2. Work and Complementary Work Equations

The deformed state $\mathbf{u}$ is assumed to be reached gradually, starting from the initial state $\mathbf{u} = 0$, due to a gradual application of some loading condition. A continuous sequence of deformed states is called a deformation path and is represented as $\mathbf{u} = \mathbf{u}(x, t)$. To a value of the parameter t corresponds a deformed shape and a point on the deformation path. It will be assumed that $\mathbf{u}(x, t)$ at any t represents an equilibrium state in which the internal forces are $\mathbf{N} = \mathbf{N}(x, t)$ and the strains $\boldsymbol{\varepsilon} = \boldsymbol{\varepsilon}(x, t)$. An incremental displacement along the deformation path is $\delta\mathbf{u} = \mathbf{u}_{,t}\,dt$. Since the member is continuously in equilibrium, the equation of virtual work, $\delta W_{\text{ext}} = \delta W_{\text{int}}$, may be applied to the actual incremental work along the deformation path. Integration along the path from the initial to the final state yields the work equation

$$W_{\text{ext}} = W_{\text{int}} \tag{5.2–1}$$

Equation (5.2–1) may be applied to a beam slice dx considered as a member. The internal work density, defined per unit dx, is

$$U = \int_0^t \mathbf{N}^T \delta\boldsymbol{\varepsilon} \tag{5.2–2}$$

For the member of length l

$$W_{\text{int}} = \int_0^l U\,dx \tag{5.2–3}$$

The complementary internal work density is defined as the integral, along the deformation path,

$$U^* = \int_0^t \boldsymbol{\varepsilon}^T \delta \mathbf{N} \tag{5.2–4}$$

and the complementary internal work as

$$W^*_{\text{int}} = \int_0^l U^* \, dx \tag{5.2–5}$$

Since $\boldsymbol{\varepsilon}^T \delta \mathbf{N} + \mathbf{N}^T \delta \boldsymbol{\varepsilon} = \delta(\mathbf{N}^T \boldsymbol{\varepsilon})$, we can write, noting that $\boldsymbol{\varepsilon}$ is zero at $t = 0$,

$$U + U^* = \mathbf{N}^T \boldsymbol{\varepsilon} \tag{5.2–6}$$

and, by integration along the member,

$$W_{\text{int}} + W^*_{\text{int}} = \int_0^l \mathbf{N}^T \boldsymbol{\varepsilon} \, dx \tag{5.2–7}$$

These equations hold for any material. Elastic materials are considered next.

5.3. Constitutive Equations for Elastic Beams

An elastic member is defined as one for which the external work vanishes for any closed deformation path, that is, for any path that brings the member back to its initial state. Since $W_{\text{ext}} = W_{\text{int}}$, the definition may be stated equivalently in terms of internal work. The definition also applies to a member of any length and thus to the internal work density U.

Since all open paths from the initial state, $\boldsymbol{\varepsilon} = 0$, to a deformed state $\boldsymbol{\varepsilon}$, may be closed by a single path, it follows that U is path independent and depends only on the final state $\boldsymbol{\varepsilon}$. The function $U(\boldsymbol{\varepsilon})$ is called the strain energy density function. The increment of internal work, $\delta U = \mathbf{N}^T \delta \boldsymbol{\varepsilon}$, is then the exact differential of U. The general form of elastic constitutive equations is thus

$$\mathbf{N} = \frac{\partial U}{\partial \boldsymbol{\varepsilon}} \tag{5.3–1}$$

The integral defining U, Eq. (5.2–2), may now be expressed in terms of $\boldsymbol{\varepsilon}$ as integration variable, or

$$U = \int_0^{\boldsymbol{\varepsilon}} \mathbf{N}^T \, d\boldsymbol{\varepsilon} \tag{5.3–2}$$

An alternate form of elastic constitutive properties may be derived in terms of U^*.

Since U and $\mathbf{N} = \partial U/\partial \boldsymbol{\varepsilon}$ are path independent, then from Eq. (5.2–6) U^* is also path independent. An argument similar to that followed in the case of $U(\boldsymbol{\varepsilon})$ allows us to consider U^* as a function of the state variables $\mathbf{N}$ with the property $\boldsymbol{\varepsilon}^T \delta \mathbf{N} = \delta U^*$. This leads to constitutive equations of the form

$$\boldsymbol{\varepsilon} = \frac{\partial U^*}{\partial \mathbf{N}} \tag{5.3–3}$$

and to the defining equation for U^*

$$U^* = \int_{\mathbf{N}_I}^{\mathbf{N}} \boldsymbol{\varepsilon}^T \delta \mathbf{N} \tag{5.3–4}$$

where $\mathbf{N}_I$ are initial stress resultants in state $\boldsymbol{\varepsilon} = 0$. $\mathbf{N}_I$ would occur, for example, if there is a change in temperature and deformation is prevented. U and U^* satisfy Eq. (5.2–6) which may serve to determine one if the other is known. For example, if U is known, Eq. (5.3–1) needs to be solved for $\boldsymbol{\varepsilon}$ in terms of $\mathbf{N}$. The result is then substituted in Eq. (5.2–6) to determine U^* as a function of $\mathbf{N}$.

The preceding derivation of the form of elastic constitutive relations relies on the property that the internal incremental work density is $\mathbf{N}^T \delta \boldsymbol{\varepsilon}$. It remains valid for any other component representation of stress resultants and strains that has the aforementioned property. The representation $\mathbf{N} = \{N_s \ N_s \ M\}$ and $\boldsymbol{\varepsilon} = \{\varepsilon_n \ \varepsilon_s \ \chi\}$ will be called normal and will be denoted $\mathbf{N}_n$ and $\boldsymbol{\varepsilon}_n$ when needed for identification purposes. If the axial strain ε_1 is chosen as a state variable instead of the normal strain ε_n, a transformation on $\mathbf{N}_n$ is needed in order to conserve the

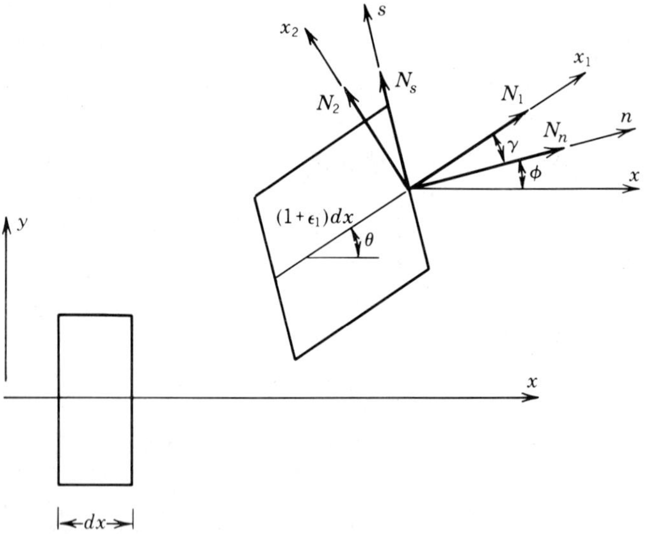

FIG. 5.3–1. Normal and axial representations of stress resultants.

form of Eq. (4.2–1). The shear strain could also be chosen as the angle γ, Fig. 2.3–1, rather than ε_s. To the representation $\boldsymbol{\varepsilon}_1 = \{\varepsilon_1\ \gamma\ \chi\}$ corresponds $\mathbf{N}_1 = \{N_1\ N_\gamma\ \chi\}$ such that

$$\mathbf{N}_n{}^T\delta\boldsymbol{\varepsilon}_n = \mathbf{N}_1{}^T\delta\boldsymbol{\varepsilon}_1 \tag{5.3–5}$$

$\delta\boldsymbol{\varepsilon}_n$ and $\delta\boldsymbol{\varepsilon}_1$ are related by a relation of the form

$$\delta\boldsymbol{\varepsilon}_n = \mathbf{T}\delta\boldsymbol{\varepsilon}_1 \tag{5.3–6}$$

Thus $\mathbf{N}_n^T\delta\boldsymbol{\varepsilon}_n = \mathbf{N}_n^T\mathbf{T}\delta\boldsymbol{\varepsilon}_1$, and consequently

$$\mathbf{N}_1 = \mathbf{T}^T\mathbf{N}_n \tag{5.3–7}$$

To determine the transformation $\mathbf{T}$, we have, using Fig. 2.3–1,

$$1 + \varepsilon_n = (1 + \varepsilon_1)\cos\gamma \tag{5.3–8a}$$

$$\varepsilon_s = (1 + \varepsilon_1)\sin\gamma \tag{5.3–8b}$$

and

$$\mathbf{T} = \frac{\partial\boldsymbol{\varepsilon}_n}{\partial\boldsymbol{\varepsilon}_1{}^T} = \begin{bmatrix} \cos\gamma & -(1+\varepsilon_1)\sin\gamma & 0 \\ \sin\gamma & (1+\varepsilon_1)\cos\gamma & 0 \\ 0 & 0 & 1 \end{bmatrix} \tag{5.3–9}$$

Except for the coefficient $(1 + \varepsilon_1)$, $\mathbf{T}$ is a vector transformation matrix. Transformation of the vector components (N_n, N_s) to components (N_1, N_2) on axes (x_1, x_2), as shown in Fig. 5.3–1, yields

$$N_1 = N_n\cos\gamma + N_s\sin\gamma \tag{5.3–10a}$$

$$N_2 = -N_n\sin\gamma + N_s\cos\gamma \tag{5.3–10b}$$

From Eq. (5.3–7), N_γ is related to N_2 through

$$N_\gamma = (1 + \varepsilon_1)N_2 \tag{5.3–11}$$

If U is assumed expressed in terms of ε_1 and γ, constitutive equations for N_1 and N_2 have the form

$$N_1 = \frac{\partial U}{\partial\varepsilon_1} \tag{5.3–12a}$$

$$N_2 = \frac{1}{1+\varepsilon_1}\frac{\partial U}{\partial\gamma} \tag{5.3–12b}$$

$\mathbf{N}_1$ and $\boldsymbol{\varepsilon}_1$ will be referred to as the axial representation

The simplest constitutive relations are linear. However, linear relations between $\mathbf{N}_n$ and $\boldsymbol{\varepsilon}_n$ are nonlinear between $\mathbf{N}_1$ and $\boldsymbol{\varepsilon}_1$, and vice versa. The distinction between N_n and N_1 and between ε_n and ε_1 disappears in a theory neglecting shear deformation, that is, if $\gamma = 0$. The shear force in such a theory is governed only by an equilibrium equation. In a theory including shear deformation, it will be shown that only in the geometrically linear theory (Chapter 3) is it consistent not to distinguish between the normal and axial representations.

Linear constitutive relations for axial force and bending moment, established in elementary beam theory for small strains, are based on linear axial stress-strain relations. They may be applied in the present theory, under the same assumption of small strain, to N_1 and M. The constitutive equation for shear is derived from Eq. (5.3–3) by evaluating the complementary strain energy due to shear in terms of the shear force. It is applied to N_2, which for $\varepsilon_1 \ll 1$, is equivalent to N_γ. With axes centered at the centroid of the cross section, the linear constitutive equations have the form

$$N_1 = EA\,\varepsilon_1 \tag{5.3–13a}$$

$$N_2 = GA_s\gamma \tag{5.3–13b}$$

$$M = EI\chi \tag{5.3–13c}$$

where E is Young's modulus, G the shear modulus, A the area of the cross section, A_s the shear area, and I the moment of inertia, or second moment of area, with respect to the centroidal z axis.

Equations (5.3–13) may be generalized to a nonhomogeneous material through the concept of transformed area and to the case where initial stresses are present. They may also be transformed to a reference point other than the centroid.

The strain energy density associated with linear constitutive relations is a second-degree polynomial in $\boldsymbol{\varepsilon}$. The general form of such a polynomial is

$$U = \tfrac{1}{2}\boldsymbol{\varepsilon}^T \mathscr{K} \boldsymbol{\varepsilon} + \boldsymbol{\varepsilon}^T \mathbf{N}_I \tag{5.3–14}$$

where $\mathscr{K}$ is symmetric. The constitutive equations are then

$$\mathbf{N} = \mathscr{K}\boldsymbol{\varepsilon} + \mathbf{N}_I \tag{5.3–15}$$

$\mathscr{K}$ is called the rigidity matrix. In the case of Eqs. (5.3–13),

$$\mathscr{K} = \lceil EA \;\; GA_s \;\; EI \rfloor.$$

$\mathscr{K}$ must be assumed nonsingular, so that Eq. (5.3–15) may be solved for $\boldsymbol{\varepsilon}$ in the form

$$\boldsymbol{\varepsilon} = \boldsymbol{\Phi}\mathbf{N} + \boldsymbol{\varepsilon}_I \tag{5.3–16}$$

where

$$\mathbf{\Phi} = \mathscr{X}^{-1} \tag{5.3–17a}$$

$$\boldsymbol{\varepsilon}_I = -\mathbf{\Phi}\mathbf{N}_I \tag{5.3–17b}$$

U^* is then found from Eq. (5.2–6) in the form

$$U^* = \tfrac{1}{2}\mathbf{N}^T\mathbf{\Phi}\mathbf{N} + \mathbf{N}^T\boldsymbol{\varepsilon}_I + C \tag{5.3–18}$$

where C is a constant that has no effect on the constitutive equations. Its value is such that $U^* = 0$ in the initial state $\boldsymbol{\varepsilon} = 0$, $\mathbf{N} = \mathbf{N}_I$. If $\mathbf{N}_I = 0$, then $\boldsymbol{\varepsilon}_I$ and C also vanish, and U and U^* are equal in value to $\tfrac{1}{2}\mathbf{N}^T\boldsymbol{\varepsilon}$.

6. SUMMARY AND COMMENTS

Governing differential equations of the structural member consist of strain-displacement relations, equilibrium equations, and constitutive equations. The strain-displacement relations may be reduced to a system of 3 differential equations for the three displacements by substituting for $\boldsymbol{\varepsilon}$ its constitutive relation in terms of $\mathbf{N}$ and by substituting for $\mathbf{N}$ the solution of the equilibrium equations $\mathbf{N} = \mathbf{bV} + \mathbf{N}_p$. It is convenient to formulate the equations in terms of displacement $\mathbf{u}_r$ relative to a member-bound reference. If $\mathbf{V}$ is assumed given, or is taken as a parameter, the support conditions in the member-bound reference form the boundary conditions for $\mathbf{u}_r$. Exact solutions are generally impossible to obtain because of the nonlinearity of the system of equations, but numerical solutions are feasible. Assuming a solution for $\mathbf{u}_r$ is found in terms of $\mathbf{V}$ and of the span load, the motion of the member-bound reference and $\mathbf{V}$ remain generally unknown if the member is part of a structure. What is needed from the solution of the member governing equations are discrete equations relating deformations $\mathbf{v}$ to member redundants $\mathbf{V}$ or end displacements $\mathbf{s}$ to end forces $\mathbf{S}$. Such relations allow us to formulate the analysis of an assemblage of members in terms of discrete rather than differential equations. The purpose of Chapters 5 and 6 is to establish such discrete relations within the context of simplified theories of the structural member.

EXERCISES

Section 2

1. Consider a rigid body displacement of the member as a whole defined by the displacements $\mathbf{s}^a$ at end a. Determine $\mathbf{s}^b$ in terms of $\mathbf{s}^a$, and verify that the deformations vanish.

2. For the rigid body displacement of Exercise 1, determine the displacements $\{u, v, \varphi\}$ at a general cross section x, and verify that the strains vanish.

3. Outline on Fig. 2.2–1 the deformations of the alternate type, and on Fig. 2.2–2 the deformations of the cantilever type.

4. In a theory neglecting shear deformation $\varepsilon_s = 0$, and φ is determined in terms of u and v by means of Eq. (2.3–1b). Derive in this case the expression of the curvature strain χ in terms of u and v, and verify that it coincides with the formula for curvature established in analytic geometry.

5. A straight line segment ab of length l along the x axis is deformed into a semicircle whose diameter ab' is on the y axis. Assume that line ab is unstrained so that a point initially at x is displaced into a point on the semicircle at the arc length distance x from point a. Determine the displacements u and v as functions of x, and verify that ε_1 vanishes.

6. Show that if $\varepsilon_1 = 0$, then

$$u = s_1^a - x + \int_0^x \sqrt{1 - (v_{,x})^2}\, dx$$

Section 3

7. Outline on Fig. 3.1–1 the end displacements, and derive the static transformation matrix T_b^a in Eq. (3.1–1b).

8. Obtained the detailed expressions of $\mathbf{T}_x^a$ and $\mathbf{P}^a$ in Eq. (3.1–1c) by specializing the expression of $\mathbf{T}_b^a$ to a segment of length x. Obtain the same result directly by means of a free body diagram.

9. Consider a member in a given deformed state subjected only to end forces and referred to a bound reference of the alternate type as shown in Fig. 3.1–3. Complete the force system and derive the static matrix $\mathbf{B}$ of the alternate type, row partitioned into $\mathbf{B}^a$ and $\mathbf{B}^b$. Ascertain that $\mathbf{B}^a = -\mathbf{T}_b^a\mathbf{B}^b$.

10. In Exercise 9 let $\mathbf{V} = 0$, and consider a span load whose resultants at point a are $\mathbf{P}^a = \{P_x^a \; P_y^a \; C^a\}$. Determine the reactions in the bound reference; then determine $\mathbf{S}_p$ by transformation to the fixed axes.

11. Derive the differential equations of equilibrium (3.2–4) by making the free body diagram of a beam slice as outlined in the text.

12. Obtain the detailed expressions of the static matrix $\mathbf{b}$ in Eqs. (3.2–3) for $\mathbf{V}$ of the cantilever and alternate types, respectively.

Section 4

13. Consider a member subjected only to end forces so that $\mathbf{S}^a = -\mathbf{T}_b^a\mathbf{S}^b$. Apply the equation of virtual work using a virtual rigid body displacement, and

deduce that $\delta \mathbf{s}^b = (\mathbf{T}_b^a)^T \delta \mathbf{s}^a$. Verify this result with the expression of $\mathbf{s}^b$ found in Exercise 1.

14. Verify that the static matrix **B**, Eqs. (3.1–3), is obtained by differentiation of the deformation-displacement relations (2.2–1) as found in Eq. (4.1–6b). Do the same for **B** and **v** of the alternate type using the results of Exercise 9 and Eq. (2.2–3).

15. Form the first variations of the strains Eqs. (2.3–1), and show that the result has the form of Eqs. (4.3–2a) with $\boldsymbol{\beta}$ and $\boldsymbol{\alpha}$ as obtained statically in Section 3.2.

16. Write Eq. (4.2–3) in detail for incremental strains and deformations taking place from the initial underformed state.

Section 5

17. a. Using Eqs. (2.3–2a) and (5.3–8), show that $\delta\varepsilon_1 = \cos\gamma\,\delta\varepsilon_n + \sin\gamma\,\delta\varepsilon_s$, and interpret this result geometrically on Fig. 2.3–1.

b. Let N_1' and N_s' be the stress resultants that are work-coupled with ε_1 and ε_s, that is, $N_1'\delta\varepsilon_1 + N_s'\delta\varepsilon_s = N_n\delta\varepsilon_n + N_s\delta\varepsilon_s$. Obtain N_1' and N_s' in terms of N_n and N_s, and show that they are the nonorthogonal components of the stress resultant vector on axes x_1 and s in Fig. 5.3–1.

3

SIMPLIFIED THEORIES OF THE STRUCTURAL MEMBER

1. INTRODUCTION

The static and kinematic equations developed in the preceding chapter may be greatly simplified if translational strains and cross-sectional rotations are assumed small compared with unity. Such assumptions are justified in most practical applications. Two levels of approximations will be implemented, leading to what is called beam-column theory (Section 2) and geometrically linear theory (Section 3).

Geometric approximations also involve static approximations. Conservation of the principles of virtual work will be taken as the basis for developing consistent approximate equations.

2. BEAM-COLUMN, OR SECOND-ORDER, THEORY

2.1 Deformation Analysis

The basic approximation in beam-column theory pertains to the orders of magnitude of rotations and axial deformations. It will be introduced first in the deformation-displacement relations and will then be specialized to a beam segment of infinitesimal length or beam slice.

The axial deformation v_1 is assumed small compared with member length l, and v_1/l is assumed to be comparable in order of magnitude to squares and products of member end rotations s_3^a and s_3^b and of member chord rotation ρ.

Equations (2/2.2–1) are simplified by developing $\cos s_3^a$ and $\sin s_3^a$ in powers of s_3^a and deleting third- and higher-order terms. Further $(s_1^b - s_1^a)$ is seen to be comparable to v_1 and must therefore be neglected in comparison with l in the expression of v_2. Deformations of the cantilever type are thus obtained as follows:

$$v_1 = s_1^b - s_1^a - \tfrac{1}{2}l(s_3^a)^2 + s_3^a(s_2^b - s_2^a) \qquad (2.1\text{–}1a)$$

$$v_2 = s_2^b - s_2^a - l s_3^a \qquad (2.1\text{–}1b)$$

$$v_3 = s_3^b - s_3^a \qquad (2.1\text{–}1c)$$

The preceding assumptions also imply that the axial deformation of the alternate type is small compared with l. With $\rho^2 \ll 1$, there comes from Eq. (2/2.2–2)

$$\rho = \frac{1}{l}(s_2^b - s_2^a) \qquad (2.1\text{–}2)$$

and deformations of the alternate type are approximated by

$$v_1 = s_1^b - s_1^a + \frac{1}{2l}(s_2^b - s_2^a)^2 \qquad (2.1\text{–}3a)$$

$$v_2 = s_3^a - \frac{1}{l}(s_2^b - s_2^a) \qquad (2.1\text{–}3b)$$

$$v_3 = s_3^b - \frac{1}{l}(s_2^b - s_2^a) \qquad (2.1\text{–}3c)$$

It is noted that for both types of deformations, v_1 is nonlinear in the displacements and that v_2 and v_3 are linear.

Strains

The approximations for a member of length l are now applied to a beam slice considered as a member of length dx. With $l = dx$, $\mathbf{s}^a = \mathbf{u}$, and $\mathbf{s}^b = \mathbf{u} + d\mathbf{u}$ in Eqs. (2.1–1), $\mathbf{v}$ turns into $\boldsymbol{\varepsilon} dx = \{\varepsilon_n \;\; \varepsilon_s \;\; \chi\} dx$, with

$$\varepsilon_n = u_{,x} - \tfrac{1}{2}\varphi^2 + \varphi v_{,x} \qquad (2.1\text{–}4a)$$

$$\varepsilon_s = v_{,x} - \varphi \qquad (2.1\text{–}4b)$$

$$\chi = \varphi_{,x} \qquad (2.1\text{–}4c)$$

For a beam slice, ρ tends to the rotation θ of the axial material fibre, and v_1/dx of the alternate type tends to the axial strain ε_1. There comes

$$\theta = v_{,x} \tag{2.1–5}$$

and

$$\varepsilon_1 = u_{,x} + \tfrac{1}{2}(v_{,x})^2 \tag{2.1–6a}$$

$$\gamma = \theta - \varphi = v_{,x} - \varphi \tag{2.1–6b}$$

The two representations of translational strains are related through

$$\varepsilon_1 = \varepsilon_n + \tfrac{1}{2}\varepsilon_s^2 \tag{2.1–7a}$$

$$\gamma = \varepsilon_s \tag{2.1–7b}$$

It is noted that nonlinearity in the strain-displacement relations now occurs only in ε_1 or ε_n.

2.2 Statical Analysis

From the statical analysis of Chapter 2 equilibrium of the member is satisfied in terms of member statical redundants **V** by equations of the form

$$\mathbf{S} = \mathbf{BV} + \mathbf{S}_p \tag{2.2–1}$$

$$\mathbf{N} = \mathbf{bV} + \mathbf{N}_p \tag{2.2–2}$$

The stress resultant components on the local axes, $\mathbf{S}^x$, are related to **N** through

$$\mathbf{S}^x = \boldsymbol{\beta}\mathbf{N} \tag{2.2–3}$$

and **N** satisfies the differential equation of equilibrium

$$(\boldsymbol{\beta}\mathbf{N})_{,x} - \boldsymbol{\alpha}\mathbf{N} + \mathbf{p} = 0 \tag{2.2–4}$$

If the considerations of virtual work seen in Chapter 2 are to remain valid, **B**, $\boldsymbol{\beta}$, and $\boldsymbol{\alpha}$ must appear in the kinematic equations:

$$\delta\mathbf{v} = \mathbf{B}^T\delta\mathbf{s} \tag{2.2–5}$$

$$\delta\boldsymbol{\varepsilon} = \boldsymbol{\beta}^T\delta\mathbf{u}_{,x} + \boldsymbol{\alpha}^T\delta\mathbf{u} \tag{2.2–6}$$

For **v** of the cantilever type, $\mathbf{B}^T$ is formed by differentiation of Eqs. (2.1–1). The corresponding static equations (2.2–1) are thus found to be

$$\begin{Bmatrix} S_1^a \\ S_2^a \\ S_3^a \\ S_1^b \\ S_2^b \\ S_3^b \end{Bmatrix} = \begin{bmatrix} -1 & 0 & 0 \\ -s_3^a & -1 & 0 \\ s_2^b - s_2^a - ls_3^a & -l & -1 \\ 1 & 0 & 0 \\ s_3^a & 1 & 0 \\ 0 & 0 & 1 \end{bmatrix} \begin{Bmatrix} V_1 \\ V_2 \\ V_3 \end{Bmatrix} + \begin{Bmatrix} S_{1p}^a \\ S_{2p}^a \\ S_{3p}^a \\ S_{1p}^b \\ S_{2p}^b \\ S_{3p}^b \end{Bmatrix} \tag{2.2–7}$$

For the alternate type, $\mathbf{B}^T$ is formed similarly using Eq. (2.1–3). There comes

$$\begin{Bmatrix} S_1^a \\ S_2^a \\ S_3^a \\ S_1^b \\ S_2^b \\ S_3^b \end{Bmatrix} = \begin{bmatrix} -1 & 0 & 0 \\ -\frac{1}{l}(s_2^b - s_2^a) & \frac{1}{l} & \frac{1}{l} \\ 0 & 1 & 0 \\ 1 & 0 & 0 \\ \frac{1}{l}(s_2^b - s_2^a) & -\frac{1}{l} & -\frac{1}{l} \\ 0 & 0 & 1 \end{bmatrix} \begin{Bmatrix} V_1 \\ V_2 \\ V_3 \end{Bmatrix} + \begin{Bmatrix} S_{1p}^a \\ S_{2p}^a \\ S_{3p}^a \\ S_{1p}^b \\ S_{2p}^b \\ S_{3p}^b \end{Bmatrix} \tag{2.2–8}$$

$\boldsymbol{\beta}^T$ and $\boldsymbol{\alpha}^T$ are determined by forming $\delta\boldsymbol{\varepsilon}$. From Eqs. (2.1–6) and (2.1–4c), there comes for the axial representation

$$\begin{Bmatrix} \delta\varepsilon_1 \\ \delta\gamma \\ \delta\chi \end{Bmatrix} = \begin{bmatrix} 1 & v_{,x} & 0 \\ 0 & 1 & 0 \\ 0 & 0 & 1 \end{bmatrix} \begin{Bmatrix} \delta u_{,x} \\ \delta v_{,x} \\ \delta\varphi_{,x} \end{Bmatrix} + \begin{bmatrix} 0 & 0 & 0 \\ 0 & 0 & -1 \\ 0 & 0 & 0 \end{bmatrix} \begin{Bmatrix} \delta u \\ \delta v \\ \delta\varphi \end{Bmatrix} \tag{2.2–9}$$

The component transformation $\mathbf{S}^x = \boldsymbol{\beta}\mathbf{N}$ is thus

$$\begin{Bmatrix} N_x \\ N_y \\ M \end{Bmatrix} = \begin{bmatrix} 1 & 0 & 0 \\ \theta & 1 & 0 \\ 0 & 0 & 1 \end{bmatrix} \begin{Bmatrix} N_1 \\ N_2 \\ M \end{Bmatrix} \tag{2.2–10}$$

where $\theta = v_{,x}$. The inverse relation $\mathbf{N} = \boldsymbol{\beta}^{-1}\mathbf{S}^x$ is

$$\begin{Bmatrix} N_1 \\ N_2 \\ M \end{Bmatrix} = \begin{bmatrix} 1 & 0 & 0 \\ -\theta & 1 & 0 \\ 0 & 0 & 1 \end{bmatrix} \begin{Bmatrix} N_x \\ N_y \\ M \end{Bmatrix} \tag{2.2–11}$$

Similar equations derived for the normal representation involve the cross-sectional rotation φ instead of θ.

The differential equations of equilibrium (2.2–4) in terms of $\{N_1 \;\; N_2 \;\; M\}$ take the form

$$N_{1,x} + p_x = 0 \tag{2.2–12a}$$

$$(N_1 v_{,x} + N_2)_{,x} + p_y = 0 \tag{2.2–12b}$$

$$M_{,x} + N_2 + m = 0 \tag{2.2–12c}$$

The preceding equations and the force equations in Eqs. (2.2–7, 8, 10, 11) contain a characteristic approximation in beam-column theory concerning force component transformation. Axial components N_x, N_n, and N_1 are equivalent and may all be called axial or normal forces. Thus the contribution to the axial force in one system of axes, of the transverse component in another system of axes is neglected. By contrast, the transverse force component in one system is obtained by adding the projections of both axial and transverse components in the other system.

The moment equation (2.2–7c) shows another aspect of beam-column theory, namely the contribution of the axial force V_1 to the bending moment because of the lever arm provided by the transverse displacements, but the effect of axial displacements on the lever arm of transverse forces is neglected. This statical property must hold in a static transformation between any pair of points on the member. The static transformation matrix from point ξ to point x, in the local axes (x, y, z), is thus

$$\mathbf{T}_{\xi}^{x} = \begin{bmatrix} 1 & 0 & 0 \\ 0 & 1 & 0 \\ v - v^{\xi} & \xi - x & 1 \end{bmatrix} \tag{2.2–13a}$$

where v and v^{ξ} are the displacement v at points x and ξ, respectively. The transformation matrix to components on the (x_1, x_2) axes at point x is

$$\boldsymbol{\beta}^{-1}\mathbf{T}_{\xi}^{x} = \begin{bmatrix} 1 & 0 & 0 \\ -\theta & 1 & 0 \\ v - v^{\xi} & \xi - x & 1 \end{bmatrix} \tag{2.2–13b}$$

where $\boldsymbol{\beta}^{-1}$ is the coefficient matrix in Eq. (2.2–11). The static matrix $\mathbf{b}$ in Eq. (2.2–2) is the product of three transformations, from $\mathbf{V}$ to $\mathbf{S}^b$, from $\mathbf{S}^b$ to $\mathbf{S}^x$, and from $\mathbf{S}^x$ to $\mathbf{N}$, or $\mathbf{b} = \boldsymbol{\beta}^{-1}\mathbf{T}_b^x\mathbf{B}^b$. For $\mathbf{V}$ of the cantilever type there comes

$$\begin{Bmatrix} N_1 \\ N_2 \\ M \end{Bmatrix} = \begin{bmatrix} 1 & 0 & 0 \\ s_3^a - \theta & 1 & 0 \\ v - s_2^b + s_3^a(l - x) & l - x & 1 \end{bmatrix} \begin{Bmatrix} V_1 \\ V_2 \\ V_3 \end{Bmatrix} + \begin{Bmatrix} N_{1p} \\ N_{2p} \\ M_p \end{Bmatrix} \tag{2.2–14}$$

and for the alternate type

$$\begin{Bmatrix} N_1 \\ N_2 \\ M \end{Bmatrix} = \begin{bmatrix} 1 & 0 & 0 \\ \rho - \theta & -\dfrac{1}{l} & -\dfrac{1}{l} \\ v - s_2^b + \rho(l - x) & \dfrac{x}{l} - 1 & \dfrac{x}{l} \end{bmatrix} \begin{Bmatrix} V_1 \\ V_2 \\ V_3 \end{Bmatrix} + \begin{Bmatrix} N_{1p} \\ N_{2p} \\ M_p \end{Bmatrix} \tag{2.2–15}$$

To determine $\mathbf{S}_p$ and $\mathbf{N}_p$, a fundamental loading case is first treated in the following examples before a general member load is considered.

Example 1. Cantilever Type

Consider concentrated actions $\mathbf{P} = \{P_x \; P_y \; C\}$ applied at point ξ, Fig. 2.2–1. $\mathbf{S}_p$ and $\mathbf{N}_p$ are in equilibrium, in the deformed state of the member, with the span

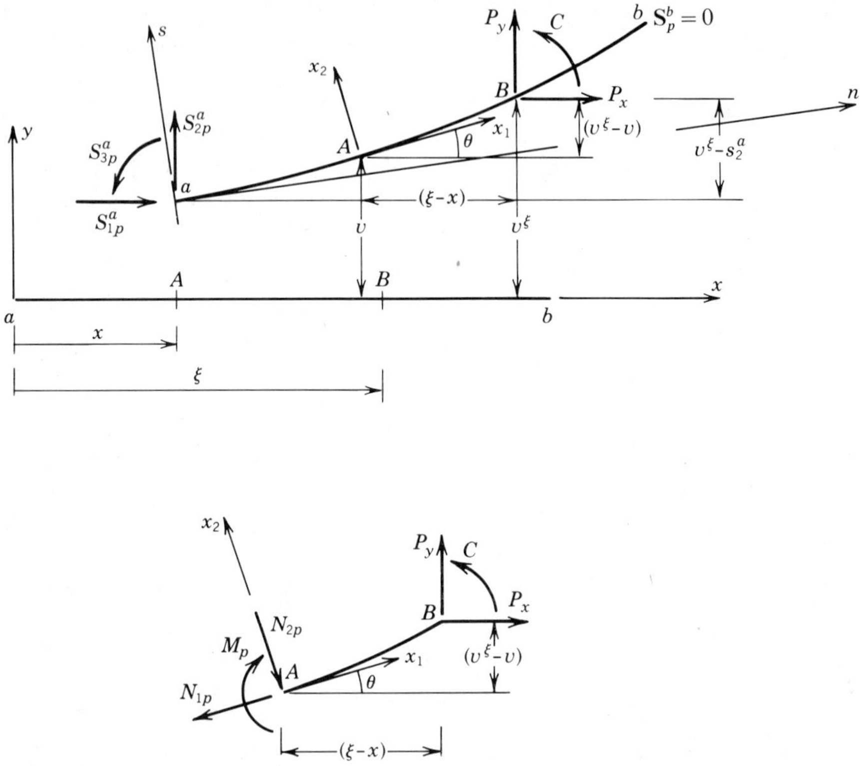

FIG. 2.2–1. Particular equilibrium solution, cantilever type.

load **P** and with $\mathbf{V} = 0$. Thus $\mathbf{S}_p^b = 0$ and $\mathbf{S}_p^a = -\mathbf{T}_\xi^a \mathbf{P}$. Also $\mathbf{N}_p = 0$ for $x > \xi$, and $\mathbf{N}_p = \boldsymbol{\beta}^{-1}\mathbf{T}_\xi^x \mathbf{P}$ for $x < \xi$. In detail,

$$\begin{Bmatrix} S_{1p}^a \\ S_{2p}^a \\ S_{3p}^a \end{Bmatrix} = \begin{bmatrix} -1 & 0 & 0 \\ 0 & -1 & 0 \\ v^\xi - s_2^a & -\xi & -1 \end{bmatrix} \begin{Bmatrix} P_x \\ P_y \\ C \end{Bmatrix}$$

and

$$\begin{Bmatrix} N_{1p} \\ N_{2p} \\ M_p \end{Bmatrix} = \begin{bmatrix} 1 & 0 & 0 \\ -\theta & 1 & 0 \\ v - v^\xi & \xi - x & 1 \end{bmatrix} \begin{Bmatrix} P_x \\ P_y \\ C \end{Bmatrix}, \quad x < \xi$$

Example 2. Alternate type

The same loading as in the previous example is shown in Fig. 2.2–2 with the member-bound reference of the alternate type. The reaction at the roller is

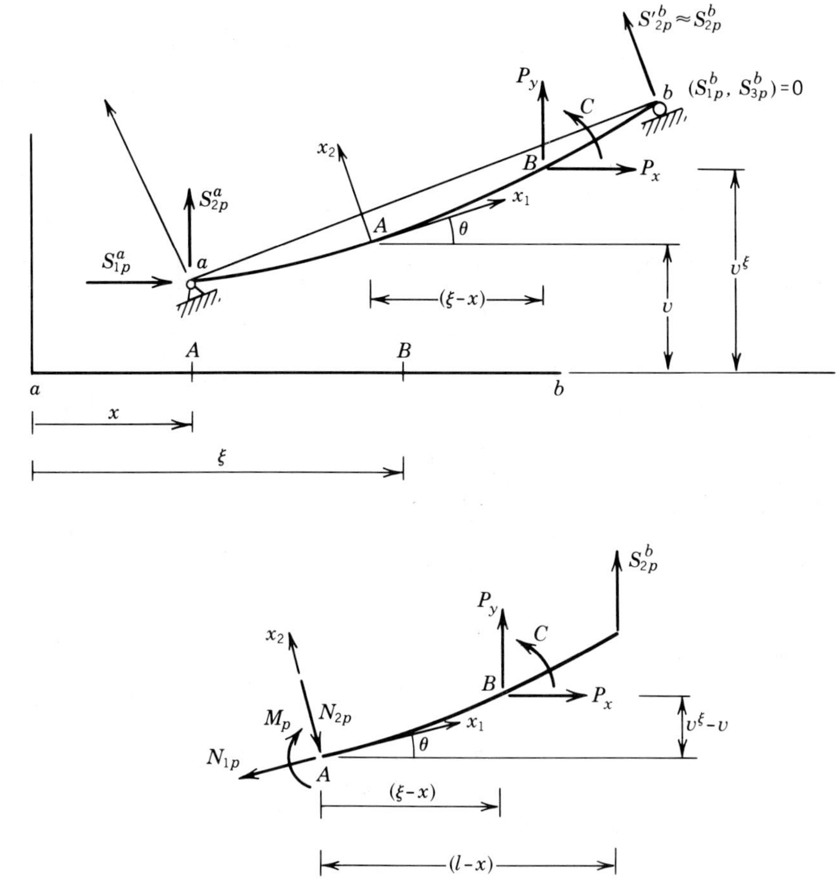

FIG. 2.2–2. Particular equilibrium solution, alternate type.

obtained by means of the moment equilibrium equation at point a. This reaction is equivalent to the y component S^b_{2p}. Thus

$$S^b_{1p} = S^b_{3p} = 0$$

$$S^b_{2p} = \tfrac{1}{l}[P_x(v^\xi - s^a_2) - P_y\xi - C]$$

and $\mathbf{S}^a_p = -\mathbf{T}^a_b\mathbf{S}^b_p - \mathbf{T}^a_\xi\mathbf{P}$, or

$$S^a_{1p} = -P_x$$

$$S^a_{2p} = -S^b_{2p} - P_y$$

$$S^a_{3p} = 0$$

For $x > \xi$, $\mathbf{N}_p = \boldsymbol{\beta}^{-1}\mathbf{T}_b^x\mathbf{S}_p^b$, or

$$N_{1p} = 0$$

$$N_{2p} = S_{2p}^b$$

$$M_p = (l - x)S_{2p}^b$$

and for $x < \xi$, $\mathbf{N}_p = -\boldsymbol{\beta}^{-1}\mathbf{T}_a^x\mathbf{S}_p^a$, or

$$N_{1p} = P_x$$

$$N_{2p} = \frac{P_x}{l}(v^\xi - s_2^a - l\theta) + P_y\left(1 - \frac{\xi}{l}\right) - \frac{C}{l}$$

$$M_p = P_x\left[v - \frac{x}{l}v^\xi - \left(1 - \frac{x}{l}\right)s_2^a\right]$$

$$- P_y x\left(1 - \frac{\xi}{1}\right) + C\frac{x}{l}$$

A general load intensity function $\mathbf{p} = \{p_x \ p_y \ m\}$ may be treated as formed of elementary concentrated actions $\mathbf{p}(\xi)d\xi$. $\mathbf{S}_p$ and $\mathbf{N}_p$ may then be expressed as integrals with respect to ξ of the preceding expressions of $\mathbf{S}_p$ and $\mathbf{N}_p$ in which $\mathbf{P}$ is replaced with $\mathbf{p}(\xi)d\xi$.

Further developments of beam-column theory are carried out in Chapter 6 where constitutive properties are used to determine the deformed shape and to derive stiffness and flexibility relations for the member.

3. GEOMETRICALLY LINEAR THEORY

3.1. Deformation Analysis

The geometrically linear theory derives its name from the property that deformation-displacement relations and strain-displacement relations are linear in the displacements and that consistently equilibrium equations are independent of the displacements.

The basic approximation leading to linear deformation-displacement $(\mathbf{v} - \mathbf{s})$ relations is that member end rotations s_3^a and s_3^b, member chord rotation ρ, and the ratio of axial deformation to member length v_1/l are all negligible with respect to unity. The $(\mathbf{v} - \mathbf{s})$ relations may be obtained from the exact relations of Chapter 2 by replacing cosines with unity and sines with their angles, and neglecting second-order terms in the displacements. A direct geometric derivation using Figs. 2/2.2–1 and 2/2.2–2 may also be made without the need to go through the exact relations.

For deformations of the cantilever type, Fig. 2/2.2–1, there comes

$$\begin{Bmatrix} v_1 \\ v_2 \\ v_3 \end{Bmatrix} = \begin{bmatrix} -1 & 0 & 0 & 1 & 0 & 0 \\ 0 & -1 & -l & 0 & 1 & 0 \\ 0 & 0 & -1 & 0 & 0 & 1 \end{bmatrix} \begin{Bmatrix} s_1^a \\ \vdots \\ s_3^b \end{Bmatrix} \tag{3.1–1}$$

and for deformations of the alternate type, Fig. 2/2.2–2

$$\rho = \frac{1}{l}(s_2^b - s_2^a) \tag{3.1–2}$$

and

$$\begin{Bmatrix} v_1 \\ v_2 \\ v_3 \end{Bmatrix} = \begin{bmatrix} -1 & 0 & 0 & 1 & 0 & 0 \\ 0 & \frac{1}{l} & 1 & 0 & -\frac{1}{l} & 0 \\ 0 & \frac{1}{l} & 0 & 0 & -\frac{1}{l} & 1 \end{bmatrix} \begin{Bmatrix} s_1^a \\ \vdots \\ s_3^b \end{Bmatrix} \tag{3.1–3}$$

Equations (3.1–1, 3) are linear in $\mathbf{s}$ and are of the form

$$\mathbf{v} = \mathbf{B}^T \mathbf{s} \tag{3.1–4}$$

Strains

The approximations for a member of length l are now applied to a beam slice considered as a member of length dx. Thus cross-sectional rotation φ, the rotation of the axial material fiber θ, and the axial strain ε_1 are assumed negligible with respect to unity. The strain-displacement ($\boldsymbol{\varepsilon} - \mathbf{u}$) relations may be obtained directly from Fig. 2/2.3–1, or by linearization of the exact relations of Chapter 2. It is found that the normal and axial representations of $\boldsymbol{\varepsilon}$ coincide. The single notation, $\boldsymbol{\varepsilon} = \{\varepsilon \;\; \gamma \;\; \chi\}$, will thus be used. The $\boldsymbol{\varepsilon} - \mathbf{u}$ relations are found to be

$$\begin{aligned} \varepsilon &= u_{,x} \\ \gamma &= \theta - \varphi = v_{,x} - \varphi \\ \chi &= \varphi_{,x} \end{aligned} \tag{3.1–5}$$

3.2. Statical Analysis

Because of the linearity of the ($\mathbf{v} - \mathbf{s}$) and ($\boldsymbol{\varepsilon} - \mathbf{u}$) relations, the statical approximations that maintain the principles of virtual work consist in neglecting displacements in statical equations. Statical analysis is thus performed in the undeformed geometry. The stress resultant representations $\mathbf{S}^x$, $\mathbf{N}_n$, $\mathbf{N}_1$, Fig.

2/5.3–1, coincide and are denoted $\mathbf{N} = \{N \ S \ M\}$. Equilibrium of the member is represented by the equations

$$\mathbf{S} = \mathbf{BV} + \mathbf{S}_p \tag{3.2–1}$$

$$\mathbf{N} = \mathbf{bV} + \mathbf{N}_p \tag{3.2–2}$$

B is the static-kinematic matrix whose transpose occurs in Eq. (3.1–4) and which is written out explicitly in Eqs. (3.1–1) and (3.1–3) for the cantilever and alternate types, respectively. Derivation of **B** by statics, using Figs. 3.2–1 and 3.2–2, yields the same result.

N is obtained in terms of **V** by evaluating the resultants at x of the external forces in the interval (x, l). For **V** of the cantilever type

$$\begin{Bmatrix} N \\ S \\ M \end{Bmatrix} = \begin{bmatrix} 1 & 0 & 0 \\ 0 & 1 & 0 \\ 0 & l-x & 1 \end{bmatrix} \begin{Bmatrix} V_1 \\ V_2 \\ V_3 \end{Bmatrix} + \begin{Bmatrix} N_p \\ S_p \\ M_p \end{Bmatrix} \tag{3.2–3}$$

and for **V** of the alternate type

$$\begin{Bmatrix} N \\ S \\ M \end{Bmatrix} = \begin{bmatrix} 1 & 0 & 0 \\ 0 & -\dfrac{1}{l} & -\dfrac{1}{l} \\ 0 & \dfrac{x}{l}-1 & \dfrac{x}{l} \end{bmatrix} \begin{Bmatrix} V_1 \\ V_2 \\ V_3 \end{Bmatrix} + \begin{Bmatrix} N_p \\ S_p \\ M_p \end{Bmatrix} \tag{3.2–4}$$

To determine $\mathbf{S}_p$ and $\mathbf{N}_p$, the member is considered supported in the member-bound reference and subjected to the span load **p** with $\mathbf{V} = 0$.

What is needed from member theory for the formulation of structural analysis are constitutive equations for the member as a whole, that is, relations between **V** and **v**. This is done in Chapter 5.

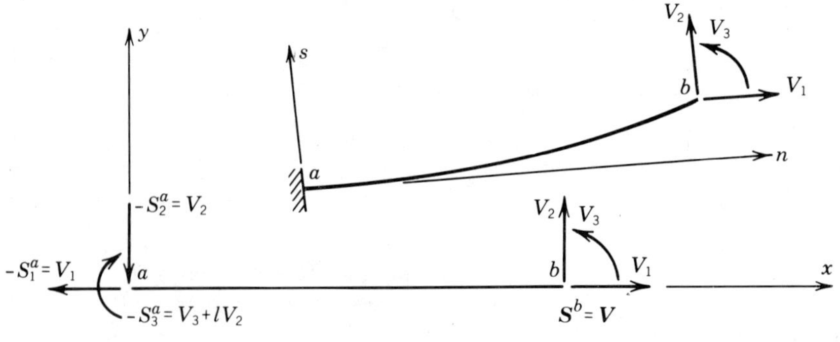

FIG. 3.2–1. Equilibrium solution $\mathbf{S} = \mathbf{BV}$, cantilever type.

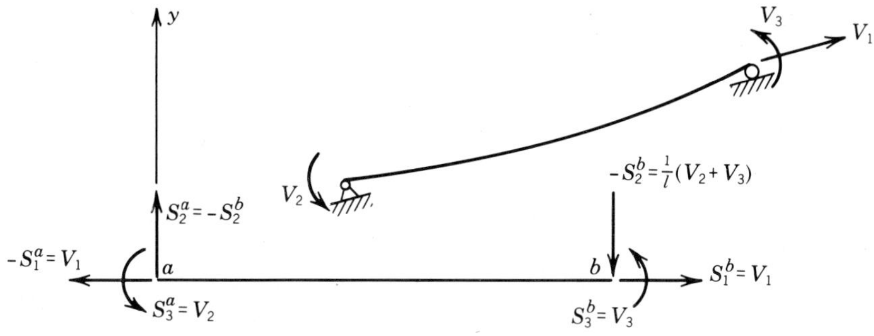

FIG. 3.2–2. Equilibrium solution $\mathbf{S} = \mathbf{BV}$, alternate type.

EXERCISES

Section 2.1

1. Do Exercises 1 and 2 of Chapter 2 within the context of the second-order theory.

2. Sketch six deformed shapes of a member corresponding in turn to one nonzero member end displacement while the remaining end displacements are zero. Outline on each sketch the deformations of the cantilever type. Repeat the exercise with deformations of the alternate type.

3. Show that if the centerline of a member is unstrained, that is, if $\varepsilon_1 = 0$, then $s_1^b - s_1^a = -\int_0^l \frac{1}{2}(v_{,x})^2\,dx$. Evaluate $s_1^b - s_1^a$ for $v = a \sin \pi x/l$, where a is a constant.

Section 2.2

4. Consider a member subjected to end forces only. Choosing statical redundants $\mathbf{V}$ of the cantilever type, complete the force system and derive the statical relations $\mathbf{S}^a = \mathbf{B}^a\mathbf{V}$ and $\mathbf{S}^b = \mathbf{B}^b\mathbf{V}$, making the approximations of the second-order theory. Ascertain that $\mathbf{B}^a$ and $\mathbf{B}^b$ are as obtained in Eq. (2.2–7)

5. Do Exercise 4 for $\mathbf{V}$ of the alternate type. Check $\mathbf{B}^a$ and $\mathbf{B}^b$ with Eq. (2.2–8).

6. Within the framework of the second-order theory (a) derive the transformation relations for the components (N_x, N_y) and (N_1, N_2) of the stress resultant vector (see Figs. 2/3.2–1 and 2/5.3–1). Check the result with Eqs. (2.2–10) and (2.2–11). (b) Derive transformation relations for (N_1, N_2) and (N_n, N_s). Check that $N_1\,\delta\varepsilon_1 + N_2\,\delta\gamma = N_n\,\delta\varepsilon_n + N_s\,\delta\varepsilon_s$. (c) Show that the nonorthogonal components (N_1', N_s') of the stress resultant vector on axes (x_1, s) are equivalent to (N_1, N_2).

7. Derive the differential equations of equilibrium (2.2–12) using a free body diagram of a beam slice. Verify that the homogeneous equations are solved by $\mathbf{N} = \mathbf{bV}$, where $\mathbf{b}$ is the coefficient matrix in Eq. (2.2–14) or Eq. (2.2–15) and $\mathbf{V}$ is arbitrary.

Section 3

8. Do Exercise 3 of Chapter 2.

9. Do Exercises 1 and 2 of Chapter 2 in the geometrically linear theory.

10. Derive by geometry the linear deformation-displacement relations (3.1–1) and (3.1–3) as suggested in Section 3.1.

11. Do Exercise 2 for the geometrically linear theory.

12. Use the statical results shown in Figs. 3.2–1 and 3.2–2 to form the static-kinematic matrix $\mathbf{B}$ of the cantilever and alternate types, respectively. Ascertain that $\mathbf{B}^T$ is the coefficient matrix in the deformation-displacement relations.

13. Derive the statical equations (3.2–3) and (3.2–4) using a free body diagram. Obtain expressions for (N_p, S_p, M_p) in the following cases:
    a. Concentrated actions $\{P_x \ \ P_y \ \ C\}$ applied at point $x = \xi$.
    b. Uniform load components $\{p_x \ \ p_y \ \ m\}$.

14. Derive the differential equations of equilibrium in the geometrically linear theory by statics of a beam slice dx. Show that the homogeneous equations are solved by $\mathbf{N} = \mathbf{bV}$, where $\mathbf{V}$ is arbitrary and $\mathbf{b}$ is the coefficient matrix in Eq. (3.2–3) or Eq. (3.2–4).

15. Establish the principles of virtual displacements and virtual forces (Chapter 2, Section 4.3) for the geometrically linear theory, starting with the strain-displacement relations (3.1–5) and the differential equations of equilibrium derived in Exercise 14.

16. In the geometrically linear theory the compatibility equation (2/4.2–3) becomes $\mathbf{v} = \int_0^l \mathbf{b}^T \boldsymbol{\varepsilon}\, dx$. Using Eq. (3.1–5) for $\boldsymbol{\varepsilon}$ and Eq. (3.2–3) or (3.2–4) for $\mathbf{b}$, show by integration that the compatibility equation yields the deformation-displacements relations.

4

CONSTITUTIVE PROPERTIES

1. INTRODUCTION: CONSTITUTIVE EQUATIONS WITHOUT SHEAR DEFORMATION

The general form of constitutive relations for elastic beams was established in Chapter 2, Section 5, in terms of strain energy and complementary strain energy density functions. A more detailed study is made in this chapter which includes shear deformation and inelastic behavior.

Elastic constitutive equations could in theory be postulated and established directly with the help of experiments conducted on beams. However, it is more instructive to connect a self-contained beam theory with three-dimensional continuum behavior. In this context the kinematic theory of Chapter 2 is considered as an adequate approximation for describing the distribution of the axial strain ε_1' over the cross section. Further a beam slice dx is considered as a bundle of axial material fibers whose axial stress σ is related to ε_1' by the uniaxial stress-strain relation of the material. Shear stresses are obtained by equilibrium equations, and shear deformation is used only to improve the accuracy of member flexibility equations. In line with this the shear strain ε_s, which is constant over a cross section, is treated as a deformation measure for the beam slice as a whole.

As a first approximation shear deformation is neglected so that the basic assumption becomes the Euler-Bernoulli assumption by which a cross section remains undeformed and perpendicular to the deformed axis. In that case normal and axial strains coincide and have the distribution

$$\varepsilon' = \varepsilon - y\chi \tag{1–1}$$

For an elastic material results of a uniaxial stress-strain test have the form

$$\sigma = f(\varepsilon') \tag{1–2}$$

where f is a single-valued function as represented in Fig. 1–1. The assumption of an elastic material implies that the stress-strain relation is independent of the manner in which a state (σ, ε') is reached. Any static loading or unloading process is represented by a point (σ, ε') remaining on the stress-strain curve.

The resultant force and resultant moment of σ over a cross section are by definition the stress resultants N and M, respectively. Thus

$$N = \int_A f(\varepsilon - y\chi)\, dA \tag{1–3a}$$

$$M = \int_A -yf(\varepsilon - y\chi)\, dA \tag{1–3b}$$

Equations (1–3) define the desired constitutive relations for the effective matrices $\mathbf{N} = \{N \;\; M\}$ and $\boldsymbol{\varepsilon} = \{\varepsilon \;\; \chi\}$. The explicit relations depend on the function f.

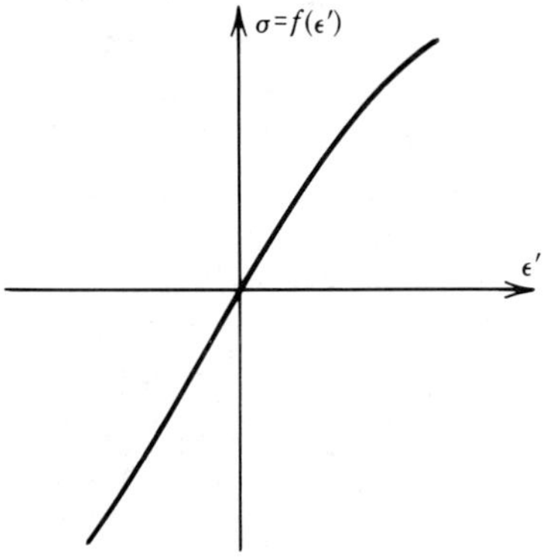

FIG. 1–1. Uniaxial stress-strain relation.

For an external load lying in the (x, y) plane the moment of σ about the y axis is zero. Thus a condition for such a loading to cause plane deformation is

$$M_y = \int_A z f(\varepsilon - y\chi)\, dA = 0 \tag{1–4}$$

Equation (1–4) holds whenever the cross section is symmetric about the y axis. In the usual case of a linear stress-strain relation, to be seen next, Eq. (1–4) requires only that the y and z axes be the principal axes of inertia. However, if the y axis is not an axis of symmetry, another condition is also required to have plane deformation and will be seen in Section 5, Example 2.

Equations (1–3) may be expressed in terms of a strain energy density function $U = U(\varepsilon, \chi)$. Letting

$$U'(\varepsilon') = \int_0^{\varepsilon'} f(\varepsilon')\, d\varepsilon' \tag{1–5}$$

and

$$U(\varepsilon, \chi) = \int_A U'(\varepsilon - y\chi)\, dA \tag{1–6}$$

we can write $\sigma = \partial U'/\partial \varepsilon', \partial \varepsilon'/\partial \varepsilon = 1, \partial \varepsilon'/\partial \chi = -y$, and, consequently

$$\frac{\partial U}{\partial \varepsilon} = \int_A \frac{\partial U'}{\partial \varepsilon'} \frac{\partial \varepsilon'}{\partial \varepsilon} = \int_A \sigma\, dA = N \tag{1–7a}$$

$$\frac{\partial U}{\partial \chi} = \int_A \frac{\partial U'}{\partial \varepsilon'} \frac{\partial \varepsilon'}{\partial \chi} = \int_A - y\sigma\, dA = M \tag{1–7b}$$

2. LINEARLY ELASTIC HOMOGENEOUS BEAMS

For a linearly elastic stress-strain relation having as Young's modulus E, we have

$$\sigma = E\varepsilon' = E\varepsilon - Ey\chi \tag{2–1}$$

Consider first a homogeneous beam slice, and let the origin of coordinates in a cross section be at the centroid. Then E is constant and $\int_A y\,dA = 0$. Equations (1–3) yield

$$N, \quad M = EA\varepsilon, \quad EI\chi \tag{2–2}$$

where $A = \int_A dA$ is the area of the cross section and $I = \int_A y^2\, dA$ is the second moment of area about the centroidal z axis. Equation (1–4) requires that $\int_A yz dA = 0$, that is, that the axes be principal axes of inertia.

In matrix notation Eqs. (2–2) are written

$$\mathbf{N} = \mathscr{E}\boldsymbol{\varepsilon} \tag{2–3}$$

where

$$\mathscr{E} = \lceil EA \;\; EI \rfloor \tag{2–4}$$

The strain energy function is as given in Eq. (2/5.3–14) without the initial stress term $\mathbf{N}_I$.

To determine σ from N and M, Eqs. (2–2) are solved for $E\varepsilon$ and $E\chi$, and the result is substituted into Eq. (2–1). This gives

$$\sigma = \frac{N}{A} - y\frac{M}{I} \tag{2–5}$$

3. LINEARLY ELASTIC NONHOMOGENEOUS BEAMS

Nonhomogeneous beams occur usually as beams of two materials such as reinforced concrete or beams formed of bounded layers of several materials. If the individual materials are linearly elastic E is piecewise constant over the cross section. In general, E is a function of position in the cross section. Applying Eqs. (1–3), the result has the form $\mathbf{N} = \mathscr{E}\boldsymbol{\varepsilon}$, where

$$\mathscr{E} = \begin{bmatrix} E_0 A^* & -E_0 Q^* \\ -E_0 Q^* & E_0 I^* \end{bmatrix} \tag{3–1}$$

In Eq. (3–1) E_0 is an arbitrary reference value for Young's modulus, and

$$A^* = \int_A dA^* = \int_A \frac{E}{E_0} dA \tag{3–2a}$$

$$Q^* = \int_A y dA^* \tag{3–2b}$$

$$I^* = \int_A y^2 dA^* \tag{3–2c}$$

Equations (3–2) are interpretable through the concept of the transformed cross section by which $dA^* = (E/E_0)dA$ is the transformed area element. A^* is then the transformed area, and Q^* and I^* are the first and second moments, respectively, of the transformed area about the z axis. The y axis is a principal axis of the transformed cross section, but the position of the origin on it may be chosen arbitrarily. By taking the origin at the centroid C^* of the transformed cross section, we can write $Q^* = 0$ and

$$\mathscr{K} = \lceil E_0 A^* \quad E_0 I^* \rfloor \tag{3–3}$$

It should be noted, however, that the locus of C^* is not necessarily a straight line and could not therefore be chosen in general as the x axis.

To determine σ from N and M, it is convenient to use C^* as reference point in the cross section. Following the same procedure as for a homogeneous material, we obtain

$$\sigma = \frac{E}{E_0}\left(\frac{N}{A^*} - y\frac{M}{I^*}\right) \tag{3–4}$$

For a given material fiber, σ is E/E_0 times the stress in a homogeneous beam slice of properties $A = A^*$ and $I = I^*$.

4. EFFECT OF INITIAL STRAINS

The uniaxial $\sigma - \varepsilon'$ relation assumed previously implies that ε' is caused only by σ. A cause of deformation other than σ is, for example, a temperature change ΔT which causes a stress-free strain

$$\varepsilon'_I = \alpha \Delta T \tag{4–1}$$

where α is the coefficient of thermal expansion of an axial fiber. A strain such as ε'_I occurring without change in stress is called initial strain. The stress σ is then assumed to cause the strain $(\varepsilon' - \varepsilon'_I)$ where ε' is the total strain as defined geometrically in terms of the displacements. For a linearly elastic material

$$\sigma = E(\varepsilon - y\chi - \varepsilon'_I) \tag{4–2}$$

Integrating σ and $-y\sigma$ over the cross section, we obtain for a homogeneous beam

$$N, \quad M = EA(\varepsilon - \varepsilon_I), \quad EI(\chi - \chi_I) \tag{4–3}$$

where

$$\varepsilon_I, \quad \chi_I = \frac{1}{A}\int_A \varepsilon'_I \, dA, \quad \frac{1}{I}\int -y\varepsilon'_I \, dA \tag{4–4}$$

In matrix form Eqs. (4–3) are written

$$\mathbf{N} = \mathscr{K}(\boldsymbol{\varepsilon} - \boldsymbol{\varepsilon}_I) = \mathscr{K}\boldsymbol{\varepsilon} + \mathbf{N}_I \tag{4–5}$$

The strain energy function is given in Eq. (2/5.3–14).

To determine σ from N and M, Eqs. (4–3) are solved for $E\varepsilon$ and $E\chi$ and are then substituted into Eq. (4–2). It is usually assumed that ε_I' is distributed linearly in y so that a stress-free deformation obeys the general theory of deformation. If that is not exactly the case, but a linear approximation to ε_I' is appropriate, then such an approximation is $\varepsilon_I' = \varepsilon_I - y\chi_I$, and ε_I and χ_I are as defined through Eqs. (4–4). The expression of σ in terms of N and M is thus found to be independent of initial strains.

For a nonhomogeneous beam a similar procedure leads to the definitions

$$\varepsilon_I, \quad \chi_I = \frac{1}{A^*}\int_A \varepsilon_I' \, dA^*, \quad \frac{1}{I^*}\int_A -y\varepsilon_I' \, dA^* \tag{4–6}$$

and to $(\mathbf{N} - \boldsymbol{\varepsilon})$ relations of the form of Eqs. (4–5), where $\mathscr{x}$ is as obtained earlier.

5. SHEAR STRESSES

The distribution of shear stresses over a cross section will be needed to derive constitutive equations for shear deformation and also to complete the study of plane deformation in the case where the (x, y) plane is not a plane of symmetry. In the derivation to follow, equilibrium equations are written in the undeformed geometry. It will be seen in Section 7 how these equations could be applied in the deformed configuration.

Consider a beam slice dx, and let it be separated into two parts by a longitudinal plane parallel to the x axis and normal to an axis η in the plane of the cross section. The part having η as outward normal is isolated, and a free body diagram of axial forces is drawn as shown in Fig. 5–1a. The longitudinal shear force qdx produces an average shear stress $\bar{\tau}_{\eta x} = q/b$. This is also equal to the

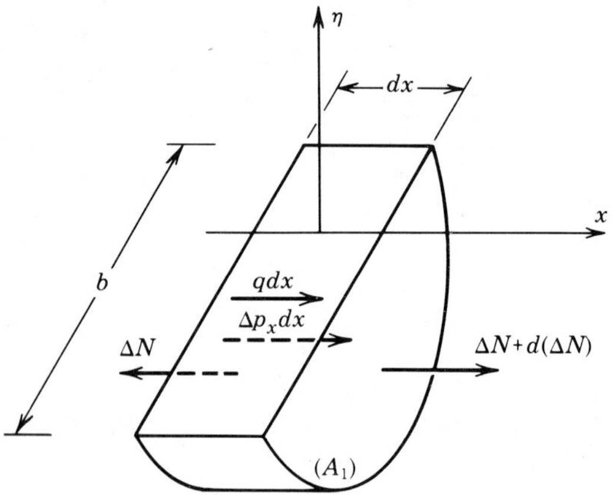

FIG. 5–1a. Free body diagram of axial forces on portion of a beam slice.

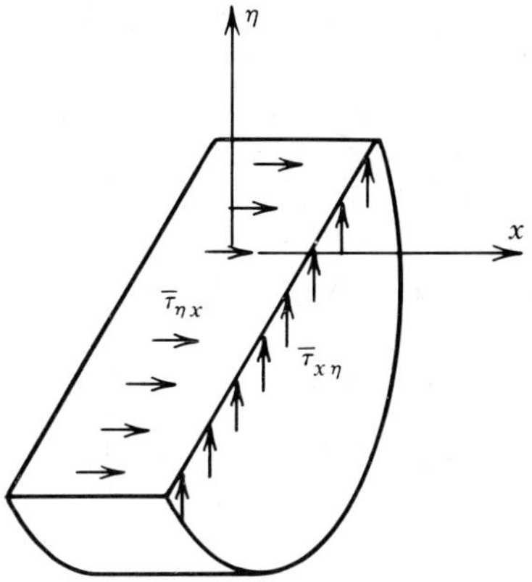

FIG. 5–1*b*. Average longitudinal and transverse shear stresses.

average over length b of the transverse shear stress $\bar{\tau}_{x\eta}$, Fig. 5–1*b*. q is sometimes called the shear flow and is determined by the axial equilibrium equation

$$q = -\frac{\mathrm{d}}{\mathrm{d}x}(\Delta N) - \Delta p_x \tag{5–1}$$

where Δp_x is the axial external load intensity acting on the isolated body. ΔN is determined as the integral $\int \sigma \, dA$ over the isolated portion A_1 of the cross section. Using Eq. (2–5) for σ, we obtain

$$\Delta N = \frac{A_1}{A} N - \frac{Q}{I} M \tag{5–2}$$

where

$$Q = \int_{A_1} y \, dA \tag{5–3}$$

Differentiation of Eq. (5–2), assuming a uniform cross section and using the differential equilibrium equations for N and M, yields

$$\frac{d(\Delta N)}{dx} = -\frac{A_1}{A} p_x + \frac{Q}{I}(N_2 + m) \tag{5–4}$$

where N_2 is the shear force, p_x is the distributed axial load, and m the distributed moment load. q may now be evaluated using Eq. (5–1). We obtain

$$\bar{\tau}_{x\eta} = \frac{q}{b} = -\frac{Q}{bI}(N_2 + m) + \frac{1}{b}\left(\frac{A_1}{A} p_x - \Delta p_x\right) \tag{5–5}$$

Usually shear stresses are due only to N_2. The other terms in Eq. (5–5) will be illustrated in Example 1. Equation (5–5) is useful for obtaining the shear stress distribution in cases where the total shear stress τ has a dominant component in a direction η and remains constant along segment b perpendicular to η. For a rectangular cross section η is the y direction. For thin-walled cross sections the shear stress has a dominant component parallel to the wall and may be considered constant along the thickness.

Example 1. Rectangular Cross Section

For a rectangular cross section the ratio Q/I for the area A_1 shown in Fig. 5–2a is

$$\frac{Q}{I}=\frac{12}{bh^3}\left[b\left(\frac{h}{2}-y\right)\frac{1}{2}\left(\frac{h}{2}+y\right)\right]=\frac{6}{h^3}\left(\frac{h^2}{4}-y^2\right)$$

The axis η that points out of the area A_1 is opposite to the y axis. Thus $\tau_{x\eta}=-\tau_{xy}\cdot\tau_{xy}$ may be considered constant along the width b in the present case. Assuming load terms other than N_2 are zero in Eq. (5–5), we obtain

$$\tau_{xy}=\frac{6N_2}{bh^3}\left(\frac{h^2}{4}-y^2\right) \qquad (5\text{–}6)$$

τ_{xy} is parabolic in y, with a maximum value at $y=0$ equal to $(3/2)(N_2/bh)$ (i.e., to 3/2 the average shear stress over the cross section). It is represented in Fig. 5–2b.

Consider now a constant axial load p_x uniformly distributed on the top face of the rectangular beam. In applying Eq. (5–5), we have $m=-p_x(h/2)$, and for the

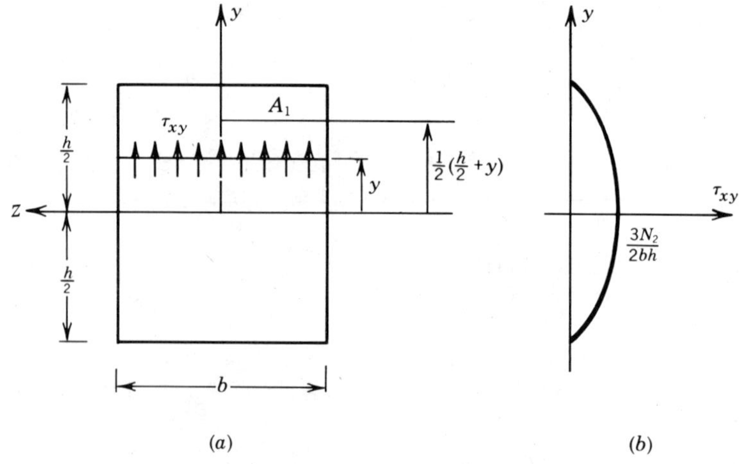

FIG. 5–2. Shear stress in rectangular cross section.

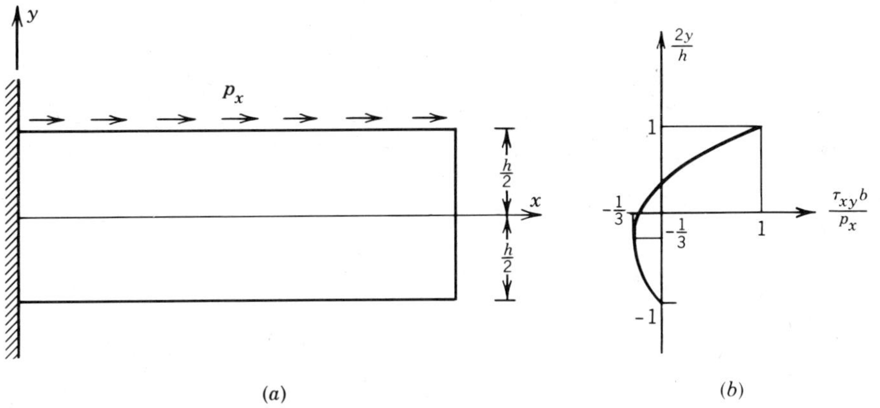

FIG. 5–3. Shear stress caused by axial surface load.

area A_1 indicated in Fig. 5–2a, $\Delta p_x = p_x$. Then, noting that $\tau_{xy} = -\tau_{x\eta}$,

$$\tau_{xy} = -\frac{q}{b} = \frac{6}{bh^3}\left(N_2 - p_x\frac{h}{2}\right)\left(\frac{h^2}{4} - y^2\right) + \frac{p_x}{bh}\left(\frac{h}{2} + y\right) \tag{5–7}$$

To consider the effect of p_x separately from that of N_2, assume the beam is fixed at $x = 0$ with a free end at $x = l$, Fig. 5–3a. Then $N_2 = 0$ and

$$\tau_{xy} = \frac{p_x}{b}\left(\frac{1}{2} + \frac{y}{h}\right)\left(\frac{3y}{h} - \frac{1}{2}\right) \tag{5–8}$$

A plot of τ_{xy} is shown in Fig. 5–3b.

Example 2. Thin-Walled Cross Sections

In a thin-walled cross section the shear stress acts predominantly in the direction parallel to the wall and may be considered constant along the thickness. It will be denoted τ instead of $\bar{\tau}_{x\eta}$. At reentrant corners there are stress concentrations that cannot be described by the present procedure. These stresses are, however, of a local character and are not within the present subject of interest.

A T cross section is shown in Fig. (5–4) with the expressions of the area moment Q and of the shear stress τ at general points in the flange and web. A diagram of the shear flow is also shown. The expressions for the left portion of the flange are valid up to the face of the web and those for the web are valid up to the lower face of the flange. The shear stress in the right portion of the flange is symmetric of the stress in the left portion with respect to the axis of symmetry y.

Other examples of shear flow for open thin-walled sections are shown in Fig. 5–5. The case of the channel cross section needs further comment because the

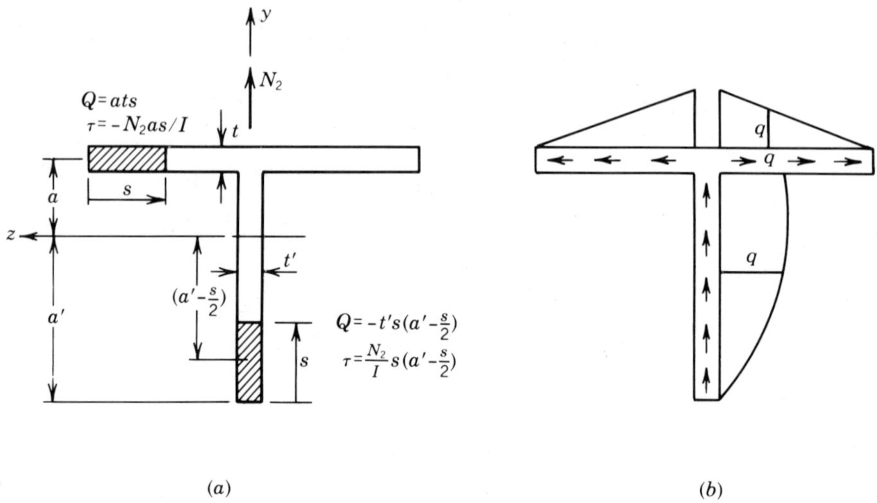

FIG. 5–4. Shear stress in T cross section.

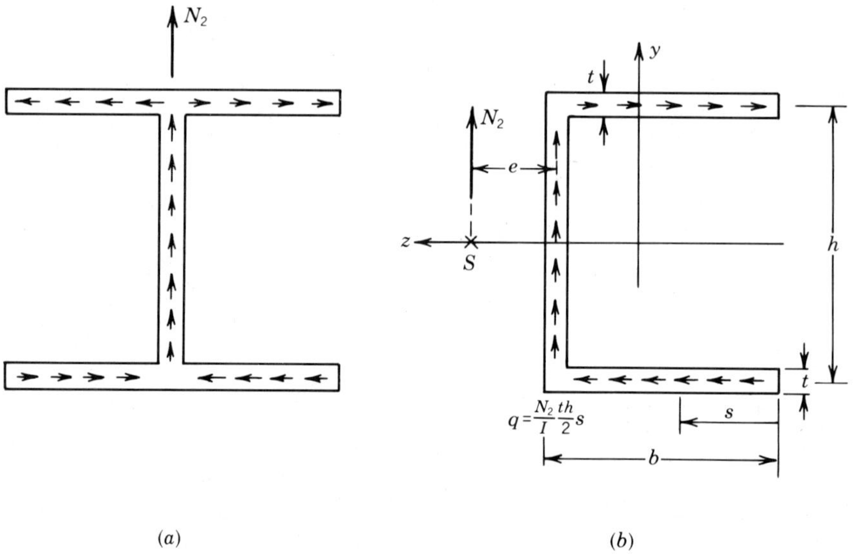

FIG. 5–5. Shear flow in I and channel cross sections.

(x, y) plane is not a plane of symmetry. The z axis is assumed to be an axis of symmetry so that the (x, y) axes are principal axes. The shear flow q as determined by means of Eq. (5–5) has a resultant force N_2 with a well-defined line of action. It is necessary that the external load yield a shear force having that same line of action if plane deformation is to occur without twisting. The resultant shear in the web may be taken equal to N_2 if the flange thickness is treated as infinitesimal.

Letting F be the resultant shear in one flange, the location of the resultant shear force is obtained by moment equivalence as

$$e = \frac{Fh}{N_2} \tag{5–9}$$

where

$$F = \int_0^b q\,ds = \frac{N_2}{I}\frac{1}{4}thb^2$$

Equation (5–9) yields

$$e = \frac{th^2b^2}{4I} \tag{5–10}$$

Point S determined by e in Fig. 5–5b is called the shear center. The case where the shear force does not pass through the shear center is discussed in Chapter 14.

For a closed thin-walled cross section the area A_1 required for evaluating the shear flow is limited by the outer and inner walls and by two lines in the direction of the thickness at two separate points, Fig. 5–6. The formula for q then evaluates the net flow out of the area A_1. If the y axis is an axis of symmetry, the shear flow at points on that axis is zero. By choosing the area A_1 to start at the axis of symmetry, the shear flow and shear stress can be evaluated at any point around the cross section. In the general case where the y axis is not an axis of symmetry, and in the case of multicelled cross sections, the condition for bending without twisting is $\oint q\,ds = 0$, where the integral is performed around the middle line of each cell and ds is the arc length along that middle line [2].

Evaluation of shear stresses and the associated question of bending without twisting are treated in more detail in texts on strength of materials and with more rigor in texts on the theory of elasticity [3]. For present purposes the shear stress distribution will be needed in deriving constitutive properties including shear deformation.

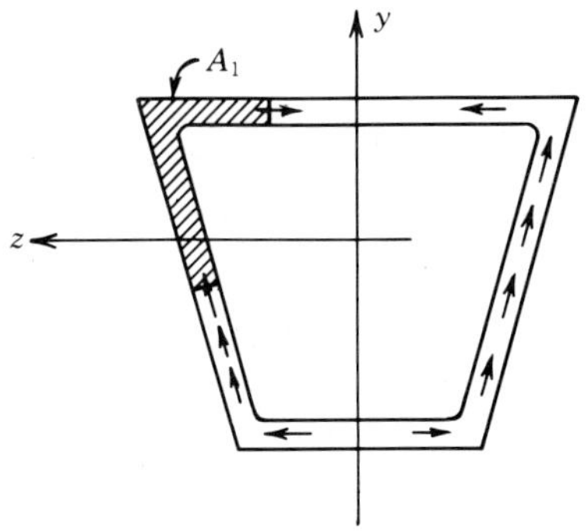

FIG. 5–6. Shear flow in closed thin-walled cross section.

6. SHEAR DEFORMATION: GEOMETRICALLY LINEAR THEORY

Constitutive equations with shear deformation are established in what follows based on a strain energy density function of the form

$$U = U_1(\varepsilon_1, \chi) + U_2(\gamma) \tag{6–1}$$

In a geometrically linear theory, the axial and normal strains coincide, and U_1 is the same as the function U determined in the case without shear deformation. U_2 is not determined directly but by means of the complementary strain energy:

$$U_2^* = \int_A U_2'^* \, dA \tag{6–2}$$

where $U_2'^*$ is the shear complementary strain energy density per unit of volume. The shear stress τ_{yz} is neglected in beam theory so that $U_2'^*$ is a function of the shear stresses τ_{xy} and τ_{xz}. The shear stress distribution over the cross section is assumed determined in terms of the shear force N_2. Equation (6–2) determines then U_2^* as a function of N_2, and the shear constitutive equation is

$$\gamma = \frac{dU_2^*}{dN_2} \tag{6–3}$$

The solution of the equation for N_2 in terms of γ has the form

$$N_2 = \frac{dU_2}{d\gamma} \tag{6–4}$$

and U_2 and U_2^* satisfy the relation (2/5.2–6),

$$U_2 + U_2^* = N_2 \gamma \tag{6–5}$$

For a linear material having a shear modulus G

$$U_2'^* = \frac{1}{2G}(\tau_{xy}^2 + \tau_{xz}^2) = \frac{\tau^2}{2G} \tag{6–6}$$

τ_{xy} and τ_{xz} are linear in N_2 and contain additional terms if there is a distributed moment load m or an axial load p_x. Such a case was seen in Section 5, Example 1. The general form of U_2^* is thus

$$U_2^* = \frac{1}{2} \frac{N_2^2}{GA_s} + \gamma_I N_2 + c \tag{6–7}$$

where γ_I and c are independent of N_2. A_s is called the shear area, and the ratio A_s/A depends only on the shape of the cross section. Applying Eq. (6–3), the constitutive shear equation is

$$\gamma = \frac{N_2}{GA_s} + \gamma_I \tag{6–8}$$

The energy function U_2 is obtained as

$$U_2 = \tfrac{1}{2}GA_s(\gamma - \gamma_I)^2 - c \tag{6–9}$$

The constant γ_I plays the role of an initial strain.

Example 1

For the rectangular cross section of Section 5, Example 1, the shear stress caused by N_2 is $\tau = \tau_{xy}$ and is given in Eq. (5–6). Integration of $\tau^2/2G$ over the cross section yields

$$U_2^* = \frac{18N_2^2}{Gbh^2}\int_{-h/2}^{h/2}\left(\frac{1}{4} - \frac{y^2}{h^2}\right)^2 dy = \frac{3}{5}\frac{N_2^2}{Gbh}$$

Identifying U_2^* with $\tfrac{1}{2}N_2^2/GA_s$, we obtain

$$A_s = \frac{5}{6}A \tag{6–10}$$

Consider now the beam of Fig. 5–3. The shear stress is given in Eq. (5–8). For this beam, $N_2 = 0$, but $\gamma = \gamma_I$. To obtain γ_I, the stress caused by both N_2 and the load p_x must be considered. This is given in Eq. (6–7). The term $\gamma_I N_2$ in Eq. (6–7) is due only to the term $p_x N_2$ in the expression of τ^2. Thus

$$\gamma_I N_2 = -\frac{18}{Gbh}N_2 p_x \int \left(\frac{1}{4} - \frac{y^2}{h^2}\right)^2 dy$$

whence

$$\gamma_I = -\frac{3}{5}\frac{p_x}{Gb}$$

For thin-walled cross sections such as shown in Figs. 5–4 and 5–5, the integral of τ^2 over the cross section yields a value of A_s that is usually close to the area of the web.

7. SHEAR DEFORMATION: GEOMETRICALLY NONLINEAR THEORY

In a geometrically nonlinear theory the choice of strain variables in the strain energy function determines the type of stress resultants that occur in the constitutive equations. The choice of ε_1 and γ leads to relations $N_1 = \partial U/\partial \varepsilon_1$ and $N_\gamma = \partial U/\partial \gamma$. It was shown in Chapter 2, Section 5.3, that N_1 and $N_2 = N_\gamma/(1+\varepsilon_1)$ are components of the stress resultant vector $\vec{N}$ on axes (x_1, x_2) as shown in Fig. 7–1. It may be shown, based on Eqs. (2/5.3–8), that if ε_1 and ε_s are chosen as strain variables, the corresponding stress resultants in the work expression $N'_1 \delta\varepsilon_1 + N'_s \delta\varepsilon_s$ are the nonorthogonal components on axes x_1 and s shown in Fig. 7–1.

We consider shear deformation within the context of small strains such that $\varepsilon_1 \ll 1$ and $\gamma^2 \ll 1$. The two representations of strain, $\boldsymbol{\varepsilon}_n = \{\varepsilon_n \ \ \varepsilon_s\}$ and $\boldsymbol{\varepsilon}_1 = \{\varepsilon_1 \ \ \gamma\}$, are then related through Eqs. (3/2.1–7), or

$$\varepsilon_1 = \varepsilon_n + \tfrac{1}{2}\varepsilon_s^2 \tag{7–1a}$$

$$\gamma = \varepsilon_s \tag{7–1b}$$

At an arbitrary point of the cross section, $\varepsilon'_n = \varepsilon_n - y\chi$, $\varepsilon'_s = \varepsilon_s$, and ε'_1 is related to ε'_n by Eq. (7–1a). Thus

$$\varepsilon'_1 = \varepsilon_1 - y\chi \tag{7–2}$$

The strain energy density associated with the axial strain ε'_1 is obtained from the uniaxial stress-strain relation as in the case without shear deformation. Thus $U_1(\varepsilon_1, \chi)$ is the same as the strain energy function in the case without shear deformation.

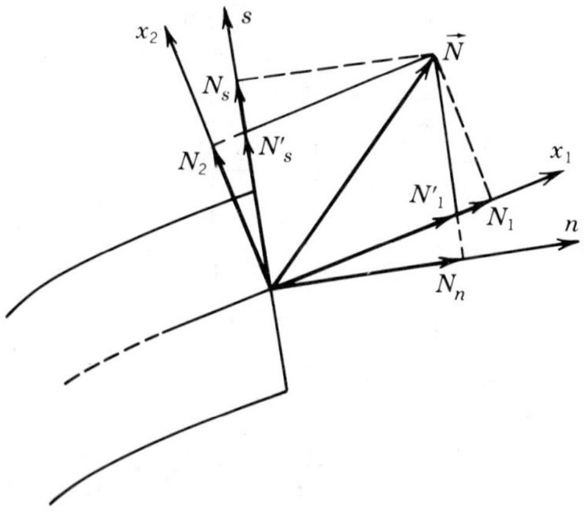

FIG. 7–1. Various choices of stress resultants.

Equality of the work expressions $N_1\delta\varepsilon_1 + N_\gamma\delta\gamma = N'_1\delta\varepsilon_1 + N'_s\delta\varepsilon_s = N_n\delta\varepsilon_n + N_s\delta\varepsilon_s$ yields

$$N_1 = N'_1 = N_n \tag{7–3a}$$

$$N_\gamma = N'_s = N_s - \gamma N_n \tag{7–3b}$$

Thus the relation $N_1 = \partial U/\partial\varepsilon_1$ may be considered to apply equivalently to N_n and N'_1 as defined in Fig. 7–1. For shear we have equivalence between N'_s, N_γ and $N_2 = N_\gamma/(1+\varepsilon_1)$, but not with N_s.

The shear constitutive equation is established by the same procedure as in the geometrically linear case. The equilibrium equations used in Section 5 to determine the shear stress distribution may be applied in the present case by resolving forces into nonorthogonal components in the x_1 and s directions. The shear force component is thus N'_s and is equivalent to N_2 and N_γ. We will adopt the notation N_2.

8. INELASTIC CONSTITUTIVE RELATIONS

Constitutive properties in inelastic behavior are based as in elastic behavior on the uniaxial stress-strain relation. We will consider materials such as structural steel whose uniaxial behavior may be idealized as being either linearly elastic or perfectly plastic as represented in Fig. 8–1. The idealized material is characterized by a yield stress σ_y and yield strain ε_y which define the linearly elastic range of behavior. It will be assumed that yield in compression occurs at the same absolute values of stress and strain as in tension.

If σ is increased, starting from a state of zero stress and zero strain, the loading path is represented by line OAB in Fig. 8–1. Plastic flow along AB is

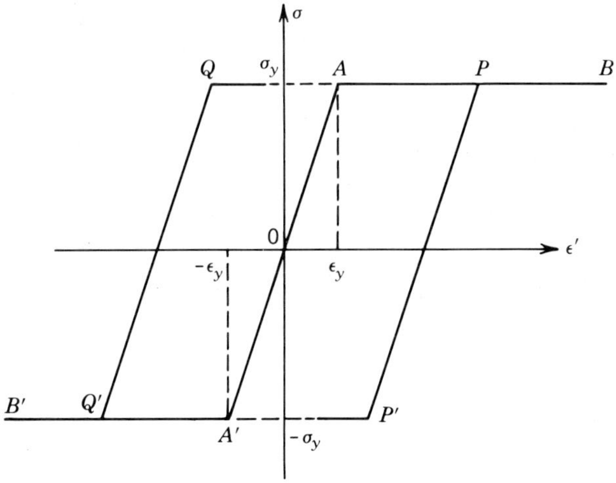

FIG. 8–1. Linearly elastic–perfectly plastic stress-strain relation.

characterized by an increasing strain $\varepsilon' > \varepsilon_y$ at the constant stress σ_y. Loading reversal from a point such as P is assumed to take place along line PP' parallel to AA' and continues on line $P'A'$ at the constant stress $-\sigma_y$. Loading reversal from a point such as Q' takes place along line $Q'Q$ and continues on line QAB. Loading paths such as PP' and QQ' are reversible. Thus as long as plastic flow does not take place, the material is incrementally linear and is characterized by the modulus of elasticity $E = \sigma_y/\varepsilon_y$.

An arbitrarily given state (σ, ε') with $|\sigma| \leqslant \sigma_y$ may be reached by different deformation paths. Thus a given strain does not determine a corresponding stress unless a definite deformation path is assumed.

Constitutive properties for the case of no load reversal are considered in what follows and are based on the Euler-Bernoulli deformation theory according to which the strain distribution over the cross section has the form

$$\varepsilon' = \varepsilon - y\chi \tag{8–1}$$

The shear stresses acting on the cross section are assumed to be small enough so that their effect on the axial yield stress may be neglected.

The defining formulas for the axial force N and bending moment M are

$$N = \int_A \sigma dA \tag{8–2a}$$

$$M = \int_A -y\sigma dA \tag{8–2b}$$

For given ε and χ, ε' varies within a range $(\varepsilon_1,\ \varepsilon_2)$ which occupies a certain segment on the ε' axis in Fig. 8–1. Since it is assumed that no prior load reversal has taken place, the segment on line $B'A'AB$ corresponding to the interval $(\varepsilon_1,\ \varepsilon_2)$ represents the stress σ to be used in Eqs. (8–2). Consider for simplicity a rectangular cross section of width b and depth h, Fig. 8–2*a*. To evaluate N and M through Eqs. (8–2), we let $dA = bdy$ and make a change of integration variable from y to ε' defined by Eq. (8–1). Equations (8–2) becomes

$$N = \frac{b}{\chi}\int_{\varepsilon_2}^{\varepsilon_1} \sigma d\varepsilon' \tag{8–3a}$$

$$M = \frac{b}{\chi^2}\int_{\varepsilon_2}^{\varepsilon_1} \sigma(\varepsilon' - \varepsilon) d\varepsilon' \tag{8–3b}$$

where

$$\varepsilon_1 = \varepsilon + \frac{h}{2}\chi \tag{8–4a}$$

$$\varepsilon_2 = \varepsilon - \frac{h}{2}\chi \tag{8–4b}$$

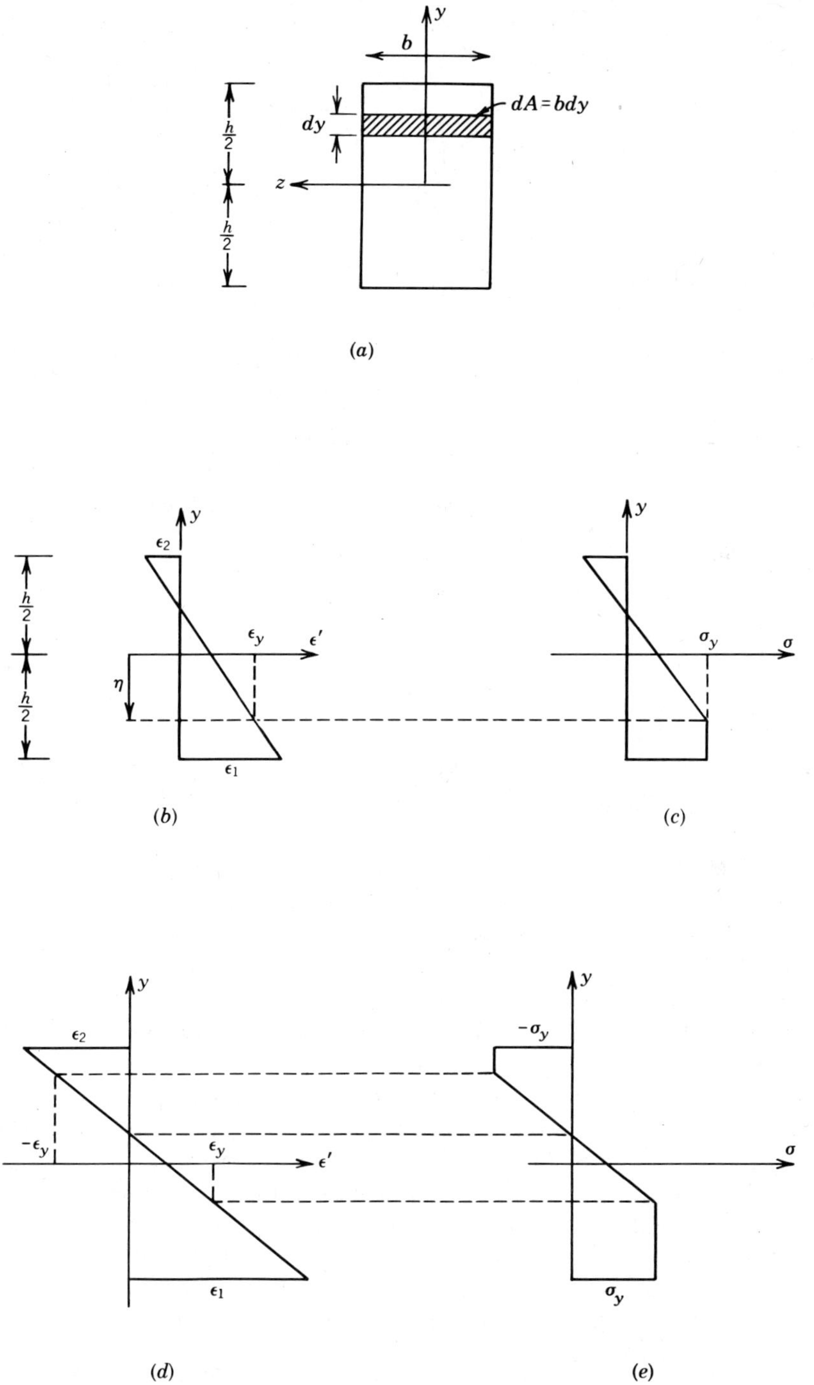

FIG. 8–2. Strain and stress diagrams in single yield (*b*), (*c*), and combined yield (*d*), (*e*).

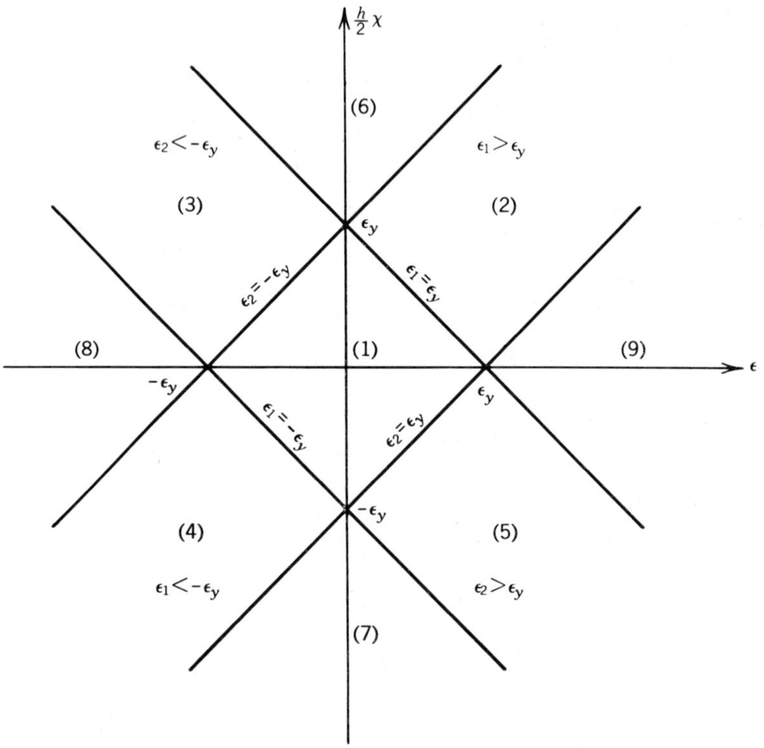

FIG. 8–3. Representation of the deformed state of a beam slice.

The various cases to be considered depend on whether $|\varepsilon_1|$ or $|\varepsilon_2|$ exceed ε_y. A graphical representation is shown in Fig. 8–3 in which a state of strain is represented by a point of coordinates $(\varepsilon, \frac{h}{2}\chi)$. There are nine regions delimited by four straight lines whose equations are $\varepsilon_1 = \pm\varepsilon_y$ and $\varepsilon_2 = \pm\varepsilon_y$. In region 1, $-\varepsilon_y \leqslant \varepsilon' \leqslant \varepsilon_y$, and the behavior is linearly elastic. In Region 2, $\varepsilon_1 > \varepsilon_y$ and $-\varepsilon_y < \varepsilon_2 < \varepsilon_y$. There is a value η of y at which $\varepsilon' = \varepsilon_y$ as shown in Fig. 8–2*b*. The behavior is linearly elastic for $y > \eta$ and is purely plastic for $y < \eta$, Fig. 8–2*c*. This case is referred to as bottom tension yield. Regions 3, 4, and 5 are similar to Region 2 in that there is single yield, that is, yield only on one side of the cross section. They correspond, respectively, to top compression yield, bottom compression yield, and top tension yield. Regions 6 and 7 are regions of combined yield. In Region 6 there is top compression yield and bottom tension yield, and in Region 7, top tension yield and bottom compression yield. In these last two cases there is an inner interval along the y axis along which ε' varies from $-\varepsilon_y$ to ε_y, and σ varies from $-\sigma_y$ to σ_y, Fig. 8–2*d* and *e*. Beyond the ends of this interval $|\varepsilon'| > \varepsilon_y$, and $\sigma = -\sigma_y$ on one side and $\sigma = \sigma_y$ on the other.

Points on the χ axis of Fig. 8–3 represent states of pure bending in which $\varepsilon = 0$ and $N = 0$.

In Region 8, $\varepsilon_1 < -\varepsilon_y$ and $\varepsilon_2 < -\varepsilon_y$, and the cross section as a whole is in compression in the plastic range. Similarly Region 9 corresponds to tension in the plastic range.

In the following, constitutive properties for single yield and combined yield will be derived corresponding to Regions 2 and 6, respectively, in Fig. 8–3. The results for Region 2 are readily applicable to the other cases of single yield by an appropriate change of the yield side or a change of tension into compression, as the case may be. Similarly results for Region 6 are applicable to Region 7. In Region 8, $N = -N_y$ and $M = 0$, and in Region 9, $N = N_y$ and $M = 0$, where N_y is the yield axial force,

$$N_y = bhE\varepsilon_y = bh\sigma_y \tag{8–5}$$

For the case of bottom tension yield, Region 2 in Fig. 8–3, Eqs. (8–3) take the form

$$N = \frac{Eb}{\chi}\int_{\varepsilon_2}^{\varepsilon_y} \varepsilon'\,d\varepsilon' + \frac{Eb\varepsilon_y}{\chi}\int_{\varepsilon_y}^{\varepsilon_2 + h\chi} d\varepsilon' \tag{8–6a}$$

$$M = \frac{Eb}{\chi^2}\int_{\varepsilon_2}^{\varepsilon_y} \varepsilon'(\varepsilon' - \varepsilon)\,d\varepsilon' + \frac{Eb\varepsilon_y}{\chi^2}\int_{\varepsilon_y}^{\varepsilon_2 + h\chi} (\varepsilon' - \varepsilon)\,d\varepsilon' \tag{8–6b}$$

The integrals in Eqs. (8–6) may be readily evaluated. It is convenient, however, to take as integration variable $e = \varepsilon' - \varepsilon_2$ and to rewrite Eqs. (8–6) so that the result is expressed in terms of $(\varepsilon_y - \varepsilon_2)$ and $h\chi$. We thus obtain

$$N = \frac{Eb}{\chi}\int_0^{\varepsilon_y - \varepsilon_2} [e - (\varepsilon_y - \varepsilon_2)]\,de + \frac{Eb}{\chi}\varepsilon_y \int_0^{h\chi} de \tag{8–7a}$$

$$M = \frac{Eb}{\chi^2}\int_0^{\varepsilon_y - \varepsilon_2} \left[e^2 - \left(\varepsilon_y - \varepsilon_2 + \frac{h\chi}{2}\right)e + \frac{h\chi}{2}(\varepsilon_y - \varepsilon_2)\right]de + \frac{Eb}{\chi^2}\varepsilon_y \int_0^{h\chi}\left(e - \frac{h\chi}{2}\right)de \tag{8–7b}$$

The second integral in the expression of M is zero. Performing the remaining integrations, there comes

$$N = -\frac{Eb}{2\chi}(\varepsilon_y - \varepsilon_2)^2 + Ebh\varepsilon_y \tag{8–8a}$$

$$M = \frac{Eb}{\chi^2}\left[-\frac{1}{6}(\varepsilon_y - \varepsilon_2)^3 + \frac{h\chi}{4}(\varepsilon_y - \varepsilon_2)^2\right] \tag{8–8b}$$

Equations (8–8) may be expressed in nondimensional form in terms of N/N_y, M/M_y, $\varepsilon/\varepsilon_y$, and χ/χ_y where χ_y and M_y are the values of χ and M, respectively, at first yield in pure bending; that is,

$$\chi_y = \frac{2}{h}\varepsilon_y \tag{8–9a}$$

$$M_y = \frac{Ebh^3}{12}\chi_y = \frac{bh^2}{6}\sigma_y \tag{8–9b}$$

There comes after replacing ε_2 with $\varepsilon - \dfrac{h\chi}{2}$ in Eqs. (8–8).

$$\frac{N}{N_y} = 1 - \frac{1}{4}\frac{\chi_y}{\chi}\left(1 - \frac{\varepsilon}{\varepsilon_y} + \frac{\chi}{\chi_y}\right)^2 \tag{8–10a}$$

$$\frac{M}{M_y} = -\frac{1}{4}\left(\frac{\chi_y}{\chi}\right)^2\left(1 - \frac{\varepsilon}{\varepsilon_y} + \frac{\chi}{\chi_y}\right)^3 + \frac{3}{4}\frac{\chi_y}{\chi}\left(1 - \frac{\varepsilon}{\varepsilon_y} + \frac{\chi}{\chi_y}\right)^2 \tag{8–10b}$$

Equations (8–10) are the constitutive equations in the case of bottom tension yield. They give N/N_y and M/M_y as functions of $\varepsilon/\varepsilon_y$ and χ/χ_y. They are applicable to the other cases of single yield as discussed earlier.

We now consider combined yield with top compression and bottom tension. Equations (8–3) take the form

$$N = \frac{Eb}{\chi}\left[-\varepsilon_y\int_{\varepsilon_2}^{-\varepsilon_y} d\varepsilon' + \int_{-\varepsilon_y}^{\varepsilon_y} \varepsilon'\, d\varepsilon' + \varepsilon_y\int_{\varepsilon_y}^{\varepsilon_2 + h\chi} d\varepsilon'\right] \tag{8–11a}$$

$$M = \frac{Eb}{\chi^2}\left[-\varepsilon_y\int_{\varepsilon_2}^{-\varepsilon_y} (\varepsilon' - \varepsilon)d\varepsilon' + \int_{-\varepsilon_y}^{\varepsilon_y} \varepsilon'(\varepsilon' - \varepsilon)d\varepsilon' + \varepsilon_y\int_{\varepsilon_y}^{\varepsilon_2 + h\chi} (\varepsilon' - \varepsilon)d\varepsilon'\right] \tag{8–11b}$$

The results of the integrations may be put in the form

$$\frac{N}{N_y} = \frac{\varepsilon}{\varepsilon_y}\frac{\chi_y}{\chi} \tag{8–12a}$$

$$\frac{M}{M_y} = \frac{3}{2}\left[1 - \left(\frac{1}{3} + \frac{\varepsilon^2}{\varepsilon_y^2}\right)\frac{\chi_y^2}{\chi^2}\right] \tag{8–12b}$$

Equations (8–12) are also applicable to the case of combined yield with top tension and bottom compression.

It is also of interest to derive the moment-curvature relation for a given axial force. It is possible to eliminate ε form Eqs. (8–10) by eliminating $(1 - \varepsilon/\varepsilon_y + \chi/\chi_y)$. Similarly it is possible to eliminate $\varepsilon/\varepsilon_y$ from Eqs. (8–12). We thus obtain for the case of single yield

$$\frac{M}{M_y} = -2\left(1 - \frac{N}{N_y}\right)^{3/2}\sqrt{\frac{\chi_y}{\chi}} + 3\left(1 - \frac{N}{N_y}\right) \tag{8–13}$$

and for combined yield

$$\frac{M}{M_y}=\frac{3}{2}\left(1-\frac{N^2}{N_y^2}\right)-\frac{1}{2}\frac{\chi_y^2}{\chi^2} \tag{8–14}$$

Equation (8–13) should yield the same value for M/M_y as in the linearly elastic case at the boundary between Region 1 and 2 in Fig. 8–3, that is, when $\varepsilon_1=\varepsilon_y$. Letting subscript o refer to the case $\varepsilon_1=\varepsilon_y$, we obtain from Eq. (8–4a)

$$\frac{\varepsilon_0}{\varepsilon_y}+\frac{\chi_0}{\chi_y}=1 \tag{8–15a}$$

then from Eqs. (8–10)

$$\frac{\chi_o}{\chi_y}=1-\frac{\varepsilon_o}{\varepsilon_y}=1-\frac{N}{N_y} \tag{8–15b}$$

$$\frac{M_o}{M_y}=\frac{\chi_o}{\chi_y}=1-\frac{N}{N_y} \tag{8–15c}$$

Similarly Eqs. (8–13) and (8–14) should yield the same value for M/M_y at the boundary between Regions 2 and 6 in Fig. 8–3, that is, when $\varepsilon_2=-\varepsilon_y$. Letting a prime and subscript o refer to this case, we obtain from Eq. (8–4b)

$$\frac{\varepsilon_o'}{\varepsilon_y}-\frac{\chi_o'}{\chi_y}+1=0 \tag{8–16a}$$

then from Eqs. (8–10) or (8–12)

$$\frac{\chi_y}{\chi_o'}=1-\frac{N}{N_y} \tag{8–16b}$$

$$\frac{M_o'}{M_y}=\frac{\chi_y}{\chi_o'}\left(3-2\frac{\chi_y}{\chi_o'}\right)=\left(1-\frac{N}{N_y}\right)\left(1+\frac{2N}{N_y}\right) \tag{8–16c}$$

Fig. 8–4 shows a typical plot of M/M_y versus χ/χ_y for constant N/N_y. The curve is made of three parts corresponding to linearly elastic behavior, single yield, and combined yield, respectively. It may be verified that the three parts connect smoothly together, that is, with the same tangent at their two common points A and B. These two boundary points occur at $\chi_0/\chi_y=1-(N/N_y)$ and $\chi_0'/\chi_y=[1-(N/N_y)]^{-1}$ with corresponding values of M/M_y as defined in Eqs. (8–15c) and (8–16c). For $N=0$ the middle part of the curve vanishes. From Eq. (8–14) it is seen that the maximum bending moment capacity of the cross section occurs as

$\chi \to \infty$ and for $N = 0$, and is equal to M_p such that

$$\frac{M_P}{M_y} = \frac{3}{2} \tag{8-17}$$

M_P is called the plastic or ultimate bending moment. The effect of the axial force is to reduce M_P to M'_P such that

$$\frac{M'_P}{M_y} = \frac{3}{2}\left(1 - \frac{N^2}{N_y^2}\right) \tag{8-18}$$

For nonrectangular cross sections the derivation of constitutive relations in the inelastic range is similar to the preceding one, although it is algebraically more complicated due to the variable width of the cross section. An approximate procedure consists in representing M/M_y by expressions similar to Eqs. (8–13) and (8–14) in their dependence on χ/χ_y but with coefficients, functions of N/N_y, that depend on the particular shape of the cross section. To do this, Eqs. (8–13) and (8–14) are expressed in terms of the coordinates of points A and B and of M'_P/M_y. These quantities are then determined for the particular cross section [4].

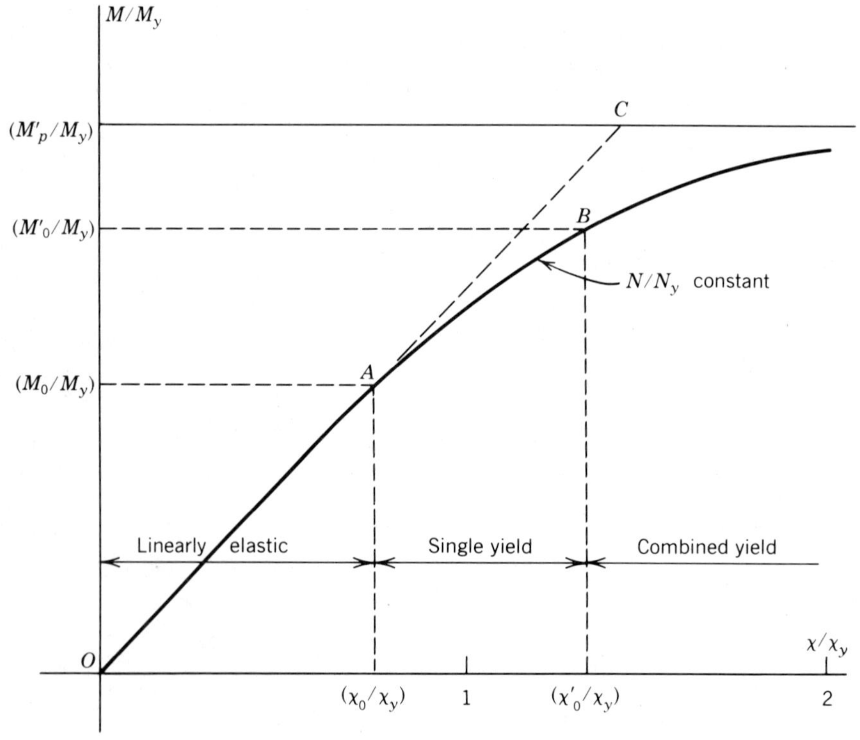

FIG. 8–4. Moment-curvature relation at constant axial force.

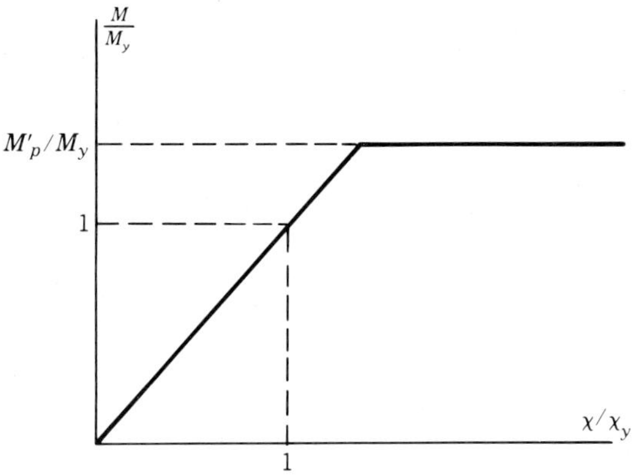

FIG. 8–5. Idealized moment-curvature relation.

In structural analysis a constitutive relation such as represented in Fig. 8–4 is sometimes idealized into a linearly elastic–perfectly plastic relation represented by line OC and the line of constant moment. For a general cross section the idealization is shown in Fig. 8–5 in which M'_P/M_y depends on the shape of the cross section and on N/N_y. In general, M'_P/M_y decreases if the area of the cross section away from the centroid increases relative to the remaining area. For I shapes and $N = 0$, typical values of M_P/M_y vary between 1.10 and 1.30 so that the transition from M_y to M_P occurs over a shorter moment interval than for a rectangular cross section.

To the linearly elastic–perfectly plastic idealization of Fig. 8–5 corresponds a physical model of behavior known as the plastic hinge. Behavior is linearly elastic until M reaches M'_P, after which the curvature may increase indefinitely at constant moment. This local kinematic property is representable as a pin connection allowing independent rotations of the cross sections on either side but transmitting an internal moment M'_P.

9. INCREMENTAL CONSTITUTIVE RELATIONS

Incremental constitutive relations are relations between differential increments $d\mathbf{N}$ and $d\boldsymbol{\varepsilon}$ taking place from a state $(\mathbf{N}, \boldsymbol{\varepsilon})$. For general material behavior, whether elastic or inelastic, incremental constitutive relations may be established from an incremental uniaxial stress-strain relation of the form

$$d\sigma = E_t(d\varepsilon' - d\varepsilon'_I) \tag{9–1}$$

where E_t is called the tangent modulus of elasticity. For a nonlinearly elastic material E_t is a function of σ or ε', whereas for inelastic behavior E_t depends also on the incremental path $d\sigma$ or $d\varepsilon'$. For example, for the idealized material of Fig. 8–1, $E_t = E$ if $|\sigma| < \sigma_y$, or if $|\sigma| = \sigma_y$, and loading reversal occurs. Otherwise

$E_t = 0$. Incrementing Eq. (1–1), we can write

$$d\varepsilon' = d\varepsilon - y\,d\chi \tag{9–2}$$

The distribution of E_t over a cross section depends on the stress distribution and, in inelastic behavior, on the path $(d\varepsilon, d\chi)$. Evaluation of the resultants of $d\sigma$ over the cross section yields the relations

$$\begin{Bmatrix} dN \\ dM \end{Bmatrix} = \begin{bmatrix} \int E_t\,dA & \int -yE_t\,dA \\ \int -yE_t\,dA & \int y^2 E_t\,dA \end{bmatrix} \begin{Bmatrix} d\varepsilon \\ d\chi \end{Bmatrix} + \begin{Bmatrix} \int -E_t\,d\varepsilon_I'\,dA \\ \int yE_t\,d\varepsilon_I'\,dA \end{Bmatrix} \tag{9–3}$$

It may be noted that the coefficient matrix of $\{d\varepsilon \;\; d\chi\}$ in the preceding relations is symmetric. In matrix form we write

$$d\mathbf{N} = \mathcal{K}_t\,d\boldsymbol{\varepsilon} + d\mathbf{N}_I \tag{9–4}$$

$\mathcal{K}_t$ is a tangent rigidity matrix. It coincides with $\mathcal{K}$ in the case of a linearly elastic material.

In the case of a general elastic material incremental constitutive relations may be expressed in terms of the strain energy density function U or the complementary strain energy density function U^*. Differentiation of Eq. (2/5.3–1) allows us to write

$$\mathcal{K}_t = \frac{\partial^2 U}{\partial \boldsymbol{\varepsilon}^T \partial \boldsymbol{\varepsilon}} \tag{9–5}$$

The inverted form of Eqs. (9–4) is

$$d\boldsymbol{\varepsilon} = \boldsymbol{\Phi}_t\,d\mathbf{N} + d\boldsymbol{\varepsilon}_I \tag{9–6}$$

where

$$\boldsymbol{\Phi}_t = \mathcal{K}_t^{-1} \tag{9–7a}$$

and $d\boldsymbol{\varepsilon}_I$ and $d\mathbf{N}_I$ are related through

$$d\mathbf{N}_I = -\mathcal{K}_t\,d\boldsymbol{\varepsilon}_I \tag{9–7b}$$

Differentiation of Eq. (2/5.3–3) allows us to write

$$\boldsymbol{\Phi}_t = \frac{\partial^2 U^*}{\partial \mathbf{N}^T \partial \mathbf{N}} \tag{9–8}$$

In the case where the deformation path in inelastic behavior is known and is described by analytic constitutive relations, incremental relations may be derived by differentiation of the finite relations. Examples of such relations are Eqs. (8–10) and (8–12) whose differentiation with respect to ε and χ presents no difficulty. It is noted, however, that the nondimensionalization used in these equations does not preserve the symmetry of the incremental rigidity matrix. An incremental bending constitutive relation at constant axial force is of particular importance in describing column behavior under increasing transverse load. It is obtained by differentiation of equations such as Eqs. (8–13) and (8–14). The resulting tangent rigidity may be expressed in terms of either M or χ. From Eq. (8–13), which applies for single yield, and a rectangular cross section, we obtain

$$\mathscr{X}_t = \frac{\partial M}{\partial \chi} = EI\left(\frac{1 - N/N_y}{\chi/\chi_y}\right)^{3/2} = EI\left[\frac{3}{2} - \frac{M/M_y}{2(1 - N/N_y)}\right]^3 \qquad (9\text{–}9a)$$

For combined yield we obtain from Eq. (8–14)

$$\mathscr{X}_t = \frac{\partial M}{\partial \chi} = \frac{EI}{(\chi/\chi_y)^3} = EI\left[3\left(1 - \frac{N^2}{N_y^2}\right) - \frac{2M}{M_y}\right]^{3/2} \qquad (9\text{–}9b)$$

The limits of single yield and combined yield are defined in Fig. 8–4 and Eqs. (8–15) and (8–16).

10. MATERIAL STABILITY

Consider a beam slice dx in a deformed equilibrium state with one face fixed and the other face acted upon by given stress resultants $\mathbf{N}$. Let the beam slice be deformed into a neighboring equilibrium state $\boldsymbol{\varepsilon} + \Delta\boldsymbol{\varepsilon}$, and let $\Delta\mathbf{N}$ be the final value of the additional stress resultants. The beam slice is said to be in a stable equilibrium if the work of the additional stress resultants along the equilibrium path from state $\boldsymbol{\varepsilon}$ to state $\boldsymbol{\varepsilon} + \Delta\boldsymbol{\varepsilon}$ is positive for any $\Delta\boldsymbol{\varepsilon}$ in some small neighborhood of $\boldsymbol{\varepsilon}$. For an infinitesimal $\Delta\boldsymbol{\varepsilon} = d\boldsymbol{\varepsilon}$ the work of the additional forces is an infinitesimal whose principal part is $\frac{1}{2}d\mathbf{N}^T d\boldsymbol{\varepsilon}$. The local stability criterion is thus

$$d\mathbf{N}^T d\boldsymbol{\varepsilon} > 0 \qquad (10\text{–}1)$$

If there is some $d\boldsymbol{\varepsilon}$ for which

$$d\mathbf{N}^T d\boldsymbol{\varepsilon} = 0 \qquad (10\text{–}2)$$

the equilibrium state is said to be neutral, and if some $d\boldsymbol{\varepsilon}$ exists for which

$$d\mathbf{N}^T d\boldsymbol{\varepsilon} < 0 \qquad (10\text{–}3)$$

the equilibrium state is said to be unstable. It is also proper to characterize the path $d\varepsilon$ as stable, neutral, or unstable, as the case may be.

To illustrate the stability criteria consider a beam slice dx subjected to an axial force N, and let the axial constitutive relation in the absence of initial strain be as shown in Fig. 10–1. At a point on the rising part of the curve, dN and $d\varepsilon$ have the same sign so that $dN d\varepsilon > 0$, and the equilibrium is stable. At a point on the falling part of the curve $dN d\varepsilon < 0$, and the equilibrium is unstable. If an experiment is conducted to establish the curve of Fig. 10–1, it is necessary to restrain the specimen in the unstable portion by controlling the displacements. At the point of the horizontal tangent in Fig. 10–1, $dN d\varepsilon = 0$, and the equilibrium state is locally neutral. At that point a state $\varepsilon + d\varepsilon$ may be considered to be an alternate equilibrium state to within an infinitesimal of at least second order in the axial force at state ε.

In the absence of initial strains the stability criterion characterizes the incremental rigidity matrix $\mathscr{K}_t$ and its inverse $\mathbf{\Phi}_t$. We can write, for $\mathscr{K}_t = \mathscr{K}_t^T$ and $d\boldsymbol{\varepsilon}_I = 0$,

$$d\mathbf{N}^T d\boldsymbol{\varepsilon} = d\boldsymbol{\varepsilon}^T \mathscr{K}_t d\boldsymbol{\varepsilon} \tag{10–4}$$

A stable equilibrium is characterized by a positive quadratic form $d\boldsymbol{\varepsilon}^T \mathscr{K}_t d\boldsymbol{\varepsilon}$ for arbitrary $d\boldsymbol{\varepsilon} \neq 0$, that is, by a positive definite $\mathscr{K}_t$. The same stability criterion holds for $\mathbf{\Phi}_t$, since $\mathbf{\Phi}_t = \mathscr{K}_t^{-1}$ and

$$d\mathbf{N}^T d\boldsymbol{\varepsilon} = d\mathbf{N}^T \mathbf{\Phi}_t d\mathbf{N} \tag{10–5}$$

If at some point of a deformation path $\mathscr{K}_t$ becomes singular, then there is some $d\boldsymbol{\varepsilon} \neq 0$ for which $d\mathbf{N} = \mathscr{K}_t d\boldsymbol{\varepsilon} = 0$. Such a point represents a neutral equilibrium for the path $d\boldsymbol{\varepsilon}$. Beyond this point equilibrium may remain neutral, become unstable, or revert to being stable again. In the case of Fig. 10–1 equilibrium becomes

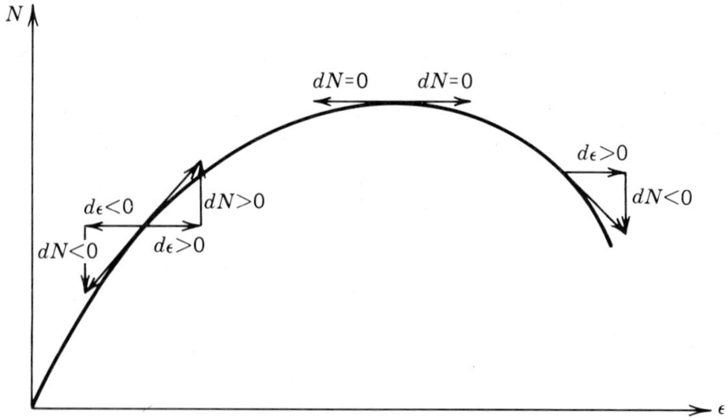

FIG. 10–1. Material stability.

unstable beyond the point of neutral equilibrium. In a case such as represented in Fig. 8–1, deformation along line AB takes place in a neutral equilibrium state, whereas unloading from point P takes place along line PP' and is formed of stable equilibrium states. In a continuous passage from stable to unstable equilibrium the quadratic form of Eq. (10–4) changes from being positive to negative by going through a neutral equilibrium position in which $\mathscr{K}_t$ is singular.

Strain increment $d\varepsilon = d\varepsilon_I$ which is purely an initial strain occurs with $d\mathbf{N} = 0$ and thus along a neutral equilibrium path.

EXERCISES

Sections 1 to 6

1. Consider constitutive equations in which the reference point for ε and M is arbitrary. The rigidity matrix is then as obtained in Eq. (3–1).

a. Show that the neutral axis, that is, the axis in the cross section on which $\varepsilon' = 0$, is located at

$$y_0 = \frac{IN + QM}{QN + AM}$$

b. Show that if $N = 0$, y_0 is the coordinate of the centroid.

c. Show that if $\chi = 0$, the bending moment at the centroidal axis is zero.

2. Determine the rigidity matrix $\mathscr{K}$ for a rectangular cross section made of two homogeneous materials occupying each half the depth and having as Young's moduli E_0 and E_1. Choose the reference point for ε and M at mid-depth.

3. Determine the rigidity matrix $\mathscr{K}$ for a cross section having two axes of symmetry and a material having the stress-strain law

$$\sigma = E\left[\varepsilon' - \frac{1}{2}\frac{(\varepsilon')^2}{\varepsilon_0}\right]$$

where E and ε_0 are constant.

4. Determine the strain energy density function U for the material of Exercise 3.

5. Consider a temperature change that varies linearly with y and is equal to T_t at the top and T_b at the bottom of the cross section. Determine the initial strains ε_I and χ_I, assuming a constant coefficient of thermal expansion α.

6. Determine the shear area A_s for a thin-walled I cross section, and compare the result to the area of the web.

7. Consider the nonlinear stress-strain relation, known as the Ramber-Osgood relation:

$$\varepsilon' = \frac{\sigma}{E}\left[1 + a\left(\frac{\sigma}{\sigma_0}\right)^{n-1}\right]$$

where E, a, σ_0, and n are constants. Prepare plots of σ/σ_0 versus $E\varepsilon'/\sigma_0$ for $n = 1, 2, 3$, and $n \to \infty$. Assume $a = 3/7$.

8. Apply Eqs. (8–3) to a nonlinearly elastic material having a stress-strain relation of the form $\varepsilon' = g(\sigma)$ where $g(\sigma)$ is a given function of σ. Letting $\sigma_1 = \sigma$ at $y = -h/2$, $\sigma_2 = \sigma$ at $y = h/2$, $\varepsilon_1 = g(\sigma_1)$, and $\varepsilon_2 = g(\sigma_2)$ show that

$$N = \frac{b}{\chi}\left(\sigma_1\varepsilon_1 - \sigma_2\varepsilon_2 + \int_{\sigma_1}^{\sigma_2} g(\sigma)\, d\sigma\right)$$

$$M = -\frac{N}{2\chi}(\varepsilon_1 + \varepsilon_2) + \frac{b}{2\chi^2}\left(\sigma_1\varepsilon_1^2 - \sigma_2\varepsilon_2^2 + \int_{\sigma_1}^{\sigma_2} g^2\, d\sigma\right)$$

Sections 8, 9

9. Show that the neutral axis in the plastic state $M = M_p$ and $N = 0$ divides the cross-sectional area A into two equal parts. Deduce that $M_p = \sigma_y(A/2)d$, where d is the distance between the centroids of the areas on the two sides of the neutral axis.

10. A beam slice of rectangular cross section of width b and depth h is subjected to an axial force N and to a bending moment $M = Nh/6$. Using Section 8, describe on Fig. 8–3 the deformation path as N increases monotonically from 0. Determine the maximum possible value of N and the deformation state as N reaches its maximum.

11. Derive the shape factor formula shown here for an I cross section of depth h, flange width b, flange thickness t, and web thickness t':

$$\frac{M_P}{M_y} = bht\left(1 - \frac{t}{h}\right) + \frac{t'h^2}{4}\left(1 - \frac{2t}{h}\right)^2$$

12. For the I cross section of the preceding exercise derive the $(M\text{–}\chi)$ relation for the case $N = 0$ and for increasing M. Consider the three stages (a) elastic behavior, (b) yield in the flanges, and (c) yield in the web.

Note: A simple procedure is to use results for a rectangular cross section and to form the I cross section by a combination of rectangles.

Answers:

a. $M = EI\chi$

b. $\dfrac{M}{M_y} = \dfrac{bh^3}{8I}\left[1 - \dfrac{1}{3}\left(\dfrac{\chi_y}{\chi}\right)^2\right] - \left(\dfrac{bh^3}{12I} - 1\right)\dfrac{\chi}{\chi_y}, \quad 1 - \dfrac{2t}{h} < \dfrac{\chi_y}{\chi} < 1$

c. $\dfrac{M}{M_y} = \dfrac{hZ}{2I} - \dfrac{t'h^3}{24I}\left(\dfrac{\chi_y}{\chi}\right)^2$

13. Derive the incremental bending rigidity $dM/d\chi$ for regimes (a), (b), and (c) of Exercise 12, and verify that $dM/d\chi$ is continuous at the transition points.

5

STIFFNESS AND FLEXIBILITY PROPERTIES OF THE STRUCTURAL MEMBER

Geometrically Linear Theory

1. INTRODUCTION

Relations between member statical redundants **V** and deformations **v** form one of the governing set of equations of structural analysis to be formulated in Chapter 8. Such relations are also useful in solving beam problems.

The methods to be developed are based on the equation of virtual work specialized to a geometrically linear theory. They are applicable with any finite or incremental constitutive equations. The linearly elastic case is developed in detail.

2. FLEXIBILITY RELATIONS BY THE METHOD OF VIRTUAL FORCES

In the geometrically linear theory, strains $\boldsymbol{\varepsilon}$ are related to displacements **u** by the same relations as the variations $\delta\boldsymbol{\varepsilon}$ and $\delta\mathbf{u}$. The equation of virtual work

(2/4.2–2a) may thus be written for an actual deformed state in the form

$$\mathbf{V}'^T\mathbf{v} + \int_0^l \mathbf{p}'^T\mathbf{u}\,dx = \int_0^l \mathbf{N}'^T\boldsymbol{\varepsilon}\,dx \tag{2–1}$$

where $(\mathbf{V}', \mathbf{p}', \mathbf{N}')$ is an arbitrary statically admissible virtual force system, and $\mathbf{u}$ is here the displacement relative to a member bound reference. The static condition is $\mathbf{N}' = \mathbf{b}\mathbf{V}' + \mathbf{N}'_p$. The compatibility equation for $\mathbf{v}$ (Chapter 2, Section 4.2) is obtained by choosing $\mathbf{p}' = 0$ and $\mathbf{V}'$ arbitrary. Then $\mathbf{N}'_p = 0$ and $\mathbf{N}' = \mathbf{b}\mathbf{V}'$. Equating coefficients of $\mathbf{V}'$ in Eq. (2–1), there comes

$$\mathbf{v} = \int_0^l \mathbf{b}^T\boldsymbol{\varepsilon}\,dx \tag{2–2}$$

For a linearly elastic material, $\boldsymbol{\varepsilon} = \boldsymbol{\Phi}\mathbf{N} + \boldsymbol{\varepsilon}_I$, and for a member in equilibrium, $\mathbf{N} = \mathbf{b}\mathbf{V} + \mathbf{N}_p$. Substitution of these relations into Eq. (2–2) yields the flexibility relations

$$\mathbf{v} = \mathbf{f}\mathbf{V} + \mathbf{v}_0 \tag{2–3}$$

where

$$\mathbf{f} = \int_0^l \mathbf{b}^T\boldsymbol{\Phi}\mathbf{b}\,dx \tag{2–4a}$$

$$\mathbf{v}_0 = \mathbf{v}_p + \mathbf{v}_I \tag{2–4b}$$

$$\mathbf{v}_p = \int_0^l \mathbf{b}^T\boldsymbol{\Phi}\mathbf{N}_p\,dx \tag{2–4c}$$

$$\mathbf{v}_I = \int_0^l \mathbf{b}^T\boldsymbol{\varepsilon}_I\,dx \tag{2–4d}$$

$\mathbf{f}$ is a member property called flexibility matrix. It has a type that corresponds to the type of deformations and member redundants. The terms $\mathbf{v}_p$ and $\mathbf{v}_I$ are deformations caused, respectively, by the span load $\mathbf{p}$ and by initial strains $\boldsymbol{\varepsilon}_I$. Their sum $\mathbf{v}_0$ will be referred to as a matrix of initial deformations.

Assuming a uniform member for which

$$\boldsymbol{\Phi} = \left\lceil \frac{1}{EA}\ \frac{1}{GA_s}\ \frac{1}{EI} \right\rfloor \tag{2–5}$$

and using the expressions of $\mathbf{b}$ derived in Eqs. (3/3.2–3) and (3/3.2–4) we obtain for the cantilever type

$$\mathbf{f} = \begin{bmatrix} \dfrac{l}{EA} & 0 & 0 \\ 0 & \dfrac{l^3}{3EI}\left(1 + \dfrac{\beta}{2}\right) & \dfrac{l^2}{2EI} \\ 0 & \dfrac{l^2}{2EI} & \dfrac{l}{EI} \end{bmatrix} \tag{2–6}$$

and for the alternate type

$$\mathbf{f} = \begin{bmatrix} \dfrac{l}{EA} & 0 & 0 \\ 0 & \dfrac{l}{3EI}\left(1 + \dfrac{\beta}{2}\right) & -\dfrac{l}{6EI}(1 - \beta) \\ 0 & -\dfrac{l}{6EI}(1 - \beta) & \dfrac{l}{3EI}\left(1 + \dfrac{\beta}{2}\right) \end{bmatrix} \tag{2–7}$$

where

$$\beta = \frac{6EI}{GA_s l^2} \tag{2–8}$$

Elements of **f** are called flexibility coefficients. Element f_{ij} in row i and column j is the deformation v_i caused by $V_j = 1$. Equation (2–3) gives **v** as a superposition of the effects of **V** and of the member loading condition, consisting in general of the span load **p** and of initial strains $\boldsymbol{\varepsilon}_I$, Fig. 2–1.

f is symmetric as may be deduced from Eq. (2–4a) and from the symmetry of $\boldsymbol{\Phi}$. Thus deformation v_i caused by a unit V_j is equal to deformation v_j caused by a unit V_i. This is an instance of what is called Betti's reciprocal theorem (see Section 5).

For a nonhomogeneous but uniform member, Eqs. (2–6, 7) remain valid in terms of properties of the transformed cross section. For a nonuniform member, $\boldsymbol{\Phi}$ in Eq. (2–4a) is represented by some function of x. The case of a stepped member is a simple one for which $\boldsymbol{\Phi}$ is piecewise constant.

β in Eq. (2–8) represents the effect of shear deformation. The ratio $I/A_s l^2$ may be written r_s^2/l^2, where r_s is a radius of gyration associated with the shear area. β may be significant for short members or for members much softer in shear than in axial deformation. For usual slender members, β is negligible compared with 1 and may be set to zero in Eqs. (2–6, 7).

For neglecting shear deformation in the general formulation, one may use effective matrices $\boldsymbol{\varepsilon} = \{\varepsilon \ \ \chi\}$, $\mathbf{N} = \{N \ \ M\}$ and corresponding effective $\boldsymbol{\Phi}$ and **b**.

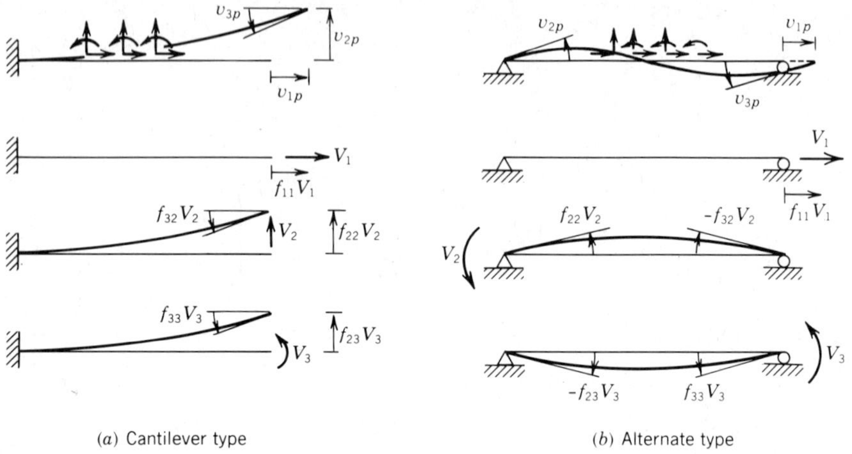

FIG. 2–1. Flexibility relations $\mathbf{v} = \mathbf{fV} + \mathbf{v}_p$: (*a*) cantilever type; (*b*) alternate type.

Evaluation of $\mathbf{v}_p$*, Fig. 2–1*

To evaluate $\mathbf{v}_p$, the stress resultants $\mathbf{N}_p = \{N_p \; S_p \; M_p\}$ caused by the span load, with $\mathbf{V} = 0$, are determined as functions of x and are substituted into Eq. (2–4c).

Example 1

Consider a set of concentrated actions $\mathbf{P} = \{P_x \; P_y \; C\}$ applied at point ξ, and let the member be referred to the bound reference of the cantilever type. Then

$$\begin{Bmatrix} N_p \\ S_p \\ M_p \end{Bmatrix} = \begin{bmatrix} 1 & 0 & 0 \\ 0 & 1 & 0 \\ 0 & \xi - x & 1 \end{bmatrix} \begin{Bmatrix} P_x \\ P_y \\ C \end{Bmatrix}, \quad 0 \leqslant x < \xi \tag{2–9}$$

and $\mathbf{N}_p = 0$ for $x > \xi$.

Evaluation of $\mathbf{v}_p$ for a uniform member by means of Eq. (2–4c) yields

$$\begin{Bmatrix} v_{1p} \\ v_{2p} \\ v_{3p} \end{Bmatrix} = \begin{bmatrix} \dfrac{\xi}{EA} & 0 & 0 \\ 0 & \dfrac{l^3}{EI}\left(\dfrac{\xi^2}{2l^2} - \dfrac{\xi^3}{6l^3} + \beta\dfrac{\xi}{6l}\right) & \dfrac{l^2}{EI}\left(\dfrac{\xi}{l} - \dfrac{\xi^2}{2l^2}\right) \\ 0 & \dfrac{\xi^2}{2EI} & \dfrac{\xi}{EI} \end{bmatrix} \begin{Bmatrix} P_x \\ P_y \\ C \end{Bmatrix} \tag{2–10}$$

If the bound reference of the alternate type is chosen, we obtain for $x < \xi$,

$$\begin{Bmatrix} N_p \\ S_p \\ M_p \end{Bmatrix} = \begin{bmatrix} 1 & 0 & 0 \\ 0 & 1 - \dfrac{\xi}{l} & -\dfrac{1}{l} \\ 0 & x\left(\dfrac{\xi}{l} - 1\right) & \dfrac{x}{l} \end{bmatrix} \begin{Bmatrix} P_x \\ P_y \\ C \end{Bmatrix} \tag{2–11a}$$

and for $x > \xi$,

$$\begin{Bmatrix} N_p \\ S_p \\ M_p \end{Bmatrix} = \begin{bmatrix} 0 & 0 & 0 \\ 0 & -\dfrac{\xi}{l} & -\dfrac{1}{l} \\ 0 & \xi\left(\dfrac{x}{l} - 1\right) & \dfrac{x}{l} - 1 \end{bmatrix} \begin{Bmatrix} P_x \\ P_y \\ C \end{Bmatrix} \tag{2–11b}$$

Evaluation of $\mathbf{v}_p$ yields

$$\begin{Bmatrix} v_{1p} \\ v_{2p} \\ v_{3p} \end{Bmatrix} = \begin{bmatrix} \dfrac{\xi}{EA} & 0 & 0 \\ 0 & \dfrac{l}{6EI}\eta\left(1 - \dfrac{\eta^2}{l^2}\right) & \dfrac{l}{6EI}\left(-1 + \dfrac{3\eta^2}{l^2} + \beta\right) \\ 0 & -\dfrac{l}{6EI}\xi\left(1 - \dfrac{\xi^2}{l^2}\right) & \dfrac{l}{6EI}\left(-1 + \dfrac{3\xi^2}{l^2} + \beta\right) \end{bmatrix} \begin{Bmatrix} P_x \\ P_y \\ C \end{Bmatrix} \tag{2–12}$$

where

$$\eta = l - \xi \tag{2–13}$$

The solution for a set of point loads $\mathbf{P}$ just obtained may be used to formulate a general solution for $\mathbf{v}_p$. In matrix notation, let Eq. (2–9) or (2–11) be written in the form

$$\mathbf{N}_p = \mathbf{b}_\xi^x \mathbf{P} \tag{2–14}$$

then Eq. (2–4c) yields $\mathbf{v}_p$ in the form

$$\mathbf{v}_p = \mathbf{G}_\xi \mathbf{P} \tag{2–15}$$

where

$$\mathbf{G}_\xi = \int_0^l \mathbf{b}^T \mathbf{\Phi} \mathbf{b}_\xi^x \, dx \tag{2–16}$$

$\mathbf{G}_\xi$ is the coefficient matrix in Eq. (2–10) or (2–12), depending on the choice of member-bound reference.

An arbitrary load intensity function $\mathbf{p} = \{p_x \; p_y \; m\}$ may be considered formed of elementary concentrated actions whose effects may be superimposed because of the linearity of the governing equations. The effect of an elementary load $\mathbf{p}(\xi)d\xi$ is obtained by replacing $\mathbf{P}$ with $\mathbf{p}(\xi)d\xi$ in the preceding example. Superposition yields the integrals

$$\mathbf{N}_p = \int_0^l \mathbf{b}_\xi^x \mathbf{p}(\xi)d\xi \tag{2–17}$$

$$\mathbf{v}_p = \int_0^l \mathbf{G}_\xi \mathbf{p}(\xi)d\xi \tag{2–18}$$

Example 2

Consider a uniform transverse load $p_y = p$. Then $\mathbf{p} = \{0 \; p \; 0\}$, and only the second column of G_ξ in Eq. (2–18) is effective. For the alternate type, Eq. (2–18) yields

$$v_{3p} = \int_0^l -\frac{l}{6EI}\xi\left(1 - \frac{\xi^2}{l^2}\right)pd\xi = -\frac{pl^3}{24EI}$$

The formula for v_{2p} is similar in terms of $\eta = l - \xi$. A change of integration variable from ξ to η yields $v_{2p} = -v_{3p}$. Formulas for $\mathbf{v}_p$ are given in Figs. 2–2 and 2–3 for typical loading conditions.

The method for evaluating $\mathbf{v}_p$ through Eq. (2–18) is called Green's function method, and the elements of $\mathbf{G}_\xi$, which represent deformations caused by a unit concentrated action applied at ξ, are called Green functions.

An interesting property of $\mathbf{G}_\xi$ appears on considering the displacements $\mathbf{u}^\xi = \{u^\xi \; v^\xi \; \varphi^\xi\}$ at point ξ caused by $\mathbf{V}$. The compatibility equation for $\mathbf{u}^\xi$ is obtained from Eq. (2–1) by using a virtual $\mathbf{P}'$ applied at point ξ. There comes

$$\mathbf{u}^\xi = \int_0^l \mathbf{b}_\xi^{xT} \boldsymbol{\varepsilon}\, dx \tag{2–19}$$

Substituting $\boldsymbol{\varepsilon} = \boldsymbol{\Phi}\mathbf{N}$ and $\mathbf{N} = \mathbf{b}\mathbf{V}$ yields

$$\mathbf{u}^\xi = \mathbf{G}_\xi^T \mathbf{V} \tag{2–20}$$

The displacements $\{u^\xi \; v^\xi \; \varphi^\xi\}$ at point ξ, caused by $V_i = 1$, form the ith column of $\mathbf{G}_\xi^T$, which is also the ith row of $\mathbf{G}_\xi$. They are thus equal, respectively, to the deformation v_i caused by unit concentrated actions P_x, P_y, and C applied in turn at point ξ. This is another instance of Betti's reciprocal theorem.

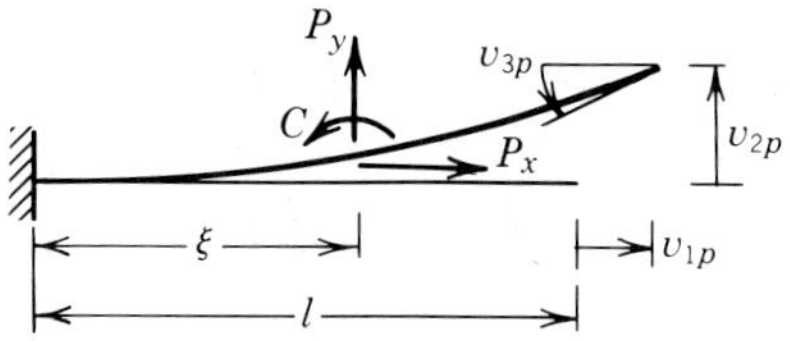

$$v_{1p} = \frac{P_x}{EA}\xi$$

$$v_{2p} = \frac{P_y l^3}{EI}\left(\frac{\xi^2}{2l^2} - \frac{\xi^3}{6l^3} + \frac{\beta\xi}{6l}\right) + \frac{Cl^2}{EI}\left(\frac{\xi}{l} - \frac{\xi^2}{2l^2}\right)$$

$$v_{3p} = \frac{P_y \xi^2}{2EI} + \frac{C\xi}{EI}$$

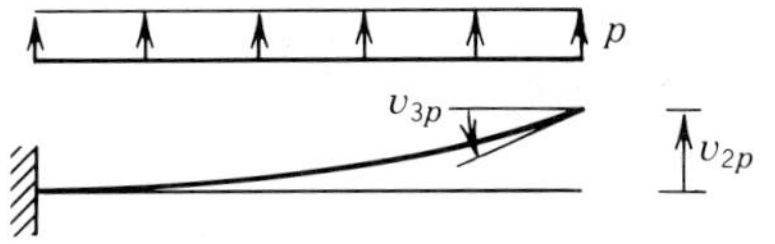

$$v_{2p} = \frac{pl^4}{8EI}\left(1 + \frac{2}{3}\beta\right)$$

$$v_{3p} = \frac{pl^3}{6EI}$$

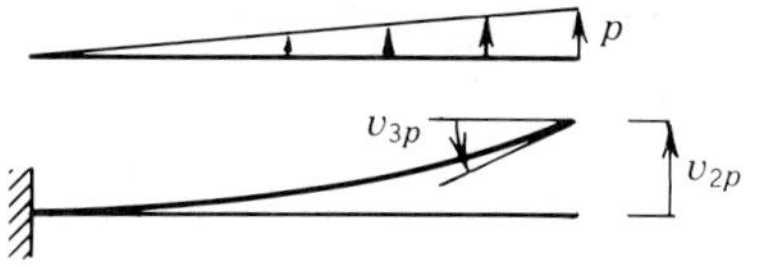

$$v_{2p} = \frac{11pl^4}{120EI}\left(1 + \frac{20}{33}\beta\right)$$

$$v_{3p} = \frac{pl^3}{8EI}$$

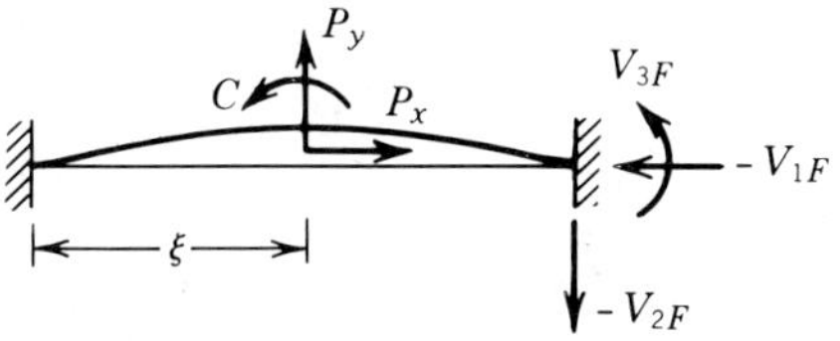

$$V_{1F} = -\frac{\xi}{l}P_x$$

$$V_{2F} = -\frac{P_y}{1 + 2\beta}\left(\frac{3\xi^2}{l^2} - \frac{2\xi^3}{l^3} + 2\beta\frac{\xi}{l}\right) - \frac{6C}{(1 + 2\beta)l}\left(\frac{\xi}{l} - \frac{\xi^2}{l^2}\right)$$

$$V_{3F} = \frac{P_y l}{1 + 2\beta}\left[(1 - \beta)\frac{\xi^2}{l^2} - \frac{\xi^3}{l^3} + \beta\frac{\xi}{l}\right] + \frac{2C}{1 + 2\beta}\left[(1 - \beta)\frac{\xi}{l} - \frac{3\xi^2}{2l^2}\right]$$

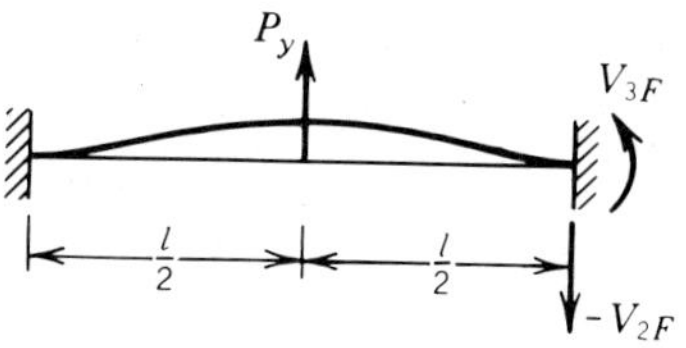

$$V_{2F} = -\frac{P_y}{2}$$

$$V_{3F} = \frac{P_y l}{8}$$

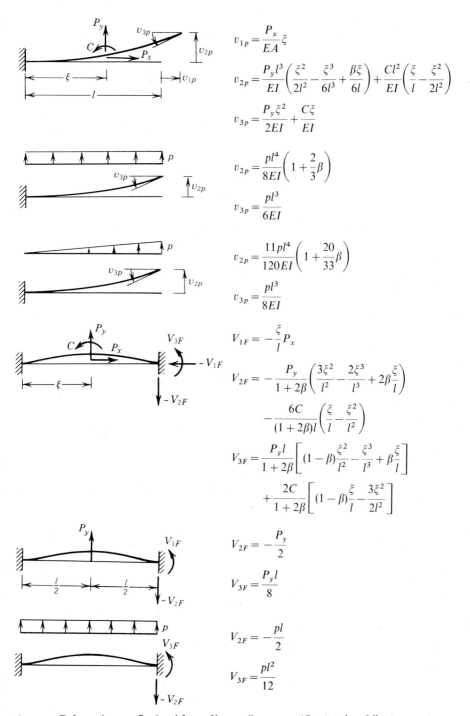

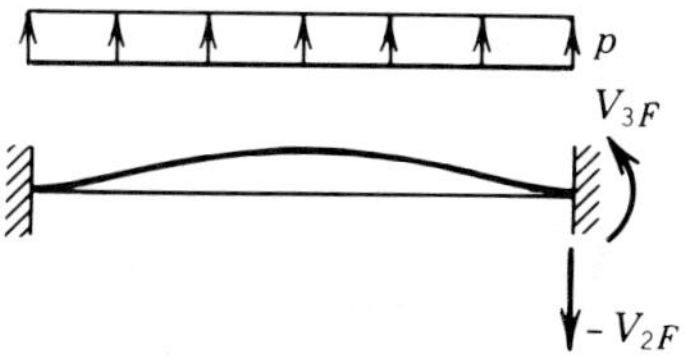

$$V_{2F} = -\frac{pl}{2}$$

$$V_{3F} = \frac{pl^2}{12}$$

Fig. 2–2. Deformations $\mathbf{v}_p$, fixed end forces $\mathbf{V}_F$, cantilever type. (*Continued on following page.*)

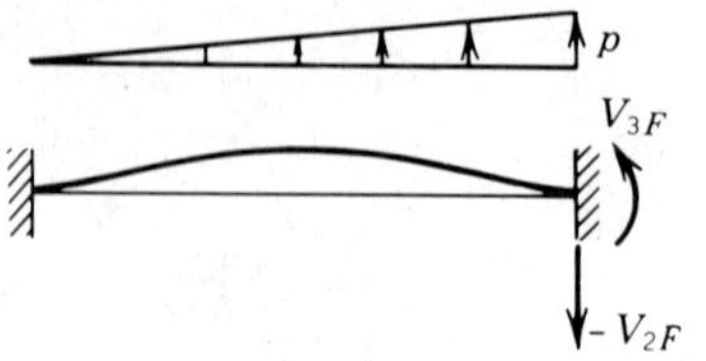

$$V_{2F} = -\frac{7pl}{20(1+2\beta)}\left(1 + \frac{40}{21}\beta\right)$$

$$V_{3F} = \frac{pl^2}{20(1+2\beta)}\left(1 + \frac{5}{3}\beta\right)$$

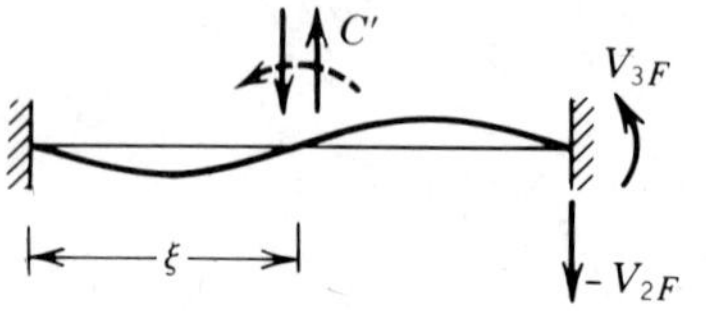

$$V_{2F} = -\frac{6C'}{(1+2\beta)l}\left(\frac{\xi}{l} - \frac{\xi^2}{l^2} + \frac{\beta}{3}\right)$$

$$V_{3F} = \frac{C'}{1+2\beta}\left[2(1-\beta)\xi - \frac{3\xi^2}{l^2} + \beta\right]$$

FIG. 2–2. *(continued)*

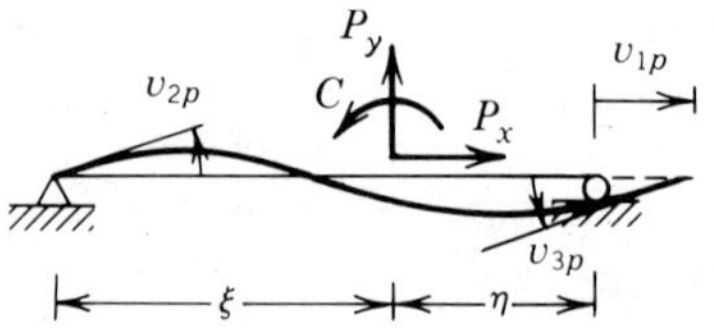

$$v_{1p} = \frac{P_x \xi}{EA}$$

$$v_{2p} = \frac{P_y l}{6EI}\eta\left(1 - \frac{\eta^2}{l^2}\right) + \frac{Cl}{6EI}\left(-1 + \frac{3\eta^2}{l^2} + \beta\right)$$

$$v_{3p} = -\frac{P_y l}{6EI}\xi\left(1 - \frac{\xi^2}{l^2}\right) + \frac{Cl}{6EI}\left(-1 + \frac{3\xi^2}{l^2} + \beta\right)$$

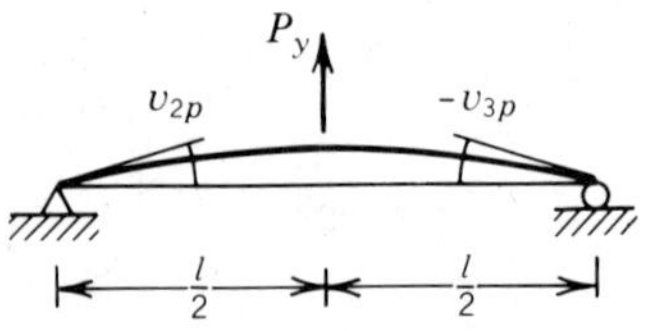

$$v_{2p} = -v_{3p} = \frac{P_y l}{16EI}$$

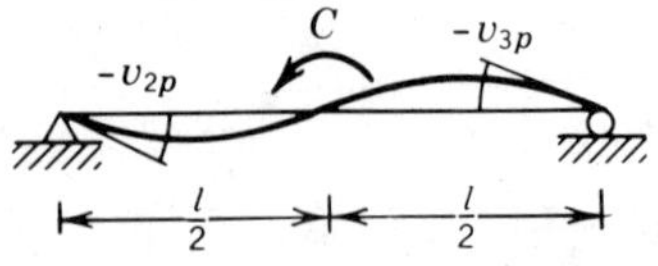

$$v_{2p} = v_{3p} = -\frac{Cl}{24EI}(1 - 4\beta)$$

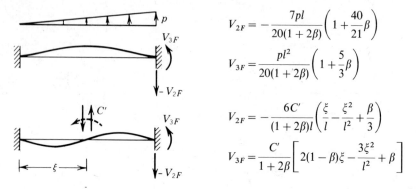

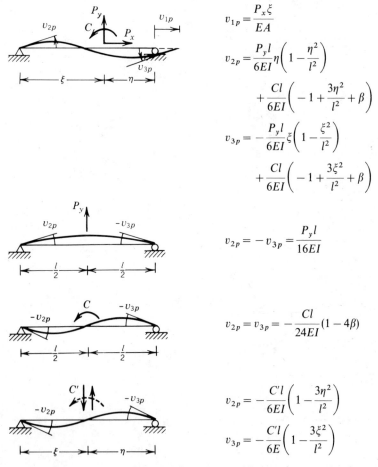

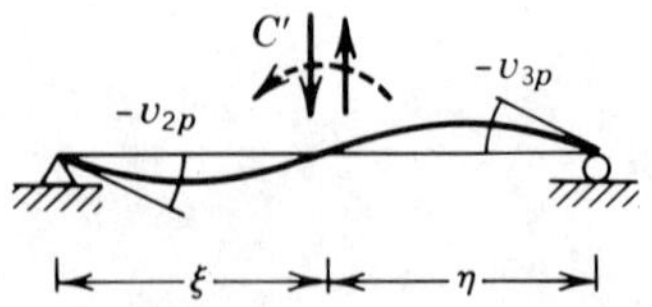

$$v_{2p} = -\frac{C'l}{6EI}\left(1 - \frac{3\eta^2}{l^2}\right)$$

$$v_{3p} = -\frac{C'l}{6E}\left(1 - \frac{3\xi^2}{l^2}\right)$$

FIG. 2–3. Deformations $\mathbf{v}_p$, alternate type. (*Continued on next page.*)

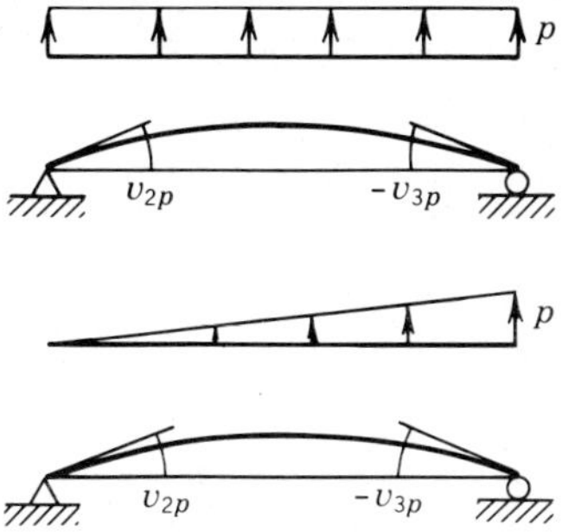

$$v_{2p} = -v_{3p} = \frac{pl^3}{24EI}$$

$$v_{2p} = \frac{7pl^3}{360EI}$$

$$v_{3p} = -\frac{pl^3}{45EI}$$

FIG. 2–3. (*continued*)

Example 3

A concentrated moment may be applied on a member in two ways as shown in Fig. 2–4. These two ways are equivalent only in a theory neglecting shear deformation. The moment C considered in Example 1 is of the first type. A moment C' applied as shown in Fig. 2–4b involves a shear force in the interval $\Delta\xi$, which tends to infinity as $\Delta\xi \to 0$. This is clearly an idealization, but it does allow us to take into account the effect of shear deformation in a finite by small interval $\Delta\xi$. By superposition, $\mathbf{v}_p$ is the sum of the effects of the loads $P_y = -C'/\Delta\xi$ applied at ξ and $P_y = C'/\Delta\xi$ applied at $\xi + \Delta\xi$. The deformations $\mathbf{v}_p$, caused by a unit P_y applied at ξ, form the second column $\mathbf{G}_{2\xi}$ of $\mathbf{G}_\xi$, Eq. (2–15). Thus

$$\mathbf{v}_p = \frac{C'}{\Delta\xi}[\mathbf{G}_{2\xi}(\xi + \Delta\xi) - \mathbf{G}_{2\xi}(\xi)].$$

In the limit

$$\mathbf{v}_p = C'\frac{d\mathbf{G}_{2\xi}}{d\xi} \tag{2–21}$$

For the cantilever, from Eq. (2–10),

$$v_{2p} = \frac{C'l^2}{EI}\left(\frac{\xi}{l} - \frac{\xi^2}{2l^2} + \frac{\beta}{6}\right)$$

$$v_{3p} = \frac{C'}{EI}\xi$$

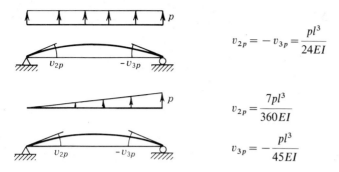

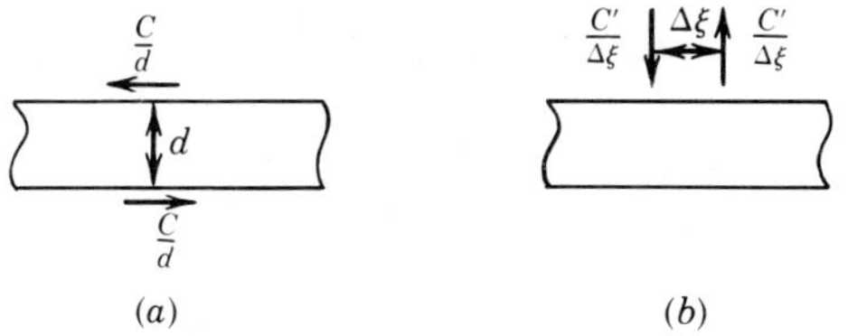

FIG. 2–4. Concentrated moments: (a) without transverse shear; (b) with transverse shear.

and for the alternate type, from Eq. (2–12),

$$v_{2p} = \frac{C'l}{6EI}\left(\frac{3\eta^2}{l^2} - 1\right)$$

$$v_{3p} = \frac{C'l}{6EI}\left(\frac{3\xi^2}{l^2} - 1\right)$$

3. STIFFNESS RELATIONS

If the flexibility equation (2–3) is solved for **V** in terms of **v**, the result is called a stiffness equation. It has the form

$$\mathbf{V} = \mathbf{k}\mathbf{v} + \mathbf{V}_F \tag{3–1}$$

where

$$\mathbf{k} = \mathbf{f}^{-1} \tag{3–2}$$

and

$$\mathbf{V}_F = -\mathbf{k}\mathbf{v}_0 \tag{3–3}$$

Inversion of **f** in Eqs. (2–6, 7) yields for the cantilever type

$$\mathbf{k} = \begin{bmatrix} \dfrac{EA}{l} & 0 & 0 \\ 0 & \dfrac{12EI}{l^3(1+2\beta)} & -\dfrac{6EI}{l^2(1+2\beta)} \\ 0 & -\dfrac{6EI}{l^2(1+2\beta)} & \dfrac{4EI}{l}\dfrac{1+\beta/2}{1+2\beta} \end{bmatrix} \tag{3–4}$$

and for the alternate type

$$\mathbf{k} = \begin{bmatrix} \dfrac{EA}{l} & 0 & 0 \\ 0 & \dfrac{4EI}{l}\dfrac{1+\beta/2}{1+2\beta} & \dfrac{2EI}{l}\dfrac{1-\beta}{1+2\beta} \\ 0 & \dfrac{2EI}{l}\dfrac{1-\beta}{1+2\beta} & \dfrac{4EI}{l}\dfrac{1+\beta/2}{1+2\beta} \end{bmatrix} \tag{3–5}$$

k is a stiffness matrix. It is symmetric because **f** is. Its elements are called stiffness

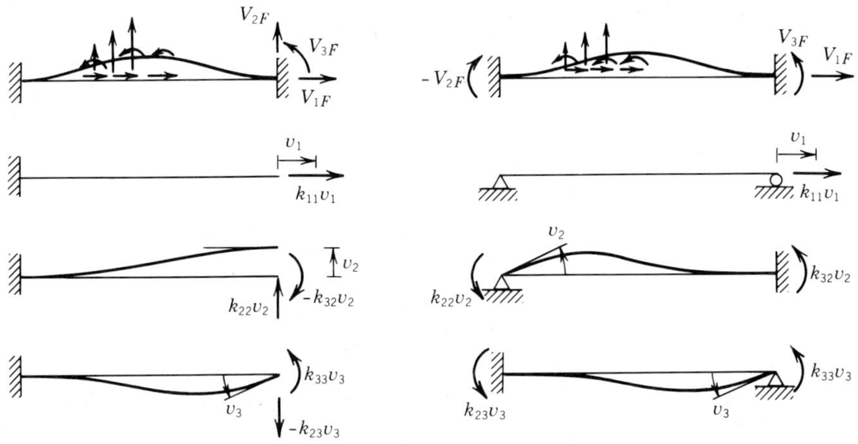

(a) Cantilever type (b) Alternate type

FIG. 3–1. Stiffness relations $\mathbf{V} = \mathbf{kv} + \mathbf{V}_F$: ($a$) cantilever type; ($b$) alternate type.

coefficients. Element k_{ij}, in row i and column j, is the force V_i required (with other end forces) to cause a unit deformation v_j, while other deformations are zero. It is equal to k_{ji}, that is, to the force V_j required to cause a unit deformation v_i. Equation (3–1) defines $\mathbf{V}$ as a superposition of forces $\mathbf{kv}$ required to cause $\mathbf{v}$ and of forces $\mathbf{V}_F$ caused by the member loading condition in the state $\mathbf{v} = 0$, Fig. 3–1. Elements of $\mathbf{V}_F$ are called fixed end forces. They are evaluated by means of Eq. (3–3), in which $\mathbf{v}_0$ is evaluated by the method of virtual forces. Results for typical loadings are shown in Fig. 2–2. Note that the Green function method may be applied to evaluate $\mathbf{V}_F$ similarly to its application to $\mathbf{v}_p$ in the preceding section.

Complete Stiffness Equations

The complete stiffness equations are relations giving the member end forces $\mathbf{S}$ in terms of member end displacements $\mathbf{s}$ and of the member loading. They are obtained from the (reduced) stiffness equations relating $\mathbf{V}$ to $\mathbf{v}$ by means of the equilibrium equation, $\mathbf{S} = \mathbf{BV} + \mathbf{S}_p$, (Chapter 3, Section 3.2), and of the deformation-displacement relations, $\mathbf{v} = \mathbf{B}^T\mathbf{s}$ (Chapter 3, Section 3–1). There comes

$$\mathbf{S} = \mathbf{k}_c\mathbf{s} + \mathbf{S}_F \tag{3–6}$$

where

$$\mathbf{k}_c = \mathbf{BkB}^T \tag{3–7}$$

$$\mathbf{S}_F = \mathbf{S}_p + \mathbf{BV}_F \tag{3–8}$$

$\mathbf{k}_c$ is called the complete stiffness matrix, and $\mathbf{S}_F$ the complete set of fixed end forces. Either type of $\mathbf{B}$ and $\mathbf{k}$ could be used in Eq. (3–7). The result is given here for a uniform member:

$$
\mathbf{k}_c = \begin{bmatrix}
\frac{EA}{l} & & & -\frac{EA}{l} & & \\
& \frac{12EI}{l^3}\frac{1}{1+2\beta} & \frac{6EI}{l^2}\frac{1}{1+2\beta} & & -\frac{12EI}{l^3}\frac{1}{1+2\beta} & \frac{6EI}{l^2}\frac{1}{1+2\beta} \\
& \frac{6EI}{l^2}\frac{1}{1+2\beta} & \frac{4EI}{l}\frac{1+\beta/2}{1+2\beta} & & -\frac{6EI}{l^2}\frac{1}{1+2\beta} & \frac{2EI}{l}\frac{1-\beta}{1+2\beta} \\
-\frac{EA}{l} & & & \frac{EA}{l} & & \\
& -\frac{12EI}{l^3}\frac{1}{1+2\beta} & -\frac{6EI}{l^2}\frac{1}{1+2\beta} & & \frac{12EI}{l^3}\frac{1}{1+2\beta} & -\frac{6EI}{l^2}\frac{1}{1+2\beta} \\
& \frac{6EI}{l^2}\frac{1}{1+2\beta} & \frac{2EI}{l}\frac{1-\beta}{1+2\beta} & & -\frac{6EI}{l^2}\frac{1}{1+2\beta} & \frac{4EI}{l}\frac{1+\beta/2}{1+2\beta}
\end{bmatrix} \tag{3–9}
$$

Evaluation of $\mathbf{S}_F$ by Eq. (3–8) is equivalent to comleting the free body diagram of the member subjected to its loading condition and to $\mathbf{V}_F$.

Elements of $\mathbf{k}_c$ have an interpretation similar to the elements of $\mathbf{k}$. Column j of $\mathbf{k}_c$ is formed of the six member end forces required to cause the displacement s_j, while other member end displacements are zero, Fig. 3–2.

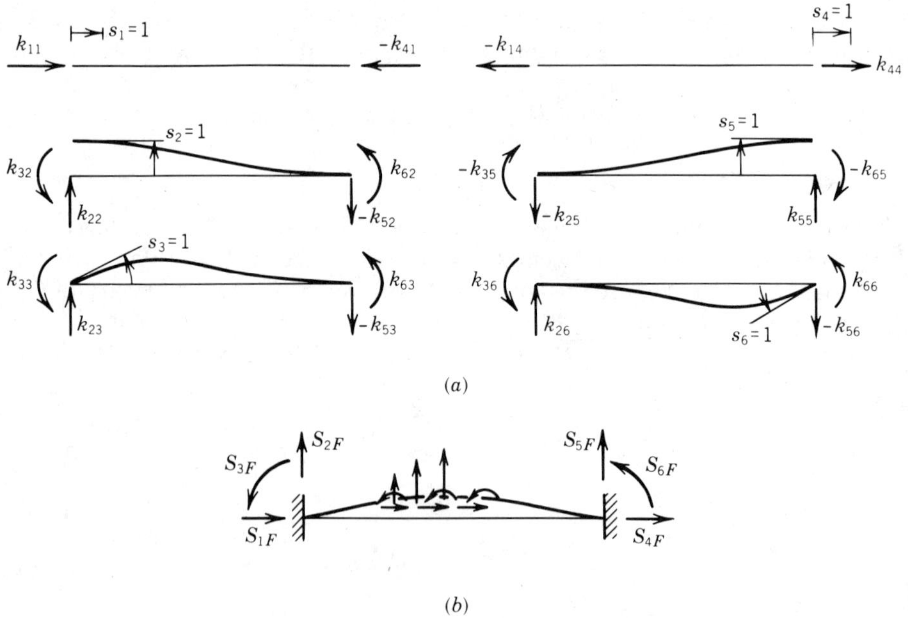

FIG. 3–2. Stiffness coefficients (*a*); fixed end forces (*b*).

Some properties of $\mathbf{k}_c$ are the following:

1. $\mathbf{k}_c$ is symmetric. This follows from Eq. (3–7) and from the symmetry of $\mathbf{k}$.
2. Any column of $\mathbf{k}_c$ is a set of self-equilibriating member end forces. Consequently its elements satisfy identically rigid body equilibrium equations. This is because $\mathbf{k}_c\mathbf{s}$ expresses $\mathbf{BV}$, and $\mathbf{BV}$ is a set of self-equilibriating end forces for any $\mathbf{V}$. The property applies thus to $\mathbf{k}_c\mathbf{s}$ for any $\mathbf{s}$ and so to each column of $\mathbf{k}_c$.
3. $\mathbf{k}_c$ is singular, and the nonzero displacements for which $\mathbf{k}_c\mathbf{s} = 0$ derive from any rigid body displacement of the member. This property is another aspect of the preceding one. For $\mathbf{k}_c\mathbf{s}$ to vanish, $\mathbf{kv}$ must vanish, and this in turn requires that $\mathbf{v}$ vanishes. Deformations vanish if and only if $\mathbf{s}$ derives from a rigid body displacement.
4. If three elements of $\mathbf{s}$ defining a rigid body displacement are constrained to be zero, the remaining elements of $\mathbf{s}$ may be considered as deformations $\mathbf{v}$, and the corresponding elements of $\mathbf{S}$ as member statical redundants $\mathbf{V}$. The corresponding reduced stiffness $\mathbf{k}$ is the subset of $\mathbf{k}_c$ consisting of the elements at the intersection of the rows and columns associated, respectively, with $\mathbf{V}$ and $\mathbf{v}$.

4. STIFFNESS RELATIONS BY THE METHOD OF VIRTUAL DISPLACEMENTS

The force system of the member satisfies the equation of virtual work (2/4.2–2b)

$$\mathbf{S}^T\delta\mathbf{s} + \int_0^l \mathbf{p}^T\delta\mathbf{u}\,dx = \int_0^l \mathbf{N}^T\delta\boldsymbol{\varepsilon}\,dx \tag{4–1}$$

for arbitrary virtual displacements. Stiffness equations are derived from Eq. (4–1) by a discretization procedure outlined in what follows. Let $\delta\mathbf{u}$ be represented in terms of a discrete set of parameters $\delta\mathbf{q}$ in the form

$$\delta\mathbf{u} = \boldsymbol{\psi}\delta\mathbf{q} \tag{4–2}$$

where $\boldsymbol{\psi}$ is function of x. For example, $\boldsymbol{\psi}\delta\mathbf{q}$ could be a polynomial expression whose coefficients are $\delta\mathbf{q}$, or it could be a truncated Fourier series. $\delta\boldsymbol{\varepsilon}$ and $\delta\mathbf{s}$ are determined from their geometric relations to $\delta\mathbf{u}$ in the form

$$\delta\boldsymbol{\varepsilon} = \mathbf{c}\delta\mathbf{q} \tag{4–3}$$

$$\delta\mathbf{s} = \mathbf{a}\delta\mathbf{q} \tag{4–4}$$

With $\delta\mathbf{u}$, $\delta\boldsymbol{\varepsilon}$, and $\delta\mathbf{s}$ from the preceding equations, the two sides of Eq. (4–1) take the form

$$\delta W_{\text{ext}} = \left[\mathbf{a}^T\mathbf{S} + \int_0^l \boldsymbol{\psi}^T\mathbf{p}\,dx\right]^T \delta\mathbf{q} \equiv \mathbf{Q}_{\text{ext}}^T\delta\mathbf{q} \tag{4–5a}$$

$$\delta W_{\text{int}} = \left[\int_0^l \mathbf{c}^T\mathbf{N}\,dx\right]^T \delta\mathbf{q} \equiv \mathbf{Q}_{\text{int}}^T\delta\mathbf{q} \tag{4–5b}$$

The equation of virtual work holds for arbitrary $\delta\mathbf{q}$, whence

$$\mathbf{Q}_{\text{ext}} = \mathbf{Q}_{\text{int}} \tag{4–6}$$

Elements of $\delta\mathbf{q}$ are called generalized virtual displacements. Corresponding elements of $\mathbf{Q}_{\text{ext}}$, as defined in Eq. (4–5a), are called generalized external forces, and those of $\mathbf{Q}_{\text{int}}$ generalized internal forces. Equations (4–6) are called generalized equilibrium equations. An individual equation in (4–6) is generated by letting one element of $\delta\mathbf{q}$ be arbitrary and all others be zero. To each element of $\delta\mathbf{q}$ corresponds what is called a degree of freedom and a generalized equilibrium equation.

Equations (4–6) turn into stiffness equations upon expressing $\mathbf{N}$ in terms of actual displacement parameters. If the same shape functions $\boldsymbol{\psi}$ are used for both virtual and actual displacements, then $\mathbf{u} = \boldsymbol{\psi}\mathbf{q}$, $\boldsymbol{\varepsilon} = \mathbf{c}\mathbf{q}$, and, for a linearly elastic material, $\mathbf{N} = \boldsymbol{\mathscr{x}}\boldsymbol{\varepsilon} + \mathbf{N}_I$, or $\mathbf{N} = \boldsymbol{\mathscr{x}}\mathbf{c}\mathbf{q} + \mathbf{N}_I$. The expression of $\mathbf{Q}_{\text{int}}$ as defined in Eq. (4–5b) takes the form

$$\mathbf{Q}_{\text{int}} = \mathbf{k}_q\mathbf{q} + \mathbf{Q}_I \tag{4–7}$$

where

$$\mathbf{k}_q = \int_0^l \mathbf{c}^T \boldsymbol{\mathscr{x}} \mathbf{c}\, dx \tag{4–8}$$

$$\mathbf{Q}_I = \int_0^l \mathbf{c}^T \mathbf{N}_I\, dx \tag{4–9}$$

Equation (4–6) turns then into the generalized stiffness equation

$$\mathbf{Q}_{\text{ext}} = \mathbf{k}_q\mathbf{q} + \mathbf{Q}_I \tag{4–10}$$

The preceding method may also be applied using the equation of virtual work relative to a member-bound reference. In that case $\mathbf{V}^T\delta\mathbf{v}$ replaces $\mathbf{S}^T\delta\mathbf{s}$, and the representation $\boldsymbol{\psi}\mathbf{q}$ must satisfy the support conditions in the member-bound reference.

Some examples are treated in what follows before discussing additional aspects of the method.

Example 1. Stiffness Equations of the Cantilever Type

Consider a uniform cantilever subjected to $\mathbf{V} = \{V_1 \;\; V_2 \;\; V_3\}$ at the free end. A discretization of $\mathbf{u}$, in terms of deformations $\mathbf{v}$ as generalized displacements, will lead to stiffness relations of the form $\mathbf{V} = \mathbf{k}\mathbf{v}$. The shape functions $\boldsymbol{\psi}$ may be determined exactly in the present example. The axial and shear forces are constant, and the bending moment is linear in x. Since the member is uniform, it follows that $\varepsilon = u_{,x}$ and $\gamma = v_{,x} - \varphi$ are constant and that $\chi = \varphi_{,x}$ is linear in x. Thus u is linear in x, φ is quadratic, and v is cubic, such that $v_{,x} - \varphi = \gamma =$ constant. Further u, v, and φ are relative to the bound reference and must thus satisfy $u = v = \varphi = 0$ at $x = 0$. To choose deformations as generalized displacements, u, v, and φ must satisfy $u = v_1, v = v_2$, and $\varphi = v_3$ at $x = l$. Starting with polynomials in terms of undetermined coefficients and letting $t = x/l$, the preceding conditions lead to the four degree-of-freedom representation

$$\begin{aligned} u &= v_1 t \\ v &= v_2(3t^2 - 2t^3) + lv_3(t^3 - t^2) + l\gamma(t - 3t^2 + 2t^3) \\ \varphi &= \frac{6v_2}{l}(t - t^2) + v_3(3t^2 - 2t) + 6\gamma(t^2 - t) \end{aligned} \tag{4–11}$$

The external virtual work is $\mathbf{V}^T \delta\mathbf{v}$. Since there is no term in $\delta\gamma$, the corresponding generalized force is zero and

$$\mathbf{Q}_{\text{ext}} = \{V_1 \ V_2 \ V_3 \ 0\} \tag{4–12}$$

Applying the strain-displacement relations, there comes

$$\begin{Bmatrix} \varepsilon \\ \gamma \\ \chi \end{Bmatrix} = \begin{Bmatrix} \frac{1}{l}u_{,t} \\ \frac{1}{l}v_{,t} - \varphi \\ \frac{1}{l}\varphi_{,t} \end{Bmatrix} = \begin{bmatrix} \frac{1}{l} & 0 & 0 & 0 \\ 0 & 0 & 0 & 1 \\ 0 & \frac{6}{l^2}(1-2t) & \frac{1}{l}(6t-2) & \frac{6}{l}(2t-1) \end{bmatrix} \begin{Bmatrix} v_1 \\ v_2 \\ v_3 \\ \gamma \end{Bmatrix} \tag{4 – 13}$$

The coefficient matrix is $\mathbf{c}$. Evaluating $\mathbf{k}_q$ using Eq. (4–8) with $\mathscr{E} = \lceil EA \ \ GA_s \ \ EI \rfloor$, the stiffness equations take the form

$$\begin{Bmatrix} V_1 \\ V_2 \\ V_3 \\ 0 \end{Bmatrix} = \begin{bmatrix} \frac{EA}{l} & 0 & 0 & 0 \\ 0 & \frac{12EI}{l^3} & -\frac{6EI}{l^2} & -\frac{12EI}{l^2} \\ 0 & -\frac{6EI}{l^2} & \frac{4EI}{l} & \frac{6EI}{l} \\ 0 & -\frac{12EI}{l^2} & \frac{6EI}{l} & GA_s l + \frac{12EI}{l} \end{bmatrix} \begin{Bmatrix} v_1 \\ v_2 \\ v_3 \\ \gamma \end{Bmatrix} \tag{4–14}$$

The last of these equations determines γ in terms of v_2 and v_3 in the form

$$\gamma = \frac{\beta}{1 + 2\beta}\left(\frac{2}{l}v_2 - v_3\right) \tag{4–15}$$

Substitution of γ into the second and third of Eqs. (4–14) yields the stiffness equations $\mathbf{V} = \mathbf{kv}$, which may be verified to coincide with those derived earlier by the method of virtual forces.

Comments

Discretization by the method of virtual displacements and the associated concepts of generalized forces and generalized equilibrium equations is due to Lagrange [5]. With minor variations in procedure, the method is also known as the Ritz method or the potential energy method. Also the generalized stiffness equations express what is referred to as Castigliano's first theorem (Section 5).

A key part of the method is the choice of shape functions. In the example treated earlier, the shape functions could be determined exactly, and consequently the exact stiffness matrix was obtained. The main advantage of the method, however, is to be applicable to problems where the exact shape functions are unknown or too complex to be used practically. The assumed shape functions are then considered as an approximation that yields an approximate stiffness matrix. With a large enough number of degrees of freedom and appropriate shape functions, it is possible to approximate any deformed shape to an arbitrary degree of accuracy.

In a geometrically linear theory the method of virtual forces is usually preferable because there is no need for shape functions. However, for nonuniform members and for nonlinearly elastic materials, the constitutive properties may be more amenable mathematically in one method than in the other, and this may be the determining factor on which of the two methods to use.

The following is a comment on the method of virtual displacements in a theory neglecting shear deformation. In such a theory $\gamma = 0$, and φ ceases to be an independent displacement but is $\varphi = v_{,x}$. In evaluating $\int_0^l \mathbf{N}^T \delta\varepsilon\, dx$, $\delta\gamma = 0$, and the shear force does not do any virtual work. The effective displacement, strain, and stress matrices are thus $\mathbf{u} = \{u \;\; v\}$, $\boldsymbol{\varepsilon} = \{\varepsilon \;\; \chi\}$, and $\mathbf{N} = \{N \;\; M\}$. If shear deformation were neglected from the start in the preceding example, the terms in γ in Eqs. (4–11), and the rows or columns corresponding to γ in Eqs. (4–12), (4–13), and (4–14) would all vanish.

Example 2

Consider the member of the preceding example subjected to $\mathbf{V}$ and to a uniformly distributed transverse load $p_y = p$. Using the same shape functions as earlier, $\mathbf{Q}_{\text{ext}}$ is now formed of two parts, one due to $\mathbf{V}$, as found in Eq. (4–12), and one due to the load p_y. This latter part, denoted $\mathbf{Q}_p$, has the definition

$$\mathbf{Q}_p^T \delta\mathbf{q} = \int_0^l p_y \delta v\, dx = p \int_0^l \delta v\, dx \tag{4–16}$$

where δv is obtained from the second of Eq. (4–11) by replacing v_2, v_3, and γ by

δv_2, δv_3, and $\delta\gamma$, respectively. We obtain

$$\mathbf{Q}_p = \left\{0 \quad p\frac{l}{2} \quad -\frac{pl^2}{12} \quad 0\right\} \tag{4–17}$$

The stiffness equations take the form $\mathbf{V} = \mathbf{kv} + \mathbf{V}_F$, where $\mathbf{V}_F = \{0 \; -pl/2 \; pl^2/12\}$. This solution for the fixed end forces turns out to be the same as the one obtained by the method of virtual forces and is thus exact. This is a remarkable result because the shape functions used in Eq. (4–16) are exact only if the member span load is zero. An explanation and a generalization of this result may be seen in Section 5 as an application of Betti's reciprocal theorem.

5. ENERGY THEOREMS

The methods outlined in previous sections lead to results that may be set in the form of theorems applicable in the wider field of solid mechanics. These and other theorems of general applicability are presented in what follows.

Castigliano's Second Theorem

Evaluation of deformations by the method of virtual forces is expressed by a theorem known as Castigliano's second theorem. From the statical equation $\mathbf{N} = \mathbf{bV} + \mathbf{N}_p$, we can write $\mathbf{b}^T = \partial\mathbf{N}^T/\partial\mathbf{V}$, and for an elastic material, $\boldsymbol{\varepsilon} = \partial U^*/\partial\mathbf{N}$. Equation (2–2) takes the form

$$\mathbf{v} = \int_0^l \frac{\partial\mathbf{N}^T}{\partial\mathbf{V}}\frac{\partial U^*}{\partial\mathbf{N}}\,dx = \frac{\partial}{\partial\mathbf{V}}\int_0^l U^*\,dx = \frac{\partial\bar{U}^*}{\partial\mathbf{V}} \tag{5–1}$$

where $\bar{U}^*$ is the complementary strain energy, expressed as a function of $\mathbf{V}$ by means of the equilibrium equation $\mathbf{N} = \mathbf{bV} + \mathbf{N}_p$. In the case of a linearly elastic material $\bar{U}^*$ takes the form

$$\bar{U}^* = \tfrac{1}{2}\mathbf{V}^T\mathbf{f}\mathbf{V} + \mathbf{V}^T\mathbf{v}_0 \tag{5–2}$$

In a similar way displacements $\mathbf{u}$, relative to a member-bound reference, may be expressed in the form

$$\mathbf{u} = \frac{\partial\bar{U}^*}{\partial\mathbf{P}} \tag{5–3}$$

where $\mathbf{P} = \{P_x \;\; P_y \;\; C\}$ is a set of concentrated actions applied at the point where $\mathbf{u}$ is evaluated. Equation (5–3) includes Eq. (5–1) as a particular case. It expresses Castigliano's second theorem. From a computational point view Eq. (5–3) offers no advantage over the method of virtual forces. In evaluating $\bar{U}^*$, $\mathbf{P}$ is included as a parameter in the member loading and is given its actual value after the partial derivative is taken. In particular, this value is zero if no concentrated actions are actually applied at the point where **u** is desired.

Castigliano's second theorem is valid for general elastic materials but only for geometrically linear compatibility equations.

Castigliano's First Theorem

The generalized stiffness equations obtained by the method of virtual displacements express what is known as Castigliano's first theorem.

The strain energy $\bar{U} = \int_0^l U\,dx$ expressed in terms of generalized displacements $\mathbf{q}$ has the property (Chapter 2, Section 5.3)

$$\delta W_{\text{int}} = \delta \bar{U} = \frac{\partial \bar{U}}{\partial \mathbf{q}^T} \delta \mathbf{q} \tag{5–4}$$

The coefficient of $\delta\mathbf{q}$ is thus $\mathbf{Q}_{\text{int}}^T$, and the generalized equilibrium equations are

$$\mathbf{Q}_{\text{ext}} = \frac{\partial \bar{U}}{\partial \mathbf{q}} \tag{5–5}$$

Equation (5–5) expresses Castigliano's first theorem. It applies in the general geometrically and elastically nonlinear theory. In the case of the linear theory $\bar{U}$ takes the form

$$\bar{U} = \int_0^l U\,dx = \tfrac{1}{2}\mathbf{q}^T \mathbf{k}_q \mathbf{q} + \mathbf{Q}_I^T \mathbf{q} \tag{5–6}$$

Betti's Reciprocal Theorem

Consider two deformed states of the same member, and assume that there are no initial stresses. In the geometrically linear theory, displacements and strains of any actual deformed state may be used as virtual quantities in the equation of virtual work. If the displacements of State (2) are used as virtual displacements with the forces of State (1), the internal virtual work, for a linearly elastic material, is the integral of $\mathbf{N}^{(1)T}\boldsymbol{\varepsilon}^{(2)}$ in which $\mathbf{N}^{(1)} = \boldsymbol{\mathscr{K}}\boldsymbol{\varepsilon}^{(1)}$. Since $\boldsymbol{\mathscr{K}} = \boldsymbol{\mathscr{K}}^T$, and a scalar is equal to its transpose, we can write

$$\int_0^l \boldsymbol{\varepsilon}^{(1)T} \boldsymbol{\mathscr{K}} \boldsymbol{\varepsilon}^{(2)}\,dx = \int_0^l \boldsymbol{\varepsilon}^{(2)T} \boldsymbol{\mathscr{K}} \boldsymbol{\varepsilon}^{(1)}\,dx \tag{5–7}$$

Equation (5–7) expresses the equality, or reciprocity, of the internal virtual work of the forces of one state through the strains of the other state. Since both states are in equilibrium, each side of Eq. (5–7) is equal to the associated external virtual work. This is Betti's reciprocal theorem by which the external virtual work of the forces of an equilibrium state through the displacements of another is equal to the external virtual work of the forces of the second state through the displacements of the first.

Application

Let State (1) be that of a member subjected only to end forces $\mathbf{S}^{(1)}$, and let $\mathbf{u}^{(1)} = \boldsymbol{\psi}\mathbf{s}^{(1)}$ be the exact deformed shape in that state, expressed in terms of the end displacements $\mathbf{s}^{(1)}$ and shape functions $\boldsymbol{\psi}$. Let State (2) be that of the same member subjected to a span load $\mathbf{p}$, with fixed end conditions $\mathbf{s}^{(2)} = 0$. Let $\mathbf{S}_F$ be the fixed end forces in State (2). Equating the external virtual works of the forces of one state through the displacements of the other, there comes

$$\mathbf{S}_F^T\mathbf{s}^{(1)} + \left[\int_0^l \mathbf{p}^T\boldsymbol{\psi}\,dx\right]\mathbf{s}^{(1)} = \mathbf{S}^{(1)T}\mathbf{s}^{(2)} = 0$$

Since $\mathbf{s}^{(1)}$ is arbitrary,

$$\mathbf{S}_F = -\int_0^l \boldsymbol{\psi}^T\mathbf{p}\,dx \tag{5–8}$$

Without the minus sign the right-hand side of the equation is by definition the generalized external forces $\mathbf{Q}_p$ due to the member load $\mathbf{p}$, when the shape functions $\boldsymbol{\psi}$ are used for discretization by the method of virtual displacements. Equation (5–8) shows that $-\mathbf{Q}_p$ are the exact fixed end forces due to $\mathbf{p}$. This is a remarkable result because the shape functions $\boldsymbol{\psi}$ cannot describe exactly a deformed state in which the member load $\mathbf{p}$ is not zero.

In particular, the fixed end forces $\mathbf{S}_F$, due to concentrated actions $\mathbf{P}$ applied at point ξ, are

$$\mathbf{S}_F = -\boldsymbol{\psi}_\xi^T\mathbf{P} \tag{5–9}$$

where $\boldsymbol{\psi}_\xi$ is $\boldsymbol{\psi}$ evaluated at $x = \xi$.

Clapeyron's Relation

It was established in Chapter 2, Section 5.3, that the strain energy in a deformed state $\mathbf{u}$ is equal to the external work done along any deformation path followed to reach state $\mathbf{u}$ from the initial state $\mathbf{u} = 0$. In the case of no initial stresses and a linear theory, the strain energy as obtained in Eq. (5–6) is equal to $\frac{1}{2}\mathbf{Q}_{\text{ext}}^T\mathbf{q}$. This is Clapeyron's relation which states that the external work done along a deformation path is equal to half the product of the final forces by the final displacements.

6. INCREMENTAL STIFFNESS AND FLEXIBILITY RELATIONS

Incremental stiffness and flexibility relations are useful in methods of analysis for nonlinear materials. They are relations between differential increments $d\mathbf{V}$ and $d\mathbf{v}$ taking place from an equilibrium state. Incremental constitutive properties have the form

$$d\mathbf{N} = æ_t d\boldsymbol{\varepsilon} + d\mathbf{N}_I \tag{6–1}$$

or

$$d\boldsymbol{\varepsilon} = \boldsymbol{\Phi}_t d\mathbf{N} + d\boldsymbol{\varepsilon}_I \tag{6–2}$$

where $æ_t$ and $\boldsymbol{\Phi}_t = æ_t^{-1}$ are tangent rigidity and compliance matrices, respectively. The methods of virtual forces and virtual displacements remain formally the same in terms of incremental quantities if $\boldsymbol{\Phi}_t$ replaces $\boldsymbol{\Phi}$ and $æ_t$ replaces $æ$. The difference in applications is that $æ_t$ and $\boldsymbol{\Phi}_t$ depend on the current deformed state, and in inelastic behavior, they depend also on the sense of the incremental strains. In the case of nonlinear but elastic materials, Eq. (6–1) is obtained by differentiating Eq. (2/5.3–1). Thus

$$æ_t = \frac{\partial^2 U}{\partial \boldsymbol{\varepsilon}^T \partial \boldsymbol{\varepsilon}} \tag{6–3}$$

Similarly from Eqs. (5–1) and (5–5) incremental or tangent flexibility and stiffness matrices satisfy the relations

$$\mathbf{f}_t = \frac{\partial^2 \bar{U}^*}{\partial \mathbf{V}^T \partial \mathbf{V}} \tag{6–4}$$

$$\mathbf{k}_t = \frac{\partial^2 \bar{U}}{\partial \mathbf{v}^T \partial \mathbf{v}} \tag{6–5}$$

EXERCISES

Section 2

1. Write in detail the compatibility equations (2–2), and derive the flexibility matrices, Eqs. (2–6) and (2–7).

2. A beam is formed of two uniform segments of properties (l_1, EA_1, EI_1) and (l_2, EA_2, EI_2) and having colinear centroidal axes. Derive the flexibility matrix of the beam using Eq. (2–4a). Neglect shear deformation, and consider in turn the cantilever and alternate types.

3. Formulate the solution of Exercise 2 if the centroidal axes of the two segments are parallel. Choose one of these two axes as reference axis for the member.

4. Show that the case where $1/EI$ is a polynomial in x presents no analytical difficulty in determining the bending flexibility coefficients. Compare this to the case where EI is a polynomial in x.

5. A haunched beam of rectangular cross section has a horizontal top face and vertical cross sections of linearly varying depth. The x axis is horizontal and passes through the centroid at the right end. Determine t'.e compliance matrix $\mathbf{\Phi}$ to be used in Eq. (2–4a). Devise a numerical procedure for evaluating the flexibility coefficients.

6. Obtain the result of Example 2, Section 2, by a direct application of the method of virtual forces.

Section 3

7. In Fig. 3–2*a* fill in the values of the stiffness coefficients for a uniform member. Ascertain that the force system in each deformation mode is in equilibrium.

8. Verify, using Eq. (3–9), that $\mathbf{k}_c\mathbf{s} = 0$ for any rigid body displacement.

9. Determine $\mathbf{v}_p$ of alternate type for a uniform member subjected to a parabolic transverse load having zero values at the ends and the value p at midspan. Determine $\mathbf{V}_F$.

10. Determine $\mathbf{v}_I$ of alternate type for a uniform thermal loading ε_I, χ_I. Determine $\mathbf{V}_F$, assuming a uniform member.

11. Do Exercises 9 and 10 for the cantilever type.

12. Derive bending flexibility coefficients of the alternate type for a uniform member whose moment-curvature relation has the form

$$\frac{\chi}{\chi_0} = \frac{M}{M_0} + a\left(\frac{M}{M_0}\right)^n$$

where $\chi_0, M_0, a,$ and n are constants.

Section 4

13. Using the method of Section 4, and the assumed axial displacement $u = v_1 x/l$, derive the generalized axial stiffness relation for a simply supported uniform member subjected to an arbitrary axial load. Solve the stiffness relation for v_1 in the following loading cases:
   a. Axial force V_1 applied at the free end.

b. Uniform axial load $p_x = p$.
c. Concentrated axial load P applied at $x = a$.
Verify that the solution for v_1 is exact in all cases but not the solution for u.

14. a. Obtain Eq. (4–11b) following the method outlined in the text in the case $\gamma = 0$.
b. Obtain similarly, for a bound reference of the alternate type,

$$v = lt(1-t)^2 v_2 - lt^2(1-t)v_3$$

Derive the bending stiffness matrix based on this representation, for a uniform member without shear deformation. Check the result with Eq. (3–5).

15. In part (b) of Exercise 14, obtain $\mathbf{Q}_p$ for a concentrated transverse force P applied at $x = a$. Verify that elements of $-\mathbf{Q}_p$ are equal to the fixed end moments $\mathbf{V}_F$.

16. Use the shape functions in bound axes of Exercises 13 and 14 to obtain the shape functions corresponding to the six member end displacements in the local reference frame.

17. Superimpose on the displacement representation of Exercise 14 part (b) the term $wt^2(1-t)^2$, where w is an additional generalized displacement, and derive the generalized stiffness equations. Solve these equations for v_2, v_3, and w in the following cases:
a. Uniform load.
b. Concentrated load at midspan.
In each case compare the results to the exact solution.

18. A member has linearly varying cross-sectional properties EA and EI defined by means of their respective values at the member ends. Derive the approximate stiffness matrix of the alternate type based on the shape functions of Exercises 13 and 14(b).

19. Let the displacement v of a simply supported member be represented by the truncated Fourier series $v = \sum_n a_n \sin n\pi x/l$. Considering a_n as generalized displacements and assuming a uniform member without shear deformation, obtain the following:
a. The generalized stiffness matrix. Show that it is a diagonal matrix.
b. The generalized external forces due to a uniform load. Solve the stiffness equations, and show that a_1 is a good approximation to the central deflection.
c. The generalized external forces due to end moments V_2 and V_3. Using only the first two terms of the series, compare the end rotations to their exact values.

6

STIFFNESS AND FLEXIBILITY PROPERTIES OF THE STRUCTURAL MEMBER

Geometrically Nonlinear Theory

1. INTRODUCTION

Geometric nonlinearity becomes important in structural analysis when displacements cannot be ignored in equilibrium equations. In particular, it is needed for investigating stability problems. The second-order, or beam-column, theory formulated in Chapter 3, Section 2, is adequate for treating most structures and is the basis for the development of the geometrically nonlinear theory treated in this chapter.

A major purpose of this chapter is to derive member flexibility and stiffness relations toward the view of formulating in Chapter 8 governing equations for geometrically nonlinear analysis. The desired relations are obtained first by integration of governing differential equations and then by variational or energy methods. The problem of instability and buckling occurs naturally in this material and is treated in separate sections.

The analysis of a beam-column, in the sense of determining stresses and displacements, is an important part of the design of a member but is not central to the present purpose. Some material and exercises are devoted to that topic, but developments remain short of providing general design information.

It will be helpful to have an overall view of the manner in which geometric nonlinearity occurs in beam-column theory. The governing equations have the following characteristics:

1. Transverse deformation-displacement relations and axial equilibrium equations are those of the linear theory.
2. Transverse equilibrium equations are bilinear in terms of axial forces and transverse displacements.
3. Axial deformations are of second order in terms of transverse displacements.

2. STIFFNESS AND FLEXIBILITY RELATIONS BY SOLUTION OF DIFFERENTIAL EQUATIONS

A linearly elastic uniform member subjected only to end forces is considered in this section and is referred to a member-bound reference. The constitutive equations $\boldsymbol{\varepsilon} = \boldsymbol{\Phi}\mathbf{N}$ become governing differential equations for the displacements upon expressing $\mathbf{N}$ by the equilibrium equation $\mathbf{N} = \mathbf{bV}$ and $\boldsymbol{\varepsilon}$ by the strain-displacement relations (3/2.1–6, 4c).

FIG. 2.1–1. Deformations and forces, alternate type.

2.1. Flexibility and Stiffness Relations of a Simply Supported Member (Alternate Type), Fig. 2.1–1

The preceding procedure, with **b** from Eq. (3/2.2–15), yields the system of differential equations

$$u_{,x} + \frac{1}{2}(v_{,x})^2 = \frac{V_1}{EA} \tag{2.1–1a}$$

$$\left(1 + \frac{V_1}{GA_s}\right)v_{,x} - \varphi = \frac{V_2 + V_3}{GA_s l} \tag{2.1–1b}$$

$$\varphi_{,x} - \frac{V_1}{EI}v = \frac{1}{EI}\left[\left(\frac{x}{l} - 1\right)V_2 + \frac{x}{l}V_3\right] \tag{2.1–1c}$$

The last two equations form a linear system in v and φ with constant coefficients. A particular solution in which φ is constant may be obtained without difficulty. The homogeneous equations reduce by elimination of v to

$$\varphi_{,xx} + \frac{\lambda^2}{l^2}\varphi = 0 \tag{2.1–2}$$

where

$$\lambda^2 = -\frac{V_1 l^2}{\alpha EI} \tag{2.1–3}$$

and

$$\alpha = 1 + \frac{V_1}{GA_s} \tag{2.1–4}$$

From the two preceding equations V_1 is related to λ^2 through

$$V_1 = -\alpha\frac{\lambda^2 EI}{l^2} = -\frac{\lambda^2 EI}{l^2}\frac{1}{1 + (EI/GA_s)(\lambda^2/l^2)} \tag{2.1–5}$$

The behavior of the solution depends on whether V_1 is tensile or compressive. The last case, for which $V_1 < 0$ and $\lambda^2 > 0$, is considered first. Superposition of the general solution of the homogeneous equations and of the particular solution yields

$$\varphi = -\frac{V_2 + V_3}{lV_1} + A\sin\frac{\lambda x}{l} + B\cos\frac{\lambda x}{l} \tag{2.1–6a}$$

$$v = -\frac{1}{V_1}\left[\left(\frac{x}{l} - 1\right)V_2 + \frac{x}{l}V_3\right] + \frac{EI\lambda}{lV_1}\left(A\cos\frac{\lambda x}{l} - B\sin\frac{\lambda x}{l}\right) \tag{2.1–6b}$$

The boundary conditions, $v = 0$ at $x = 0$ and $x = l$, determine the integration constants A and B in terms of V_2 and V_3. We thus obtain after simplification of trigonometric expressions and use of Eq. (2.1–3)

$$\begin{Bmatrix} v \\ \varphi \end{Bmatrix} = \frac{l^2}{EI\alpha\lambda^2} \begin{bmatrix} \dfrac{\sin(\lambda - \lambda x/l)}{\sin\lambda} - 1 + \dfrac{x}{l} & -\dfrac{\sin(\lambda x/l)}{\sin\lambda} + \dfrac{x}{l} \\ -\dfrac{\alpha\lambda}{l\sin\lambda}\cos\left(\lambda - \lambda\dfrac{x}{l}\right) + \dfrac{1}{l} & -\dfrac{\alpha\lambda}{l\sin\lambda}\cos\dfrac{\lambda x}{l} + \dfrac{1}{l} \end{bmatrix} \begin{Bmatrix} V_2 \\ V_3 \end{Bmatrix} \tag{2.1–7}$$

The bending deformations v_2 and v_3 are the values of φ at $x = 0$ and $x = l$, or

$$\begin{Bmatrix} v_2 \\ v_3 \end{Bmatrix} = \frac{l}{EI\alpha\lambda^2 \sin\lambda} \begin{bmatrix} \sin\lambda - \alpha\lambda\cos\lambda & \sin\lambda - \alpha\lambda \\ \sin\lambda - \alpha\lambda & \sin\lambda - \alpha\lambda\cos\lambda \end{bmatrix} \begin{Bmatrix} V_2 \\ V_3 \end{Bmatrix} \tag{2.1–8}$$

The stiffness relations obtained by inverting these relations are

$$\begin{Bmatrix} V_2 \\ V_3 \end{Bmatrix} = \frac{EI\lambda}{4l\sin(\lambda/2)[\sin(\lambda/2) - (\alpha\lambda/2)\cos(\lambda/2)]} \begin{bmatrix} \sin\lambda - \alpha\lambda\cos\lambda & \alpha\lambda - \sin\lambda \\ \alpha\lambda - \sin\lambda & \sin\lambda - \alpha\lambda\cos\lambda \end{bmatrix} \begin{Bmatrix} v_2 \\ v_3 \end{Bmatrix} \tag{2.1–9}$$

It will be convenient to have a separate notation for axial quantities and to let $\mathbf{v} = \{v_2 \ \ v_3\}$ and $\mathbf{V} = \{V_2 \ \ V_3\}$. $\mathbf{f}$ and $\mathbf{k} = \mathbf{f}^{-1}$ are then the bending flexibility and stiffness matrices appearing in Eqs. (2.1–8) and (2.1–9), respectively.

The axial displacement u is determined from Eq. (2.1–1a) by integration of $u_{,x}$. Thus

$$u = \frac{V_1}{EA}x - \int_0^x \frac{1}{2}(v_{,x})^2\,dx \tag{2.1–10}$$

and at $x = l$,

$$v_1 = \frac{V_1 l}{EA} - \int_0^l \frac{1}{2}(v_{,x})^2\,dx \tag{2.1–11}$$

The last term represents the effect of geometric nonlinearity on the axial flexibility relation. It may be evaluated using the first of Eqs. (2.1–7). The result is the quadratic form

$$\int_0^l \frac{1}{2}(v_{,x})^2\,dx = \frac{1}{2lV_1^2}\mathbf{V}^T\mathbf{\Lambda}\mathbf{V} \tag{2.1–12}$$

where

$$\mathbf{\Lambda} = \begin{bmatrix} \dfrac{\lambda(2\lambda + \sin 2\lambda)}{4\sin^2\lambda} - 1 & \dfrac{\lambda(\sin\lambda + \lambda\cos\lambda)}{2\sin^2\lambda} - 1 \\ \swarrow & \searrow \end{bmatrix} \tag{2.1–13}$$

A representation in terms of $\mathbf{v} = \{v_2 \ \ v_3\}$ is also useful and is obtained by substituting **kv** for **V**. The result has the form

$$\int_0^l \frac{1}{2}(v_{,x})^2\,dx = \frac{1}{2}\mathbf{v}^T\mathbf{g}\mathbf{v} \tag{2.1–14}$$

and the axial stiffness relation is then

$$V_1 = \frac{EA}{l}\left(v_1 + \frac{1}{2}\mathbf{v}^T\mathbf{g}\mathbf{v}\right) \tag{2.1–15}$$

g may be evaluated through the formulas

$$\mathbf{g} = \begin{bmatrix} a+b & b-a \\ b-a & a+b \end{bmatrix} \tag{2.1–16a}$$

$$a = \frac{l}{8\lambda}\frac{\lambda - \sin\lambda}{\alpha^2 \sin^2(\lambda/2)} \tag{2.1–16b}$$

$$b = \frac{l}{32}\frac{\lambda(\lambda+\sin\lambda) - 8\sin^2(\lambda/2)}{[\sin(\lambda/2) - (\alpha\lambda/2)\cos(\lambda/2)]^2} \tag{2.1–16c}$$

The effect of shear deformation is represented in the preceding solution by $\alpha = 1 + V_1/GA_s$. Except for special members much softer in shear than in axial compression, $V_1/GA_s \ll 1$, and shear deformation is negligible. Built-up members are sometimes modeled as a beam-column with uniform properties, and shear deformation may be nonnegligible for such members [6, p. 135].

The main property of member behavior represented in Eqs. (2.1–8, 9) is that for a fixed axial force the bending relations are linear, but the influence coefficients (flexibility or stiffness) are functions of the axial force. By contrast, the axial deformation v_1 is related nonlinearly to all three forces V_1, V_2, and V_3.

The bending influence coefficients are tabulated in nondimensional form in Table 2.1–1 by dividing them by their respective values in the linear theory. As $V_1 \to 0$, these coefficients take the indeterminate form 0/0, but the limits could be shown by means of series developments in powers of λ^2 to be the values in the linear theory. Shear deformation is neglected in the tabulated values.

In Eq. (2.1–15) the nonlinear term represents a shortening of the member chord due to the deflected shape of the member. This term does not vanish for $V_1 = 0$ but is neglected in the linear theory.

The parameter λ, Eq. (2.1–3), is called the stability parameter. An examination of Table 2.1–1 shows that f_{22} and f_{23} are increasing functions of λ—that is, the member becomes more flexible in bending as the compressive force increases. Further f_{22} and f_{23} tend to infinity as λ tends to the critical value $\lambda_{cr} = \pi$, for which

$$-V_{1,cr} = \frac{\pi^2 EI}{l^2}\frac{1}{1 + (\pi^2/l^2)(EI/GA_s)} \tag{2.1–17}$$

TABLE 2.1–1. Flexibility and stiffness coefficients, alternate type.

λ	$\dfrac{V_1}{V_{1,cr}}=\dfrac{\lambda^2}{\pi^2}$	$\dfrac{3EI}{l}f_{22}$	$-\dfrac{6EI}{l}f_{23}$	$\dfrac{l}{4EI}k_{22}$	$\dfrac{l}{2EI}k_{23}$
0.	0.	1.	1.	1.	1.
0.2	0.00405	1.00268	1.00469	0.99867	1.00067
0.4	0.01621	1.01083	1.01899	0.99466	1.00268
0.6	0.03648	1.02485	1.04366	0.98794	1.00607
0.8	0.06485	1.04545	1.08006	0.97849	1.01088
1.0	0.10132	1.07372	1.13037	0.96622	1.01720
1.2	0.14590	1.11138	1.19792	0.95107	1.02511
1.4	0.19859	1.16102	1.28777	0.93292	1.03476
1.6	0.25938	1.22665	1.40785	0.91164	1.04631
1.8	0.32828	1.31477	1.57100	0.88708	1.05996
2.0	0.40528	1.43649	1.79925	0.85903	1.07596
2.2	0.49039	1.61242	2.13360	0.82726	1.09465
2.4	0.58361	1.88544	2.65950	0.79147	1.11640
2.6	0.68493	2.36176	3.58902	0.75131	1.14172
2.8	0.79436	3.39626	5.63151	0.70635	1.17123
3.0	0.91189	7.34859	13.5057	0.65605	1.20573
π	1.	∞	∞	$\pi^2/16$	$\pi^2/8$
3.2	1.0375	−15.7398	−32.7063	0.5997	1.2462
4.0	1.6211	−0.4603	−2.3570	0.2933	1.5019
4.5	2.0518	0.0044	−1.6603	−0.0048	1.8070
5.0	2.5330	0.2975	−1.4914	−0.4772	2.3923
6.0	3.6476	1.8015	−3.7456	−5.1594	10.7270
2π	4.	∞	$-\infty$	$-\infty$	∞

$V_{1,cr}$ is called Euler's buckling load. To interpret $V_{1,cr}$, consider a member in pure compression. Thus $\mathbf{V}=0$, and the bending stiffness equations are $\mathbf{kv}=0$. For $|V_1|<|V_{1,cr}|$, $\mathbf{k}$ is nonsingular, and $\mathbf{v}=0$. For $V_1=V_{1,cr}$, $\mathbf{k}$ is singular, and consequently the homogeneous equations $\mathbf{kv}=0$ have a nonzero solution. A bent or buckled shape becomes thus possible at $V_1=V_{1,cr}$ and is called an alternate equilibrium state, Fig. 2.1–2. In the solution of the differential equation the integration constant A is zero, and B remains indeterminate. The buckled shape for v is $\sin(\pi x/l)$. The indeterminate amplitude must be interpreted physically as infinitesimal. A more refined theory is needed to analyze postbuckling behavior [6, p. 76]. For $|V_1|>|V_{1,cr}|$, the straight configuration of the member becomes unstable. The question of stability will be treated further in Section 5.

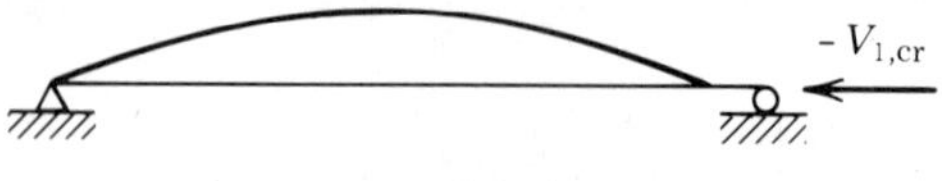

FIG. 2.1–2. Euler buckling.

For a beam-column subjected to a given $\mathbf{V}$ and an increasing compressive force $(-V_1)$, deformations increase nonlinearly in terms of V_1. The flexibility relations remain valid as long as the limit of linear elasticity or the geometric limit of beam-column theory are not exceeded. For usual structural members the elastic limit is reached first. The ratio $V_1/V_{1,\mathrm{cr}}$ is a useful parameter for assessing geometric nonlinearity of a beam-column in combined bending and compression.

The case where V_1 is tensile is now examined briefly. λ^2, as defined in Eq. (2.1–3), is then negative. The algebra of the solution remains valid with complex variables if the substitutions $\lambda \to i\lambda$, $A \to iA$, $B \to B$, $\sin(\lambda x/l) \to i \sinh(\lambda x/l)$, $\cos(\lambda x/l) \to \cosh(\lambda x/l)$, and $i^2 = -1$ are made. It is found that the member becomes stiffer in bending as the tensile force increases.

2.2 Flexibility and Stiffness Relations of Cantilever Type, Fig. 2.2–1

Following a similar procedure as in the previous case, using the equilibrium equation (3/2.2–14), the deformed shape is found to be

$$\begin{Bmatrix} v \\ \varphi \end{Bmatrix} = \frac{l^3}{EI} \begin{bmatrix} \dfrac{\sin\lambda - \sin(\lambda - \lambda x/l) - \alpha(x/l)\lambda\cos\lambda}{\alpha^2\lambda^3\cos\lambda} & \dfrac{1-\cos(\lambda x/l)}{\alpha\lambda^2 l\cos\lambda} \\ \dfrac{\cos(\lambda - \lambda x/l) - \cos\lambda}{\alpha\lambda^2 l\cos\lambda} & \dfrac{\sin(\lambda x/l)}{l^2\lambda\cos\lambda} \end{bmatrix} \begin{Bmatrix} V_2 \\ V_3 \end{Bmatrix} \tag{2.2–1}$$

The bending deformations are $v_2 = (v)_{x=l}$ and $v_3 = (\varphi)_{x=l}$. Thus

$$\begin{Bmatrix} v_2 \\ v_3 \end{Bmatrix} = \begin{bmatrix} \dfrac{l^3}{EI}\dfrac{1}{\alpha\lambda^2}\left(\dfrac{\tan\lambda}{\alpha\lambda} - 1\right) & \dfrac{l^2}{EI}\dfrac{1-\cos\lambda}{\alpha\lambda^2\cos\lambda} \\ \dfrac{l^2}{EI}\dfrac{1-\cos\lambda}{\alpha\lambda^2\cos\lambda} & \dfrac{l}{EI}\dfrac{\tan\lambda}{\lambda} \end{bmatrix} \begin{Bmatrix} V_2 \\ V_3 \end{Bmatrix} \tag{2.2–2}$$

The inverse relations are found to be

$$\begin{Bmatrix} V_2 \\ V_3 \end{Bmatrix} = \frac{EI\alpha\lambda^2}{2l^2[\tan(\lambda/2) - (\alpha\lambda/2)]} \begin{bmatrix} \dfrac{\alpha\lambda}{l} & -\tan\dfrac{\lambda}{2} \\ -\tan\dfrac{\lambda}{2} & \dfrac{l}{\alpha\lambda}\left(1 - \dfrac{\alpha\lambda}{\tan\lambda}\right) \end{bmatrix} \begin{Bmatrix} v_2 \\ v_3 \end{Bmatrix} \tag{2.2–3}$$

and v_1 is put in the form of Eq. (2.1–11, 12) with

$$\mathbf{\Lambda} = \begin{bmatrix} l^2\left[1 + \left(\dfrac{1}{2\alpha^2} - \dfrac{2}{\alpha}\right)\dfrac{\tan\lambda}{\lambda} + \dfrac{1}{2\alpha^2\cos^2\lambda}\right] & \dfrac{l}{\cos\lambda}\left(\dfrac{\lambda\tan\lambda}{2\alpha} + \cos\lambda - 1\right) \\ \text{symmetric} & \dfrac{\lambda^2}{2\cos^2\lambda}\left(1 - \dfrac{\sin 2\lambda}{2\lambda}\right) \end{bmatrix} \tag{2.2–4}$$

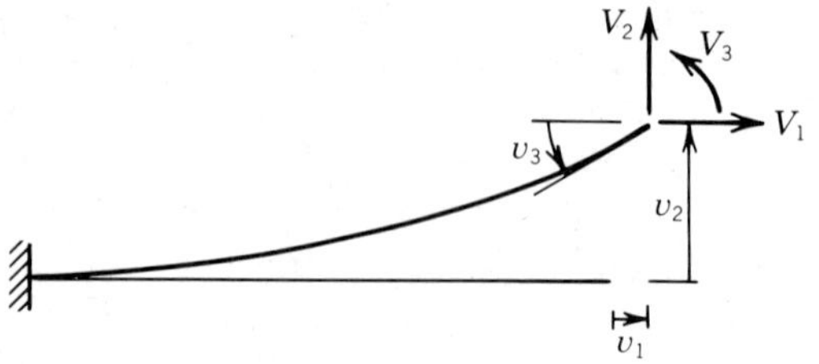

FIG. 2.2–1. Deformations and forces, cantilever type.

TABLE 2.2–1. Flexibility coefficients, cantilever type.

λ	$\dfrac{V_1}{V_{1,\mathrm{cr}}} = \dfrac{4\lambda^2}{\pi^2}$	$\dfrac{3EI}{l^3} f_{22}$	$\dfrac{2EI}{l^2} f_{23}$	$\dfrac{EI}{l} f_{33}$
0.	0.	1.	1.	1.
0.1	0.00405	1.00402	1.00418	1.00335
0.2	0.01621	1.01626	1.01694	1.01355
0.3	0.03648	1.03736	1.03892	1.03112
0.4	0.06485	1.06843	1.07131	1.05698
0.5	0.10132	1.11126	1.11595	1.09260
0.6	0.14590	1.16857	1.17571	1.14023
0.7	0.19859	1.24450	1.25494	1.20327
0.8	0.25938	1.34554	1.36039	1.28705
0.9	0.32828	1.48213	1.50303	1.40018
1.0	0.40528	1.67222	1.70163	1.55741
1.1	0.49039	1.94912	1.99108	1.78615
1.2	0.58361	2.38221	2.44403	2.14346
1.3	0.68493	3.14352	3.24063	2.77084
1.4	0.79436	4.80818	4.98315	4.14135
1.5	0.91189	11.2013	11.67718	9.40095
$\pi/2$	1.	∞	∞	∞
2.5	2.5330	−0.6234	−0.7194	−0.2988
π	4.	$-3/\pi^2$	$-4/\pi^2$	0.
4.	6.4846	−0.1332	−0.3162	0.2895
$3\pi/2$	9.	∞	$-\infty$	∞
5.	10.1321	−0.2011	0.2020	−0.6761
2π	16.	$-3/(4\pi^2)$	0.	0.

FIG. 2.2–2. Cantilever buckling.

TABLE 2.2–2. Stiffness coefficients, cantilever type.

λ	$\dfrac{V_1}{V_{1,cr}} = \dfrac{4\lambda^2}{\pi^2}$	$\dfrac{l^3}{12EI} k_{22}$	$-\dfrac{l^2}{6EI} k_{23}$	$\dfrac{l}{4EI} k_{33}$
0.	0.	1.	1.	1.
0.1	0.00405	0.99900	0.99983	0.99967
0.2	0.01621	0.99600	0.99933	0.99867
0.3	0.03648	0.99100	0.99850	0.99700
0.4	0.06485	0.98400	0.99733	0.99466
0.5	0.10132	0.97499	0.99583	0.99164
0.6	0.14590	0.96398	0.99398	0.98794
0.7	0.19859	0.95097	0.99180	0.98356
0.8	0.25938	0.93595	0.98928	0.97849
0.9	0.32828	0.91892	0.98642	0.97271
1.0	0.40528	0.89988	0.98321	0.96622
1.1	0.49039	0.87882	0.97966	0.95901
1.2	0.58361	0.85575	0.97575	0.95107
1.3	0.68493	0.83065	0.97149	0.94237
1.4	0.79436	0.80353	0.96687	0.93292
1.5	0.91189	0.77438	0.96188	0.92268
$\pi/2$	1.	0.75251	0.95813	0.91495
2.5	2.5330	0.3700	0.8908	0.7720
π	4.	0.	$\pi^2/12$	$\pi^2/16$
4.	6.4846	−0.6372	0.6961	0.2933
$3\pi/2$	9.	−1.2992	0.5514	−0.1755
5.	10.1321	−1.6040	0.4793	−0.4772
2π	16.	$-\pi^2/3$	0.	$-\infty$

The influence coefficients are tabulated in nondimensional form in Tables 2.2–1 and 2.2–2, with shear deformation neglected. The role of the compressive axial force is qualitatively similar to that discussed earlier. The critical value of λ is found here to be $\lambda_{cr} = \pi/2$. The corresponding $V_{1,cr}$ is thus one-fourth that of the simply supported member, Fig. 2.2–2. Critical loads for additional restraints at the member ends are seen in Section 5.

2.3. Flexibility and Stiffness Relations with Member Transverse load

Consider now a member referred to a bound reference and subjected to $\{V_1 \;\; V_2 \;\; V_3\}$ and to a transverse span load $\{p_y \;\; m\}$. Since $p_x = 0$, the axial force N is constant and equal to V_1. A basic property of the bending governing equations is that they are linear and their coefficients depend linearly on V_1 as a parameter. Solutions corresponding to different transverse loading conditions may thus be superimposed, provided each is obtained with V_1 applied. The bending flexibility relations have then the form

$$\mathbf{v} = \mathbf{f}\mathbf{V} + \mathbf{v}_p \tag{2.3–1}$$

where $\mathbf{f}$ is the flexibility matrix determined earlier and $\mathbf{v}_p = \{\mathbf{v}_{2p} \ \mathbf{v}_{3p}\}$ are the deformations due to the span load $\mathbf{p}$ and to V_1, with $V_2 = V_3 = 0$. $\mathbf{v}_p$ will be determined first for a point load, $\mathbf{P} = \{P_y \ C\}$, applied at point ξ. Green's function method (Chapter 5, Section 2) will then be applied to obtain the solution for any transverse loading. $\mathbf{v}_p$ may be obtained by solving the governing differential equations, but a much shorter procedure is to use Betti's reciprocal theorem (Chapter 5, Section 5) which applies to the linear bending equations. Betti's reciprocal theorem is applied to two states. State (1) refers to the problem to be solved, namely the member subjected to $\mathbf{P}$ at point ξ and to V_1. The deformations in that state are $\mathbf{v}_p$. State (2) refers to the member subjected only to $\mathbf{V}$ and V_1. The deformed shape $\mathbf{u} = \{v \ \varphi\}$ in that state was determined earlier, Eqs. (2.1–7) and (2.2–1). At point ξ, $\mathbf{u}$ has the form $\mathbf{u}^\xi = \mathbf{G}_\xi^T \mathbf{V}$. Equating the external virtual works of the transverse forces of one state through the transverse displacements of the other, there comes $\mathbf{V}^T \mathbf{v}_p = \mathbf{P}^T \mathbf{u}^\xi = \mathbf{P}^T \mathbf{G}_\xi^T \mathbf{V}$. The last term is a scalar equal to its transpose. $\mathbf{V}$ being arbitrary, we obtain

$$\mathbf{v}_p = \mathbf{G}_\xi \mathbf{P} \tag{2.3–2}$$

For the cantilever $\mathbf{G}_\xi$ is the transpose of the coefficient matrix in Eq. (2.2–1), evaluated at $x = \xi$. Thus

$$\mathbf{G}_\xi = \frac{l^3}{EI} \left[\begin{array}{c|c} \dfrac{\sin\lambda - \sin(\lambda - \lambda\xi/l) - \alpha(\xi/l)\lambda\cos\lambda}{\alpha^2\lambda^3\cos\lambda} & \dfrac{\cos(\lambda - \lambda\xi/l) - \cos\lambda}{\alpha\lambda^2 l\cos\lambda} \\ \hline \dfrac{1 - \cos(\lambda\xi/l)}{\alpha\lambda^2 l\cos\lambda} & \dfrac{\sin(\lambda\xi/l)}{l^2\lambda\cos\lambda} \end{array} \right] \tag{2.3–3}$$

For the simply supported member, we obtain similarly from Eq. (2.1–7),

$$\mathbf{G}_\xi = \frac{l^2}{EI}\frac{1}{\alpha\lambda^2} \begin{bmatrix} \dfrac{\sin(\lambda - \lambda\xi/l)}{\sin\lambda} - 1 + \dfrac{\xi}{l} & -\dfrac{\alpha\lambda}{l\sin\lambda}\cos\left(\lambda - \dfrac{\lambda\xi}{l}\right) + \dfrac{1}{l} \\ -\dfrac{\sin(\lambda\xi/l)}{\sin\lambda} + \dfrac{\xi}{l} & -\dfrac{\alpha\lambda}{l\sin\lambda}\cos\dfrac{\lambda\xi}{l} + \dfrac{1}{l} \end{bmatrix} \tag{2.3–4}$$

For a general loading function $\mathbf{p}(\xi) = \{p_y \ m\}$, Green's function method allows to evaluate $\mathbf{v}_p$ by means of the integral

$$\mathbf{v}_p = \int_0^l \mathbf{G}_\xi \mathbf{p}(\xi)\, d\xi \tag{2.3–5}$$

Results for typical loadings are given in Figs. 2.3–1 and 2.3–2.

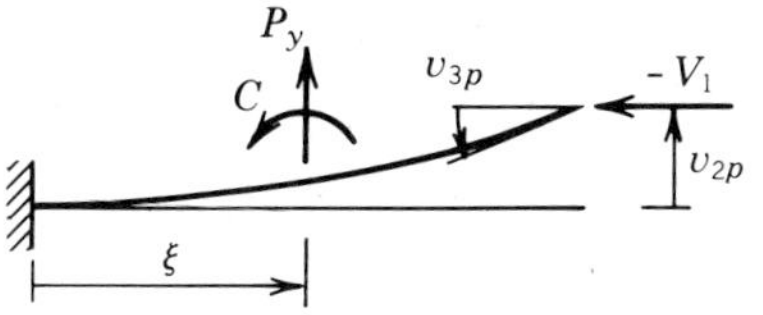

$$\begin{Bmatrix} v_{2p} \\ v_{3p} \end{Bmatrix} = [\mathbf{G}_\xi] \begin{Bmatrix} P_y \\ C \end{Bmatrix}, \quad \text{see Eq. (6/2.3–3)}$$

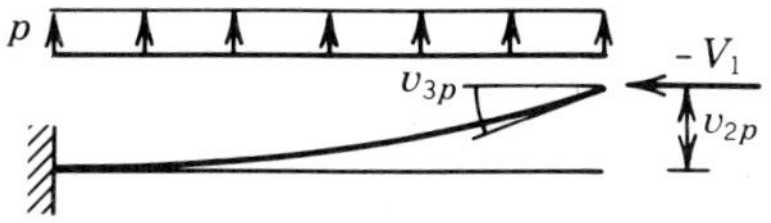

$$v_{2p} = \frac{pl^4}{EI\alpha^2\lambda^4}\left(\lambda \tan\lambda - \frac{1}{\cos\lambda} + 1 - \frac{\alpha\lambda^2}{2}\right)$$

$$v_{3p} = \frac{pl^3}{EI\alpha\lambda^3}\left(\frac{\lambda}{\cos\lambda} - \tan\lambda\right)$$

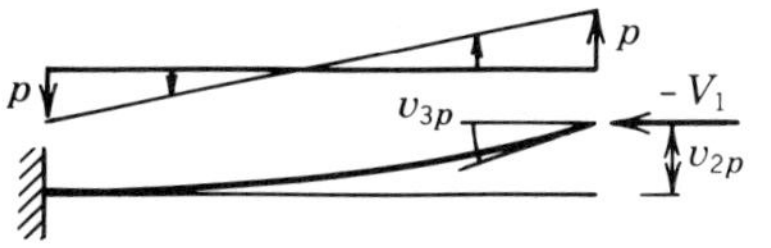

$$v_{2p} = \frac{pl^4}{EI\alpha^2\lambda^5}\left[-\frac{\alpha\lambda^3}{6} - \lambda\left(1 + \frac{1}{\cos\lambda}\right) + 2\tan\lambda\right]$$

$$v_{3p} = \frac{2pl^3}{EI\alpha\lambda^4}\left(-\frac{\lambda}{2}\tan\lambda - 1 + \frac{1}{\cos\lambda}\right)$$

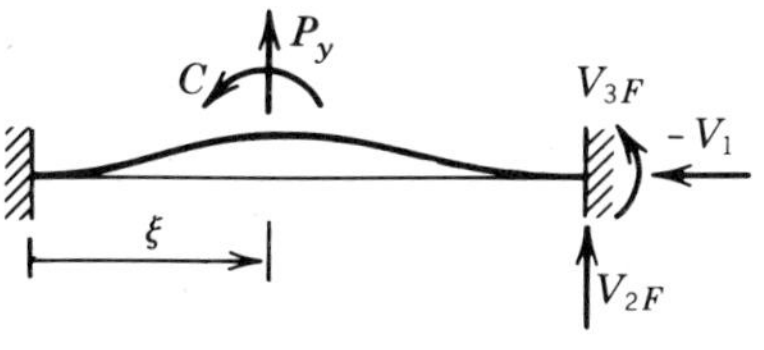

$$\begin{Bmatrix} V_{2F} \\ V_{3F} \end{Bmatrix} = [-\boldsymbol{\psi}_\xi^T] \begin{Bmatrix} P_y \\ C \end{Bmatrix}, \quad \text{see Eq. (6/2.3–10)}$$

$$\boldsymbol{\psi}_\xi = \mathbf{k}\mathbf{G}_\xi$$

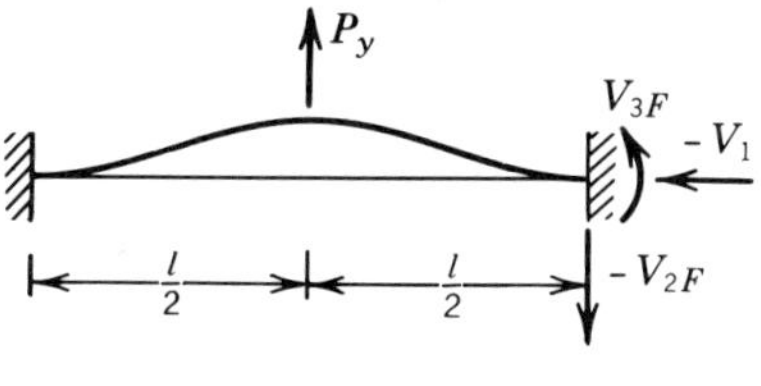

$$V_{2F} = -\frac{P_y}{2}$$

$$V_{3F} = \frac{P_y l}{2\alpha\lambda}\tan\frac{\lambda}{4}$$

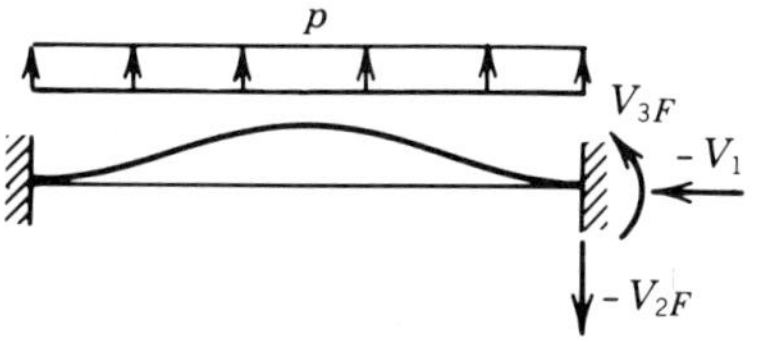

$$V_{2F} = -\frac{pl}{2}$$

$$V_{3F} = \frac{pl^2}{\alpha\lambda^2}\left(1 - \frac{\lambda}{2}\cot\frac{\lambda}{2}\right)$$

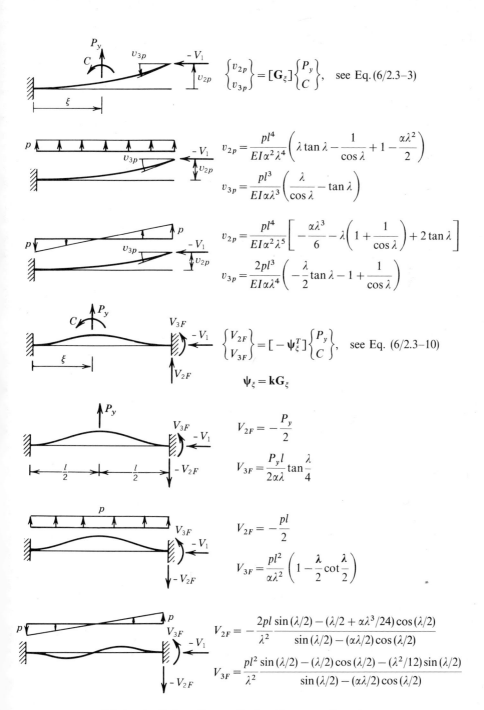

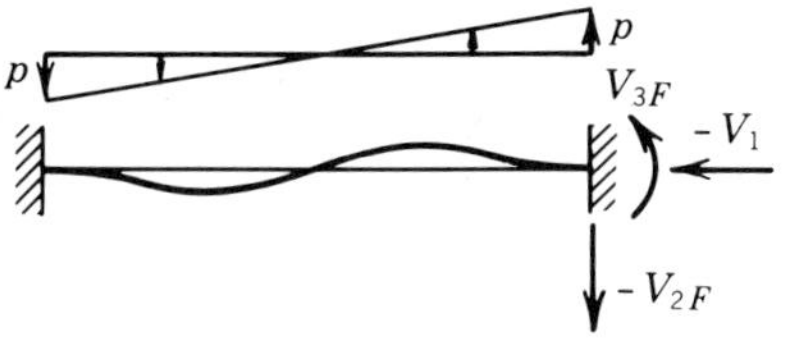

$$V_{2F} = -\frac{2pl}{\lambda^2}\,\frac{\sin(\lambda/2) - (\lambda/2 + \alpha\lambda^3/24)\cos(\lambda/2)}{\sin(\lambda/2) - (\alpha\lambda/2)\cos(\lambda/2)}$$

$$V_{3F} = \frac{pl^2}{\lambda^2}\,\frac{\sin(\lambda/2) - (\lambda/2)\cos(\lambda/2) - (\lambda^2/12)\sin(\lambda/2)}{\sin(\lambda/2) - (\alpha\lambda/2)\cos(\lambda/2)}$$

FIG. 2.3–1. Deformations $\mathbf{v}_p$, fixed end forces, cantilever type.

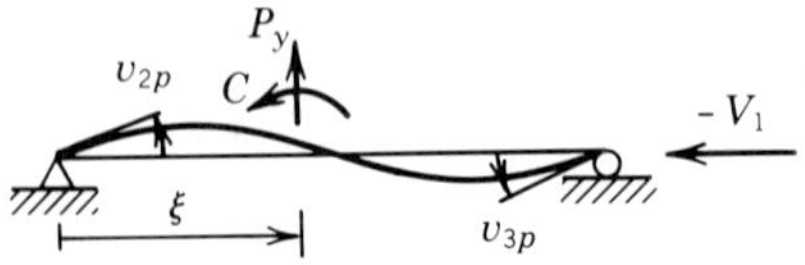

$$\begin{Bmatrix} v_{2p} \\ v_{3p} \end{Bmatrix} = [\mathbf{G}_\xi] \begin{Bmatrix} P_y \\ C \end{Bmatrix}, \quad \text{see Eq. (6/2.3–4)}$$

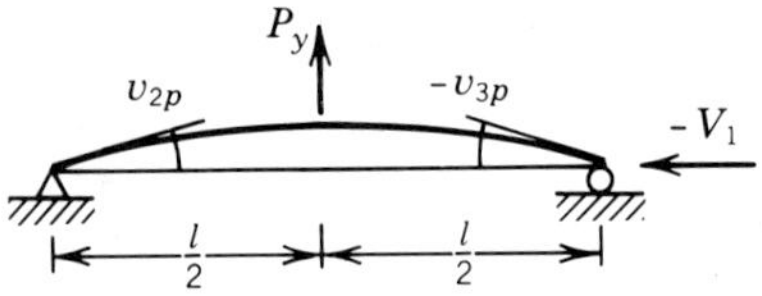

$$v_{2p} = -v_{3p} = \frac{P_y l^2}{2EI\alpha\lambda^2}\left(\frac{1}{\cos\lambda/2} - 1\right)$$

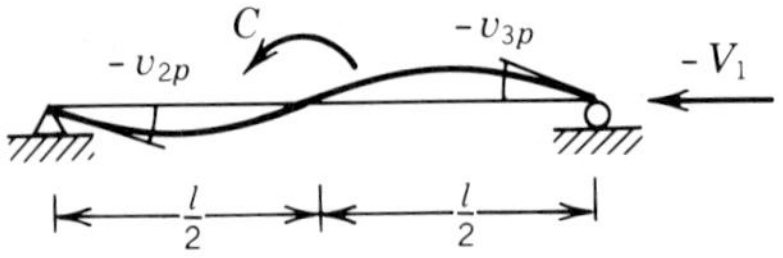

$$v_{2p} = v_{3p} = -\frac{Cl}{EI\alpha\lambda^2}\left(\frac{\alpha\lambda}{2}\frac{1}{\sin\lambda/2} - 1\right)$$

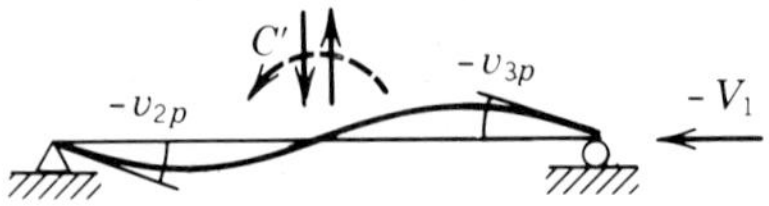

$$v_{2p} = v_{3p} = -\frac{C'l}{EI\alpha\lambda^2}\left(\frac{\lambda}{2}\frac{1}{\sin\lambda/2} - 1\right)$$

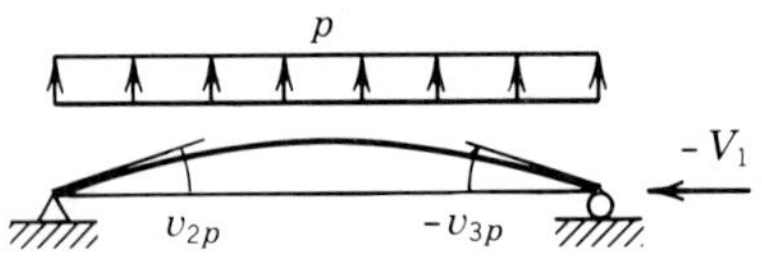

$$v_{2p} = -v_{3p} = \frac{pl^3}{EI\alpha\lambda^3}\left(\tan\frac{\lambda}{2} - \frac{\lambda}{2}\right)$$

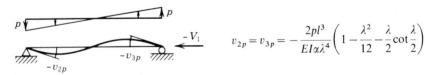

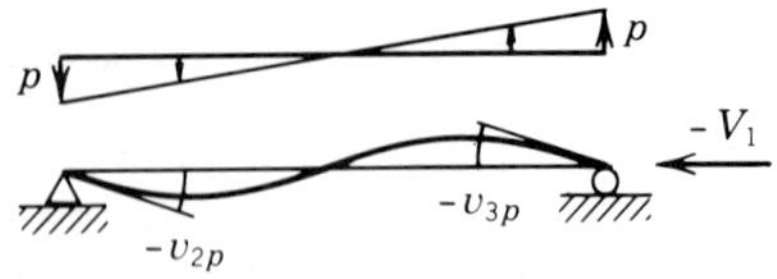

$$v_{2p} = v_{3p} = -\frac{2pl^3}{EI\alpha\lambda^4}\left(1 - \frac{\lambda^2}{12} - \frac{\lambda}{2}\cot\frac{\lambda}{2}\right)$$

$$\frac{\alpha EI\lambda^2}{l^2} = -V_1, \quad \alpha = 1 + \frac{V_1}{GA_s}$$

FIG. 2.3–2. Deformations $\mathbf{v}_p$, alternate type.

If a concentrated moment is applied by means of transverse forces, and shear deformation is to be taken into account, the procedure outlined in Example 3 of Chapter 5, Section 2, may be used. $\mathbf{v}_p$ is then obtained by means of Eq. (5/2–21) in which $\mathbf{G}_{2\xi}$ is the first column of $\mathbf{G}_\xi$ as obtained here.

The axial flexibility relation is modified by the transverse span load because its nonlinear term depends on the deformed shape. Instead of a tedious exact calculation, the nonlinear term may be approximated by Eq. (2.1–14), in which $\mathbf{v}$ is now due to the total loading. The basis for this approximation is to represent the deformed shape v in the form

$$v = \boldsymbol{\psi}\mathbf{v} \tag{2.3–6}$$

and to adopt for $\boldsymbol{\psi}$ the shape functions for the member subjected to end actions only, then

$$\mathbf{g} = \int_0^l \boldsymbol{\psi}_{,x}^T \boldsymbol{\psi}_{,x}\, dx \tag{2.3–7}$$

$\mathbf{g}$ is given for the alternate type of relations in Eq. (2.1–16). Explicit expressions for $\boldsymbol{\psi}$ are encountered in the next topic.

Stiffness Relations with Member Transverse Load

Bending stiffness relations with member transverse load are obtained by inverting the flexibility relations. Thus

$$\mathbf{V} = \mathbf{k}\mathbf{v} + \mathbf{V}_F \tag{2.3–8}$$

where $\mathbf{k} = \mathbf{f}^{-1}$ was determined earlier and

$$\mathbf{V}_F = -\mathbf{k}\mathbf{v}_p \tag{2.3–9}$$

Elements of $\mathbf{V}_F$ are interpreted to be fixed end forces, as in Chapter 5, Section 3. For the case of a point load $\mathbf{P} = \{P_y \;\; C\}$ applied at ξ, $\mathbf{V}_F = -\mathbf{k}\mathbf{G}_\xi\mathbf{P}$. For the cantilever, with $\eta = (l/2) - \xi$,

$$V_{2F} = -\frac{1}{2}\left[1 + \frac{(\alpha\eta\lambda/l)\cos(\lambda/2) - \sin(\lambda\eta/l)}{\sin(\lambda/2) - (\alpha\lambda/2)\cos(\lambda/2)}\right]P_y - \frac{\alpha\lambda}{2l}\frac{\cos(\eta\lambda/l) - \cos(\lambda/2)}{\sin(\lambda/2) - (\alpha\lambda/2)\cos(\lambda/2)}C \tag{2.3–10a}$$

$$V_{3F} = l\left\{\frac{\cos(\eta\lambda/l) - \cos(\lambda/2)}{2\alpha\lambda\sin(\lambda/2)} + \frac{(2\eta/l)\sin(\lambda/2) - \sin(\eta\lambda/l)}{4[\sin(\lambda/2) - (\alpha\lambda/2)\cos(\lambda/2)]}\right\}P_y + \frac{1}{2}\left[\frac{\sin(\eta\lambda/l)}{\sin(\lambda/2)} - \frac{\sin(\lambda/2) - (\alpha\lambda/2)\cos(\eta\lambda/l)}{\sin(\lambda/2) - (\alpha\lambda/2)\cos(\lambda/2)}\right]C \tag{2.3–10b}$$

For the simply supported member,

$$V_{2F} = l\left\{ -\frac{\cos(\eta\lambda/2) - \cos(\lambda/2)}{2\alpha\lambda\sin(\lambda/2)} + \frac{(2\eta/l)\sin(\lambda/2) - \sin(\eta\lambda/l)}{4[\sin(\lambda/2) - (\alpha\lambda/2)\cos(\lambda/2)]} \right\} P_y + \frac{1}{2}\left[-\frac{\sin(\eta\lambda/l)}{\sin(\lambda/2)} - \frac{\sin(\lambda/2) - (\alpha\lambda/2)\cos(\eta\lambda/l)}{\sin(\lambda/2) - (\alpha\lambda/2)\cos(\lambda/2)} \right] C \tag{2.3–11}$$

V_{3F} is the same as in Eq. (2.3–10b).

For a general loading $\mathbf{V}_F$ may be obtained by means of Eq. (2.3–9) or by direct application of Green's function method. Some results are given in Fig. 2.3–1. A property of Green's functions for $\mathbf{V}_F$ was established in Chapter 5, Section 5, as an application of Betti's reciprocal theorem. This property is that the deformed shape due to end forces only, expressed in the form $\mathbf{u} = \boldsymbol{\psi}\mathbf{v}$, where $\boldsymbol{\psi} = \boldsymbol{\psi}(x)$, allows us to evaluate $\mathbf{V}_F$ due to a point load $\mathbf{P}$ applied at point ξ, through $\mathbf{V}_F = -\boldsymbol{\psi}_\xi^T\mathbf{P}$ where $\boldsymbol{\psi}_\xi = \boldsymbol{\psi}(\xi)$. The coefficient matrices in Eqs. (2.3–10, 11) are two forms of $-\boldsymbol{\psi}_\xi^T$. The corresponding shape functions $\boldsymbol{\psi}$ are obtained from $\boldsymbol{\psi}_\xi$ by replacing ξ with x, that is, η with $(l/2) - x$.

2.4. Deformed Shape Caused by a Span Load

The deformed shape in a beam-column is an important part of member analysis because it is needed for the evaluation of the bending moment and the shear force. It may be determined by superposition of the solutions due, respectively, to $\{V_1 \ V_2 \ V_3\}$ and to the span load acting together with V_1. The case of a point force P_y applied at ξ is fundamental because it allows us to obtain by superposition the solution due to a general loading. The deformed shape of a simply supported uniform member subjected to P_y and V_1, Fig. 2.4–1, is determined in what follows. Instead of using the differential equation method, $\{v \ \varphi\}$ will be determined by using the stiffness properties of portion AB of the member, referred to the bound reference (x_r, y_r), Fig. 2.4–1. The deformation at A is

$$v_2' = v_{2p} - \frac{v}{x} \tag{2.4–1}$$

In the equation v_{2p} is given the expression found in Section 2.3, and v_2' is expressed by the flexibility relation for a member of length x, with $\alpha' = \alpha$, and

$$\lambda'^2 = -\frac{V_1 x^2}{\alpha EI} = \lambda^2\frac{x^2}{l^2} \tag{2.4–2}$$

Equation (2.4–1) is then solved for v. The rotation φ may be determined from the flexibility relation for the deformation at B, or equivalently, from the stress-

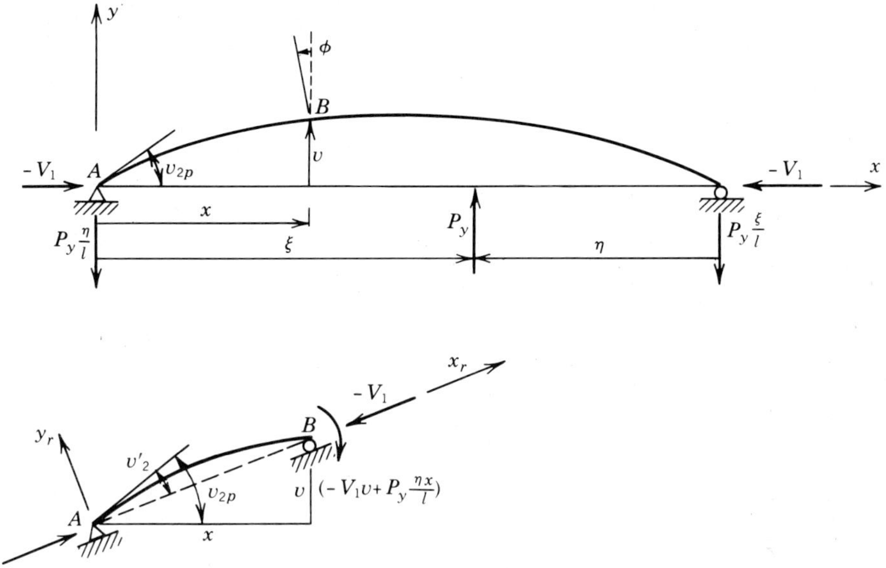

FIG. 2.4–1. Evaluation of v.

displacement relation for shear

$$v_{,x} - \varphi = \frac{1}{GA_s}\left(P_y\frac{\eta}{l} - V_1 v_{,x}\right) \tag{2.4–3}$$

The result may be put in the form

$$v = -\frac{P_y}{V_1}\left(\frac{l}{\alpha\lambda \sin\lambda}\sin\frac{\lambda\eta}{l}\sin\frac{\lambda x}{l} - \frac{\eta x}{l}\right), \qquad x \leqslant l - \eta \tag{2.4–4a}$$

$$\varphi = -\frac{P_y}{V_1}\left(\frac{1}{\sin\lambda}\sin\frac{\lambda\eta}{l}\cos\frac{\lambda x}{l} - \frac{\eta}{l}\right), \quad x \leqslant l - \eta \tag{2.4–4b}$$

For $x > l - \eta$, η is replaced by ξ, x by $l - x$, and the sign of φ is changed in these equations.

For a load function $p_y = p_y(\xi)$, application of the Green function method yields for the deflection function $v(x)$,

$$v = -\frac{1}{V_1}\int_0^x \left[\frac{l}{\alpha\lambda\sin\lambda}\sin\frac{\lambda\xi}{l}\sin\left(\lambda - \frac{\lambda x}{l}\right) - \xi\left(1 - \frac{x}{l}\right)\right] p_y\, d\xi$$
$$-\frac{1}{V_1}\int_0^{l-x}\left(\frac{l}{\alpha\lambda\sin\lambda}\sin\frac{\lambda\eta}{l}\sin\frac{\lambda x}{l} - \frac{\eta x}{l}\right)p_y\, d\eta \tag{2.4–5}$$

For a uniform load $p_y = p$, the result of the integration may be put in the form

$$v = -\frac{p}{V_1}\left[\frac{2l^2}{\alpha\lambda^2 \cos(\lambda/2)} \sin\frac{\lambda x}{2l} \sin\left(\frac{\lambda}{2} - \frac{\lambda x}{2l}\right) - \frac{1}{2}x(l-x)\right] \qquad (2.4\text{–}6)$$

2.5. Deformed Shape with Initial Imperfections, Fig. 2.5–1

Consider an initially curved member whose shape is $v_I = v_I(x)$. When subjected to an axial force V_1, the member deforms into the shape $v = v(x)$. Neglecting shear deformation, the change in curvature is $\chi = (v - v_I)_{,xx}$. For a simply supported member, the bending moment is $M = V_1 v$. The equation, $\chi = M/EI$, takes the form

$$v_{,xx} - \frac{V_1}{EI} v = v_{I,xx} \qquad (2.5\text{–}1)$$

Assuming $V_1 < 0$, EI constant, and the initial shape

$$v_I = a \sin\frac{\pi x}{l} \qquad (2.5\text{–}2)$$

The solution of Eq. (2.5–1), which satisfies the end conditions $v = 0$, is readily found to be

$$v = \frac{a}{1 - (V_1/V_{1,\mathrm{cr}})} \sin\frac{\pi x}{l} \qquad (2.5\text{–}3)$$

where $V_{1,\mathrm{cr}} = \pi^2 EI/l^2$.

The quantity $1/(1 - V_1/V_{1,\mathrm{cr}})$ is referred to as a magnification factor. It has a wide range of applications in approximate methods of analysis and will be seen again in Section 3.2. The effect of the compressive axial force on the deflection is qualitatively similar to that encountered earlier with the flexibility coefficients.

2.6. Complete Stiffness Relations in a Fixed Reference

The stiffness relations derived previously are now considered to apply in a (moving) member-bound reference and will be transformed into relations in the local reference. In the present theory the axial load component p_x is the same in

FIG. 2.5–1. Initial imperfection.

both references and is thus also assumed to vanish in the local reference. Practically, a small variation in the axial force along the member, relative to its average, may be neglected. The notation separating axial from transverse components used previously is applied here, with $\mathbf{s}_1 = \{s_1^a \ s_1^b\}$, $\mathbf{s} = \{s_2^a \ s_3^a \ s_2^b \ s_3^b\}$, and similar definitions of $\mathbf{S}_1$ and $\mathbf{S}$. The transformation to be made consists in expressing $\mathbf{S}_1$ and $\mathbf{S}$ in terms of V_1 and $\mathbf{V}$ by statics and expressing v_1 and $\mathbf{v}$ in terms of $\mathbf{s}_1$ and $\mathbf{s}$ by geometry. These expressions are found in Chapter 3, Sections 2.2 and 2.1, for either type of bound reference. They have the form

$$\mathbf{S}_1 = \mathbf{B}_1 V_1 = \{-1 \ \ 1\} V_1 \tag{2.6–1a}$$

$$\mathbf{S} = \mathbf{B}\mathbf{V} + \mathbf{c}_1 \mathbf{s} V_1 + \mathbf{S}_p \tag{2.6–1b}$$

and

$$v_1 = \mathbf{B}_1^T \mathbf{s}_1 + \tfrac{1}{2}\mathbf{s}^T \mathbf{c}_1 \mathbf{s} \tag{2.6–2a}$$

$$\mathbf{v} = \mathbf{B}^T \mathbf{s} \tag{2.6–2b}$$

The only terms that differ from those of the linear theory are those involving $\mathbf{c}_1$ which is a 4×4 constant matrix. Substitution of the stiffness relations for V_1 and $\mathbf{V}$ into Eqs. (2.6–1), and use of Eqs. (2.6–2), yields

$$\mathbf{S}_1 = \mathbf{k}_{11}\mathbf{s}_1 + \frac{1}{2}\frac{EA}{l}\mathbf{B}_1 \mathbf{s}^T \mathbf{g}_s \mathbf{s} \tag{2.6–3a}$$

$$\mathbf{S} = \mathbf{k}_s \mathbf{s} + \mathbf{S}_F \tag{2.6–3b}$$

where

$$\mathbf{k}_{11} = \frac{EA}{l}\mathbf{B}_1 \mathbf{B}_1^T = \frac{EA}{l}\begin{bmatrix} 1 & -1 \\ -1 & 1 \end{bmatrix} \tag{2.6–4a}$$

$$\mathbf{g}_s = \mathbf{B}\mathbf{g}\mathbf{B}^T + \mathbf{c}_1 \tag{2.6–4b}$$

$$\mathbf{S}_F = \mathbf{S}_p + \mathbf{B}\mathbf{V}_F \tag{2.6–4c}$$

$$\mathbf{k}_s = \mathbf{B}\mathbf{k}\mathbf{B}^T + V_1 \mathbf{c}_1 \tag{2.6–4d}$$

These matrix operations could be carried out with either type of bound reference. For the alternate type, $\mathbf{c}_1$ will have further applications. It is obtained from Eq. (3/2.2–8), or Eq. (3/2.1–3a), as

$$\mathbf{c}_1 = (\mathbf{c}_1)_a = \frac{1}{l}\begin{bmatrix} 1 & 0 & -1 & 0 \\ 0 & 0 & 0 & 0 \\ -1 & 0 & 1 & 0 \\ 0 & 0 & 0 & 0 \end{bmatrix} \tag{2.6–5}$$

$\mathbf{k}_s$ is the complete bending stiffness matrix and is found, with $\alpha = 1$, to be

$$\mathbf{k}_s = \frac{EI\lambda}{4l\sin(\lambda/2)[\sin(\lambda/2) - (\lambda/2)\cos(\lambda/2)]}$$

$$\begin{bmatrix} \frac{\lambda^2}{l^2}\sin\lambda & \frac{2\lambda}{l}\sin^2\frac{\lambda}{2} & -\frac{\lambda^2}{l^2}\sin\lambda & \frac{2\lambda}{l}\sin^2\frac{\lambda}{2} \\ & \sin\lambda - \lambda\cos\lambda & -\frac{2\lambda}{l}\sin^2\frac{\lambda}{2} & \lambda - \sin\lambda \\ \text{symmetric} & & \frac{\lambda^2}{l^2}\sin\lambda & -\frac{2\lambda}{l}\sin^2\frac{\lambda}{2} \\ & & & \sin\lambda - \lambda\cos\lambda \end{bmatrix} \tag{2.6–6}$$

Portion $V_1\mathbf{c}_1$ of $\mathbf{k}_s$ in Eq. (2.6–4d) is the only one that need be considered for a member in pure compression or tension. Such a "two-force member" is typical of a pin-jointed truss subjected to joint loads. The nonzero terms in $V_1\mathbf{c}_1$ are transverse components S_2^a and S_2^b due to the rotation of the member axis. Accordingly $V_1\mathbf{c}_1$ is called the string geometric stiffness matrix, Fig. 2.6–1*b*.

Note that the submatrix of $\mathbf{k}_s$ associated with (S_2^b, s_2^b) and (S_3^b, s_3^b) is the stiffness of the cantilever type, and the submatrix associated with (S_3^a, s_3^a) and (S_3^b, s_3^b) is the stiffness of the alternate type. The remaining elements of $\mathbf{k}_s$ satisfy the equilibrium equations of the member end forces and the symmetry of $\mathbf{k}_s$.

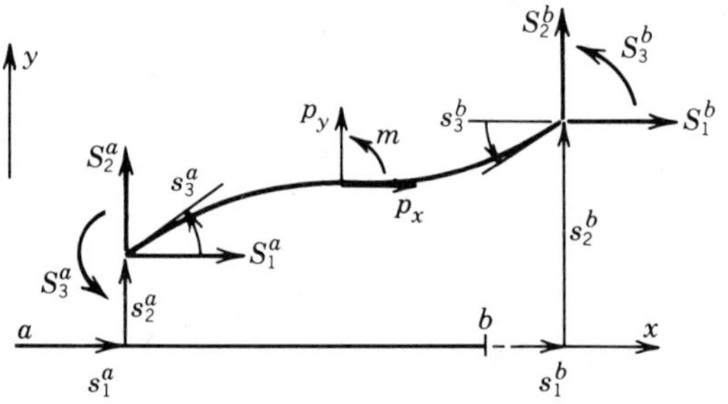

(*a*)

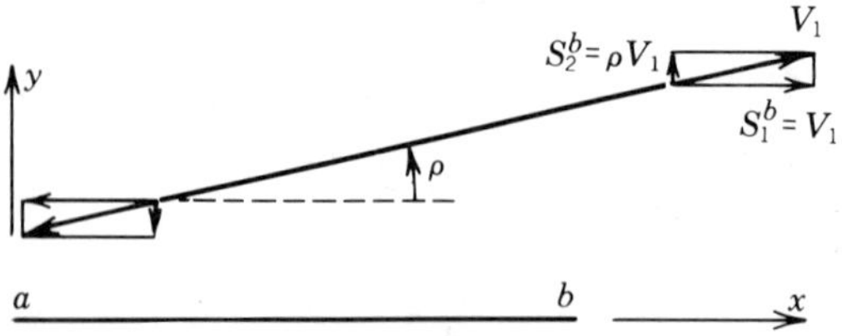

(*b*)

FIG. 2.6–1. Member variables in local axes (*a*); string geometric stiffness (*b*).

Evaluation of $\mathbf{g}_s$ is made using the results obtained for $\mathbf{g}$ in Eqs. (2.1–16). There comes, with the same notation for a and b,

$$\mathbf{g}_s = \begin{bmatrix} \frac{4b}{l^2}+\frac{1}{l} & \frac{2b}{l} & -\frac{4b}{l^2}-\frac{1}{l} & \frac{2b}{l} \\ & a+b & -\frac{2b}{l} & b-a \\ \text{symmetric} & & \frac{4b}{l^2}+\frac{1}{l} & -\frac{2b}{l} \\ & & & a+b \end{bmatrix} \tag{2.6–7}$$

3. STIFFNESS EQUATIONS BY THE METHOD OF VIRTUAL DISPLACEMENTS

3.1. Stiffness Equations in the Local Reference

The method of virtual displacements in a geometrically nonlinear theory has the same basis as the method presented and applied in Chapter 5, Section 4, within the context of a linear theory. Its advantage lies in its ability to deal analytically as well as numerically with problems where exact solutions are not possible or are too cumbersome to obtain. It also applies to tensile as well as compressive axial forces. The method is conveniently formulated in the form of Castigliano's first theorem;

$$\mathbf{Q}_{\text{ext}} = \frac{\partial \bar{U}}{\partial \mathbf{q}} \tag{3.1–1}$$

For simplicity, shear deformation is neglected, and a linearly elastic material without initial strains is considered. The strain energy has then the expression

$$\bar{U} = \int_0^l (U_a + U_b)\,dx \tag{3.1–2}$$

where

$$U_a = \tfrac{1}{2}EA\varepsilon^2 = \tfrac{1}{2}EA[u_{,x} + \tfrac{1}{2}(v_{,x})^2]^2 \tag{3.1–3a}$$
$$U_b = \tfrac{1}{2}EI\chi^2 = \tfrac{1}{2}EI(v_{,xx})^2 \tag{3.1–3b}$$

A representation of u and v in terms of generalized displacements $\mathbf{q}$ makes $\bar{U}$ a function of $\mathbf{q}$. $\mathbf{Q}_{\text{ext}}$ is evaluated by identifying the external virtual work δW_{ext} with $\mathbf{Q}_{\text{ext}}^T \delta\mathbf{q}$. For deriving stiffness equations in the local axes,

$$\delta W_{\text{ext}} = \mathbf{S}_1^T \delta\mathbf{s}_1 + \mathbf{S}^T \delta\mathbf{s} + \int_0^l (p_x \delta u + p_y \delta v + m\delta v_{,x})\,dx \tag{3.1–4}$$

Equation (3.1–1) could be used to derive both axial and bending stiffness equations by means of a representation of both u and v in terms of generalized displacements $\mathbf{q}$. However, it will be easier to derive the axial relations separately, by integrating the differential equation for u as done in Section 2. The problem is made more general here by including an axial span load p_x and a variable axial rigidity EA. The differential equation for u is Eq. (2.1–1a), in which V_1 is replaced by the axial force

$$N = V_1 + N_p = V_1 + \int_0^l p_x\,dx \tag{3.1–5}$$

In the simplest application $\mathbf{q}$ is chosen as the end displacements $\mathbf{s}$, and v is represented by means of shape functions $\boldsymbol{\psi}$ as $v = \boldsymbol{\psi}\mathbf{s}$. Integration of $u_{,x}$ in the interval $(0, x)$, and use of the condition $u = s_1^a$ at $x = 0$, yields

$$u = s_1^a + V_1 \int_0^x \frac{dx}{EA} + \int_0^x \frac{N_p}{EA}\,dx - \frac{1}{2}\mathbf{s}^T\mathbf{g}_x\mathbf{s} \tag{3.1–6}$$

where

$$\mathbf{g}_x = \int_0^x \boldsymbol{\psi}_{,x}^T \boldsymbol{\psi}_{,x}\,dx \tag{3.1–7}$$

At $x = l, u = s_1^b$, and Eq. (3.1–6) is solved for V_1 in the form

$$V_1 = k_{11}(s_1^b - s_1^a) + V_{1F} + \tfrac{1}{2}k_{11}\mathbf{s}^T\mathbf{g}_s\mathbf{s} \tag{3.1–8}$$

where

$$k_{11} = \left[\int_0^l \frac{dx}{EA}\right]^{-1} \tag{3.1–9a}$$

$$V_{1F} = -k_{11}\int_0^l \frac{N_p}{EA}\,dx \tag{3.1–9b}$$

$$\mathbf{g}_s = \int_0^l \boldsymbol{\psi}_{,x}^T \boldsymbol{\psi}_{,x}\,dx \tag{3.1–9c}$$

In the case of a uniform member, $k_{11} = EA/l$.

Axial stiffness relations for S_1^a and S_1^b are obtained through the equilibrium equations,

$$\begin{Bmatrix} S_1^a \\ S_1^b \end{Bmatrix} = \begin{Bmatrix} -1 \\ 1 \end{Bmatrix} V_1 + \begin{Bmatrix} -\int_0^l p_x\,dx \\ 0 \end{Bmatrix} \tag{3.1–10}$$

which, in matrix notation, are $\mathbf{S}_1 = \mathbf{B}_1 V_1 + \mathbf{S}_{1p}$. There comes

$$\mathbf{S}_1 = \mathbf{k}_{11}\mathbf{s}_1 + \mathbf{S}_{1F} + \tfrac{1}{2}k_{11}\mathbf{B}_1\mathbf{s}^T\mathbf{g}_s\mathbf{s} \tag{3.1–11}$$

where

$$\mathbf{k}_{11} = k_{11}\begin{bmatrix} 1 & -1 \\ -1 & 1 \end{bmatrix} \tag{3.1–12}$$

$$\mathbf{S}_{1F} = \mathbf{S}_{1p} + \mathbf{B}_1 V_{1F} \tag{3.1–13}$$

Equation (3.1–11) may be seen as a generalization of Eq. (2.6–3a). It now includes the fixed end forces $\mathbf{S}_{1F}$ due to the axial load.

In deriving the bending stiffness relations, δu is now constrained by the condition $\varepsilon = N/EA$, which makes $\delta\varepsilon = 0$—that is, the virtual displacements are inextensional. Thus $\delta U_a = 0$ and only the bending strain energy need be evaluated. With $v = \boldsymbol{\psi}\mathbf{s}$, Eq. (3.1–3b) yields

$$\bar{U}_b = \tfrac{1}{2}\mathbf{s}^T\mathbf{k}_{s0}\mathbf{s} \tag{3.1–14}$$

where

$$\mathbf{k}_{s0} = \int_0^l EI\boldsymbol{\psi}^T_{,xx}\boldsymbol{\psi}_{,xx}\,dx \tag{3.1–15}$$

In evaluating δW_{ext}, δv is $\boldsymbol{\psi}\delta\mathbf{s}$, δu is obtained by varying only s_1^a and $\mathbf{s}$ in Eq. (3.1–6), and δs_1^b is δu at $x = l$. The resulting expression contains the term $(S_1^a + S_1^b + \int_0^l p_x\,dx)\delta s_1^a$, which is zero because of axial equilibrium, Eq. (3.1–10). Identifying δW_{ext} with $\mathbf{Q}^T_{\text{ext}}\,\delta\mathbf{s}$, $\mathbf{Q}_{\text{ext}}$ is found in the form

$$\mathbf{Q}_{\text{ext}} = \mathbf{S} - (V_1\mathbf{g}_s + \mathbf{k}_{Gp})\mathbf{s} - \mathbf{S}_F \tag{3.1–16}$$

where

$$\mathbf{k}_{Gp} = \int_0^l \mathbf{g}_x p_x\,dx \tag{3.1–17}$$

$$\mathbf{S}_F = -\int_0^l (\boldsymbol{\psi}^T p_y + \boldsymbol{\psi}^T_{,x} m)\,dx \tag{3.1–18}$$

Solving the equation $\mathbf{Q}_{\text{ext}} = \partial\bar{U}_b/\partial\mathbf{s}$ for $\mathbf{S}$, there comes

$$\mathbf{S} = \mathbf{k}_s\mathbf{s} + \mathbf{S}_F \tag{3.1–19}$$

where

$$\mathbf{k}_s = \mathbf{k}_{s0} + V_1\mathbf{g}_s + \mathbf{k}_{Gp} \tag{3.1–20}$$

The last two terms are referred to as geometric stiffness matrices. Their sum may be expressed, by integration by parts in Eq. (3.1–17), as a single integral involving the axial force N, as

$$\mathbf{k}_{sG} = V_1\mathbf{g}_s + \mathbf{k}_{Gp} = \int_0^l N\boldsymbol{\psi}^T_{,x}\boldsymbol{\psi}_{,x}\,dx \tag{3.1–21}$$

The preceding outline is readily specialized to the derivation of reduced stiffness equations for a cantilever or for a simply supported member by deleting the zero displacements and the associated shape functions. The notation to be adopted in that case is that used in earlier sections; that is,

$$V_1 = k_{11}v_1 + \tfrac{1}{2}k_{11}\mathbf{v}^T\mathbf{g}\mathbf{v} + V_{1F} \qquad (3.1\text{–}22a)$$

$$\mathbf{V} = \mathbf{k}\mathbf{v} + \mathbf{V}_F \qquad (3.1\text{–}22b)$$

and

$$\mathbf{k} = \mathbf{k}_0 + V_1\mathbf{g} + \mathbf{k}_{gp} \qquad (3.1\text{–}23)$$

The stiffness relations of the preceding sections could all be established by the present method if the exact shape functions are used. What is of interest, however, is that these same shape functions, and even the simpler ones established in the linear theory, could be used in the general context of variable axial force and a nonuniform member to derive approximate stiffness relations.

Example 1. Cantilever

The method outlined is now applied to a uniform cantilever, using the shape functions of the linear theory, Eq. (5/4–11). With shear deformation neglected,

$$v = [3t^2 - 2t^3 \quad l(-t^2 + t^3)]\{v_2 \ v_3\} \qquad (3.1\text{–}24)$$

where $t = x/l$. Performing the integrations in Eq. (3.1–7), there comes

$$\mathbf{g}_x = \frac{1}{l}\begin{bmatrix} \frac{6}{5}(10t^3 - 15t^4 + 6t^5) & \frac{l}{10}(-40t^3 + 75t^4 - 36t^5) \\ \text{Symmetric} & \frac{l^2}{15}(20t^3 - 45t^4 + 27t^5) \end{bmatrix} \qquad (3.1\text{–}25)$$

$\mathbf{g}$ is equal to $\mathbf{g}_x$ at $t = 1$, or

$$\mathbf{g} = \begin{bmatrix} \frac{6}{5l} & -\frac{1}{10} \\ -\frac{1}{10} & \frac{2l}{15} \end{bmatrix} \qquad (3.1\text{–}26)$$

Considering the case of a uniform axial load p_x, the associated geometric stiffness is obtained using Eq. (3.1–17), as

$$\mathbf{k}_{gp} = p_x l \begin{bmatrix} \frac{3}{5l} & -\frac{1}{10} \\ -\frac{1}{10} & \frac{l}{30} \end{bmatrix} \qquad (3.1\text{–}27)$$

$\mathbf{k}_0$ is the stiffness matrix in the linear theory, based on the assumed shape. For a uniform member it is obtained in Eq. (5/3–4), in which shear deformation is neglected by letting $\beta = 0$. The stiffness matrix $\mathbf{k}$, Eq. (3.1–23), may be put in the form

$$\mathbf{k} = EI\begin{bmatrix} \frac{12}{l^3} - \frac{6}{5}\frac{\lambda^2}{l^3} - \frac{3}{5}\frac{\mu^2}{l^3} & -\frac{6}{l^2} + \frac{1}{10}\frac{\lambda^2}{l^2} + \frac{1}{10}\frac{\mu^2}{l^2} \\ \text{Symmetric} & \frac{4}{l} - \frac{2}{15}\frac{\lambda^2}{l} - \frac{1}{30}\frac{\mu^2}{l} \end{bmatrix} \tag{3.1–28}$$

where

$$\lambda^2 = -\frac{V_1 l^2}{EI} \tag{3.1–29a}$$

$$\mu^2 = -\frac{p_x l^3}{EI} \tag{3.1–29b}$$

To assess the accuracy of the method, consider first the case $p_x = 0$. Elements of $\mathbf{k}$ are compared with the exact values listed in Table 2.2–2. A good accuracy is achieved for a range of values of λ^2 starting at 0 and exceeding the critical value $\lambda_{\text{cr}}^2 = \pi^2/4$. At this critical value the relative error on the elements of $\mathbf{k}$ is less than 1%. However, at λ_{cr} the exact $\mathbf{k}$ is singular, and the flexibility coefficients tend to infinity. The critical value, as obtained from the approximate $\mathbf{k}$, is the lowest root of the determinantal equation $|\mathbf{k}| = 0$ and is found to be $\lambda^2 = 2.486$. This exceeds the exact value by only 0.77%.

It may be verified that $\mathbf{k}_0 + V_1\mathbf{g}$ coincides with the truncated series development in powers of λ^2 of the exact expression, Eq. (2.2–3).

A comparable accuracy is found for the case $V_1 = 0$ and p_x constant. This loading will be discussed further within the context of instability analysis.

Example 2

This example deals with a simply supported member and is otherwise entirely similar to the preceding one. The representation of v is

$$v = l[t(1-t)^2 \quad -t^2(1-t)]\{v_2 \ v_3\} \tag{3.1–30}$$

from which the following results are obtained

$$\mathbf{g}_x = l\begin{bmatrix} t - 4t^2 + \frac{22}{3}t^3 - 6t^4 + \frac{9}{5}t^5 & -t^2 + \frac{11}{3}t^3 - \frac{9}{2}t^4 + \frac{9}{5}t^5 \\ \text{Symmetric} & \frac{4}{3}t^3 - 3t^4 + \frac{9}{5}t^5 \end{bmatrix} \tag{3.1–31}$$

$$\mathbf{g} = l\begin{bmatrix} \frac{2}{15} & -\frac{1}{30} \\ -\frac{1}{30} & \frac{2}{15} \end{bmatrix} \tag{3.1–32}$$

For p_x constant

$$\mathbf{k}_{gp} = p_x l^2 \begin{bmatrix} \frac{1}{10} & -\frac{1}{60} \\ -\frac{1}{60} & \frac{1}{30} \end{bmatrix} \tag{3.1–33}$$

and

$$\mathbf{k} = \frac{EI}{l} \begin{bmatrix} 4 - \frac{2\lambda^2}{15} - \frac{\mu^2}{10} & 2 + \frac{\lambda^2}{30} + \frac{\mu^2}{60} \\ \text{Symmetric} & 4 - \frac{2\lambda^2}{15} - \frac{\mu^2}{30} \end{bmatrix} \tag{3.1–34}$$

As in the case of the cantilever, comparison of **k** with the exact matrix in the case of a constant axial force shows a good accuracy for low values of λ^2. In the present case, however, the approximate λ_{cr} is found equal to 12, whereas the exact value is $\pi^2 = 9.87$. For $V_1/V_{1,cr} = 0.8$, the relative error on the stiffness coefficients remains at about 5%.

To improve the approximation, the assumed shape must include additional degrees of freedom, or more appropriate shape functions should be used. (See Example 4.)

Example 3

In this example the complete bending stiffness matrix $\mathbf{k}_s$ is established based on the shape functions of the linear theory. A direct derivation would include a repetition of the calculations done in the two previous examples. Instead, $\mathbf{k}_s$ may be formed by recognizing that it contains as submatrices the reduced stiffness matrices of both the cantilever and alternate types. $\mathbf{k}_s$ is completed by symmetry and by requiring each column to consist of forces in equilibrium, including the moment of axial forces. The result for the case of a uniform axial load p_x is

$$\mathbf{k}_s = \frac{EI}{l^3} \begin{bmatrix} 12 - \frac{6}{5}\lambda^2 - \frac{3}{5}\mu^2 & l\left(6 - \frac{\lambda^2}{10}\right) & -12 + \frac{6}{5}\lambda^2 + \frac{3}{5}\mu^2 & l\left(6 - \frac{\lambda^2}{10} - \frac{\mu^2}{10}\right) \\ & l^2\left(4 - \frac{2\lambda^2}{15} - \frac{\mu^2}{10}\right) & -l\left(6 - \frac{\lambda^2}{10}\right) & l^2\left(2 + \frac{\lambda^2}{30} + \frac{\mu^2}{60}\right) \\ \text{Symmetric} & & 12 - \frac{6}{5}\lambda^2 - \frac{3}{5}\mu^2 & -l\left(6 - \frac{\lambda^2}{10} - \frac{\mu^2}{10}\right) \\ & & & l^2\left(4 - \frac{2\lambda^2}{15} - \frac{\mu^2}{30}\right) \end{bmatrix} \tag{3.1–35}$$

As an example, the equilibrium diagram by which column 3 of $\mathbf{k}_s$ is formed is shown in Fig. 3.1–1. The ordering (1, 2, 3, 4) corresponds to $\{s_2^a \; s_3^a \; s_2^b \; s_3^b\}$. k_{33}

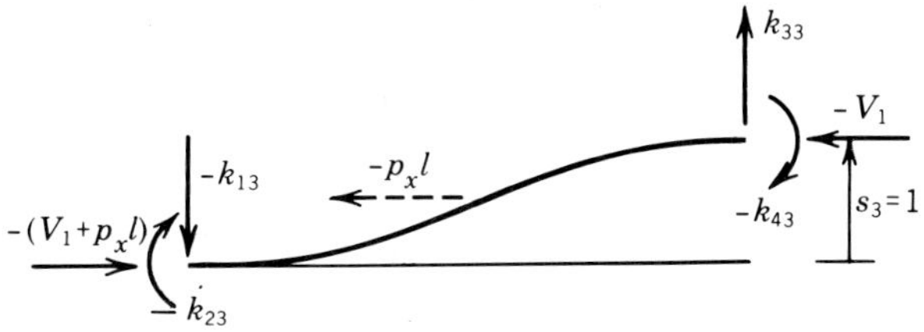

FIG. 3.1–1. Stiffness coefficients due to $\mathbf{s}_3 = 1$.

and k_{43} are obtained from the case of the cantilever. Then $k_{13} = -k_{33}$, and k_{23} is found by moment equilibrium.

Example 4

To obtain a better approximation for the simply supported member of Example 2, v is represented by means of three generalized displacements in the form

$$v = \boldsymbol{\psi}\mathbf{q} = [\boldsymbol{\psi}_v \;\; \Psi_w]\{\mathbf{v} \;\; w\} \tag{3.1–36}$$

where $\boldsymbol{\psi}_v$ is the shape function matrix of Example 2. At the beam ends v must remain zero, and $v_{,x}$ must remain equal to the deformations $\mathbf{v}$. Thus Ψ_w must satisfy the conditions $\Psi_w = \Psi_{w,x} = 0$ at $x = 0$ and $x = l$. The simplest polynomial satisfying these conditions is proportional to $t^2(1-t)^2$. If wl is chosen to be the value of this polynomial at midspan, then

$$\Psi_w = 16lt^2(1-t)^2 \tag{3.1–37}$$

The bending strain energy takes the form $\frac{1}{2}\mathbf{q}^T\mathbf{k}_0'\mathbf{q}$, and $\mathbf{k}_0'$ is found as

$$\mathbf{k}_0' = \frac{EI}{l}\begin{bmatrix} 4 & 2 & 0 \\ 2 & 4 & 0 \\ 0 & 0 & \dfrac{1024}{5} \end{bmatrix} \tag{3.1–38}$$

The geometric stiffness matrix is obtained similarly to Eq. (3.1–21) as

$$V_1\mathbf{g}' + \mathbf{k}_{gp}' = \int_0^l N\boldsymbol{\psi}_{,x}^T\boldsymbol{\psi}_{,x}\,dx \tag{3.1–39}$$

The results are

$$V_1\mathbf{g}' = \frac{V_1 l}{30}\begin{bmatrix} 4 & -1 & 16 \\ -1 & 4 & -16 \\ 16 & -16 & \dfrac{1024}{7} \end{bmatrix} \tag{3.1–40}$$

and for p_x constant,

$$\mathbf{k}'_{gp} = p_x l^2 \begin{bmatrix} \dfrac{1}{10} & -\dfrac{1}{60} & \dfrac{32}{105} \\ & \dfrac{1}{30} & -\dfrac{24}{105} \\ \text{Symmetric} & & \dfrac{256}{105} \end{bmatrix} \tag{3.1–41}$$

The bending stiffness equations have the form

$$\mathbf{Q} = (\mathbf{k}'_0 + V_1 \mathbf{g}' + \mathbf{k}'_{gp})\mathbf{q} \tag{3.1–42}$$

where $\mathbf{Q}$ is the set of generalized forces due to $\mathbf{V}$ and to the transverse span load $\{p_y \ m\}$. Identifying the virtual work of these forces with $\mathbf{Q}^T \delta \mathbf{q}$, there comes

$$\mathbf{Q} = \{\mathbf{V} \ 0\} + \int_0^l (\boldsymbol{\psi}^T p_y + \boldsymbol{\psi}_{,x}^T m)\, dx \tag{3.1–43}$$

The axial stiffness relation is derived following the general procedure outlined earlier and takes the form

$$V_1 = k_{11}(v_1 + \tfrac{1}{2}\mathbf{q}^T \mathbf{g}' \mathbf{q}) - k_{11} \int_0^l \frac{1}{EA} N_p\, dx \tag{3.1–44}$$

Equation (3.1–42) is a generalized stiffness equation that may be reduced to an equation relating $\mathbf{V}$ to $\mathbf{v}$ by a procedure called static condensation. The procedure is to solve the third of Eqs. (3.1–42) for w and to substitute the result in the first two. The substitution for w is also made in the axial stiffness relation.

The condensed bending stiffness matrix in the case $p_x = 0$ is found to be

$$\mathbf{k} = \frac{EI}{l} \begin{bmatrix} 4 - \dfrac{2\lambda^2}{15} - \dfrac{7\lambda^4}{120(42 - \lambda^2)} & 2 + \dfrac{\lambda^2}{30} + \dfrac{7\lambda^4}{120(42 - \lambda^2)} \\ \swarrow & \searrow \end{bmatrix} \tag{3.1–45}$$

A comparison with the exact $\mathbf{k}$ shows an excellent accuracy for a range of λ exceeding Euler's critical value. In particular, Euler's load is overestimated by only 0.06%.

For the case p_x constant and $V_1 = 0$, the condensed $\mathbf{k}$ is found as

$$\mathbf{k} = \frac{EI}{l} \begin{bmatrix} 4 - \dfrac{\mu^2}{10} - \dfrac{\mu^4}{2205(1 - \mu^2/84)} & 2 + \dfrac{\mu^2}{60} + \dfrac{\mu^4}{2940(1 - \mu^2/84)} \\ \text{Symmetric} & 4 - \dfrac{\mu^2}{30} - \dfrac{\mu^4}{3920(1 - \mu^2/84)} \end{bmatrix} \tag{3.1–46}$$

Additional results that may be found for this example, such as formulas for fixed end forces and geometric stiffness matrices for different axial loads, are left to exercises.

Example 5

Consider a pin-ended tie rod subjected to a transverse load p_y, and let the problem be to determine the axial force caused by p_y.

Starting with stiffness equations of the form, $\mathbf{V} = \mathbf{kv} + \mathbf{V}_F$, the condition of zero end moments $\mathbf{V}$ gives

$$\mathbf{v} = -\mathbf{k}^{-1}\mathbf{V}_F \tag{3.1–47}$$

The condition $v_1 = 0$ reduces the axial stiffness equation to

$$V_1 = \tfrac{1}{2}k_{11}\mathbf{v}^T\mathbf{g}\mathbf{v} = \tfrac{1}{2}k_{11}\mathbf{V}_F^T\mathbf{k}^{-1}\mathbf{g}\mathbf{k}^{-1}\mathbf{V}_F \tag{3.1–48}$$

The right-hand side of Eq. (3.1–48) is a function of V_1, whose form depends on the assumed shape used to obtain $\mathbf{k}, \mathbf{g}$, and $\mathbf{V}_F$. If the shape of Example 2 is used, only $\mathbf{k}$ depends on V_1, and (3.1–48) may be reduced to a cubic equation.

For the tie rod problem a more appropriate assumed shape is a half sine wave,

$$v = a \sin\frac{\pi x}{l}$$

which may be considered as the first term of a Fourier series representation of v. The axial stiffness relation is obtained by integrating the axial stress-displacement relation in the interval $(0, l)$. With $v_1 = 0$, there comes

$$\frac{V_1 l}{EA} = \int_0^l \frac{1}{2}(v_{,x})^2\,dx = \frac{\pi^2 a^2}{4l} \tag{3.1–49}$$

The bending strain energy is

$$\bar{U}_b = \int_0^l \frac{1}{2}EI(v_{,xx})^2\,dx = \frac{\pi^4 EIa^2}{4l^3}$$

In applying Castigliano's theorem with inextensional virtual displacements, the virtual δv_1 is

$$\delta v_1 = -\delta\int_0^l \frac{1}{2}(v_{,x})^2\,dx = -\frac{\pi^2 a\delta a}{2l}$$

The external virtual work defines Q_{ext} as

$$Q_{\text{ext}}\delta a = V_1\delta v_1 + \delta a\int_0^l p_y \sin\frac{\pi x}{l}\,dx$$

Assuming a uniform p_y, and equating Q_{ext} to $\partial \bar{U}_b/\partial a$, obtain

$$\frac{2lp_y}{\pi} - \frac{\pi^2 a}{2l} V_1 = \frac{\pi^4 EI}{2l^3} a \tag{3.1–50}$$

Equations (3.1–49, 50) are now solved for V_1 and a. Elimination of the unknown a yields a cubic equation for V_1 which may be put in the form

$$x(1+x)^2 = \frac{1}{4}\left(\frac{v_0}{r}\right)^2 \tag{3.1–51a}$$

where

$$x = V_1 l^2/\pi^2 EI \tag{3.1–51b}$$

$$v_0 = 4p_y l^4/\pi^5 EI \tag{3.1–51c}$$

$$r = \sqrt{I/A} \tag{3.1–51d}$$

3.2. Magnification Factor Method

In analyzing a beam-column, determination of the internal forces could be simplified by using an approximate deflected shape. A simple procedure that is often accurate enough is based on assuming that the deflected shape v is proportional to the shape in the geometrically linear theory v_L, or

$$v = qv_L \tag{3.2–1}$$

q is referred to as magnification factor. It will be determined in what follows using Castigliano's first theorem.

Letting a subscript L indicate a quantity based on the shape v_L, the bending strain energy is $\bar{U}_b = q^2 \bar{U}_{bL}$, and by Castigliano's first theorem

$$Q_{\text{ext}} = \frac{\partial \bar{U}_b}{\partial q} = 2q\bar{U}_{bL} \tag{3.2–2}$$

To evaluate Q_{ext}, inextensional virtual displacements are assumed, in which $\delta v = v_L \delta q$. The virtual chord shortening is

$$\delta v_1 = -\int_0^l v_{,x}\delta v_{,x} dx = 2v_{1L} q\delta q \tag{3.2–3}$$

where

$$v_{1L} = -\int_0^l \frac{1}{2}(v_{L,x})^2 dx \tag{3.2–4}$$

The external virtual work is

$$\delta W_{\text{ext}} = V_1 \delta v_1 + \mathbf{V}^T \delta \mathbf{v} + \int_0^l (p_y \delta v + m \delta v_{,x}) dx \tag{3.2–5}$$

Expressing δW_{ext} in terms of v_L, q, and δq, the coefficient of δq is found to be

$$Q_{\text{ext}} = 2qV_1 v_{1L} + \mathbf{V}^T \mathbf{v}_L + \int_0^l (p_y v_L + m v_{L,x})\, dx \qquad (3.2\text{–}6)$$

By Clapeyron's relation the last two terms are equal to $2\bar{U}_{bL}$. Solving Eq. (3.2–2) for q, there comes

$$q = \frac{1}{1 - V_1 v_{1L}/\bar{U}_{bL}} \qquad (3.2\text{–}7)$$

The denominator in this equation may be interpreted by considering the problem of buckling of the member, in a state of pure compression, with certain homogeneous kinematic end conditions. It will be seen in Section 5.4 that if v_L is a good approximation to the buckled shape, then the ratio $\bar{U}_{bL}/v_{1L}$ is a good approximation to the buckling load $V_{1,\text{cr}}$. Assuming this to be the case, Eq. (3.2–7) becomes

$$q = \frac{1}{1 - (V_1/V_{1,\text{cr}})} \qquad (3.2\text{–}8)$$

In using Eq. (3.2–8), it is not necessary to evaluate $V_{1,\text{cr}}$ by means of Eq. (3.2–7), since it may be known by other means. For consistency, however, the denominator of Eq. (3.2–7) should yield a good approximation to $V_{1,\text{cr}}$. For example, the shape caused by a uniform load, and to a lesser extent, the shape caused by a concentrated force at midspan, would give a good approximation to $V_{1,\text{cr}}$ for the various homogeneous kinematic end conditions. Also the shape caused by end moments V_2 and $V_3 = -V_2$ would give a good approximation to $V_{1,\text{cr}}$ for a simply supported member. Equation (3.2–8) is then applicable to all these cases.

Equation (3.2–8) may also be applied to the case where v_L is not caused by any loading but represents an initially curved member. A case where such an application is exact was treated in Section 2.5.

3.3. Transformation from Bound to Local Reference

Derivation of stiffness equations in a bound reference, and their subsequent transformation to the local reference, may be an easier procedure than a direct derivation in the local reference. It also allows a separation of geometric nonlinearity in the bound reference from the effect of rigid body motion of that reference. The static-kinematic transformation from the bound to the local reference may then be carried out using the exact equations derived in Chapter 2.

Within the context of the second-order theory, in both the bound and the fixed references, the transformation from bound to local reference was carried out in Section 2.6 in the case of no axial span load. If there is such as load, additional displacement-dependent terms occur in the transformation and are considered in what follows.

It is assumed that the load components p_x and p_y are given in the local reference. In the bound reference $p_{xr} = p_x$ and $p_{yr} = p_y - \rho_0 p_x$, where ρ_0 is the rotation of the bound reference. For the cantilever type, $\rho_0 = s_3^a$, and for the alternate type, $\rho_0 = (s_2^b - s_2^a)/l$. These linear relations are put in the form

$$\rho_0 = \mathbf{B}_\rho^T \mathbf{s} \tag{3.3–1}$$

The reduced stiffness equations (3.1–22) are now considered relative to the bound reference, with the load component p_y replaced by $p_{yr} = p_y - \rho_0 p_x$. The transverse load term $-\rho_0 p_x$ causes a displacement-dependent term in the fixed end forces $\mathbf{V}_F$, which should be brought out explicitly in the stiffness equations. We can write

$$\mathbf{V}_F = \mathbf{V}'_F - \rho_0 \mathbf{V}_{Fx} = \mathbf{V}'_F - \mathbf{V}_{Fx} \mathbf{B}_\rho^T \mathbf{s} \tag{3.3–2}$$

$\mathbf{V}'_F$ is due to the load components $\{p_x \ \ p_y \ \ m\}$ applied along the bound axes, and $\mathbf{V}_{Fx}$ is due to p_x applied transversely, that is, in the bound y direction. Thus

$$\mathbf{V}_{Fx} = -\int_0^l \boldsymbol{\psi}_r^T p_x \, dx \tag{3.3–3}$$

where $\boldsymbol{\psi}_r$ is a shape function matrix representing the transverse displacement v_r relative to the bound reference, in the form

$$v_r = \boldsymbol{\psi}_r \mathbf{v} \tag{3.3–4}$$

The transformation from bound to local axes leads to Eqs. (3.1–11, 19), in which $\mathbf{S}_F$ hides a displacement-dependent term. To bring this term out, let

$$\mathbf{S}_p = \mathbf{S}_{pL} + \mathbf{S}_{px} \tag{3.3–5}$$

where $\mathbf{S}_{pL}$ is due to the transverse load, and is evaluated in the initial geometry, and $\mathbf{S}_{px}$ is due to p_x. Elements of $\mathbf{S}_{px}$ are statically equivalent to a couple, in equilibrium with the moment of the load p_x about member end a. This moment is $C = \int_0^l -(v_r + \rho_0 x) p_x \, dx$, or, with $v_r = \boldsymbol{\psi}_r \mathbf{v}$,

$$C = \mathbf{V}_{Fx}^T \mathbf{v} - \rho_0 \int_0^l x p_x \, dx \tag{3.3–6}$$

The equilibrium equation relating $\mathbf{S}_{px}$ to C is established by means of the principle of virtual displacements for a rigid body: for any virtual rotation $\delta\rho_0$, $\mathbf{S}_{px}^T \delta\mathbf{s} + C\delta\rho_0 = 0$. This virtual work equation remains true, however, for any $\delta\mathbf{s}$ because the force system $\mathbf{S}_{px}$ has zero $\mathbf{V}$, and therefore $\mathbf{V}^T \delta\mathbf{v} = 0$. With

$\delta\rho_0 = \mathbf{B}_\rho^T \delta\mathbf{s}$, there comes

$$\mathbf{S}_{px} = -\mathbf{B}_\rho C = -\mathbf{B}_\rho \mathbf{V}_{Fx}^T \mathbf{v} + \mathbf{B}_\rho \rho_0 \int_0^l x p_x \, dx \tag{3.3–7}$$

The displacement-dependent term in $(\mathbf{S}_p + \mathbf{B}\mathbf{V}_F)$ is evaluated using Eqs. (3.3–7) and (3.3–2). With $\mathbf{v} = \mathbf{B}^T\mathbf{s}$, this term takes the form

$$\mathbf{S}_{px} + \mathbf{B}(-\rho_0 \mathbf{V}_{Fx}) = \mathbf{k}_{GF}\mathbf{s} \tag{3.3–8}$$

where

$$\mathbf{k}_{GF} = -(\mathbf{B}_\rho \mathbf{V}_{Fx}^T \mathbf{B}^T + \mathbf{B}\mathbf{V}_{Fx}\mathbf{B}_\rho^T) + \left(\int_0^l x p_x \, dx\right)\mathbf{B}_\rho \mathbf{B}_\rho^T \tag{3.3–9}$$

The transformation of the reduced stiffness equations to the local axes results in the bending stiffness equations $\mathbf{S} = \mathbf{k}_s\mathbf{s} + \mathbf{S}_F$, where

$$\mathbf{k}_s = \mathbf{B}\mathbf{k}\mathbf{B}^T + V_1\mathbf{c}_1 + \mathbf{k}_{GF} \tag{3.3–10}$$

$$\mathbf{S}_F = \mathbf{S}_{pL} + \mathbf{B}\mathbf{V}_F' \tag{3.3–11}$$

Equations (3.3–10) and (3.1–20) are two alternate approaches to forming $\mathbf{k}_s$. In the preceding expression $\mathbf{B}$, $\mathbf{k}$, $\mathbf{c}_1$, and $\mathbf{k}_{GF}$ depend on the type of bound reference, but not $\mathbf{k}_s$.

Example 1

As an example of $\mathbf{k}_{GF}$, consider a bound reference of the alternate type, a uniform member, a constant p_x, and the shape functions of the linear theory. Then

$$\rho_0 = \mathbf{B}_\rho^T\mathbf{s} = \begin{bmatrix} -\frac{1}{l} & 0 & \frac{1}{l} & 0 \end{bmatrix} \{s_2^a \;\; s_3^a \;\; s_2^b \;\; s_3^b\} \tag{3.3–12}$$

$$\mathbf{V}_{Fx} = \frac{p_x l^2}{12}\{-1 \;\; 1\} \tag{3.3–13}$$

$$\int_0^l x p_x \, dx = \frac{p_x l^2}{2}$$

and $\mathbf{B}$ may be seen in Eq. (3/2.2–8). Performing the operations in Eq. (3.3–9), there comes

$$\mathbf{k}_{GF} = p_x \begin{bmatrix} \frac{1}{2} & -\frac{l}{12} & -\frac{1}{2} & \frac{l}{12} \\ & 0 & \frac{l}{12} & 0 \\ \text{Symmetric} & & \frac{1}{2} & -\frac{l}{12} \\ & & & 0 \end{bmatrix} \tag{3.3–14}$$

4. INCREMENTAL STIFFNESS EQUATIONS

Incremental stiffness relations have been defined in Chapter 5, Section 6, within the context of the geometrically linear theory. In the present theory they are useful in solution procedures for geometrically nonlinear problems that may also be materially nonlinear. For a linear material, incremental relations may be established by differentiation of finite relations such as Eqs. (3.1–11, 19) or (3.1–22). Assuming the shape functions to be independent of the axial forces, the differential of a term such as $\mathbf{s}^T\mathbf{g}_s\mathbf{s}$ is $2\mathbf{s}^T\mathbf{g}_s d\mathbf{s}$ because $\mathbf{g}_s$ is symmetric and independent of $\mathbf{s}$. The differential of $\mathbf{k}_s\mathbf{s}$ is $\mathbf{k}_s d\mathbf{s} + (d\mathbf{k}_s)\mathbf{s}$, where $d\mathbf{k}_s$ is due to increments dV_1 and dp_x. From Eq. (3.1–8), and with the notation $\mathbf{B}_1 = \{-1 \;\; 1\}$,

$$dV_1 = k_{11}(\mathbf{B}_1^T d\mathbf{s}_1 + \mathbf{s}^T\mathbf{g}_s d\mathbf{s}) + dV_{1F} \tag{4–1}$$

then by incrementing Eqs. (3.1–11) and (3.1–19),

$$\begin{Bmatrix} d\mathbf{S}_1 \\ d\mathbf{S} \end{Bmatrix} = \begin{bmatrix} \mathbf{k}_{11} & k_{11}\mathbf{B}_1\mathbf{s}^T\mathbf{g}_s \\ k_{11}\mathbf{g}_s\mathbf{s}\mathbf{B}_1^T & \mathbf{k}_s + k_{11}\mathbf{g}_s\mathbf{s}\mathbf{s}^T\mathbf{g}_s \end{bmatrix} \begin{Bmatrix} d\mathbf{s}_1 \\ d\mathbf{s} \end{Bmatrix} + \begin{Bmatrix} d\mathbf{S}_{1F} \\ d\mathbf{S}_0 \end{Bmatrix} \tag{4–2}$$

where

$$d\mathbf{S}_0 = d\mathbf{S}_F + (d\mathbf{k}_{Gp})\mathbf{s} + \mathbf{g}_s\mathbf{s}\,dV_{1F} \tag{4–3}$$

These incremental matrices may be readily obtained by incrementing the corresponding defining formulas. The incremental stiffness matrix may be seen to be symmetric.

Similar incremental equations may be derived in a bound reference, noting, however, that $d\mathbf{V}_F$ hides incremental displacements if there is an axial load p_x (see Section 3.3).

For a nonlinear material, incremental stiffness relations may be derived by means of an incremental equation of virtual displacements. If axial and bending constitutive equations are uncoupled, the result is formally similar to Eqs. (4–2) except that k_{11}, Eq. (3.1–9a), and $\mathbf{k}_{s0}$, Eq. (3.1–15), are defined in terms of incremental rigidities.

Generalized Tangent Stiffness Matrix

The general form of incremental stiffness equations in terms of generalized displacements and forces is

$$d\mathbf{Q} = \mathbf{k}_t d\mathbf{q} + d\mathbf{Q}_0 \tag{4–4}$$

$\mathbf{k}_t$ is called incremental or tangent stiffness matrix.

For a general elastic material, $\mathbf{k}_t$ may be defined in terms of energy functions.

By incrementing Eq. (3.1–1), there comes

$$d\mathbf{Q}_{\text{ext}} = \frac{\partial^2 \bar{U}}{\partial \mathbf{q}^T \partial \mathbf{q}} d\mathbf{q} + d\mathbf{Q}_I \tag{4–5}$$

where $d\mathbf{Q}_I$ is due to an increment in initial stress resultants $\mathbf{N}_I$. In the general case $d\mathbf{Q}_{\text{ext}}$ depends on $d\mathbf{q}$. The case of conservative external forces is of particular interest. Such forces have a potential $\bar{U}_{\text{ext}}$, functional of the displacements, with the property

$$\delta W_{\text{ext}} = -\,\delta \bar{U}_{\text{ext}} \tag{4–6}$$

In a discrete formulation $\bar{U}_{\text{ext}}$ is a function of $\mathbf{q}$, and

$$\delta W_{\text{ext}} = -\frac{\partial \bar{U}_{\text{ext}}}{\partial \mathbf{q}^T} \delta \mathbf{q}$$

Thus

$$\mathbf{Q}_{\text{ext}} = -\frac{\partial \bar{U}_{\text{ext}}}{\partial \mathbf{q}} \tag{4–7}$$

and

$$d\mathbf{Q}_{\text{ext}} = -\frac{\partial^2 \bar{U}_{\text{ext}}}{\partial \mathbf{q}^T \partial \mathbf{q}} d\mathbf{q} + d\mathbf{Q} \tag{4–8}$$

where $d\mathbf{Q}$ is independent of $d\mathbf{q}$. The tangent stiffness matrix is then

$$\mathbf{k}_t = \frac{\partial^2 \Pi}{\partial \mathbf{q}^T \partial \mathbf{q}} \tag{4–9}$$

where Π is the total potential energy,

$$\Pi = \bar{U} + \bar{U}_{\text{ext}} \tag{4–10}$$

Example 1

Consider a simply supported member subjected to $V_1, \mathbf{V}$, and to a span load $\{p_x,\ p_y,\ m\}$. Neglecting shear deformation,

$$\delta W_{\text{ext}} = V_1 \delta v_1 + \mathbf{V}^T \delta \mathbf{v} + \int_0^l (p_x \delta u + p_y \delta v + m \delta v_{,x})\, dx$$

where $\delta v_1 = (\delta u)_{x=l}$ and $\delta \mathbf{v} = \delta\{(v_{,x})_{x=0}(v_{,x})_{x=l}\}$. Then

$$\bar{U}_{\text{ext}} = -V_1 v_1 - \mathbf{V}^T\mathbf{v} - \int_0^l (p_x u + p_y v + m v_{,x})\, dx \tag{4–11a}$$

The term in p_x may be integrated by parts and combined with the term in V_1 to express $\bar{U}_{\text{ext}}$ in the form

$$\bar{U}_{\text{ext}} = -\mathbf{V}^T\mathbf{v} - \int_0^l (N u_{,x} + p_y v + m v_{,x})\, dx \tag{4–11b}$$

where N is the axial force,

$$N = V_1 + N_p = V_1 + \int_x^l p_x\, dx \tag{4–12}$$

The procedure of Section 3.1, by which the bending stiffness relations are derived, is represented here by restricting $u_{,x}$ to satisfy the axial stress-displacement relation. Then v remains as the only independent displacement, and $\bar{U}_{\text{ext}}$ takes the form

$$\bar{U}_{\text{ext}} = -\mathbf{V}^T\mathbf{v} + \int_0^l \frac{1}{2}N(v_{,x})^2\, dx - \int_0^l (p_y v + m v_{,x})\, dx - \int_0^l \frac{N^2}{EA}\, dx \tag{4–13}$$

The last term is independent of v and may be dropped without affecting $\delta\bar{U}_{\text{ext}}$. Similarly the axial strain energy is now expressed in terms of N and may be dropped from $\bar{U}$. The representation $v = \boldsymbol{\psi}\mathbf{v}$ yields, with the notation used in Section 3.1

$$\bar{U}_{\text{ext}} = -(\mathbf{V} - \mathbf{V}_F)^T\mathbf{v} + \tfrac{1}{2}\mathbf{v}^T(V_1\mathbf{g} + \mathbf{k}_{gp})\mathbf{v} \tag{4–14}$$

and

$$\bar{U} = \bar{U}_b = \tfrac{1}{2}\mathbf{v}^T\mathbf{k}_0\mathbf{v} \tag{4–15}$$

The bending stiffness relations are

$$\frac{\partial(\bar{U} + \bar{U}_{\text{ext}})}{\partial \mathbf{v}} = (\mathbf{k}_0 + V_1\mathbf{g} + \mathbf{k}_{gp})\mathbf{v} - \mathbf{V} + \mathbf{V}_F = 0 \tag{4–16}$$

The incremental stiffness matrix at fixed axial forces is here the same as the finite stiffness matrix. The effect of increments in axial forces is described in Eq. (4–2).

5. STABILITY

5.1. Buckling of Columns

The problem of column buckling, which came up within the context of flexibility and stiffness relations, was discussed in Section 2.1 for Euler's column and in Section 2.2 for a cantilever. This study is now extended to other end conditions, Fig. 5.1–1, before presenting more general ideas and methods for stability analysis.

Proped Cantilever, Fig. 5.1–1d

For the proped cantilever of Fig. 5.1–1*d*, the end condition is $v_2 = 0$. The effective bending stiffness relation is $V_3 = k_{33} v_3$. For $V_3 = 0$, a bent shape is possible if $k_{33} = 0$, or

$$\tan \lambda = \alpha \lambda \tag{5.1–1}$$

For the case without shear deformation, $\alpha = 1$, the smallest nonzero root in λ is found by trial as

$$\lambda_{cr} = 4.4934 \ldots \approx \frac{\pi}{0.70} \tag{5.1–2}$$

Guided Cantilever, Fig. 5.1–1c

A guided cantilever has the end condition $v_3 = 0$. The condition $k_{22} = 0$ yields the same critical load as Euler's column.

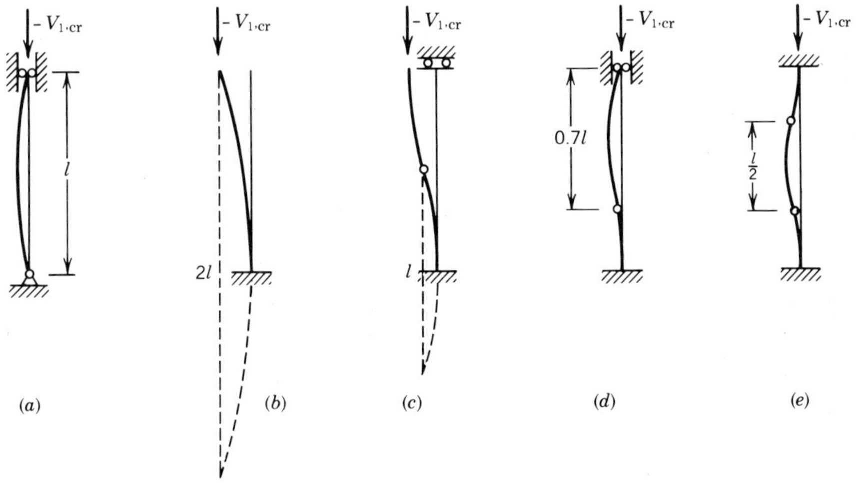

FIG. 5.1–1. Column buckling, effective length.

Both Ends Restrained, Fig. 5.1–1e

If $\mathbf{v} = 0$, a bent shape requires nonzero reactions $\mathbf{V}$, and $\mathbf{V} = \mathbf{k}\mathbf{v}$. Thus elements of $\mathbf{k}$ must tend to infinity. Using Eq. (2.1–9), this occurs if $\sin(\lambda/2) = 0$, or

$$\lambda_{cr} = 2\pi \tag{5.1–3}$$

The concept of effective length Kl reduces all formulas for $V_{1,cr}$ to an Euler's type formula, or

$$-V_{1,cr} = \frac{\pi^2 EI}{(Kl)^2} \tag{5.1–4}$$

Kl is the distance between two inflexion points of the buckled shape. This portion of the member is equivalent to Euler's column, Fig. 5.1–1.

5.2. Column Buckling with Nonlinear Material

Consider a material having a nonlinear compressive stress-strain relation as shown in Fig. 5.2–1a. A column in pure compression has a uniform stress σ and a uniform strain ε represented by point A. If the column takes a bent shape infinitesimally close to the straight one, the changes $d\sigma$ and $d\varepsilon$ at any point in the beam are related by an incremental relation $d\sigma = E_t d\varepsilon$. If the material is elastic, the tangent modulus E_t is the same whether $d\sigma$ is positive or negative and is equal to the slope at point A. Investigation of the bent shape as an alternate equilibrium shape is entirely similar to that of the linearly elastic case, if E is replaced with E_t. Equation (5.1–4) is thus generalized to

$$-V_{1,cr} = \frac{\pi^2 E_t I}{(Kl)^2} \tag{5.2–1}$$

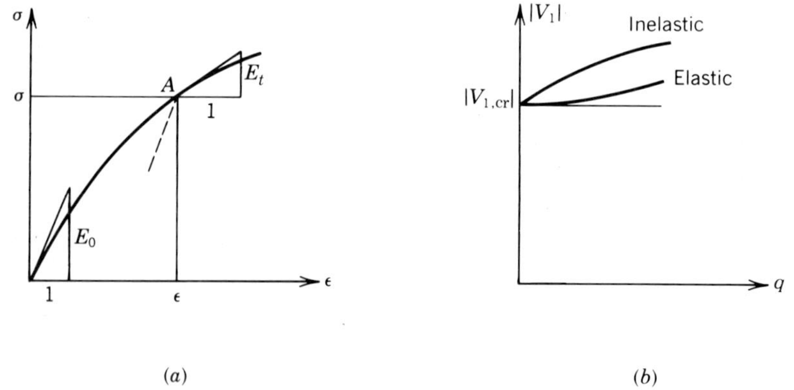

FIG. 5.2–1. Inelastic buckling.

If the material is inelastic, E_t depends on whether $d\sigma$ is an increase or a decrease in compressive stress. For an increase, E_t is the same as defined earlier. For a decrease, or unloading, E_t is defined on the unloading path and has a larger value E_0. If an alternate equilibrium state is sought at the same V_1, it is found that unloading must take place on the convex side of the member, and the resulting formula differs from Eq. (5.2–1) by replacing E_t with an equivalent modulus $E_r > E_t$, which depends on both E_t and E_0. However, if an increment $|dV_1|$ in compressive force is considered in investigating the alternate equilibrium state, it is found that the unloading stress is an infinitesimal of second order relative to the first-order bending displacements. Graphically point A in Fig. 5.2–1*a* moves differently for different points of the member, but it either moves upward along the tangent or remains stationary to within infinitesimals of second order. The conclusion is that the alternate equilibrium state becomes possible at the critical value of V_1 given by the tangent modulus theory, Eq. (5.2–1), with unloading taking place subsequently as $|V_1|$ increases. This explanation is essentially due to Shanley [7]. A graphical representation of V_1 versus a transverse generalized displacement q is shown in Fig. 5.2–1*b* for two different columns. In the elastic case the incremental stiffness at the initiation of buckling is zero.

5.3. Work and Potential Energy Criteria of Stability

Stability of a general system in equilibrium is a property that describes the response of the system to small disturbances. An equilibrium state $\mathbf{u}$ is locally stable, if positive external work ΔW is required to displace the system into any kinematically admissible neighboring equilibrium state, $\mathbf{u} + \delta\mathbf{u}$. If ΔW_{ext} is the work done by the loading of the system and ΔW_{int} is the internal work, then, by the work equation (Chapter 2, Section 5.2)

$$\Delta W = \Delta W_{\text{int}} - \Delta W_{\text{ext}} \tag{5.3–1}$$

The stability criterion is $\Delta W > 0$ for any kinematically admissible $\delta\mathbf{u}$. For a conservative system, that is, a system having a total potential energy, ΔW_{int} is the change $\Delta\bar{U}$ in strain energy $\bar{U}$, and $-\Delta W_{\text{ext}}$ is the change $\Delta\bar{U}_{\text{ext}}$ in external potential $\bar{U}_{\text{ext}}$. ΔW is thus the change, or total variation, of the potential energy, and the stability criterion is

$$\Delta\Pi = \Pi(\mathbf{u} + \delta\mathbf{u}) - \Pi(\mathbf{u}) > 0 \tag{5.3–2}$$

It is concluded that, in a stable equilibrium state, Π is minimum over all kinematically admissible neighboring states.

In a discrete formulation in terms of generalized displacements $\mathbf{q}$, Π is a function of $\mathbf{q}$, and by Taylor's series development,

$$\Delta\Pi = \frac{\partial\Pi}{\partial\mathbf{q}^T}\delta\mathbf{q} + \frac{1}{2}\delta\mathbf{q}^T\frac{\partial^2\Pi}{\partial\mathbf{q}^T\partial\mathbf{q}}\delta\mathbf{q} + \cdots \tag{5.3–3}$$

In an equilibrium state $\partial\Pi/\partial\mathbf{q} = 0$. For small $\delta\mathbf{q}$ the sign of $\Delta\Pi$ is governed by that of the quadratic form which is the second variation $\delta^2\Pi$. The stability criterion becomes

$$\delta^2\Pi = \delta\mathbf{q}^T\mathbf{k}_t\delta\mathbf{q} > 0 \tag{5.3–4}$$

It is thus found that the criterion of stability is that the tangent stiffness matrix be positive definite.

If some $\delta\mathbf{q}$ exists for which $\delta\mathbf{q}^T\mathbf{k}_t\delta\mathbf{q} < 0$, the equilibrium state $\mathbf{q}$ is unstable, it must be restrained in displacing it through $\delta\mathbf{q}$.

A stable deformation path ceases to be stable if at some point the tangent stiffness matrix, with respect to all degrees of freedom, ceases to be positive definite. For a continuous $\mathbf{k}_t$ this may occur only if $\mathbf{k}_t$ becomes singular. In that case there is some $\delta\mathbf{q} \neq 0$ for which $\mathbf{k}_t\delta\mathbf{q} = 0$. Thus the increment in external forces, required for equilibrium in state $\mathbf{q} + \delta\mathbf{q}$, is at least of second order in $\delta\mathbf{q}$. State $\mathbf{q}$ is called a neutral equilibrium state, and state $\mathbf{q} + \delta\mathbf{q}$ is called an alternate equilibrium state adjacent to state $\mathbf{q}$. An investigation of stability may be carried out by seeking the conditions under which $\mathbf{k}_t$ is singular or by directly seeking an alternate equilibrium state.

As a point on a deformation path a neutral equilibrium state may be of two types, as represented in Fig. 5.3–1. In one case, referred to as Euler-type buckling, there is a bifurcation of equilibrium at point A. The dotted line represents the unstable path. In the other case there is no bifurcation but a decreasing incremental stiffness reaching zero at point B. The descending part of the curve is unstable as it has a negative incremental stiffness. This type of instability is usually due to material inelastic behavior, but it can also occur solely because of geometric nonlinearity (see Chapter 12, Fig. 5–4).

To illustrate the criterion $|\mathbf{k}_t| = 0$, consider the incremental stiffness equations (4–2) specialized to a cantilever or to a simply supported member. Here $\mathbf{q}$ is $\{v_1 \ \ \mathbf{v}\}$, and the equation $\mathbf{k}_t\delta\mathbf{q} = 0$ takes the form

$$\begin{bmatrix} k_{11} & k_{11}\mathbf{v}^T\mathbf{g} \\ k_{11}\mathbf{g}\mathbf{v} & \mathbf{k} + k_{11}\mathbf{g}\mathbf{v}\mathbf{v}^T\mathbf{g} \end{bmatrix} \begin{Bmatrix} \delta v_1 \\ \delta\mathbf{v} \end{Bmatrix} = 0 \tag{5.3–5}$$

The first of Eqs. (5.3–5) expresses $dV_1 = 0$ and determines δv_1 in terms of $\delta\mathbf{v}$.

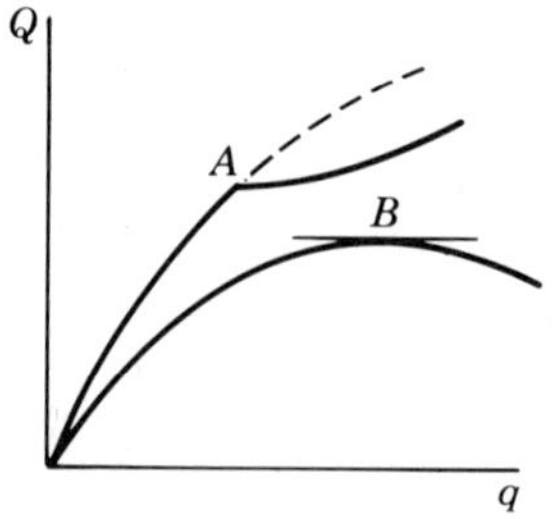

FIG. 5.3–1. Types of instability.

Elimination of δv_1 reduces the second equation to $\mathbf{k}\delta\mathbf{v} = 0$. Instability may thus be investigated by means of the bending stiffness $\mathbf{k}$, as done in Sections 2 and 3. For an elastic material, $\mathbf{k}$ does not depend on the transverse loading. If such loading is applied, or if the member has initial bending deformations, the axial load cannot reach a level at which $\mathbf{k}$ becomes singular, as seen in Section 2.1.

The incremental stiffness relations may be applied to inelastic buckling if k_{11} and $\mathbf{k}$ are defined in terms of tangent rigidities. In the straight configuration, $\mathbf{v} = 0$. An alternate equilibrium state $\delta\mathbf{v} \neq 0$ requires $|\mathbf{k}| = 0$. However, dV_1 needs not be zero.

5.4. Buckling as a Characteristic-Value Problem

Consider a linear material and a generalized bending stiffness $\mathbf{k}$, established for a given axial load by means of shape functions $\boldsymbol{\psi}$ and generalized displacements $\mathbf{q}$. The prime symbol used in Section 3.1 to distinguish between $\mathbf{v}$ and a general choice of $\mathbf{q}$ is here deleted. $\mathbf{k}$ has the form

$$\mathbf{k} = \mathbf{k}_0 + \mathbf{k}_g = \mathbf{k}_0 + V_1\mathbf{g} + \mathbf{k}_{gp} \tag{5.4–1}$$

$\mathbf{k}_0$ and $\mathbf{k}_g$ are determined by evaluating the bending strain energy $\bar{U}_b$, and the external potential of the axial load $\bar{U}_{\text{ext}}$, restricted to inextensional displacements. This results in the quadratic forms

$$\bar{U}_b = \tfrac{1}{2}\mathbf{q}^T\mathbf{k}_0\mathbf{q} \tag{5.4–2}$$

$$\bar{U}_{\text{ext}} = \tfrac{1}{2}\mathbf{q}^T\mathbf{k}_g\mathbf{q} \tag{5.4–3}$$

The axial load is assumed proportional to a parameter γ. The axial force N is then of the form, $N = \gamma N_0$, where N_0 is a fixed function of x. $\mathbf{k}_g$ is linear in N and takes the form $\mathbf{k}_g = \gamma\mathbf{k}_{g0}$. The critical axial load is the lowest load for which the homogeneous equations

$$(\mathbf{k}_0 + \gamma\mathbf{k}_{g0})\mathbf{q} = 0 \tag{5.4–4}$$

have a nonzero solution. Equation (5.4–4) represents a linear characteristic value problem, or eigenvalue problem. The condition for a nonzero solution is the characteristic equation, $|\mathbf{k}_0 + \gamma\mathbf{k}_{g0}| = 0$, whose lowest root is γ_{cr}. The associated characteristic shape $\mathbf{q}$ is the buckled shape. γ_{cr} has a minimum property that is important in assessing the accuracy of a solution. The potential energy associated with a shape $\mathbf{q}$ is here the change $\Delta\Pi$ that takes place from the equilibrium state $\mathbf{q} = 0$, and $\mathbf{q}$ plays the role of $\delta\mathbf{q}$ in the energy criterion of stability. γ_{cr} is the minimum load level at which $\Delta\Pi = 0$. The solution of $\Delta\Pi = 0$ for γ is

$$\gamma = -\frac{\mathbf{q}^T\mathbf{k}_0\mathbf{q}}{\mathbf{q}^T\mathbf{k}_{g0}\mathbf{q}} \tag{5.4–5}$$

and minimization of γ leads back to the characteristic value problem of Eq. (5.4–4).

If two discretizations are used such that the generalized displacements of one are a subset of the other, then the minimum obtained by the more refined discretization cannot be less than that obtained by the coarser one, and both are not less than the exact solution for γ_{cr}. A complete discretization is one that is capable of representing the potential energy functional to an arbitrary degree of accuracy. A typical example is a truncated Fourier series representation in which the number of terms may be increased indefinitely. By including more terms, a decreasing sequence of approximations is obtained whose lower limit is the exact γ_{cr}.

Example 1. Nonuniform Member

The energy method is applied here to determine the critical axial force for a nonuniform member. Assuming simple supports and a bending rigidity symmetric about midspan, the buckled shape will also be symmetric. A sine Fourier series in the interval $(0, l)$ satisfies the kinematic end conditions. Keeping only the symmetric terms, the buckled shape is sought in the form

$$v = a_1 \sin\frac{\pi x}{l} + a_3 \sin\frac{3\pi x}{l} + \cdots \tag{5.4–6}$$

The bending rigidity is assumed to be

$$EI = EI_0\left(\frac{1}{2} + \frac{x}{l}\right), \quad 0 \leqslant x \leqslant \frac{l}{2}$$

Evaluation of $\bar{U}_b$ and $\bar{U}_{ext}$, using only the first two terms of the series, yields

$$\bar{U}_b = 2 \times \frac{1}{2}\int_0^{l/2} EI(v_{,xx})^2\, dx = \frac{\pi^4}{4}\frac{EI_0}{l^3}\left[\left(\frac{3}{4} + \frac{1}{\pi^2}\right)a_1^2 + 81\left(\frac{3}{4} + \frac{1}{9\pi^2}\right)a_3^2 + \frac{18}{\pi^2}a_1 a_3\right]$$

$$\bar{U}_{ext} = V_1 \int_0^l \frac{1}{2}(v_{,x})^2\, dx = V_1 \frac{\pi^2}{4l}(a_1^2 + 9a_3^2)$$

Then

$$\mathbf{k}_0 = \frac{\partial^2 \bar{U}_b}{\partial \mathbf{q}^T \partial \mathbf{q}} = \frac{\pi^4 EI_0}{4l^3}\begin{bmatrix} \dfrac{3}{2} + \dfrac{2}{\pi^2} & \dfrac{18}{\pi^2} \\ \dfrac{18}{\pi^2} & 81\left(\dfrac{3}{2} + \dfrac{2}{9\pi^2}\right) \end{bmatrix} \tag{5.4–7}$$

$$V_1 \mathbf{g} = \frac{\partial^2 \bar{U}_{ext}}{\partial \mathbf{q}^T \partial \mathbf{q}} = V_1 \frac{\pi^2}{4l}\begin{bmatrix} 2 & 0 \\ 0 & 18 \end{bmatrix} \tag{5.4–8}$$

The characteristic equation $|\mathbf{k}_0 + V_1\mathbf{g}| = 0$ is quadratic in V_1. The lower root for $(-V_1 l^2/EI_0)$ is found to be 8.25. If only the first term of the series is used the root is 8.40. The improvement of about 2% from a one-term to a two-term representation indicates that a sufficiently accurate answer has been reached.

Example 2. Cantilever Subjected to a Uniform Axial Load

The assumed shape for cantilever buckling may be obtained from the sine Fourier series of the preceding example by shifting the origin to midspan and replacing l with $l/2$. The result is the cosine series

$$v = -(a_1 + a_2 + \cdots) + a_1 \cos\frac{\pi x}{2l} + a_3 \cos\frac{3\pi x}{2l} + \cdots \tag{5.4–9}$$

For EI constant, use of three terms yields

$$\bar{U}_b = \frac{1}{2}\int_0^l EI(v_{,xx})^2\, dx = \frac{\pi^4 EI}{64 l^3}(a_1^2 + 81 a_3^2 + 625 a_5^2)$$

and thus

$$\mathbf{k}_0 = \frac{\pi^4 EI}{32 l^3}\begin{bmatrix} 1 & 81 & 625 \end{bmatrix} \tag{5.4–10}$$

For a uniform compressive axial load p, $N = -p(l - x)$. Evaluation of $\mathbf{k}_{gp}$, using Eq. (3.1–21), yields

$$\mathbf{k}_{gp} = -p\begin{bmatrix} \dfrac{\pi^2}{16} - \dfrac{1}{4} & \dfrac{3}{4} & -\dfrac{5}{36} \\ \dfrac{3}{4} & \dfrac{9\pi^2}{16} - \dfrac{1}{4} & \dfrac{15}{4} \\ -\dfrac{5}{36} & \dfrac{15}{4} & \dfrac{25}{16}\pi^2 - \dfrac{1}{4} \end{bmatrix} \tag{5.4–11}$$

The characteristic equation is cubic in p. The lowest root is found by trial and error to be

$$p_{\text{cr}} = 7.838\frac{EI}{l^3} \tag{5.4–12}$$

A two-term representation yields the same answer rounded to four digits. A sufficient accuracy is thus achieved.

6. BUCKLING BY THE METHOD OF VIRTUAL FORCES

6.1. Governing Integral Equations

The equation of virtual forces, written in the geometrically linear theory, Eq. (5/2–1), is a condition of geometric compatibility for linear strain-displacement relations. It applies therefore to the bending part of these relations in beam-column theory. Inclusion of shear deformation in the method of virtual forces presents no particular difficulties. It will be omitted here, however, for simplicity of presentation. Use of a unit virtual force yields the compatibility equation for the displacement $v(x)$,

$$v(x) = \int_0^l b_x^\xi \chi \, d\xi \tag{6.1–1}$$

where b_x^ξ is the bending moment at ξ due to $P_y' = 1$ applied at x and χ is the curvature function of ξ. Use of a virtual $\mathbf{V}' = \mathbf{I}$ yields the compatibility equations for deformations

$$\mathbf{v} = \int_0^l \mathbf{b}^T \chi \, dx \tag{6.1–2}$$

The reference axes in Eqs. (6.1–1, 2) are member-bound axes, whose type is chosen by convenience in any particular application. The alternate type is convenient for a member whose support conditions prevent displacement of the member chord. The cantilever type is convenient if a member end is fixed. If both of these end conditions apply, then both types may be equally convenient.

For a member subjected to a general loading condition, the bending moment has the expression

$$M = \mathbf{bV} + \mathbf{b}_v V_1 + M_p \tag{6.1–3}$$

The displacement v occurs linearly in Eq. (6.1–3) as a lever arm of axial forces. Thus, if $\chi = M/EI$ is substituted into Eq. (6.1–1), the result is a linear integral equation for v. Equation (6.1–2) provides boundary conditions if elements of $\mathbf{v}$ are prescribed. Equations (6.1–1, 2) may be used in general as a basis for developing flexibility and stiffness relations. They will be applied here only in the restricted context of buckling.

Consider a straight member subjected only to axial forces and having homogeneous kinematic end conditions. In the buckled configuration elements of $\mathbf{V}$ are either zero or are reactions to restrained deformations. Elimination of these reactions from the integral equations (6.1–1, 2) yields a linear homogeneous integral equation for $v(x)$. If the axial load is poportional to one load parameter λ^2, the integral equation takes the form

$$v = \lambda^2 \mathscr{L}(v) \tag{6.1–4}$$

where $\mathscr{L}$ is a linear integral operator. Equation (6.1–4) has nontrivial solutions for characteristic values of λ^2, the lowest of which, λ^2_{cr}, defines the critical load.

Approximate methods for determining λ_{cr} are based on assuming a buckled shape. Some of these methods are outlined in what follows through examples.

6.2. Collocation Method

Example 1. Simply Supported Member

Consider a simply supported member subjected only to V_1. The compatibility equation (6.1–1) is

$$v(x) = -\left(1 - \frac{x}{l}\right)\int_0^x \xi\chi\,d\xi - \frac{x}{l}\int_0^l (l-\xi)\chi\,d\xi \tag{6.2–1}$$

In the buckled configuration $M = V_1 v$ and $\chi = V_1 v/EI$. If EI is variable, the load parameter λ^2 is defined in terms of a reference EI_0, as

$$\lambda^2 = -\frac{V_1 l^2}{EI_0} \tag{6.2–2}$$

Substituting $\chi = V_1 v/EI$ into Eq. (6.2–1), the result takes the form $v = \lambda^2 \mathscr{L}(v)$, where

$$\mathscr{L}(v) = \frac{1}{l^2}\left(1 - \frac{x}{l}\right)\int_0^l \frac{EI_0}{EI}\xi v\,d\xi + \frac{1}{l^2}\frac{x}{l}\int_0^l \frac{EI_0}{EI}(l-\xi)v\,d\xi \tag{6.2–3}$$

The collocation method consists in assuming a buckled shape v and of evaluating

$$\lambda^2 = \frac{v}{\mathscr{L}(v)} \tag{6.2–4}$$

at discrete or collocation points. In the case $EI = EI_0$ the assumed shape $v = a\sin \pi x/l$ yields a constant ratio $v/\mathscr{L}(v) = \pi^2$. To illustrate how an approximate solution is obtained, let $v = ax(l-x)$ be an assumed shape for Euler's column. Substituting $v = a\xi(l-\xi)$ in Eq. (6.2–3), the ratio $v/\mathscr{L}(v)$ is found to be

$$\frac{v}{\mathscr{L}(v)} = \frac{1}{t^2/3 - t^3/4 + (1-t)^2/3 - (1-t)^3/4} \tag{6.2–5}$$

where $t = x/l$. Because the assumed shape is approximate, the preceding ratio is not constant but is a variable approximation to λ_{cr}^2. A study of that ratio as a function of t shows that

$$\frac{48}{5} \leqslant \frac{v}{\mathscr{L}(v)} < 12 \tag{6.2–6}$$

The lower limit occurs at midspan, $t = \frac{1}{2}$, and the upper limit at the member ends. The lower limit, $\lambda^2 = 9.6$, is a good approximation from below to the exact value $\pi^2 = 9.87$. In general, a value of $v/\mathscr{L}(v)$ in a region of high flexibility such as midspan may be expected to be a better approximation than in a region of high stiffness such as near the supports.

It will be shown now that for simple supports, the minimum and maximum of $v/\mathscr{L}(v)$ are, respectively, lower and upper bounds to λ_{cr}^2. The proof will be given for a variable EI.

The computed shape $\mathscr{L}(v)$ may be considered as the exact buckling shape of a member having a variable bending rigidity EI'. $\mathscr{L}(v)$ satisfies exactly the moment curvature relation

$$\frac{d^2\mathscr{L}(v)}{dx^2} = \frac{V_1 v}{EI\lambda^2} = -\frac{EI_0}{EI}\frac{v}{l^2} \tag{6.2–7}$$

and is thus the exact shape for the variable rigidity

$$EI' = -\frac{V_1 l^2}{EI_0} EI \frac{\mathscr{L}(v)}{v} \tag{6.2–8}$$

If V_1 is given the value

$$|V_1|_{\min} = \frac{EI_0}{l^2} \min\left(\frac{v}{\mathscr{L}(v)}\right) \tag{6.2–9}$$

then EI' does not exceed EI anywhere along the beam, and $|V_1|_{\min}$ is necessarily a lower bound to $|V_{1,cr}|$. Similarly the maximum of $v/\mathscr{L}(v)$ gives an upper bound. Thus

$$\left(\frac{v}{\mathscr{L}(v)}\right)_{\min} < \lambda_{cr}^2 < \left(\frac{v}{\mathscr{L}(v)}\right)_{\max} \tag{6.2–10}$$

The procedure followed in Example 1 may be iterated by taking $\mathscr{L}(v)$ as an assumed shape.

The existence of a lower bound to λ_{cr} by the present method should be contrasted with the potential energy method which gives only upper bounds. The existence of both bounds is a useful property in assessing the accuracy of a solution.

Example 2. Restrained Cantilever

The method of Example 1 is now applied to the buckling of a restrained cantilever subjected to V_1. In the buckled configuration there is a moment V_2 at the fixed end, and

$$M = V_1 v - V_2\left(1 - \frac{x}{l}\right) \tag{6.2–11}$$

The condition $v_2 = 0$ is obtained from Eq. (6.1–2) in the form

$$v_2 = \int_0^l -\left(1 - \frac{x}{l}\right)\frac{M}{EI}dx = 0 \tag{6.2–12}$$

For EI constant Eq. (6.2–12) yields

$$V_2 = \frac{3V_1}{l}\int_0^l \left(1 - \frac{x}{l}\right) v\,dx \tag{6.2–13}$$

Taking as the assumed shape $v = (ax^2/l^3)(l - x)$, Eq. (6.2–13) yields $V_2 = V_1 a/10$. Substituting V_2 in Eq. (6.2–11) and $\chi = M/EI$ in Eq. (6.1–1), the ratio $v/\mathscr{L}(v)$ is found to be

$$\frac{v}{\mathscr{L}(v)} = \frac{20}{1 + 2t/3 - t^2} \tag{6.2–14}$$

This ratio varies from 20 at $t = 0$ to 30 at $t = 1$, with a minimum at $t = 1/3$ equal to $\lambda_{\min}^2 = 18$. The exact value is $\lambda_{cr}^2 = 20.19$. In this example the maxima of assumed and computed shapes occur at different but close locations, and $\lambda_{\min}^2$ occurs at yet another location more toward the fixed end. A good estimate of λ_{cr}^2 may be expected at the maxima of v and $\mathscr{L}(v)$. $v_{\max}$ occurs at $t = 2/3$, where $\lambda^2 = 20$, and $[\mathscr{L}(v)]_{\max}$ occurs at $t = 0.623$, where $\lambda^2 = 19.5$.

The collocation method may be applied by assuming a shape depending on several parameters, $\mathbf{q} = \{q_1\ q_2 \cdots q_n\}$. By equating v to $\lambda^2\mathscr{L}(v)$ at n collocation points, a linear homogeneous system of equations is obtained whose lowest characteristic value is an approximation to λ_{cr}^2. However, methods based on equating weighted averages of v and $\mathscr{L}(v)$ are generally preferable and are presented next.

6.3. Method of Weighted Averages

Instead of satisfying Eq. (6.1–4) at discrete points, Eq. (6.1–4) may be multiplied through by a weight function $p' = p'(x)$ and integrated in the interval $(0, l)$. This yields

$$\int_0^l p' v\,dx = \lambda^2 \int_0^l p' \mathscr{L}(v)\,dx \tag{6.3–1}$$

The procedure may be handled more simply, however, without determining $\mathscr{L}(v)$ by applying the equation of virtual forces with a virtual load function $p'_y = p'(x)$. The virtual bending moment has the expression, $M' = \mathbf{b}\mathbf{V}' + M'_p$, in which M'_p is the statically determinate part, and $\mathbf{V}'$ are virtual statical redundants. The equation of virtual forces is

$$\mathbf{V}'^T\mathbf{v} + \int_0^l p'v\,dx = \int_0^l M'_p\chi\,dx + \mathbf{V}'^T\int_0^l \mathbf{b}^T\chi\,dx \tag{6.3–2}$$

Equation (6.3–2) allows to use weight factors $\mathbf{V}'$ for deformations, together with the weight function p' for the displacement function. If an element of $\mathbf{v}$ is restrained, then the corresponding compatibility equation in $\int_0^l \mathbf{b}^T\chi\,dx = 0$ is satisfied separately.

Example 1

Consider the restrained cantilever of Example 2 of the last section and the same assumed shape. The fixed end moment V_2 is determined as in that example, and the final expression of the bending moment is

$$M = V_1 a\left[\frac{x^2}{l^2}\left(1 - \frac{x}{l}\right) - \frac{1}{10}\left(1 - \frac{x}{l}\right)\right] \tag{6.3–3}$$

Now Eq. (6.3–2) satisfies $v_2 = 0$ and only V'_3 need be kept. Choosing $V'_3 = 0$, Eq. (6.3–2) reduces to

$$\int_0^l p'v\,dx = \frac{1}{EI}\int_0^l M'_p M\,dx \tag{6.3–4}$$

For p' constant, $M'_p = -p'x(l-x)/2$. Equation (6.3–4) yields $\lambda^2 = -V_1 l^2/EI = 20$. The same answer is found if p' is chosen proportional to the assumed shape. It is recalled that $\lambda^2_{cr} = 20.19$.

6.4. Complementary Energy Method

The principle of virtual forces applied to an elastic material may be transformed into a complementary energy principle, which is a generalization of Castigliano's second theorem seen in the geometrically linear theory. This will be carried out in what follows for the buckling problem of a straight member in uniform compression supported at both ends. The case of an axial distributed load will be seen subsequently.

In the buckled configuration the bending moment has the expression

$$M = V_1 v - V_2\left(1 - \frac{x}{l}\right) + V_3\frac{x}{l} \tag{6.4–1}$$

Homogeneous end conditions are considered so that V_2 and V_3 are either zero or are reactions to restrained deformations. The virtual bending moment is now considered as an arbitrary variation δM of M, generated by virtual δv, δV_2, and δV_3. Note that δV_2 and δV_3 occur only if V_2 and V_3 occur in M.

For a linear material with $\chi = M/EI$, the internal virtual work becomes

$$\int_0^l \delta M \frac{M}{EI} dx = \delta \bar{U}_b^* \tag{6.4–2}$$

where

$$\bar{U}_b^* = \int_0^l \frac{1}{2} \frac{M^2}{EI} dx \tag{6.4–3}$$

The virtual load δp is obtained through the equilibrium equation of the linear theory, namely

$$\delta p = \delta M_{,xx} = V_1 \delta v_{,xx} \tag{6.4–4}$$

In the external virtual work, $\mathbf{v}^T \delta \mathbf{V}$ vanishes because deformations are zero where virtual redundants are not. The virtual work expression of δp is now integrated by parts, and the end conditions, $v = 0$, are enforced. There comes

$$\int_0^l v \delta p \, dx = V_1 \int_0^l v \delta v_{,xx} dx = -V_1 \int_0^l v_{,x} \delta v_{,x} dx \tag{6.4–5}$$

The case of a variable axial force is treated similarly. The differential equation of equilibrium for δM is

$$\delta M_{,x} = -\delta N_2 = N \delta v_{,x} + \text{constant} \tag{6.4–6}$$

where N_2 is the shear force. The integration by parts of $\int_0^l (\delta M)_{,xx} v \, dx$ is carried out as before with the end conditions $v = 0$. The external virtual work takes the variational form $-\delta \bar{U}_{\text{ext}}$, where

$$\bar{U}_{\text{ext}} = \int_0^l \frac{1}{2} N (v_{,x})^2 \, dx \tag{6.4–7}$$

In $\bar{U}_b^*$ the expression of M includes a term depending on the axial span load p_x and on v.

The equation of virtual forces takes the form

$$\delta \Pi^* = 0 \tag{6.4–8}$$

where

$$\Pi^* = \bar{U}_b^* + \bar{U}_{\text{ext}} \tag{6.4–9}$$

It may be stated then that the condition of geometric compatibility of an admissible buckled shape is that Π^* be stationary with respect to arbitrary but admissible variations δv, δV_2, and δV_3. The admissibility conditions are the kinematic support conditions on v and the static ones on $\mathbf{V}$.

Equation (6.4–8) is consistent with the work criterion of stability presented in Section 5.3. For a linear material, with $\chi = M/EI$, the value of $\bar{U}_b^*$ in a true equilibrium state is equal to that of the bending strain energy $\bar{U}_b$, and the value of Π^* is that of the potential energy Π. Note that Π^* and Π are in fact changes $\Delta\Pi^*$ and $\Delta\Pi$ from the straight to the buckled configuration. This change is zero to within third-order terms in the displacements. Here Π^* is of second order, and thus in the buckled state

$$\Pi^* = 0 \tag{6.4–10}$$

This last property is obtained directly by applying the equation of virtual forces to the buckled state, with $\delta v = v$, $\delta V_2 = V_2$, and $\delta V_3 = V_3$, and thus with $\delta M = M$.

Example 1

Consider buckling of Euler's column and the assumed shape

$$v = \frac{ax}{l^2}(l - x)$$

then

$$\bar{U}_{\text{ext}} = V_1 \int_0^l \frac{1}{2}(v_{,x})^2\,dx = \frac{1}{6}\frac{V_1 a^2}{l} \tag{6.4–11}$$

$$\bar{U}_b^* = \int_0^l \frac{1}{2EI}(V_1 v)^2\,dx = \frac{1}{60}\frac{V_1^2 a^2 l}{EI} \tag{6.4–12}$$

The condition $\Pi^* = 0$ yields $\lambda^2 = -V_1 l^2/EI = 10$. By comparison, the potential energy method based on the same shape yields $\lambda^2 = 12$. The higher accuracy of the complementary energy method is due to the property that Π^* depends on first derivatives of v, which are usually better approximated than the second derivatives which appear in the potential energy.

Example 2

Consider the buckling problem of a simply supported member subjected to a uniformly distributed compressive axial load p. The equilibrium solution in the

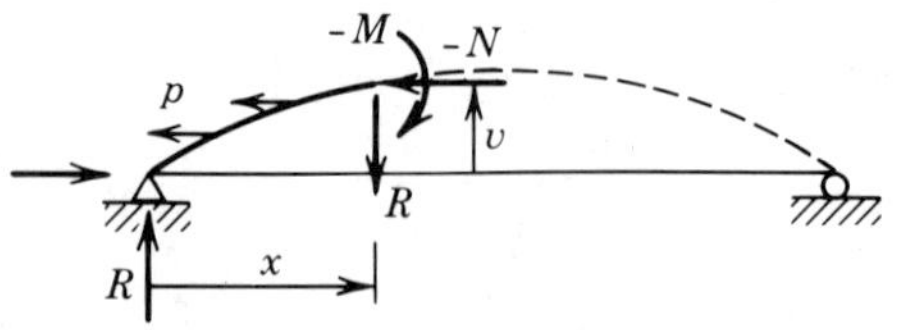

FIG. 6.4–1. Buckling under uniform compressive load.

buckled state, Fig. 6.4–1, is

$$N = -p(l - x) \tag{6.4–13}$$

and

$$M = Rx - plv + p\int_0^x [v(x) - v(\xi)]\,d\xi \tag{6.4–14}$$

where $R = (p\int_0^l v\,dx)/l$. M may be written in the form

$$M = Nv - p\int_0^x v\,dx + \frac{px}{l}\int_0^l v\,dx \tag{6.4–15}$$

The buckled shape is given the sine Fourier series representation

$$v = \sum_{i=1}^{n} a_i \sin\frac{i\pi x}{l} \tag{6.4–16}$$

which satisfies identically the admissibility conditions $v = 0$ at $x = 0$ and $x = l$. Then

$$v_{,x} = \sum a_i \frac{i\pi}{l}\cos\frac{i\pi x}{l} \tag{6.4–17}$$

$$M = pl\sum a_i M_i \tag{6.4–18}$$

and M_i is found as

$$M_i = \frac{1}{i\pi}\left(1 - \cos\frac{i\pi x}{l}\right) + \left(1 - \frac{x}{l}\right)\sin\frac{i\pi x}{l} - [1 + (-1)^{i+1}]\frac{x}{i\pi l} \tag{6.4–19}$$

Evaluation of Π^* is feasible analytically and results in an expression of the form

$$\Pi^* = \frac{p^2 l^3}{EI}\mathbf{a}^T\mathbf{F}\mathbf{a} - p\mathbf{a}^T\mathbf{G}\mathbf{a} \tag{6.4–20}$$

where $\mathbf{a} = \{a_1\ a_2 \ldots a_n\}$ and $\mathbf{F}$ and $\mathbf{G}$ are square symmetric matrices. The condition $\delta\Pi^* = 0$ for arbitrary $\delta\mathbf{a}$ yields

$$(\mathbf{G} - \mu^2\mathbf{F})\mathbf{a} = 0 \tag{6.4–21}$$

where

$$\mu^2 = \frac{pl^3}{EI} \tag{6.4–22}$$

Equation (6.4–21) represents a linear characteristic value problem, and **G** and **F** are positive definite matrices, as is readily deduced from the definition of Π^*. This characterises the lowest eigenvalue as the minimum μ^2 for which $\Pi^* = 0$, or

$$\mu_{cr}^2 = \min \frac{\mathbf{a}^T \mathbf{G} \mathbf{a}}{\mathbf{a}^T \mathbf{F} \mathbf{a}} \tag{6.4–23}$$

The sine Fourier series allows us to obtain a decreasing sequence of approximations to μ_{cr}^2. Use of one-, two-, and three-term series yields the sequence $\mu^2 = 18.78, 18.570, 18.569$. The critical value to four significant digits may thus be taken as

$$\mu_{cr}^2 = 18.57 \tag{6.4–24}$$

EXERCISES

Section 2

1. Using a free body diagram of a portion of the member in Fig. 2.1–1, obtain the statical solution $\mathbf{N} = \mathbf{bV}$, then derive Eqs. (2.1–1). Obtain similarly the statical solution for the cantilever of Fig. 2.2–1 and then the governing differential equations for v and φ and their solution, Eq. (2.2–1).
2. Prepare plots of the flexibility and stiffness coefficients of the alternate type as functions of λ. Do the same for the cantilever type.
3. Obtain the first two terms of series developments of the stiffness coefficients in powers of λ^2. Determine the ranges of λ in which the two-term series differ by less than 5% from the exact expressions.
4. Determine the limits as $\lambda \to 0$ of Eqs. (2.1–12) and (2.1–14).
5. Establish the flexibility and stiffness matrices of the alternate type in the case of a tensile axial force by performing the substitutions indicated in Section 2.1.
6. Consider a simply supported member subjected at its ends to equal bending moments $V_2 = -V_3$ and to a compressive axial force. The displacements are obtained in Eq. (2.1–7). Show the following:
   a. The deflection at midspan is

$$v_m = \frac{V_3}{V_1}\left(\frac{1}{\cos \lambda/2} - 1\right)$$

   b. The bending moment at midspan is

$$M_m = \frac{V_3}{\cos \lambda/2}$$

c. The maximum compressive stress takes the form, known as the secant formula,

$$\sigma_m = \frac{V_1}{A}\left(1 + \frac{ec}{r^2 \cos \lambda/2}\right)$$

where $e = |V_3/V_1|$, $r^2 = I/A$, and c is the distance to the neutral axis.

d. The ratio of v_m to the deflection v_{mL} in the geometrically linear theory, or deflection magnification factor, is

$$\frac{v_m}{v_{mL}} = \frac{8}{\lambda^2}\left(\frac{1}{\cos \lambda/2} - 1\right)\left(1 + \frac{EI}{GA_s}\frac{\lambda^2}{l^2}\right)$$

e. $v_3 = -v_2 = v_{3L} \tan(\lambda/2)/(\lambda/2)$
where $v_{3L} = V_3 l/2EI$ is the deformation in the linear theory.

7. A simply supported member with an I cross section having the properties $A = 40$ in^2, $I = 568$ in^4, and $d =$ depth $= 14$ in is subjected to a compressive axial force causing a stress of 17 ksi. For what value of equal end bending moments ($V_2 = -V_3$) does the maximum stress reach the value 34 ksi assuming linearly elastic behavior? (Use results of Exercise 6.)

8. Apply Eqs. (2.3–4) and (2.3–5) to determine $\mathbf{v}_p$ of the alternate type for a sinusoidal transverse load $p_y = p \sin \pi x/l$. Determine $\mathbf{V}_F$ using Eq. (2.3–9).

9. Consider a simply supported member without shear deformation subjected to V_1, V_2, V_3 and to an arbitrary transverse load. Let M_L be the bending moment in the linear theory. Show that the differential equation for v is

$$\frac{d^2v}{dx^2} + \frac{\lambda^2}{l^2}v = \frac{M_L}{EI}$$

Determine M_L for the loading of Exercise 8, and show that v is a sinusoidal function.

10. Verify that Eq. (2.6–4d) yields the same stiffness matrix with either type of bound reference.

Section 3.1

11. Obtain the complete bending stiffness matrix in Eq. (3.1–35) from the reduced matrices by making equilibrium diagrams similar to that of Fig. 3.1–1.

12. Derive the geometric matrices $\mathbf{k}_{gp}$ in Section 3.1, Examples 1 and 2, for an axial load $p_x = p(x/l)$.

13. For the member of Section 3.1, Example 2, subjected only to end actions, derive generalized bending stiffness relations based on the assumed shape

$$v = q_1 \sin\frac{\pi x}{l} + q_2 \sin\frac{2\pi x}{l}$$

Transform these stiffness relations into relations between **V** and **v**. Compare the resulting stiffness matrix for accuracy with the one based on Eq. (3.1–30).

14. In Section 3.1, Example 4, obtain **Q** in Eq. (3.1–43) for the case $m = 0$ and p_y constant; then use the stiffness equations to obtain the fixed end moments. Compare the result to the exact one.

Answer:

$$V_{3F} = -V_{2F} = \frac{pl^2}{12}\left(1 + \frac{1}{60}\frac{\lambda^2}{1 - \lambda^2/42}\right)$$

15. Do the preceding exercise for a concentrated transverse force P applied at midspan.

Answer:

$$V_{3F} = -V_{2F} = \frac{Pl}{8}\left(1 + \frac{1}{48}\frac{\lambda^2}{1 - \lambda^2/42}\right)$$

16. Formulate the tie-rod problem of Section 3.1, Example 5, based on the assumed shape of Example 2. Reduce the problem to the cubic equation

$$x(1 + x)^2 = \frac{32}{225}\left(\frac{v_0}{r}\right)^2$$

where $x = -\lambda^2/12$ and $v_0 = 5pl^4/384EI$

17. Use Section 3.1, Example 4, to formulate the tie-rod problem with fixed ends and a transverse load p_y. Reduce the problem to the cubic equation

$$x(1 + x)^2 = \frac{128}{2205}\left(\frac{v_0}{r}\right)^2$$

where

$$x = -\frac{\lambda^2}{42}, \quad v_0 = \frac{5}{1024}\frac{Q_w l^2}{EI}, \quad \text{and} \quad Q_w = \int_0^l \Psi_w p_y \, dx$$

Sections 3.2, 3.3

18. Show that the ratio found in Exercise 6 (part d) is well approximated by $(1 - V_1/V_{1,\text{cr}})^{-1}$. Neglect shear deformation.

19. A 20 ft long steel beam has a cross section with two axes of symmetry and the properties $I = 10$ in^4, $A = 3$ in^2, and depth $= 10/3$ in. Determine the maximum deflection and the maximum stress in the two following cases:
a. Simply supported beam subjected to a concentrated force of 1000 lb at midspan and a compressive axial force of 30,000 lb.
b. The beam is clamped at the ends and loaded as in (a). Compare the results obtained through Eq. (3.2–8) to those of an exact analysis.

20. Determine the magnification factors relative to the linear theory for v_2 and v_3 of the cantilever type in the loading condition $V_3 = 0$, $(V_1, V_2) \neq 0$, and neglecting shear deformation. Show that the magnification factor for v_2 is well approximated by $(1 - V_1/V_{1,cr})^{-1}$ where $V_{1,cr}$ pertains to the cantilever. Determine the magnification factor for the bending moment at the fixed end.

21. Verify that the stiffness matrix in Eq. (3.1–35) may be obtained by application of Eq. (3.3–10).

Section 5.1

22. Figure P22 shows the buckled shape of a cantilever having a rigid part of length a. Determine P_{cr} by applying either the flexibility or stiffness relations for the flexible part of the member.

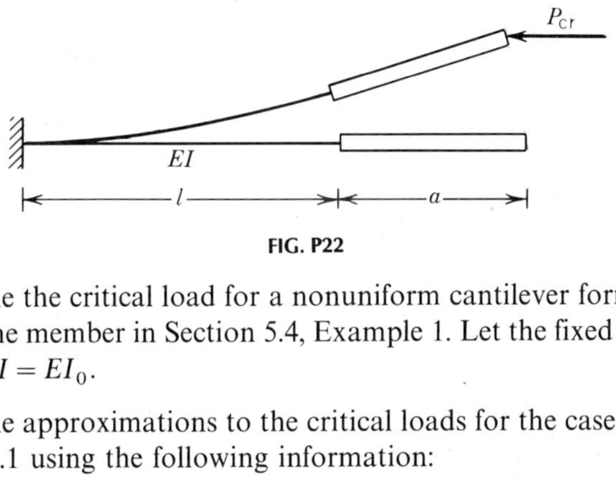

FIG. P22

23. Determine the critical load for a nonuniform cantilever formed of half the span of the member in Section 5.4, Example 1. Let the fixed end be the one having $EI = EI_0$.

24. Determine approximations to the critical loads for the cases considered in Section 5.1 using the following information:
a. The stiffness matrix $\mathbf{k}_s$, Eq. (3.1–35), with $\mu = 0$.
b. The improved reduced stiffness matrix of Eq. (3.1–46).
Comment on the accuracy of the results.

25. Using the stiffness matrix of Eqs. (3.1–28), determine an approximation to the critical value of a uniform compressive load p_x acting on a cantilever. Compare the result with the one obtained in Section 5.4, Example 2.

26. Do the preceding exercise for a simply supported member using the stiffness matrices in Eqs. (3.1–34) and (3.1–46), respectively. A more accurate value is obtained in Eq. (6.4–24).

Sections 5.3, 5.4

27. Use the energy method and the admissible shapes given here to determine the critical axial force for a uniform member without shear deformation in the following cases:
a. Proped cantilever, $v = ax^2(l - x)$.
b. Member with restrained ends, $v = ax^2(l - x)^2$ and $v = a(1 - \cos 2\pi x/l)$.
c. Simply supported member, origin of x at midspan, $v = -(a_1 + a_2) + a_1(2x/l)^2 + a_2(2x/l)^4$.

28. Use the energy method to determine the critical axial force for a member having $EI = EI_1(1 - x/l) + EI_2(x/l)$, assuming simple supports and a two-term sine series representation of the buckled shape.

29. Apply the energy method to Euler's column including shear deformation and using the assumed shape $v = a \sin \pi x/l$. Obtain φ in terms of v using the shear constitutive equation, and find

$$\varphi = \left(1 + \frac{V_1}{GA_s}\right)\frac{dv}{dx}$$

30. Do the preceding exercise keeping the amplitudes of the assumed shapes for v and φ independent.

31. Consider Euler's column, but let the ends be elastically restrained against rotation by springs of stiffness k so that the end moments are related to the end rotations through $V_2 = -kv_2$ and $V_3 = -kv_3$. Express the potential energy of the column, assuming inextensional buckling. Determine the critical axial force based on the assumed shape $v = a \sin \pi x/l + b \sin 3\pi x/l$, and study its behavior as k varies from 0 to ∞.

Section 6

32. Apply the method of Section 6.2 to determine bounds on the critical axial force of a uniform cantilever using the following assumed shapes:
a. $v = a(x/l)^2$.
b. $v = a(1 - \cos \pi x/2l)$

33. Do the preceding exercise with the shape function (b) for a cantilever having $1/EI = (1/EI_1)(1 - x/2l)$.

34. Apply the method of Section 6.3 to the case of a uniform cantilever using the assumed shapes of Exercise 32 and the virtual distributed loads (a) $p' = 1$ and (b) $p' = x/l$.

35. Apply the complementary energy method to Euler's column using as the assumed shape two straight line segments originating from the member ends and meeting at midspan.

36. Apply thc complementary energy method to the problems of Exercise 27.

37. Apply the complementary energy method to determine the critical axial force of a member for which

$$\frac{1}{EI} = \frac{1}{EI_1}\left(1 - \frac{x}{l}\right) + \frac{1}{EI_2}\frac{x}{l}$$

and for the following cases:

a. Cantilever, assumed shape $v = a(x/l)^2 + b(x/l)^3$.

b. Simply supported member, assumed shape $v = a \sin \pi x/l + b \sin 2\pi x/l$.

7

TRANSFORMATIONS

Effective Member Equations

1. INTRODUCTION: STATIC-KINEMATIC PAIRS

To be suitable for a general formulation of structural analysis, the static and kinematic equations of the structural member, and its stiffness and flexibility equations, derived in preceding chapters, need to be subjected to certain transformations. The basic need is for a transformation from the local reference, or a member-bound reference, to the global reference. Other transformations pertain to the formulation of effective equations that take into account particular static or kinematic conditions prescribed at the member ends.

In addition to its usefulness in specific applications, the concept of transformation provides a theoretical basis for generalization of governing equations.

The member state variables occur in static-kinematic pairs, such as $(\mathbf{V}, \mathbf{v})$ and $(\mathbf{S}, \mathbf{s})$ and, at the differential level, $(\mathbf{N}, \boldsymbol{\varepsilon})$. If $(\mathbf{Q}, \mathbf{q})$ denotes any static-kinematic pair, a common property is that $\mathbf{Q}^T \delta\mathbf{q}$ is a virtual work expression. Transformation of member equations may start with a kinematic transformation on $\mathbf{q}$ or with a static transformation on $\mathbf{Q}$. In either case invariance of the expression $\mathbf{Q}^T \delta\mathbf{q}$ (i.e., conservation of its value and form) will be required in order that the transformed equations as well as the principles of virtual work conserve the same form as in the original formulation. Thus either type of transformation will determine that of both kinematic and static variables and of associated flexibility and stiffness equations.

Transformations in the linear theory are treated first. In the geometrically nonlinear theory, both linear and nonlinear transformations are considered. In the last two subsections large rigid body rotations are considered and use is made of the exact static-kinematic equations of Chapter 2.

2. TRANSFORMATION TO GLOBAL COORDINATES IN LINEAR THEORY

2.1. Member Variables in Global Coordinates

In a plane structure the orientation of the local axes of a member with respect to the global axes is defined by means of the angle θ between the global X axis and

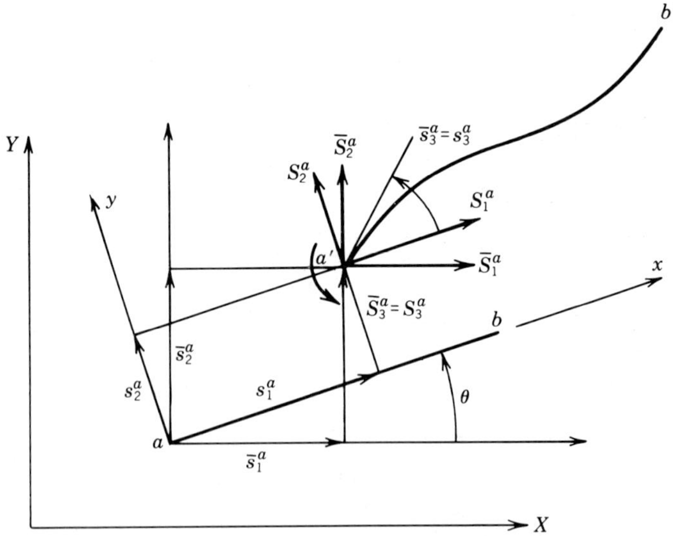

FIG. 2.1–1. Transformation from local to global coordinates.

the local x axis. To define θ algebraically, the member ends are labeled arbitrarily as a and b, and the x axis is oriented from a to b. A rotation of the global axes about the Z axis through the angle θ brings them into coincidence with the local axes. A positive θ is defined according to the right-hand rule. θ is thus defined to within $\pm 2\pi$, and sin θ and cos θ have unambiguous signed values.

In Fig. 2.1–1, s_1^a and s_2^a are the local translational displacements at end a and $\bar{s}_1^a$ and $\bar{s}_2^a$ are the global ones. Both sets are components of the vector $\overrightarrow{aa'}$. s_1^a is obtained as the sum of the orthogonal projections of $\bar{s}_1^a$ and $\bar{s}_2^a$ on the x axis. s_2^a is obtained similarly by projection on the y axis. s_3^a is the same in both coordinate systems. Thus

$$\begin{Bmatrix} s_1^a \\ s_2^a \\ s_3^a \end{Bmatrix} = \begin{bmatrix} \cos\theta & \sin\theta & 0 \\ -\sin\theta & \cos\theta & 0 \\ 0 & 0 & 1 \end{bmatrix} \begin{Bmatrix} \bar{s}_1^a \\ \bar{s}_2^a \\ \bar{s}_3^a \end{Bmatrix} \tag{2.1–1a}$$

or

$$\mathbf{s}^a = \boldsymbol{\lambda}^a \bar{\mathbf{s}}^a \tag{2.1–1b}$$

At end b the same transformation applies. Thus $\mathbf{s}^b = \boldsymbol{\lambda}^b \bar{\mathbf{s}}^b$ and $\boldsymbol{\lambda}^b = \boldsymbol{\lambda}^a$. For $\mathbf{s} = \{\mathbf{s}^a \ \mathbf{s}^b\}$,

$$\mathbf{s} = \boldsymbol{\lambda} \bar{\mathbf{s}} \tag{2.1–2a}$$

where

$$\boldsymbol{\lambda} = \begin{bmatrix} \boldsymbol{\lambda}^a & \\ & \boldsymbol{\lambda}^b \end{bmatrix} \tag{2.1–2b}$$

The transformation of force components in the present case obeys the same vector law. For a treatment that will remain valid in more general cases, however, the relation between local components $\mathbf{S}$, and global components $\bar{\mathbf{S}}$ is derived from the condition of virtual work invariance presented in the introduction. This is the equality of the scalar products

$$\mathbf{S}^T \delta\mathbf{s} = \bar{\mathbf{S}}^T \delta\bar{\mathbf{s}} \tag{2.1–3}$$

Substituting $\delta\mathbf{s} = \boldsymbol{\lambda}\delta\bar{\mathbf{s}}$ in Eq. (2.1–3), and equating the coefficients of $\delta\bar{\mathbf{s}}$ on both sides, there comes

$$\bar{\mathbf{S}} = \boldsymbol{\lambda}^T \mathbf{S} \tag{2.1–4}$$

In detail, $\bar{\mathbf{S}}^a = \boldsymbol{\lambda}^{aT}\mathbf{S}^a$, and $\bar{\mathbf{S}}^b = \boldsymbol{\lambda}^{bT}\mathbf{S}^b$. In the present case $\boldsymbol{\lambda}^a$ is an orthogonal matrix with the property that $\boldsymbol{\lambda}^{aT} = \boldsymbol{\lambda}^{a-1}$. Thus $\boldsymbol{\lambda}^T = \boldsymbol{\lambda}^{-1}$, and Eq. (2.1–4) may be inverted into $\mathbf{S} = \boldsymbol{\lambda}\bar{\mathbf{S}}$. Equation (2.1–2) and (2.1–4) form what is called a pair of contragradient relations, or together, a contragradient transformation.

2.2. Static-Kinematic Equations in Global Coordinates

The deformation-displacement relations and the member equilibrium solution are

$$\mathbf{v} = \mathbf{B}^T\mathbf{s} \tag{2.2–1a}$$

$$\mathbf{S} = \mathbf{B}\mathbf{V} + \mathbf{S}_p \tag{2.2–1b}$$

Substituting $\mathbf{s} = \boldsymbol{\lambda}\bar{\mathbf{s}}$ into Eq. (2.2–1a) and evaluating $\bar{\mathbf{S}}$ by means of Eq. (2.1–4), there comes

$$\mathbf{v} = \bar{\mathbf{B}}^T\bar{\mathbf{s}} = (\mathbf{B}^T\boldsymbol{\lambda})\bar{\mathbf{s}} \tag{2.2–2a}$$

$$\bar{\mathbf{S}} = \bar{\mathbf{B}}\mathbf{V} + \bar{\mathbf{S}}_p = (\boldsymbol{\lambda}^T\mathbf{B})\mathbf{V} + \boldsymbol{\lambda}^T\mathbf{S}_p \tag{2.2–2b}$$

Note that if Eq. (2.2–1b) is written $\mathbf{S} - \mathbf{S}_p = \mathbf{B}\mathbf{V}$, then Eqs. (2.2–1) form a contragradient pair, and this property is conserved in the transformed equations. In more detail, $\bar{\mathbf{B}}^T = [\bar{\mathbf{B}}^{aT} \;\; \bar{\mathbf{B}}^{bT}]$, $\bar{\mathbf{B}}^{aT} = \mathbf{B}^{aT}\boldsymbol{\lambda}^a$, and $\bar{\mathbf{B}}^{bT} = \mathbf{B}^{bT}\boldsymbol{\lambda}^b$. With λ^a and $\boldsymbol{\lambda}^b$ from Eq. (2.1–1a), and $\mathbf{B}^T$ from Eqs. (3/3.1–1) and (3/3.1–3), the expression of $\bar{\mathbf{B}}^T$ for the cantilever type is found as

$$(\bar{\mathbf{B}}^T)_c = [\bar{\mathbf{B}}^{aT} \;\; \bar{\mathbf{B}}^{bT}] = \left[\begin{array}{ccc|ccc} -\cos\theta & -\sin\theta & 0 & \cos\theta & \sin\theta & 0 \\ \sin\theta & -\cos\theta & -l & -\sin\theta & \cos\theta & 0 \\ 0 & 0 & -1 & 0 & 0 & 1 \end{array}\right] \tag{2.2–3}$$

and, for the alternate type

$$(\bar{\mathbf{B}}^T)_a = [\bar{\mathbf{B}}^{aT} \;\; \bar{\mathbf{B}}^{bT}] = \left[\begin{array}{ccc|ccc} -\cos\theta & -\sin\theta & 0 & \cos\theta & \sin\theta & 0 \\ -\dfrac{\sin\theta}{l} & \dfrac{\cos\theta}{l} & 1 & \dfrac{\sin\theta}{l} & -\dfrac{\cos\theta}{l} & 0 \\ -\dfrac{\sin\theta}{l} & \dfrac{\cos\theta}{l} & 0 & \dfrac{\sin\theta}{l} & -\dfrac{\cos\theta}{l} & 1 \end{array}\right] \tag{2.2–4}$$

A suggested exercise is to derive $\bar{\mathbf{B}}$ directly by statics.

2.3. Stiffness Equations in Global Coordinates

Stiffness equations in a bound reference have the form $\mathbf{V} = \mathbf{k}\mathbf{v} + \mathbf{V}_F$. Substituting this expression of $\mathbf{V}$ into Eq. (2.2–2b) and letting $\mathbf{v} = \bar{\mathbf{B}}^T\bar{\mathbf{s}}$, there comes

$$\bar{\mathbf{S}} = \bar{\mathbf{k}}_c\bar{\mathbf{s}} + \bar{\mathbf{S}}_F \tag{2.3–1}$$

where

$$\bar{\mathbf{k}}_c = \bar{\mathbf{B}}\mathbf{k}\bar{\mathbf{B}}^T \tag{2.3–2a}$$

$$\bar{\mathbf{S}}_F = \bar{\mathbf{S}}_p + \bar{\mathbf{B}}\mathbf{V}_F \tag{2.3–2b}$$

This derivation is similar to that perfomed in Chapter 5, Section 3, for the stiffness $\mathbf{k}_c$ in local coordinates. It is also possible to perform the transformation $\bar{\mathbf{S}} = \boldsymbol{\lambda}^T\mathbf{S}$, $\mathbf{S} = \mathbf{k}_c\mathbf{s} + \mathbf{S}_F$ and $\mathbf{s} = \boldsymbol{\lambda}\bar{\mathbf{s}}$. There results

$$\bar{\mathbf{k}}_c = \boldsymbol{\lambda}^T\mathbf{k}_c\boldsymbol{\lambda} \qquad (2.3\text{–}3a)$$

$$\bar{\mathbf{S}}_F = \boldsymbol{\lambda}^T\mathbf{S}_F \qquad (2.3\text{–}3b)$$

Equations (2.3–2a, 3a) define what is called a congruent transformation.

For a general member the stiffness properties are stated in most concise form in the 3 × 3 reduced stiffness matrix **k**. Also the axial stiffness k_{11} is uncoupled from the bending stiffnesses. Because of symmetry, these are described by three stiffness coefficients k_{22}, k_{33} and k_{23}. Equation (2.3–2a) allows then to express the 36 elements of $\bar{\mathbf{k}}_c$ in terms of four independent stiffness coefficients, θ and l. If $\bar{\mathbf{B}}$ is partitioned by columns,

$$\bar{\mathbf{B}} = [\bar{\mathbf{B}}_1 \;\; \bar{\mathbf{B}}_2 \;\; \bar{\mathbf{B}}_3] \qquad (2.3\text{–}4)$$

then

$$\bar{\mathbf{k}}_c = \sum_{i=1}^{3}\sum_{j=1}^{3} \bar{\mathbf{B}}_i k_{ij} \bar{\mathbf{B}}_j^T \qquad (2.3\text{–}5a)$$

or

$$\bar{\mathbf{k}}_c = k_{11}\bar{\mathbf{B}}_1\bar{\mathbf{B}}_1^T + k_{22}\bar{\mathbf{B}}_2\bar{\mathbf{B}}_2^T + k_{33}\bar{\mathbf{B}}_3\bar{\mathbf{B}}_3^T + k_{23}(\bar{\mathbf{B}}_2\bar{\mathbf{B}}_3^T + \bar{\mathbf{B}}_3\bar{\mathbf{B}}_2^T) \qquad (2.3\text{–}5b)$$

3. TRANSFORMATION LAWS

3.1. Contragradient Transformation

The transformation formulas established in the preceding section are applications of general transformation laws that govern the elements of a static-kinematic pair $(\mathbf{Q}, \mathbf{q})$, the constitutive equations relating these elements, and static-kinematic relations between two static-kinematic pairs.

Only linear transformations of linear equations are considered here. Nonlinearity will be treated separately.

A transformation has a type: kinematic or static. A linear kinematic transformation has the form of Eq. (3.1–2a), in which $\mathbf{q}'$ is a new set of independent variables, **T** is a constant transformation matrix, and $\mathbf{q}_0$ is a fixed set of values. $\mathbf{q}_0$ is the value of **q** in the initial state $\mathbf{q}' = 0$. If $\mathbf{q}_0 = 0$ the transformation is called homogeneous. **T** is constant in the sense that it does not depend on **q** or $\mathbf{q}'$. The transformation of the static variables is deduced from the requirement of virtual work invariance

$$\mathbf{Q}'^T\delta\mathbf{q}' = \mathbf{Q}^T\delta\mathbf{q} \qquad (3.1\text{–}1)$$

Since $\mathbf{q}_0$ is fixed, $\delta q = \mathbf{T}\delta q'$, and $\delta q'$ is arbitrary. Equation (3.1–1) defines $\mathbf{Q}'$ as the second of the next equations

$$\mathbf{q} = \mathbf{T}\mathbf{q}' + \mathbf{q}_0 \tag{3.1–2a}$$
$$\mathbf{Q}' = \mathbf{T}^T\mathbf{Q} \tag{3.1–2b}$$

Equations (3.1–2) are said to form a contragradient transformation, or a contragradient pair of relations.

A linear static transformation has the form $\mathbf{Q} = \mathbf{L}\mathbf{Q}' + \mathbf{Q}_0$. If $\mathbf{Q}_0 \neq 0$, it will be convenient to define the transformation for $(\mathbf{Q} - \mathbf{Q}_0)$. Consider then a homogeneous static transformation, $\mathbf{Q} = \mathbf{L}\mathbf{Q}'$. The invariance requirement (3.1–1) defines $\delta\mathbf{q}' = \mathbf{L}^T\delta\mathbf{q}$. This last relation may be integrated since $\mathbf{L}$ is constant, and the result is the pair of contragradient relations

$$\mathbf{Q} = \mathbf{L}\mathbf{Q}' \tag{3.1–3a}$$
$$\mathbf{q}' = \mathbf{L}^T\mathbf{q} + \mathbf{q}_0' \tag{3.1–3b}$$

The integration constants $\mathbf{q}_0'$ are determined by prescribing the initial state for $\mathbf{q}'$ at $\mathbf{q} = 0$.

A contragradient pair of relations such as Eqs. (3.1–2) transforms into contragradient relations if $\mathbf{q}'$ is transformed kinematically or $\mathbf{Q}$ is transformed statically. For example, if $\mathbf{q}' = \mathbf{T}'\mathbf{q}''$, then the transformation from $\mathbf{q}$ to $\mathbf{q}''$ has the matrix $\mathbf{T}\mathbf{T}'$ and is called a product of the transformations, from $\mathbf{q}$ to $\mathbf{q}'$ and from $\mathbf{q}'$ to $\mathbf{q}''$, respectively.

Example 1

Examples of contragradient transformations encountered previously are the two pairs of equations ($\mathbf{s} = \boldsymbol{\lambda}\bar{\mathbf{s}}, \bar{\mathbf{S}} = \boldsymbol{\lambda}^T\mathbf{S}$) and ($\mathbf{v} = \mathbf{B}^T\mathbf{s}, \mathbf{S} - \mathbf{S}_p = \mathbf{B}\mathbf{V}$). The second pair of relations also holds in global coordinates, as seen in the preceding section.

Example 2

The static transformation, $\mathbf{Q} = \mathbf{T}_b^a\mathbf{S}^b$ defines $\mathbf{Q}$ as the statical equivalents at a of the forces $\mathbf{S}^b$ applied at b. The contragradient relation $\mathbf{s}^b = \mathbf{T}_b^{aT}\mathbf{q}$ is then a rigid body kinematic transformation.

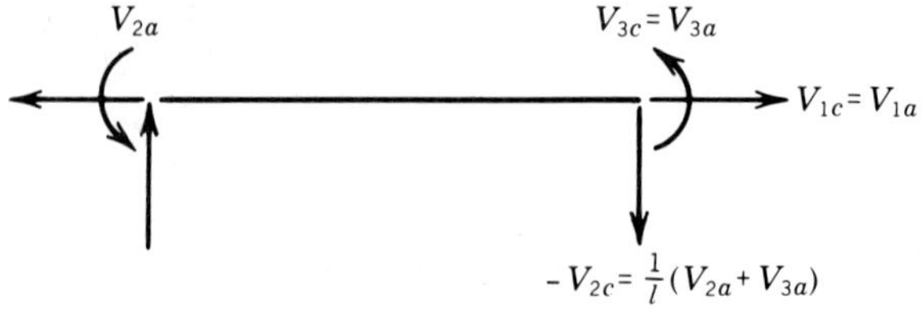

FIG. 3.1–1. Static transformation of bound reference.

Example 3

Consider a transformation from one bound reference to another. Let $(\mathbf{V}_c, \mathbf{v}_c)$ refer to the cantilever type and $(\mathbf{V}_a, \mathbf{v}_a)$ refer to the alternate type. The transformation can be defined kinematically or statically. For example, by statics, Fig. 3.1–1,

$$\begin{Bmatrix} V_{1,c} \\ V_{2,c} \\ V_{3,c} \end{Bmatrix} = \begin{bmatrix} 1 & 0 & 0 \\ 0 & -\dfrac{1}{l} & -\dfrac{1}{l} \\ 0 & 0 & 1 \end{bmatrix} \begin{Bmatrix} V_{1,a} \\ V_{2,a} \\ V_{3,a} \end{Bmatrix} \tag{3.1–4a}$$

The contragradient law yields

$$\begin{Bmatrix} v_{1,a} \\ v_{2,a} \\ v_{3,a} \end{Bmatrix} = \begin{bmatrix} 1 & 0 & 0 \\ 0 & -\dfrac{1}{l} & 0 \\ 0 & -\dfrac{1}{l} & 1 \end{bmatrix} \begin{Bmatrix} v_{1,c} \\ v_{2,c} \\ v_{3,c} \end{Bmatrix} \tag{3.1–4b}$$

A reverse approach could have been used by starting with the kinematic transformation, Fig. 3.1–2. Also the inverse transformations could be defined similarly.

Example 4

This example generalizes the contragradient law as presented earlier. If a member is viewed as formed of differential elements, the homogeneous equilibrium solution $\mathbf{N} = \mathbf{bV}$, where $\mathbf{N}$ and $\mathbf{b}$ are functions of x, may be viewed as a set of equations for the differential elements. The contragradient law in this case is based on the virtual work equation for the set of differential elements,

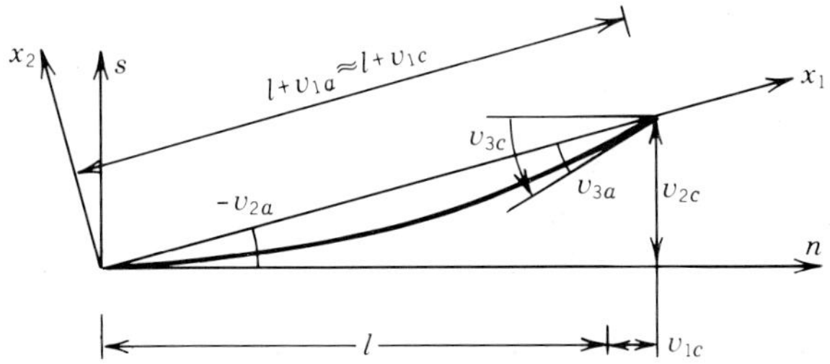

FIG. 3.1–2. Kinematic transformation of bound reference.

geometrically constrained to remain continuous, or

$$\int_0^l \mathbf{N}^T \delta\boldsymbol{\varepsilon}\, dx = \mathbf{V}^T \delta\mathbf{v}$$

With $\mathbf{N} = \mathbf{b}\mathbf{V}$, and $\mathbf{V}$ arbitrary, there comes

$$\delta\mathbf{v} = \int_0^l \mathbf{b}^T \delta\boldsymbol{\varepsilon}\, dx$$

Since $\mathbf{b}$ is constant in the linear theory, the equation is integrated into

$$\mathbf{v} = \int_0^l \mathbf{b}^T \boldsymbol{\varepsilon}\, dx$$

where the integration constants are chosen as zero.

3.2. Effective Variables

As part of a structure a member may have constraints placed on some of its state variables. Effective variables are a reduced set in terms of which member governing equations are formulated.

Constraints arise usually from construction conditions, but they could also represent mathematical approximations. Typically kinematic constraints represent support conditions and prescribe values, zero or not, to certain displacements. Static constraints that prescribe values to internal forces arise from structural connections that enforce continuity of only certain displacements while releasing that of the remaining displacements. For example, a structural member rigidly connected at end a to a structural joint and pin-connected at end b to another joint has a released rotation at end b and a prescribed end moment. If an external moment is applied at end b, there is a choice in defining the internal moment. The applied moment may be considered as a joint load. The bending moment at the member end is then equal to the applied moment. The other choice is to consider the applied moment as a span load and the bending moment at the pin as zero. This latter choice will be assumed in what follows.

An effective static-kinematic pair $(\mathbf{Q}', \mathbf{q}')$ is formed by applying the contragradient law to a pair $(\mathbf{Q}, \mathbf{q})$. In the case of a kinematic constraint, let $\mathbf{q} = \{\mathbf{q}_1 \;\; \mathbf{q}_2\}$ and $\mathbf{Q} = \{\mathbf{Q}_1 \;\; \mathbf{Q}_2\}$, where $\mathbf{q}_2$ is prescribed. The constraint $\delta\mathbf{q}_2 = 0$ is enforced so that $\mathbf{Q}^T \delta\mathbf{q} = \mathbf{Q}_1^T \delta\mathbf{q}_1$. In the case of a kinematic release, let $\mathbf{Q}_2 = 0$ be the corresponding static conditions. Then $\mathbf{Q}^T \delta\mathbf{q} = \mathbf{Q}_1^T \delta\mathbf{q}_1$. In both cases the pair $(\mathbf{Q}_1, \mathbf{q}_1)$ may be chosen as effective. Effective contragradient relations are considered following some examples.

Example 1. Pin-Connected Member

A member pin-connected at both ends has the static conditions $V_2 = V_3 = 0$, where V_2 and V_3 are the end moments. Thus effective variables are $\mathbf{V} = \{V_1\}$ and

$\mathbf{v} = \{v_1\}$. The static conditions are also $S_3^a = S_3^b = 0$ and $\bar{S}_3^a = \bar{S}_3^b = 0$. Thus effective variables are $\mathbf{S} = \{S_1^a \; S_2^a \; S_1^b \; S_2^b\}$, $\mathbf{s} = \{s_1^a \; s_2^a \; s_1^b \; s_2^b\}$, and $\bar{\mathbf{S}}$ and $\bar{\mathbf{s}}$ have similar components. If the member has no transverse load it becomes a two-force member. A further reduction to $\mathbf{S} = \{S_1^a \; S_1^b\}$ and $\mathbf{s} = \{s_1^a \; s_1^b\}$ is then possible. $\bar{\mathbf{S}}$ and $\bar{\mathbf{s}}$ retain in general their respective four elements.

Example 2

The structure of Fig. 3.2–1 is used to illustrate possible choices of effective variables. The kinematic constraints and releases are identified for each member, for the pairs (**S**, **s**) and (**V**, **v**), in Fig. 3.2–2. The alternate type is assumed for **V** and **v**. The effective variables are chosen by deleting any pair of static-kinematic variables that has one constrained element. Two members may have the same effective variables for different types of constraints. Such members will have different constitutive equations, as will be seen in the next section. For member 7

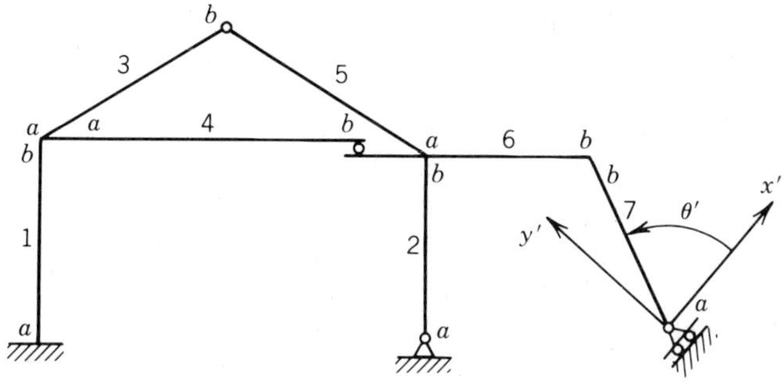

FIG. 3.2–1. Choice of effective variables, Example 2.

Member	Kinematic Constraints	Kinematic Releases	Effective variables	
1	$\mathbf{s}^a = 0$		$\mathbf{S} = \mathbf{S}^b$	$\mathbf{s} = \mathbf{s}^b$
			$\mathbf{V} = \{V_1 \; V_2 \; V_3\}$	$\mathbf{v} = \{v_1 \; v_2 \; v_3\}$
2	$s_1^a = s_2^a = 0$	$S_3^a = 0$	$\mathbf{S} = \mathbf{S}^b$	$\mathbf{s} = \mathbf{s}^b$
		$V_2 = 0$	$\mathbf{V} = \{V_1 \; V_3\}$	$\mathbf{v} = \{v_1 \; v_3\}$
3		$S_3^b = 0$	$\mathbf{S} = \{\mathbf{S}^a \; S_1^b \; S_2^b\}$	$\mathbf{s} = \{\mathbf{s}^a \; s_1^b \; s_2^b\}$
		$V_3 = 0$	$\mathbf{V} = \{V_1 \; V_2\}$	$\mathbf{v} = \{v_1 \; v_2\}$
4		$S_1^b = S_3^b = 0$	$\mathbf{S} = \{\mathbf{S}^a \; S_2^b\}$	$\mathbf{s} = \{\mathbf{s}^a \; s_2^b\}$
		$V_1 = V_3 = 0$	$\mathbf{V} = \{V_2\}$	$\mathbf{v} = \{v_2\}$
5		$S_3^a = 0$	$\mathbf{S} = \{S_1^a \; S_2^a \; \mathbf{S}^b\}$	$\mathbf{s} = \{s_1^a \; s_2^a \; \mathbf{s}^b\}$
		$V_2 = 0$	$\mathbf{V} = \{V_1 \; V_3\}$	$\mathbf{v} = \{v_1 \; v_3\}$
6			$\mathbf{S} = \{\mathbf{S}^a \; \mathbf{S}^b\}$	$\mathbf{s} = \{\mathbf{s}^a \; \mathbf{s}^b\}$
			$\mathbf{V} = \{V_1 \; V_2 \; V_3\}$	$\mathbf{v} = \{v_1 \; v_2 \; v_3\}$
7	$s_2'^a = 0$	$S_1'^a = S_3^a = 0$	$\mathbf{S} = \mathbf{S}^b$	$\mathbf{s} = \mathbf{s}^b$
		$V_1' = V_3' = 0$	$\mathbf{V}' = \{V_2'\}$	$\mathbf{v}' = \{v_2'\}$

FIG. 3.2–2. Effective variables for structure of Fig. 3.2–1.

the constraint that applies to an element of $\mathbf{S}$ is $S_3^a = 0$. To reduce further $\mathbf{s}$ and $\mathbf{S}$, $\mathbf{s}^a$ and $\mathbf{S}^a$ are transformed to axes (x', y'), as shown. Then $s'^a_2 = 0$ and $S'^a_1 = 0$, as indicated in the figure. A similar transformation may be made for member 7 from $\mathbf{V}$ to $\mathbf{V}' = \mathbf{S}'^a$. $\mathbf{S}'^a$ is thereby considered as a set of member redundants of cantilever type relative to axes bound to end b and oriented in the x', y' directions. $\mathbf{v}'$ is then the set of displacements at end a relative to the bound axes at b. Because of the two kinematic releases at end a, the effective variables are V'_2 and v'_2.

In global coordinates, effective pairs $(\bar{\mathbf{S}}, \bar{\mathbf{s}})$ are formed similarly to $(\mathbf{S}, \mathbf{s})$ whenever a force or a displacement component in global coordinates is constrained. For members 1 through 6 all constraints may be expressed in the same form in local or global axes. For member 7 a transformation from $\mathbf{s}^a$ to $\mathbf{s}'^a$ allows us to have $\mathbf{S}^b$ and $\mathbf{s}^b$ as effective variables.

To effective variables correspond effective static-kinematic equations and effective contragradient relations. The effective form of Eqs. (3.1–2) is obtained most simply either if an element of $\mathbf{q}'$ is prescribed or if an element of $\mathbf{Q}$ is prescribed. The two other possibilities are usually less simple. Consider, for example, Eqs. (2.2–2). If $\bar{\mathbf{s}}_i = 0$, the ith column of $\bar{\mathbf{B}}^T$, which multiplies $\bar{s}_i$ in the product $\bar{\mathbf{B}}^T\bar{\mathbf{s}}$, is removed together with the removal of $\bar{s}_i$ from $\bar{\mathbf{s}}$. In the static equation (2.2–2b) the equation for $\bar{S}_i$ is removed altogether, leaving the same effective $\bar{\mathbf{B}}$ as in the kinematic equation. In the other simple case an element V_i of $\mathbf{V}$ is zero. Then in the static equation the ith column of $\bar{\mathbf{B}}$ is removed together with the removal of V_i from $\mathbf{V}$. In the kinematic equation the equation for v_i is removed altogether, leaving the same effective $\bar{\mathbf{B}}$.

The two less simple cases alluded to earlier are ones in which either an element of $\mathbf{v}$ or an element of $\bar{\mathbf{S}}$ is prescribed. In the first case elements of $\bar{\mathbf{s}}$ cease to be kinematically independent, and in the second, elements of $\mathbf{V}$ cease to be statically independent. These cases belong to the general areas of kinematic and static transformations, respectively. A simplification occurs, however, if the prescribed element of $\bar{\mathbf{S}}$ is also an element of $\mathbf{V}$. This occurs typically for rotational releases and is dealt with similarly in local or global coordinates. For example, if $\bar{S}_i = V_j = 0$, the effective $\bar{\mathbf{S}}$, $\bar{\mathbf{s}}$, $\mathbf{V}$ and $\mathbf{v}$ are formed following the general procedure. The effective $\bar{\mathbf{B}}$ is then formed by deleting both the ith row and jth column because the ith static equation is removed and because the jth column multiplies $V_j = 0$. The effective kinematic equations $\mathbf{v} = \bar{\mathbf{B}}^T\bar{\mathbf{s}}$ are thereby determined. Consistency is ensured by virtual work invariance. In the present case $\bar{s}_i \neq 0$ and $v_j \neq 0$, but the elements of the ith column of $\mathbf{B}^T$ that pertain to the effective deformations are zero. The nonzero element of this column corresponds to v_j whose equation is removed.

3.3. Transformation of Stiffness Relations: Effective Stiffness

A constitutive stiffness equation for a static-kinematic pair $(\mathbf{Q}, \mathbf{q})$ has the form

$$\mathbf{Q} = \mathbf{k}\mathbf{q} + \mathbf{Q}_F \tag{3.3–1}$$

A kinematic transformation from $\mathbf{q}$ to $\mathbf{q}'$, Eq. (3.1–2a), defines $\mathbf{Q}' = \mathbf{T}^T\mathbf{Q}$ and leads to

$$\mathbf{Q}' = \mathbf{k}'\mathbf{q}' + \mathbf{Q}'_F \tag{3.3–2}$$

where

$$\mathbf{k}' = \mathbf{T}^T\mathbf{k}\mathbf{T} \tag{3.3–3a}$$

$$\mathbf{Q}'_F = \mathbf{T}^T(\mathbf{Q}_F + \mathbf{k}\mathbf{q}_0) \tag{3.3–3b}$$

A static transformation $\mathbf{Q} = \mathbf{L}\mathbf{Q}'$ cannot be applied in the same way to the stiffness equations. This problem will be addressed in the particular context of effective stiffness equations.

Example 1

Examples of the congruent transformation law, Eq. (3.3–3a), may be seen in Eqs. (2.3–2a) and (2.3–3a).

Example 2

Equation (3.1–4b) defines a kinematic transformation from deformations of alternate type to deformations of cantilever type. If the stiffness of alternate type is assumed given, Eq. (3.3–3a) determines the stiffness of cantilever type.

Effective Stiffness Relations due to Kinematic Constraints

Let Eq. (3.3–1) be partitioned in the form

$$\begin{Bmatrix} \mathbf{Q}_1 \\ \mathbf{Q}_2 \end{Bmatrix} = \begin{bmatrix} \mathbf{k}_{11} & \mathbf{k}_{12} \\ \mathbf{k}_{21} & \mathbf{k}_{22} \end{bmatrix} \begin{Bmatrix} \mathbf{q}_1 \\ \mathbf{q}_2 \end{Bmatrix} + \begin{Bmatrix} \mathbf{Q}_{1F} \\ \mathbf{Q}_{2F} \end{Bmatrix} \tag{3.3–4}$$

and assume that $\mathbf{q}_2$ is prescribed. Effective variables are $(\mathbf{Q}_1, q_1)$. The effective stiffness relations are the first subset in which $\mathbf{q}_2$ is placed on the right-hand side, or

$$\mathbf{Q}_1 = \mathbf{k}_{11}\mathbf{q}_1 + \mathbf{Q}'_{1F} \tag{3.3–5a}$$

where

$$\mathbf{Q}'_{1F} = \mathbf{Q}_{1F} - \mathbf{k}_{12}\mathbf{q}_2 \tag{3.3–5b}$$

A formal application of the transformation formulas would yield the same result.

If the effective stiffness relations exclude rigid body degrees of freedom, their inverse yields effective flexibility relations.

Example 3

For member 1 of Fig. 3.2–1, effective $\mathbf{S}$ and $\mathbf{s}$ are $\mathbf{S} = \mathbf{S}^b$ and $\mathbf{s} = \mathbf{s}^b$. The effective stiffness equations are extracted from the complete set as the subset at end b, in which $\mathbf{s}^a = 0$, or

$$\mathbf{S}^b = \mathbf{k}_c^{bb}\mathbf{s}^b + \mathbf{S}_F^b$$

This equation is the same as the reduced stiffness equation of the cantilever type.

If now end a has a prescribed support settlement $\mathbf{s}^a$, the effective stiffness equations become

$$\mathbf{S}^b = \mathbf{k}_c^{bb}\mathbf{s}^b + \mathbf{S}_F^b + \mathbf{k}_c^{ba}\mathbf{s}^a \tag{3.3–6}$$

The last two terms form effective fixed end forces $\mathbf{S}_F'^b$ at end b. For a uniform member with a uniform load p_y and a prescribed

$$\mathbf{s}^a = \{0 \;\; s_2^a \;\; 0\}$$

we have

$$\mathbf{S}_F^b = \left\{0 \quad -\frac{p_y l}{2} \quad \frac{p_y l^2}{12}\right\}$$

$$\mathbf{k}_c^{ba}\mathbf{s}^a = \left\{0 \quad -\frac{12EI}{l^3}s_2^a \quad \frac{6EI}{l^2}s_2^a\right\}$$

In these relations s_2^a and p_y obey their sign conventions. The local y axis for member 1 is oriented to the left.

Effective Stiffness Relations due to Kinematic Releases: Static Condensation

Consider the partitioned stiffness equation (3.3–4), and let kinematic releases cause the condition

$$\mathbf{Q}_2 = \mathbf{k}_{21}\mathbf{q}_1 + \mathbf{k}_{22}\mathbf{q}_2 + \mathbf{Q}_{2F} = 0 \tag{3.3–7a}$$

The effective variables are then $\mathbf{Q}_1$ and q_1. To obtain the effective stiffness relations, $\mathbf{q}_2$ must be determined in terms of $\mathbf{q}_1$ by solving Eq. (3.3–7a), or

$$\mathbf{q}_2 = -\mathbf{k}_{22}^{-1}\mathbf{k}_{21}\mathbf{q}_1 - \mathbf{k}_{22}^{-1}\mathbf{Q}_{2F} \tag{3.3–7b}$$

Equation (3.3–7b) defines the kinematic transformation

$$\begin{Bmatrix}\mathbf{q}_1\\ \mathbf{q}_2\end{Bmatrix} = \begin{bmatrix}\mathbf{I}\\ -\mathbf{k}_{22}^{-1}\mathbf{k}_{21}\end{bmatrix}\mathbf{q}_1 + \begin{Bmatrix}0\\ -\mathbf{k}_{22}^{-1}\mathbf{Q}_{2F}\end{Bmatrix} \tag{3.3–8}$$

Application of the general transformation formulas is equivalent to substitution of the expression of $\mathbf{q}_2$ into the stiffness equations for $\mathbf{Q}_1$. The process is called static condensation and results in

$$\mathbf{Q}_1 = \mathbf{k}_{11}'\mathbf{q}_1 + \mathbf{Q}_{1F}' \tag{3.3–9}$$

where

$$\mathbf{k}'_{11} = \mathbf{k}_{11} - \mathbf{k}_{12}\mathbf{k}_{22}^{-1}\mathbf{k}_{21} \tag{3.3–10a}$$

$$\mathbf{Q}'_{1F} = \mathbf{Q}_{1F} - \mathbf{k}_{12}\mathbf{k}_{22}^{-1}\mathbf{Q}_{2F} \tag{3.3–10b}$$

It will be seen in Section 3.4 that if the pair $(\mathbf{Q}, \mathbf{q})$ has flexibility relations, then kinematic releases are implemented by deletion of rows and columns from the flexibility relations. The effective stiffness relations may then be obtained by inversion of effective flexibility relations.

Example 4

Consider member 2 in Fig. 3.2–1. A first set of effective stiffness equations in local axes may be formed by deleting from the complete set the rows and columns associated with the kinematic support conditions. For a uniform member without shear deformation,

$$\begin{Bmatrix} S_3^a \\ S_1^b \\ S_2^b \\ S_3^b \end{Bmatrix} = \begin{bmatrix} \frac{4EI}{l} & 0 & -\frac{6EI}{l^2} & \frac{2EI}{l} \\ 0 & \frac{EA}{l} & 0 & 0 \\ -\frac{6EI}{l^2} & 0 & \frac{12EI}{l^3} & -\frac{6EI}{l^2} \\ \frac{2EI}{l} & 0 & -\frac{6EI}{l^2} & \frac{4EI}{l} \end{bmatrix} \begin{Bmatrix} s_3^a \\ s_1^b \\ s_2^b \\ s_3^b \end{Bmatrix} + \begin{Bmatrix} S_{3F}^a \\ S_{1F}^b \\ S_{2F}^b \\ S_{3F}^b \end{Bmatrix} \tag{3.3–11}$$

The fixed end forces depend on the member span load, which is left arbitrary in this example.

To further reduce the stiffness equations, a static condensation may be performed, based on the static condition $S_3^a = 0$ which yields

$$s_3^a = \frac{3}{2l}s_2^b - \frac{1}{2}s_3^b - \frac{l}{4EI}S_{3F}^a \tag{3.3–12}$$

Substitution into the equations for S_2^b and S_3^b yields the effective stiffness relations

$$\begin{Bmatrix} S_1^b \\ S_2^b \\ S_3^b \end{Bmatrix} = \begin{bmatrix} \frac{EA}{l} & 0 & 0 \\ 0 & \frac{3EI}{l^3} & -\frac{3EI}{l^2} \\ 0 & -\frac{3EI}{l^2} & \frac{3EI}{l} \end{bmatrix} \begin{Bmatrix} s_1^b \\ s_2^b \\ s_3^b \end{Bmatrix} + \begin{Bmatrix} S_{1F}^b \\ S_{2F}^b + \frac{3}{2l}S_{3F}^a \\ S_{3F}^b - \frac{1}{2}S_{3F}^a \end{Bmatrix} \tag{3.3–13}$$

Example 5

For member 2 of Fig. 3.2–1 effective **V** and **v** of the alternate type are $\mathbf{V} = \{V_1 \ V_3\}$ and $\mathbf{v} = \{v_1 \ v_3\}$ because of the static condition $V_2 = 0$, or

$$V_2 = \frac{4EI}{l} v_2 + \frac{2EI}{l} v_3 + V_{2F} = 0 \tag{3.3–14}$$

Solving for v_3 and substituting into the stiffness equation for V_3, there comes

$$\begin{Bmatrix} V_1 \\ V_3 \end{Bmatrix} = \begin{bmatrix} \frac{EA}{l} & 0 \\ 0 & \frac{3EI}{l} \end{bmatrix} \begin{Bmatrix} v_1 \\ v_3 \end{Bmatrix} + \begin{Bmatrix} V_{1F} \\ V_{3F} - \frac{1}{2} V_{2F} \end{Bmatrix} \tag{3.3–15}$$

These stiffness equations may now be transformed to local axes to find the same result as in the previous example. The transformation is made by means of the effective static-kinematic equations

$$\mathbf{S} = \mathbf{B}\mathbf{V} + \mathbf{S}_p$$

$$\mathbf{v} = \mathbf{B}^T \mathbf{s}$$

The static equations are

$$\begin{Bmatrix} S_1^b \\ S_2^b \\ S_3^b \end{Bmatrix} = \begin{bmatrix} 1 & 0 \\ 0 & -\frac{1}{l} \\ 0 & 1 \end{bmatrix} \begin{Bmatrix} V_1 \\ V_3 \end{Bmatrix} + \begin{Bmatrix} S_{1p}^b \\ S_{2p}^b \\ S_{3p}^b \end{Bmatrix} \tag{3.3–16}$$

With **B** as just obtained and **k** as in Eq. (3.3–15), the congruent law yields

$$\mathbf{k}_c = \mathbf{B}\mathbf{k}\mathbf{B}^T \tag{3.3–17}$$

and the result may be verified to coincide with that of Example 4.

Example 6

Consider member 3 of Fig. 3.2–1. The member is pinned at b. Effective variables in local axes are $\mathbf{S} = \{S_1^a \ S_2^a \ S_3^a \ S_1^b \ S_2^b\}$ and $\mathbf{s} = \{s_1^a \ s_2^a \ s_3^a \ s_1^b \ s_2^b\}$. The effective stiffness relations may be obtained by a static condensation of the complete set, based on the condition $S_3^b = 0$. The procedure is similar to that of Example 1.

A more interesting procedure is to start with effective reduced stiffness equations which are of the same form as those of member 2, except for the name of the member ends. For member 3

$$\begin{Bmatrix} V_1 \\ V_2 \end{Bmatrix} = \begin{bmatrix} \frac{EA}{l} & 0 \\ 0 & \frac{3EI}{l} \end{bmatrix} \begin{Bmatrix} v_1 \\ v_2 \end{Bmatrix} + \begin{Bmatrix} V_{1F} \\ V_{2F} - \frac{1}{2} V_{3F} \end{Bmatrix} \tag{3.3–18}$$

Transformation to local axes obeys the congruent law $\mathbf{k}_c = \mathbf{B}\mathbf{k}\mathbf{B}^T$, in which $\mathbf{B}$ is obtained by deleting the third column and sixth row from the complete matrix. The corresponding static equations are

$$\begin{Bmatrix} S_1^a \\ S_2^a \\ S_3^a \\ S_1^b \\ S_2^b \end{Bmatrix} = \begin{bmatrix} -1 & 0 \\ 0 & \frac{1}{l} \\ 0 & 1 \\ 1 & 0 \\ 0 & -\frac{1}{l} \end{bmatrix} \begin{Bmatrix} V_1 \\ V_2 \end{Bmatrix} + \begin{Bmatrix} S_{1p}^a \\ S_{2p}^a \\ S_{3p}^a \\ S_{1p}^b \\ S_{2p}^b \end{Bmatrix} \tag{3.3–19}$$

$\mathbf{k}_c$ may be evaluated as

$$\mathbf{k}_c = \frac{EA}{l}\mathbf{B}_1\mathbf{B}_1^T + \frac{3EI}{l}\mathbf{B}_2\mathbf{B}_2^T \tag{3.3–20}$$

where $\mathbf{B}_1$ and $\mathbf{B}_2$ are the first and second columns of $\mathbf{B}$, respectively.

The axial part of $\mathbf{k}_c$ is well known. The bending part is

$$(\mathbf{k}_c)_b = \frac{3EI}{l}\mathbf{B}_2\mathbf{B}_2^T = \frac{3EI}{l}\begin{bmatrix} 0 & 0 & 0 & 0 & 0 \\ 0 & \frac{1}{l^2} & \frac{1}{l} & 0 & -\frac{1}{l^2} \\ 0 & \frac{1}{l} & 1 & 0 & -\frac{1}{l} \\ 0 & 0 & 0 & 0 & 0 \\ 0 & -\frac{1}{l^2} & -\frac{1}{l} & 0 & \frac{1}{l^2} \end{bmatrix} \tag{3.3–21}$$

In global coordinates

$$\bar{\mathbf{k}}_c = \frac{EA}{l}\bar{\mathbf{B}}_1\bar{\mathbf{B}}_1^T + \frac{3EI}{l}\bar{\mathbf{B}}_2\bar{\mathbf{B}}_2^T \tag{3.3–22}$$

where

$$\bar{\mathbf{B}}_1^T = [-\cos\theta \quad -\sin\theta \quad 0 \quad \cos\theta \quad \sin\theta] \tag{3.3–23}$$

$$\bar{\mathbf{B}}_2^T = \left[-\frac{\sin\theta}{l} \quad \frac{\cos\theta}{l} \quad 1 \quad \frac{\sin\theta}{l} \quad -\frac{\cos\theta}{l} \right]$$

Example 7

Consider member 7 in Fig. 3.2–1. A choice of effective variables consists of $\mathbf{S}^b$ and $\mathbf{s}^b$, provided a coordinate transformation is made to axes (x', y') at a. This transformation, $\mathbf{s}^a = \boldsymbol{\lambda}\mathbf{s}'^a$, is

$$\begin{Bmatrix} s_1^a \\ s_2^a \\ s_3^a \end{Bmatrix} = \begin{bmatrix} \cos\theta' & \sin\theta' & 0 \\ -\sin\theta' & \cos\theta' & 0 \\ 0 & 0 & 1 \end{bmatrix} \begin{Bmatrix} s_1'^a \\ s_2'^a \\ s_3'^a \end{Bmatrix} \tag{3.3–24}$$

Starting with the complete $\mathbf{k}_c$ partitioned by member ends, the congruent transformation yields

$$\mathbf{k}_c' = \mathbf{T}^T\mathbf{k}_c\mathbf{T} = \begin{bmatrix} \boldsymbol{\lambda}^T\mathbf{k}^{aa}\boldsymbol{\lambda} & \boldsymbol{\lambda}^T\mathbf{k}^{ab} \\ \mathbf{k}^{ba}\boldsymbol{\lambda} & \mathbf{k}^{bb} \end{bmatrix} \tag{3.3–25}$$

A first set of effective variables is formed of $\mathbf{S}' = \{S_1'^a \ S_3'^a \ S_1^b \ S_2^b \ S_3^b\}$ and the corresponding $\mathbf{s}'$. The effective stiffness for this set is obtained by deleting from $\mathbf{k}_c'$ the second row and column. For the final effective set a static condensation is performed, corresponding to the conditions $S_1'^a = 0$ and $S_3'^a = 0$. The procedure for this follows that of an earlier example.

3.4. Transformation of Flexibility Relations: Effective Flexibility

For a static-kinematic pair to have flexibility relations, rigid body degrees of freedom must be absent from $\mathbf{q}$. Flexibility relations have then the form

$$\mathbf{q} = \mathbf{f}\mathbf{Q} + \mathbf{q}_0 \tag{3.4–1}$$

A static transformation $\mathbf{Q} = \mathbf{L}\mathbf{Q}'$ defines $\mathbf{q}' = \mathbf{L}^T\mathbf{q}$ and leads to

$$\mathbf{q}' = \mathbf{f}'\mathbf{Q}' + \mathbf{q}_0' \tag{3.4–2}$$

where

$$\mathbf{f}' = \mathbf{L}^T\mathbf{f}\mathbf{L} \tag{3.4–3a}$$

$$\mathbf{q}_0' = \mathbf{L}^T\mathbf{q}_0 \tag{3.3–4b}$$

A kinematic transformation of $\mathbf{q}$ cannot be applied in the same manner to the flexibility equations. This problem will be addressed in the particular context of effective flexibility equations.

Example 1

Equation (3.1–4a) defines a static transformation from reduced forces of the cantilever type to reduced forces of the alternate type. If the flexibility relations of cantilever type are assumed given, Eqs. (3.4–3) determine the relations of the alternate type.

Effective Flexibility Relations due to Kinematic Constraints

Let Eq. (3.4–1) be partitioned in the form

$$\begin{Bmatrix} \mathbf{q}_1 \\ \mathbf{q}_2 \end{Bmatrix} = \begin{bmatrix} \mathbf{f}_{11} & \mathbf{f}_{12} \\ \mathbf{f}_{21} & \mathbf{f}_{22} \end{bmatrix} \begin{Bmatrix} \mathbf{Q}_1 \\ \mathbf{Q}_2 \end{Bmatrix} + \begin{Bmatrix} \mathbf{q}_{1,0} \\ \mathbf{q}_{2,0} \end{Bmatrix} \tag{3.4–4}$$

and assume that $\mathbf{q}_2$ is prescribed. To obtain the effective relations for $\mathbf{q}_1$, the second part of Eq. (3.4–4) is solved for $\mathbf{Q}_2$, and the result is substituted into the first part. There comes

$$\mathbf{Q}_2 = -\mathbf{f}_{22}^{-1}\mathbf{f}_{21}\mathbf{Q}_1 - \mathbf{f}_{22}^{-1}(\mathbf{q}_{2,0} - \mathbf{q}_2) \tag{3.4–5}$$

$$\mathbf{q}_1 = \mathbf{f}'_{11}\mathbf{Q}_1 + \mathbf{q}'_{1,0} \tag{3.4–6}$$

where

$$\mathbf{f}'_{11} = \mathbf{f}_{11} - \mathbf{f}_{12}\mathbf{f}_{22}^{-1}\mathbf{f}_{21} \tag{3.4–7a}$$

$$\mathbf{q}'_{1,0} = \mathbf{q}_{1,0} - \mathbf{f}_{12}\mathbf{f}_{22}^{-1}(\mathbf{q}_{2,0} - \mathbf{q}_2) \tag{3.4–7b}$$

Equation (3.4–5) defines a nonhomogeneous static transformation. Section 3.1 outlines how this may be reduced to a homogeneous one. The congruent law applies and yields Eq. (3.4–7a).

The process just described for kinematic constraints parallels the static condensation of stiffness equations due to kinematic releases. The effective stiffness relations for kinematic constraints are simply obtained by deletion of rows and columns. Inversion of these relations is another procedure to derive effective flexibility relations.

Effective Flexibility Relations due to Kinematic Releases

Consider Eqs. (3.4–4), and let kinematic releases cause the conditions $\mathbf{Q}_2 = 0$. The effective flexibility relations are then

$$\mathbf{q}_1 = \mathbf{f}_{11}\mathbf{Q}_1 + \mathbf{q}_{1,0} \tag{3.4–8}$$

If an external load is applied in the released directions, its effect on Eq. (3.4–8) is to add the term $\mathbf{f}_{12}\mathbf{Q}_2$ to $\mathbf{q}_{1,0}$.

The process described for kinematic releases parallels the one applied to obtain stiffness equations due to kinematic constraints. The flexibility relations obtained here may be inverted to obtain effective stiffness relations due to kinematic releases. Rigid body degrees of freedom may then be added by means of member static-kinematic equations.

Example 2. Pin-Connected Member

For a pin-connected member the effective flexibility relation is the axial one, or

$$v_1 = f_{11}V_1 + v_{1,0}$$

The effective stiffness relation is the inverse relation,

$$V_1 = k_{11}v_1 + V_{1F}$$

The static-kinematic equations in local coordinates are

$$v_1 = \mathbf{B}^T\mathbf{s} = [-1 \;\; 0 \;\; 1 \;\; 0]\{s_1^a \;\; s_2^a \;\; s_1^b \;\; s_2^b\} \tag{3.4–8a}$$

and

$$\mathbf{S} = \mathbf{B}V_1 + \mathbf{S}_p \tag{3.4–8b}$$

The stiffness equations in local coordinates are then

$$\mathbf{S} = \mathbf{k}_c\mathbf{s} + \mathbf{S}_F \tag{3.4–9a}$$

where

$$\mathbf{k}_c = k_{11}\mathbf{B}\mathbf{B}^T = k_{11}\begin{bmatrix} 1 & 0 & -1 & 0 \\ 0 & 0 & 0 & 0 \\ -1 & 0 & 1 & 0 \\ 0 & 0 & 0 & 0 \end{bmatrix} \tag{3.4–9b}$$

and

$$\mathbf{S}_F = \mathbf{S}_p + \mathbf{B}V_{1F} \tag{3.4–9c}$$

The four elements of $\mathbf{S}_F$ are the end forces caused by the member loading condition with the member ends pinned.

If there is no transverse load, the member becomes a two-force member and $\mathbf{S}_F$ contains only axial components. In that case effective $\mathbf{S}$ and $\mathbf{s}$ may be reduced to the axial components and $\mathbf{k}_c$ to a 2×2 matrix.

For stiffness equations in global coordinates,

$$v_1 = \bar{\mathbf{B}}^T\bar{\mathbf{s}} = [-\cos\theta \;\; -\sin\theta \;\; \cos\theta \;\; \sin\theta]\{\bar{s}_1^a \;\; \bar{s}_2^a \;\; \bar{s}_1^b \;\; \bar{s}_2^b\} \tag{3.4–10}$$

and

$$\bar{\mathbf{k}}_c = k_{11}\bar{\mathbf{B}}\bar{\mathbf{B}}^T = k_{11}\left[\begin{array}{cc|c} c^2 & cs & \\ & & \longrightarrow(-1) \\ cs & s^2 & \\ \hline \downarrow & \searrow & \\ (-1) & & (+1) \end{array}\right] \tag{3.4–11}$$

$$\bar{\mathbf{S}}_F = \bar{\mathbf{S}}_p + \bar{\mathbf{B}}V_{1F} \tag{3.4–12}$$

where

$$c = \cos\theta \tag{3.4–13a}$$

$$s = \sin\theta \tag{3.4–13b}$$

Example 3

Consider member 4 of Fig. 3.2–1. It has two releases at end b which cause the static conditions $V_1 = 0$ and $V_3 = 0$. The effective flexibility is obtained by deleting the first and third rows and columns from the complete matrix. For a uniform member there remains

$$v_2 = \frac{l}{3EI}V_2 + v_{2,0} \tag{3.4–14}$$

The effective stiffness equation is then

$$V_2 = \frac{3EI}{l}v_2 + V'_{2F} \tag{3.4–15a}$$

where

$$V'_{2F} = -\frac{3EI}{l}v_{2,0} \tag{3.4–15b}$$

V'_{2F} is the fixed end moment at a caused by the member loading, with end b released as shown in the figure.

Effective stiffness relations in local coordinates are obtained through the transformation formulas:

$$\mathbf{k}_c = \frac{3EI}{l}\mathbf{B}\mathbf{B}^T \tag{3.4–16a}$$

$$\mathbf{S}_F = \mathbf{S}_p + \mathbf{B}V'_{2F} \tag{3.4–16b}$$

The effective $\mathbf{S}$ is formed by deleting S_1^b and S_3^b because the corresponding displacements are released. The effective $\mathbf{B}$ is then the second column of the complete $\mathbf{B}$ from which the rows corresponding to S_1^b and S_3^b are deleted. The resulting $\mathbf{k}_c$ is the same as the bending part of the stiffness matrix of member 3, determined in Section 3.3, Example 6. The difference here is that the axial stiffness for S_1^a is zero, which is as should be expected because of the axial release at end b. The axial stiffness relation reduces to the statical equation $S_1^a = S_{1p}^a$.

Example 4

Assume that member 4 in Fig. 3.2–1 is now pin connected at end a. It has then three releases which make it statically determinate. Effective $\mathbf{V}$ and $\mathbf{v}$ become empty—that is, they have zero dimensions. Effective $\mathbf{S}$ is $\mathbf{S} = \{S_1^a \;\; S_2^a \;\; S_2^b\}$ because the kinematic releases make $S_3^a = S_1^b = S_3^b = 0$. There is no stiffness associated with $\mathbf{S}$, but the effective stiffness relations reduce to the equilibrium solution, $\mathbf{S} = \mathbf{S}_p$.

4. TRANSFORMATION TO GLOBAL COORDINATES IN SECOND-ORDER THEORY

The definition of member variables $\mathbf{s}$ and $\mathbf{S}$, the coordinate transformation $\mathbf{s} = \boldsymbol{\lambda}\bar{\mathbf{s}}$, and the contragradient law $\bar{\mathbf{S}} = \boldsymbol{\lambda}^T\mathbf{S}$ seen in Section 2.1, are all valid in any theory of plane behavior. The stiffness relations in local axes, (6/3.1–11, 19) or (6/2.6–3), are partitioned into axial relations for $\mathbf{S}_1$ and bending relations for $\mathbf{S}$. Adopting this notation here, the transformation formulas become

$$\begin{Bmatrix} \mathbf{s}_1 \\ \mathbf{s} \end{Bmatrix} = \begin{bmatrix} \boldsymbol{\lambda}_1 \\ \boldsymbol{\lambda} \end{bmatrix} \bar{\mathbf{s}} \tag{4–1a}$$

$$\bar{\mathbf{S}} = \boldsymbol{\lambda}_1^T \mathbf{S}_1 + \boldsymbol{\lambda}^T \mathbf{S} \tag{4–1b}$$

Expression of $\mathbf{S}$ in terms of $\mathbf{s}$ takes the form

$$\bar{\mathbf{S}} = \bar{\mathbf{k}}_s \bar{\mathbf{s}} + \bar{\mathbf{S}}_F + \bar{\mathbf{S}}_N \tag{4–2}$$

where

$$\bar{\mathbf{k}}_s = \boldsymbol{\lambda}_1^T \mathbf{k}_{11} \boldsymbol{\lambda}_1 + \boldsymbol{\lambda}^T \mathbf{k}_s \boldsymbol{\lambda} \tag{4–3a}$$

$$\bar{\mathbf{S}}_F = \boldsymbol{\lambda}_1^T \mathbf{S}_{1F} + \boldsymbol{\lambda}^T \mathbf{S}_F \tag{4–3b}$$

$$\bar{\mathbf{S}}_N = \tfrac{1}{2} k_{11} \boldsymbol{\lambda}_1^T \mathbf{B}_1 \bar{\mathbf{s}}^T \bar{\mathbf{g}}_s \bar{\mathbf{s}} \tag{4–3c}$$

$$\bar{\mathbf{g}}_s = \boldsymbol{\lambda}^T \mathbf{g}_s \boldsymbol{\lambda} \tag{4–3d}$$

With axial forces treated as parameters, nonlinearity in Eq. (4–2) is separated into the term $\mathbf{S}_N$. It is noted that the congruent law applies to the stiffness matrix and to the geometric matrix $\bar{\mathbf{g}}_s$.

Transformation from member-bound axes to the local axes was seen in Chapter 6, Sections 2.6 and 3.1. Transformation from bound to global axes is similar and could be formulated either directly or via the local axes.

5. EFFECTIVE STIFFNESS AND FLEXIBILITY RELATIONS IN SECOND-ORDER THEORY

The basis and rules for choosing effective variables remain the same in a geometrically nonlinear theory as in the linear theory. Effective stiffness equations arising from prescribed displacements and effective flexibility relations arising from kinematic releases are also formed, respectively, by the procedures outlined in the linear theory. Particular considerations for the nonlinear term in the axial flexibility relation will appear through an example.

To obtain effective stiffness equations due to kinematic releases, a start with flexibility followed by inversion and transformation to local or global axes may be analytically simpler than a static condensation of the stiffness equations.

Example 1. Flexibility and Stiffness Relations with Rotational Release

Consider member 3 in Fig. 3.2–1. Using member redundants of the alternate type, the rotational release at b causes $V_3 = 0$. The effective bending flexibility relation is that for v_2. For a uniform member without span load, from Eq. (6/2.1–8),

$$v_2 = f_{22} V_2 = \frac{l}{EI} \frac{1 - \alpha\lambda \cot \lambda}{\alpha\lambda^2} V_2 \tag{5–1}$$

With $V_3 = 0$, the axial flexibility relation (6/2.1–11, 12) becomes

$$v_1 = \frac{l}{EA} V_1 - \frac{V_2^2}{2lV_1^2} \left[\frac{\lambda(2\lambda + \sin 2\lambda)}{4 \sin^2 \lambda} - 1 \right] \tag{5–2}$$

The effective stiffness relations are obtained by inverting the flexibility relations. If the nonlinear term in Eq. (5–2) is expressed in terms of v_2, there comes

$$V_2 = k'_{22} v_2 = \frac{EI}{l} \frac{\alpha\lambda^2}{1 - \alpha\lambda \cot \lambda} v_2 \tag{5–3}$$

$$V_1 = \frac{EA}{l}(v_1 + \tfrac{1}{2} g'_{22} v_2^2) \tag{5–4a}$$

where

$$g'_{22} = l \frac{\lambda(2\lambda + \sin 2\lambda) - 4 \sin^2 \lambda}{4(\sin \lambda - \alpha\lambda \cos \lambda)^2} \tag{5–4b}$$

The other and less simple procedure to obtain the effective stiffness relations involves a static condensation of the bending relations and a transformation of the term $\mathbf{v}^T \mathbf{g} \mathbf{v}$.

If the member has a transverse load, Eqs. (5–1) and (5–3) become $v_2 =$

$f_{22}V_2 + v_{2p}$ and $V_2 = k'_{22}v_2 + V'_{2F}$, respectively, where $V'_{2F} = -k'_{22}v_{2p}$. The nonlinear term in the axial stiffness relation is modified in a more fundamental way, however, because it depends on the deformed shape. Here static condensation offers an approximate procedure that consists in using the expression $\mathbf{v}^T\mathbf{g}\mathbf{v}$ with v_3 determined in terms of v_2 and of the transverse load.

Equations (5–3) and (5–4) are now transformed into effective stiffness equations in the local axes. The member static-kinematic equations, separated into axial and bending equations are obtained from Eqs. (6/2.6–1, 2). Since $V_3 = 0$, the corresponding column of $\mathbf{B}$ is removed from $\mathbf{BV}$, and the relation for v_3 is deleted. Also $\mathbf{S}_p = 0$ in the present example. Thus

$$\mathbf{S}_1 = \mathbf{B}_1 V_1 = \{-1 \;\; 1\} V_1 \tag{5–5a}$$

$$\mathbf{S} = \mathbf{B}_2 V_2 + (V_1 \mathbf{c}_1)\mathbf{s} \tag{5–5b}$$

and

$$v_1 = \mathbf{B}_1^T \mathbf{s}_1 + \tfrac{1}{2}\mathbf{s}^T \mathbf{c}_1 \mathbf{s} \tag{5–5c}$$

$$v_2 = \mathbf{B}_2^T \mathbf{s} \tag{5–5d}$$

The static condition $V_3 = 0$ is also $S_3^b = 0$, and this is satisfied by Eq. (5–5b) because the rows and columns of $\mathbf{c}_1$ associated with the rotational displacements are zero. Effective variables and matrices are thus $\mathbf{S} = \{S_2^a \;\; S_3^a \;\; S_2^b\}$, $\mathbf{s} = \{s_2^a \;\; s_3^a \;\; s_2^b\}$ and

$$\mathbf{B}_2^T = \begin{bmatrix} \frac{1}{l} & 1 & -\frac{1}{l} \end{bmatrix} \tag{5–6}$$

$$\mathbf{c}_1 = \frac{1}{l}\begin{bmatrix} 0 & 0 & -1 \\ 0 & 0 & 0 \\ -1 & 0 & 1 \end{bmatrix} \tag{5–7}$$

Equations (5–5) turn into the desired stiffness equations upon substituting in them the stiffness relations for V_1 and V_2 and the geometric relations for v_1 and v_2. There comes

$$\mathbf{S}_1 = \frac{EA}{l}[\mathbf{B}_1\mathbf{B}_1^T\mathbf{s}_1 + \tfrac{1}{2}\mathbf{B}_1\mathbf{s}^T(g'_{22}\mathbf{B}_2\mathbf{B}_2^T + \mathbf{c}_1)\mathbf{s}] \tag{5–8}$$

$$\mathbf{S} = (k'_{22}\mathbf{B}_2\mathbf{B}_2^T + V_1\mathbf{c}_1)\mathbf{s} \tag{5–9}$$

With $V_1 = -\alpha\lambda^2 EI/l^2$, the bending stiffness matrix, coefficient of $\mathbf{s}$, is

$$\mathbf{k}_s = \frac{EI\alpha\lambda^2}{l^3(1-\alpha\lambda\cot\lambda)}\begin{bmatrix} 1 & l & -1 \\ l & l^2 & -l \\ -1 & -l & 1 \end{bmatrix} - \frac{EI\alpha\lambda^2}{l^3}\begin{bmatrix} 1 & 0 & -1 \\ 0 & 0 & 0 \\ -1 & 0 & 1 \end{bmatrix} \tag{5–10a}$$

or

$$\mathbf{k}_s = \frac{EI}{l^3}\frac{\alpha\lambda^2}{1-\alpha\lambda\cot\lambda}\begin{bmatrix} \alpha\lambda\cot\lambda & l & -\alpha\lambda\cot\lambda \\ l & l^2 & -l \\ -\alpha\lambda\cot\lambda & -l & \alpha\lambda\cot\lambda \end{bmatrix} \tag{5–10b}$$

Equations (5–8, 9) may be transformed to global coordinates by means of Eqs. (4–3). For a direct transformation from bound to global coordinates, Eqs. (5–5) become

$$\bar{\mathbf{S}} = \bar{\mathbf{B}}_1 V_1 + \bar{\mathbf{B}}_2 V_2 + V_1 \bar{\mathbf{c}}_1 \bar{\mathbf{s}} \tag{5–11a}$$

$$v_1 = \bar{\mathbf{B}}_1^T \bar{\mathbf{s}} + \tfrac{1}{2}\bar{\mathbf{s}}^T \bar{\mathbf{c}}_1 \bar{\mathbf{s}} \tag{5–11b}$$

$$v_2 = \bar{\mathbf{B}}_2^T \bar{\mathbf{s}} \tag{5–11c}$$

where

$$\bar{\mathbf{B}}_1 = \boldsymbol{\lambda}_1^T \mathbf{B}_1 = \{-\cos\theta \quad -\sin\theta \quad 0 \quad \cos\theta \quad \sin\theta \quad 0\} \tag{5–12a}$$

$$\bar{\mathbf{B}}_2 = \boldsymbol{\lambda}^T \mathbf{B}_2 = \left\{-\frac{\sin\theta}{l} \quad \frac{\cos\theta}{l} \quad 1 \quad -\frac{\sin\theta}{l} \quad -\frac{\cos\theta}{l} \quad 0\right\} \tag{5–12b}$$

$$\bar{\mathbf{c}}_1 = \boldsymbol{\lambda}^T \mathbf{c}_1 \boldsymbol{\lambda} = \frac{1}{l}\left[\begin{array}{ccc|c} \sin^2\theta & -\sin\theta\cos\theta & 0 & \\ -\sin\theta\cos\theta & \cos^2\theta & 0 & \rightarrow(-1) \\ 0 & 0 & 0 & \\ \hline & \downarrow(-1) & & \searrow(+1) \end{array}\right] \tag{5–12c}$$

The stiffness equations take the form

$$\bar{\mathbf{S}} = \bar{\mathbf{k}}_s \bar{\mathbf{s}} + \bar{\mathbf{S}}_N$$

where

$$\bar{\mathbf{k}}_s = \frac{EA}{l}\bar{\mathbf{B}}_1 \bar{\mathbf{B}}_1^T + k'_{22}\bar{\mathbf{B}}_2 \bar{\mathbf{B}}_2^T + V_1 \bar{\mathbf{c}}_1 \tag{5–13a}$$

$$\bar{\mathbf{S}}_N = \frac{1}{2}\frac{EA}{l}\bar{\mathbf{B}}_1 \bar{\mathbf{s}}^T (g'_{22}\bar{\mathbf{B}}_2 \bar{\mathbf{B}}_2^T + \bar{\mathbf{c}}_1)\bar{\mathbf{s}} \tag{5–13b}$$

It is noted that there is no effective stiffness for $\bar{S}_3^b$ as should be expected from the condition $\bar{S}_3^b = V_3 = 0$. An effective $\bar{\mathbf{S}}$ could have been formed by deleting S_3^b as was done in local coordinates.

Example 2. Two-Force Member

For a two-force member, the effective stiffness relations in the bound reference reduce to the axial relation $V_1 = EAv_1/l$. For transformation to local coordinates, Eqs. (5–5) of the previous example may be made effective by implementing the condition $S_3^a = V_2 = 0$. Thus $\mathbf{S}_1 = \{S_1^a \ S_1^b\}$, and $\mathbf{S} = \{S_2^a \ S_2^b\}$. Equations (5–8, 9) are specialized to the present case by deleting the terms in g'_{22} and k'_{22} and reducing $\mathbf{c}_1$ to *a* 2×2 matrix. There comes

$$\mathbf{S}_1 = \frac{EA}{l}(\mathbf{B}_1\mathbf{B}_1^T\mathbf{s}_1 + \tfrac{1}{2}\mathbf{B}_1\mathbf{s}^T\mathbf{c}_1\mathbf{s}) \qquad (5\text{–}14a)$$

$$\mathbf{S} = V_1\mathbf{c}_1\mathbf{s} \qquad (5\text{–}14b)$$

The relation for $\mathbf{S}$ is purely geometric, and $V_1\mathbf{c}_1$ is called the string geometric stiffness matrix. In global coordinates, $\bar{\mathbf{S}} = \{\bar{S}_1^a \ \bar{S}_2^a \ \bar{S}_1^b \ \bar{S}_2^b\}$, and

$$\bar{\mathbf{S}} = \left[\frac{EA}{l}\bar{\mathbf{B}}_1\bar{\mathbf{B}}_1^T + V_1\bar{\mathbf{c}}_1\right]\bar{\mathbf{s}} + \frac{1}{2}\frac{EA}{l}\bar{\mathbf{B}}_1\bar{\mathbf{s}}^T\bar{\mathbf{c}}_1\bar{\mathbf{s}} \qquad (5\text{–}15)$$

where $\bar{\mathbf{c}}_1$ and $\bar{\mathbf{B}}_1$ are obtained from the equations of the previous example by deleting the empty rows and columns associated with $\bar{s}_3^a$ and $\bar{s}_3^b$.

6. NONLINEAR TRANSFORMATIONS

6.1. Contragradient Law

A nonlinear kinematic transformation from $\mathbf{q}$ to $\mathbf{q}'$ is a general functional relation $\mathbf{q} = \mathbf{f}(\mathbf{q}')$. The transformation for virtual generalized displacements $\delta\mathbf{q}$ is obtained by differentiation of $\mathbf{f}$ and has the form

$$\delta\mathbf{q} = \mathbf{T}\,\delta\mathbf{q}' \qquad (6.1\text{–}1a)$$

where

$$\mathbf{T} = \frac{\partial\mathbf{q}}{\partial\mathbf{q}'^T} \qquad (6.1\text{–}1b)$$

The contragradient law is based on invariance of the virtual work expression $\mathbf{Q}^T\delta\mathbf{q}$. Thus, as in the linear theory,

$$\mathbf{Q}' = \mathbf{T}^T\mathbf{Q} \qquad (6.1\text{–}2)$$

except that here $\mathbf{T}$ depends on the displacements $\mathbf{q}'$.

If a static transformation of the form

$$\mathbf{Q} = \mathbf{L}\mathbf{Q}' \qquad (6.1\text{–}3a)$$

is considered, the contragradient law defines

$$\delta\mathbf{q}' = \mathbf{L}^T\delta\mathbf{q} \qquad (6.1\text{–}3b)$$

but it does not necessarily allow a definition of $\mathbf{q}'$ unless $\mathbf{L}^T\delta\mathbf{q}$ is an exact differential.

Example 1

An example of a nonlinear transformation consists of the deformation-displacement relations, written here symbolically as $\mathbf{v} = \mathbf{v}(\mathbf{s})$. The contragradient law applies to the self-equilibriating part of the member end forces and is $\mathbf{S} - \mathbf{S}_p = \mathbf{B}\mathbf{V}$, where

$$\mathbf{B} = \frac{\partial \mathbf{v}}{\partial \mathbf{s}^T} \tag{6.1–4}$$

This relation is valid in the exact theory of Chapter 2 as well as in second-order and linear theories.

Example 2. Transformation of Bound Reference

In the exact theory of Chapter 2, deformations of the cantilever type $\{v_1 \;\; v_2 \;\; v_3\}$ are related to deformations of the alternate type $\{v_1' \;\; v_2' \;\; v_3'\}$ through the relations

$$v_1 = -l + (l + v_1')\cos v_2' \tag{6.1–5a}$$

$$v_2 = -(l + v_1')\sin v_2' \tag{6.1–5b}$$

$$v_3 = v_3' - v_2' \tag{6.1–5c}$$

$\mathbf{V}'$ is thus related to $\mathbf{V}$ through

$$\mathbf{V}' = \mathbf{T}^T\mathbf{V} \tag{6.1–6a}$$

where

$$\mathbf{T} = \frac{\partial \mathbf{v}}{\partial \mathbf{v}'^T} = \begin{bmatrix} \cos v_2' & -(l + v_1')\sin v_2' & 0 \\ -\sin v_2' & -(l + v_1')\cos v_2' & 0 \\ 0 & -1 & 1 \end{bmatrix} \tag{6.1–6b}$$

6.2. Transformation of Bound Reference: Second-Order Theory

Relations expressing deformations of the cantilever type $\{v_1 \;\; \mathbf{v}\}$ in terms of deformations of the alternate type $\{v_1' \;\; \mathbf{v}'\}$ may be obtained by specialization of the general $\mathbf{v}$–$\mathbf{s}$ relations by letting $s_1^a = s_2^a = s_2^b = 0$ and $s_3^a = v_2, s_1^b = v_1$, and $s_3^b = v_3$. In the second-order theory

$$v_1 = v_1' - \frac{l}{2}(v_2')^2 \tag{6.2–1}$$

$$\begin{Bmatrix} v_2 \\ v_3 \end{Bmatrix} = \begin{bmatrix} -l & 0 \\ -1 & 1 \end{bmatrix} \begin{Bmatrix} v_2' \\ v_3' \end{Bmatrix} \tag{6.2–2}$$

Only the first equation is nonlinear. We obtain by differentiation

$$\delta v_1 - \delta v_1' = -lv_2'\delta v_2' = -\frac{1}{l}v_2\delta v_2 \tag{6.2–3}$$

The contragradient law may be applied by requiring the equality $\mathbf{V}^T\delta\mathbf{v} + V_1\delta v_1 = \mathbf{V}'^T\delta\mathbf{v}' + V_1'\delta v_1'$. There comes

$$V_1' = V_1 \tag{6.2–4}$$

$$\begin{Bmatrix} V_2' \\ V_3' \end{Bmatrix} = \begin{bmatrix} -l & -1 \\ 0 & 1 \end{bmatrix} \begin{Bmatrix} V_2 \\ V_3 \end{Bmatrix} + V_1 \begin{Bmatrix} -lv_2' \\ 0 \end{Bmatrix} \tag{6.2–5}$$

The stiffness equations

$$V_1 = k_{11}(v_1 + \tfrac{1}{2}\mathbf{v}^T\mathbf{g}\mathbf{v}) + V_{1F} \tag{6.2–6a}$$

$$\mathbf{V} = \mathbf{k}\mathbf{v} + \mathbf{V}_F \tag{6.2–6b}$$

transform into similar equations with primed symbols in which $k_{11}' = k_{11}$, $V_{1F}' = V_{1F}$, and

$$\mathbf{k}' = \mathbf{T}^T\mathbf{k}\mathbf{T} + V_1\begin{bmatrix} -l & 0 \\ 0 & 0 \end{bmatrix} \tag{6.2–7a}$$

$$\mathbf{g}' = \mathbf{T}^T\mathbf{g}\mathbf{T} + \begin{bmatrix} -l & 0 \\ 0 & 0 \end{bmatrix} \tag{6.2–7b}$$

$$\mathbf{V}_F' = \mathbf{T}^T\mathbf{V}_F \tag{6.2–7c}$$

where $\mathbf{T}$ is the coefficient matrix in Eq. (6.2–2).

Equations (6.2–1, 2, 4, 5) allow us to transform the flexibility relations from the alternate to the cantilever type. The bending flexibility relation $\mathbf{v}' = \mathbf{f}'\mathbf{V}' + \mathbf{v}_p'$ transforms by means of Eqs. (6.2–2) and (6.2–5) into

$$\mathbf{v} = (\mathbf{T}\mathbf{f}'\mathbf{T}^T)\mathbf{V} + V_1\mathbf{T}\mathbf{f}'\begin{Bmatrix} v_2 \\ 0 \end{Bmatrix} + \mathbf{T}\mathbf{v}_p' \tag{6.2–8}$$

The equation remains to be solved for $\mathbf{v}$ because v_2 appears on the right-hand side.

6.3. Two-Force Member: General Nonlinear Theory

The effective force variables for a two-force member are $\mathbf{V} = \{V_1\}$ and $\mathbf{S} = \{S_1^a \; S_2^a \; S_1^b \; S_2^b\}$. The equilibrium equation is $\mathbf{S} = \mathbf{B}_1 V_1$ in which

$$\mathbf{B}_1 = [-\cos\rho \quad -\sin\rho \quad \cos\rho \quad \sin\rho] \tag{6.3–1}$$

and ρ is the member rotation. The linear axial stiffness relation $V_1 = k_{11}v_1$ transforms into the nonlinear relation

$$\mathbf{S} = \mathbf{B}_1 k_{11} v_1 \tag{6.3–2}$$

v_1 and ρ are related to $\mathbf{s}$ through the relations

$$(l + v_1)\cos\rho = l + s_1^b - s_1^a \tag{6.3–3a}$$

$$(l + v_1)\sin\rho = s_2^b - s_2^a \tag{6.3–3b}$$

from which

$$v_1 = -l + [(l + s_1^b - s_1^a)^2 + (s_2^b - s_2^a)^2]^{1/2} \tag{6.3–4}$$

In methods of nonlinear analysis there is need for incremental stiffness equations in fixed axes. They are obtained by incrementing Eq. (6.3–2). It is recalled that $dv_1 = \mathbf{B}_1^T\, d\mathbf{s}$, and therefore

$$\mathbf{B}_1 = \frac{\partial v_1}{\partial \mathbf{s}} \tag{6.3–5}$$

In incrementing Eq. (6.3–2), the term $d\mathbf{B}_1 k_{11} v_1$ contributes the geometric stiffness term

$$d\mathbf{B}_1 V_1 = V_1 \mathbf{c}_1\, d\mathbf{s} \tag{6.3–6}$$

where

$$\mathbf{c}_1 = \frac{\partial \mathbf{B}_1}{\partial \mathbf{s}^T} = \frac{\partial^2 v_1}{\partial \mathbf{s}\,\partial \mathbf{s}^T} \tag{6.3–7}$$

The incremental stiffness equation is then

$$d\mathbf{S} = (k_{11}\mathbf{B}_1\mathbf{B}_1^T + V_1\mathbf{c}_1)d\mathbf{s} \tag{6.3–8}$$

Derivation of $\mathbf{c}_1$ and a kinematic rederivation of $\mathbf{B}_1$ is made conveniently in terms of ρ. If the total differentials of Eqs. (6.3–3) are solved for dv_1 and $d\rho$, there comes

$$dv_1 = (ds_1^b - ds_1^a)\cos\rho + (ds_2^b - ds_2^a)\sin\rho \tag{6.3–9a}$$

$$d\rho = \frac{1}{l + v_1}[(ds_2^b - ds_2^a)\cos\rho - (ds_1^b - ds_1^a)\sin\rho] \tag{6.3–9b}$$

Equation (6.3–9a) yields the statically derivable Eq. (6.3–1). Equation (6.3–9b) allows us to evaluate the partial derivatives of ρ. We thus find

$$\mathbf{c}_1 = \frac{\partial^2 v_1}{\partial \mathbf{s} \partial \mathbf{s}^T} = \frac{1}{l + v_1} \left[\begin{array}{cc|c} \sin^2 \rho & -\sin \rho \cos \rho & (-1) \\ -\sin \rho \cos \rho & \cos^2 \rho & \rightarrow \\ \hline \downarrow (-1) & & \searrow (+1) \end{array} \right] \tag{6.3.10}$$

In global coordinates, Eq. (6.3–8) retains the same form in terms of $\bar{\mathbf{B}}_1$ and $\bar{\mathbf{c}}_1$. These matrices are obtained from $\mathbf{B}_1$ and $\mathbf{c}_1$, respectively, by replacing ρ with $(\theta + \rho)$.

6.4. Beam-Column with Large Rotations

If the second-order theory is assumed to hold only in a member-bound reference, transformation from bound to fixed axes is based on the exact static-kinematic equations of Chapter 2. The general static solution, $\mathbf{S} = \mathbf{BV} + \mathbf{S}_p$, becomes a nonlinear stiffness relation upon substituting the stiffness relations for $\mathbf{V}$ in terms of $\mathbf{v}$ and the kinematic relations for $\mathbf{v}$ in terms of $\mathbf{s}$. $\mathbf{S}$ may thus be evaluated for a given deformed state. In methods of analysis there is need for incremental equations. Transformation of incremental stiffness equations from a bound reference to a fixed reference is now treated for an arbitrary rigid body motion of the bound reference. Section 3.3 of Chapter 6 deals with a similar topic in the second-order theory.

Let (x_r, y_r) be the bound axes and ρ_0 be their angle of rotation from the local axes. For bound axes of the cantilever type, $\rho_0 = s_3^a$, and for the alternate type, ρ_0 is the chord rotation ρ. The increment $d\rho_0$ is expressed as

$$d\rho_0 = \mathbf{B}_\rho^T d\mathbf{s} \tag{6.4–1}$$

The load components in the bound axes, p_{xr} and p_{yr}, are related to p_x and p_y through

$$p_{xr} = p_x \cos \rho_0 + p_y \sin \rho_0 \tag{6.4–2a}$$

$$p_{yr} = -p_x \sin \rho_0 + p_y \cos \rho_0 \tag{6.4–2b}$$

p_x and p_y are assumed given. Increments in p_{xr} and p_{yr} are due to dp_x and dp_y and to the geometric effect of the rotation $d\rho_0$. Since the second-order theory is assumed to govern the member in the bound axes, the geometric effect on dp_{xr} must be neglected for the same reason that p_x is equivalent to p_{xr} when the second-order theory applies in the local axes. Differentiation of Eqs. (6.4–2) yields then

$$dp_{xr} = (dp_{xr})_0 \tag{6.4–3a}$$

$$dp_{yr} = (dp_{yr})_0 - p_{xr} d\rho_0 \tag{6.4–3b}$$

where $(dp_{xr})_0$ and $(dp_{yr})_0$ are the nongeometric increments. The geometric term in dp_{yr} contributes a geometric term to the incremental stiffness equations in the bound reference. These equations are obtained by incrementing Eqs. (6/3.1–22) as

$$\begin{Bmatrix} dV_1 \\ d\mathbf{V} \end{Bmatrix} = \begin{bmatrix} k_{11} & k_{11}\mathbf{v}^T\mathbf{g} \\ k_{11}\mathbf{g}\mathbf{v} & \mathbf{k} + k_{11}\mathbf{g}\mathbf{v}\mathbf{v}^T\mathbf{g} \end{bmatrix} \begin{Bmatrix} dv_1 \\ d\mathbf{v} \end{Bmatrix} + \begin{Bmatrix} dV_{1F} \\ d\mathbf{V}_F \end{Bmatrix} + \begin{Bmatrix} 0 \\ dV_{1F}\mathbf{g}\mathbf{v} + (d\mathbf{k}_{gp})\mathbf{v} \end{Bmatrix} \tag{6.4–4}$$

where, by specialization of Eq. (6/3.1–18),

$$d\mathbf{V}_F = -\int_0^l (\boldsymbol{\psi}_r^T dp_{yr} + \boldsymbol{\psi}_{r,x}^T dm)\, dx \tag{6.4–5}$$

The geometric term appears only in $d\mathbf{V}_F$ and is due to the term $-p_{xr}d\rho_0$ in dp_{yr}. Letting

$$\mathbf{V}_{Fxr} = -\int_0^l \boldsymbol{\psi}_r^T p_{xr}\, dx \tag{6.4–6}$$

the geometric term is designated $d\mathbf{V}_{FG}$ and is

$$d\mathbf{V}_{FG} = -\mathbf{V}_{Fxr}\mathbf{B}_\rho^T d\mathbf{s} \tag{6.4–7}$$

We now revert to the unpartitioned notation and write Eqs. (6.4–4) in the form

$$d\mathbf{V} = \mathbf{k}_t d\mathbf{v} + d\mathbf{V}_0 - \mathbf{V}_{Fxr}\mathbf{B}_\rho^T d\mathbf{s} \tag{6.4–8}$$

where

$$\mathbf{V}_{Fxr} = \{0 \quad -\int_0^l \boldsymbol{\psi}_r^T p_{xr}\, dx\} \tag{6.4–9}$$

Transformation to the local axes is made by means of the incremental kinematic and static equations

$$d\mathbf{v} = \mathbf{B}^T d\mathbf{s} \tag{6.4–10a}$$

$$d\mathbf{S} = \mathbf{B}d\mathbf{V} + (d\mathbf{B})\mathbf{V} + d\mathbf{S}_p \tag{6.4–10b}$$

The last two terms contribute geometric terms which are now obtained.

First, $(d\mathbf{B})\mathbf{V}$ is written as $\sum V_i(d\mathbf{B}_i)$, where $\mathbf{B}_i$ is the ith column of $\mathbf{B}$. Since $\mathbf{B}_i = \partial v_i/\partial \mathbf{s}$, there comes

$$(d\mathbf{B})\mathbf{V} = \mathbf{k}_G d\mathbf{s} \tag{6.4–11}$$

where

$$\mathbf{k}_G = \sum V_i \mathbf{c}_i \tag{6.4–12a}$$

and

$$\mathbf{c}_i = \frac{\partial \mathbf{B}_i}{\partial \mathbf{s}^T} = \frac{\partial^2 v_i}{\partial \mathbf{s} \partial \mathbf{s}^T} \tag{6.4–12b}$$

Next, $d\mathbf{S}_p$ is considered. To express $\mathbf{S}_p$ in terms of the span load, let R_x and R_y be the resultant force components on the local axes and C^a be the moment at end a of the span load. The force system $(\mathbf{S}_p, R_x, R_y, C^a)$ is in equilibrium and thus satisfies the virtual work equation for virtual rigid body displacements:

$$\mathbf{S}_p^T \delta\mathbf{s} + R_x \delta s_1^a + R_y \delta s_2^a + C^a \delta\rho_0 = 0 \tag{6.4–13}$$

Equation (6.4–13) is in fact satisfied for any $\delta\mathbf{s}$ because the force system $\mathbf{S}_p$ has zero $\mathbf{V}$ and the term $\mathbf{V}^T\delta\mathbf{v}$ is zero for an arbitrary $\delta\mathbf{s}$. Equation (6.4–13) is now incremented with respect to $\mathbf{s}$. R_x and R_y are independent of $\mathbf{s}$, and the increments in $\mathbf{S}_p$ and $\mathbf{C}^a$ are the desired geometric terms $d\mathbf{S}_{pG}$ and dC_G^a. Thus, with $\delta\rho_0 = \mathbf{B}_\rho^T \delta\mathbf{s}$, there comes

$$d\mathbf{S}_{pG} = -\mathbf{B}_\rho dC_G^a - (d\mathbf{B}_\rho)C^a \tag{6.4–14}$$

C^a is evaluated in terms of quantities in the bound axes as

$$C^a = \int_0^l (xp_{yr} - v_r p_{xr} + m)\, dx \tag{6.4–15}$$

and, with use of eqs. (6.4–3),

$$dC_G^a = d\rho_0 \int_0^l -xp_{xr}\, dx - \int_0^l dv_r p_{xr}\, dx$$

In this equation $d\rho_0 = \mathbf{B}_\rho^T d\mathbf{s}$, and $dv_r = \boldsymbol{\psi}_r d\mathbf{v}$. Noting the defining equation of $\mathbf{V}_{Fxr}$, dC_G^a takes the form

$$dC_G^a = \left(\int_0^l -xp_{xr}\, dx \right) \mathbf{B}_\rho^T d\mathbf{s} + \mathbf{V}_{Fxr}^T \mathbf{B}^T d\mathbf{s} \tag{6.4–16}$$

Finally,

$$(d\mathbf{B}_\rho)C^a = C^a d\left(\frac{\partial \rho_0}{\partial \mathbf{s}} \right) = C^a \frac{\partial^2 \rho_0}{\partial \mathbf{s} \partial \mathbf{s}^T} d\mathbf{s} = -C^a \mathbf{c}_3 d\mathbf{s} \tag{6.4–17}$$

$\mathbf{c}_3$, as defined in terms of ρ_0, is the same as in Eq. (6.4–12b) for $i = 3$ because, in either type of bound reference, v_3 differs from $-\rho_0$ by a linear term in $\mathbf{s}$.

The expression of $d\mathbf{S}_{pG}$ in Eq. (6.4–14) is now determined by means of Eqs. (6.4–16) and (6.4–17). The total geometric term contributed to Eq. (6.4–10b) by $d\mathbf{S}_p$ and $d\mathbf{V}_F$ is

$$d\mathbf{S}_{pG} + \mathbf{B}d\mathbf{V}_{FG} = \mathbf{k}_{GF}d\mathbf{s} \tag{6.4–18a}$$

where $\mathbf{k}_{GF}$ is the symmetric matrix

$$\mathbf{k}_{GF} = \left(\int_0^l xp_{xr}\,dx\right)\mathbf{B}_\rho\mathbf{B}_\rho^T - \mathbf{B}_\rho\mathbf{V}_{Fxr}^T\mathbf{B}^T - \mathbf{B}\mathbf{V}_{Fxr}\mathbf{B}_\rho^T + C^a\mathbf{c}_3 \tag{6.4–18b}$$

The incremental stiffness equations have the form

$$d\mathbf{S} = \mathbf{k}_{st}\,d\mathbf{s} + d\mathbf{S}_0 \tag{6.4–19a}$$

where

$$\mathbf{k}_{st} = \mathbf{B}\mathbf{k}_t\mathbf{B}^T + \mathbf{k}_G + \mathbf{k}_{GF} \tag{6.4–19b}$$

For equations in the global axes, $\mathbf{k}_{st}$ and $d\mathbf{S}_0$ transform according to the congruent and contragradient laws, respectively.

Expressions of $\mathbf{B}$, $\mathbf{B}_\rho$, and $\mathbf{k}_G$ are derived in what follows for a bound reference of the alternate type. v_1 is given in Eq. (6.3–4), and $v_2 = s_3^a - \rho$ and $v_3 = s_3^b - \rho$. From the expressions of dv_1 and $d\rho$ obtained in Eqs. (6.3–9), there comes

$$\mathbf{B}^T = \frac{\partial \mathbf{v}}{\partial \mathbf{s}^T} = \begin{bmatrix} -\cos\rho & -\sin\rho & 0 & \cos\rho & \sin\rho & 0 \\ -\dfrac{\sin\rho}{l'} & \dfrac{\cos\rho}{l'} & 1 & \dfrac{\sin\rho}{l'} & -\dfrac{\cos\rho}{l'} & 0 \\ -\dfrac{\sin\rho}{l'} & \dfrac{\cos\rho}{l'} & 0 & \dfrac{\sin\rho}{l'} & -\dfrac{\cos\rho}{l'} & 0 \end{bmatrix} \tag{6.4–20}$$

where $l' = l + v_1 \approx l$, and

$$\mathbf{B}_\rho^T = \frac{\partial\rho}{\partial\mathbf{s}^T} = \frac{1}{l'}[\sin\rho \quad -\cos\rho \quad 0 \quad -\sin\rho \quad \cos\rho \quad 0] \tag{6.4–21}$$

$\mathbf{k}_G$ is the sum $V_i\mathbf{c}_i$ as defined in Eq. (6.4–12). $\mathbf{c}_1$ is determined in reduced form in Eq. (6.3–10), and $\mathbf{c}_2$ and $\mathbf{c}_3$ are

$$c_2 = c_3 = -\frac{\partial^2\rho}{\partial\mathbf{s}\partial\mathbf{s}^T} = \frac{1}{(l')^2}\left[\begin{array}{ccc|c} -\sin 2\rho & \cos 2\rho & 0 & \\ \cos 2\rho & \sin 2\rho & 0 \rightarrow (-1) & \\ 0\ \downarrow & 0 & 0\ \searrow & \\ \hline (-1) & & & (+1) \end{array}\right] \tag{6.4–22}$$

EXERCISES

Section 2

1. Let members 3 and 5 in Fig. 3.2–1 be inclined, respectively, at 30° and 45° to the horizontal. Determine the coordinate transformation matrix $\boldsymbol{\lambda}$ for members 1, 3, 4, and 5. How does $\boldsymbol{\lambda}$ change if the member axes are reversed in orientation?

2. Obtain by statics the static-kinematic matrix in global coordinates $\bar{\mathbf{B}}$, and check the result with Eq. (2.2–3) or Eq. (2.2–4).

3. Obtain $\bar{\mathbf{B}}^T$ in Exercise 2 by geometry.

4. Obtain the detailed expression of the complete stiffness matrix using Eq. (2.3–5b) in terms of the reduced stiffness coefficients of either the cantilever or the alternate type.

Section 3

5. Obtain the contragradient relations in Section 3.1, Example 2, by statics and kinematics, respectively.

6. Express by geometry deformations of the cantilever type in terms of deformations of the alternate type, Fig. 3.1–2; then write the contragradient transformation law for the member statical redundants.

7. Consider the static transformation $V_2 = V_s + V_a$, $V_3 = -V_s + V_a$ in which V_2 and V_3 are the member end moments. Determine the contragradient transformation defining $\{v_s \; v_a\}$. Interpret the transformed static and kinematic quantities.

8. Obtain the transformed stiffness equations arising from the kinematic transformation of Exercise 6. Obtain also the transformed flexibility equations arising from the statical transformation.

9. Obtain the transformed stiffness relations arising from the transformation of Exercise 7.

10. A simply supported member is formed of three consecutive segments of lengths a, l, and b, respectively. Segments a and b are infinitely rigid. Let $\mathbf{v}'$ be the deformations of alternate type for segment l and $\mathbf{v}$ be the deformations for the member. Express by geometry $\mathbf{v}'$ in terms of $\mathbf{v}$, verify by statics the contragradient relation, and obtain the stiffness matrix relating $\mathbf{V}$ to $\mathbf{v}$.

11. Derive the flexibility matrix for the member of Exercise 10, starting with a static transformation.

12. Apply either a kinematic or a static transformation to perform a given reordering with a possible sign change of the elements of $\mathbf{s}$ and $\mathbf{S}$ and to obtain the reordered stiffness matrix.

13. Consider member 1 in Fig. 3.2–1, and let s_2^a be a prescribed support displacement. Derive the effective stiffness relations between $\mathbf{S}^b$ and $\mathbf{s}^b$.

14. Do Exercise 11 for member 2 in Fig. 3.2–1.

15. Let member 3 in Section 3.3, Example 6, be subjected to a vertical uniformly distributed load of intensity p per unit horizontal length. Let l be the member length and θ its inclination. Obtain the effective fixed end forces in the stiffness relations in local and in global axes.

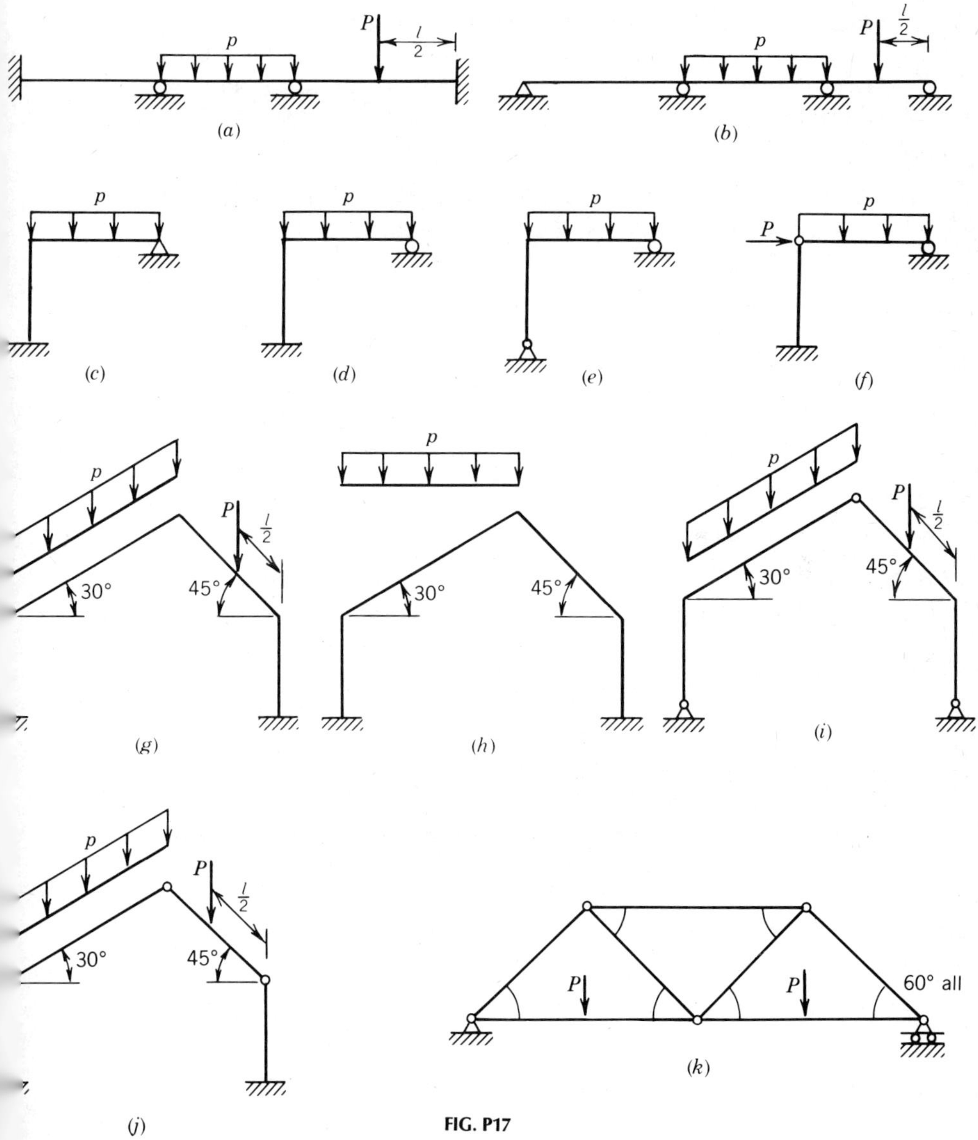

FIG. P17

16. Obtain effective flexibility relations for member 3 of Fig. 3.2–1. Assume a uniform member of length l and inclination θ and a uniform vertical load of intensity p per unit horizontal length.

17. For the structures shown in Fig. P17, define for each member the effective relations:
 a. $\mathbf{v} = \mathbf{f}\mathbf{V} + \mathbf{v}_0$
 b. $\mathbf{V} = \mathbf{k}\mathbf{v} + \mathbf{V}_F$
 c. $\mathbf{S} = \mathbf{k}_c\mathbf{s} + \mathbf{S}_F$
 d. $\bar{\mathbf{S}} = \bar{\mathbf{k}}_c\bar{\mathbf{s}} + \bar{\mathbf{S}}_F$
 Identify by means of sketches the elements of the static-kinematic pairs and, by formulas or by reference to formulas in the text, the elements of the flexibility and stiffness matrices and the load-dependent terms. Assume a linear theory and uniform member of properties l, EA, and EI without shear deformation.

18. An inextensible member is one that is subjected to the kinematic constraint $v_1 = 0$. Define the effective $\mathbf{V}$ and $\mathbf{v}$ as outlined in Section 3.2.

19. For an inextensible member as defined in Exercise 18, the independent displacements in local coordinates are $\mathbf{s}' = \{q \;\; s_2^a \;\; s_3^a \;\; s_2^b \;\; s_3^b\}$, where $q = s_1^a = s_1^b$ is the axial translation of the member. Consider the transformation from $\mathbf{s}$ to $\mathbf{s}'$; then obtain the contragradient statical transformation and the transformed stiffness equations.

Sections 5, 6

20. Obtain the effective stiffness relations of Section 5, Example 1, using the approximate stiffness coefficients established in Chapter 6.

21. Show that under a general kinematic transformation $q = f(q')$, Castigliano's stiffness relation $\mathbf{Q} = \partial\bar{U}/\partial\mathbf{q}$ transforms into $\mathbf{Q}' = \partial\bar{U}/\partial\mathbf{q}'$.

22. Verify that the member stiffness matrices in beam-column theory established in Chapter 6 satisfy the transformation (6.2–7a). Do this for the exact as well as for approximate matrices. Verify similarly that the flexibility matrices satisfy the transformation (6.2–8).

23. A cantilever is formed of two consecutive segments of lengths l and a, respectively. Segment a is infinitely rigid and is subjected at its free end to $\mathbf{V} = \{V_1 \;\; V_2 \;\; V_3\}$. Derive the stiffness matrix that relates $\mathbf{V}$ and $\mathbf{v}$ in beam-column theory from the stiffness matrix of segment l using a kinematic transformation for deformations.

24. Derive the stiffness matrix of the member of Exercise 10 in beam-column theory using a kinematic transformation.

25. A static transformation in geometrically nonlinear theory of the form

$\mathbf{Q}' = \mathbf{T}^T\mathbf{Q}$ leads to the contragradient relation $\delta q = \mathbf{T}\delta q'$. Determine by statics the matrix $\mathbf{T}^T$ in Exercise 23; then verify that the contragradient relation is the differential of the finite kinematic transformation.

26. Specialize the incremental stiffness relations of Section 6.3 to the second-order theory of Section 5, Example 2.

8

GOVERNING EQUATIONS OF STRUCTURAL ANALYSIS

1. STRUCTURE IDEALIZATION: THE PROBLEM OF STRUCTURAL ANALYSIS

A structure is an assemblage of members connected at their ends into joints. It is represented as a graph of line elements, Fig. 1–1. The members are governed by engineering beam theory, and the joints have static-kinematic properties that must be defined.

A joint enforces geometric continuity and transmits internal forces at the member ends. A member end could be connected to a joint in various ways which are idealized into standard types. A rigid connection establishes full continuity of

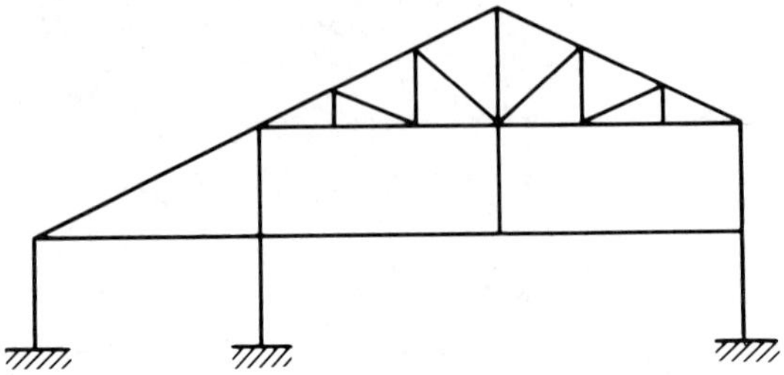

FIG. 1–1. Structure representation.

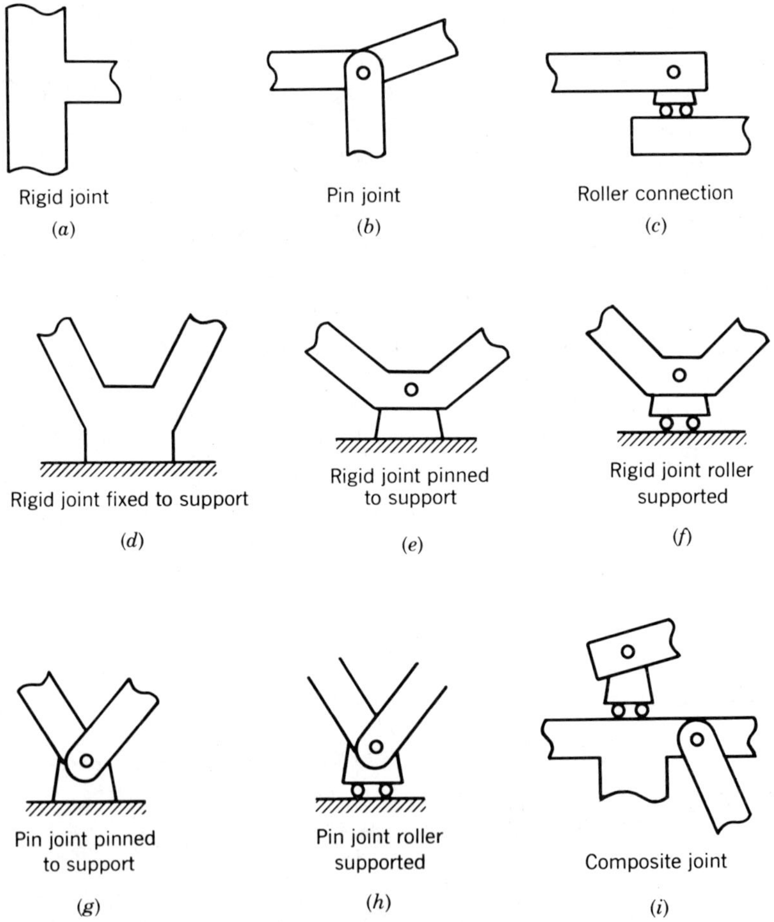

FIG. 1–2. Types of joints.

translational and rotational displacements and is capable of transmitting all internal force and moment components. A pin connection establishes the samc continuity except for the possibility of a relative rotational displacement between the connected parts about the axis of the pin. If the pin is frictionless, it cannot transmit an internal moment about its axis. The member end rotation about the axis of the pin is said to be released. Other typical connections are the universal joint and the roller. The former releases completely the member end rotations but maintains continuity of translational displacements. In plane behavior it is equivalent to a pin. A roller connection in plane behavior releases the member end rotation and one translational displacement. Connections such as these described are also used at supports. In analysis, a joint is isolated as a free body. It is then an assemblage of member stubs. Various types are shown in Fig. 1–2. The rigid and pin joints are the most common idealizations. The axes of members connected at a joint are usually made to intersect at a point. The joint is then idealized into that material point, keeping, however, its specific kinematic properties as an assemblage. This idealization is assumed in examples, unless stated otherwise. It needs not be assumed in the general theory.

There are also idealizations in which the internal force corresponding to a released displacement is not zero. In an elastic connection the internal force is a function of the released relative displacement. Such a stiffness relationship may be ascribed to a member of infinitesimal size. In a plastic connection the internal force remains constant as the released deformation increases.

Finally, two unusual types of joints are shown in Fig. 1–3. In (*a*) and (*b*) member ends are rigidly connected by pairs, but the joint is a mechanism. In (*c*) ABC represents a continuous cable that may slip over a smooth surface at the end B of member BD. The cable may be treated in this case as formed of two members AB' and CC', and the joint is the free body enclosed in the dotted circle.

The given data in a typical problem of structural analysis describes the geometry of the structure in the initial undeformed state, the support conditions, the material properties, and the loading condition. In its general sense a loading condition includes any cause that deforms the structure. In addition to prescribed external forces it may include prescribed displacements, thermal variations, and geometric errors relative to an initial ideal configuration. A loading condition may be subdivided into joint and member loading conditions. The former

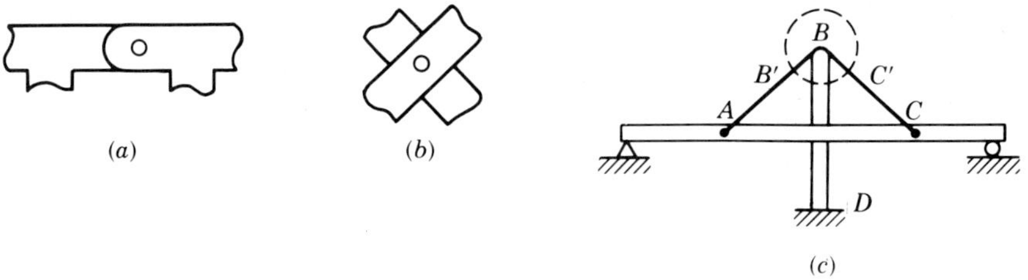

FIG. 1–3. Special types of joints.

consists of prescribed forces or displacements at the joints. The latter consists of member span loads, including thermal variations and geometric errors.

The main problem of structural analysis is to determine member end forces and displacements caused by a given loading condition. With static and kinematic end conditions determined, each member could then be isolated and analyzed further for stresses and displacements. The object of this chapter is to formulate a set of governing equations for the general problem. The linear theory is developed first and forms a necessary background to the geometrically nonlinear theory. Three alternate formulations are developed that correspond to different choices of coordinates for describing member constitutive equations.

2. GOVERNING EQUATIONS OF THE LINEAR THEORY

2.1. Joint Displacements: Notation

Governing equations for a structure are formulated by applying the three basic concepts of geometric continuity, equilibrium, and constitutive equations. The constitutive equations are those of the members. Equilibrium must be satisfied for each member and for each joint. Geometric continuity at each joint must be satisfied by the member end displacements according to the specific types of connections.

Member equilibrium equations are either identically satisfied by member stiffness equations or are solved in terms of member statical redundants **V**, in the form $\mathbf{S} = \mathbf{BV} + \mathbf{S}_p$. The task that remains is then to formulate geometric continuity equations at the joints and joint equilibrium equations.

The member end displacements must satisfy the geometric continuity imposed by the joints. The independent variables that determine kinematically the member end displacements are the joint displacements **r**. A choice of **r** is a basic first step in formulating a problem.

A rigid joint j displaces as a rigid body. Its displacements are denoted $\mathbf{r}^j = \{r_1^j\, r_2^j\, r_3^j\}$. r_1^j and r_2^j are translations in the global X and Y directions, and r_3^j is the rigid body rotation. At a composite joint a complete set of joint displacements $\mathbf{r}^j$ must determine the complete set of member end displacements at that joint. Thus, if the joint is formed of a rigid part and of articulated member stubs, then $\mathbf{r}^j = \{r_1^j\;\; r_2^j\;\; r_3^j\;\; r_4^j \ldots\}$. The first three displacements describe the motion of the rigid part, and $r_4^j \ldots$ are independent released displacements of the articulated member stubs. An example is shown in Fig. 2.1–1.

A member having kinematic releases at its ends may be described by effective variables that exclude the released displacements. Thus, in general, joint displacements are defined as an independent set that determines kinematically the effective member end displacements. For the composite joint of Fig. 2.1–1 an effective **r** may exclude the rotation r_4. For a pinned joint effective displacements may be limited to the two translational components.

The notation to be adopted uses a superscript to refer to a joint and subscripts

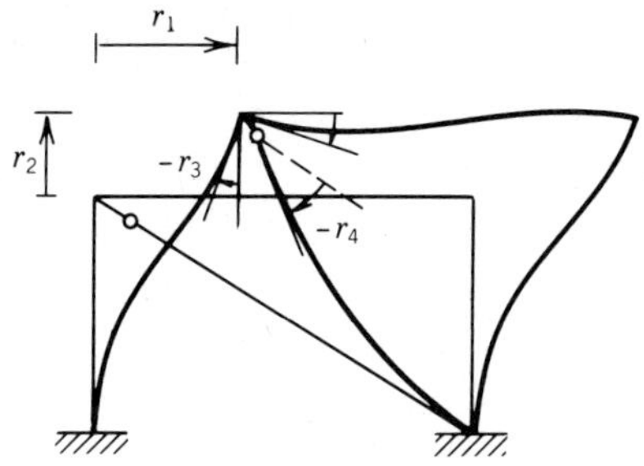

FIG. 2.1–1. Complete set of joint displacements.

to refer to particular displacement components. The joint set is ordered arbitrarily. Displacements of joint j are $\mathbf{r}^j = \{r_1^j \; r_2^j \; \ldots\}$, and the set of joint displacements is ordered by joints as

$$\mathbf{r} = \{\mathbf{r}^1 \; \mathbf{r}^2 \; \ldots \; \mathbf{r}^j \; \ldots \; \mathbf{r}^{NJ}\}$$

where NJ is the number of joints.

The notation for member matrices used earlier in the text is now adopted for the member set. A superscript is used to refer to a particular member, and another superscript refers to a particular end of that member. Thus $\mathbf{s}^{m,a}$ refers to the displacements of end a of member m, in local coordinates. Ends a and b of member m may be given the names of the joints at the member ends. Thus for member m spanning joints i and j,

$$\mathbf{s}^m = \{\mathbf{s}^{m,i} \; \mathbf{s}^{m,j}\} = \{s_1^{m,i} \; s_2^{m,i} \; s_3^{m,i} \; s_1^{m,j} \; s_2^{m,j} \; s_3^{m,j}\}$$

When referring to a specific member, and no ambiguity is possible, superscript m may be deleted to make reading easier, and the original member notation is used.

The members are ordered arbitrarily, and their displacement matrices are ordered accordingly into

$$\mathbf{s} = \{\mathbf{s}^1 \; \mathbf{s}^2 \; \ldots \; \mathbf{s}^m \; \ldots \; \mathbf{s}^{NM}\}$$

where NM is the number of members.

A similar notation is adopted for member end forces $\mathbf{S}$, for displacements and forces in global coordinates $\bar{\mathbf{s}}$ and $\bar{\mathbf{S}}$, and for deformations and member statical redundants $\mathbf{v}$ and $\mathbf{V}$. Note, however, that $\mathbf{v}$ and $\mathbf{V}$ are partitioned by members, not by member ends.

2.2. Governing Equations in Global Coordinates

Connectivity Equations

Member end displacements $\bar{\mathbf{s}}^{m,j}$ are determined kinematically by the joint displacements $\mathbf{r}^j$. The relationship is linear of the form

$$\bar{\mathbf{s}}^{m,j} = \bar{\mathbf{a}}^{m,j}\mathbf{r}^j \tag{2.2–1}$$

$\bar{\mathbf{a}}^{m,j}$ is called a connectivity matrix. At a rigid joint idealized into a point, $\bar{\mathbf{s}}^{m,j} = \mathbf{r}^j$ and $\bar{\mathbf{a}}^{m,j}$ is the 3×3 identity matrix. Similarly, at a pinned joint an effective $\bar{\mathbf{a}}^{m,j}$ is the 2×2 identity matrix. Equation (2.2–1) is written for each member at ends a and b, and the total set of equations ordered by the members has the form

$$\bar{\mathbf{s}} = \bar{\mathbf{a}}\mathbf{r} \tag{2.2–2}$$

Example 1a

A rigid frame with ordered joints and members is shown in Fig. 2.2–1. Support conditions are not yet prescribed. The connectivity equations partitioned by member ends and joints are

$$\begin{Bmatrix} \bar{\mathbf{s}}^{1,1} \\ \bar{\mathbf{s}}^{1,2} \\ \bar{\mathbf{s}}^{2,2} \\ \bar{\mathbf{s}}^{2,3} \\ \bar{\mathbf{s}}^{3,3} \\ \bar{\mathbf{s}}^{3,4} \\ \bar{\mathbf{s}}^{4,4} \\ \bar{\mathbf{s}}^{4,5} \end{Bmatrix} = \begin{bmatrix} \mathbf{I} & & & & \\ & \mathbf{I} & & & \\ & \mathbf{I} & & & \\ & & \mathbf{I} & & \\ & & \mathbf{I} & & \\ & & & \mathbf{I} & \\ & & & \mathbf{I} & \\ & & & & \mathbf{I} \end{bmatrix} \begin{Bmatrix} \mathbf{r}^1 \\ \mathbf{r}^2 \\ \mathbf{r}^3 \\ \mathbf{r}^4 \\ \mathbf{r}^5 \end{Bmatrix}$$

The general scheme for forming $\bar{\mathbf{a}}$ is shown in Fig. 2.2–2. $\bar{\mathbf{a}}$ is row-partitioned by members and column-partitioned by joints. The mth row is further partitioned by member ends. In each of these two rows the connectivity matrix is placed in the

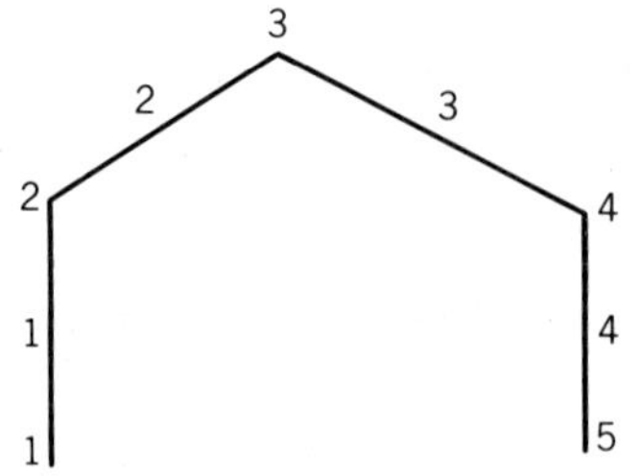

FIG. 2.2–1. Connectivity, Example 1.

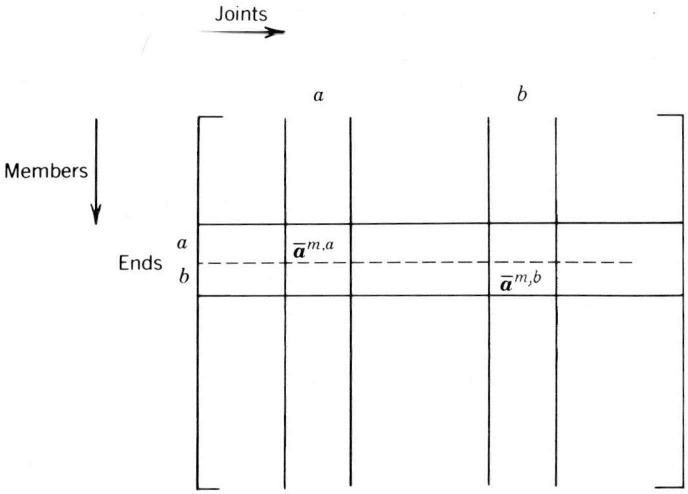

FIG. 2.2–2. Scheme for forming $\bar{\mathbf{a}}$ or $\mathbf{a}$.

column associated with the joint at the corresponding member end. All other submatrices remain empty.

Matrix $\bar{\mathbf{a}}$ is sparse and would waste storage space in both computer and hand methods. One should distinguish, however, between the matrix representation of equations and operational methods of programming or hand calculations. Such aspects are left for later consideration in Chapter 10.

Joint Equilibrium

External forces and moments applied at joint j form the joint load matrix $\mathbf{R}^j = \{R_1^j \; R_2^j \ldots\}$. Elements of $\mathbf{R}^j$ are defined to correspond in order, number, and sign convention to the joint displacements $\mathbf{r}^j = \{r_1^j \; r_2^j \ldots\}$. Thus for a rigid joint, R_1^j and R_2^j are force components in the direction of the global axes, and R_3^j is a moment. If $\mathbf{r}^j$ contains a released displacement, $\mathbf{R}^j$ contains a corresponding force (moment) applied in the released direction at the member end involved. This is illustrated in Fig. 2.2–3 for the joint displacements of Fig. 2.1–1. Since a physical

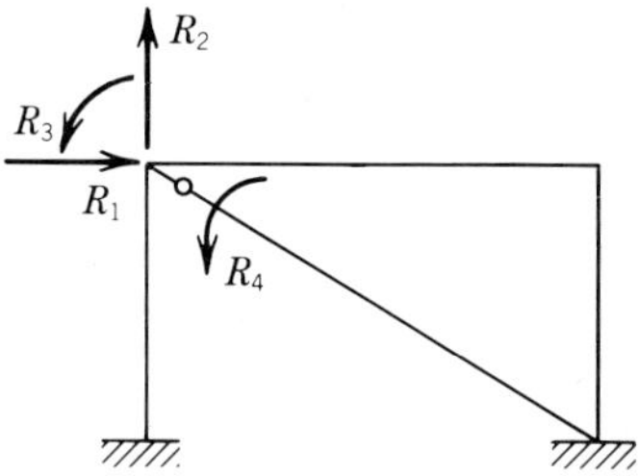

FIG. 2.2–3. Joint loads corresponding to displacements of Fig. 2.1–1.

joint has some finite size, elements of $\mathbf{R}^j$ are resultants of the actual force system applied at the physical joint.

The correspondance between $\mathbf{R}^j$ and $\mathbf{r}^j$ is such that $\mathbf{R}^{jT}\delta\mathbf{r}^j$ is the virtual work of the external forces applied at joint j through an arbitrary infinitesimal virtual displacement $\delta\mathbf{r}^j$.

A joint, isolated from a structure in an equilibrium state, is acted on by the joint load $\mathbf{R}^j$ and by the member end forces, reversed in sense, $-\mathbf{S}^{m,j}$. The joint is an assemblage of member stubs whose equilibrium equations may be derived by the principle of virtual displacements: a necessary and sufficient condition for the force system acting on the isolated joint to be in equilibrium is that its virtual work vanishes for any kinematically admissible virtual displacement $\delta\mathbf{r}^j$. The virtual member end displacements are $\delta\bar{\mathbf{s}}^{m,j} = \bar{\mathbf{a}}^{m,j}\delta\mathbf{r}^j$. The condition for joint equilibrium is then

$$\mathbf{R}^{jT}\delta\mathbf{r}^j + \sum_m -\bar{\mathbf{S}}^{m,jT}\bar{\mathbf{a}}^{m,j}\delta\mathbf{r}^j = 0 \tag{2.2–3}$$

$\delta\mathbf{r}^j$ being arbitrary, there comes

$$\sum_m \bar{\mathbf{a}}^{m,jT}\bar{\mathbf{S}}^{m,j} = \mathbf{R}^j \tag{2.2–4}$$

The sum in this equation ranges over the member ends at joint j. The total set of joint equilibrium equations may be derived as a whole by applying the principle of virtual displacements to the set of joints isolated as free bodies. Letting $\mathbf{R} = \{\mathbf{R}^1\ \mathbf{R}^2 \ldots \mathbf{R}^j \ldots \mathbf{R}^{NJ}\}$, the equation of virtual work takes the form

$$\mathbf{R}^T\delta\mathbf{r} - \bar{\mathbf{S}}^T\delta\bar{\mathbf{s}} = 0 \tag{2.2–5}$$

in which $\delta\bar{\mathbf{s}} = \bar{\mathbf{a}}\delta\mathbf{r}$ and $\delta\mathbf{r}$ is arbitrary. There comes

$$\bar{\mathbf{a}}^T\bar{\mathbf{S}} = \mathbf{R} \tag{2.2–6}$$

Equation (2.2–6) represents the set (2.2–4) ordered by joints.

Derivation of equilibrium equations by the principle of virtual displacements brings out the static-kinematic character of the connectivity matrix $\bar{\mathbf{a}}$. It also brings out the important concept of associating each equilibrium equation to a specific degree of freedom. An element δr_i of $\delta\mathbf{r}$ is a kinematically possible displacement from the equilibrium state $\mathbf{r}$ and represents what is called a degree of freedom. The static equilibrium equation generated by the virtual displacement δr_i is the condition that this kinematically possible displacement does not in fact take place. This concept may be seen in a generalized context in Section 4, Example 1.

Governing Equations

Governing equations in global coordinates are obtained by appending member stiffness relations to the geometric and static equations established earlier, or

$$\bar{\mathbf{s}} = \bar{\mathbf{a}}\mathbf{r} \tag{2.2–7a}$$

$$\bar{\mathbf{a}}^T\bar{\mathbf{S}} = \mathbf{R} \tag{2.2–7b}$$

$$\bar{\mathbf{S}} = \bar{\mathbf{k}}_c\bar{\mathbf{s}} + \bar{\mathbf{S}}_F \tag{2.2–7c}$$

$\bar{\mathbf{k}}_c$ is formed by a diagonal ordering of member stiffness matrices $\bar{\mathbf{k}}_c^m$.

Equations (2.2–7) form a linear set of simultaneous equations which, as a governing set, should have as many equations as unknowns. Assume, first, that support conditions prescribe zero values to certain joint displacements that are removed from $\mathbf{r}$. Accordingly the unknown reactions at support joints are not part of $\mathbf{R}$, and $\mathbf{R}$ consists of prescribed joint loads. The fixed end forces $\bar{\mathbf{S}}_F^m$ are determined for each member from its loading condition. The unknowns in Eqs. (2.2–7) are then $\bar{\mathbf{s}}$, $\bar{\mathbf{S}}$, and $\bar{\mathbf{r}}$. If NS is the size of $\bar{\mathbf{s}}$ and $\bar{\mathbf{S}}$, and NR is the size of $\mathbf{r}$ and $\mathbf{R}$, the number of unknowns is $2(NS) + NR$. There are exactly as many equations: NR joint equilibrium equations, NS connectivity equations, and NS stiffness equations.

Let now $\mathbf{r}$ be a complete set that includes a subset $\mathbf{r}_s$ of prescribed displacements. To $\mathbf{r}_s$ corresponds in $\mathbf{R}$ a subset $\mathbf{R}_s$ of unknown joint forces. By comparison with the preceding set of equations, the present one contains as many additional equilibrium equations as there are unknown reactions in $\mathbf{R}_s$.

To form Eqs. (2.2–7) requires the following steps:

1. Identify joint displacements $\mathbf{r}$ and member end displacements $\bar{\mathbf{s}}$. These are, in general, effective displacements.
2. Form $\bar{\mathbf{k}}_c^m$ and $\bar{\mathbf{S}}_F^m$ for each member. In typical cases this requires only an identification of available formulas.
3. Form $\mathbf{R}$ from the prescribed joint loads.
4. Form the connectivity matrix $\bar{\mathbf{a}}$.

Example 1b. Rigid Frame

The frame of Fig. 2.2–1 is now supported and loaded as shown in Fig. 2.2–4*a*. Effective joint displacements are $\mathbf{r} = \{\mathbf{r}^2\ \mathbf{r}^3\ \mathbf{r}^4\}$. The connectivity matrix $\bar{\mathbf{a}}$ is obtained by deleting the first and fifth column of the complete $\mathbf{a}$ formed in Example 1a. Member displacements $\bar{\mathbf{s}}^{1,1}$ and $\bar{\mathbf{s}}^{4,5}$ could also be deleted, but this is not necessary for consistency. The connectivity equations enforce the conditions $\bar{\mathbf{s}}^{1,1} = \bar{\mathbf{s}}^{4,5} = 0$. For the joint load shown, $\mathbf{R}^3 = \mathbf{R}^4 = 0$, and $\mathbf{R}^2 = \{P\ 0\ 0\}$. For member 2, assumed to be uniform, the fixed end forces are shown in Fig. 2.2–4*b*. Thus

$$\bar{\mathbf{S}}_F^2 = \left\{0\ \ \frac{pl'}{2}\ \ \frac{pl'^2}{12}\ \ 0\ \ \frac{pl'}{2}\ \ -\frac{pl'^2}{12}\right\}$$

For the remaining members $\bar{\mathbf{S}}_F^m = 0$.

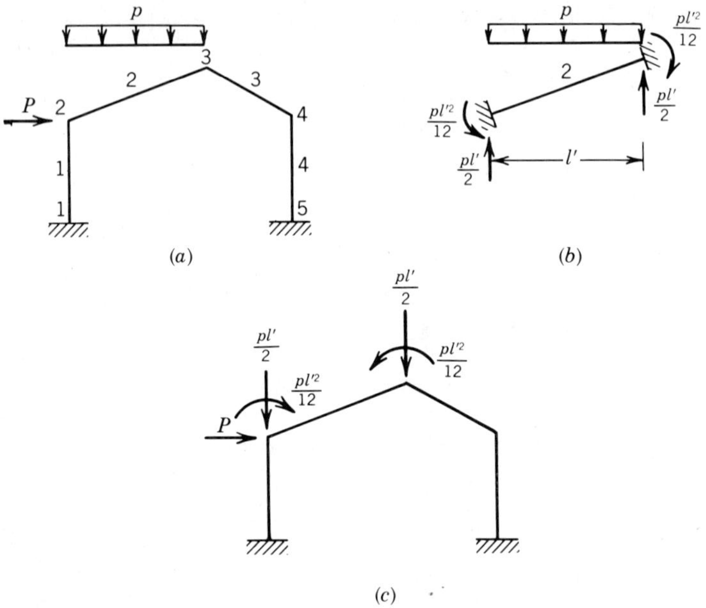

FIG. 2.2–4. Fixed end forces, equivalent joint load, Example 1.

To complete the governing equations, the member stiffness matrices $\bar{\mathbf{k}}_c^m$ are placed on the main diagonal of the member-set matrix $\bar{\mathbf{k}}_c$, that is,

$$\bar{\mathbf{k}}_c = \lceil \bar{\mathbf{k}}_c^1 \;\; \bar{\mathbf{k}}_c^2 \;\; \bar{\mathbf{k}}_c^3 \;\; \bar{\mathbf{k}}_c^4 \rfloor$$

$\bar{\mathbf{k}}_c^m$ is a 6×6 matrix obtained from the 3×3 reduced stiffness matrix as seen in Chapter 7, Section 2.3.

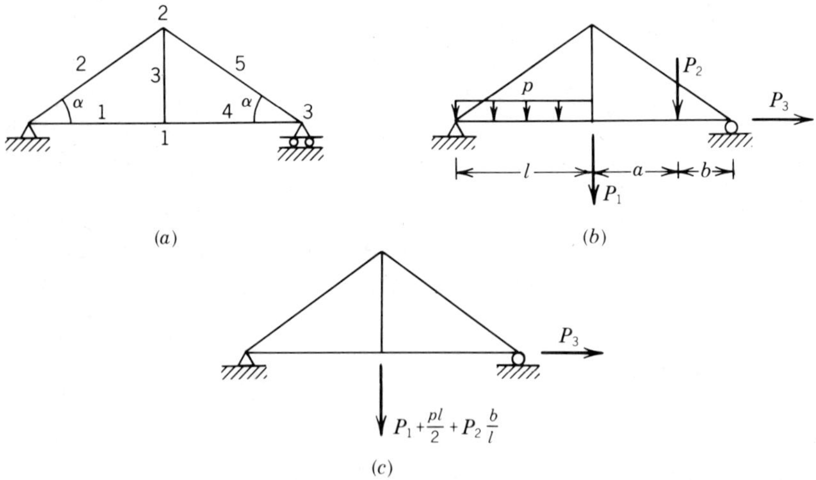

FIG. 2.2–5. Truss of Example 2 (*a*), (*b*); equivalent joint load (*c*).

Example 2. Pin-Jointed Truss

For a pin-jointed truss effective displacements consist only of joint translations. For a typical member $\bar{\mathbf{s}}^m = \{\bar{s}_1^a \;\; \bar{s}_2^a \;\; \bar{s}_1^b \;\; \bar{s}_2^b\}$, and for a typical joint $\mathbf{r}^j = \{r_1^j \;\; r_2^j\}$. For the truss of Fig. 2.2–5*a*, $\mathbf{r}$ is reduced by deleting the prescribed joint displacements. Thus $\mathbf{r} = \{\mathbf{r}^1 \;\; \mathbf{r}^2 \;\; \mathbf{r}^3\}$, and $\mathbf{r}^3 = \{r_1^3\}$. A similar reduction is made from member displacements and member forces, although this is not required for consistency. The connectivity equations are formed according to the general scheme of Fig. 2.2–2. Connectivity matrices $\bar{\mathbf{a}}^{m,j}$ are 2×2 identity matrices, except at joint 3 where $\bar{\mathbf{a}}^{4,3} = \bar{\mathbf{a}}^{5,3} = [1]$. This is shown as

$$\begin{Bmatrix} \bar{\mathbf{s}}^{1,1} \\ \hline \bar{\mathbf{s}}^{2,2} \\ \hline \bar{\mathbf{s}}^{3,1} \\ \hline \bar{\mathbf{s}}^{3,2} \\ \hline \bar{\mathbf{s}}^{4,1} \\ \hline \bar{\mathbf{s}}^{4,3} \\ \hline \bar{\mathbf{s}}^{5,2} \\ \hline \bar{\mathbf{s}}^{5,3} \end{Bmatrix} = \left[\begin{array}{c|c|c} \mathbf{I} & & \\ \hline & \mathbf{I} & \\ \hline \mathbf{I} & & \\ \hline & \mathbf{I} & \\ \hline \mathbf{I} & & \\ \hline & & 1 \\ \hline & \mathbf{I} & \\ \hline & & 1 \end{array}\right] \begin{Bmatrix} \mathbf{r}^1 \\ \mathbf{r}^2 \\ r_1^3 \end{Bmatrix}$$

For the loading shown in Fig. 2.2–5*b*, the joint load is formed of $\mathbf{R}^1 = \{0 \;\; -P_1\}$, $\mathbf{R}^2 = 0$, and $R_1^3 = P_3$.

The fixed member end forces are effective in that they are obtained by fixing the effective displacements while letting released displacements be free. Thus $\bar{\mathbf{S}}_F^m$ are the force reactions of member m pinned at both ends and subjected to its loading condition. Elements of $\bar{\mathbf{S}}_F^m$ that correspond to deleted member displacements are also deleted. Thus for members 1 and 4, $\bar{\mathbf{S}}_F^1 = \bar{\mathbf{S}}_F^{1,1} = \{0 \;\; pl/2\}$, and $\bar{\mathbf{S}}_F^4 = \{\bar{\mathbf{S}}_F^{4,1} \;\; \bar{\mathbf{S}}_F^{4,3}\} = \{0 \;\; P_2 b/l \;\; 0\}$.

The effective stiffness matrix for a pin-connected member is given in Eq. (7/3.4–11). Here it is made further effective for members 1, 2, 4, and 5 by deletion of the rows and columns associated with the prescribed displacements. The resulting matrices correspond then to the effective $\bar{\mathbf{s}}^m$ and $\bar{\mathbf{S}}^m$. Letting $k_m = EA/l$ be the axial stiffness of member m, we obtain

$$\bar{\mathbf{k}}_c^1 = k_1 \begin{bmatrix} 1 & 0 \\ 0 & 0 \end{bmatrix}$$

$$\bar{\mathbf{k}}_c^2 = k_2 \begin{bmatrix} \cos^2\alpha & \cos\alpha\sin\alpha \\ \cos\alpha\sin\alpha & \sin^2\alpha \end{bmatrix}$$

$$\bar{\mathbf{k}}_c^3 = k_3 \left[\begin{array}{cc|c} 0 & 0 & \\ 0 & 1 & (-1) \\ \hline & (-1) & (+1) \end{array}\right]$$

$\bar{\mathbf{k}}_c^4$ is similar to $\bar{\mathbf{k}}_c^1$, and $\bar{\mathbf{k}}_c^5$ is similar to $\bar{\mathbf{k}}_c^2$, with α replaced by $-\alpha$. Finally,

$$\bar{\mathbf{k}}_c = \lceil \bar{\mathbf{k}}_c^1 \;\; \bar{\mathbf{k}}_c^2 \;\; \bar{\mathbf{k}}_c^3 \;\; \bar{\mathbf{k}}_c^4 \;\; \bar{\mathbf{k}}_c^5 \rfloor$$

Example 3. Finite Composite Joint

Fig. 2.2–6*a* shows a finite joint having a rigid part of dimensions a and b which is connected rigidly to member 1 and pinned to member 2. The complete set of joint displacements consists of the displacements r_1, r_2, and r_3 of the rigid part and of the independent rotation r_4 of member stub 2. Letting j be the name of the joint, we can write the connectivity relations

$$\begin{Bmatrix} \bar{s}_1^{1,j} \\ \bar{s}_2^{1,j} \\ \bar{s}_3^{1,j} \end{Bmatrix} = \begin{bmatrix} 1 & 0 & -b & 0 \\ 0 & 1 & 0 & 0 \\ 0 & 0 & 1 & 0 \end{bmatrix} \begin{Bmatrix} r_1 \\ r_2 \\ r_3 \\ r_4 \end{Bmatrix}$$

$$\begin{Bmatrix} \bar{s}_1^{2,j} \\ \bar{s}_2^{2,j} \\ \bar{s}_3^{2,j} \end{Bmatrix} = \begin{bmatrix} 1 & 0 & 0 & 0 \\ 0 & 1 & a & 0 \\ 0 & 0 & 0 & 1 \end{bmatrix} \begin{Bmatrix} r_1 \\ r_2 \\ r_3 \\ r_4 \end{Bmatrix}$$

These coefficient matrices are the connectivity matrices $\bar{\mathbf{a}}^{1,j}$ and $\bar{\mathbf{a}}^{2,j}$, respectively. An effective choice of displacements may delete $\bar{s}_3^{2,j}$ and r_4, with corresponding deletions from $\bar{\mathbf{a}}^{1,j}$ and $\bar{\mathbf{a}}^{2,j}$. If this is done, the stiffness matrix of member 2 must be transformed to take into account the released rotation (see Chapter 7, Sections 3.3 and 3.4). Deletion of r_4 is accompanied by a deletion of the moment equilibrium equation at the end of member 2. With reference to Fig. 2.2–6*b*, the deleted equation is $\bar{S}_3^{2,j} = R_4$. R_4, if applied, becomes part of the span load and affects the effective fixed end forces in the transformed stiffness equation of member 2.

Structure Stiffness Equations

Equations (2.2–7) are suitable for choosing $\mathbf{r}$ as a basic set of unknowns. In the displacement method of analysis, $\bar{\mathbf{s}}$ and $\bar{\mathbf{S}}$ are eliminated and Eqs. (2.2–7) are

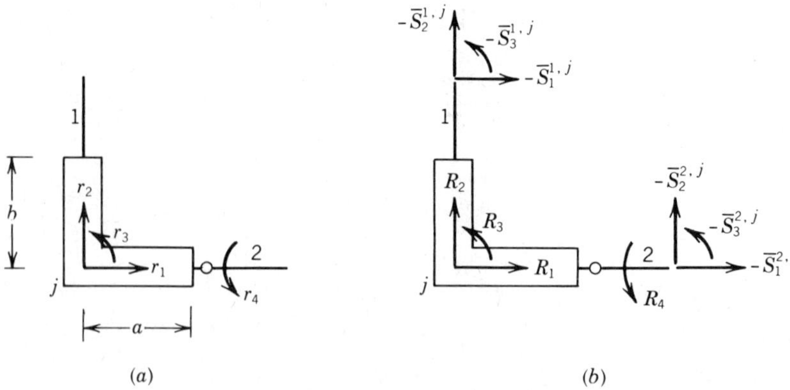

FIG. 2.2–6. Finite composite joint, Example 3.

reduced to the equilibrium equations expressed in terms of $\mathbf{r}$. This yields the structure stiffness equations

$$\mathbf{Kr} = \mathbf{R} - \mathbf{R}_F \tag{2.2–8a}$$

where

$$\mathbf{K} = \bar{\mathbf{a}}^T \bar{\mathbf{k}}_c \bar{\mathbf{a}} \tag{2.2–8b}$$

$$\mathbf{R}_F = \bar{\mathbf{a}}^T \bar{\mathbf{S}}_F \tag{2.2–8c}$$

$\mathbf{K}$ is the structure stiffness matrix, and $\mathbf{R} - \mathbf{R}_F$ is called the total equivalent joint load.

The term $(-\bar{\mathbf{a}}^T \bar{\mathbf{S}}_F)$ is equivalent to the member loading condition in the following sense: if the joint displacements are fixed in the initial state $\mathbf{r} = 0$, and the members are subjected to their loading condition, their end forces are $\bar{\mathbf{S}}_F$, and the load required to fix the joints is $\mathbf{R}_F = \bar{\mathbf{a}}^T \bar{\mathbf{S}}_F$. This load, reversed in sense, is the equivalent joint load which, together with $\mathbf{R}$, causes the joint displacements $\mathbf{r}$.

Examples of the evaluation of $(\mathbf{R} - \mathbf{R}_F)$ are represented in Figs. 2.2–4*c* and 2.2–5*c*. $-\mathbf{R}_F$ is obtained by adding the member fixed end forces at a joint and reversing the result.

Developments of the stiffness or displacement method are continued in Chapter 10.

2.3. Governing Equations in Local Coordinates

By governing equations in local coordinates is meant equations similar to Eqs. (2.2–7) in which member end displacements and forces are resolved on the member local axes. The derivation is entirely similar to the preceding one. The connectivity matrix is denoted $\mathbf{a}$. In applying the principle of virtual displacements, the virtual work of the member end forces is $\mathbf{S}^T \delta \mathbf{s}$. We thus obtain

$$\mathbf{s} = \mathbf{ar} \tag{2.3–1a}$$

$$\mathbf{a}^T \mathbf{S} = \mathbf{R} \tag{2.3–1b}$$

$$\mathbf{S} = \mathbf{k}_c \mathbf{s} + \mathbf{S}_F \tag{2.3–1c}$$

The partitioning of $\mathbf{a}$ by member ends and by joints is entirely similar to that of $\bar{\mathbf{a}}$, Fig. 2.2–2. At end j of member m, $\mathbf{a}^{m,j}$ involves the coordinate transformation $\mathbf{s}^{m,j} = \boldsymbol{\lambda}^{m,j} \bar{\mathbf{s}}^{m,j}$. The connectivity relation for $\bar{\mathbf{s}}^{m,j}$ yields

$$\mathbf{a}^{m,j} = \boldsymbol{\lambda}^{m,j} \bar{\mathbf{a}}^{m,j} \tag{2.3–2a}$$

For the structure as a whole, $\mathbf{s} = \boldsymbol{\lambda} \bar{\mathbf{s}}$, and

$$\mathbf{a} = \boldsymbol{\lambda} \bar{\mathbf{a}} \tag{2.3–2b}$$

The transformation $\mathbf{s} = \boldsymbol{\lambda} \bar{\mathbf{s}}$ is accompanied by the contragradient relation $\bar{\mathbf{S}} =$

$\boldsymbol{\lambda}^T\mathbf{S}$. If $\boldsymbol{\lambda}\bar{\mathbf{a}}$ is substituted for $\mathbf{a}$ into Eq. (2.3–1b), the result is the equation in global coordinates. The structure stiffness equations (2.2–8a) may be derived from the present governing equations. $\mathbf{K}$ and $\mathbf{R}_F$ are then obtained as

$$\mathbf{K} = \mathbf{a}^T\mathbf{k}_c\mathbf{a} \tag{2.3–3a}$$

$$\mathbf{R}_F = \mathbf{a}^T\mathbf{S}_F \tag{2.3–3b}$$

The transformation $\mathbf{a} = \boldsymbol{\lambda}\bar{\mathbf{a}}$ transforms $\mathbf{k}_c$ into $\bar{\mathbf{k}}_c$, $\mathbf{S}_F$ into $\bar{\mathbf{S}}_F$, and leaves $\mathbf{K}$ and $\mathbf{R}_F$ unchanged. If $\bar{\mathbf{a}}$ is easier to form than $\mathbf{a}$, by contrast, $\mathbf{k}_c$ is more readily formed than $\bar{\mathbf{k}}_c$.

The main advantage of Eqs. (2.3–1) is that the member end forces are obtained in local coordinates and are thus in the appropriate form for member stress analysis. Another advantage may occur with use of effective matrices. For example, in a pin-jointed truss it is possible to use an effective $\mathbf{s}^m$ consisting of the two axial displacements at the member ends, whereas four translational displacements constitute $\bar{\mathbf{s}}^m$. It may be noted, however, that individual member equations could be written independently in local or global coordinates in the same set of governing equations.

Example 1. Rigid Frame

For a rigid frame $\bar{\mathbf{a}}^{m,j} = \mathbf{I}$, and $\mathbf{a}^{m,j}$ is the coordinate transformation matrix $\boldsymbol{\lambda}^{m,j}$, Eq. (7/2.1–1). For the frame of Fig. 2.3–1, $\mathbf{a}$ is shown partitioned by member ends

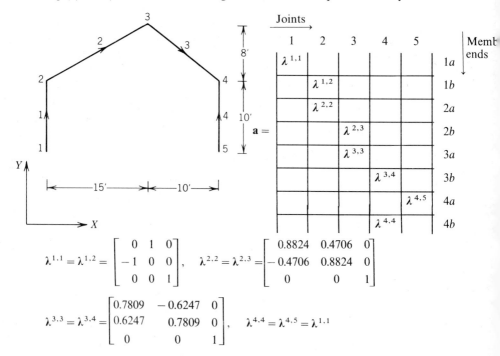

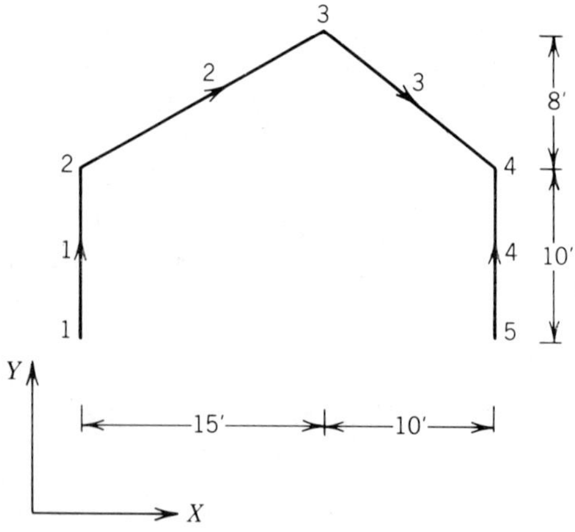

$\mathbf{a} =$ (Joints → columns; Member ends ↓ rows)

Member ends	1	2	3	4	5
1a	$\lambda^{1,1}$				
1b		$\lambda^{1,2}$			
2a		$\lambda^{2,2}$			
2b			$\lambda^{2,3}$		
3a			$\lambda^{3,3}$		
3b				$\lambda^{3,4}$	
4a					$\lambda^{4,5}$
4b				$\lambda^{4,4}$	

$$\lambda^{1,1} = \lambda^{1,2} = \begin{bmatrix} 0 & 1 & 0 \\ -1 & 0 & 0 \\ 0 & 0 & 1 \end{bmatrix}, \quad \lambda^{2,2} = \lambda^{2,3} = \begin{bmatrix} 0.8824 & 0.4706 & 0 \\ -0.4706 & 0.8824 & 0 \\ 0 & 0 & 1 \end{bmatrix}$$

$$\lambda^{3,3} = \lambda^{3,4} = \begin{bmatrix} 0.7809 & -0.6247 & 0 \\ 0.6247 & 0.7809 & 0 \\ 0 & 0 & 1 \end{bmatrix}, \quad \lambda^{4,4} = \lambda^{4,5} = \lambda^{1,1}$$

FIG. 2.3–1. Connectivity matrix **a**, Example 1.

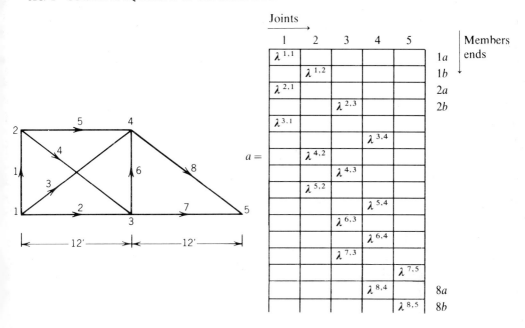

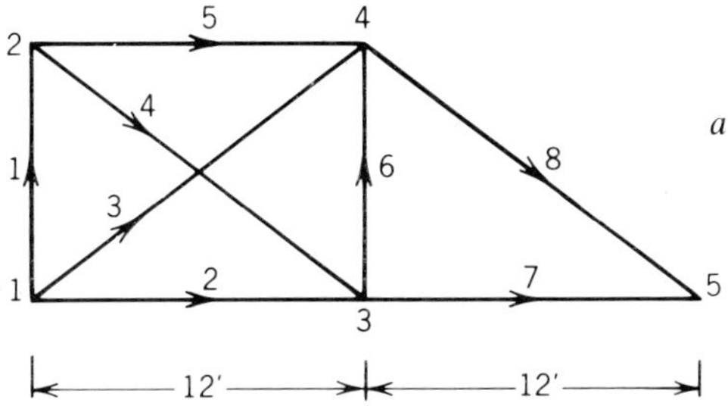

Joints →

Members ends ↓

$a =$

	1	2	3	4	5	
	$\lambda^{1,1}$					1a
		$\lambda^{1,2}$				1b
	$\lambda^{2,1}$					2a
			$\lambda^{2,3}$			2b
	$\lambda^{3,1}$					
				$\lambda^{3,4}$		
		$\lambda^{4,2}$				
			$\lambda^{4,3}$			
		$\lambda^{5,2}$				
				$\lambda^{5,4}$		
			$\lambda^{6,3}$			
				$\lambda^{6,4}$		
			$\lambda^{7,3}$			
					$\lambda^{7,5}$	
				$\lambda^{8,4}$		8a
					$\lambda^{8,5}$	8b

$$\boldsymbol{\lambda}^{1,1} = \boldsymbol{\lambda}^{1,2} = \boldsymbol{\lambda}^{6,3} = \boldsymbol{\lambda}^{6,4} = [0 \quad 1]$$
$$\boldsymbol{\lambda}^{2,1} = \boldsymbol{\lambda}^{2,3} = \boldsymbol{\lambda}^{5,2} = \boldsymbol{\lambda}^{5,4} = \boldsymbol{\lambda}^{7,3} = \boldsymbol{\lambda}^{7,5} = [1 \quad 0]$$
$$\boldsymbol{\lambda}^{3,1} = \boldsymbol{\lambda}^{3,4} = [0.8 \quad 0.6]$$
$$\boldsymbol{\lambda}^{4,2} = \boldsymbol{\lambda}^{4,3} = \boldsymbol{\lambda}^{8,4} = \boldsymbol{\lambda}^{8,5} = [0.8 \quad -0.6]$$

FIG. 2.3–2. Connectivity matrix **a**, Example 2.

and joints. The oriented members of the frame define the local x axes. The local y axis is obtained by a counterclockwise rotation of 90° of the local axis. The orientation angle θ used in $\boldsymbol{\lambda}^{m,j}$ is $\pi/2$ for members 1 and 4. For member 2, $\theta = \tan^{-1}(8/15)$ and for member 3, $\theta = -\tan^{-1}(0.8)$.

Example 2. Pin-Jointed Truss

For a pin-joined truss subjected only to joint loads, each member is a two-force member. Effective $\mathbf{s}^m$ and $\mathbf{S}^m$ contain only the axial components. The typical coordinate transformation matrix $\boldsymbol{\lambda}^{m,j}$ is

$$\boldsymbol{\lambda}^{m,j} = [\cos\theta \quad \sin\theta]$$

where θ is the orientation angle of member m. An example is shown in Fig. 2.3–2. The members are oriented in order to define θ.

2.4. Governing Equations in Member-Bound Axes

Governing equations in member-bound axes are formulated with member deformations $\mathbf{v}$ and member statical redundants $\mathbf{V}$ as state variables. Member equilibrium equations are solved in terms of member statical redundants (or reduced forces) in the form

$$\mathbf{S} = \mathbf{B}\mathbf{V} + \mathbf{S}_p \tag{2.4–1}$$

A similar equation holds in global coordinates.

Member connectivity equations are derived from the deformation-displacement relations

$$\mathbf{v} = \mathbf{B}^T\mathbf{s} = \bar{\mathbf{B}}^T\bar{\mathbf{s}} \tag{2.4–2}$$

and the connectivity relations $\mathbf{s} = \mathbf{a}\mathbf{r}$, or $\bar{\mathbf{s}} = \bar{\mathbf{a}}\mathbf{r}$. There comes

$$\mathbf{v} = \mathbf{a}_v\mathbf{r} \tag{2.4–3}$$

$$\mathbf{a}_v = \mathbf{B}^T\mathbf{a} = \bar{\mathbf{B}}^T\bar{\mathbf{a}} \tag{2.4–4}$$

For an individual member connected to joints a and b

$$\mathbf{v}^m = [\mathbf{a}_v^a \ \mathbf{a}_v^b]\{\mathbf{r}^a \ \mathbf{r}^b\} \tag{2.4–5}$$

$$\mathbf{a}_v^a = \mathbf{B}^{aT}\mathbf{a}^a = \bar{B}^{aT}\bar{\mathbf{a}}^a \tag{2.4–6a}$$

$$\mathbf{a}_v^b = \mathbf{B}^{bT}\mathbf{a}^b = \bar{\mathbf{B}}^{bT}\bar{\mathbf{a}}^b \tag{2.4–6b}$$

Equation (2.4–3) is the set of Eqs. (2.4–5) ordered by members. The general scheme for forming $\mathbf{a}_v$ from member matrices $\mathbf{a}_v^a$ and $\mathbf{a}_v^b$ is shown in Fig. 2.4–1, where a superscript m is added to $\mathbf{a}_v^a$ and $\mathbf{a}_v^b$. Joint equilibrium equations may be derived by substituting $\mathbf{S} = \mathbf{B}\mathbf{V} + \mathbf{S}_p$ into $\mathbf{a}^T\mathbf{S} = \mathbf{R}$, or by a similar procedure in global coordinates. For a direct derivation by the principle of virtual displacements, the equation of virtual work at the joints (2.2–5) takes the form

$$\mathbf{V}^T\delta\mathbf{v} + \mathbf{S}_p^T\delta\mathbf{s} = \mathbf{R}^T\delta\mathbf{r} \tag{2.4–7}$$

With $\delta\mathbf{v} = \mathbf{a}_v\delta\mathbf{r}$ and $\delta\mathbf{s} = \mathbf{a}\delta\mathbf{r}$, the result is Eq. (2.4–8b). Member flexibility or

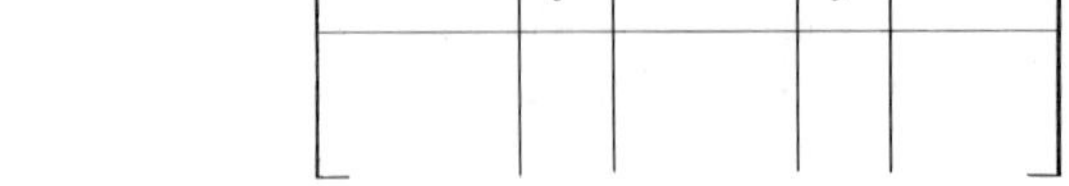

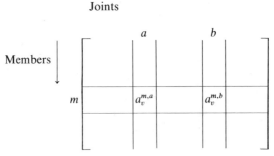

FIG. 2.4–1. Scheme for formin $\mathbf{a}_v$.

stiffness relations complete the governing set. Thus, altogether,

$$\mathbf{v} = \mathbf{a}_v \mathbf{r} \tag{2.4–8a}$$

$$\mathbf{a}_v^T \mathbf{V} = \mathbf{R} - \mathbf{R}_p \tag{2.4–8b}$$

$$\mathbf{v} = \mathbf{f}\mathbf{V} + \mathbf{v}_0 \tag{2.4–8c}$$

or

$$\mathbf{V} = \mathbf{k}\mathbf{v} + \mathbf{V}_F \tag{2.4–8d}$$

with

$$\mathbf{R}_p = \mathbf{a}^T \mathbf{S}_p = \bar{\mathbf{a}}^T \bar{\mathbf{S}}_p \tag{2.4–9}$$

$$\mathbf{V}_F = -\mathbf{k}\mathbf{v}_0 \tag{2.4–10}$$

Among the three governing sets of equations, the present one is the most fundamental. It may be taken as the starting point for developing both the stiffness and the flexibility methods of analysis. In particular, it is the appropriate set for studying statically determinate structures. By contrast, the two previous sets are directly appropriate only for the stiffness method. The member flexibility and stiffness matrices, in a bound reference, are the basic finite constitutive properties, whereas $\mathbf{k}_c$ and $\bar{\mathbf{k}}_c$ are both obtainable by transformation of $\mathbf{k}$. $\mathbf{k}$ is a 3×3 matrix, whereas $\mathbf{k}_c$ and $\bar{\mathbf{k}}_c$ are each 6×6. The connectivity matrix $\mathbf{a}_v$ is sparse, but to a lesser degree than $\mathbf{a}$ and $\bar{\mathbf{a}}$, and has a smaller row size. Member equilibrium equations are not part of the present governing set but are solved in terms of the member statical redundants. Accordingly the joint equilibrium equations in the present set involve both the joint load $\mathbf{R}$ and the members span loads that determine $\mathbf{R}_p$. In the two previous sets of governing equations, member equilibrium is incorporated into the member stiffness equations, and the joint equilibrium equations involve only the loads applied at the joints.

Before treating some examples, the definition of $\mathbf{S}_p$ and $\mathbf{v}_0$ will be generalized in order to make application of Eqs. (2.4–8) more flexible. $\mathbf{S}_p$ and $\mathbf{v}_0$ are determined independently for each member. As defined until now, $\mathbf{S}_p^m$ is the set of end forces in equilibrium with the span load and with $\mathbf{V}^m = 0$. This particular, or initial, equilibrium state is statically determinate. In it $\mathbf{v}_0^m$ is the set of deformations caused by the total member loading condition. $\mathbf{v}_0^m$ may thus include thermal deformations or geometric errors $\mathbf{v}_I^m$, in addition to the deformations $\mathbf{v}_p^m$ caused by the span loads. The choice of a statically determinate state as an initial equilibrium state is theoretically fundamental but not necessary in applications. For usual uniform members and typical loads, statically indeterminate solutions are readily available and may be more convenient to use. It is thus advantageous to generalize the definition of $\mathbf{S}_p^m$ and $\mathbf{v}_0^m$ as pertaining to any particular equilibrium state of the member subjected to its span loading condition. Equation (2.4–9) defines $\mathbf{R}_p$ as an equivalent joint load in equilibrium with $\mathbf{S}_p$. The initial equilibrium state of the members may or may not be geometrically compatible at the joints. This is of no concern because in the final equilibrium state the deformations satisfy the connectivity equation. If necessary in order to avoid confusion, a prime is used with symbols for which the initial equilibrium state has a nonzero $\mathbf{V}$.

The initial equilibrium state that is the most direct for the stiffness method (Chapter 10) is the fixed state $\bar{\mathbf{s}} = 0$. It is also a compatible state in which $\mathbf{r} = 0$. In this initial state $\mathbf{v}'_0 = 0$, and $\mathbf{S}'_p$ become the fixed end forces $\mathbf{S}_F$. The forces $\mathbf{V}'$ are now additional to those of the fixed state, and the term $\mathbf{V}'_F = -\mathbf{k}\mathbf{v}'_0$ in the member stiffness equations becomes zero. In the two previous governing sets, the fixed initial state is represented by the terms $\mathbf{S}_F$ and $\bar{\mathbf{S}}_F$.

Another choice of initial equilibrium state in the present formulation, which is often simple enough, is one where the member is pinned at both ends. The member is then statically indeterminate axially but is determinate for transverse forces (see Example 1b).

Example 1a. Rigid Frame

For a rigid frame with point joints, $\bar{\mathbf{a}}^a = \bar{\mathbf{a}}^b = \mathbf{I}$, and consequently

$$\mathbf{v}^m = [\bar{\mathbf{B}}^{aT} \ \ \bar{\mathbf{B}}^{bT}]\{\mathbf{r}^a \ \ \mathbf{r}^b\} \tag{2.4–11}$$

The choice of type for $\mathbf{v}$ and $\mathbf{V}$ is arbitrary. For the alternate type, from Eqs. (7/2.2–4).

$$[\bar{\mathbf{B}}^{aT}|\bar{\mathbf{B}}^{bT}] = \left[\begin{array}{ccc|ccc} -\cos\theta & -\sin\theta & 0 & \cos\theta & \sin\theta & 0 \\ -\dfrac{\sin\theta}{l} & \dfrac{\cos\theta}{l} & 1 & \dfrac{\sin\theta}{l} & -\dfrac{\cos\theta}{l} & 0 \\ -\dfrac{\sin\theta}{l} & \dfrac{\cos\theta}{l} & 0 & \dfrac{\sin\theta}{l} & -\dfrac{\cos\theta}{l} & 1 \end{array}\right] \tag{2.4–12}$$

Consider the rigid frame of Fig. 2.3–1 with as yet unspecified supports. The preceding equation is applied for each member, and $\bar{\mathbf{B}}^{aT}$ and $\bar{\mathbf{B}}^{bT}$ are placed in $\mathbf{a}_v$, as shown in Fig. 2.4–2. Ends a and b are at the tail and tip of the oriented member, respectively.

To illustrate how $\mathbf{R}_p$ and $\mathbf{v}_0$ are formed in a given loading condition, consider the loading shown in Fig. 2.4–3. Only members 2 and 3, which have a span load, need be considered. First, a statically determinate choice of $\mathbf{S}_p$ will be illustrated. Since $\mathbf{V}$ is chosen to be of the alternate type, each member is simply supported so that $\mathbf{V} = 0$. The reactions are determined in global coordinates and form $\bar{\mathbf{S}}_p$. $\mathbf{R}_p$ is evaluated by adding at each joint the forces contributed by the member ends. The components of the equivalent joint load $-\mathbf{R}_p$ are shown in Fig. 2.4–3f. Evaluation of $\mathbf{v}_p$ for members 2 and 3 is shown in (c) and (e), assuming uniform members. If a thermal loading were part of the loading condition, the free thermal deformations $\mathbf{v}_I$ would be added to $\mathbf{v}_p$ to form the total initial deformation term $\mathbf{v}_0$. A simpler choice of $\mathbf{S}_p$ and $\mathbf{v}_p$ is made in Example 1b.

In solving the governing equations, the support conditions are enforced by letting $\mathbf{r}^1 = \mathbf{r}^5 = 0$. Accordingly $\mathbf{R}^1$ and $\mathbf{R}^5$ are unknown reactions. A count of

Joint →

$\mathbf{a}_v =$

	1	2	3	4	5	Members ↓
	$\bar{\mathbf{B}}^{1,1T}$	$\bar{\mathbf{B}}^{1,2T}$				1
		$\bar{\mathbf{B}}^{2,2T}$	$\bar{\mathbf{B}}^{2,3T}$			2
			$\bar{\mathbf{B}}^{3,3T}$	$\bar{\mathbf{B}}^{3,4T}$		3
				$\bar{\mathbf{B}}^{4,4T}$	$\bar{\mathbf{B}}^{4,5T}$	4

$$\bar{\mathbf{B}}^{1,1T} = \begin{bmatrix} 0 & -1 & 0 \\ -0.1 & 0 & 1 \\ -0.1 & 0 & 0 \end{bmatrix}, \quad \bar{\mathbf{B}}^{1,2T} = \begin{bmatrix} 0 & 1 & 0 \\ 0.1 & 0 & 0 \\ 0.1 & 0 & 1 \end{bmatrix}$$

$$\bar{\mathbf{B}}^{2,2T} = \begin{bmatrix} -0.8824 & -0.4706 & 0 \\ -0.02768 & 0.05190 & 1 \\ -0.02768 & 0.05190 & 0 \end{bmatrix}, \quad \bar{\mathbf{B}}^{2,3T} = \begin{bmatrix} 0.8824 & 0.4706 & 0 \\ 0.02768 & -0.05190 & 0 \\ 0.02768 & -0.05190 & 1 \end{bmatrix}$$

$$\bar{\mathbf{B}}^{3,3T} = \begin{bmatrix} -0.7809 & 0.6247 & 0 \\ 0.04878 & 0.06098 & 1 \\ 0.04878 & 0.06098 & 0 \end{bmatrix}, \quad \bar{\mathbf{B}}^{3,4T} = \begin{bmatrix} 0.7809 & -0.6247 & 0 \\ -0.04878 & -0.06098 & 0 \\ -0.04878 & -0.06098 & 1 \end{bmatrix}$$

$$\bar{\mathbf{B}}^{4,4T} = \bar{\mathbf{B}}^{1,2T}, \quad \bar{\mathbf{B}}^{4,5T} = \bar{\mathbf{B}}^{1,1T}$$

FIG. 2.4–2. Connectivity matrix $\mathbf{a}_v$ for frame of Fig. 2.3–1, Example 1*a*.

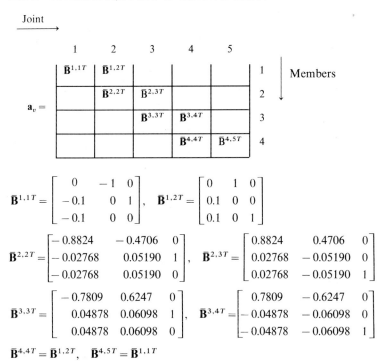

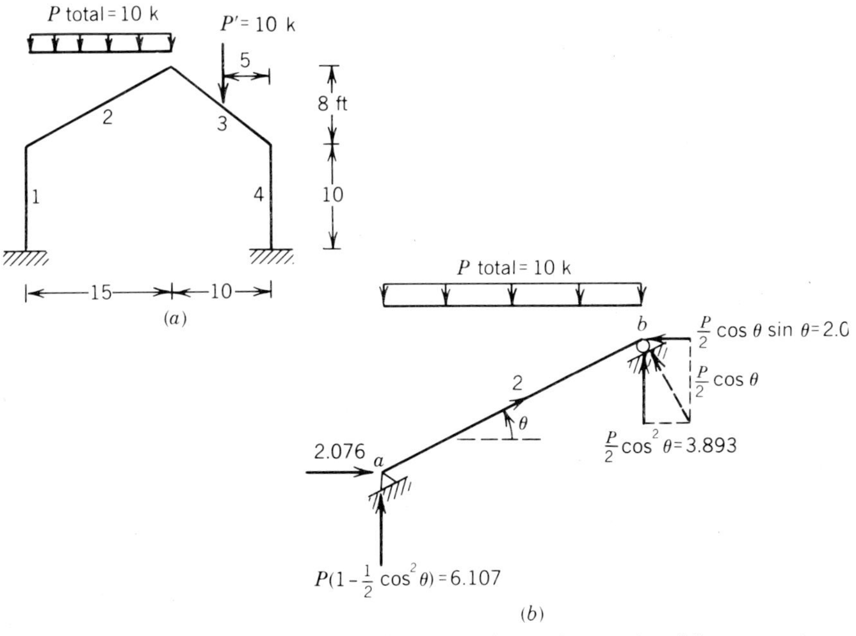

FIG. 2.4–3. Evaluation of $\mathbf{R}_p$ and $\mathbf{v}_0$, Example 1*a*. (*Continued on following page.*)

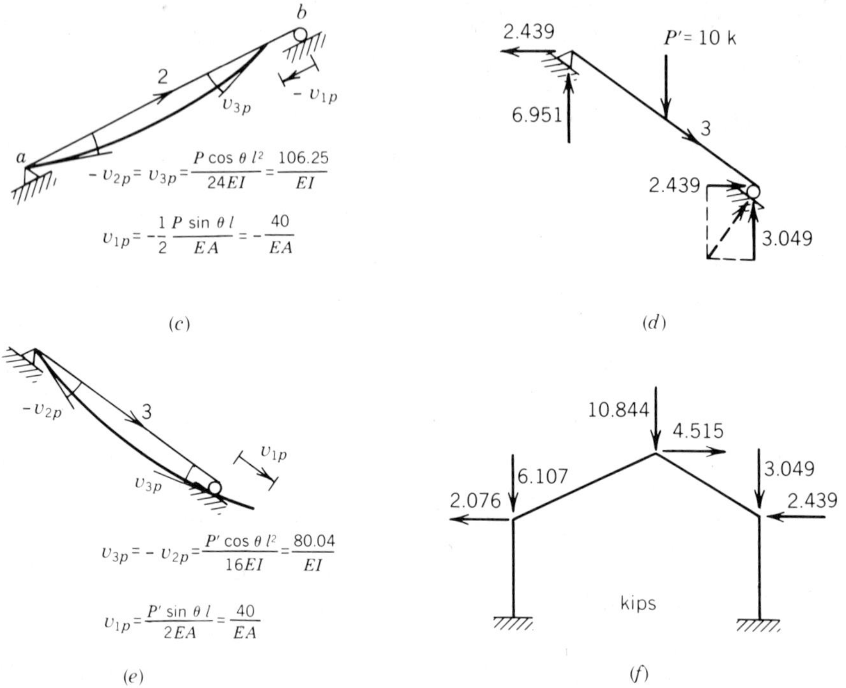

FIG. 2.4–3. *(continued)*

equations gives 12 for connectivity, 15 for equilibrium, and 12 for member flexibility, for a total of 39. Count of unknowns gives 12 deformations **v**, 12 forces **V**, 9 joint displacements $\{\mathbf{r}^2\ \mathbf{r}^3\ \mathbf{r}^4\}$, and 6 support reactions $\{\mathbf{R}^1\ \mathbf{R}^5\}$, for a total of 39.

The stiffness equations, obtained by elimination of **v** and **V**, form a set of 15 equilibrium equations. Nine of these correspond to the unrestrained degrees of freedom and form a subsystem in the 9 joint displacements. The remaining 6 equations serve to evaluate the support reactions.

In the flexibility method of analysis to be developed in Chapter 11, 15 out of the 18 force unknowns are eliminated from the 15 equilibrium equations, and the 21 unknowns **v** and **r** are eliminated from the 24 remaining equations. The result consists of 3 compatibility equations in 3 force unknowns.

The present problem could be formulated in terms of the effective joint displacements $\mathbf{r} = \{\mathbf{r}^2\ \mathbf{r}^3\ \mathbf{r}^4\}$. Then $\mathbf{a}_v$ is obtained from the preceding one by deleting the first and last sets of three columns. Accordingly **R** is reduced to $\{\mathbf{R}^2\ \mathbf{R}^3\ \mathbf{R}^4\}$ and thus excludes the reactions. The preceding comments on the stiffness method remain applicable except that the support reactions are not obtained in the solution. In this problem the reactions also appear as member end forces and are thus readily obtained. In the flexibility method the reduction to a system of three equations in three force unknowns remains, but these unknowns are now necessarily elements of **V**.

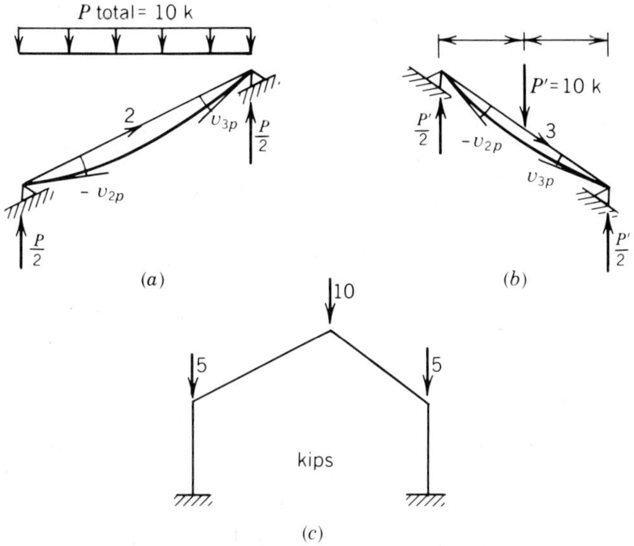

FIG. 2.4–4. Evaluation of $\mathbf{R}_p$ and $\mathbf{v}_0$, Example 1*b*.

Example 1b. Choice of Initial Equilibrium State

For the frame of Fig. 2.4–3 an initial equilibrium state in which each member is pin supported at both ends is shown in Fig. 2.4–4. Now $\mathbf{v}'_p = \{0 \;\; v_{2p} \;\; v_{3p}\}$ and $\mathbf{V}'_p = \{V_{1F} \;\; 0 \;\; 0\}$, where $V_{1F} = -(EA/l)v_{1p}$ and v_{1p} is the axial deformation for the statically determinate member as determined in Example 1*a*. The final result for the reactions is that they are vertical and are equal each to half the vertical load applied on the member. The equivalent joint load $-\mathbf{R}_p$ is shown in the figure.

Note that if the member is not uniform, the reactions in the pinned state may contain a couple of horizontal forces and unequal vertical reactions. Also this statically indeterminate problem is solved by determining v_{1p} for the determinate member, and thus by going through the fundamental equilibrium state of the previous example.

Example 1c

Refer to Example 1*a* and Fig. 2.4–3, and assume now that the support conditions are modified into a pin and a roller. The number of support reactions is now 3 and the number of force unknowns is $12 + 3 = 15$. This is equal to the number of joint equilibrium equations that form then a statically determinate system. The solution for $\mathbf{V}$ and for the support reactions is obtained by solving the joint equilibrium equations, without making use either of connectivity or of material properties. Statically determinate structures are seen in Chapters 9 and 11.

Note that if the choice of initial equilibrium state of Example 1*b* is used in the present problem, the axial stiffness of the members is introduced in forming $\mathbf{R}_p$, but the final solution for the member end forces is independent of material properties. There is thus something artificial about using a statically indeterminate initial state in the analysis of a statically determinate structure.

Example 1d. Forming $\mathbf{a}_v$ by Columns

This example is intended to bring out the "influence character" of a coefficient matrix such as $\mathbf{a}_v$ in the relation $\mathbf{v} = \mathbf{a}_v\mathbf{r}$. A given column of $\mathbf{a}_v$, say column i, multiplies the ith displacement r_i. Thus column i is the set of member deformations due to $r_i = 1$, while all other elements of $\mathbf{r}$ are zero. This provides a method for forming $\mathbf{a}_v$ by individual columns or by sets of columns. Individual columns will be considered for illustration.

The deformation modes due to $r_1^2 = 1$ and to $r_2^3 = 1$ are shown in Fig. 2.4–5. The axial deformations are evaluated by projecting the joint displacement on the initial position of the members. The rotational deformations (alternate type) are the rotations from the member chords. The absolute values of the deformations are shown in the figure, but in forming $\mathbf{a}_v$, counterclockwise deformations are considered positive. Since there are 3 displacements per joint, r_1^2 occupies position 4 in $\mathbf{r}$, and the corresponding 4th column of $\mathbf{a}_v$ consists of ordered member deformations due to $r_1^2 = 1$, or

$$\{0 \quad 0.1 \quad 0.1 \quad -0.8824 \quad -0.02768 \quad -0.02768 \quad 0 \quad 0 \quad 0 \quad 0 \quad 0 \quad 0\}$$

Similarly the 8th column of $\mathbf{a}_v$ corresponds to r_2^3 and is found, using Fig. 2.4–5*b*, to be

$$\{0 \quad 0 \quad 0 \quad 0.4706 \quad -0.0519 \quad -0.0519 \quad 0.6247 \quad 0.06098 \quad 0.06098 \quad 0 \quad 0 \quad 0\}$$

Columns of $\mathbf{a}_v$ corresponding to joint rotations are the easiest to form. The deformation mode due to $r_3^3 = 1$ is shown in Fig. 2.4–5*c*. The member orientations determine the ordering of the deformations as shown in Fig. 2.4–5*d*. To r_3^3 corresponds the 9th column of $\mathbf{a}_v$,

$$\{0 \; 0 \; 0 \; 0 \; 0 \; 1 \; 0 \; 1 \; 0 \; 0 \; 0 \; 0\}$$

Completion of $\mathbf{a}_v$ is left as an exercise and the result should coincide with that in Fig. 2.4–2.

Example 2. Pin-Jointed Truss

For a pin-jointed truss the effective $\mathbf{V}$ consists of the axial forces, and the effective $\mathbf{v}$ consists of the axial deformations. The member connectivity equation is

$$v_1^m = [\bar{\mathbf{B}}^{aT} \; \bar{\mathbf{B}}^{bT}]\begin{Bmatrix}\mathbf{r}^a \\ \mathbf{r}^b\end{Bmatrix} = [-\boldsymbol{\lambda}^a \; \boldsymbol{\lambda}^b]\begin{Bmatrix}\mathbf{r}^a \\ \mathbf{r}^b\end{Bmatrix} \qquad (2.4\text{–}14a)$$

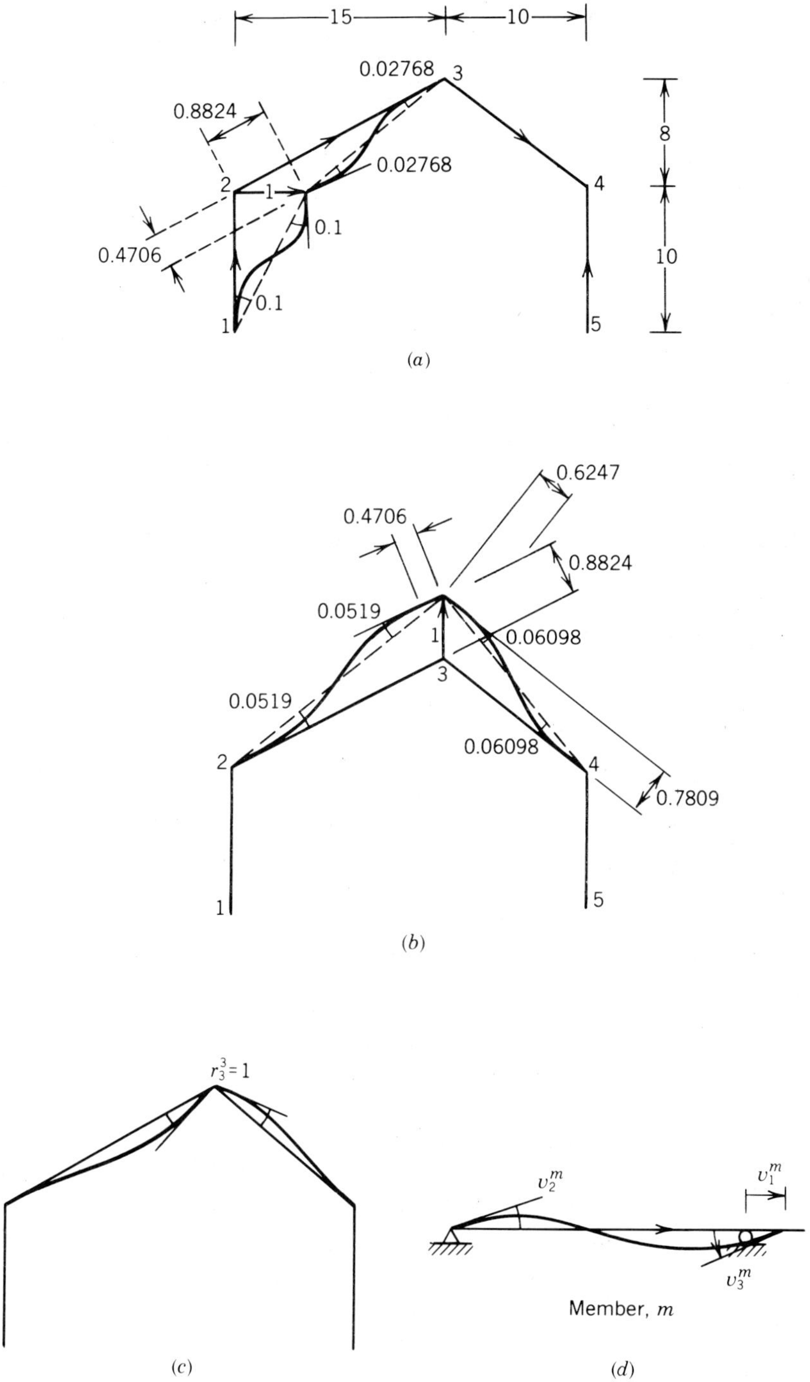

FIG. 2.4–5. Deformation modes, Example 1*d*.

Joints →

$\mathbf{a}_v =$

	1	2	3	4	5	Members ↓
	$-\lambda^{1,1}$	$\lambda^{1,2}$				1
	$-\lambda^{2,1}$		$\lambda^{2,3}$			2
	$-\lambda^{3,1}$			$\lambda^{3,4}$		3
		$-\lambda^{4,2}$	$\lambda^{4,3}$			4
		$-\lambda^{5,2}$		$\lambda^{5,4}$		5
			$-\lambda^{6,3}$	$\lambda^{6,4}$		6
			$-\lambda^{7,3}$		$\lambda^{7,5}$	7
				$-\lambda^{8,4}$	$\lambda^{8,5}$	8

FIG. 2.4–6. Connectivity matrix $\mathbf{a}_v$ for truss of Fig. 2.3–2, Example 2.

where

$$[-\boldsymbol{\lambda}^a \; \boldsymbol{\lambda}^b] = [-\cos\theta \quad -\sin\theta \quad \cos\theta \quad \sin\theta] \tag{2.4–14b}$$

$\mathbf{a}_v$ is formed for the truss of Fig. 2.3–2 using the matrices $\boldsymbol{\lambda}^{m,a}$ and $\boldsymbol{\lambda}^{m,b}$ given in the figure. The result is shown in Fig. 2.4–6.

The matrix $\mathbf{a}_v$ could be formed also by columns. Column i contains the set of deformations due to $r_i = 1$. For example, for $r_1^1 = 1$, the deformations of members 2 and 3 are -1 and -0.8, respectively. Member 1 has zero deformation because in the linear theory the deformation is evaluated from the orthogonal projection of the displaced joints on the initial member axis. The remaining members are not incident on joint 1 and are not affected. The first column of $\mathbf{a}_v$ is thus

$$\{0 \quad -1 \quad -0.8 \quad 0 \quad 0 \quad 0 \quad 0 \quad 0\}$$

2.5. Support Members

The statical unknowns in the governing set (2.4–8) consist of the elements of $\mathbf{V}$ and of the joint forces $\mathbf{R}_s$ corresponding to prescribed displacements $\mathbf{r}_s$ at support joints. A device for representing all the statical unknowns in $\mathbf{V}$ is to replace a support by a member connected to the support joint in the same manner as the actual support and supported at the other end. The prescribed displacements $\mathbf{r}_s$ are included in $\mathbf{r}$ in order that their degrees of freedom generate the corresponding equilibrium equations. Support members are given zero flexibilities so that $\mathbf{r}_s$ is enforced to be zero. Nonzero prescribed displacements and elastic supports may also be modeled by this device.

Example 1

The pin and roller supports of the pin-jointed truss in Fig. 2.5–1*a* may be modeled as shown in (*b*). Because two-force members are typical of truss

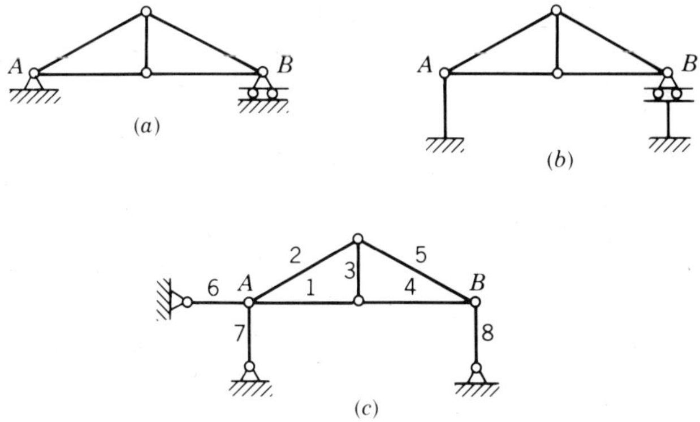

FIG. 2.5–1. Support members, Example 1.

structures, the support member at A may be replaced by two support links as in (c). The support member at B in (b) is exactly equivalent to the roller in (a). The modeling in (c) is also valid in the geometrically linear theory because the effect of displacements on the orientation of forces and the lengthening of the support link due to a horizontal motion at B are both absent from the linear theory.

If the three support links in (c) are numbered 6, 7, and 8, the corresponding connectivity equations are $v_6 = r_1^A$, $v_7 = r_2^A$, and $v_8 = r_2^B$. These three equations add three rows to $\mathbf{a}_v$ which becomes an 8×8 matrix. The joint equilibrium equations in this case are statically determinate. In the solution for $\mathbf{V}$ due to a given loading condition, a positive V_6 represents tension and thus a reaction force at A acting to the left. V_7 and V_8 are interpreted similarly.

Support members for the frame of Fig. 2.5–2a are shown in (b) as AA' and BB'. Joint A is rigid and fixed, and joint B is rigid and pinned to the support. These types of connections are maintained for the support members.

Because support members have zero flexibilities, the formulation that uses flexibility equations (2.4–8c) rather than stiffness equations is the theoretically appropriate one. If stiffness equations are used, zero flexibilities can only be simulated with large stiffness coefficients, and numerical problems that this may cause have to be avoided.

The device of support members may be readily used for representing prescribed nonzero displacements and for representing elastic support conditions. Prescribed support displacements are initial deformations $\mathbf{v}_0$ of the

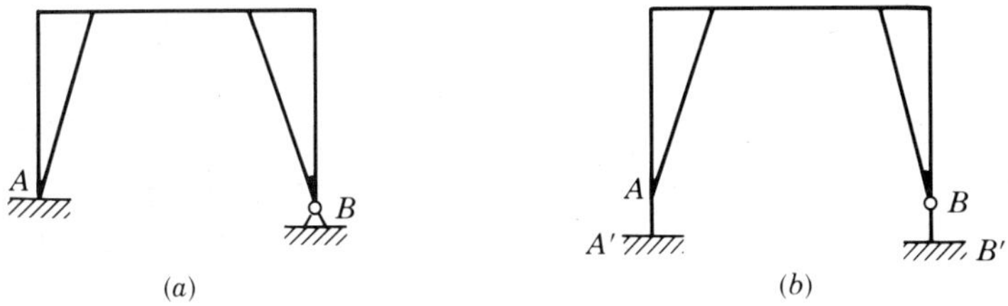

FIG. 2.5–2. Support members, Example 1.

support member whose constitutive equation in the set (2.4–8c) is $\mathbf{v} = \mathbf{v}_0$. Elastic support conditions are modeled into a flexibility or a stiffness matrix for the support member.

3. CHOICE OF EFFECTIVE VARIABLES

The property that a governing set of equations may be formulated with effective variables allows more efficiency in treating certain problems and more flexibility in applications, while remaining within the same theoretical framework. The main advantage is a reduced number of unknowns, but other considerations may make it more appropriate not to bring this number to the minimum possible.

The topic to be addressed here is the process by which a choice of effective variables is made. The process depends on whether the possibility arises from kinematic constraints or from kinematic releases. For kinematic releases the process starts at the member ends. If a released displacement is deleted from member variables, it is also deleted from the joint displacements. A released displacement also allows us to delete the released deformation. What remains to be done is to obtain member constitutive equations that take into account the kinematic release involved and the span loading condition. This is treated in Chapter 7.

Kinematic constraints may be of two types. A simple constraint prescribes a joint displacement and thereby some member end displacements. Here the process of choosing effective variables starts with the joint. A prescribed joint displacement may always be deleted from $\mathbf{r}$. The effect of this is to delete a corresponding equilibrium equation involving the reaction to the prescribed displacement. If the prescribed joint displacement is removed, the member displacements prescribed by that same condition usually are also removed, although this is not required for consistency. In either case what remains to be done is to obtain the corresponding constitutive equations. Note that if nonzero prescribed displacements are removed, their effect remains in the effective constitutive equations.

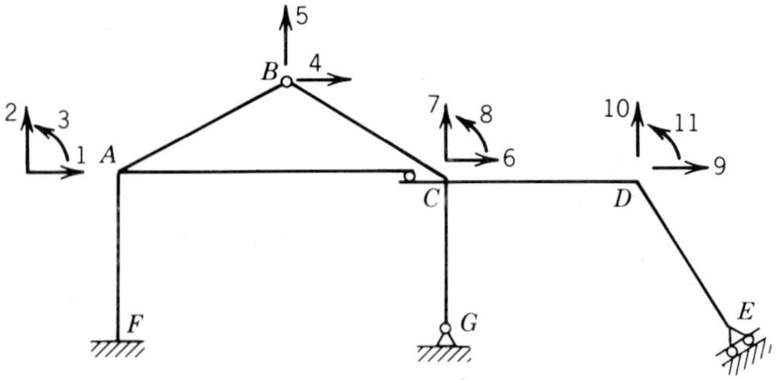

FIG. 3–1. Effective joint displacements, Example 1.

Example 1

Refer to Fig. 7/3.2–1 and to the choice of effective variables shown in Fig. 7/3.2–2. A corresponding choice of 11 joint displacements is shown in Fig. 3–1. If released displacements were kept, there would be two additional displacements at each of joints *B*, *C*, and *E*, and one rotation at *G*, for a total of 7. If it is also desired to include the reactions at supports in the set of unknowns, the complete set of 24 joint displacements is obtained.

The other type of kinematic constraint, which may be called complex, takes the form of a constraint equation on the joint displacements. This type of constraint is treated subsequently within the context of transformations and is applied to the formulation of inextensional analysis of frames.

4. TRANSFORMATION OF GOVERNING EQUATIONS: INEXTENSIBLE FRAMES

The concept of static-kinematic pair introduced for the member in Chapter 7, Section 1, applies here to the set of members and to the additional static-kinematic pair $(\mathbf{R}, \mathbf{r})$. The material presented in Chapter 7, Section 3, on contragradient and congruent transformations due either to kinematic or to static transformations remains also applicable.

Any one of the governing set of equations derived in Section 2 may be considered as the result of a kinematic transformation performed on the set of unassembled members. For example, the transformation $\mathbf{s} = \mathbf{ar}$ assembles kinematically the member ends. The contragradient law yields the joint equilibrium equation $\mathbf{a}^T\mathbf{S} = \mathbf{R}$, and the congruent law yields the structure stiffness, or assembled stiffness matrix, $\mathbf{K} = \mathbf{a}^T\mathbf{k}_c\mathbf{a}$.

The three governing sets of equations may be transformed one into another by a static or by a kinematic transformation. In each set the connectivity and equilibrium equations form a contragradient pair, and this property is maintained under kinematic or static transformation. For example, the static transformation $\bar{\mathbf{S}} = \boldsymbol{\lambda}^T\mathbf{S}$ transforms the set in global coordinates into the set in local coordinates. The kinematic transformation $\mathbf{v} = \mathbf{B}^T\mathbf{s}$ transforms the set in bound axes into the set in local coordinates.

The choice of joint displacements to describe the degrees of freedom of a structure is generalized by means of a linear kinematic transformation of the form

$$\mathbf{r} = \mathbf{T}\mathbf{q} \tag{4–1a}$$

in which $\mathbf{q}$ is a set of independent generalized displacements. Generalized external forces are defined by virtual work equivalence which yields the contragradient law. Thus $\mathbf{R}$ transforms into

$$\mathbf{Q} = \mathbf{T}^T\mathbf{R} \tag{4–1b}$$

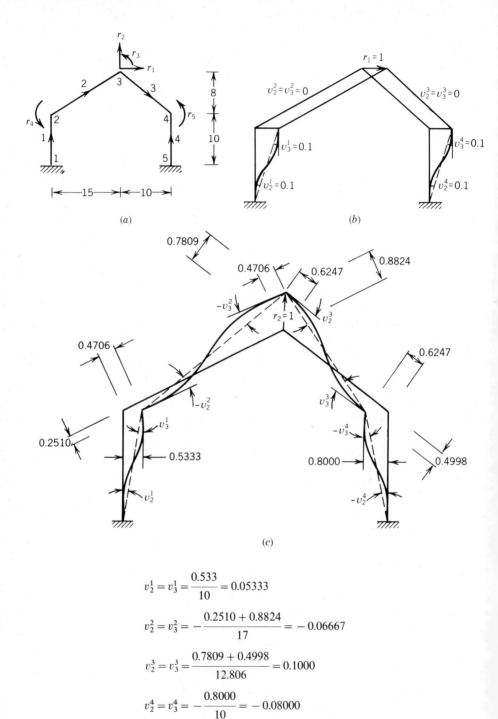

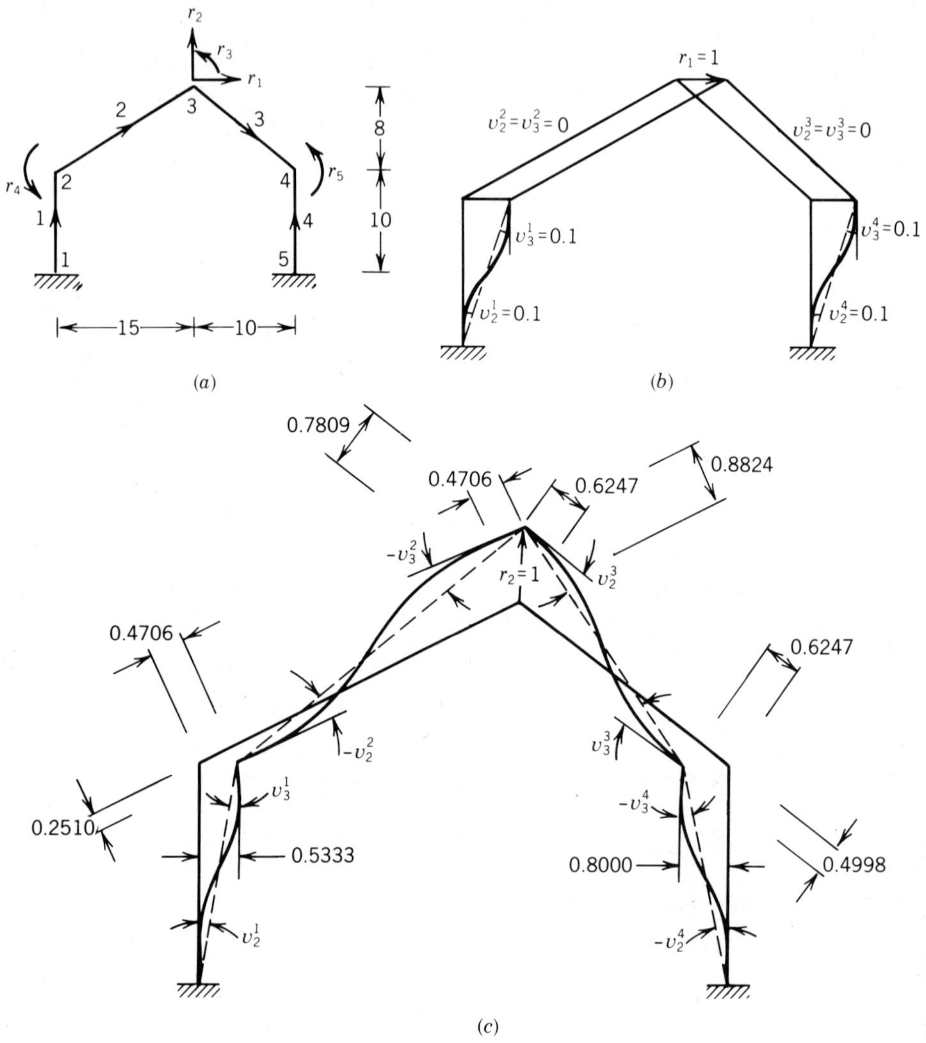

$$v_2^1 = v_3^1 = \frac{0.533}{10} = 0.05333$$

$$v_2^2 = v_3^2 = -\frac{0.2510 + 0.8824}{17} = -0.06667$$

$$v_2^3 = v_3^3 = \frac{0.7809 + 0.4998}{12.806} = 0.1000$$

$$v_2^4 = v_3^4 = -\frac{0.8000}{10} = -0.08000$$

FIG. 4–1. Inextensional deformation, Example 1*a*.

Governing equations in terms of $\mathbf{q}$ and $\mathbf{Q}$ are similar to those derived in terms of $\mathbf{r}$ and $\mathbf{R}$. For example, the set in local coordinates becomes $\mathbf{s} = \boldsymbol{\alpha}\mathbf{q}$, $\boldsymbol{\alpha}^T\mathbf{S} = \mathbf{Q}$, and there is no change in the constitutive equations. The transformed connectivity matrix $\boldsymbol{\alpha}$ is

$$\boldsymbol{\alpha} = \mathbf{aT} \tag{4–2}$$

Connectivity matrices $\bar{\boldsymbol{\alpha}}$ and $\boldsymbol{\alpha}_v$ are obtained similarly for $\bar{\mathbf{s}}$ and $\mathbf{v}$. The transformed structure stiffness matrix is obtained by the congruent law into any one of the three equivalent forms

$$\mathbf{K}_q = \boldsymbol{\alpha}^T\mathbf{k}_c\boldsymbol{\alpha} = \bar{\boldsymbol{\alpha}}^T\bar{\mathbf{k}}_c\bar{\boldsymbol{\alpha}} = \boldsymbol{\alpha}_v^T\mathbf{k}\boldsymbol{\alpha}_v \tag{4–3}$$

The concept of transformation has a wide range of applications. One application to be seen now is the treatment of inextensible frames.

Inextensible Frames

An approximate analysis of rigid frames is based on assuming that certain or all members are inextensible. The approximation is in turn based on the property that the axial stiffness is much larger than a transverse stiffness. A detailed analytical study of this topic is made in Chapters 9 and 10. Examples are treated here that also illustrate the concept of transformation.

Example 1a

If the four members of the frame in Fig. 4–1*a* are constrained to be inextensible, there are four independent conditions, of the form $v_1^m = 0$, that reduce the number of independent joint displacements from nine to five. These include necessarily the joint rotations. The remaining two represent translational displacement modes and may be chosen in various ways. One such choice is shown in (*b*) and (*c*).

The condition $v_1^m = 0$ requires that the orthogonal projection of the deformed member on the initial member axis has the same original length. Thus joints 2 and 4 can move only horizontally. In (*b*) r_1 represents a translation of the two top members. In (*c*) r_2 is accompanied by horizontal displacements at joints 2 and 4 whose values must satisfy the inextensibility condition and which are thus determined in terms of r_2. The transformation from the unconstrained joint displacements to the inextensional ones is read off the figure and has the form

$$\begin{Bmatrix} r_1^2 \\ r_2^2 \\ r_3^2 \\ r_1^3 \\ r_2^3 \\ r_3^3 \\ r_1^4 \\ r_2^4 \\ r_3^4 \end{Bmatrix} = \begin{bmatrix} 1 & 0.5333 & 0 & 0 & 0 \\ 0 & 0 & 0 & 0 & 0 \\ 0 & 0 & 0 & 1 & 0 \\ 1 & 0 & 0 & 0 & 0 \\ 0 & 1 & 0 & 0 & 0 \\ 0 & 0 & 1 & 0 & 0 \\ 1 & -0.8 & 0 & 0 & 0 \\ 0 & 0 & 0 & 0 & 0 \\ 0 & 0 & 0 & 0 & 1 \end{bmatrix} \begin{Bmatrix} r_1 \\ r_2 \\ r_3 \\ r_4 \\ r_5 \end{Bmatrix} \tag{4–4}$$

Each column of this transformation matrix represents a displacement mode of the structure due to a unit value of the corresponding displacement.

If the transformation (4–4) is written $\mathbf{r} = \mathbf{T}\mathbf{q}$, the generalized joint loads are $\mathbf{Q} = \mathbf{T}^T\mathbf{R}$. $\mathbf{Q}$ is a set of Lagrangian generalized forces: element Q_i of $\mathbf{Q}$ is equal to the virtual work of the actual joint loads $\mathbf{R}$ through the displacement mode $q_i = 1$, that is, through the ith column of $\mathbf{T}$. Thus Q_3, Q_4, and Q_5 coincide with the joint moment loads at joints 3, 2, and 4, respectively. Q_1 and Q_2 are equal, respectively, to the virtual work of $\mathbf{R}$ through the joint displacements of Fig. 4–1*b* and 4–1*c*. For example, a joint load consisting of the two forces R_1^2 and R_1^4 transforms into $Q_1 = R_1^2 + R_1^4$ and $Q_2 = 0.5333\ R_1^2 - 0.8\ R_1^4$.

The governing equations in bound axes have the form

$$\mathbf{v} = \boldsymbol{\alpha}_v \mathbf{q} \qquad (4\text{–}5a)$$

$$\boldsymbol{\alpha}_v^T \mathbf{V} = \mathbf{Q} - \mathbf{Q}_p \qquad (4\text{–}5b)$$

$$\mathbf{v} = \mathbf{f}\mathbf{V} + \mathbf{v}_p \qquad (4\text{–}5c)$$

or

$$\mathbf{V} = \mathbf{k}\mathbf{v} + \mathbf{V}_F \qquad (4\text{–}5d)$$

$\boldsymbol{\alpha}_v$ is the transform of $\mathbf{a}_v$ and is $\mathbf{a}_v\mathbf{T}$. Equation (4–5b) is the generalized joint equilibrium equation. $\mathbf{Q}_p$ is the transform of $\mathbf{R}_p$ in the same manner as $\mathbf{Q}$ is the transform of $\mathbf{R}$.

Each generalized equilibrium equation corresponds to a degree of freedom represented by a virtual displacement δq_i. The following procedure helps to bring out the meaning of these equations. A free body diagram of joints 2, 3, and 4 is made. The external joint forces are $\mathbf{R}$, and the internal joint forces, which are the member end forces reversed in sense, are expressed in terms of the member reduced forces $\mathbf{V}$ and of the forces $\mathbf{S}_p$ of the particular equilibrium state. The virtual work of these forces through the displacement mode $\delta q_1 = 1$, Fig. 4–1b, is set to zero to obtain the first generalized equilibrium equation. The second equation is obtained similarly through the displacement mode $\delta q_2 = 1$, Fig. 4–1*c*. The remaining equations coincide with the original moment equilibrium equations.

Since virtual displacements are inextensional, the axial forces in $\mathbf{V}$ do no virtual work and are absent from the equations. This is the basis for forming effective $\mathbf{V}$ and $\mathbf{v}$ that exclude the axial components and corresponding effective $\mathbf{f}$ and $\mathbf{k}$.

To illustrate this point, consider the equivalent joint loads ($-\mathbf{R}_p$) shown in Figs. 2.4–3 and 2.4–4. These two sets of loads are obtained on a different basis. In the first the particular equilibrium state is chosen with each member simply supported. In the second the members are pinned. The two sets differ only by pairs of equal and opposite axial forces for each loaded member. It should be verified that both sets transform into the same generalized joint load ($-\mathbf{Q}_p$).

Note that the generalized equilibrium equations are exact but are only a subset of the total number of equations that would be obtained with a complete set of degrees of freedom.

The connectivity matrix $\boldsymbol{\alpha}_v$ may be formed by direct kinematic analysis, as shown in Fig. 4–1. The effective $\mathbf{v}^m = \{v_2^m \ v_3^m\}$ is chosen to be of the alternate type. v_2^m and v_3^m are end rotations relative to the displaced member chord and are positive if counterclockwise. Thus a clockwise chord rotation produces positive deformations. We thus obtain

$$\begin{Bmatrix} v_2^1 \\ v_3^1 \\ v_2^2 \\ v_3^2 \\ v_2^3 \\ v_3^3 \\ v_2^4 \\ v_3^4 \end{Bmatrix} = \begin{bmatrix} 0.1 & 0.05333 & 0 & 0 & 0 \\ 0.1 & 0.05333 & 0 & 1 & 0 \\ 0 & -0.06667 & 0 & 1 & 0 \\ 0 & -0.06667 & 1 & 0 & 0 \\ 0 & 0.1 & 1 & 0 & 0 \\ 0 & 0.1 & 0 & 0 & 1 \\ 0.1 & -0.08 & 0 & 0 & 0 \\ 0.1 & -0.08 & 0 & 0 & 1 \end{bmatrix} \begin{Bmatrix} r_1 \\ r_2 \\ r_3 \\ r_4 \\ r_5 \end{Bmatrix} \tag{4–6}$$

The transformed stiffness matrix of the structure is obtained as the last expression in Eq. (4–3), in which $\mathbf{k} = \lceil \mathbf{k}^1 \ \mathbf{k}^2 \ \mathbf{k}^3 \ \mathbf{k}^4 \rfloor$, and, for uniform members without shear deformation,

$$\mathbf{k}^m = \frac{EI}{l}\begin{bmatrix} 4 & 2 \\ 2 & 4 \end{bmatrix}$$

A direct formulation, as done here, is convenient for hand methods and small-size problems. A more systematic kinematic analysis of deformations is presented next.

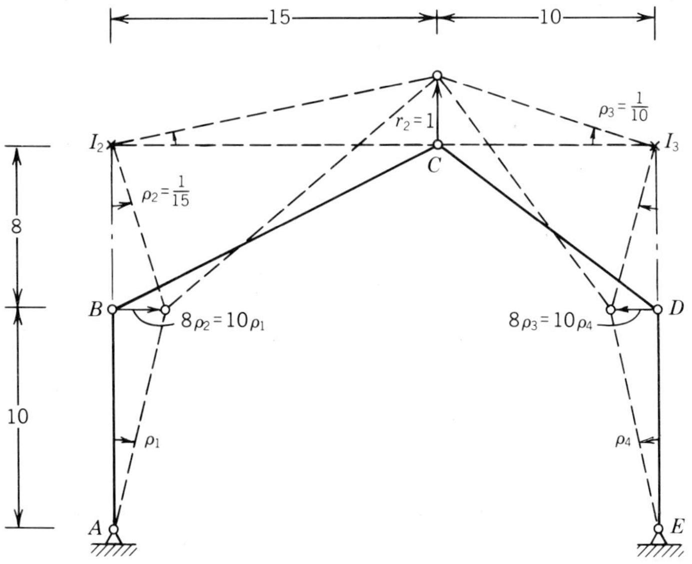

FIG. 4–2. Kinematics of inextensional displacements, Example 1b.

Example 1b

The rotational deformations of an inextensible member due to translational displacements are equal to the rotation ρ of the member chord but have the opposite sense. The geometry of the member chords is that of a pin-connected mechanism. In a mechanism motion each member has an instantaneous center of rotation located at the intersection of the perpendiculars to the displacements at the member's ends. For the structure of Fig. 4–1 the mechanism mode corresponding to r_2 is shown in Fig. 4–2, with the instantaneous centers of members BC and CD labeled I_2 and I_3, respectively. The displacement at joint B is due both to the rotation ρ_1 about A and to the rotation ρ_2 about I_2. Thus $8\rho_2 = 10\rho_1$, as indicated in the figure. A similar equation relates ρ_3 and ρ_4 at D. The rotations ρ_2 about I_2 and ρ_3 about I_3 are both determined from the unit displacement $r_2 = 1$ at C, as shown. There comes

$$(\rho_1, \rho_2, \rho_3, \rho_4) = (0.05333,\ 0.06667,\ 0.1,\ 0.08)$$

These numbers define the second column of $\boldsymbol{\alpha}_v$ in Eq. (4–6), noting that deformations are positive if ρ is clockwise.

Example 2

The rectangular frame of Fig. 4–3 has 60 unrestrained joints and thus 180 joint displacements. If all 110 members are assumed inextensible, the number of independent displacements is reduced to $180 - 110 = 70$. Sixty of these are identified as joint rotations, and the remaining 10 as floor translations. The joints in a given floor are thus constrained to move only horizontally and together as a rigid body.

The problem may be formulated partially by means of effective variables by excluding from $\mathbf{r}$ the zero joint displacements whose number is 60. The remaining horizontal displacements, grouped six to a floor, are set equal to the corresponding floor translation. Connectivity equations in terms of the new set of

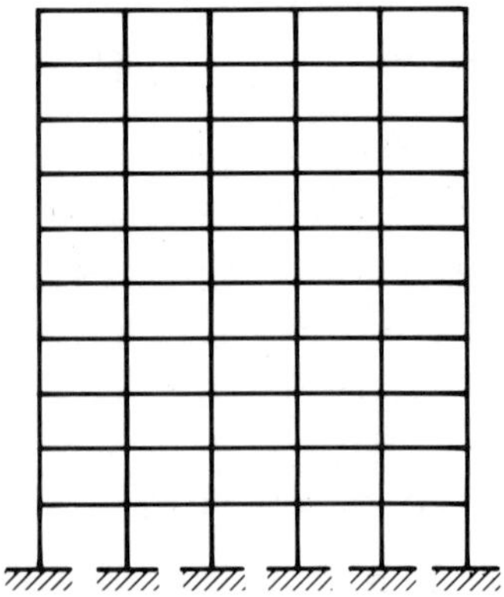

FIG. 4–3. Rectangular frame, Example 2.

displacements may be written directly without the need for a formal transformation.

For tall rectangular frames significant errors may occur if columns are assumed to be inextensible. In that case only the horizontal displacements are constrained.

5. GOVERNING EQUATIONS OF GEOMETRICALLY NONLINEAR THEORY

5.1. Introduction and Notation

The governing equations to be developed in this section are based on the second-order theory developed in Chapters 3 and 6 for the structural member. The reader is referred to the introduction to Chapter 6 for a summary statement about the need for geometrically nonlinear analysis and about the mathematical nature of this nonlinearity in second-order theory. In Section 6 a more general geometric nonlinearity is examined.

The notation to be adopted for member-set variables is the one adopted in Chapter 6 for member variables. Axial quantities are separated from transverse quantities and are referred to by subscripting the general matrix symbol with 1. The unsubscripted symbol refers to the transverse quantities. Thus member end displacements in local coordinates are $\{\mathbf{s}_1 \ \mathbf{s}\}$, where $\mathbf{s}_1$ is the set of axial member end displacements and $\mathbf{s}$ is the set of transverse displacements (and rotations). $\mathbf{s}_1$ and $\mathbf{s}$ are partitioned by members. For member m with ends a and b, $\mathbf{s}_1^m = \{s_1^a \ s_1^b\}$, and $\mathbf{s}^m = \{s_2^a \ s_3^a \ s_2^b \ s_3^b\}$. The notation for $\{\mathbf{S}_1 \ \mathbf{S}\}$, $\{\mathbf{v}_1 \ \mathbf{v}\}$, and $\{\mathbf{V}_1 \ \mathbf{V}\}$ is similar. If distinction is needed between two member ends at the same joint, the notation $s_1^{m,a}$ and $\mathbf{s}^{m,a} = \{s_2^{m,a} \ s_3^{m,a}\}$ is used.

The formulations to follow are made in terms of effective variables and equations. Member effective equations in the geometrically nonlinear theory are treated in Chapter 7.

5.2. Governing Equations in Fixed Axes: Second-Order Theory

In order to avoid an initial difficulty, joints are assumed idealized into points, and releases at member ends are limited to rotational ones. The connectivity equations and the joint equilibrium equations in either local or global coordinates are the same as in the linear theory. Geometric nonlinearity occurs only in the member constitutive equations. With the notation described in Section 5.1, the governing equations in local coordinates are partitioned in the form

$$\begin{Bmatrix} \mathbf{s}_1 \\ \mathbf{s} \end{Bmatrix} = \begin{bmatrix} \mathbf{a}_1 \\ \mathbf{a} \end{bmatrix} \mathbf{r} \tag{5.2–1a}$$

$$\mathbf{a}_1^T \mathbf{S}_1 + \mathbf{a}^T \mathbf{S} = \mathbf{R} \tag{5.2–1b}$$

$$\begin{Bmatrix} \mathbf{S}_1 \\ \mathbf{S} \end{Bmatrix} = \begin{bmatrix} \mathbf{k}_{11} & \\ & \mathbf{k}_s \end{bmatrix} \begin{Bmatrix} \mathbf{s}_1 \\ \mathbf{s} \end{Bmatrix} + \begin{Bmatrix} \mathbf{S}_{1F} \\ \mathbf{S}_F \end{Bmatrix} + \begin{Bmatrix} \mathbf{S}_{1N} \\ 0 \end{Bmatrix} \tag{5.2–1c}$$

where $\mathbf{S}_{1N}$ is nonlinear in the displacements and is formed by a column ordering of member forces $\mathbf{S}_{1N}^m$ as shown next:

$$\mathbf{S}_{1N} = \{\mathbf{S}_{1N}^m\} = \{\tfrac{1}{2}\ k_{11}^m\ \mathbf{B}_1^m \mathbf{s}^{mT} \mathbf{g}_s^m \mathbf{s}^m\} \tag{5.2–2}$$

Forming of the member stiffness equations is described in Chapter 6, Sections 2.6 and 3.1. A summary overview may be appropriate.

The term k_{11}^m is the member axial stiffness in the linear theory to which correspond the 2×2 axial stiffness matrix $\mathbf{k}_{11}^m$. The 4×4 bending stiffness matrix $\mathbf{k}_s^m$ depends on the axial force. $\mathbf{S}_{1F}^m$ and $\mathbf{S}_F^m$ are fixed end forces; that is, they are the end forces caused by the member loading condition in the fixed state $\mathbf{s}_1^m = 0$, $\mathbf{s}^m = 0$.

Exact formulas for $\mathbf{k}_s^m$ are available for a uniform member and a constant axial force. The corresponding exact $\mathbf{S}_F^m$ depends on the axial force, and the exact $\mathbf{S}_{1F}^m$ depends on the transverse load.

For general members, with axial span loads, $\mathbf{k}_s^m$ is established by approximate methods. If the variation in the axial force due to the span load is small, a constant axial force is a good approximation. In the simplest approximate methods, $\mathbf{k}_s^m$ is obtained by adding a geometric stiffness matrix to the matrix of the linear theory, and $\mathbf{S}_{1F}$ and $\mathbf{S}_F$ are the same as in the linear theory. The accuracy of an approximate $\mathbf{k}_s^m$ decreases for an increasing compressive axial force. The eigenvalues of $\mathbf{k}_s^m$ are approximations to critical values of the axial force corresponding to various homogeneous end conditions. The largest of these values, which is also an acceptable approximation, provides a range of axial force within which $\mathbf{k}_s^m$ is an acceptable approximation.

A basic method of solution of Eqs. (5.2–1) to be described in Chapter 12 deals with incremental equations. Since Eqs. (5.2–1a, b) are linear, their coefficient matrices remain the same in the incremental equations. The incremental stiffness equation is obtained in Eq. (6/4–2), assuming $\mathbf{k}_s$ is based on shape functions independent of the axial force. The argument may then be made that the exact $\mathbf{k}_s$ may also be used in Eq. (6/4–2) by neglecting the effect of variation in axial force on the exact shape functions. It will be seen that this latter approximation does not prevent obtaining an exact solution as a limit of an iterative process known as the Newton–Raphson method.

5.3. Governing Equations in Member-Bound Axes: Second-Order Theory

To derive governing equations in terms of member deformations $\{\mathbf{v}_1\ \ \mathbf{v}\}$ and member reduced forces $\{\mathbf{V}_1\ \ \mathbf{V}\}$, we start with the member static and kinematic equations (6/2.6–1) and (6/2.6–2) and the member stiffness equations (6/3.1–22). Written for the set of members, these equations take the form

$$\mathbf{S}_1 = \mathbf{B}_1 \mathbf{V}_1 \tag{5.3–1a}$$

$$\mathbf{S} = \mathbf{B}\mathbf{V} + \mathbf{k}_G \mathbf{s} + \mathbf{S}_p \tag{5.3–1b}$$

$$\mathbf{v}_1 = \mathbf{B}_1^T \mathbf{s}_1 + \mathbf{v}_{1N} \tag{5.3–2a}$$

$$\mathbf{v} = \mathbf{B}^T \mathbf{s} \tag{5.3–2b}$$

and

$$\mathbf{V}_1 = \lceil k_{11}^m \rfloor \mathbf{v}_1 + \mathbf{V}_{1F} + \mathbf{V}_{1N} \tag{5.3–3a}$$

$$\mathbf{V} = \mathbf{k}\mathbf{v} + \mathbf{V}_F \tag{5.3–3b}$$

where

$$\mathbf{k}_G = \lceil \mathbf{k}_G^m \rfloor = \lceil V_1^m \mathbf{c}_1^m \rfloor \tag{5.3–4a}$$

$$\mathbf{v}_{1N} = \{ v_{1N}^m \} = \{ \tfrac{1}{2}\, \mathbf{s}^{mT} \mathbf{c}_1^m \mathbf{s}^m \} \tag{5.3–4b}$$

$$\mathbf{V}_{1N} = \{ V_{1N}^m \} = \{ \tfrac{1}{2}\, k_{11}^m \mathbf{v}^{mT} \mathbf{g}^m \mathbf{v}^m \} \tag{5.3–4c}$$

$\mathbf{k}_G$ is formed of a diagonal ordering of the member string geometric stiffness. $\mathbf{v}_{1N}$ and $\mathbf{V}_{1N}$ are formed of a column ordering of nonlinear terms, as indicated.

If there are member axial span loads, the terms $\mathbf{S}_p$ and $\mathbf{V}_F$ hide displacement-dependent terms that have been made explicit in Chapter 6, Section 3.3. The result is

$$\mathbf{S}_p + \mathbf{B}\mathbf{V}_F = \mathbf{S}_{pL} + \mathbf{B}\mathbf{V}_F' + \mathbf{k}_{GF}\mathbf{s} = \mathbf{S}_F + \mathbf{k}_{GF}\mathbf{s} \tag{5.3–5}$$

where $\mathbf{k}_{GF}$ is a symmetric geometric stiffness associated with the rotation of the bound reference and the axial span load, Eq. (6/3.3–9). It is of the same type as the string geometric stiffness $\mathbf{k}_G$.

We are thus lead to the change of variable

$$\mathbf{V} = \mathbf{V}' - \mathbf{V}_{Fx} \mathbf{B}_p^T \mathbf{s} \tag{5.3–6}$$

where the last term is the portion of $\mathbf{V}_F$ due to the axial span load. Governing equations in terms of $\mathbf{v}_1$, $\mathbf{v}$, $\mathbf{V}_1$, and $\mathbf{V}'$ are now formulated. Equations (5.3–2) become structure connectivity equations on substituting $\mathbf{s}_1 = \mathbf{a}_1\mathbf{r}$ and $\mathbf{s} = \mathbf{ar}$. To derive joint equilibrium equations, we may use the principle of virtual displacements in the form

$$\mathbf{V}_1^T \delta\mathbf{v}_1 + \mathbf{V}^T \delta\mathbf{v} + \mathbf{S}_p^T \delta\mathbf{s} = \mathbf{R}^T \delta\mathbf{r} \tag{5.3–7}$$

or, equivalently, substitute Eqs. (5.3–1) into the joint equilibrium equations derived earlier in terms of $\mathbf{S}_1$ and $\mathbf{S}$. The two procedures are equivalent because the expression $(\mathbf{S}_1^T \delta\mathbf{s}_1 + \mathbf{S}^T \delta\mathbf{s})$ turns into the left-hand side of Eq. (5.3–7) on use of the member equilibrium solution (5.3–1). With the change of variable from $\mathbf{V}$ to $\mathbf{V}'$, we obtain

$$\begin{Bmatrix} \mathbf{v}_1 \\ \mathbf{v} \end{Bmatrix} = \begin{bmatrix} \mathbf{a}_{1v} \\ \mathbf{a}_v \end{bmatrix} \mathbf{r} + \begin{Bmatrix} \mathbf{v}_{1N} \\ \mathbf{0} \end{Bmatrix} \tag{5.3–8a}$$

$$\mathbf{a}_{1v}^T \mathbf{V}_1 + \mathbf{a}_v^T \mathbf{V}' + (\mathbf{K}_G + \mathbf{K}_{GF})\mathbf{r} = \mathbf{R} - \mathbf{a}_1^T \mathbf{S}_{1p} - \mathbf{a}^T \mathbf{S}_{pL} \tag{5.3–8b}$$

$$\begin{Bmatrix} \mathbf{V}_1 \\ \mathbf{V}' \end{Bmatrix} = \begin{bmatrix} \lceil k_{11}^m \rfloor & \\ & \mathbf{k} \end{bmatrix} \begin{Bmatrix} \mathbf{v}_1 \\ \mathbf{v} \end{Bmatrix} + \begin{Bmatrix} \mathbf{V}_{1F} \\ \mathbf{V}_F' \end{Bmatrix} + \begin{Bmatrix} \mathbf{V}_{1N} \\ 0 \end{Bmatrix} \tag{5.3–8c}$$

where

$$\mathbf{K}_G = \mathbf{a}^T \mathbf{k}_G \mathbf{a} \tag{5.3–9a}$$

$$\mathbf{K}_{GF} = \mathbf{a}^T \mathbf{k}_{GF} \mathbf{a} \tag{5.3–9b}$$

By comparison with the governing equations in the local reference, the present set separates geometric stiffnesses into two parts. One part is relative to the bound axes and affects **k**. The other part is due to the rigid body rotation of the bound reference and is represented by $\mathbf{K}_G$ and $\mathbf{K}_{GF}$. This separation is used in incremental equations of a general nonlinear theory in the next section.

As in the linear theory, the present set of equations may be used as the starting point of both stiffness and flexibility methods of analysis. It will be applied in Chapter 12.

6. GOVERNING EQUATIONS WITH LARGE ROTATIONS

In this section equations are developed in which members are governed by the second-order theory in their respective bound references but in which these references may undergo arbitrary rigid body motions. With such governing equations are associated incremental equations needed in methods of solution.

Equations needed in what follows are the exact static-kinematic equations of Chapter 2. We revert here to the unpartitioned notation except where needed. Deformations **v** are defined as functions **v**(**s**) of **s**. Incremental deformations are

$$d\mathbf{v} = \mathbf{B}^T d\mathbf{s} \tag{6–1a}$$

where

$$\mathbf{B} = \frac{\partial \mathbf{v}}{\partial \mathbf{s}^T} \tag{6–1b}$$

Member equilibrium equations are solved in the form

$$\mathbf{S} = \mathbf{B}\mathbf{V} + \mathbf{S}_p \tag{6–2}$$

Equations (6–1, 2) may be written in terms of $d\bar{\mathbf{s}}$ and $\bar{\mathbf{S}}$ by the coordinate transformation $\mathbf{s} = \boldsymbol{\lambda}\bar{\mathbf{s}}$, which yields

$$\bar{\mathbf{B}} = \boldsymbol{\lambda}^T \mathbf{B} = \frac{\partial \mathbf{v}}{\partial \bar{\mathbf{s}}^T} \tag{6–3}$$

Equations in Fixed Axes

Connectivity and joint equilibrium equations in local or global axes are the same as in the linear and second-order theories. Constitutive equations are obtained by substituting into Eq. (6–2) the stiffness relations for **V** and by expressing **v** in terms of **s**.

The incremental connectivity and joint equilibrium equations have the same coefficient matrices, $\mathbf{a}$ and $\mathbf{a}^T$, as in the finite equations, since $\mathbf{a}$ is constant. The incremental stiffness equations have been derived in Chapter 7, Sections 6.3 and 6.4.

Equations in Bound Axes

Connectivity equations in bound axes are obtained by substituting $\mathbf{s} = \mathbf{ar}$ into the expressions $\mathbf{v}(\mathbf{s})$. Joint equilibrium equations are obtained by the principle of virtual displacements, or by substituting from Eq. (6–2) into $\mathbf{a}^T\mathbf{S} = \mathbf{R}$. There comes

$$\mathbf{a}_v^T\mathbf{V} = \mathbf{R} - \mathbf{R}_p \tag{6–4}$$

where

$$\mathbf{a}_v^T = \mathbf{a}^T\mathbf{B} \tag{6–5}$$

and $\mathbf{a}_v$ is the incremental connectivity matrix for deformations; that is,

$$d\mathbf{v} = \mathbf{a}_v d\mathbf{r} \tag{6–6}$$

The governing set is completed either with the flexibility or with the stiffness equations for $\mathbf{v}$ and $\mathbf{V}$.

The incremental equations consist of Eq. (6–6), of incremental equilibrium equations, and incremental constitutive equations. A review of Chapter 7, Section 6.4, yields the incremental member equilibrium equation

$$d\mathbf{S} = \mathbf{B}d\mathbf{V} + \mathbf{k}_G d\mathbf{s} + d\mathbf{S}_{pG} + d\mathbf{S}_{po} \tag{6–7}$$

the incremental constitutive equation

$$d\mathbf{V} = \mathbf{k}_t d\mathbf{v} + d\mathbf{V}_{FG} + d\mathbf{V}_0 \tag{6–8}$$

and the geometric stiffness equation

$$d\mathbf{S}_{pG} + \mathbf{B}d\mathbf{V}_{FG} = \mathbf{k}_{GF} d\mathbf{s} \tag{6–9}$$

where $\mathbf{k}_{GF}$ is symmetric and depends on the member span load. This leads to the change of variable

$$d\mathbf{V} = d\mathbf{V}' + d\mathbf{V}_{FG} \tag{6–10}$$

with incremental stiffness equations for $d\mathbf{V}'$,

$$d\mathbf{V}' = \mathbf{k}_t d\mathbf{v} + d\mathbf{V}_0 \tag{6–11}$$

The incremental equilibrium equation, $\mathbf{a}^T d\mathbf{S} = d\mathbf{R}$, is expressed in terms of $d\mathbf{V}$ and $d\mathbf{r}$ by means of Eq.(6–7) and by substituting $d\mathbf{s} = \mathbf{a}d\mathbf{r}$. There comes

$$\mathbf{a}_v^T d\mathbf{V}' + (\mathbf{K}_G + \mathbf{K}_{GF})d\mathbf{r} = d\mathbf{R} - \mathbf{a}^T d\mathbf{S}_{p0} \tag{6–12}$$

where

$$\mathbf{K}_G = \mathbf{a}^T \mathbf{k}_G \mathbf{a} \tag{6–13a}$$

$$\mathbf{K}_{GF} = \mathbf{a}^T \mathbf{k}_{GF} \mathbf{a} \tag{6–13b}$$

The incremental governing equations are (6–6), (6–11), and (6–12). Equation (6–11) may be replaced by its inverse flexibility relation.

EXERCISES

Sections 1 to 3

1. Form the governing equations in global coordinates, Eqs. (2.2–7), for the following structures and data using the steps outlined in Section 2.2 before Example 1b.

a. For the pin-jointed truss in Fig. 9/1–2e assume the following conditions:

i. The horizontal members and the two vertical members at the supports have the same length a.

ii. All members have the same axial stiffness $k = EA/l$.

iii. A distributed vertical load of intensity p acts on the horizontal members, and a vertical force P acts at the right-most joint.

b. For the rigid frame in Fig. 10/3–1a assume the members are uniform with given lengths and stiffness properties and that a uniform transverse load p acts on the beams.

2. Form the governing equations in local coordinates, Eqs. (2.3–1), for the structures of Exercise 1.

3. From the governing equations in member-bound axes, Eqs. (2.4–8), for the structures of Exercise 1.

4. For the truss of Fig. 2.3–2 establish the joint equilibrium equations $\mathbf{a}_v^T\mathbf{V} = \mathbf{R}$ by means of free body diagrams of the joints. Check the result with the geometrically derived $\mathbf{a}_v$ in Fig. 2.4–6.

5. Consider the governing equations in member-bound axes, Eqs. (2.4–8), and assume that prescribed displacements are excluded from the effective joint displacements. Outline steps for solving the governing equations if $\mathbf{a}_v$ is square and nonsingular. Apply these steps to the truss of Fig. 2.2–5, assuming $\alpha = 30°$ and all members have the same axial flexibility l/EA.

6. Form the connectivity matrices $\bar{\mathbf{a}}$, $\mathbf{a}$, and $\mathbf{a}_v$ for the pin-jointed truss of Fig. P 6.

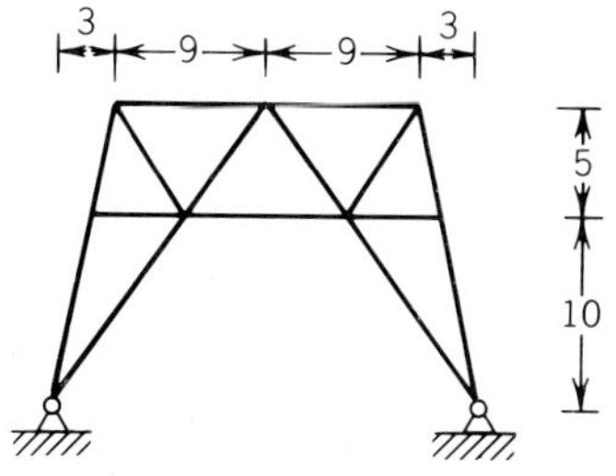

FIG. P6

7. For the structure and effective joint displacements shown in Fig. 3–1, outline how the member effective stiffness equations are formed from given member stiffness properties, lengths, and orientations.

8. Identify by means of sketches effective sets of joint displacements **r** and member end displacements **s** for the structures in Fig. 9/1–9. The numbers shown in these figures may be ignored. However, NR is a possible number of effective joint displacements. Assuming member length, orientation, and stiffness properties are given, outline how the effective member stiffness equations are formed.

Section 4

9. For the frames indicated next which are assumed to be inextensible, select independent translational displacements, and sketch the corresponding deformation modes:
   a. Frame in Fig. 10/3–1*a*.
   b. Frame in Fig. 9/1–9*h*.

10. Reduce the displacements shown in Fig. 3–1 to an independent set if all members except AC are assumed inextensible.

11. The independent translational displacements of the inextensible frame of Section 4, Example 1, Fig. 4–1*a*, may be chosen as the horizontal displacements at joints 2 and 4. Sketch the corresponding deformation modes, and form the connectivity equations for deformations of the alternate type. Determine the joint load matrix due to joint loads R_1^3 and R_2^3.

12. a. For the frames of Chapter 10, Fig. P3i, j assumed to be inextensible, select a set of generalized displacements; then form the connectivity matrix for member end displacements, and derive the structure stiffness equations. Use the least number of effective displacements.
    b. Repeat part (a) by forming the connectivity matrix for deformations of alternate type.

13. Do Exercise 12 for the frame having finite rigid joints shown in Fig. P13.

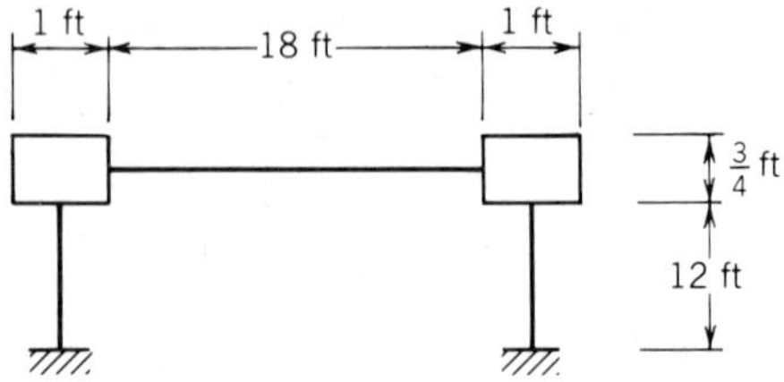

FIG. P.13

14. For any of the structures of Exercises 12 and 13 form the generalized joint equilibrium equations by making free body diagrams and applying the principle of virtual displacements. Ascertain that the static-kinematic property of the connectivity matrices is satisfied by the derived equilibrium equations.

Section 5

15. Form the joint equilibrium equations for the pin-jointed trusses of Fig. P15 within the context of the second-order theory. These equations should have the form of Eq. (5.3–8*b*) in which $(\mathbf{V}', \mathbf{K}_{GF}, \mathbf{S}_{1p}, \mathbf{S}_{pL}) = 0$.

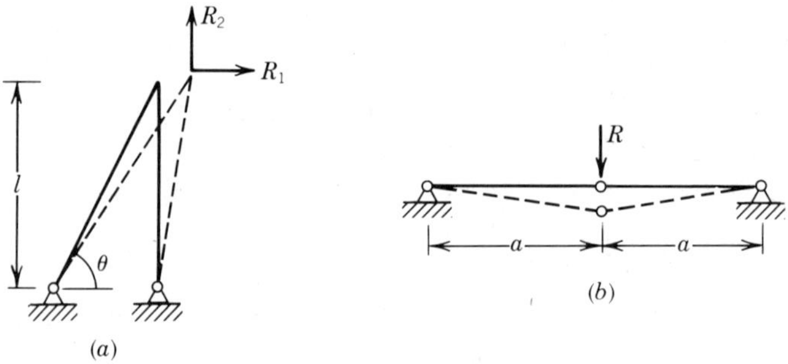

FIG. P15

16. The member in Fig. P16 is formed of two uniform segments and is treated as a two-member structure as shown. Form the joint equilibrium equations in the second-order theory.

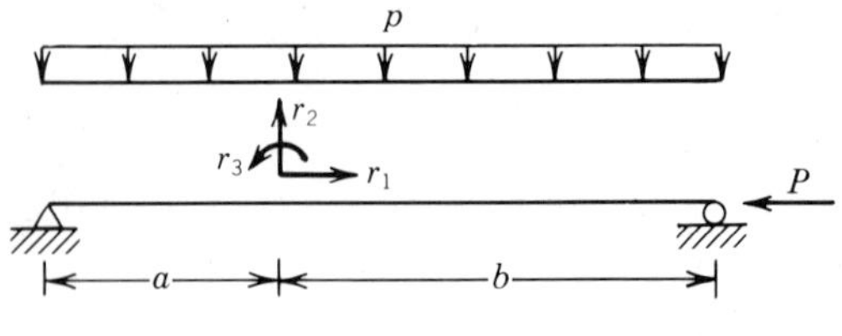

FIG. P16

17. Form the joint equilibrium equations (5.3–8b) for the frame in Fig. P17. Note that $\mathbf{V}' = \mathbf{V}$ and $(\mathbf{K}_{GF}, \mathbf{S}_{1p}, \mathbf{S}_{pL}) = 0$.

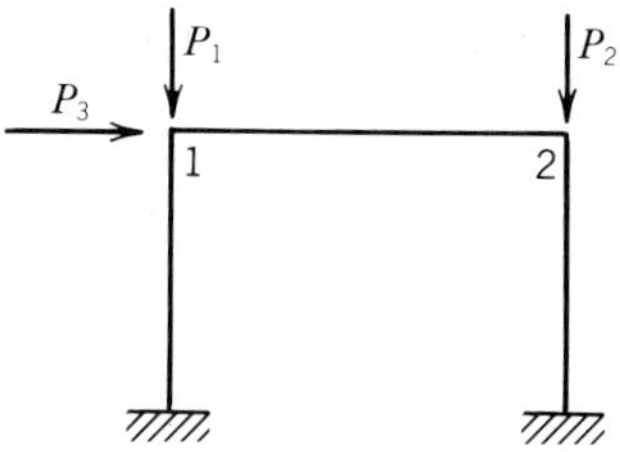

FIG. P17

18. For the frame of Exercise 17 perform a kinematic transformation from joint displacements $\mathbf{r} = \{\mathbf{r}^1 \mathbf{r}^2\}$ to $\mathbf{q} = \{\mathbf{q}_1 \mathbf{q}_2\}$, where $\mathbf{q}_1 = \{\frac{1}{2}(r_1^1 + r_1^2)\ r_3^1\ r_3^2\}$ and $\mathbf{q}_2 = \{(r_1^2 - r_1^1)\ r_2^1\ r_2^2\}$. Obtain the transformed equilibrium equations.

9

STATIC-KINEMATIC PROPERTIES OF STRUCTURES

1. KINEMATIC STABILITY

A structure is said to be kinematically or geometrically stable if rigid body or mechanism motion is impossible. Mathematically it is impossible to find displacements $\mathbf{r}$ for which all member deformations $\mathbf{v}$ vanish. In the geometrically linear theory, rigid body motion is linearized, and accordingly, rigid body rotation is considered to be infinitesimal. $\mathbf{v}$ is then related linearly to $\mathbf{r}$ through

$$\mathbf{v} = \mathbf{a}_v \mathbf{r} \tag{1-1}$$

A structure is kinematically stable in the geometrically linear theory, if the homogeneous system of equations $\mathbf{a}_v \mathbf{r} = 0$ has only the trivial solution $\mathbf{r} = 0$. From the theory of simultaneous equations, the rank ρ of $\mathbf{a}_v$ must be equal to the number of elements (NR) of $\mathbf{r}$, that is, to the number of degrees of freedom. The condition for kinematic stability is thus

$$\rho = (NR) \tag{1-2}$$

In a geometrically nonlinear theory an incremental kinematic stability is defined

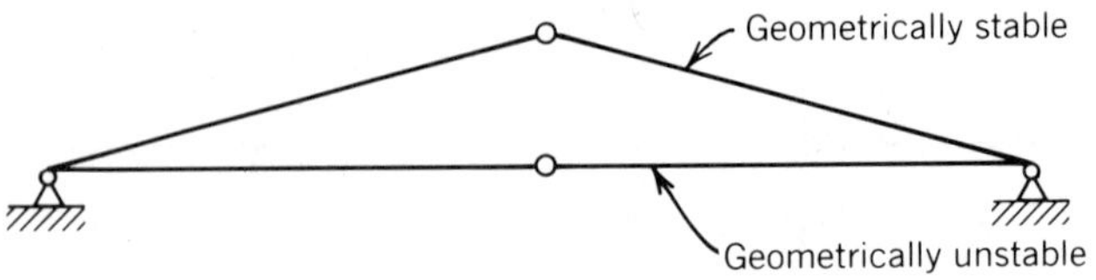

FIG. 1–1. Change in stability.

in terms of $\delta \mathbf{v}$ and $\delta \mathbf{r}$. The incremental connectivity relation is

$$\delta \mathbf{v} = \mathbf{a}_v \delta \mathbf{r} \tag{1–3}$$

where $\mathbf{a}_v$ depends on the deformed state $\mathbf{r}$. The condition for incremental stability is expressed in terms of the rank of $\mathbf{a}_v$ by Eq. (1–2). Because $\mathbf{a}_v$ depends on $\mathbf{r}$, it is possible for a structure to be incrementally unstable in state $\mathbf{r} = 0$ and to become stable as displacements take place. The reverse is also possible. An illustration is given in Fig. 1–1.

The material to be presented in this chapter applies to the geometrically linear theory and to incremental stability in the geometrically nonlinear theory. However, for simplicity, only the finite notation of Eq. (1–1) is used.

Pin-Jointed Trusses: Geometric Study

For investigating kinematic stability, members are considered as rigid bodies. If the structure as a whole, free of any supports, forms a rigid body, the assemblage of members is called rigid. The simplest rigid assemblage of more than one member consists of three members forming a triangle which is uniquely determined from the lengths of its three sides. A triangulated system is a rigid assemblage formed by starting with one triangle and by constructing additional triangles using for each new triangle one member of the existing system and two new members. Each additional triangle thus adds two members and one joint to the system. Thus $M - 3 = 2(J - 3)$, where M is the number of members and J is the number of joints, or

$$M = 2J - 3 \tag{1–4}$$

For a triangulated system to become a kinematically stable structure, it needs only to be supported against rigid body displacements. This may be done by simply supporting the structure, that is, by pinning one joint A to a fixed support and restraining the rotation about A by a roller support at another joint. Examples are shown in Fig. 1–2*a*, *d*, *e*, *f*, and *g*. The information written with each figure is discussed subsequently. In (*b*) the assemblage is triangulated, and the structure is stable, but the supports conditions exceed those of simple supports. In (*c*) the assemblage contains two more members than is needed to form a triangulated one.

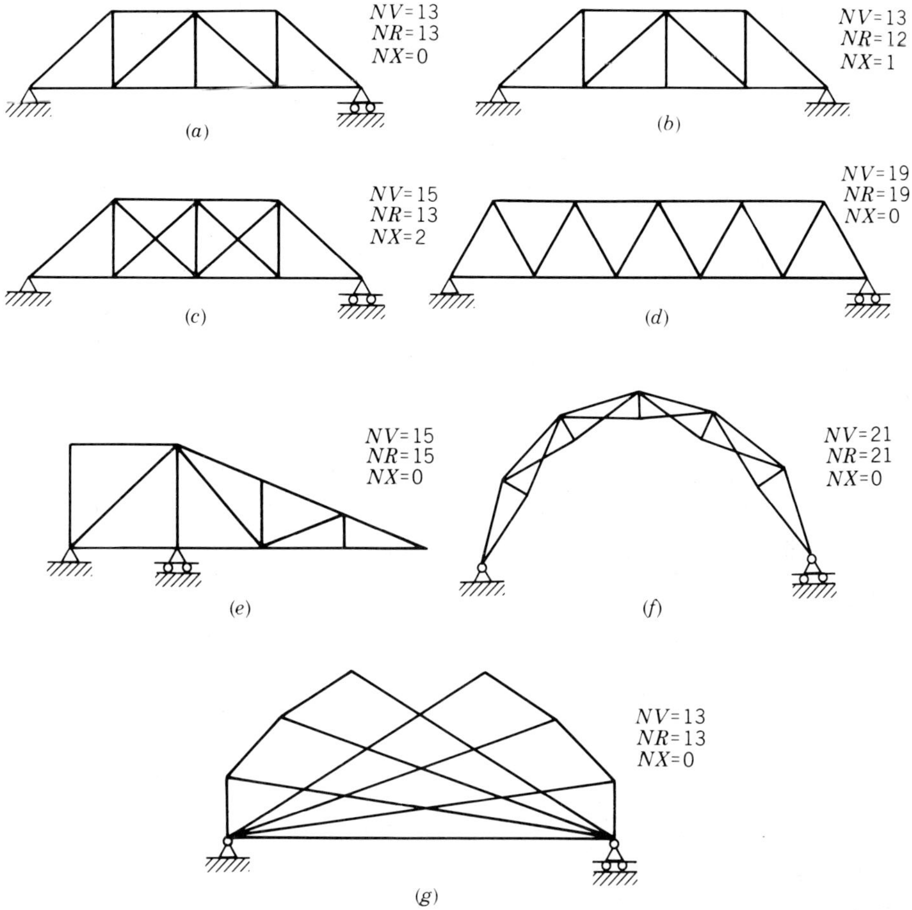

FIG. 1–2 Stable pin-jointed trusses, statically determinate ($NX = 0$); statically indeterminate ($NX > 0$).

The construction of a rigid assemblage may be generalized beyond the triangulated system. If A and B are two arbitrary joints of a rigid assemblage, distance AB is fixed. Geometrically AB may be treated similarly to a member in constructing a triangulated system, by adding two noncolinear members AC and BC pinned at a new joint C. A rigid assemblage constructed in this manner continues to satisfy Eq. (1–4) if the starting assemblage does. It is referred to as a simple assemblage. For example, in Fig. 1–3*a* portion $ADBEFG$ is a triangulated system to which CA and CB are added. Other examples of simple assemblages are shown in Fig. 1–3.

A general way of supporting a body against plane rigid body displacements is by means of three nonconcurrent and nonparallel simple support members as shown in Fig. 1–4*a*. For infinitesimal displacements these support members may be equivalently replaced with pin-ended links as shown in (*b*). This property of

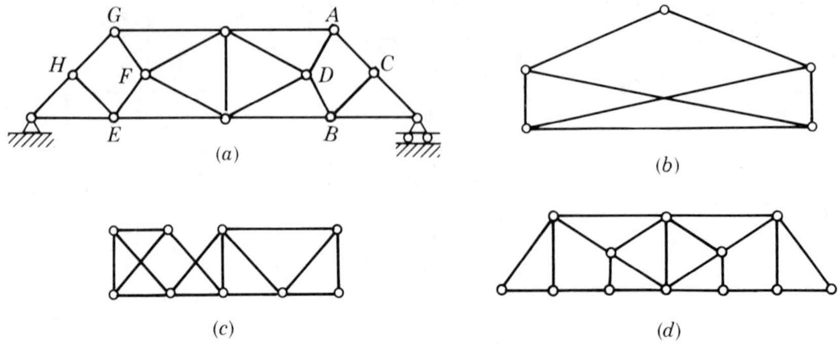

FIG. 1–3. Simple assemblages.

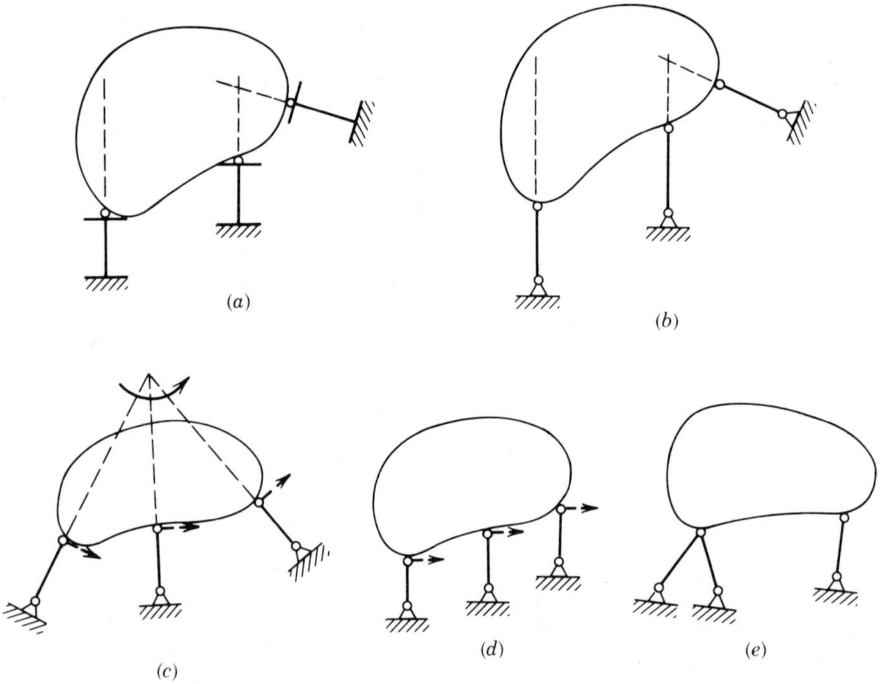

FIG. 1–4. Stable and unstable support conditions.

three nonconcurrent links is deduced from the kinematics of infinitesimal plane rigid body motion. Such a motion has an instantaneous center of rotation which is at the intersection of the perpendiculars to the displacements of all points of the rigid body. If a rigid body is supported by three concurrent links as shown in (*c*), their point of intersection is an instantaneous center of rotation, and if the three links are parallel as in (*d*), the infinitesimal rotation becomes a translation perpendicular to the links. If the links are not concurrent, an instantaneous center cannot be constructed, and thus rigid body displacement is impossible. A pin and roller supports are equivalent to the particular form of three nonconcurrent links shown in (*e*). It follows from the preceding that another general way of constructing a rigid assemblage is to connect two rigid assemblages with three nonconcurrent (and nonparallel) links. The resulting assemblage is referred to as a compound assemblage and is shown schematically in Fig. 1–5*a*. If the component rigid assemblages satisfy Eq. (1–4), then the compound assemblage also satisfies that equation. Examples of compound assemblages are shown in Fig. 1–5. Figure 1–6 shows examples of nonrigid assemblages that are, however, supported so that the structure is kinematically stable. In these examples a portion of the structure that is supported in a stable manner provides itself a support for the remaining portion. The truss of Fig. 1–6*a* is formed of three triangulated systems, two of which are simply supported and the third is in turn supported by the first two by means of pin *A* and link *BC*. In (*b*) a portion of the structure is a simply supported triangulated system, and another portion is a triangulated system connected to the first and to a support by two links and a roller. The two links and roller are equivalent to three nonconcurrent links and thus provide stable support conditions.

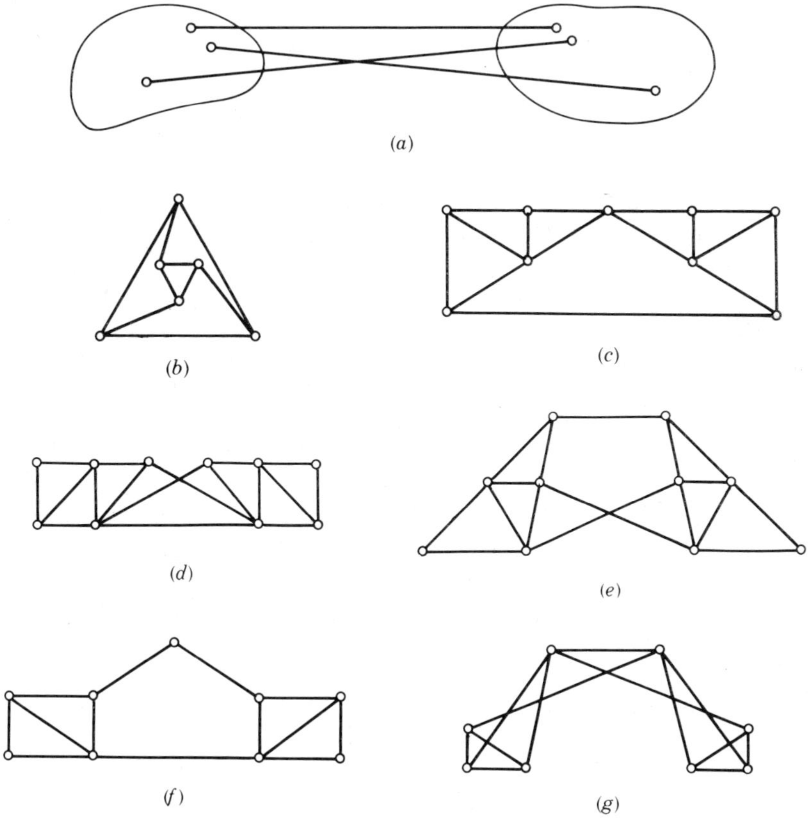

FIG. 1–5. Compound assemblages.

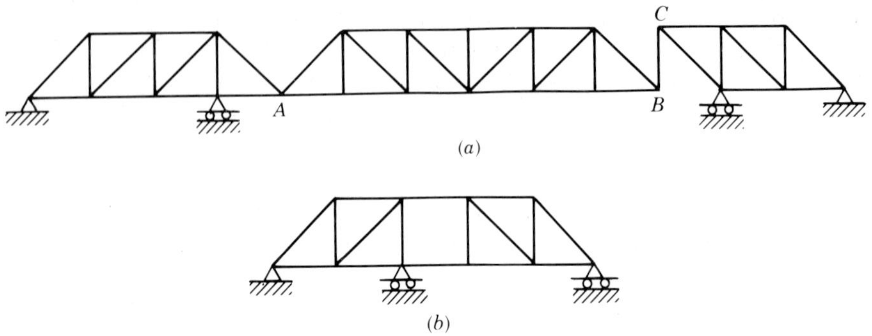

FIG. 1–6. Stable nonrigid assemblages, statically determinate.

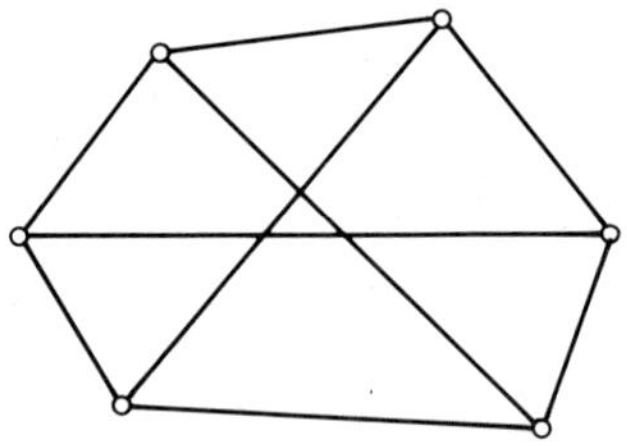

FIG. 1–7. Complex assemblage.

If an assemblage cannot be constructed by assembling stable subassemblages, it is denoted as complex. An example is shown in Fig. 1–7, and its stability is discussed subsequently.

Examples of unstable truss structures are shown in Fig. 1–8. To find the mechanism modes in (*a*), it is noted that the structure is formed of two triangulated systems AEB and BCD, pin connected at B and supported, respectively, at pin A and roller D. AEB may rotate about A, and the instantaneous center of rotation of BCD is then found at the intersection of lines AB and DC. This is the only possible motion, and there is thus only one mechanism mode. The structure in (*b*) has the same mechanism mode as the preceding one.

The structures of Fig. 1–8*a*, *b* may be made stable by changing the roller into a pin support, thus forming what is called a three-hinged arch, or by adding a member connecting the supported joints. The complex assemblage of Fig. 1–8*c* has an axis of symmetry and will be shown to have the mechanism mode shown in dotted lines. Four 4-bar mechanisms may be identified in the figure. They are (AF, AD, FC, CD), (AB, AD, BC, CD), (FE, FC, ED, CD), and (AF, AB, FE, BE). The motion of the first mechanism is defined by the angle θ which is the same for bars AD and FC because of symmetry. The rotations of bars AD and FC determine in turn the motion of the next two mechanisms. For example, the instantaneous center of bar BC is at the intersection of lines AB and FC. The geometry of the 4-bar mechanism determines the rotations of BC and of AB. The

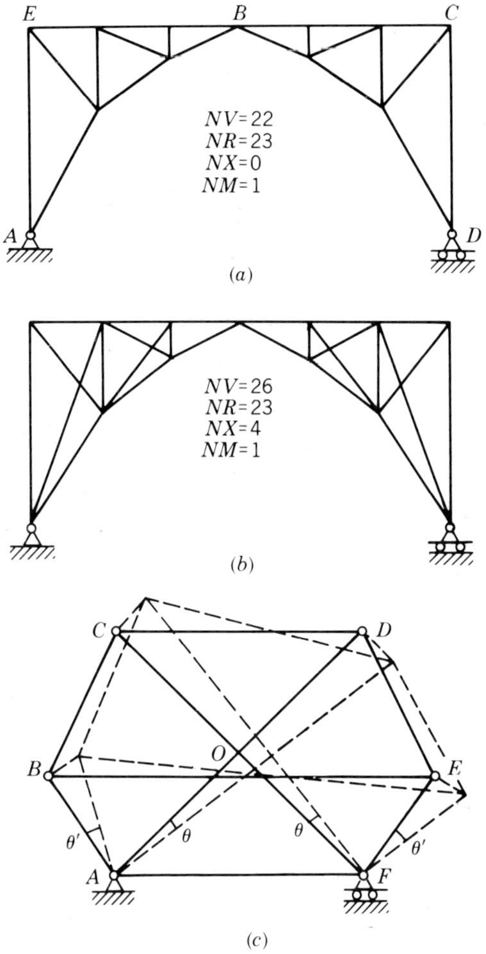

FIG. 1–8. Unstable assemblages.

third mechanism being symmetric of the second, bars AB and FE rotate by the same angle θ', and this is compatible with the motion of the fourth mechanism.

Without symmetry the motions of the second and third mechanisms would have different rotations for bars AB and FE. These in turn would be in general incompatible with the geometry of the fourth mechanism, and the assemblage would then be stable.

Frames: Geometric Study of Stability

A structure with rigid joints is by definition a rigid assemblage. For kinematic stability it is only required to prevent rigid body motion. General frames may have any types of member joint connections and include pin-jointed trusses as a particular case.

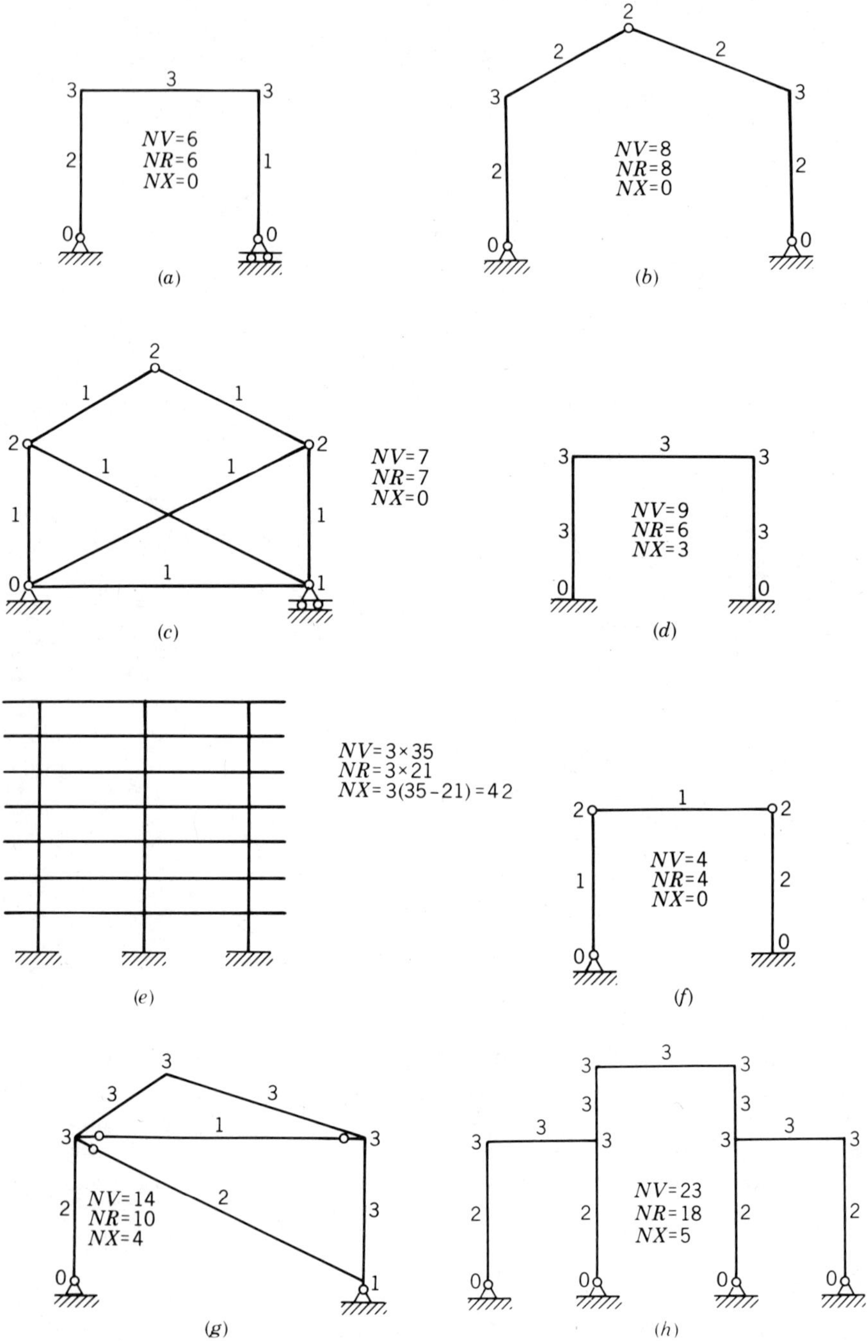

FIG. 1–9. Stable structures, statically determinate ($NX = 0$); statically indeterminate ($NX > 0$). (*Continued on next page.*)

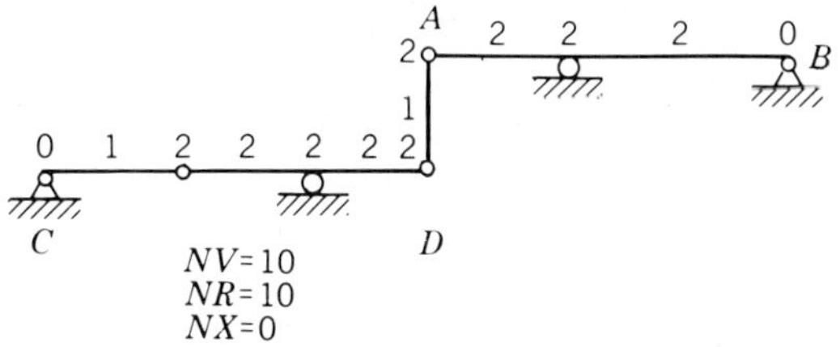

(i)

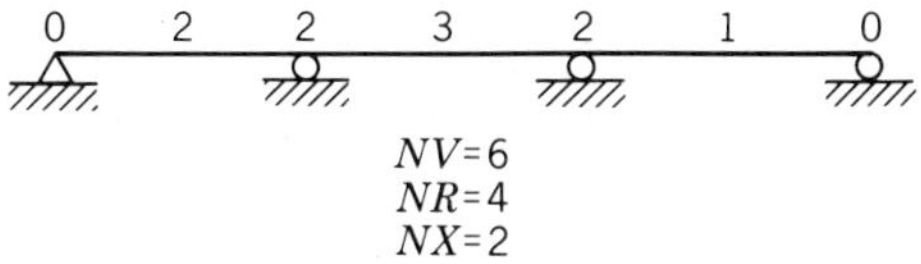

(j)

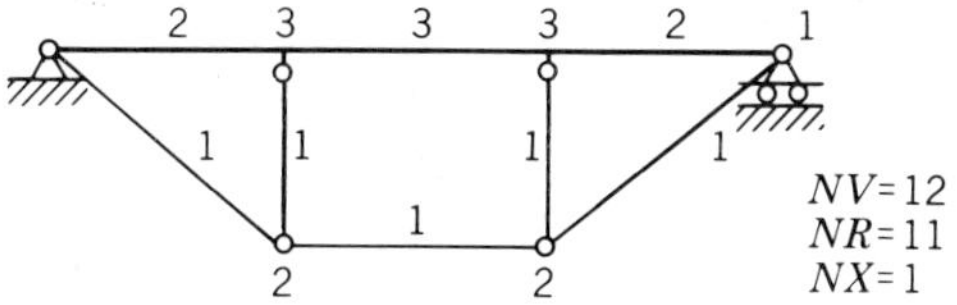

(k)

FIG. 1–9 (*continued*)

The kinematic stability of the frames in Fig. 1–9 may be readily ascertained by the geometric approach followed in the case of trusses. Rigidly joined members form a single rigid body. The structure in (i) may need further consideration. Portion AB is rigid and simply supported and is thus individually stable. Portion CD has one mechanism mode which is, however, restrained by link A which ties it to the stable portion AB.

2. STATIC-KINEMATIC PROPERTIES OF KINEMATICALLY STABLE STRUCTURES

Two questions are to be addressed in the present topic. One is statical and may be stated as follows: Given a stable structure and an arbitrary loading, is it always possible to determine internal forces and reactions at supports that satisfy member and joint equilibrium? The other question is geometric: If a set of member deformations $\mathbf{v}$ is given, what conditions, if any, must be satisfied in order that $\mathbf{v}$ be compatible with some joint displacements $\mathbf{r}$. The answers to these questions are contained in the mathematical properties of the static-kinematic pair of equations, (8/2.4–8a, b),

$$\mathbf{a}_v^T \mathbf{V} = \mathbf{R} - \mathbf{R}_p \tag{2–1a}$$

$$\mathbf{v} = \mathbf{a}_v \mathbf{r} \tag{2–1b}$$

It will be assumed that the device of support members is used, so that reactions at supports are elements of $\mathbf{V}$, and $(\mathbf{R} - \mathbf{R}_p)$ is completely known for any given loading condition.

For a kinematically stable structure the rank ρ of $\mathbf{a}_v$ is equal to the number of degrees of freedom (NR). Thus the equilibrium equations (2–1a) have a coefficient matrix whose rank is equal to the number of equations. Consequently the number of elements (NV) in $\mathbf{V}$ cannot be less than (NR). Letting

$$(NX) = (NV) - (NR) \tag{2–2}$$

The cases $(NX) = 0$ and $(NX) > 0$ correspond, respectively, to statically determinate and statically indeterminate structures; these are treated next.

Statically Determinate Structures

A statically determinate kinematically stable structure is characterized by the equalities

$$(NV) = (NR) = \rho \tag{2–3}$$

$\mathbf{a}_v$ is then square and nonsingular. Letting

$$\mathbf{b} = \mathbf{a}_v^{T-1} \tag{2–4}$$

the solution of the joint equilibrium equations is

$$\mathbf{V} = \mathbf{b}(\mathbf{R} - \mathbf{R}_p) \tag{2–5}$$

Thus to an arbitrarily given loading corresponds a unique statical solution given by Eq. (2–5). In particular, if $\mathbf{R} - \mathbf{R}_p = 0$, then $\mathbf{V} = 0$. Thus a nonmechanical loading, such as a thermal one or one consisting of prescribed support settlements, cannot cause any member forces nor support reactions. Also the solution for $\mathbf{V}$ is independent of any material properties.

Consider now the kinematic question. Given a set of deformations $\mathbf{v}$, Eq. (2–1b) may be solved uniquely for $\mathbf{r}$. $\mathbf{a}_v^{-1}$ is $\mathbf{b}^T$, and the solution is

$$\mathbf{r} = \mathbf{b}^T \mathbf{v} \tag{2–6}$$

For a given loading, $\mathbf{V}$ is determined by means of Eq. (2–5); then deformations $\mathbf{v}$ are determined from member flexibility relations, and joint displacements are determined through Eq. (2–6). This procedure is the force method for analyzing statically determinate structures and is discussed further in Chapter 11.

Having determined that a structure is kinematically stable, a test for statical determinacy is

$$(NX) = (NV) - (NR) = 0 \tag{2–7}$$

In order to make a minimum count, it is convenient to use effective degrees of freedom (NR) and effective member statical redundants (NV). Thus released member end displacements may be deleted from (NR), and the corresponding member forces, which are statically determined by the releases, are then removed from the count of (NV). Further one may choose to include reactions at supports in (NV) and the equal number of prescribed joint displacements in (NR), or to exclude both without affecting $(NV - NR)$. The physical interpretation of this procedure is that the joint equilibrium equations are those corresponding to the unrestrained degrees of freedom, and the statical unknowns in these equations exclude the reactions at supports.

For pin-jointed trusses there is one force unknown per member, namely the axial force V_1, and two displacements per joint. A triangulated, simple, or compound assemblage, which is supported against rigid body motion, has three unknown reactions at supports and a total count of $(NV) = M + 3$. The count of (NR) is $2J$. All such assemblages satisfy Eq. (1–4) and are thus statically determinate. An alternate count in which $(NV) = M$ and $(NR) = 2J - 3$ is possible and corresponds to the deletion of the support reactions from (NV) and of the prescribed displacements from (NR). Examples are shown in Fig. 1–2 for the cases having $NX = 0$. In Fig. 1–6 the structures have been ascertained earlier to be stable compound assemblages of triangulated parts. All that is needed to be established is that they are statically determinate. A count of (NV) and (NR) should yield the same conclusion.

For a general frame a member with rigid connections at both ends contributes 3 to (NV). To each kinematic release at a member end corresponds a decrease of 1 from (NV). Released member end displacements are then omitted from the effective degrees of freedom.

It is common in frames for a single member end to be considered as a joint, for example, as at the base of a column, or at the tip of an overhang. If the member end is supported, there is a reduction by 1 of the effective (NR) for every restrained degree of freedom, and to every unrestrained degree of freedom corresponds a kinematic release reducing by 1 both (NV) and (NR). The result is that in all cases of a joint at a single member end $(NR) = 0$, and (NV) for the member is reduced by the number of kinematic releases at the member ends. This applies in particular to a member with a free end that contributes zero to (NV). This manner of counting (NV) and (NR) is shown in Fig. (1–9) next to each member and each joint, respectively. All the structures in this figure are kinematically stable. Those having $NX = 0$ are thus statically determinate.

Statically Indeterminate Structures

A kinematically stable structure is statically indeterminate if the number of force unknowns (NV) exceeds the number of equilibrium equations (NR). The degree of statical indeterminacy is $(NX) = (NV) - (NR)$.

The structures of Figs. 1–2 and 1–9 may all be ascertained by geometry to be stable. The count of (NV) and (NR) is made as outlined earlier. The statically indeterminate cases are those for which $NX > 0$.

Another way of determining (NX) than by counting (NV) and (NR) is to determine the number of kinematic releases that would transform the given structure into a statically determinate structure while retaining kinematic stability. Such releases could be applied to prescribed displacements, or they could be introduced at member joint connections or at any location along the member. For example, a full cut in the beam of Fig. 1–9*d* is equivalent to three kinematic releases and yields a determinate structure formed of two cantilever parts. A similar consideration applies to the frame of Fig. 1–9*e*. A full cut through the 14 beams of the seven floors is equivalent to $3 \times 14 = 42$ releases and yields a determinate structure in three cantilever parts. There are various ways of introducing kinematic releases that transform a statically indeterminate structure into a determinate one. An alternate way for the last two examples is to modify the end connections of each beam into a pin at one end and a roller at the other. Other ways, still for the frame of Fig. 1–9*d*, correspond to the determinate frames in (*a*) and (*f*).

For pin-jointed trusses, cutting a member reduces the degree of statical undeterminacy by one, provided the cut does not make the structure unstable. A cut should be understood simply as releasing the axial continuity of the member. For example, Fig. 1–2*a* is obtained from (*c*) by cutting two diagonals.

In the case of Fig. 1–2*c* no release can be introduced in the support conditions because any such release would allow rigid body motion. The reactions at the supports are statically determinate, and the structure is classified as internally indeterminate. The same applies to Fig. 1–9*k*. The structure of Fig. 1–9*d* is externally indeterminate because releases at supports would make it a determinate structure as in (*a*). Note that this does not prevent the possibility of making the structure determinate by internal releases. The structure of Fig. 1–9*g* is both externally and internally indeterminate because only one release may be made at the support joints and three more are required to make the structure determinate.

This approach which determines the degree of statical indeterminacy will be considered again in order to interpret physically the analytical formulation of the static-kinematic properties.

The static-kinematic pair of equations (2–1) may be partitioned in the form

$$[\mathbf{a}_1^T \ \mathbf{a}_2^T] \begin{Bmatrix} \mathbf{V}_1 \\ \mathbf{V}_2 \end{Bmatrix} = \mathbf{R} - \mathbf{R}_p \tag{2–9a}$$

$$\begin{Bmatrix} \mathbf{v}_1 \\ \mathbf{v}_2 \end{Bmatrix} = \begin{bmatrix} \mathbf{a}_1 \\ \mathbf{a}_2 \end{bmatrix} \mathbf{r} \tag{2–9b}$$

where $\mathbf{V}_1$ contains (NR) elements and $\mathbf{V}_2$ (NX) elements. Elements of $\mathbf{v}_1$ and $\mathbf{v}_2$

correspond, respectively, to those of $\mathbf{V}_1$ and $\mathbf{V}_2$. Since $\rho=(NR)$, it is always possible to order the elements of $\mathbf{V}$ so that $\mathbf{a}_1$ is nonsingular. Equation (2–9a) may then be solved for $\mathbf{V}_1$ in terms of $\mathbf{V}_2$ and the load matrix, and the first of Eq. (2–9b) may be solved for $\mathbf{r}$ in terms of $\mathbf{v}_1$. We thus obtain

$$\mathbf{V}_1=\mathbf{b}_1(\mathbf{R}-\mathbf{R}_p)+\mathbf{c}_1\mathbf{V}_2 \tag{2–10}$$

$$\mathbf{r}=\mathbf{b}_1^T\mathbf{v}_1 \tag{2–11a}$$

and by substituting for $\mathbf{r}$ into the second of Eq. (2–9b),

$$[\mathbf{c}_1^T\ \mathbf{I}]\begin{Bmatrix}\mathbf{v}_1\\ \mathbf{v}_2\end{Bmatrix}=0 \tag{2–11b}$$

In the preceding equations

$$\mathbf{b}_1=\mathbf{a}_1^{T-1} \tag{2–12}$$

$$\mathbf{c}_1=-\mathbf{a}_1^{T-1}\mathbf{a}_2^T \tag{2–13}$$

It will be convenient to name the statical redundants $\mathbf{X}$ and to append the equation $\mathbf{V}_2=\mathbf{X}$ to Eq. (2–10). Thus

$$\begin{Bmatrix}\mathbf{V}_1\\ \mathbf{V}_2\end{Bmatrix}=\begin{bmatrix}\mathbf{b}_1\\ \mathbf{O}\end{bmatrix}(\mathbf{R}-\mathbf{R}_p)+\begin{bmatrix}\mathbf{c}_1\\ \mathbf{I}\end{bmatrix}\mathbf{X} \tag{2–14a}$$

This equation and Eqs. (2–11) may be written in the form

$$\mathbf{V}=\mathbf{b}(\mathbf{R}-\mathbf{R}_p)+\mathbf{c}\mathbf{X} \tag{2–14b}$$

$$\mathbf{r}=\mathbf{b}^T\mathbf{v} \tag{2–15a}$$

$$\mathbf{c}^T\mathbf{v}=0 \tag{2–15b}$$

Equations (2–14) and (2–15) are transformed static-kinematic equations that bring out the static-kinematic properties of the structure.

The general solution of the joint equilibrium equation, is formed by superposition of $\mathbf{b}(\mathbf{R}-\mathbf{R}_p)$ and of $\mathbf{cX}$, with $\mathbf{X}$ arbitrary. The first term is a particular solution. The second term is the general solution of the homogeneous equation and is thus a set of self-equilibriating member forces. The (NX) columns of $\mathbf{c}$ represent independent self-equilibriating modes, corresponding each to a unit value of a statical redundant. The linear combination of these modes with arbitrary $\mathbf{X}$ forms the general solution for self-equilibriating member forces. Elements of $\mathbf{X}$ are called statical redundants. The choice of the elements of $\mathbf{X}$ from the set $\mathbf{V}$ must be such that $\mathbf{a}_1$ is nonsingular but is otherwise arbitrary.

Kinematically a given set of deformations $\mathbf{v}$ must satisfy the geometric compatibility equation (2–11b) in order that it be compatible with some displacements $\mathbf{r}$. In that case $\mathbf{r}$ is determined in terms of the subset $\mathbf{v}_1$ by means of Eq. (2–11a).

The force method of analysis (to be presented in Chapter 11) expresses the (NX) compatibility equations as a set of simultaneous equations for determining the (NX) statical redundants by means of the member flexibility equations.

3. PRIMARY STRUCTURE

A statically determinate structure obtained from an indeterminate one by introduction of (NX) kinematic releases is called a primary structure. It provides a physical model for the description of the static and kinematic properties of the indeterminate structure.

Let $\mathbf{X}$ be a choice of statical redundants, taken from the set $\mathbf{V}$. Each element of $\mathbf{X}$ is an internal force that may be made zero by introducing a corresponding kinematic release. To a moment corresponds a rotational release, and to a force corresponds a translational release. A complete cut produces three releases and makes three force elements zero. By introducing (NX) releases which make $\mathbf{X} = 0$, there results a primary structure whose internal forces under the given loading condition form the particular solution $\mathbf{b}(\mathbf{R} - \mathbf{R}_p)$ in Eq. (2–14b). If now arbitrary values $\mathbf{X}$ are applied externally at the corresponding releases, with each element of $\mathbf{X}$ being applied as a pair of equal and opposite actions, the internal forces caused by $\mathbf{X}$ form, together with $\mathbf{X}$, the homogeneous solution $\mathbf{cX}$. The statics of the primary structure subjected to the given loading condition and to an arbitrary set of statical redundants are thus identical with the statics of the indeterminate structure.

Example 1

The truss of Fig. 3–1*a*, is indeterminate to the second degree. In order to allow for all possibilities of choosing statical redundants from the set of unknowns $\mathbf{V}$, the supports are replaced by support members in (*b*). Two releases are shown in (*c*) with the corresponding statical redundants applied each as a pair of opposite forces.

If the degrees of freedom used in formulating the equilibrium equations exclude support displacements, $\mathbf{V}$ excludes the reactions at supports. Statical redundants may always be chosen from the set of member forces. A possible choice is shown in (*d*).

A similar example is illustrated for the frame of Fig. 3–1*e*. Support members are shown in (*f*). In (*g*) the statical redundants are the three forces of the support member. They represent the reactions at E. In (*h*) the statical redundants are chosen from the member end moments. Note that at C, X_2 is chosen either from the forces of member CD or from those of member CB. The moments of the two members at C remain governed by the joint equilibrium equation.

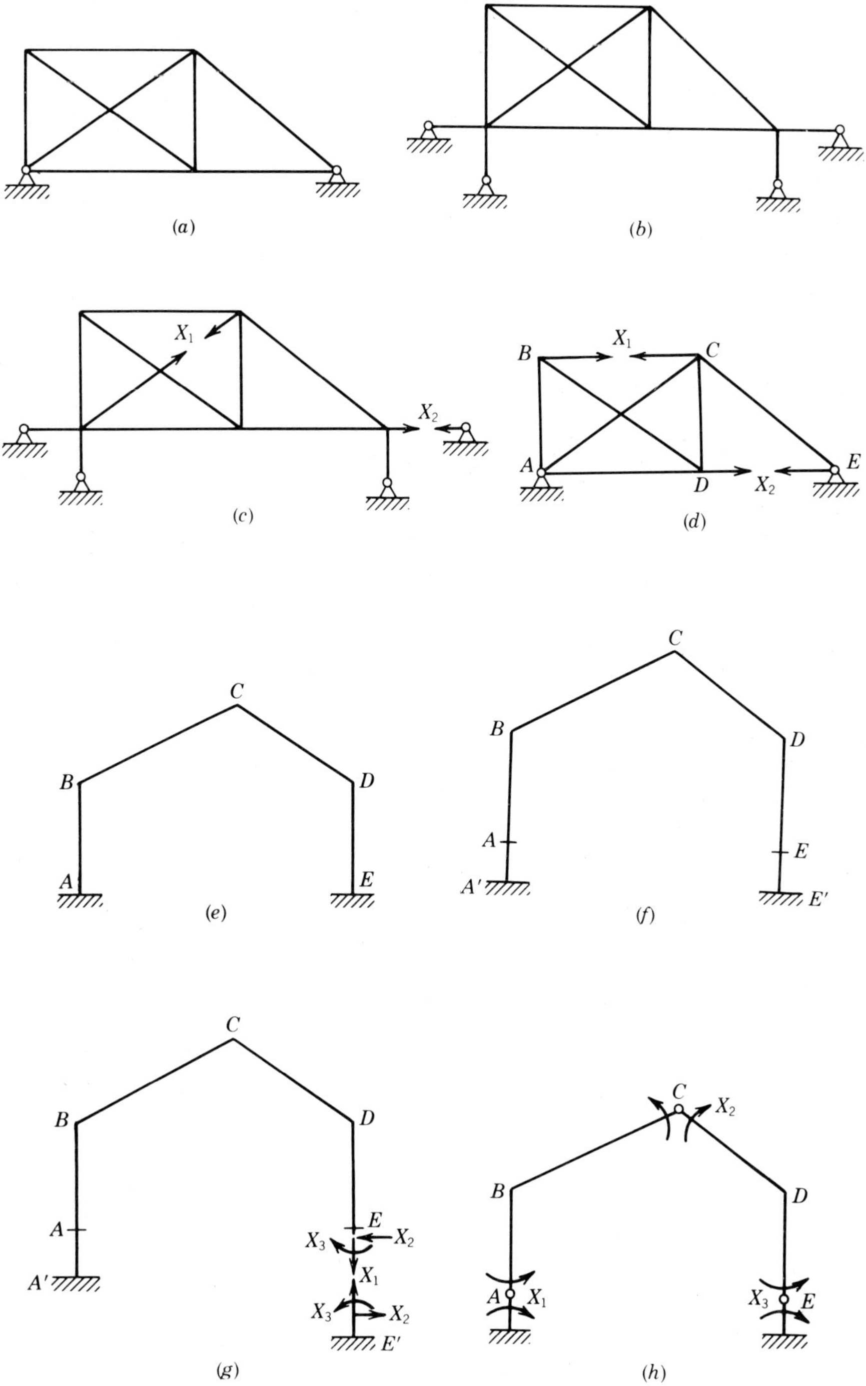

FIG. 3–1. Primary structure and statical redundants.

The statics of the primary structure are the same as those of the indeterminate structure, but the kinematics are not. An arbitrary set of member deformations in the primary structure is compatible because the structure is statically determinate. However, such member deformations would produce relative displacements, or deformations, at the releases. Let $\mathbf{\Delta} = \{\Delta_1, \Delta_2, \ldots, \Delta_{NX}\}$ be the deformations at the releases due to an arbitrarily given set of member deformations $\mathbf{v}$. The expression of $\mathbf{\Delta}$ in terms of $\mathbf{v}$ is derived by the principle of virtual forces. We will also derive in the same process the expression of the joint displacements $\mathbf{r}$ in terms of $\mathbf{v}$. Let $\mathbf{R}'$ be a virtual joint load, and $\mathbf{X}'$ be a set of virtual statical redundants applied at the releases. The external virtual work is $\mathbf{R}'^T\mathbf{r} + \mathbf{X}'^T\mathbf{\Delta}$. The expression $\mathbf{X}'^T\mathbf{\Delta}$ defines $\mathbf{\Delta}$ in a precise manner both geometrically and in sign convention. The internal virtual work is $\mathbf{V}'^T\mathbf{v}$ in which $\mathbf{V}'$ must be statically admissible with $\mathbf{R}'$ and $\mathbf{X}'$. Thus $\mathbf{V}' = \mathbf{b}\mathbf{R}' + \mathbf{c}\mathbf{X}'$. Equating external and internal virtual work for arbitrary $\mathbf{R}'$ and $\mathbf{X}'$ yields Eq. (2–15a) for $\mathbf{r}$, and

$$\mathbf{\Delta} = \mathbf{c}^T\mathbf{v} \tag{3–1}$$

The compatibility equation of the indeterminate structure is $\mathbf{\Delta} = 0$.

Behavior of the indeterminate structure may now be seen as a particular state of the primary structure in which the deformations $\mathbf{\Delta}$ caused by the loading condition and by $\mathbf{X}$ vanish. For elastic materials the primary structure may be taken through any loading path that reaches the final state $\mathbf{\Delta} = 0$. The loading condition may be applied first with $\mathbf{X} = 0$. $\mathbf{X}$ is then applied and is determined so that the final $\mathbf{\Delta}$ vanishes (see Chapter 11).

The analytical derivation of Eq. (2–14) from the equilibrium equations makes $\mathbf{X}$ necessarily a subset of $\mathbf{V}$, whereas the geometric way of choosing a primary structure allows more freedom. The theory covers this added freedom through the concept of static transformation. In such a transformation the equation of virtual work makes $\mathbf{X}$ and $\mathbf{\Delta}$ a static kinematic pair that transforms according to the contragradient law. Equations (2–14, 15) retain their form for a general choice of $\mathbf{X}$, but $\mathbf{b}$ and $\mathbf{c}$ are not necessarily constructed from $\mathbf{b}_1$ and $\mathbf{c}_1$ as in Eq. (2–14a). This generalization is useful, in particular, in manual methods of analysis.

4. STATIC-KINEMATIC PROPERTIES OF KINEMATICALLY UNSTABLE STRUCTURES

General Properties

Usual structures are built as kinematically stable, but the subject of kinematic instability has several applications. The properties of unstable structures complement the understanding of stable ones. The topic also arises in using virtual displacement modes for deriving equilibrium equations, in analyzing modes of collapse of structures and in determining inextensional displacement modes of frames. Geometric examples are treated after the following analysis.

The kinematic criterion of instability is that zero member deformations $\mathbf{a}_v\mathbf{r}$ are possible for some nonzero $\mathbf{r}$. Independent solutions of the homogeneous equations $\mathbf{a}_v\mathbf{r} = 0$ represent independent rigid body or mechanism modes of the structure. Any linear combination of such modes represents a possible mechanism mode. Mathematically the criterion is

$$\rho < (NR)$$

Let $\mathbf{a}_{11}$ be a nonsingular square submatrix of $\mathbf{a}_v$, of size equal to the rank ρ, and assume $\mathbf{v}$ and $\mathbf{r}$ are reordered such that the connectivity and equilibrium equations partition in the form

$$\begin{bmatrix} \mathbf{a}_{11} & \mathbf{a}_{12} \\ \mathbf{a}_{21} & \mathbf{a}_{22} \end{bmatrix} \begin{Bmatrix} \mathbf{r}_1 \\ \mathbf{r}_2 \end{Bmatrix} = \begin{Bmatrix} \mathbf{v}_1 \\ \mathbf{v}_2 \end{Bmatrix} \tag{4–1}$$

$$\begin{bmatrix} \mathbf{a}_{11}^T & \mathbf{a}_{21}^T \\ \mathbf{a}_{12}^T & \mathbf{a}_{22}^T \end{bmatrix} \begin{Bmatrix} \mathbf{V}_1 \\ \mathbf{V}_2 \end{Bmatrix} = \begin{Bmatrix} \mathbf{R}_1 \\ \mathbf{R}_2 \end{Bmatrix} \tag{4–2}$$

Let

$$(NM) = (NR) - \rho \tag{4–3a}$$

$$(NX) = (NV) - \rho \tag{4–3b}$$

In Eq. (4–1), $\mathbf{a}_{12}$ is $\rho \times (NM)$, $\mathbf{a}_{21}$ is $(NX) \times \rho$, and $\mathbf{a}_{22}$ is $(NX) \times (NM)$. (NX) may be zero, in which case the second row partition of $\mathbf{a}_v$ does not exist, and accordingly, $\mathbf{v}$ and $\mathbf{V}$ are not partitioned.

Since $\mathbf{a}_{11}$ is nonsingular, the first of Eq. (4–1) may be solved for $\mathbf{r}_1$. Substituting the result in the second part, the homogeneous part of the equation is satisfied identically because of the rank property of the matrix. We thus obtain

$$\mathbf{a}_{22} - \mathbf{a}_{21}\mathbf{a}_{11}^{-1}\mathbf{a}_{12} = 0 \tag{4–4}$$

and

$$\mathbf{r}_1 = -\mathbf{a}_{11}^{-1}\mathbf{a}_{12}\mathbf{r}_2 + \mathbf{a}_{11}^{-1}\mathbf{v}_1 \tag{4–5a}$$

$$[-\mathbf{a}_{21}\mathbf{a}_{11}^{-1} \quad \mathbf{I}] \begin{Bmatrix} \mathbf{v}_1 \\ \mathbf{v}_2 \end{Bmatrix} = 0 \tag{4–5b}$$

Equations. (4–5) are written in the form

$$\mathbf{r}_1 = \mathbf{M}_1\mathbf{r}_2 + \mathbf{b}_1^T\mathbf{v}_1 \tag{4–6a}$$

$$[\mathbf{c}_1^T \quad \mathbf{I}] \begin{Bmatrix} \mathbf{v}_1 \\ \mathbf{v}_2 \end{Bmatrix} = 0 \tag{4–6b}$$

The general deformed state of the structure may be described by prescribing $\mathbf{r}_2$ and $\mathbf{v}_1$ arbitrarily and then determining $\mathbf{r}_1$ and $\mathbf{v}_2$ by Eqs. (4–6), respectively. For $\mathbf{v}_1=0$, Eqs. (4–6) yield $\mathbf{v}_2=0$, and $\mathbf{r}_1=\mathbf{M}_1\mathbf{r}_2$. Appending the identity $\mathbf{r}_2=\mathbf{r}_2$, we can write $\mathbf{r}=\mathbf{M}\mathbf{r}_2$, where

$$\mathbf{M}=\begin{bmatrix}\mathbf{M}_1\\ \mathbf{I}\end{bmatrix}=\begin{bmatrix}-\mathbf{a}_{11}^{-1}\mathbf{a}_{12}\\ \mathbf{I}\end{bmatrix} \tag{4–7}$$

The (NM) columns of $\mathbf{M}$ are independent mechanism modes, corresponding, respectively, to unit values of the elements of $\mathbf{r}_2$.

Similarly the first part of Eq. (4–2) is solved for $\mathbf{V}_1$. Substituting the result into the second part, the homogeneous part of the equation is identically satisfied. The result has the form

$$\mathbf{V}_1=\mathbf{b}_1\mathbf{R}_1+\mathbf{c}_1\mathbf{V}_2 \tag{4–8a}$$

$$[\mathbf{M}_1^T\ \ \mathbf{I}]\begin{Bmatrix}\mathbf{R}_1\\ \mathbf{R}_2\end{Bmatrix}=0 \tag{4–8b}$$

For equilibrium to be possible, $\mathbf{R}$ must satisfy Eq. (4–8b). Thus the virtual work of $\mathbf{R}$ through the (NM) independent mechanism modes must vanish. This is expressed by saying that $\mathbf{R}$ must be orthogonal to these modes. If this condition is satisfied, $\mathbf{V}_2$ is a set of (NX) statical redundants, and $\mathbf{V}_1$ is determined through Eq. (4–8a).

Equations (4–8) may be derived by requiring the equation of virtual displacements, $\mathbf{V}^T\delta\mathbf{v}=\mathbf{R}^T\delta\mathbf{r}$, to hold for arbitrary $\delta\mathbf{r}_2$ and $\delta\mathbf{v}_1$, with $\delta\mathbf{r}_1$ and $\delta\mathbf{v}_2$ obtained by variation of Eqs. (4–6).

The case of a statically determinate unstable structure corresponds to $(NX)=(NV)-\rho=0$. Since $(NM)=(NR)-\rho>0$, it follows that $(NR)>(NV)$. The test for statical determinacy of a stable structure is $(NR)=(NV)$. This equality in the case of an unstable structure entails necessarily a degree of statical indeterminacy (NX) equal to the degree of kinematic instability (NM).

The degree of kinematic instability may be found geometrically by determining the number of simple constraints required to obtain a stable structure without raising the degree of statical indeterminacy. The process may also be reversed. Starting with a stable and determinate structure, introduction of kinematic releases produces an unstable structure with as many mechanism modes.

Example 1

The structure of Fig. 4–1*a* may be made stable and determinate by requiring rotational continuity at *A* and *B*. It has thus two independent mechanism modes that may be obtained by releasing one at a time these rotational continuities. To each release corresponds a mechanism with one degree of freedom. The mechanism mode obtained through the release at *B* is shown in (*b*). To define this mechanism mode, it is noted that the instantaneous centers of *EAC* and *DB* are at

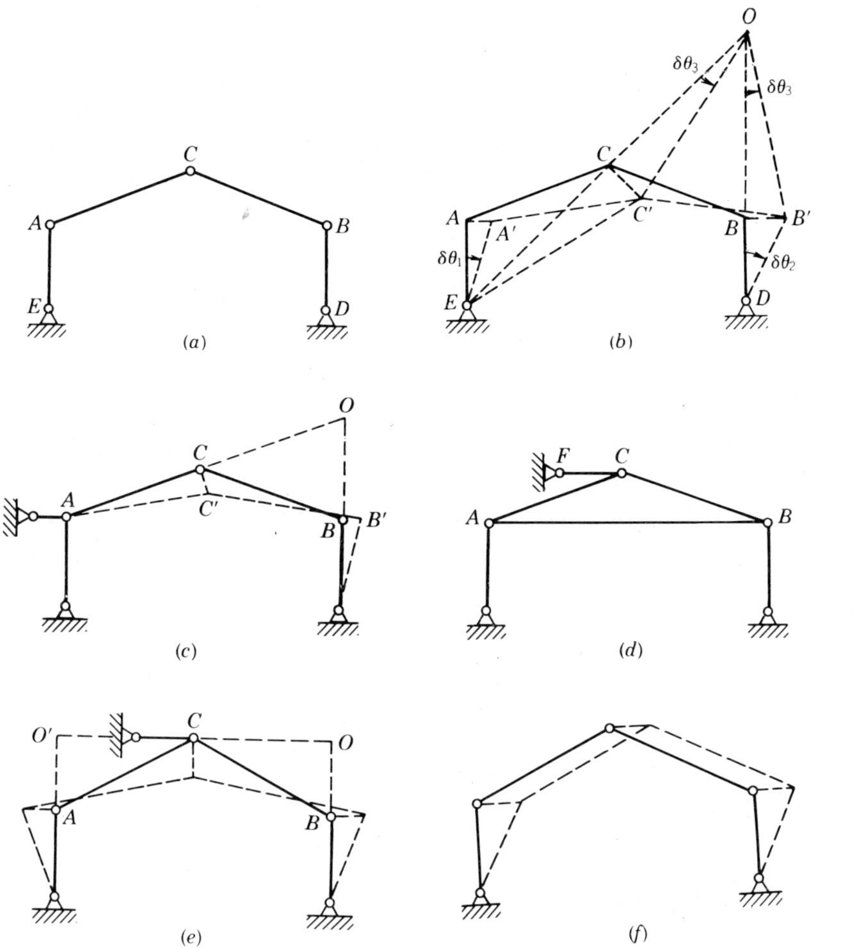

FIG. 4–1. Mechanism modes.

E and D, respectively. This determines the orientations of the infinitesimal displacements CC' and BB' as perpendicular to EC and DB, respectively. The instantaneous center of CB is thus at the intersection O of EC and DB. The infinitesimal rotations of EAC, CB and BD are related by the two conditions,

$$(CC') = (EC)\delta\theta_1 = (OC)\delta\theta_3$$
$$(BB') = (BD)\delta\theta_2 = (OB)\delta\theta_3$$

which leave one independent rotation.

The second mechanism interchanges the roles of A and B and is similar to the first.

There are various ways of transforming the structure of Fig. 4–1a into a kinematically stable one, with corresponding ways of obtaining two independent mechanism modes. For example, two diagonal links *AD* and *EB* transform the structure into a stable and determinate truss. Equivalently, lateral support links at *A* and *B* would achieve the same effect of preventing displacements of these joints. Independent mechanism modes may then be obtained by releasing one at a time these two lateral links. Fig. 4–1*c* shows the mechanism obtained by keeping *A* fixed and cutting the link at *B*.

Finally, a third way by which two independent mechanism modes may be obtained, corresponds to the structure shown in (*d*). Cutting links *AB* and *CF* yields, respectively, the mechanism modes shown in (*e*) and (*f*).

Application: Equilibrium through Virtual Mechanism Modes

Consider a statically determinate, kinematically stable structure subjected to a given loading, and let *V* be an internal force at a certain location. Let a virtual kinematic release corresponding to *V* be introduced. The structure, thus modified, has one mechanism displacement mode that may be used as virtual displacement. The only virtual deformation in the mechanism mode occurs as a relative displacement δv at the kinematic release. The internal virtual work is then $V\delta v$, and the equation of virtual work is

$$V\delta v = \delta W_{\text{ext}} \tag{4–9}$$

An example is shown in Fig. 4–2*a*, *b* for a bending moment *V* at location *x* doing virtual work through the virtual deformation δv. The virtual work of **P** is

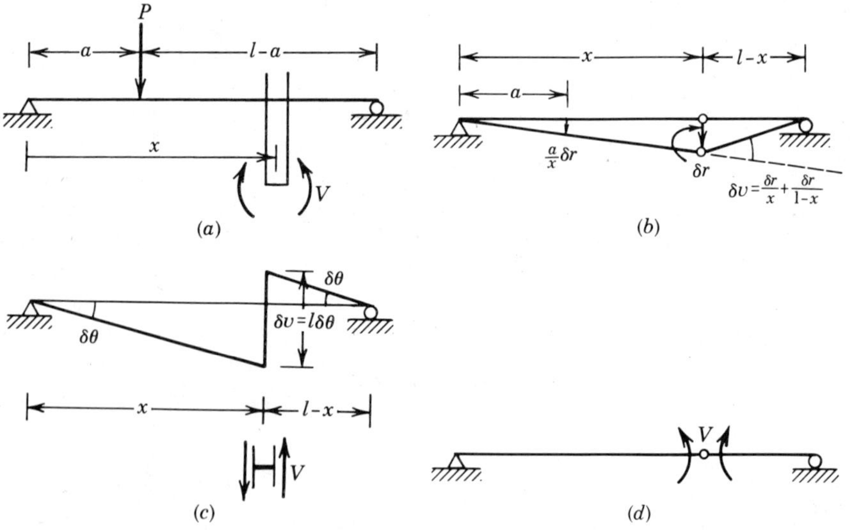

FIG. 4–2. Virtual mechanisms.

$\delta W_{\text{ext}} = (Pa/x)\delta r$, and Eq. (4–9) yields, after dividing through by δr,

$$V\left(\frac{1}{x} + \frac{1}{l-x}\right) = \frac{Pa}{x}$$

The equation simplifies to

$$V = Pa\left(1 - \frac{x}{l}\right)$$

and holds for $x \geqslant a$.

For determining the shear force at x, the virtual release is as shown in (*c*). It must maintain the rotational continuity so that only the shear force does nonzero virtual work. Letting V be the shear force, the equation of virtual work for a unit $\delta\theta$ is $Vl = Pa$ and yields $V = Pa/l$, for $x > a$.

It is possible to give Eq. (4–9) an alternate interpretation by representing V as a pair of opposite forces applied externally at the two sides of the kinematic release, Fig. 4–2*d*. In the modified structure there are no virtual member deformations, and the internal virtual work vanishes. The equation of virtual work for the modified structure equates to zero the external virtual work as a condition for equilibrium. The virtual work of V in this representation is $-V\delta v$, and the result coincides with Eq. (4–9).

In the application of Eq. (4–9) to a general determinate structure, δv and δW_{ext} are proportional to one generalized virtual displacement that cancels out, leaving V determined. It is thus always possible to determine directly any internal force in a statically determinate structure without having to solve simultaneously for other statical unknowns. This applies as well to a support reaction for which the virtual release frees the prescribed displacement. The procedure does not apply to a statically indeterminate structure except if the connectivity is such that there are some locations at which one kinematic release causes a mechanism. However, if a sufficient number of virtual releases are introduced into a statically indeterminate structure to form a one degree-of-freedom mechanism, the principle of virtual displacements allows us to derive an equilibrium equation whose only unknowns are the internal forces at the releases.

In evaluating the external virtual work in a virtual mechanism motion, one may evaluate the scalar product of each force by the virtual displacement of its point of application, or one may make use of the property of rigid body kinematics, whereby the virtual work of a force system through a rigid body displacement is equal to the virtual work of its resultant force and resultant moment taken at an arbitrary point. The virtual mechanism is formed of rigid bodies each of which has an instantaneous center of rotation I, a virtual rotation $\delta\theta$ about I, and an external moment C about I. The external virtual work is then $\sum C\delta\theta$. If a part of the mechanism has a translational motion, $C\delta\theta$ is replaced by $F\delta u$, where δu is the virtual translation and F the component in the direction of the translation of the resultant external force. In Fig. 4–3*a* the bending moment

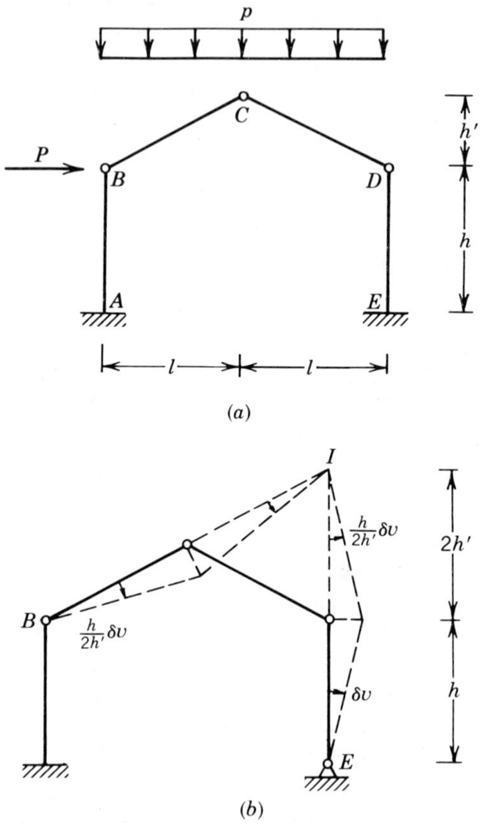

FIG. 4–3. Virtual mechanism.

at E is evaluated through the virtual mechanism shown in (b). The distributed load over BC has the moment $pl^2/2$ about the center of rotation B, and does the virtual work $(pl^2/2)\,(h/2h')\delta v$. A similar calculation is made for CD whose center of rotation is at I. P does no virtual work and thus does not contribute to the bending moment at E. The equation of virtual work for $\delta v = 1$ yields

$$V = \frac{1}{2} p l^2 \frac{h}{h'}$$

For the indeterminate frame in Fig. 4–4a, the mechanism mode in (b) allows to derive an equilibrium equation relating the bending moments at A and B. This is

$$-V_A + V_B = Ph$$

The sign convention for V_A and V_B is to produce tension on the inside of the frame if positive. In Fig. 4–5a the axial force V in bar AB is evaluated by means of the

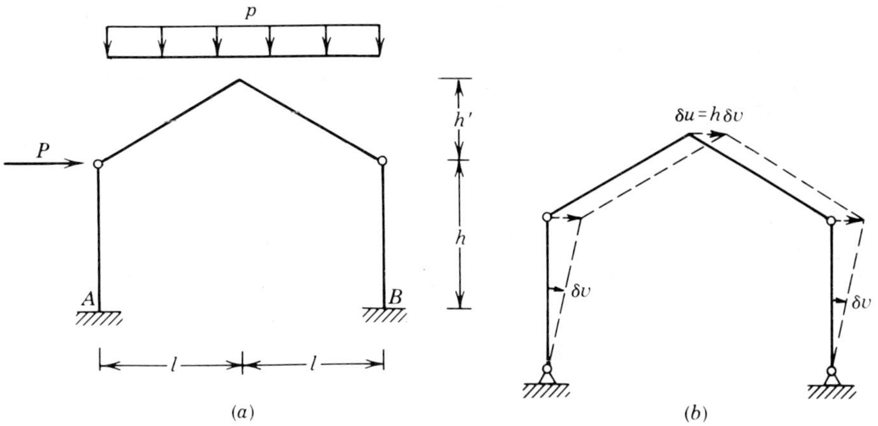

FIG. 4–4. Virtual mechanism.

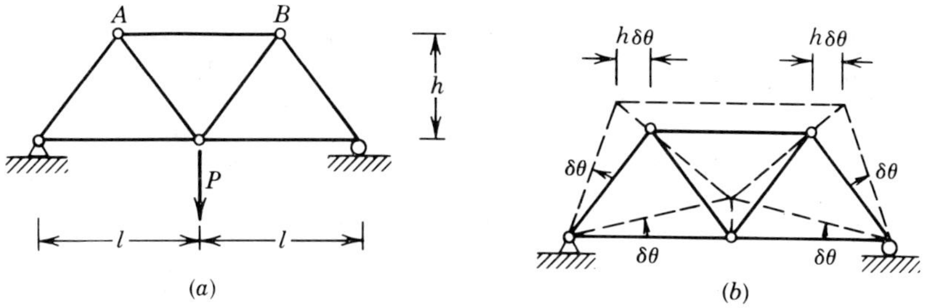

FIG. 4–5. Virtual mechanism.

virtual mechanism shown in (*b*). δv is the virtual elongation of AB and is equal to $2h\delta\theta$. The equation of virtual work for $\delta\theta = 1$ yields, $V \times 2h = -Pl$, or $V = -Pl/2h$. The negative sign indicates that the direction of V is opposite to that of δv and is thus compressive.

5. INEXTENSIONAL AND EXTENSIONAL DEFORMATION MODES

The degrees of freedom of a structure may generally be separated into two parts corresponding, respectively, to inextensional and extensional deformation modes. Behavior of rigid frames is unsually well approximated by the inextensional part. Stable pin-jointed trusses have only extensional degrees of freedom. The topic to be addressed here is that of transformation from joint displacements **r** to generalized displacements **q** of equal number, which may be separated into inextensional and extensional parts.

Inextensional degrees of freedom do not induce axial deformations and thus do not change the members chord length. Joint rotations are thus identified at the outset as inextensional. Translational inextensional displacements may be visualized by modifying the rigid joints of the given structure into pinned ones. If this modified structure is unstable, the joint translations of its mechanism modes are the inextensional translational modes of the given structure.

To separate extensional from inextensional modes, the connectivity relations, $\mathbf{v} = \mathbf{a}_v \mathbf{r}$, are reordered and partitioned into

$$\begin{Bmatrix} \mathbf{v}_a \\ \mathbf{v}_b \end{Bmatrix} = \begin{bmatrix} \mathbf{a}_{au} & 0 \\ \mathbf{a}_{bu} & \mathbf{a}_{b\varphi} \end{bmatrix} \begin{Bmatrix} \mathbf{r}_u \\ \mathbf{r}_\varphi \end{Bmatrix} \tag{5–1}$$

where $\mathbf{v}_a$ are the axial deformations, $\mathbf{v}_b$ the bending deformations, $\mathbf{r}_u$ the joint translations, and $\mathbf{r}_\varphi$ the joint rotations. The zero submatrix in Eq. (5–1) applies to structures with point joints, for which joint rotations do not affect axial deformations. The inextensional modes are such that

$$\mathbf{v}_a = \mathbf{a}_{au} \mathbf{r}_u = 0 \tag{5–2}$$

The properties of a homogeneous system of linear equations have been discussed in Section 4. Let (NU) be the size of $\mathbf{r}_u$, and (NA) the size of $\mathbf{v}_a$. For Eq. (5–2) to have nontrivial solutions, the rank ρ of $\mathbf{a}_{au}$ must be less than (NU), in which case the number of independent solutions is

$$(NM) = (NU) - \rho \tag{5–3}$$

These solutions form (NM) independent translational inextensional modes. In addition to these modes, there are $(N\Phi)$ rotations $\mathbf{r}_\varphi$.

To choose generalized displacements for the translational inextensional modes, let $\mathbf{a}_{11}$ be a square submatrix of $\mathbf{a}_{au}$ of rank ρ, and let $\mathbf{a}_{au}$ be reordered and partitioned in the form

$$\mathbf{a}_{au} = \begin{bmatrix} \mathbf{a}_{11} & \mathbf{a}_{12} \\ \mathbf{a}_{21} & \mathbf{a}_{22} \end{bmatrix} \tag{5–4}$$

$\mathbf{r}_u$, $\mathbf{v}_a$ and the connectivity equations are reordered and partitioned accordingly in the form

$$\begin{bmatrix} \mathbf{a}_{11} & \mathbf{a}_{12} \\ \mathbf{a}_{21} & \mathbf{a}_{22} \end{bmatrix} \begin{Bmatrix} \mathbf{r}_1 \\ \mathbf{r}_2 \end{Bmatrix} = \begin{Bmatrix} \mathbf{v}_1 \\ \mathbf{v}_2 \end{Bmatrix} \tag{5–5}$$

Following the same procedure as in the analysis of kinematically unstable structures, we obtain the transformation

$$\begin{Bmatrix} \mathbf{r}_1 \\ \mathbf{r}_2 \\ \mathbf{r}_\varphi \end{Bmatrix} = \begin{bmatrix} \mathbf{a}_{11}^{-1} & -\mathbf{a}_{11}^{-1}\mathbf{a}_{12} & 0 \\ 0 & \mathbf{I} & 0 \\ 0 & 0 & \mathbf{I} \end{bmatrix} \begin{Bmatrix} \mathbf{v}_1 \\ \mathbf{r}_2 \\ \mathbf{r}_\varphi \end{Bmatrix} \tag{5–6}$$

and the compatibility equation

$$\mathbf{v}_2 = \mathbf{a}_{21}\mathbf{a}_{11}^{-1}\mathbf{v}_1 \tag{5–7}$$

The extensional modes are defined by $\mathbf{v}_1$ and have $\mathbf{r}_2 = 0$ and $\mathbf{r}_\varphi = 0$. The inextensional translational modes are defined by $\mathbf{r}_2$, and the rotational ones by $\mathbf{r}_\varphi$.

If the rank of $\mathbf{a}_{au}$ is equal to the number of axial deformations, the partitioning of $\mathbf{a}_{au}$ becomes $[\mathbf{a}_{11} \ \ \mathbf{a}_{12}]$. Then $\mathbf{v}_1 = \mathbf{v}_a$, and there is no compatibility equation for the axial deformations. In this case the unstable structure obtained by transforming rigid joints into pinned joints is statically determinate.

If the rank of $\mathbf{a}_{au}$ is equal to the number of joint translations, the pin-jointed structure obtained from the actual one is kinematically stable, and the only inextensional modes are rotational.

Equation (5–6) may be written in the form

$$\mathbf{r} = \mathbf{T}\mathbf{q} = [\mathbf{T}_1 \ \ \mathbf{T}_2]\begin{Bmatrix}\mathbf{q}_1\\ \mathbf{q}_2\end{Bmatrix} \tag{5–7}$$

where

$$\mathbf{q}_1 = \{\mathbf{r}_2 \ \ \mathbf{r}_\varphi\} \tag{5–8a}$$

$$\mathbf{q}_2 = \mathbf{v}_1 \tag{5–8b}$$

$$\mathbf{T}_1 = \begin{bmatrix} -\mathbf{a}_{11}^{-1}\mathbf{a}_{12} & 0\\ \mathbf{I} & 0\\ 0 & \mathbf{I}\end{bmatrix} \tag{5–9a}$$

$$\mathbf{T}_2 = \begin{bmatrix} \mathbf{a}_{11}^{-1}\\ 0\\ 0\end{bmatrix} \tag{5–9b}$$

An example illustrating the preceding may be seen in Chapter 10, Section 12.

EXERCISES

Sections 1 to 3

1. Explain why the assemblages in Fig. 1–2*f*, *g* are triangulated, those of Fig. 1–3*b*, *c*, *d* are simple, and those of Fig. 1–5 are compound.
2. Determine the degree of static indeterminacy of the trusses shown in Fig. P2. Indicate which of these trusses are kinematically unstable.
3. Do the preceding exercise for the frames of Fig. P3.
4. Give an example of a primary structure for each stable indeterminate structure in Exercises 2 and 3.

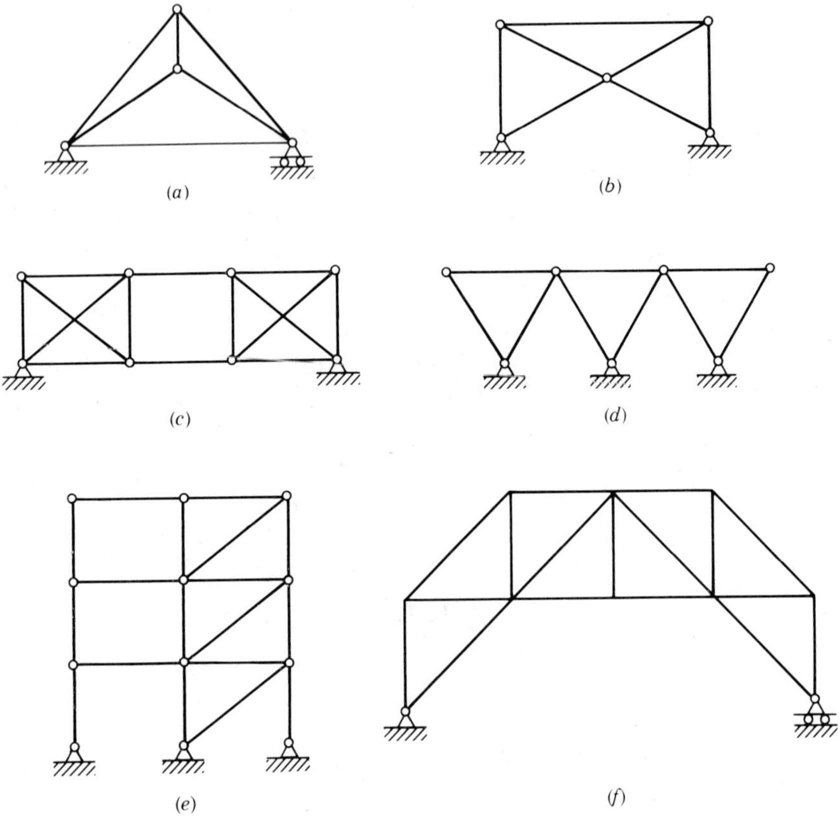

FIG. P2

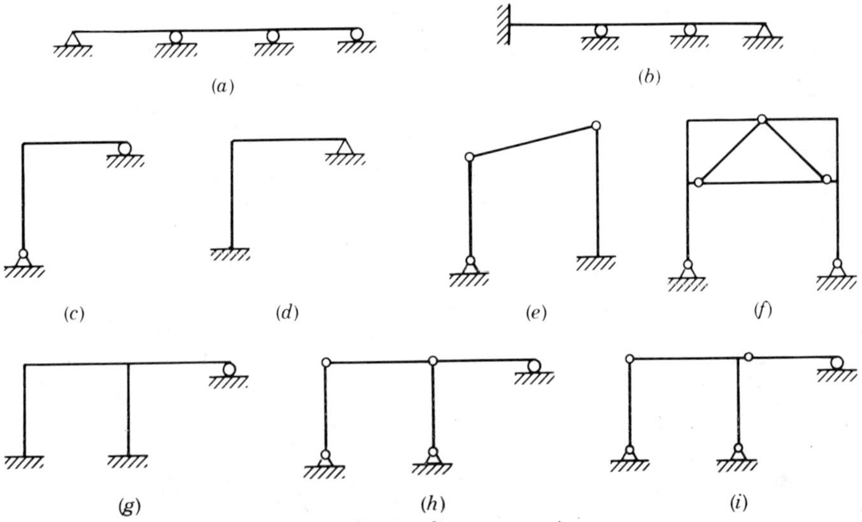

FIG. P3 (*Continued on next page.*)

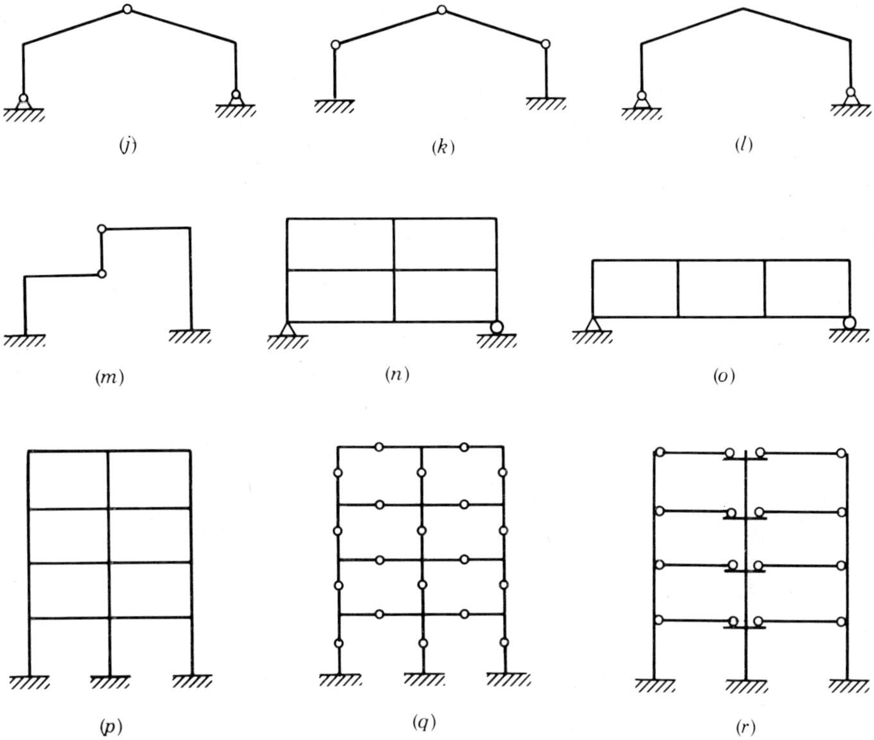

FIG. P3 (*continued*)

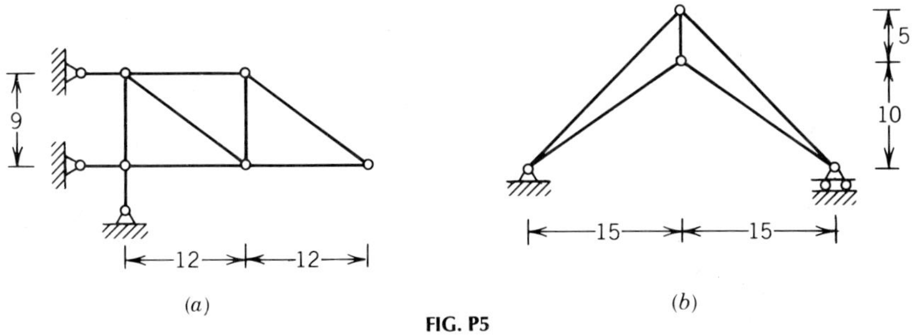

FIG. P5

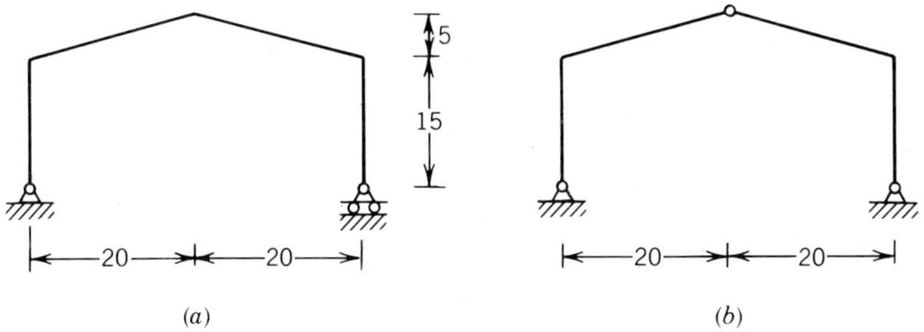

FIG. P6

5. Form by statics the matrix **b** (Eq. 2–5) for the trusses of Fig. P5. The *i*th column of **b** is the set of member forces in equilibrium with $R_i = 1$. Show the ordering chosen for **V** and **R**.

6. Do the preceding exercise for the frame of Fig. P6. Choose member redundants of alternate type.

7. In the truss of Fig. P5*a* let there be a second diagonal member. Choosing the axial force in that member as statical redundant form by statics matrices **b** and **c** (Eq. 2–14b).

8. In the frame of Fig. P6*b* let the pin connection at the top become a rigid connection. Choosing the bending moment at that location as a statical redundant, form by statics **b** and **c**.

9. Do an exercise similar to the preceding one for the frame of Fig. P6*a* by replacing the roller support by a pin.

10. For any of Exercises 5 through 9 form $\mathbf{a}_v$, and with the help of a calculator or a computer determine **b** and **c** as defined in terms of operations on $\mathbf{a}_v$.

Sections 4, 5

11. For the frames of Fig. P3 determine the number of translational degrees of freedom if the members are assumed inextensional.

12. Determine by the method of virtual displacements the axial forces in the members of the truss of Fig. P12 one by one.

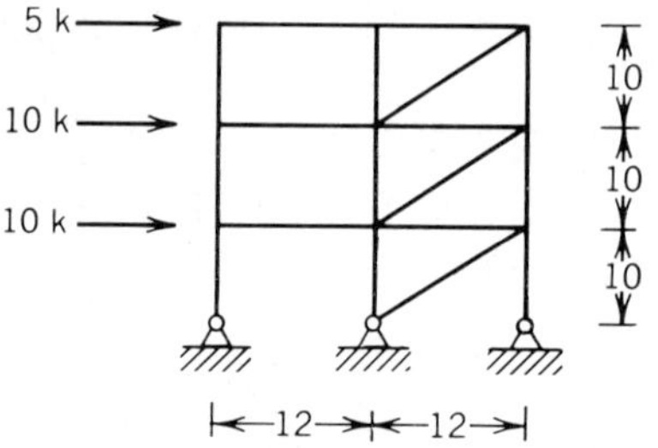

FIG. P12

13. Determine by the method of virtual displacements the bending moments at *A*, *B*, *C*, *D*, and *E* in Fig. P13 one by one.

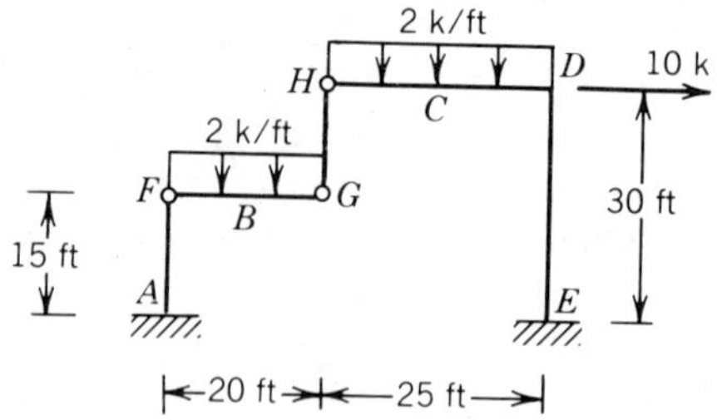

FIG. P13

14. Assume that the joints at F, G, and H of Fig. P13 are rigid. Derive by the method of virtual displacements equilibrium equations relating the bending moments at the following locations:

a. A, F, G, H
b. F, G, H, D
c. H, C, D
d. A, F, D, E,

15. Determine by the method of virtual displacements the ratio of P_1 to P_2 so that the bar systems in Fig. P15 are in equilibrium. In (*b*) the joints at A and B are as shown in Fig. 8/1–3b.

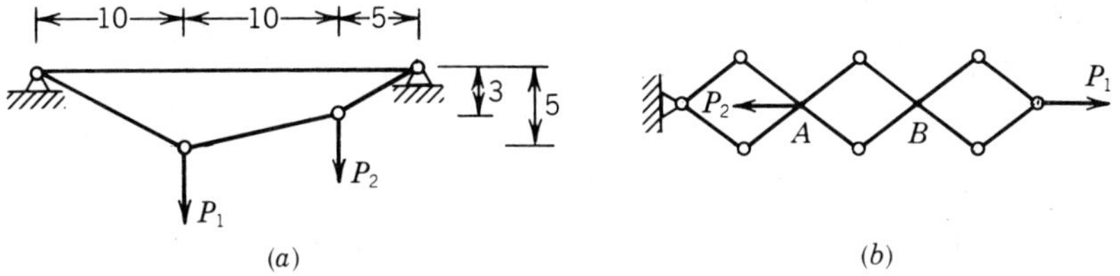

FIG. P15

16. Transform the joint displacements $\mathbf{r}$ of the rectangular frame in Fig. P3*p* to a set $\mathbf{q} = \{\mathbf{q}_1\ \mathbf{q}_2\}$ corresponding to inextensional and extensional deformation modes, respectively.

10

DISPLACEMENT (STIFFNESS) METHOD

Linear Theory

1. STIFFNESS EQUATIONS: SOLUTION STEPS

In the displacement method the governing system of equations is reduced to the joint equilibrium equations expressed in terms of the displacements. The equations thus obtained are referred to as the structure stiffness equations and are derived in Chapter 8, Section 2.2, in the form

$$\mathbf{Kr} = \mathbf{R} - \mathbf{R}_F \tag{1–1}$$

$\mathbf{K}$ and $\mathbf{R}_F$ have, respectively, three equivalent expressions, depending on the governing system used. These are

$$\mathbf{K} = \bar{\mathbf{a}}^T \bar{\mathbf{k}}_c \bar{\mathbf{a}} = \mathbf{a}^T \mathbf{k}_c \mathbf{a} = \mathbf{a}_v^T \mathbf{k} \mathbf{a}_v \tag{1–2}$$

$$\mathbf{R}_F = \bar{\mathbf{a}}^T \bar{\mathbf{S}}_F = \mathbf{a}^T \mathbf{S}_F = \mathbf{a}_v^T \mathbf{V}_F + \mathbf{a}^T \mathbf{S}_p \tag{1–3}$$

Governing equations are described with examples in Chapter 8, Sections 2.2, 2.3, 2.4, 3, and 4. $\mathbf{K}$ and $\mathbf{R}_F$ may be evaluated for any one of these examples by executing defining matrix operations in Eqs. (1–2) and (1–3). However, a procedure for forming $\mathbf{K}$ and $\mathbf{R}_F$, which is more efficient both for hand calculations and programmed procedures, is presented in Section 3.

The stiffness equations are generalized in Chapter 8, Section 4, with $\mathbf{r}$ representing generalized displacements, and $(\mathbf{R} - \mathbf{R}_F)$ generalized equivalent joint loads. This terminology is used only where needed for clarity.

If $\mathbf{r}$ includes support displacements, support conditions are yet to be enforced in the stiffness equations. $\mathbf{K}$ is then called the complete stiffness matrix, although it may be effective in some other aspect.

Let $\mathbf{r}_s$ be a subset of prescribed joint displacements, and $\mathbf{r}_f$ the remaining subset of $\mathbf{r}$, or free displacements. To the partitioning $\mathbf{r} = \{\mathbf{r}_f \ \mathbf{r}_s\}$ corresponds the partitioning $(\mathbf{R} - \mathbf{R}_F) = \{\mathbf{R}_f - \mathbf{R}_{f,F}, \mathbf{R}_s - \mathbf{R}_{s,F}\}$. The unknowns in the joint loads are the reactions $\mathbf{R}_s$. The stiffness equations partition into

$$\begin{bmatrix} \mathbf{K}_{ff} & \mathbf{K}_{fs} \\ \mathbf{K}_{sf} & \mathbf{K}_{ss} \end{bmatrix} \begin{Bmatrix} \mathbf{r}_f \\ \mathbf{r}_s \end{Bmatrix} = \begin{Bmatrix} \mathbf{R}_f - \mathbf{R}_{f,F} \\ \mathbf{R}_s - \mathbf{R}_{s,F} \end{Bmatrix} \tag{1–4}$$

The first set of the partitioned equations gives

$$\mathbf{K}_{ff} \mathbf{r}_f = \mathbf{R}_f - \mathbf{R}_{f,F} - \mathbf{K}_{fs} \mathbf{r}_s \tag{1–5}$$

It will be seen in Section 10 that $\mathbf{K}_{ff}$ is nonsingular if the supported structure is kinematically stable. Assuming this to be the case, Eq. (1–5) is solved for $\mathbf{r}_f$. The second set of equations allows us then to compute the reactions at the supports through

$$\mathbf{R}_s = \mathbf{K}_{sf} \mathbf{r}_f + \mathbf{K}_{ss} \mathbf{r}_s + \mathbf{R}_{s,F} \tag{1–6}$$

Having determined $\mathbf{r}_f$, the remaining unknowns are found using equations of the governing sets established in Chapter 8. For example, in local coordinates $\mathbf{s}$ and $\mathbf{S}$ are determined using $\mathbf{s} = \mathbf{a}\mathbf{r}$ and $\mathbf{S} = \mathbf{k}_c \mathbf{s} + \mathbf{S}_F$. Alternately, $\mathbf{v}$ and $\mathbf{V}$ are determined through $\mathbf{v} = \mathbf{a}_v \mathbf{r}$ and $\mathbf{V} = \mathbf{k}\mathbf{v} + \mathbf{V}_F$; then $\mathbf{S} = \mathbf{S}_p + \mathbf{B}\mathbf{V}$.

If support conditions at a joint i prescribe displacement components on axes different from the axes used for defining $\mathbf{r}^i$, a coordinate transformation is made from $\mathbf{r}^i$ to displacements $\mathbf{r}'^i$ which include the displacements to be prescribed. The rest of the procedure remains unchanged. The transformation from $\mathbf{r}^i$ to $\mathbf{r}'^i$ may be made on the structure stiffness equations (Chapter 8, Section 4), or on the member stiffness equations (Chapter 7, Section 3.3), prior to forming Eqs. (1–1).

Inclusion of support displacements in **r** is convenient for treating different cases of support conditions and for obtaining the reactions at supports as part of the general solution. If such considerations are secondary, $\mathbf{r}_f$ may be chosen as effective displacements, and the stiffness equations reduce then to Eq. (1–5).

Example 1

A continuous beam is a type of structure for which, with a proper choice of effective variables, the three sets of governing equations may be made to coincide. The stiffness equations for the structure of Fig. 1–1 are established in what follows, using the least number of effective degrees of freedom:

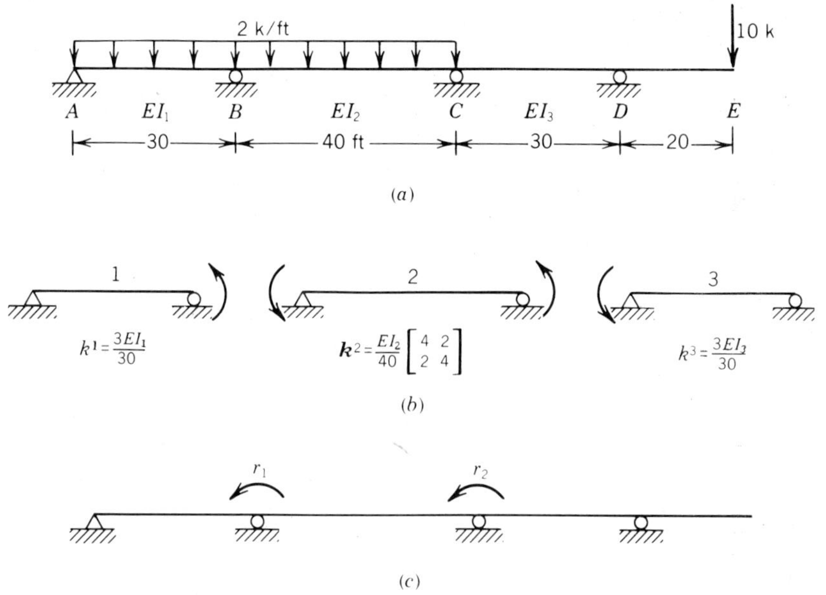

$$\mathbf{a}_v = \mathbf{a} = \bar{\mathbf{a}} = \begin{bmatrix} 1 & 0 \\ 1 & 0 \\ 0 & 1 \\ 0 & 1 \end{bmatrix} \quad (d) \qquad \mathbf{k} = \frac{E}{20}\begin{bmatrix} 2I_1 & & & \\ & 2I_2 & I_2 & \\ & I_2 & 2I_2 & \\ & & & 2I_3 \end{bmatrix} \quad (e)$$

$$\mathbf{K} = \mathbf{a}_v^T \mathbf{k} \mathbf{a}_v = \frac{E}{20}\begin{bmatrix} 2I_1 + 2I_2 & I_2 \\ I_2 & 2I_2 + 2I_3 \end{bmatrix} \quad (f)$$

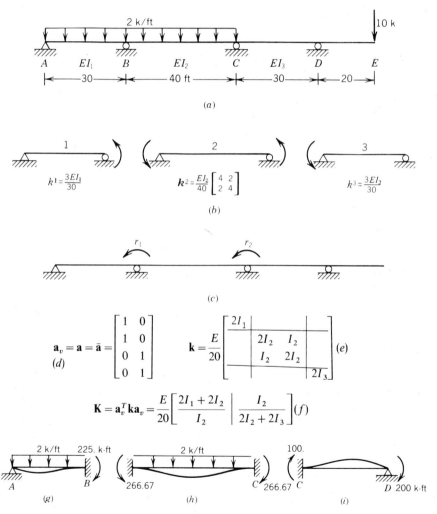

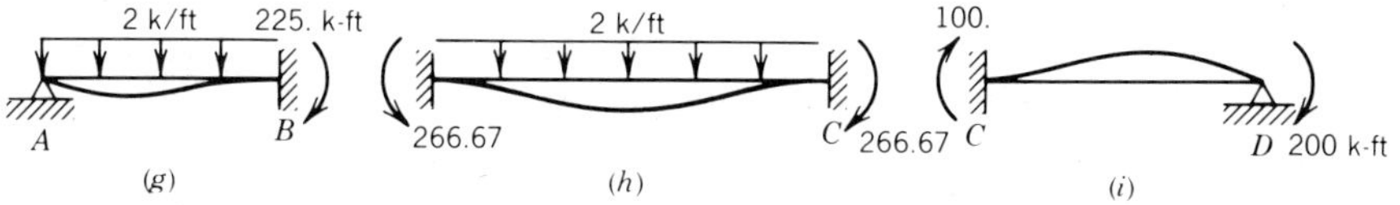

FIG. 1–1. Continuous beam, Section 1, Example 1; Section 3, Example 1. (*Continued on following page.*)

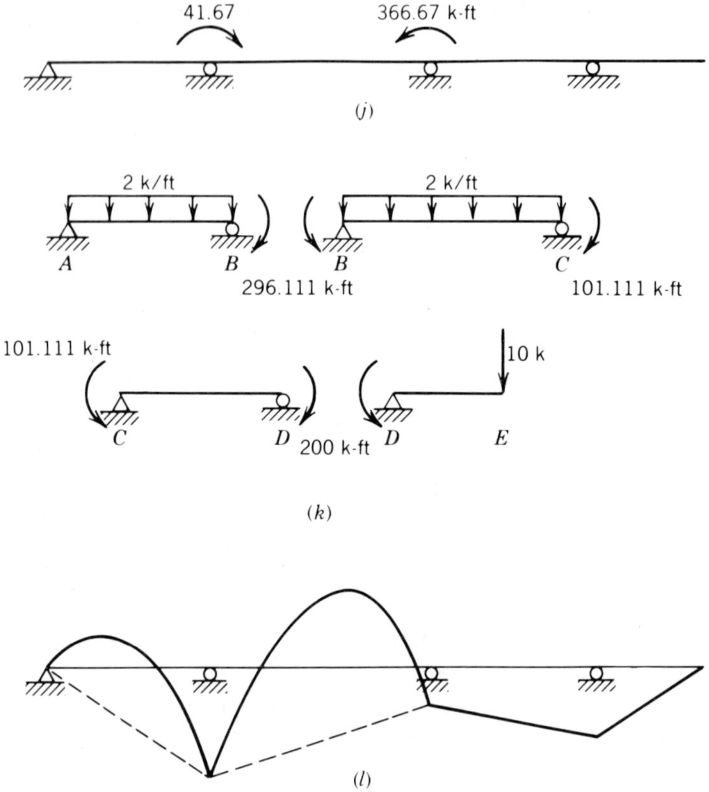

FIG. 1–1. *(continued)*

First, it is noted that in the geometrically linear theory the axial behavior is independent of the bending behavior. For the loading shown, there are no axial forces so that only a bending analysis is made. The effective member end displacements are identified in (*b*). They coincide in this case with deformations of the alternate type. Member *DE* is removed from the set of members because it is statically determinate. Its forces at *D* are applied to the rest of the structure. The effective member stiffness matrices $\mathbf{k}^m$ are shown under each member with shear deformation neglected. The effective joint displacements are the rotations $\mathbf{r}_1$ and $\mathbf{r}_2$ at the two inner joints. $\mathbf{a}_v$, $\mathbf{k}$ and $\mathbf{K}$ are shown in (*d*), (*e*) and (*f*). The effective fixed end moments are obtained in the state $\mathbf{r} = 0$ as shown in (*g*), (*h*), and (*i*). Note that the rotations at *A* and *D* must remain released in this calculation and that the applied moment at *D* is not part of the joint load matrix because the rotation at *D* is removed from $\mathbf{r}$. The fixed end moments shown in (*g*), (*h*) and (*i*) define $\mathbf{V}_F = \mathbf{S}_F = \{-225 \;\; 266.67 \;\; -266.67 \;\; -100\}$; then $-\mathbf{R}_F = -\mathbf{a}^T\mathbf{S}_F = \{-41.67 \;\; 366.67\}$ k-ft. This result is also found by summing the fixed end moments at each joint and reversing the sense of the results. In this example, $\mathbf{R} = 0$ and $-\mathbf{R}_F$ is the total equivalent joint load as shown in (*j*). Assuming for simplicity that $EI_1 = EI_2 = EI_3$, the structure stiffness equations take the form

$$\frac{EI}{20}\begin{bmatrix} 4 & 1 \\ 1 & 4 \end{bmatrix}\begin{Bmatrix} r_1 \\ r_2 \end{Bmatrix} = \begin{Bmatrix} -41.67 \\ 366.67 \end{Bmatrix}$$

The solution is found to be

$$\mathbf{r} = \frac{1}{EI}\{-711.111 \;\; 2011.111\}$$

To obtain member deformations and forces, we have

$$\mathbf{v} = \mathbf{a}_v\mathbf{r} = \{r_1 \;\; r_1 \;\; r_2 \;\; r_2\}$$

$$\mathbf{V} = \mathbf{kv} + \mathbf{V}_F = \frac{EI}{20}\{2r_1,\; 2r_1 + r_2,\; r_1 + 2r_2,\; 2r_2\} + \mathbf{V}_F$$

$$= \{-71.111 \;\; 29.444 \;\; 165.555 \;\; 201.111\} + \mathbf{V}_F$$

$$= \{-296.111 \;\; 296.111 \;\; -101.111 \;\; 101.111\}\text{k} - \text{ft}$$

These results are shown in (*k*). The analysis is concluded by a bending moment diagram that is prepared separately for each member. The dotted line in the figure represents the bending moment due to the end moments. The value measured from the dotted line is the bending moment in the simply supported member.

2. STRUCTURES WITHOUT JOINT TRANSLATIONS: CONTINUOUS BEAMS

Structures without joint translations occur typically in laterally supported frames, assumed to be inextensible, and in continuous beams. The displacements for such structures are the joint rotations, and the stiffness equations express the joint moment equilibrium equations.

The stiffness equations of continuous beams are simple and systematic enough to receive a particular treatment. The joint rotations are given the same ordering as the joints, as shown in Fig. 2–1. A general member i spans joints i and $i+1$. It is shown isolated in (*b*) The bending stiffness equations of member i are

$$V_2^i = k_{22}^i r_i + k_{23}^i r_{i+1} + V_{2F}^i \tag{2–1a}$$

$$V_3^i = k_{32}^i r_i + k_{33}^i r_{i+1} + V_{3F}^i \tag{2–1b}$$

The ith stiffness equation of the structure is the moment equilibrium equation at joint i,

$$V_3^{i-1} + V_2^i = R_i \tag{2–2}$$

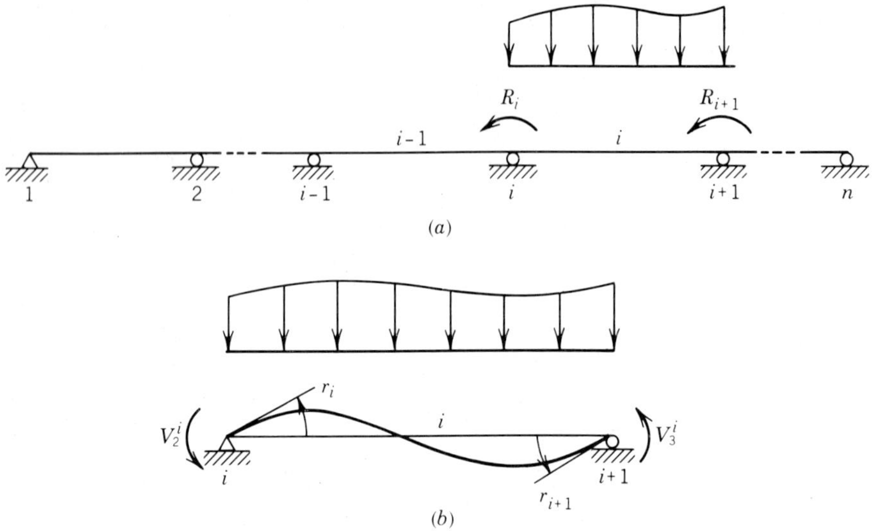

FIG. 2–1. Notation and sign convention for continuous beam.

where R_i is a concentrated moment load at joint i. The equivalent moment load at joint i is

$$R_{iT} = R_i - R_{iF} \tag{2–3a}$$

where

$$R_{iF} = V_{3F}^{i-1} + V_{2F}^{i} \tag{2–3b}$$

The ith stiffness equation at interior joints takes the form

$$k_{32}^{i-1} r_{i-1} + (k_{33}^{i-1} + k_{22}^{i}) r_i + k_{23}^{i} r_{i+1} = R_{iT} \tag{2–4}$$

If the first and last joints are unrestrained against rotation, the stiffness equation at these joints coincides with the member stiffness equation. There results n stiffness equations in n unknown joint rotations. Note that it is also possible to condense the first and last unrestrained rotations by establishing effective stiffness equations for the first and last member, thus reducing to $n-2$ the number of equations and unknowns. If end 1 is fixed, no stiffness equation is written at that end, and the condition $r_1 = 0$ is enforced in the stiffness equation at joint 2. A similar procedure is followed if joint n is fixed.

For a uniform member, or for a member having a plane of symmetry, the stiffness matrix has equal terms on the main diagonal. These are denoted k_i for member i, and the off-diagonal terms are denoted k_i'. Equation (2–4) becomes

$$k_{i-1}' r_{i-1} + (k_{i-1} + k_i) r_i + k_i' r_{i+1} = R_{iT} \tag{2–5}$$

For a uniform span without shear deformation,

$$k_i = \frac{4EI}{l} \tag{2–6a}$$

$$k_i' = \frac{2EI}{l} \tag{2–6b}$$

Example 1

Consider as an example the continuous beam of Fig. 2–2. The analysis will first be carried out without condensation of r_1 and r_4. The fixed end moments for member 1 are

$$\{V_{2F}^1 \; V_{3F}^1\} = \left\{\frac{pl^2}{12} \quad -\frac{pl^2}{12}\right\}$$

The stiffness equations take the form

$$\frac{2EI}{l}\begin{bmatrix} 2 & 1 & & \\ 1 & 4 & 1 & \\ & 1 & 4 & 1 \\ & & 1 & 2 \end{bmatrix}\begin{Bmatrix} r_1 \\ r_2 \\ r_3 \\ r_4 \end{Bmatrix} = \frac{pl^2}{12}\begin{Bmatrix} -1 \\ 1 \\ 0 \\ 0 \end{Bmatrix}$$

The first and last equations yield

$$r_1 = -\frac{1}{2}r_2 - \frac{pl^3}{24EI}$$

$$r_4 = -\frac{1}{2}r_3$$

The second and third equations become

$$\frac{EI}{l}\begin{bmatrix} 7 & 2 \\ 2 & 7 \end{bmatrix}\begin{Bmatrix} r_2 \\ r_3 \end{Bmatrix} = \frac{pl^2}{8}\begin{Bmatrix} 1 \\ 0 \end{Bmatrix}$$

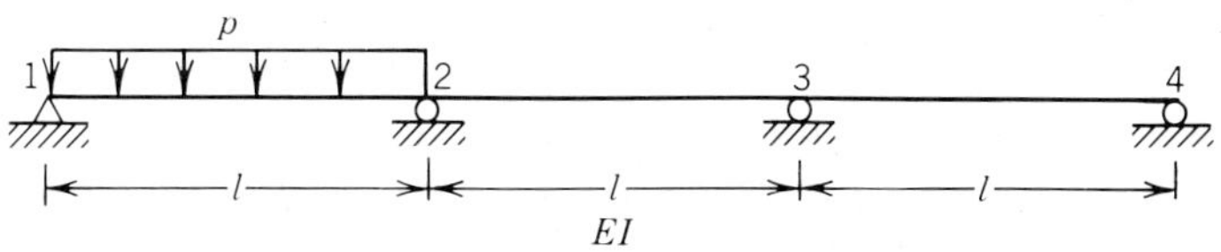

FIG. 2–2. Continuous beam, Section 2, Example 1.

The solution is found to be

$$r_2 = \frac{7pl^3}{360EI}$$

$$r_3 = -\frac{2pl^3}{360EI}$$

r_1 and r_4 are evaluated using their expressions, as given earlier, and the member end moments are found by substituting for the displacements in the member stiffness equations.

To illustrate the use of effective stiffness relations for the first and last member, we have for the first member, $V_2^1 = 0$ and

$$V_3^1 = \frac{3EI}{l} r_2 - \frac{pl^2}{8}$$

For the last member, $V_3^3 = 0$, and

$$V_2^3 = \frac{3EI}{l} r_3$$

The moment equilibrium equations at joints 2 and 3 now yield the two equations for r_2 and r_3 found here.

3. ASSEMBLY OF THE STIFFNESS EQUATIONS BY THE DIRECT STIFFNESS METHOD

Connectivity matrices are usually sparse and are more efficiently dealt with member by member. The direct stiffness method is a method of forming the stiffness matrix $\mathbf{K}$ and the load matrix $\mathbf{R}_T = \mathbf{R} - \mathbf{R}_F$ by processing member matrices one member at a time, thus avoiding the inefficiency of wasted storage space and some useless matrix operations inherent in applying the defining equations of $\mathbf{K}$ and $\mathbf{R}_T$. To outline the method, it is first assumed that there are no internal kinematic constraints such as member inextensibility and that $\mathbf{r}$ and $\mathbf{R}$ are partitioned by joints in the form $\mathbf{r} = \{\mathbf{r}^1 \ \mathbf{r}^2 \ldots \mathbf{r}^i \ldots \mathbf{r}^j\}$, and similarly for $\mathbf{R}$. The joint equilibrium equation for a general joint i is

$$\sum_m (\bar{\mathbf{a}}^{m,i})^T \bar{\mathbf{S}}^{m,i} = \mathbf{R}^i \tag{3–1}$$

In Eq. (3–1) the sum extends over all members m incident on joint i. The connectivity relation at such a member end is

$$\bar{\mathbf{s}}^{m,i} = \bar{\mathbf{a}}^{m,i} \mathbf{r}^i \tag{3–2}$$

Letting j denote the other end of member m, the stiffness equations of the member, partitioned by member ends, give for $\bar{\mathbf{S}}^{m,i}$,

$$\bar{\mathbf{S}}^{m,i} = \bar{\mathbf{k}}_c^{m,ii}\bar{\mathbf{s}}^{m,i} + \bar{\mathbf{k}}_c^{m,ij}\bar{\mathbf{s}}^{m,j} + \bar{\mathbf{S}}_F^{m,i} \tag{3–3}$$

Substituting for $\bar{\mathbf{S}}^{m,i}$ into Eq. (3–1), and using the connectivity relation (3–2) for $\bar{\mathbf{s}}^{m,i}$ and $\bar{\mathbf{s}}^{m,j}$, Eq. (3–1) turns into

$$\mathbf{K}^{ii}\mathbf{r}^i + \sum_{j \neq i} \mathbf{K}^{ij}\mathbf{r}^j = \mathbf{R}^i - \mathbf{R}_F^i \tag{3–4}$$

where

$$\mathbf{K}^{ij} = \mathbf{K}^{m,ij} = (\bar{\mathbf{a}}^{m,i})^T \bar{\mathbf{k}}_c^{m,ij} \bar{\mathbf{a}}^{m,j} \tag{3–5}$$

$$\mathbf{K}^{ii} = \sum_m \mathbf{K}^{m,ii} = \sum_m (\bar{\mathbf{a}}^{m,i})^T \bar{\mathbf{k}}_c^{m,ii} \bar{\mathbf{a}}^{m,i} \tag{3–6}$$

$$\mathbf{R}_F^i = \sum_m \mathbf{R}_F^{m,i} = \sum_m (\bar{\mathbf{a}}^{m,i})^T \bar{\mathbf{S}}_F^{m,i} \tag{3–7}$$

Equation (3–4) is the ith equation of the partitioned stiffness equations. $\mathbf{K}$ is partitioned into $J \times J$ submatrices or blocks, such that submatrix $\mathbf{K}^{ij}$ occupies the position at row i and column j. $\mathbf{R}_T$ is partitioned into J submatrices, such that $\mathbf{R}_T^i$ occupies row i.

According to Eq. (3–6) a main diagonal block of $\mathbf{K}$ such as $\mathbf{K}^{ii}$ is formed by superposition of member contributions $\mathbf{K}^{m,ii}$ for all members connected to joint i. From Eq. (3–5) an off-diagonal block such as $\mathbf{K}^{ij}$ is the contribution $\mathbf{K}^{m,ij}$ of the member connecting joints i and j and is zero if no such member exists. Thus the contribution to $\mathbf{K}$ of a given member m with ends at joints i and j is formed of four blocks $\mathbf{K}^{m,ii}$, $\mathbf{K}^{m,ij}$, $\mathbf{K}^{m,ji}$, and $\mathbf{K}^{m,jj}$ to be placed, respectively, in locations (i, i), (i, j), (j, i), and (j, j). Equation (3–5), which is written for $j \neq i$ holds in fact for all four combinations (i, i), (i, j), (j, i), and (j, j).

Similar derivations in terms of $\mathbf{k}_c^m$ and $\mathbf{k}^m$ yield the equivalent formulas

$$\mathbf{K}^{m,ij} = (\bar{\mathbf{a}}^{m,i})^T \bar{\mathbf{k}}_c^{m,ij} \bar{\mathbf{a}}^{m,j} = (\mathbf{a}^{m,i})^T \mathbf{k}_c^{m,ij} \mathbf{a}^{m,j} = (\mathbf{a}_v^{m,i})^T \mathbf{k}^m \mathbf{a}_v^{m,j} \tag{3–8}$$

The contribution of member m joining joints i and j to $\mathbf{R}_T$ consists of $-\mathbf{R}_F^{m,i}$ and $-\mathbf{R}_F^{m,j}$. From Eq. (3–7) and similar equations written in terms of $\mathbf{S}^m$ and $\mathbf{V}^m$, we can write, with i referring to either member end, the three equivalent expressions:

$$\mathbf{R}_F^{m,i} = (\bar{\mathbf{a}}^{m,i})^T \bar{\mathbf{S}}_F^{m,i} = (\mathbf{a}^{m,i})^T \mathbf{S}_F^{m,i} = (\bar{\mathbf{a}}^{m,i})^T \bar{\mathbf{S}}_p^{m,i} + (\mathbf{a}_v^{m,i})^T \mathbf{V}_F^m \tag{3–9}$$

The four submatrices contributed to $\mathbf{K}$ by member m may be evaluated together as submatrices of a matrix $\mathbf{K}^m$ partitioned by joints. Similarly the contributions of member m to $\mathbf{R}_F$ may be evaluated together, as submatrices of

$\mathbf{R}_F^m$. Letting $\mathbf{r}^m = \{\mathbf{r}^i \ \mathbf{r}^j\}$, connectivity equations for member m have the form

$$\mathbf{s}^m = \begin{Bmatrix} \mathbf{s}^{m,i} \\ \mathbf{s}^{m,j} \end{Bmatrix} = \begin{bmatrix} \mathbf{a}^{m,i} & \\ & \mathbf{a}^{m,j} \end{bmatrix} \begin{Bmatrix} \mathbf{r}^i \\ \mathbf{r}^j \end{Bmatrix} = \mathbf{a}^m \mathbf{r}^m \tag{3–10}$$

$$\mathbf{v}^m = [\mathbf{a}_v^{m,i} \ \ \mathbf{a}_v^{m,j}] \begin{Bmatrix} \mathbf{r}^i \\ \mathbf{r}^j \end{Bmatrix} = \mathbf{a}_v^m \mathbf{r}^m \tag{3–11}$$

$\mathbf{K}^m$ and $\mathbf{R}_F^m$ have, respectively, the three equivalent expressions

$$\mathbf{K}^m = \bar{\mathbf{a}}^{mT} \bar{\mathbf{k}}_c^m \bar{\mathbf{a}}^m = \mathbf{a}^{mT} \mathbf{k}_c^m \mathbf{a}^m = \mathbf{a}_v^{mT} \mathbf{k}^m \mathbf{a}_v^m \tag{3–12}$$

$$\mathbf{R}_F^m = \bar{\mathbf{a}}^{mT} \bar{\mathbf{S}}_F^m = \mathbf{a}^{mT} \mathbf{S}_F^m = \bar{\mathbf{a}}^{mT} \bar{\mathbf{S}}_p^m + \mathbf{a}_v^{mT} \mathbf{V}_F^m \tag{3–13}$$

Example 1

Consider as an example the rectangular frame shown in Fig. 3–1*a*. Let the zero support displacements be excluded from the effective displacements, and let **r** be ordered by joints, or $\mathbf{r} = \{\mathbf{r}^1 \mathbf{r}^2 \ldots \mathbf{r}^5\}$. At joint j, $\mathbf{r}^j = \{r_1^j \, r_2^j \, r_3^j\}$ as illustrated for joint 5. The ordering of member global displacements is shown for typical members in (*b*) and (*c*). In this example all connectivity matrices $\bar{\mathbf{a}}^{m,i}$ or $\bar{\mathbf{a}}^{m,j}$ are 3×3 identity matrices. Consequently the contribution $\mathbf{K}^m$ of any member m to **K** is its stiffness matrix $\bar{\mathbf{k}}_c^m$, and the contribution $\mathbf{R}_F^m$ to $\mathbf{R}_F$ is $\bar{\mathbf{S}}_F^m$. **K** and $\mathbf{R}_F$ are shown partitioned by joints in (*d*) and (*e*), respectively. For member m, joining joints i and j, $\bar{\mathbf{k}}_c^m$ is partitioned by joints into four blocks $\mathbf{K}^{m,ij}$. Each block is added into, or merged, in block (i, j) of **K**. For example, for member 5, $i = 2$ and $j = 3$. $\mathbf{K}^{5,22}$, $\mathbf{K}^{5,23}$, $\mathbf{K}^{5,32}$, and $\mathbf{K}^{5,33}$ are merged into **K**, in blocks (2, 2), (2, 3), (3, 2), and (3, 3), respectively. For member 6, the four blocks are merged into blocks (1, 1), (1, 4), (4, 1), and (4, 4), respectively. The procedure is applied to all members, with this particular note for members 1, 2, and 3: if a support joint is labeled O, and the free joint is labeled j, then the effective $\mathbf{K}^m$ is $\mathbf{K}^{m,jj}$. The three other blocks of $\bar{\mathbf{k}}_c^m$ are ignored, and $\mathbf{K}^{m,jj}$ is merged into block (j, j). **K** is thus assembled by processing one member at a time, and the result is shown in (*d*).

For the horizontal members the global coordinates coincide with the local ones so that $\bar{\mathbf{k}}_c^m = \mathbf{k}_c^m$. For the vertical members the ordering of the global coordinates in (*c*) is a reordering of the local coordinates with a possible change in sign. The correspondance is indicated as follows:

Local ordering: 1, 2, 3, 4, 5, 6.
Global ordering: 2, − 1, 3, 5, − 4, 6.

$\bar{\mathbf{k}}_c^m$ may be formed by a symmetric merge of $\mathbf{k}_c^m$ into a 6×6 matrix according to this correspondance. For example, element (1, 4) of $\mathbf{k}_c^m$ goes into location (2, 5) of $\bar{\mathbf{k}}_c^m$, and element (4, 1) goes symmetrically into location (5, 2). Element (3, 5) of $\mathbf{k}_c^m$ goes into location (3, 4) of $\bar{\mathbf{k}}_c^m$, and its sign is changed. Element (2, 5) of $\mathbf{k}_c^m$

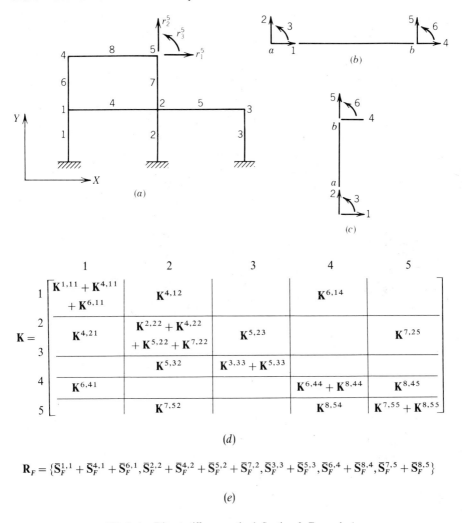

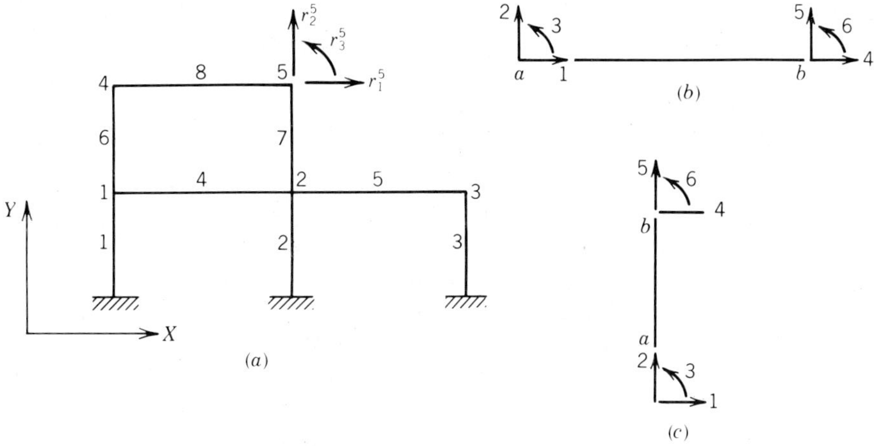

$$
\mathbf{K} = \left[\begin{array}{c|c|c|c|c}
 & 1 & 2 & 3 & 4 & 5 \\
1 & \mathbf{K}^{1,11} + \mathbf{K}^{4,11} + \mathbf{K}^{6,11} & \mathbf{K}^{4,12} & & \mathbf{K}^{6,14} & \\
\hline
2,3 & \mathbf{K}^{4,21} & \mathbf{K}^{2,22} + \mathbf{K}^{4,22} + \mathbf{K}^{5,22} + \mathbf{K}^{7,22} & \mathbf{K}^{5,23} & & \mathbf{K}^{7,25} \\
\hline
 & & \mathbf{K}^{5,32} & \mathbf{K}^{3,33} + \mathbf{K}^{5,33} & & \\
\hline
4 & \mathbf{K}^{6,41} & & & \mathbf{K}^{6,44} + \mathbf{K}^{8,44} & \mathbf{K}^{8,45} \\
\hline
5 & & \mathbf{K}^{7,52} & & \mathbf{K}^{8,54} & \mathbf{K}^{7,55} + \mathbf{K}^{8,55}
\end{array}\right]
$$

(*d*)

$$
\mathbf{R}_F = \{\bar{\mathbf{S}}_F^{1,1} + \bar{\mathbf{S}}_F^{4,1} + \bar{\mathbf{S}}_F^{6,1}, \bar{\mathbf{S}}_F^{2,2} + \bar{\mathbf{S}}_F^{4,2} + \bar{\mathbf{S}}_F^{5,2} + \bar{\mathbf{S}}_F^{7,2}, \bar{\mathbf{S}}_F^{3,3} + \bar{\mathbf{S}}_F^{5,3}, \bar{\mathbf{S}}_F^{6,4} + \bar{\mathbf{S}}_F^{8,4}, \bar{\mathbf{S}}_F^{7,5} + \bar{\mathbf{S}}_F^{8,5}\}
$$

(*e*)

FIG. 3–1. Direct stiffness method, Section 3, Example 1.

goes into location (1, 4) of $\bar{\mathbf{k}}_c^m$ without sign change because there are two sign reversals in the correspondance. The procedure described here may be applied directly to merge $\mathbf{k}_c^m$ into $\mathbf{K}$ without going through $\bar{\mathbf{k}}_c^m$. For example, the correspondance between member local ordering and $\mathbf{r}$ ordering for member 6 is as follows:

Local ordering: 1, 2, 3, 4, 5, 6.
Global ordering: r_2^1, $-r_1^1$, r_3^1, r_2^4, $-r_1^4$, r_6^4.

The global ordering is referred to as the merging sequence of the member stiffness matrix.

In forming $\mathbf{R}_T = \mathbf{R} - \mathbf{R}_F$, $\mathbf{R}$ consists of concentrated forces or moments applied directly at the joints. The forming of $\mathbf{R}_F$ by superposition of member contributions is shown in (*e*). For example, for member 5, $\bar{\mathbf{S}}_F^5 = \{\bar{\mathbf{S}}_F^{5,2}\bar{\mathbf{S}}_F^{5,3}\}$. $\bar{\mathbf{S}}_F^{5,2}$ is merged into block 2, and $\bar{\mathbf{S}}_F^{5,3}$ into block 3. Note that it is $-\bar{\mathbf{S}}_F^m$ that should be merged in forming $\mathbf{R}_T$.

Example 2

In the example of the continuous beam in Fig. 1–1*a*, the member contributions to $\mathbf{K}$ are shown in (*b*). Merging of these matrices is straightforward and leads to (*f*) without explicitly forming $\mathbf{a}$ and $\mathbf{k}$. Similarly the elements of $\mathbf{S}_F^m$ shown in Fig. 1–1*g*, *h*. *i* are merged with the proper sign in $\mathbf{R}_T$.

Example 3

For a pin-jointed truss the procedure for forming $\mathbf{K}$ and $\mathbf{R}_T$ is similar to the case of the rigid frame. The connectivity matrices $\bar{\mathbf{a}}^{m,i}$ and $\bar{\mathbf{a}}^{m,j}$ are 2×2 identity-matrices so that $\mathbf{K}^m = \bar{\mathbf{k}}_c^m$ and $\bar{\mathbf{k}}_c^m$ is the 4×4 stiffness matrix in Eq. (7/3.4–11). If support displacements are deleted, the corresponding rows and columns are deleted from $\bar{\mathbf{k}}_c^m$. For the truss in Fig. 3–2, with support displacements deleted,

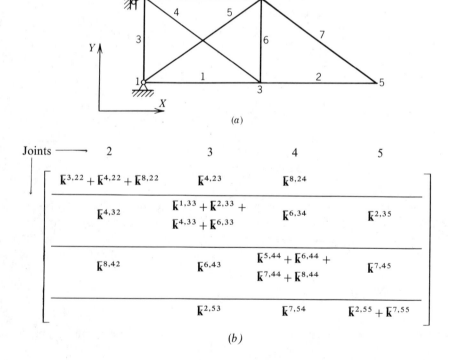

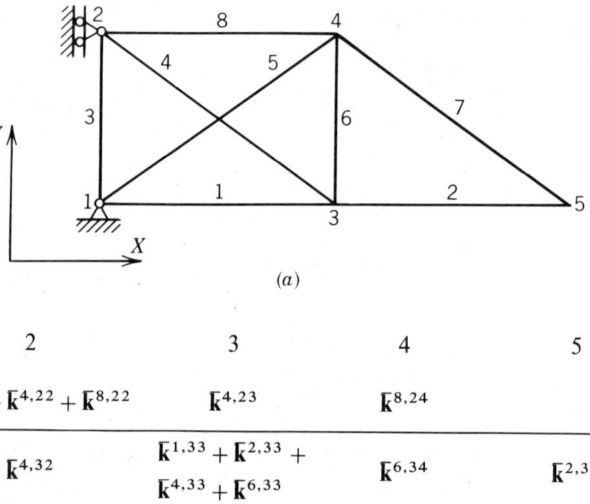

(*a*)

Joints → 2, 3, 4, 5

$$
\left[\begin{array}{c|c|c|c}
\bar{\mathbf{k}}^{3,22} + \bar{\mathbf{k}}^{4,22} + \bar{\mathbf{k}}^{8,22} & \bar{\mathbf{k}}^{4,23} & \bar{\mathbf{k}}^{8,24} & \\
\hline
\bar{\mathbf{k}}^{4,32} & \bar{\mathbf{k}}^{1,33} + \bar{\mathbf{k}}^{2,33} + \bar{\mathbf{k}}^{4,33} + \bar{\mathbf{k}}^{6,33} & \bar{\mathbf{k}}^{6,34} & \bar{\mathbf{k}}^{2,35} \\
\hline
\bar{\mathbf{k}}^{8,42} & \bar{\mathbf{k}}^{6,43} & \bar{\mathbf{k}}^{5,44} + \bar{\mathbf{k}}^{6,44} + \bar{\mathbf{k}}^{7,44} + \bar{\mathbf{k}}^{8,44} & \bar{\mathbf{k}}^{7,45} \\
\hline
 & \bar{\mathbf{k}}^{2,53} & \bar{\mathbf{k}}^{7,54} & \bar{\mathbf{k}}^{2,55} + \bar{\mathbf{k}}^{7,55}
\end{array}\right]
$$

(*b*)

FIG. 3–2. Direct stiffness method, Section 3, Example 3.

$\mathbf{r} = \{\mathbf{r}^2 \mathbf{r}^3 \mathbf{r}^4 \mathbf{r}^5\}$, $\mathbf{r}^2 = \{r_2^2\}$, and for the remaining joints, $\mathbf{r}^j = \{r_1^j r_2^j\}$. $\mathbf{K}$ is accordingly partitioned by joints into 4×4 blocks and is formed by merging member stiffness matrices as indicated in (*b*).

In partitioning $\bar{\mathbf{k}}_c^m$ by joints, either end could be labeled i and the other j; the 2×2 submatrices are

$$\bar{\mathbf{k}}^{m,ii} = \bar{\mathbf{k}}^{m,jj} = k_{11}\begin{bmatrix} c^2 & cs \\ cs & s^2 \end{bmatrix}$$

and

$$\bar{\mathbf{k}}^{m,ij} = \bar{\mathbf{k}}^{m,ji} = -\bar{\mathbf{k}}^{m,ii}$$

For a uniform member $k_{11} = EA/l$. For example, for member 7, $c = 0.8$, $s = -0.6$ (or $c = -0.8$, $s = 0.6$), and

$$\bar{\mathbf{k}}^{7,44} = \bar{\mathbf{k}}^{7,55} = -\bar{\mathbf{k}}^{7,45} = -\bar{\mathbf{k}}^{7,54} = \frac{EA}{l}\begin{bmatrix} 0.64 & -0.48 \\ -0.48 & 0.36 \end{bmatrix}$$

Member 4 has a similar stiffness matrix in terms of its own EA/l, but the first row and first column that correspond to r_1^2 are deleted. Thus

$$\bar{\mathbf{k}}^{4,22} = \frac{EA}{l}[0.36]$$

$$\bar{\mathbf{k}}^{4,23} = -\frac{EA}{l}[-0.48 \quad 0.36]$$

$$\bar{\mathbf{k}}^{4,32} = (\bar{\mathbf{k}}^{4,23})^T$$

$$\bar{\mathbf{k}}^{4,33} = \frac{EA}{l}\begin{bmatrix} 0.64 & -0.48 \\ -0.48 & 0.36 \end{bmatrix}$$

For the horizontal and vertical members the 4×4 $\bar{\mathbf{k}}_c^m$ contains two zero rows and two zero columns. Instead of forming $\bar{\mathbf{k}}_c^m$, the 2×2 stiffness in local coordinates,

$$\mathbf{k}_c^m = \frac{EA}{l}\begin{bmatrix} 1 & -1 \\ -1 & 1 \end{bmatrix}$$

may be merged directly into $\mathbf{K}$. For member 2, for example, the merging sequence is (r_1^3, r_1^5) (or equivalently r_1^5, r_1^3). For member 6 the merging sequence is (r_2^3, r_2^4). For member 3 there is one axial displacement, and $\mathbf{k}_c^3 = EA/l[1]$ is merged in location r_1^2 on the main diagonal of $\mathbf{K}$.

It is finally noted that **K** depends only on the axial stiffnesses of the members and on their orientations. Thus two geometrically similar trusses in which pairs of corresponding members have the same k_{11} have the same stiffness matrices.

To form $\mathbf{R}_T$, **R** is formed from forces applied directly at the joints, and $-\mathbf{R}_F$ is formed by merging $-\bar{\mathbf{S}}_F^m$. $\bar{\mathbf{S}}_F^m$ consists of the reaction forces at the ends of the pin-supported member subjected to its span load. For example, if all members are uniform and subjected to a uniform load p per unit member length, member 2 contributes $-6p$ to locations r_2^3 and r_2^5 of $\mathbf{R}_T$, member 4 contributes $-7.5p$ to locations r_2^3 and r_2^2, member 5 contributes $-7.5p$ to location r_2^4, and member 6 contributes $-4.5p$ to locations r_2^3 and r_2^4.

General Merging Procedure

Consider now the general case where **r** is a set of generalized displacements. Connectivity equations for member m involve only a certain subset $\mathbf{r}^m$ of **r**. They have the form $\mathbf{s}^m = \mathbf{a}^m\mathbf{r}^m$ or $\mathbf{v}^m = \mathbf{a}_v^m\mathbf{r}^m$, but $\mathbf{r}^m$ is not necessarily formed of joint displacements at the member ends, and the ordering of the elements of $\mathbf{r}^m$ in **r** is arbitrary. The generalized equilibrium equations for the structure are derived by the principle of virtual displacements, represented by the virtual work equation

$$\sum \delta\mathbf{s}^{mT}\mathbf{S}^m = \delta\mathbf{r}^T\mathbf{R} \qquad (3\text{–}14)$$

Substituting the stiffness equation for $\mathbf{S}^m$ and the connectivity equations for $\mathbf{s}^m$ and $\delta\mathbf{s}^m$, Eq. (3–14) turns into

$$\sum \delta\mathbf{r}^{mT}(\mathbf{a}^{mT}\mathbf{k}_c^m\mathbf{a}^m\mathbf{r}^m + \mathbf{a}^{mT}\mathbf{S}_F^m) = \delta\mathbf{r}^T\mathbf{R} \qquad (3\text{–}15)$$

Upon equating coefficients of $\delta\mathbf{r}^m$ on both sides of Eq. (3–15), there follows the contributions $\mathbf{K}^m = \mathbf{a}^{mT}\mathbf{k}_c^m\mathbf{a}^m$ and $\mathbf{R}_F^m = \mathbf{a}^{mT}\mathbf{S}_F^m$ of member m. The equation of virtual work written in terms of $\mathbf{S}^m$ or $\mathbf{V}^m$ would yield the other equivalent expressions in Eqs. (3–12) and (3–13).

The merging of $\mathbf{K}^m$ into **K** and of $\mathbf{S}_F^m$ into $\mathbf{R}_F$ is made according to the positions that the elements of $\mathbf{r}^m$ occupy in **r**. If the sequence $1, 2, \ldots p, \ldots q, \ldots$ represents the ordering in $\mathbf{r}^m$, the ordering in **r** of the elements of $\mathbf{r}^m$ is a sequence of the form

$$l_1, l_2, \ldots l_p, \ldots l_q, \ldots$$

Element (p, q) of $\mathbf{K}^m$ is added into, or merged, in element (l_p, l_q) of **K**. Similarly element p of $\bar{\mathbf{S}}_F^m$ is merged into element l_p of $\mathbf{R}_F$. Another description of the merging procedure is that row p of $\mathbf{K}^m$ is merged into row l_p of **K**, and successive elements of this row are merged, respectively, into locations $l_1, l_2, \ldots$. The merging procedure may be described by columns and is symmetric of the merging by rows. The sequence $l_1, l_2, \ldots$ is referred to as the merging sequence.

Example 4a. Inextensible Frame

Stiffness equations for an inextensional analysis are established by the direct stiffness method for the frame of Fig. 3–3. There are three inextensibility conditions that reduce the number of independent translational displacements from 4 to 1. A choice of translational displacement r_1 is shown in (*b*). Joint rotations r_2 and r_3 complete the choice of $\mathbf{r}$. Deformations of alternate type are chosen and are shown in (*b*) for the displacement mode r_1. Modes r_2 and r_3 are simple enough and are not shown. Adopting the notation, $\mathbf{v}^m = \{v_a^m \; v_b^m\}$, where a and b refer to the member ends as identified in the figure, the connectivity equation for member 1, $\mathbf{v}^1 = \mathbf{a}_v^1 \mathbf{r}^1$, is

$$\begin{Bmatrix} v_a^1 \\ v_b^1 \end{Bmatrix} = \begin{bmatrix} \dfrac{1}{l} & 0 \\ \dfrac{1}{l} & 1 \end{bmatrix} \begin{Bmatrix} r_1 \\ r_2 \end{Bmatrix}$$

Assuming uniform members without shear deformation, the member stiffness is

$$\mathbf{k}^1 = \frac{EI}{l_1} \begin{bmatrix} 4 & 2 \\ 2 & 4 \end{bmatrix}$$

Contribution of member 1 to the structure stiffness is

$$\mathbf{K}^1 = \mathbf{a}_v^{1T} \mathbf{k}^1 \mathbf{a}_v^1 = \frac{EI}{l_1^3} \begin{bmatrix} 12 & 6l_1 \\ 6l_1 & 4l_1^2 \end{bmatrix}$$

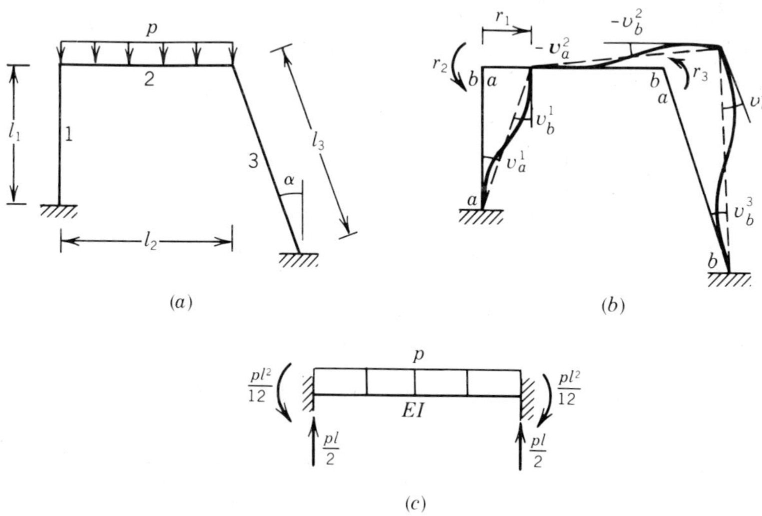

FIG. 3–3. Inextensional analysis, Section 3, Example 4.

For member 2,

$$\begin{Bmatrix} v_a^2 \\ v_b^2 \end{Bmatrix} = \begin{bmatrix} -\dfrac{\tan\alpha}{l_2} & 1 & 0 \\ -\dfrac{\tan\alpha}{l_2} & 0 & 1 \end{bmatrix} \begin{Bmatrix} r_1 \\ r_2 \\ r_3 \end{Bmatrix}$$

$$\mathbf{K}^2 = \mathbf{a}_v^{2T}\mathbf{k}^2\mathbf{a}_v^2 = \frac{EI}{l_2^3}\begin{bmatrix} 12\tan^2\alpha & -6l_2\tan\alpha & -6l_2\tan\alpha \\ & 4l_2^2 & 2l_2^2 \\ \text{Symmetric} & & 4l_2^2 \end{bmatrix}$$

For member 3,

$$\begin{Bmatrix} v_a^3 \\ v_b^3 \end{Bmatrix} = \begin{bmatrix} \dfrac{1}{l_3\cos\alpha} & 1 \\ \dfrac{1}{l_3\cos\alpha} & 0 \end{bmatrix} \begin{Bmatrix} r_1 \\ r_3 \end{Bmatrix}$$

$$\mathbf{K}^3 = \mathbf{a}_v^{3T}\mathbf{k}^3\mathbf{a}_v^3 = \frac{EI}{l_3^3}\begin{bmatrix} \dfrac{12}{\cos^2\alpha} & \dfrac{6l_3}{\cos\alpha} \\ \dfrac{6l_3}{\cos\alpha} & 4l_3^2 \end{bmatrix}$$

The merging sequences for $\mathbf{K}^1$, $\mathbf{K}^2$, and $\mathbf{K}^3$ are, respectively, (1, 2), (1, 2, 3), and (1, 3). The result is

$$\mathbf{K} = EI\begin{bmatrix} \dfrac{12}{l_1^3} + \dfrac{12}{l_2^3}\tan^2\alpha + \dfrac{12}{l_3^3\cos^2\alpha} & \dfrac{6}{l_1^2} - \dfrac{6\tan\alpha}{l_2^2} & -\dfrac{6}{l_2^2}\tan\alpha + \dfrac{6}{l_3^2\cos\alpha} \\ & \dfrac{4}{l_1} + \dfrac{4}{l_2} & \dfrac{2}{l_2} \\ \text{Symmetric} & & \dfrac{4}{l_2} + \dfrac{4}{l_3} \end{bmatrix}$$

the equivalent joint load in the present example is due entirely to the span load of member 2 and is equal to $(-\mathbf{R}_F^2)$. Evaluation of $\mathbf{R}_F^2$ by means of any one of the expressions of Eq. (3–13) is equivalent to the following procedure. The fixed end forces are determined as shown in (*c*). The virtual work of these forces through the displacement modes $r_1 = 1$, $r_2 = 1$, and $r_3 = 1$ is equal to $\mathbf{R}_F^2$. Thus $-\mathbf{R}_F = -\mathbf{R}_F^2 = \{-(pl_2/2)\tan\alpha \quad -pl_2^2/12 \quad pl_2^2/12\}$.

Example 4b

The same problem as Example 4a is treated by means of connectivity and stiffness matrices in local axes. The local x axis is oriented from end a to end b, and the y axis is obtained by a counterclockwise rotation of the x axis by $\pi/2$. For member 1 the effective connectivity equation, $\mathbf{s}^1 = \mathbf{a}^1\mathbf{r}^1$, is

$$\begin{Bmatrix} s_2^b \\ s_3^b \end{Bmatrix} = \begin{bmatrix} -1 & 0 \\ 0 & 1 \end{bmatrix} \begin{Bmatrix} r_1 \\ r_2 \end{Bmatrix}$$

The effective stiffnes $\mathbf{k}_c^1$ is the submatrix $\mathbf{k}_c^{bb}$ of the complete matrix, from which the row and column associated with axial terms are deleted. The product $\mathbf{a}^{1T}\mathbf{k}_c^1\mathbf{a}^1$ yields $\mathbf{K}^1$ as found in the previous example.

For member 2, the effective variables are $\mathbf{s}^2 = \{s_1^a \; s_3^a \; s_1^b \; s_2^b \; s_3^b\}$. However, the transformed member stiffness does not contain axial terms. It is thus possible, for forming $\mathbf{K}^2$, to choose $\mathbf{s}^2 = \{s_3^a \, s_2^b \, s_3^b\}$. The connectivity equation, $\mathbf{s}^2 = \mathbf{a}^2\mathbf{r}^2$, is

$$\begin{Bmatrix} s_3^a \\ s_2^b \\ s_3^b \end{Bmatrix} = \begin{bmatrix} 0 & 1 & 0 \\ \tan\alpha & 0 & 0 \\ 0 & 0 & 1 \end{bmatrix} \begin{Bmatrix} r_1 \\ r_2 \\ r_3 \end{Bmatrix}$$

The effective stiffness is obtained by deletion of rows and columns from the complete one. The product $\mathbf{a}^{2T}\mathbf{k}_c^2\mathbf{a}^2$ should yield the same $\mathbf{K}^2$ as in the previous example.

Contribution of member 2 to $\mathbf{R}_F$ is $\mathbf{R}_F^2 = \mathbf{a}^{2T}\mathbf{S}_F^2$. The effective $\mathbf{S}_F^2$ is obtained from Fig. 3–3c by including only the components corresponding to the elements of $\mathbf{s}^2$. Thus

$$\mathbf{S}_F^2 = \begin{Bmatrix} \dfrac{pl_2^2}{12} & \dfrac{pl_2}{2} & -\dfrac{pl_2^2}{12} \end{Bmatrix}$$

The product $\mathbf{a}^{2T}\mathbf{S}_F^2$ yields $\mathbf{R}_F^2$ as found earlier.

Note that if member 2 had a horizontal span load, evaluation of $\mathbf{R}_F^2$ by the product $\mathbf{a}^{2T}\mathbf{S}_F^2$ requires that the horizontal displacements s_1^a and s_1^b remain part of the effective member displacements.

For member 3 the effective connectivity equation is

$$\begin{Bmatrix} s_2^a \\ s_3^a \end{Bmatrix} = \begin{bmatrix} \dfrac{1}{\cos\alpha} & 0 \\ 0 & 1 \end{bmatrix} \begin{Bmatrix} r_1 \\ r_3 \end{Bmatrix}$$

The effective stiffness $\mathbf{k}_c^3$ is the submatrix $\mathbf{k}_c^{aa}$ of the complete matrix, from which the row and column associated with axial terms are deleted.

Example 5. Inextensible Rectangular Frame

The stiffness matrix of an inextensible rectangular frame may be assembled in a particularly simple manner from member stiffness matrices in local coordinates. Consider the frame of Fig. 3–4 with displacements ordered as shown. The

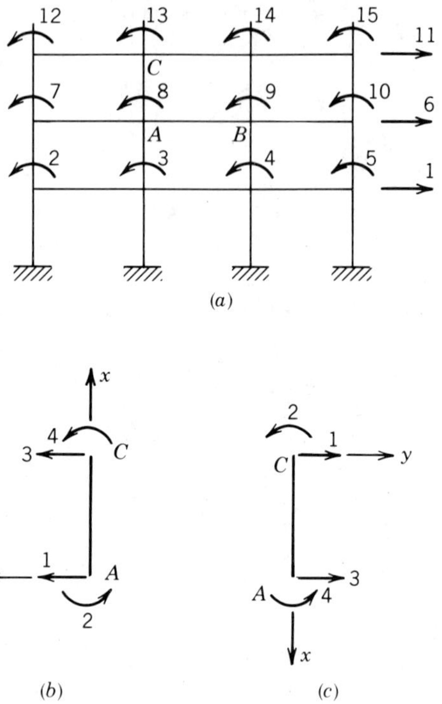

FIG. 3–4. Inextensional rectangular frame, Section 3, Example 5.

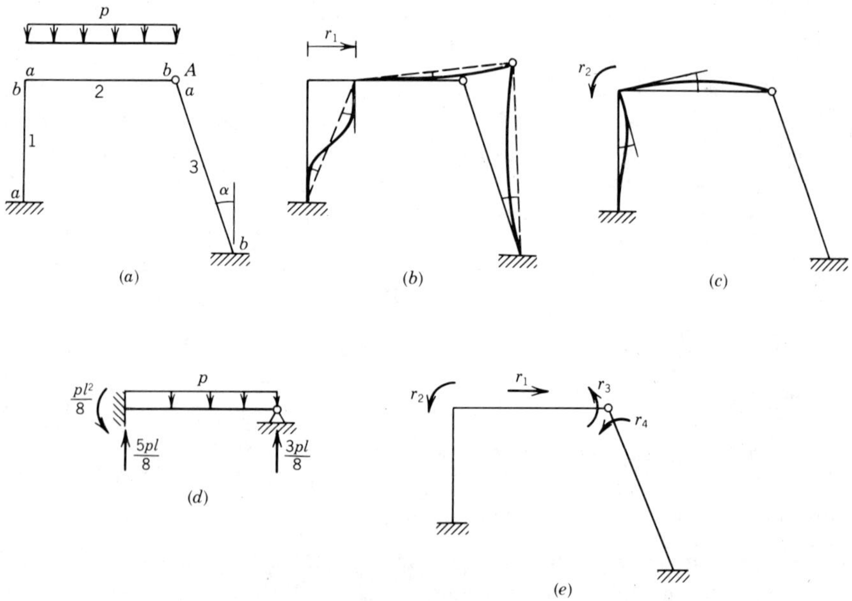

FIG. 3–5. Effective and complete variables, Section 3, Example 6.

effective stiffness of a beam is the 2×2 reduced bending stiffness for end rotations. It is merged in **K** according to the ordering of the end rotations. For example, the merging sequence for member *AB* is (8, 9). The effective stiffness matrix for a column is obtained by deleting the axial terms. For a typical column such as *AC*, an ordering of $\mathbf{k}_c$ in local coordinate is shown in (*b*). $\mathbf{k}_c$ is merged into **K** with the merging sequence (-6, 8, -11, 13). The same result can be obtained from the ordering shown in (*c*) with merging sequence (11, 13, 6, 8). The procedure is particularly simple to program because of the regularity of the grid.

Example 6

This example uses the frame of Fig. 3–5*a* to illustrate several points. The frame differs from that considered in Example 4 by the pinned joint at *A*. First, an inextensional formulation is made using the least number of displacements. These are $\{r_1 r_2\}$ as shown in (*b*)(*c*). Contribution of member 1 to **K** is the same as in Example 4a. Members 2 and 3 have each one effective deformation whose connectivity equation is the same as in Example 4a. Thus

$$\mathbf{v}^2 = v_a^2 = \begin{bmatrix} -\dfrac{\tan\alpha}{l_2} & 1 \end{bmatrix} \begin{Bmatrix} r_1 \\ r_2 \end{Bmatrix}$$

$$\mathbf{v}^3 = v_b^3 = \frac{1}{l_3 \cos\alpha} r_1$$

The effective stiffness matrices are

$$\mathbf{k}^2 = \frac{3EI}{l_2}[1]$$

$$\mathbf{k}^3 = \frac{3EI}{l_3}[1]$$

Thus

$$\mathbf{K}^2 = \frac{3EI}{l_2} \begin{bmatrix} -\dfrac{\tan\alpha}{l_2} \\ 1 \end{bmatrix} \begin{bmatrix} -\dfrac{\tan\alpha}{l_2} & 1 \end{bmatrix} = \frac{3EI}{l_2} \begin{bmatrix} \dfrac{\tan^2\alpha}{l_2^2} & -\dfrac{\tan\alpha}{l_2} \\ -\dfrac{\tan\alpha}{l_2} & 1 \end{bmatrix}$$

$$\mathbf{K}^3 = \frac{3EI}{l_3}\left(\frac{1}{l_3\cos\alpha}\right)^2$$

Merging of $\mathbf{K}^1$, $\mathbf{K}^2$, and $\mathbf{K}^3$ into **K** follows the general procedure.

To form the equivalent joint load, the procedure of Example 4a is followed, but the effective fixed end forces are those shown in (*d*). Thus

$$-\mathbf{R}_F = -\mathbf{R}_F^2\begin{Bmatrix} -\dfrac{3pl_2}{8}\tan\alpha & -\dfrac{pl_2^2}{8}\end{Bmatrix}$$

A second point to be illustrated is the inclusion in **r** of the rotations at the pinned joint. **r** consists then of four displacements as shown in (*e*). Evaluation of $\mathbf{K}^1$, $\mathbf{K}^2$ and $\mathbf{K}^3$ is exactly the same as in Example 4a, and the merging sequences remain the same for members 1 and 2. For member 3 the merging sequence now is (1, 4). The equivalent joint load is obtained by augmenting the matrix found in Example 4a with a zero.

The last point to be illustrated with the structure of Fig. 3–5*e* is a transformation to another set of displacements. Let

$$\begin{Bmatrix} r_3 \\ r_4 \end{Bmatrix} = \begin{bmatrix} 1 & \frac{1}{2} \\ 1 & -\frac{1}{2} \end{bmatrix}\begin{Bmatrix} r_3' \\ r_4' \end{Bmatrix}$$

r_3' is then the average rotation $(r_3 + r_4)/2$, and r_4' is the relative rotation, or joint angular deformation $(r_3 - r_4)$. The connectivity equations are obtained by substituting the expressions of r_3 and r_4 in the previously derived equations. The direct stiffness method follows then the general procedure using the transformed connectivity matrices. Joint loads R_3 and R_4 transform according to the contragradient law:

$$\begin{Bmatrix} R_3' \\ R_4' \end{Bmatrix} = \begin{bmatrix} 1 & 1 \\ \frac{1}{2} & -\frac{1}{2} \end{bmatrix}\begin{Bmatrix} R_3 \\ R_4 \end{Bmatrix}$$

The transformation could also be applied to the assembled stiffness equation. Letting $\mathbf{r}' = \{r_1\ r_2\ r_3'\ r_4'\}$ the transformation $\mathbf{r} = \mathbf{T}\mathbf{r}'$ is

$$\begin{Bmatrix} r_1 \\ r_2 \\ r_3 \\ r_4 \end{Bmatrix} = \begin{bmatrix} 1 & 0 & 0 & 0 \\ 0 & 1 & 0 & 0 \\ 0 & 0 & 1 & \frac{1}{2} \\ 0 & 0 & 1 & -\frac{1}{2} \end{bmatrix}\begin{Bmatrix} r_1 \\ r_2 \\ r_3' \\ r_4' \end{Bmatrix}$$

The transformed stiffness matrix is obtained by the congruent law and the transformed equivalent joint load by the contragradient law (Chapter 8, Section 4).

4. INTERPRETATION OF STIFFNESS COEFFICIENTS

If joint displacements are not kinematically constrained, so that each joint has independent degrees of freedom, the structure stiffness coefficients have a direct interpretation as joint loads required to produce a given displacement. The stiffness coefficients K_{ij}, for $i = 1, 2, \ldots, NR$, from the ith column of $\mathbf{K}$ and consist of the joint loads required to produce the displacement $r_j = 1$, with all other displacements zero. These external joint loads are balanced by internal joint forces exerted by the members. This approach leads to forming the stiffness matrix by columns and may be convenient in manual methods.

As an illustration, consider the frame shown in Fig. 4–1a with 15 degrees of freedom ordered as shown. A unit displacement $r_7 = 1$ causes the deformation shown in (b), and internal joint forces at joints B, C and D in the directions of the

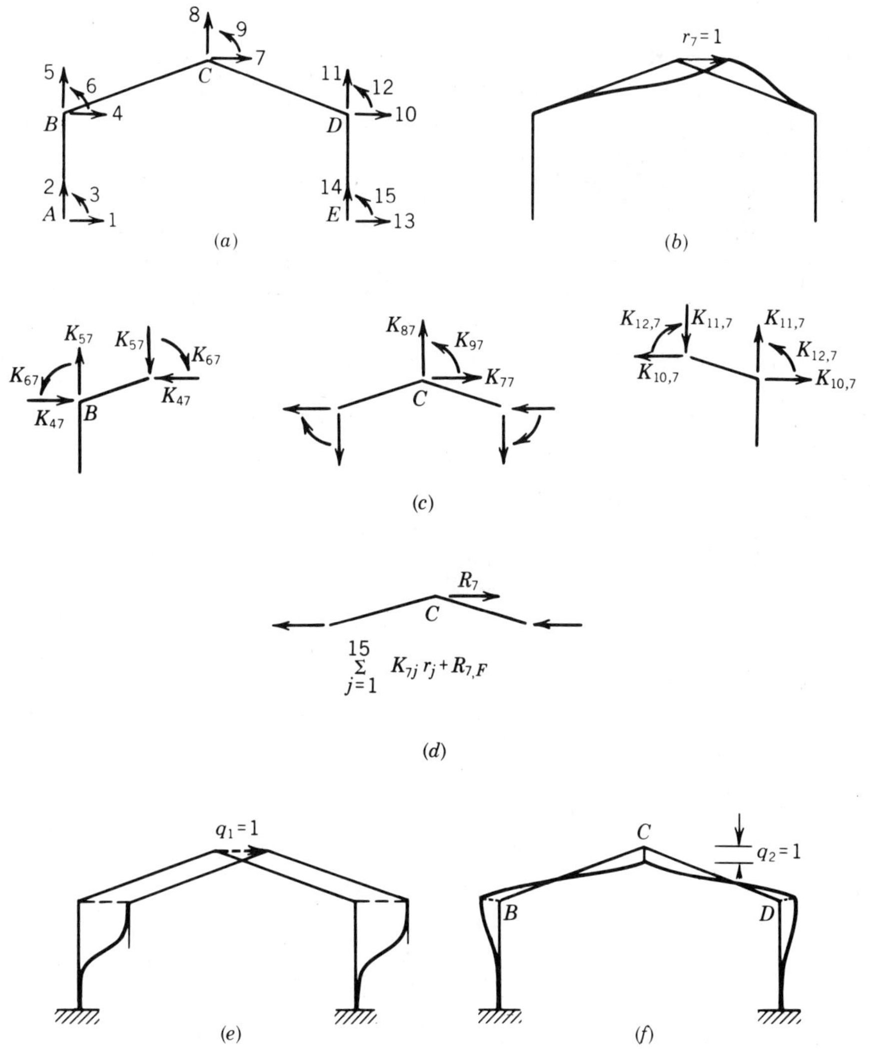

FIG. 4–1. Interpretation of stiffness coefficients, Section 4, Example 1.

three degrees of freedom of each joint. Since joints B and D remain fixed, members AB and DE are undeformed and have no end forces due to r_7. The external and internal joint forces required to produce $r_7 = 1$ are shown and labeled in (c).

The joint equilibrium equation associated with the degree of freedom δr_7 is $\sum_{j=1}^{15} K_{7j}r_j = R_7 - R_{7F}$. A corresponding free body diagram is shown in (d). The stiffness coefficients K_{7j} in this equation are equal, respectively, to the coefficients K_{j7} shown in (c) but have a different physical meaning. They are all forces at joint C, in the direction of freedom 7, caused, respectively, by unit values $r_j = 1$, for $j = 1, \ldots, 15$.

If the stiffness equations are transformed by means of a kinematic transformation from joint displacements $\mathbf{r}$ to generalized displacements $\mathbf{q}$, the interpretation of the stiffness coefficients is thereby modified from that of individual joint forces to generalized forces. A generalized displacements $q_j = 1$ induces joint displacements in a certain zone of influence and corresponding member end forces. K_{ij} is the virtual work of these forces through the virtual displacements induced by $\delta q_i = 1$.

Example 1

As an example consider the frame of Fig. 4–1, fixed at the base, and assume that all members are inextensible. A choice of translational generalized displacements is shown in (e) and (f). The stiffness coefficient K_{12} is equal to the virtual work of the member end forces caused by $q_2 = 1$ through the virtual joint displacements induced by $\delta q_1 = 1$. This is then the sum of the horizontal components of the member end forces at joints B, C, and D in the deformed state (f). Since end forces for each member are in equilibrium, this sum reduces to that of the shear forces at the top of the columns. If the frame has a vertical plane of symmetry at C, these shear forces are equal and opposite, and $K_{12} = 0$.

To interpret K_{21}, one considers the member end forces in (e) and their virtual work through the displacements in (f). It is a simple exercise to ascertain that $K_{21} = K_{12}$.

The interpretation or definition of an individual stiffness element K_{ij} in terms of virtual work is the unifying concept by which the case of independent joint displacements becomes a particular case of the general concept of generalized displacements. Section 11 deals further with this topic.

5. SYMMETRY AND ANTISYMMETRY

A structure has a plane of symmetry (S) if the part on one side of (S) is the mirror image, in geometry and material properties, of the other part. Examples are shown in Fig. 5–1.

It will be possible to formulate the analysis of a symmetric structure under a general loading condition, as the superposition of a symmetric loading and an antisymmetric loading, and to analyze in each case only half of the structure on one side of the plane of symmetry.

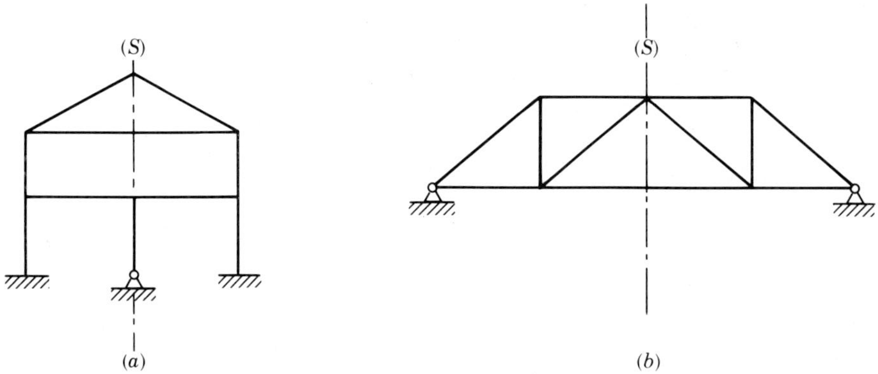

FIG. 5–1. Symmetric structures.

A symmetric deformation keeps the deformed structure symmetric and is caused by a symmetric loading. Displacements and oriented angles of rotation, and forces and moments on one side of (S), are mirror images of the quantities on the other side, as shown in Fig. 5–2*a*. Note, however, that if rotations and moments are represented by vectors, then the vectors on one side of (S) are the antimirror image of the vectors on the other side—that is, symmetric rotations and moments are represented by antisymmetric vectors. The vector representation of rotations and moments will not be used so that symmetry of deformation and loading corresponds to a mirror image representation.

An antisymmetric deformation has an antimirror image representation and is caused by an antisymmetric loading. The representation of antisymmetric displacements and rotations and of corresponding forces is shown in Fig. 5–2*b*.

The plane of symmetry may cut certain members at midspan. It may also pass through certain joints, and it may contain certain members. It will be convenient to consider a member lying in the plane of symmetry as formed of two identical members having each half the stiffness properties of the actual member and placed, respectively, on the two sides of the plane of symmetry. The ends of these members have the same connectivity as the actual member.

A load applied in the plane of symmetry may similarly be divided into two equal parts applied on the two sides of the plane of symmetry.

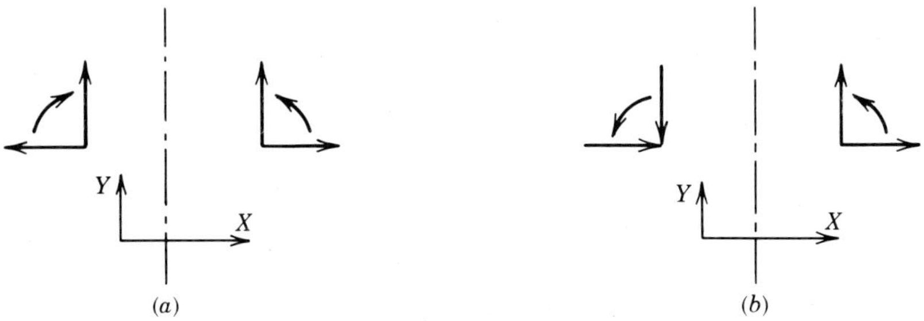

FIG. 5–2. Symmetric displacements and forces (*a*); antisymmetric displacements and forces (*b*).

The structure is now cut by the plane of symmetry only at joints, or at midspan of certain members, which may also be considered as joints. Conditions at these joints in symmetric and antisymmetric loading cases will allow us in each case to analyze only half of the structure.

Symmetric Conditions

For a rigid joint in the plane of symmetry, it is deduced from Fig. 5–2*a* that the X displacement and the rotation vanish. An external Y force may be applied freely, but the internal Y force at a cut by the plane of symmetry must vanish because it must consist of a pair of opposite forces and must also be symmetric. If the continuity of the X displacement or of the rotation is released, a pair of symmetric X forces or a pair of symmetric moments may be applied at the joint.

Symmetric conditions for the structures of Fig. 5–1 are shown in Fig. 5–3*a*, *b*. For a member cut at midspan by the plane of symmetry, effective stiffness relations avoid introducing additional degrees of freedom at midspan. Consider the bending stiffness relations of the alternate type. If the member is symmetric, then

$$k_{22} = k_{33} = k \tag{5–1a}$$

Let also

$$k_{23} = k_{32} = k' \tag{5–1b}$$

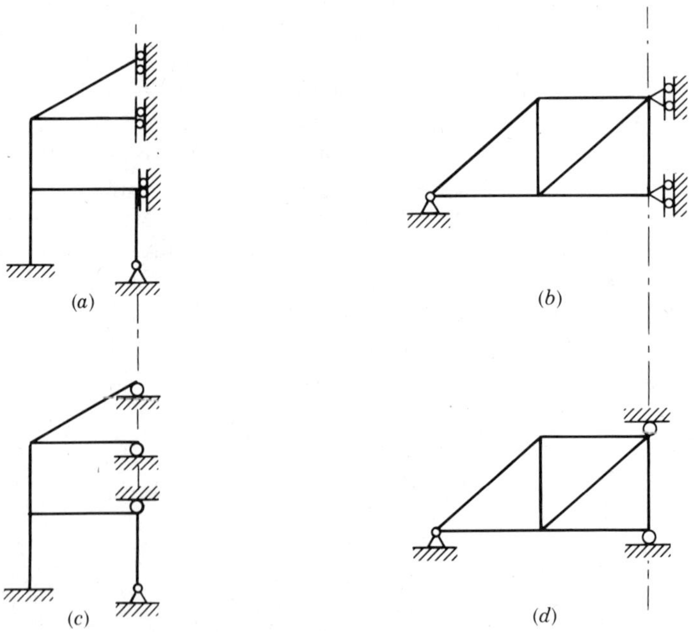

FIG. 5–3. Boundary conditions for symmetry (*a*), (*b*); for antisymmetry (*c*), (*d*).

In a symmetric deformation mode, $v_3 = -v_2$, $V_{3F} = -V_{2F}$, and effective stiffness relations in terms of V_2 are

$$V_2 = (k - k')v_2 + V_{2F} \tag{5–2a}$$

For a uniform member, neglecting shear deformation,

$$k - k' = \frac{2EI}{l} \tag{5–2b}$$

Antisymmetric Conditions

For a rigid joint in the plane of symmetry, it is deduced from Fig. 5–2*b* that the *Y* displacement vanishes. An external *X* force and an external moment may be applied freely, but the internal *X* force and the internal moment at a cut by the plane of symmetry must vanish because they must be concurrently as in (*b*) and (*a*). If the continuity of the *Y* displacement is released, a pair of antisymmetric *Y* forces may be applied at the joint.

Antisymmetric conditions for the structures of Fig. 5–1 are shown in Fig. 5–3*c*, *d*. As in the symmetric case, effective stiffness relations may be used for a member cut at midspan by the plane of symmetry. In the present case, $v_3 = v_2$, and $V_{3F} = V_{2F}$. The effective stiffness relation for V_2 is

$$V_2 = (k + k')v_2 + V_{2F} \tag{5–3a}$$

For a uniform member without shear deformation

$$k + k' = \frac{6EI}{l} \tag{5–3b}$$

It will now be shown that any loading (*L*) acting on a symmetric structure may be represented as the sum of a symmetric loading (*SL*) and of an antisymmetric loading (*AL*). Symbolically.

$$(L) = (SL) + AL)$$

Let $(L)'$ be the loading obtained from the given loading (*L*) by replacing each load in (*L*) by its mirror image with respect to the plane of symmetry, and let

$$(SL) = \tfrac{1}{2}(L) + \tfrac{1}{2}(L)'$$

$$(AL) = \tfrac{1}{2}(L) - \tfrac{1}{2}(L)'$$

(*SL*) is a symmetric loading since its mirror image changes (*L*) into $(L)'$ and $(L)'$ into (*L*) leaving (*SL*) unchanged. The mirror image of (*AL*) changes (*AL*) into

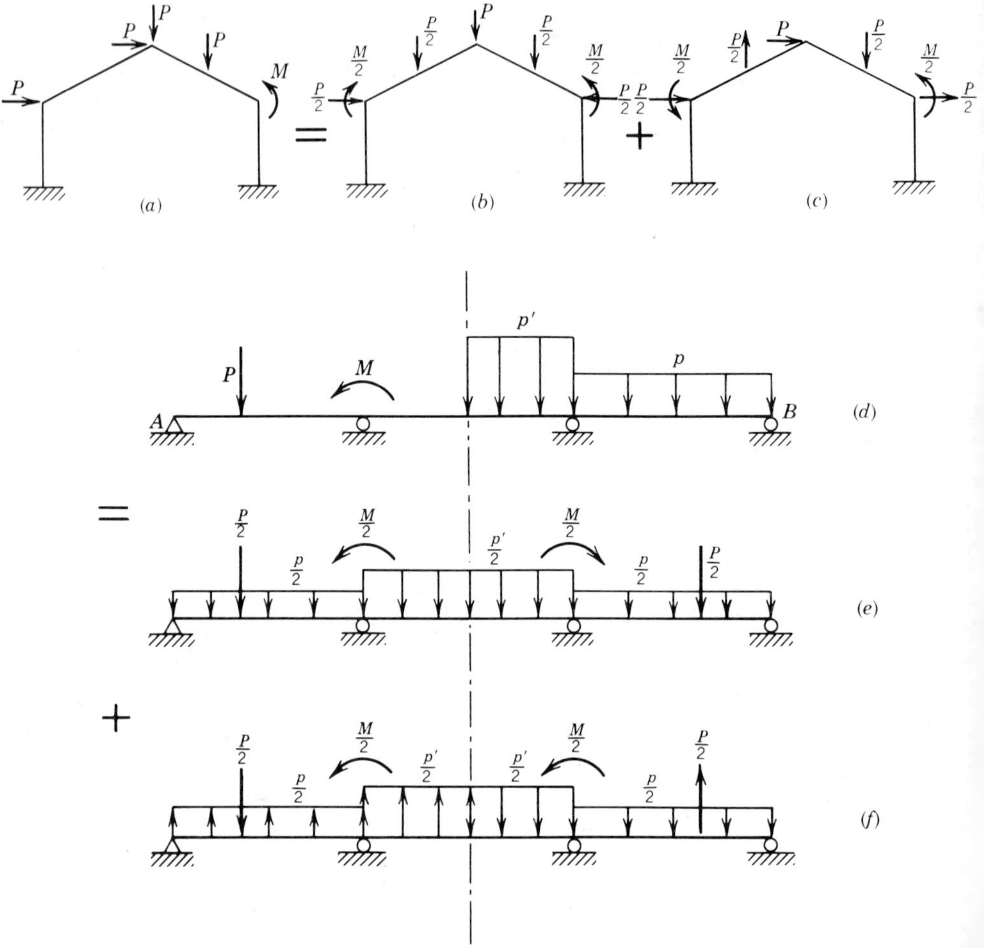

FIG. 5–4. Decomposition of a loading into symmetric and antisymmetric loadings.

$-(AL)$ so that (AL) is an antisymmetric loading. It is clear that the superposition of (SL) and (AL) as defined here yields the original loading (L). Examples are shown in Fig. 5–4. Note that in Fig. 5–4*d* the geometric symmetry of the structure is violated by the different axial support conditions at A and B. However, this has no influence on the symmetry or antisymmetry properties of bending behavior.

6. BAND WIDTH OF STIFFNESS MATRIX

Methods of solution of the stiffness equations may be made more efficient if account is taken of the zero elements of the stiffness matrix. Stiffness matrices of large structures, typically have their nonzero elements within a band running in the direction of the main diagonal.

There are various techniques for increasing the efficiency of solution procedures of simultaneous equations. The commonly used Gauss reduction method has the property that all elements outside the band remain zero through the reduction process and thus are not needed to solve the stiffness equations. Further the stiffness matrix is symmetric, and there is need to store only the main diagonal and a half band on one side of it. Minimization of the half band width through appropriate ordering of the joint displacements is thus desirable for increased solution efficiency.

Assembly of the stiffness matrix by the direct stiffness method indicates that the half band width is equal to the maximum difference of the degree-of-freedom numbers affecting any one member.

In general, the most efficient degree-of-freedom numbering is a topological problem. For a simple configuration such as a tall rectangulr frame, ordering of joint displacements in the short direction of the floors leads to the minimum band width.

7. STATIC CONDENSATION

Let $\mathbf{r}$ and $\mathbf{R}$ be partitioned into $\mathbf{r} = \{\mathbf{r}_1 \ \mathbf{r}_2\}$ and $\mathbf{R} = \{\mathbf{R}_1 \ \mathbf{R}_2\}$. In certain applications it may be desired to obtain stiffness equations relating $\mathbf{R}_1$ and $\mathbf{r}_1$ for fixed values of $\mathbf{R}_2$. This is done by eliminating $\mathbf{r}_2$ from the stiffness equations. The process is called static condensation and is formally similar to the process described for the member in Chapter 7, Section 3.3. For example, $\mathbf{r}_1$ may be the floor translations of a rectangular frame and $\mathbf{r}_2$ the remaining displacements.

Example 1

The portal frame of Fig. 7–1, if treated as inextensible, has one translational displacement r_1 and two joint rotations r_2 and r_3. Assume that a load R_1 is applied, and that it is desired to obtain the condensed stiffness relation $R_1 = kr_1$. First, it is possible to take advantage of antisymmetry. The displacements are then $\{r_1 \ r_2\}$, with $r_3 = r_2$. The antisymmetric stiffness of the beam is, from Eq. (5–

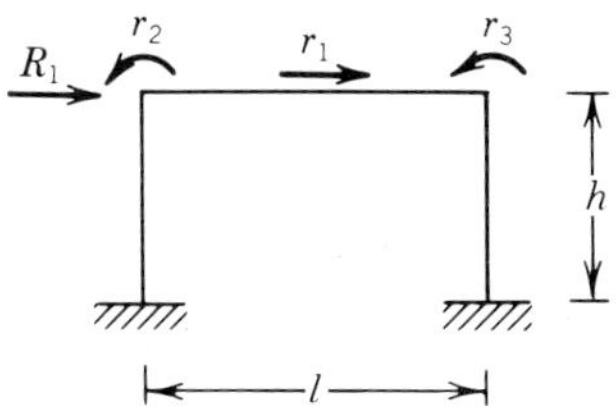

FIG. 7–1. Static condensation, Example 1.

3b), $6EI/l$. The assembled stiffness equations are then

$$\frac{EI}{h^3}\begin{bmatrix} 12 & 6h \\ 6h & 4h^2 + 6\frac{h^3}{l} \end{bmatrix}\begin{Bmatrix} r_1 \\ r_2 \end{Bmatrix} = \begin{Bmatrix} R_1 \\ 0 \end{Bmatrix}$$

Elimination of r_2 yields

$$R_1 = \frac{12EI}{h^3}\left(1 - \frac{3}{4 + 6h/l}\right)r_1$$

8. ANALYSIS BY SUBSTRUCTURES

An assemblage of members may be considered a superelement, or substructure, that may be connected to other substructures to form a final structure. The theory of the final assemblage, as formulated from the stiffness properties and connectivity matrices of the substructures, is similar to the theory developed for an assemblage of members. A particular aspect of a substructure is that its displacements are of two types: internal displacements and boundary displacements. Connectivity equations involve only the boundary displacements. To reduce the size of substructure matrices, the internal displacements may be eliminated by static condensation. The effective substructure matrices that result may be used in a direct stiffness method, as described for the simple member. All that is needed is the merging sequence that identifies the correspondance between substructure ordering and global ordering of displacements.

For example, in a rectangular frame a substructure may be formed from one or several stories bounded by horizontal planes cutting the columns above two floors. Joint displacements in the bounding planes are boundary displacements. If a substructure is formed of several stories, the displacements between the bounding planes are eliminated by static condensation.

Subdivision of a structure into substructures may be useful in fitting a large problem into limited storage capabilities. The largest stiffness matrix to be dealt with is either the largest substructure stiffness matrix or the assembled matrix of the condensed substructure matrices. An assemblage of members may also be devised as an approximation to a continuum filling the space bounded by the assemblage. Approximations to plate, shell, and three-dimensional continuum problems are thus feasible.

9. ENERGY THEOREMS

The equation of virtual work for the structure as a whole may be obtained by summing the member equations and enforcing at the joints the condition

$\mathbf{S}^T\delta\mathbf{s} = \mathbf{R}^T\delta\mathbf{r}$. Equality of external virtual work, δW_{ext}, to internal virtual work, δW_{int}, takes the form

$$\sum_m \int_0^l \mathbf{p}^T \delta\mathbf{u}\, dx + \mathbf{R}^T \delta\mathbf{r} = \sum_m \int_0^l \mathbf{N}^T \delta\boldsymbol{\varepsilon}\, dx \tag{9–1}$$

Equation (9–1) may be shown formally, as was done in Chapter 2 for the individual member, to be a necessary and sufficient condition for member differential equilibrium and joint equilibrium. Energy theorems based on the equation of virtual displacements and established for a general member in Chapter 5 are established in a similar manner for the structure.

The strain energy $\bar{U}$ is defined as the sum of the strain energies of the members. For member m the strain energy $\bar{U}^m$ is expressed in terms of member generalized displacements as obtained in Eq. (5/5–6). Taking these as member end displacements, and enforcing the connectivity equations, $\bar{U}$ takes the form

$$\bar{U} = \tfrac{1}{2}\mathbf{r}^T\mathbf{K}\mathbf{r} + \mathbf{Q}_I^T\mathbf{r} \tag{9–2}$$

$\mathbf{Q}_I$ is a set of generalized forces due to initial stress resultants $\mathbf{N}_I$.

Castigliano's first theorem is an application of the equation of virtual work. A property of the strain energy function $\bar{U}$ is $\delta W_{\text{int}} = \delta\bar{U} = (\partial\bar{U}/\partial\mathbf{r}^T)\delta\mathbf{r}$. The external virtual work is

$$\delta W_{\text{ext}} = (\mathbf{R}^T + \mathbf{Q}_p^T)\delta\mathbf{r} \tag{9–3}$$

where $\mathbf{Q}_p$ is the set of generalized forces due to the span loads $\mathbf{p}$. Castigliano's first theorem is thus obtained in the form

$$\mathbf{R} + \mathbf{Q}_p = \mathbf{K}\mathbf{r} + \mathbf{Q}_I \tag{9–4}$$

A property established for the member in Chapter 5, Section 5, is that $-(\mathbf{Q}_p - \mathbf{Q}_I)$ is equal to the fixed end forces due to the complete span loading condition. For the structure

$$\mathbf{Q}_p - \mathbf{Q}_I = -\mathbf{R}_F \tag{9–5}$$

Equation (9–4) reduces then to the stiffness equations derived earlier. A transformation from $\mathbf{r}$ to generalized displacements allows us to state Castigliano's first theorem in terms of generalized quantities.

Betti's reciprocal theorem and Clapeyron's theorem may be established for a structure in the same way as they were established for the member. Both theorems apply only in the linear theory and without initial stresses. Their statements in Chapter 5, Section 5, need only to be applied to the structure instead of a single member.

10. PROPERTIES OF THE STIFFNESS MATRIX

The structure stiffness matrix has properties similar to those of the member stiffness matrix presented in Chapter 5, Section 3. Some of these properties are the following:

1. **K** is symmetric. This property is readily deduced from a defining formula such as $\mathbf{a}^T\mathbf{k}_c\mathbf{a}$ and from symmetry of $\mathbf{k}_c$. Symmetry of **K** is conserved by a congruent transformation. The symmetry relation, $K_{ij} = K_{ji}$, is an expression of Betti's reciprocal theorem.

2. If rigid body degrees of freedom are included in **r**, each column of **K**, which is considered as a set of generalized forces, satisfies the corresponding rigid body equilibrium equations.

This property is expressed by the principle of virtual displacements in the form, $\mathbf{K}\delta\mathbf{r} = 0$, in which $\delta\mathbf{r}$ is any rigid body displacement. To show this, $\mathbf{K}\delta\mathbf{r}$ is written as $\mathbf{a}_v^T\mathbf{k}\mathbf{a}_v\delta\mathbf{r}$, and $\mathbf{a}_v\delta\mathbf{r}$ is recognized as a set of virtual deformations $\delta\mathbf{v}$ which is zero for any rigid body displacement.

In general, $\delta\mathbf{v}$ is zero for any mechanism displacement mode. The rigid body degrees of freedom are then understood in this generalized sense.

3. If **r** includes rigid body modes, **K** is singular, and the nontrivial solutions of the system of equations $\mathbf{K}\mathbf{r} = 0$ are the rigid body modes.

This was shown in Property 2, with $\delta\mathbf{r}$ replacing **r**. There remains, however, to show that **Kr** cannot be zero for nonrigid body modes. We can write $\mathbf{r}^T\mathbf{K}\mathbf{r} = \mathbf{v}^T\mathbf{k}\mathbf{v} = \sum_m \mathbf{v}^{mT}\mathbf{k}^m\mathbf{v}^m$. A nonrigid body mode causes nonzero deformations in at least one member, and $\mathbf{v}^{mT}\mathbf{k}^m\mathbf{v}^m$ for any such member is a strain energy function that is positive. Thus **Kr** cannot be zero if **r** is not zero and is not a rigid body mode.

4. A kinematically stable structure is by definition one for which **r** excludes rigid body modes. A consequence of Property 3 is that, for a kinematically stable structure, **K** is nonsingular and is also positive definite.

For a kinematically stable structure the system of equations $\mathbf{a}_v\mathbf{r} = 0$ admits only $\mathbf{r} = 0$ as a solution. The connectivity matrix $\mathbf{a}_v$ has then a rank ρ equal to the number of degrees of freedom (NR).

If **r** includes rigid body modes, their independent number (NM) is related to ρ and (NR) through

$$(NM) = (NR) - \rho$$

11. GENERALIZED STIFFNESS: NORMAL DEFORMATION MODES

The concept of generalized stiffness was introduced in Section 4. K_{ij} is caused by the displacement $r_j = 1$ and corresponds to the degree of freedom δr_i. It is equal to the virtual work of the member end forces caused by $r_j = 1$ through the virtual displacements induced by $\delta r_i = 1$. Further applications of this concept are presented in what follows within the context of transformation from one set of displacements to another.

Let the transformation $\mathbf{r} = \mathbf{Tq}$ be partitioned into individual modes $\mathbf{T}_i q_i$ in the form

$$\mathbf{r} = [\mathbf{T}_1 \mathbf{T}_2 \ldots \mathbf{T}_i \ldots \mathbf{T}_n]\{q_1 \; q_2 \ldots q_i \ldots q_n\}$$

If $\mathbf{K}'$ is the transformed stiffness matrix, its (i, j) element is

$$K'_{ij} = \mathbf{T}_i^T \mathbf{K} \mathbf{T}_j \qquad (11\text{–}1)$$

Columns $\mathbf{T}_i$ and $\mathbf{T}_j$ define joint displacement modes, $q_i = 1$ and $q_j = 1$, respectively. In Eq. (11–1) $\mathbf{T}_i$ plays the role of virtual displacements in the mode $\delta q_i = 1$, and $\mathbf{KT}_j$ are the joint forces in the mode $q_j = 1$.

The behavior of a structure may be approximated by a transformation to a lesser number of degrees of freedom. Transformation to a single degree of freedom is illustrated in the following example.

Example 1

Consider the springs in series shown in Fig. 11–1. This could be a simplified model for describing the lateral stiffness of a multistory frame. What follows is readily generalized to any number of springs. The direct stiffness method leads to the assembled $\mathbf{K}$ shown in (*b*).

As an example of a deformation mode, let r_1 be proportional to i. Taking $r_6 = q$

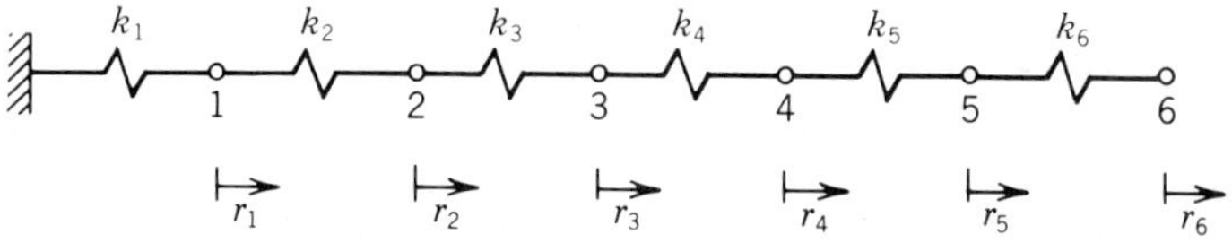

(*a*)

$$\mathbf{K} = \begin{bmatrix} k_1 + k_2 & -k_2 & & & & \\ -k_2 & k_2 + k_3 & -k_3 & & & \\ & -k_3 & k_3 + k_4 & -k_4 & & \\ & & -k_4 & k_4 + k_5 & -k_5 & \\ & & & -k_5 & k_5 + k_6 & -k_6 \\ & & & & -k_6 & k_6 \end{bmatrix}$$

(*b*)

FIG. 11–1. Springs in series, Example 1.

as generalized coordinate, the deformation mode is

$$\mathbf{T} = \frac{1}{6}\{1 \ 2 \ 3 \ 4 \ 5 \ 6\}$$

The transformation $\mathbf{r} = \mathbf{T}q$ yields

$$K' = \mathbf{T}^T\mathbf{K}\mathbf{T} = \frac{1}{36}(k_1 + k_2 + \ldots + k_6)$$

$$\mathbf{Q} = \mathbf{T}^T\mathbf{R} = \frac{1}{6}(R_1 + 2R_2 + \ldots + 6R_6)$$

Consider the case $k_1 = k_2 = \ldots = k_6 = k$, for which $K' = k/6$, and a loading consisting only of $R_6 = R$. Then $Q = R$, $q = Q/K' = 6R/k$, and

$$\mathbf{r} = \mathbf{T}q = \frac{R}{k}\{1 \ 2 \ 3 \ 4 \ 5 \ 6\}$$

This is the exact solution for the loading considered which causes a constant axial force R in the springs, a constant spring elongation R/k, and a displacement r_i equal to the total elongation of the preceding springs, or iR/k.

Consider now the loading, $R_i = R$, with $i < 6$ and all other loads zero. The exact solution consists of $r_i = r_{i+1} = \ldots = r_6 = iR/k$, and for the joints preceding joint i, the displacements increase in proportion to the joint number. The transformed stiffness equations yield $Q = iR_i/6$ and $q = Q/K' = iR_i/k$. Thus the exact solution is again obtained for r_6, but not for the remaining displacements. Since r_6 is obtained exactly for any individual load R_i, it is also obtained exactly for any combination of joint loads. This apparently curious result is similar to one found in obtaining member stiffness relations by the displacement method in Chapter 5, Section 4, Example 2. In the case of the member the deflected shape is represented by a finite number of degrees of freedom. In the present case the six degree-of-freedom spring is reduced to a one degree-of-freedom representation. In both cases the assumed representation is exact for loads applied only at the ends. This property and Betti's theorem are used in the case of the member to show that the fixed end forces are obtained exactly for any span load. A similar result may be established in the present case. Since the fixed end force is exact, and the representation is exact for a force applied at the end, the solution for r_6 is exact for any span load.

Normal Deformation Modes

A fundamental type of deformation mode $\mathbf{r} = \mathbf{\Phi}$, called normal or characteristic, has the property

$$\mathbf{K\Phi} = \lambda\mathbf{\Phi} \qquad (11\text{–}2a)$$

or

$$(\mathbf{K} - \lambda\mathbf{I})\mathbf{\Phi} = 0 \tag{11–2b}$$

where λ is a characteristic value. This property is such that the displacement mode $\mathbf{\Phi}$ is proportional to the load $\lambda\mathbf{\Phi}$. Equation (11–2) represents a linear characteristic-value problem or eigenvalue problem. For a nontrivial solution $\mathbf{\Phi} \neq 0$ to exist, λ must be a root of the determinantal, or characteristic, equation

$$|\mathbf{K} - \lambda\mathbf{I}| = 0 \tag{11–3}$$

The determinant in the equation is a polynomial in λ of degree equal to the number of degrees of freedom n. It is shown in linear algebra that for a symmetric matrix $\mathbf{K}$, which is the case for the stiffness matrix, all n roots of the determinant are real, although some roots may be multiple. It is also shown that for a positive definite matrix, which is the case for a kinematically stable structure, all roots are positive. For a structure having mechanism modes, it was seen in Section 10 that such modes satisfy Eq. (11–2) with $\lambda = 0$. If there are (NM) independent mechanism modes, $\lambda = 0$ is a root of multiplicity (NM), and the remaining roots are all positive.

The n roots of Eq. (11–3), $\lambda_1, \lambda_2, \ldots, \lambda_n$, are the characteristic values or eigenvalues of $\mathbf{K}$. They will be interpreted in what follows as generalized stiffnesses.

To λ_i corresponds a solution $\mathbf{\Phi}_i = \{\Phi_{1i} \ldots \Phi_{ni}\}$ of Eq. (11–2). $\mathbf{\Phi}_i$ is called, equivalently, a characteristic shape, a characteristic vector, an eigenvector, or a normal deformation mode. If λ_i is a simple root, the solution of Eq. (11–2) has the form $\mathbf{\Phi}_i = C\mathbf{A}$ where C is an arbitrary constant and $\mathbf{A}$ is any particular solution. $\mathbf{A}$ may be determined by giving one of the unknowns some value, say 1, and solving $(n-1)$ equations for the remaining unknowns. The nth equation will be necessarily satisfied, as this is what the determinantal equation ensures. A definite choice of C is a normalization procedure. For example, the largest element of $\mathbf{\Phi}_i$ in absolute value may be set equal to 1. A standard normalization procedure is

$$\mathbf{\Phi}_i^T \mathbf{\Phi}_i = 1 \tag{11–4}$$

Equation (11–4) yields $C^2\mathbf{A}^T\mathbf{A} = 1$. Since $\mathbf{A}^T\mathbf{A} > 0$, C^2 can always be determined. The sign of C is chosen arbitrarily.

Two distinct normal modes $\mathbf{\Phi}_i$ and $\mathbf{\Phi}_j$ satisfy the relations $\mathbf{K}\mathbf{\Phi}_i = \lambda_i\mathbf{\Phi}_i$ and $\mathbf{K}\mathbf{\Phi}_j = \lambda_j\mathbf{\Phi}_j$. Thus they also satisfy

$$\mathbf{\Phi}_j^T \mathbf{K}\mathbf{\Phi}_i = \lambda_i \mathbf{\Phi}_j^T \mathbf{\Phi}_i \tag{11–5a}$$

$$\mathbf{\Phi}_i^T \mathbf{K}\mathbf{\Phi}_j = \lambda_j \mathbf{\Phi}_i^T \mathbf{\Phi}_j \tag{11–5b}$$

The left-hand sides of Eqs. (11–5) are scalars, equal each to its transpose.

Since $\mathbf{K} = \mathbf{K}^T$, the left-hand sides are equal, and thus also the right-hand sides. For $\lambda_i \neq \lambda_j$, there comes

$$\mathbf{\Phi}_j^T \mathbf{\Phi}_i = 0 \tag{11–6}$$

Equation (11–6) represents the orthogonality property of the normal modes. It may be said that the external load of one normal mode does zero virtual work through the displacements of another mode.

Consider now the case of a double root $\lambda_i = \lambda_{i+1}$. The general solution of Eq. (7–10b) is then of the form

$$\mathbf{\Phi} = C_1 \mathbf{A}_1 + C_2 \mathbf{A}_2 \tag{11–7}$$

where C_1 and C_2 are arbitrary constants and $\mathbf{A}_1$ and $\mathbf{A}_2$ are two independent particular solutions. A particular solution $\mathbf{A}_1$ may be obtained by giving two unknowns arbitrary values and solving $(n-2)$ equations for the remaining unknowns. $\mathbf{A}_2$ is obtained similarly. It is now possible to choose two sets of constants C_1 and C_2 yielding, respectively, normalized and orthogonal modes $\mathbf{\Phi}_i$ and $\mathbf{\Phi}_{i+1}$. For example, $\mathbf{\Phi}_i$ may be determined by letting $C_2 = 0$ and normalizing $\mathbf{A}_1$. Then $\mathbf{\Phi}_{i+1}$ is sought, as in Eq. (11–7), and is required to be orthogonal to $\mathbf{\Phi}_i$. Thus

$$\mathbf{\Phi}_{i+1}^T \mathbf{\Phi}_i = C_1 + C_2 \mathbf{A}_2^T \mathbf{\Phi}_i = 0 \tag{11–8}$$

If $\mathbf{A}_2^T \mathbf{\Phi}_i = 0$, then $C_1 = 0$, and $\mathbf{A}_2$ need only to be normalized to become $\mathbf{\Phi}_{i+1}$. If $\mathbf{A}_2^T \mathbf{\Phi}_i \neq 0$, Eq. (11–8) determines C_1 in terms of C_2. The normalizing condition determines then C_2.

What has been described for a double root is readily generalized to a root of higher order multiplicity, although this is a rare occurrence.

In conclusion, there are n orthogonal normal deformation modes $\mathbf{\Phi}_1, \ldots, \mathbf{\Phi}_n$ associated with n characteristic values $\lambda_1, \ldots, \lambda_n$.

Consider now the $n \times n$ matrix

$$\mathbf{\Phi} = [\mathbf{\Phi}_1 \;\; \mathbf{\Phi}_2 \ldots \mathbf{\Phi}_n] \tag{11–9}$$

and define the diagonal matrix

$$\boldsymbol{\lambda} = \lceil \lambda_1 \;\; \lambda_2 \ldots \lambda_n \rfloor \tag{11–10}$$

$\mathbf{\Phi}$ satisfies

$$\mathbf{K\Phi} = \mathbf{\Phi}\boldsymbol{\lambda} \tag{11–11}$$

and from Eqs. (11–4) and (11–6),

$$\mathbf{\Phi}^T \mathbf{\Phi} = \mathbf{I} \tag{11–12}$$

Equation (7–12) characterizes an orthogonal matrix and shows that

$$\mathbf{\Phi}^{-1} = \mathbf{\Phi}^T \tag{11–13}$$

A transformation from $\mathbf{r}$ to generalized displacements $\mathbf{q}$ using $\mathbf{\Phi}$ as a transformation matrix,

$$\mathbf{r} = \mathbf{\Phi q} \tag{11–14}$$

is called transformation to normal coordinates. The stiffness equations $\mathbf{Kr} = \mathbf{R}$ transform into

$$\mathbf{K}^*\mathbf{q} = \mathbf{R}^* \tag{11–15}$$

where, noting Eqs. (11–11) and (11–12),

$$\mathbf{K}^* = \mathbf{\Phi}^T\mathbf{K\Phi} = \boldsymbol{\lambda} \tag{11–16}$$

$$\mathbf{R}^* = \mathbf{\Phi}^T\mathbf{R} \tag{11–17}$$

The characteristic values $\boldsymbol{\lambda}$ take thus the physical meaning of generalized stiffnesses. To mode $\mathbf{\Phi}_i$ corresponds the generalized or modal stiffness

$$K_i^* = \mathbf{\Phi}_i^T\mathbf{K\Phi}_i = \lambda_i \tag{11–18}$$

and the generalized or modal load

$$R_i^* = \mathbf{\Phi}_i^T\mathbf{R} \tag{11–19}$$

If the structure has rigid body modes, the corresponding generalized stiffnesses are zero, as seen earlier. The transformed stiffness equations (11–15) are called normal, or modal, equations. They are uncoupled into n separate equations of the form

$$\lambda_i q_i = R_i^* \tag{11–20}$$

In the absence of rigid body modes all λ's are nonzero and positive. The solution is then

$$q_i = \lambda_i^{-1} R_i^* \tag{11–21a}$$

or

$$\mathbf{q} = \boldsymbol{\lambda}^{-1}\mathbf{R}^* \tag{11–21b}$$

and from Eq. (11–14),

$$\mathbf{r} = \mathbf{\Phi q} = \mathbf{\Phi}_1 q_1 + \mathbf{\Phi}_2 q_2 + \ldots + \mathbf{\Phi}_n q_n \tag{11–22a}$$

or

$$\mathbf{r} = \mathbf{\Phi}\boldsymbol{\lambda}^{-1}\mathbf{R}^* = \lambda_1^{-1} R_1^* \mathbf{\Phi}_1 + \ldots + \lambda_n^{-1} R_n^* \mathbf{\Phi}_n \tag{11–22b}$$

If $\mathbf{\Phi}_i$ is a rigid body mode, then $\lambda_i = 0$, and Eq. (11–20) requires that $R_i^* = 0$. Thus for equilibrium to be possible, the virtual work of the load $\mathbf{R}$ through the rigid body modes should vanish. The rigid body displacements remain indeterminate.

Equation (11–22a) gives $\mathbf{r}$ as a superposition of deformation modes $\mathbf{\Phi}_i$ with amplitudes q_i. If the modal stiffnesses are ordered in ascending order,

$$\lambda_1 \leqslant \lambda_2 \leqslant \ldots \leqslant \lambda_n \tag{11–23}$$

then λ_1^{-1} is the largest modal flexibility, followed by λ_2^{-1}, and so on. This provides a basis for approximating Eq. (11–22) by deleting the contribution of the higher modes, that is, the stiffer ones.

Example 2

Consider three springs in series of equal stiffness k. The stiffness matrix is obtained by specializing that of Fig. 11–1. The stiffness equations of the normal modes are then

$$\begin{bmatrix} 2k-\lambda & -k & 0 \\ -k & 2k-\lambda & -k \\ 0 & -k & k-\lambda \end{bmatrix} \begin{Bmatrix} \Phi_1 \\ \Phi_2 \\ \Phi_3 \end{Bmatrix} = 0$$

The determinantal equation reduces to

$$x^3 - 5x^2 + 6x - 1 = 0$$

where $x = \lambda/k$. The roots are found to be

$$\boldsymbol{\lambda} = k\{0.19806 \quad 1.55496 \quad 3.24698\}$$

To determine the first normal mode, λ is given the value 0.19806k. Then, from the first stiffness equation, $\Phi_2 = 1.80194\Phi_1$. From the third, $\Phi_3 = 2.24698\Phi_1$. The normalizing condition $\Phi_1^2 + \Phi_2^2 + \Phi_3^2 = 1$ determines Φ_1^2. Choosing $\Phi_1 > 0$, we obtain $\Phi_1 = 0.32799$. The solution is renamed $\{\Phi_{11} \quad \Phi_{21} \quad \Phi_{31}\}$ and is placed in the first column of $\mathbf{\Phi}$. The second and third columns of $\mathbf{\Phi}$ are obtained similarly. The result is

$$\mathbf{\Phi} = \begin{bmatrix} 0.32799 & 0.73698 & 0.59101 \\ 0.59101 & 0.32799 & -0.73698 \\ 0.73698 & -0.59101 & 0.32799 \end{bmatrix}$$

Consider now the loading

$$\mathbf{R} = \{0 \ \ 0 \ \ R\}$$

The transformed loading according to Eq. (11–17) is

$$\mathbf{R}^* = R\{0.73698 \ \ -0.59101 \ \ 0.32799\}$$

then

$$\mathbf{q} = \boldsymbol{\lambda}^{-1}\mathbf{R}^* = \frac{R}{k}\{3.72099 \ \ -0.38008 \ \ 0.10101\}$$

and

$$\mathbf{r} = \boldsymbol{\Phi}\mathbf{q} = \frac{R}{k}\{1.0 \ \ 2.0 \ \ 3.0\}$$

The result is found to four significant digits and is the exact solution, as seen in Example 1.

An approximation to the preceding problem based on the first mode is the first column of $\boldsymbol{\Phi}$ multiplied by $q_1 = 3.72099R/k$, or

$$\mathbf{r} = \boldsymbol{\Phi}_1 q_1 = \frac{R}{k}\{1.220 \ \ 2.199 \ \ 2.742\}$$

The relative error on the largest displacement r_3 is $(3–2.742)/3 = 0.086$

As a different loading, let

$$\mathbf{R} = R\{1 \ \ 2 \ \ 3\}$$

then

$$\mathbf{R}^* = \boldsymbol{\Phi}^T\mathbf{R} = R\{3.72099 \ \ -0.38008 \ \ 0.10101\}$$

$$\mathbf{q} = \boldsymbol{\lambda}^{-1}\mathbf{R}^* = \frac{R}{k}\{18.7872 \ \ -0.2444 \ \ 0.03111\}$$

and

$$\mathbf{r} = \boldsymbol{\Phi}\mathbf{q} = \frac{R}{k}\{6.000 \ \ 11.000 \ \ 14.000\}$$

It is readily verified that this is the exact solution. An approximation based on the first mode gives

$$\mathbf{r} = \mathbf{\Phi}_1 q_1 = \frac{R}{k}\{6.162 \quad 11.103 \quad 13.846\}$$

The relative error on r_3 is $(14-13.846)/14 = 0.011$.

What favors a good approximation through the fundamental mode is the increase of modal stiffness λ_i with mode number i and a decrease of modal loads R_i^* with increasing i.

12. INEXTENSIONAL BEHAVIOR OF FRAMES

Rigid frames such as shown in Fig. 3–1 gain their kinematic stability through their rigid joints. If the member ends of such a frame were pin connected, the frame would become kinematically unstable having translational mechanism modes. These modes, modified by preventing joint rotations, are inextensional deformation modes of the rigid frame. They form with the displacement modes due to joint rotations the complete set of inextensional modes of the structure. An arbitrary deformed state of the structure may be represented as a superposition of inextensional and extensional deformation modes. The latter are often negligible because member axial stiffnesses are much larger than transverse stiffnesses.

An examination of the orders of magnitudes involved in frame behavior, and of the general conditions under which the inextensional approximation is valid, is made in what follows.

Consider a transformation

$$\mathbf{r} = [\mathbf{T}_1 \quad \mathbf{T}_2]\begin{Bmatrix} \mathbf{q}_1 \\ \mathbf{q}_2 \end{Bmatrix} \tag{12–1}$$

from joint displacements $\mathbf{r}$ to inextensional and extensional generalized displacements, $\mathbf{q}_1$ and $\mathbf{q}_2$, respectively. Such a transformation is described analytically in Chapter 9, Section 5. Connectivity and member stiffness equations are partitioned into equations for axial deformations $\mathbf{v}_a$ and bending deformations $\mathbf{v}_b$, in the form

$$\begin{Bmatrix} \mathbf{v}_a \\ \mathbf{v}_b \end{Bmatrix} = \begin{bmatrix} 0 & \boldsymbol{\alpha}_{a2} \\ \boldsymbol{\alpha}_{b1} & \boldsymbol{\alpha}_{b2} \end{bmatrix}\begin{Bmatrix} \mathbf{q}_1 \\ \mathbf{q}_2 \end{Bmatrix} \tag{12–2}$$

$$\begin{Bmatrix} \mathbf{V}_a \\ \mathbf{V}_b \end{Bmatrix} = \begin{bmatrix} \mathbf{k}_a & \\ & \mathbf{k}_b \end{bmatrix}\begin{Bmatrix} \mathbf{v}_a \\ \mathbf{v}_b \end{Bmatrix} \tag{12–3}$$

For simplicity the discussion will be limited to the effects of joint loads only. The fixed end forces are thus omitted from the stiffness equations.

The transformed equilibrium equations form a contragradient pair with Eq. (12–2) and are thus

$$\begin{bmatrix} 0 & \boldsymbol{\alpha}_{b1}^T \\ \boldsymbol{\alpha}_{a2}^T & \boldsymbol{\alpha}_{b2}^T \end{bmatrix} \begin{Bmatrix} \mathbf{V}_a \\ \mathbf{V}_b \end{Bmatrix} = \begin{Bmatrix} \mathbf{Q}_1 \\ \mathbf{Q}_2 \end{Bmatrix} \tag{12–4}$$

where

$$\begin{Bmatrix} \mathbf{Q}_1 \\ \mathbf{Q}_2 \end{Bmatrix} = \begin{bmatrix} \mathbf{T}_1^T \\ \mathbf{T}_2^T \end{bmatrix} \mathbf{R} \tag{12–5}$$

The structure stiffness equations take the form

$$\begin{bmatrix} \mathbf{K}_{11} & \mathbf{K}_{12} \\ \mathbf{K}_{21} & \mathbf{K}_{22} \end{bmatrix} \begin{Bmatrix} \mathbf{q}_1 \\ \mathbf{q}_2 \end{Bmatrix} = \begin{Bmatrix} \mathbf{Q}_1 \\ \mathbf{Q}_2 \end{Bmatrix} \tag{12–6}$$

where

$$\mathbf{K}_{11} = \boldsymbol{\alpha}_{b1}^T \mathbf{k}_b \boldsymbol{\alpha}_{b1} \tag{12–7a}$$

$$\mathbf{K}_{12} = \mathbf{K}_{21}^T = \boldsymbol{\alpha}_{b1}^T \mathbf{k}_b \boldsymbol{\alpha}_{b2} \tag{12–7b}$$

$$\mathbf{K}_{22} = \boldsymbol{\alpha}_{a2}^T \mathbf{k}_a \boldsymbol{\alpha}_{a2} + \boldsymbol{\alpha}_{b2}^T \mathbf{k}_b \boldsymbol{\alpha}_{b2} \tag{12–7c}$$

The preceding transformation is illustrated in Example 1 and is then followed by an examination of structural behavior.

Example 1

The frame of Fig. 12–1 was analyzed by the inextensional approximation in Section 3, Example 4a. Here the transformation to inextensional and extensional

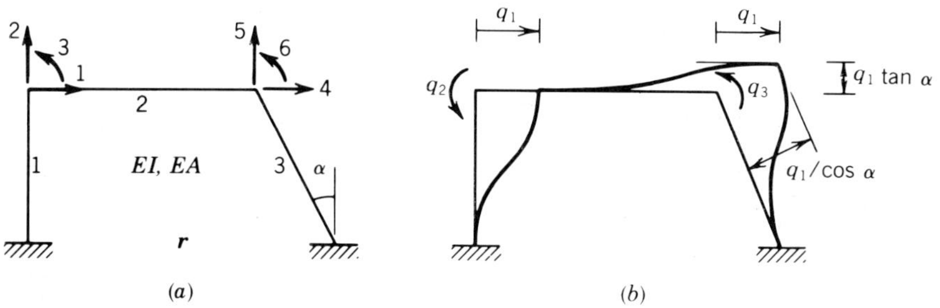

FIG. 12–1. Transformation to inextensional and extensional displacements, Example 1.

modes is illustrated. The complete set of joint displacements $\mathbf{r}$ is identified in (*a*), and the inextensional displacements $\mathbf{q}_1 = \{q_1 \; q_2 \; q_3\}$ in (*b*). The extensional generalized displacements $\mathbf{q}_2 = \{q_4 \; q_5 \; q_6\}$ are chosen as the elongation of members 1, 2, and 3, respectively. We thus have

$$\begin{aligned} r_2 &= q_4 \\ r_4 - r_1 &= q_5 \\ -r_4 \sin\alpha + r_5 \cos\alpha &= q_6 \end{aligned}$$

In relating q_1 to $\mathbf{r}$, it is possible to identify q_1 with the horizontal displacement of any point of member 2. A choice that treats r_1 and r_4 equally is

$$\tfrac{1}{2}(r_1 + r_4) = q_1$$

These four equations are readily solved for r_1, r_2, r_4, and r_5 in terms of q_1, q_4, q_5, and q_6. Adjoining the relations $r_3 = q_2$ and $r_6 = q_3$, the transformation matrix is obtained as

$$\mathbf{T} = [\mathbf{T}_1 \;\; \mathbf{T}_2] = \left[\begin{array}{ccc|ccc} 1 & 0 & 0 & 0 & -\frac{1}{2} & 0 \\ 0 & 0 & 0 & 1 & 0 & 0 \\ 0 & 1 & 0 & 0 & 0 & 0 \\ 1 & 0 & 0 & 0 & \frac{1}{2} & 0 \\ t & 0 & 0 & 0 & \frac{t}{2} & \frac{1}{c} \\ 0 & 0 & 1 & 0 & 0 & 0 \end{array}\right]$$

where

$$\begin{aligned} t &= \tan\alpha \\ c &= \cos\alpha \end{aligned}$$

Axial and bending deformations are

$$\begin{aligned} \mathbf{v}_a &= \{v_1^1 \; v_1^2 \; v_1^3\} \\ \mathbf{v}_b &= \{v_2^1 \; v_3^1 \; v_2^2 \; v_3^2 \; v_2^3 \; v_3^3\} \end{aligned}$$

Connectivity matrix $\boldsymbol{\alpha}_v$ may be formed from $\mathbf{a}_v$ by the transformation $\boldsymbol{\alpha}_v = \mathbf{a}_v\mathbf{T}$, or directly by geometry. For deformations of the alternate type, $\boldsymbol{\alpha}_v$ is obtained in the partitioned form

$$
\begin{bmatrix} 0 & \boldsymbol{\alpha}_{a2} \\ \boldsymbol{\alpha}_{b1} & \boldsymbol{\alpha}_{b2} \end{bmatrix} =
\left[\begin{array}{ccc|ccc}
 & & & 1 & & \\
 & & & & 1 & \\
 & & & & & 1 \\
\hline
\frac{1}{l_1} & & & & -\frac{1}{2l_1} & \\
\frac{1}{l_1} & 1 & & & -\frac{1}{2l_1} & \\
-\frac{t}{l_2} & 1 & & \frac{1}{l_2} & -\frac{t}{2l_2} & -\frac{1}{cl_2} \\
-\frac{t}{l_2} & & 1 & \frac{1}{l_2} & -\frac{t}{2l_2} & -\frac{1}{cl_2} \\
\frac{1}{cl_3} & & & & \frac{1}{2cl_3} & \frac{t}{l_3} \\
\frac{1}{cl_3} & & 1 & & \frac{1}{2cl_3} & \frac{t}{l_3}
\end{array}\right]
$$

The member stiffness matrices are $\mathbf{k}_b = [\mathbf{k}_b^1 \;\; \mathbf{k}_b^2 \;\; \mathbf{k}_b^3]$ and $\mathbf{k}_a = \lceil k_a^1 \;\; k_a^2 \;\; k_a^3 \rfloor$, where for $m = 1, 2, 3$,

$$
\mathbf{k}_b^m = \frac{EI}{l_m}\begin{bmatrix} 4 & 2 \\ 2 & 4 \end{bmatrix}
$$

$$
k_a^m = \frac{EA}{l_m}
$$

Applying Eqs. (12–7), the partitions of the stiffness matrix are obtained as

$$
\mathbf{K}_{11} = EI \left[\begin{array}{c|c|c}
\frac{12}{l_1^3} + \frac{12t^2}{l_2^3} + \frac{12}{l_3^3 c^2} & \frac{6}{l_1^2} - \frac{6t}{l_2^2} & \frac{6}{l_3^2 c} - \frac{6t}{l_2^2} \\
\hline
\text{symmetric} & \frac{4}{l_2} + \frac{4}{l_1} & \frac{2}{l_2} \\
\cline{2-3}
 & & \frac{4}{l_2} + \frac{4}{l_3}
\end{array}\right]
$$

$$\mathbf{K}_{12}=\mathbf{K}_{21}^T=EI\begin{bmatrix} -\dfrac{12t}{l_2^3} & \dfrac{6t^2}{l_2^3}+\dfrac{6}{l_3^3c^2}-\dfrac{6}{l_1^3} & \dfrac{t}{c}\left(\dfrac{12}{l_2^3}+\dfrac{12}{l_3^3}\right) \\ \dfrac{6}{l_2^2} & -\dfrac{3}{l_1^2}-\dfrac{3t}{l_2^2} & -\dfrac{6}{l_2^2c} \\ \dfrac{6}{l_2^2} & -\dfrac{3t}{l_2^2}+\dfrac{3}{l_3^2c} & -\dfrac{6}{l_2^2c}+\dfrac{6t}{l_3^2} \end{bmatrix}$$

$$\mathbf{K}_{22}=E\begin{bmatrix} \dfrac{A}{l_1}+\dfrac{12I}{l_2^3} & -\dfrac{6It}{l_2^3} & -\dfrac{12I}{l_2^3c} \\ & \dfrac{A}{l_2}+\dfrac{3I}{l_1^3}+\dfrac{3It^2}{l_2^3}+\dfrac{3I}{l_3^3c^2} & \dfrac{t}{c}\left(\dfrac{6I}{l_2^3}+\dfrac{6I}{l_3^3}\right) \\ \text{symmetric} & & \dfrac{A}{l_3}+\dfrac{12It^2}{l_3^3}+\dfrac{12I}{l_2^3c^2} \end{bmatrix}$$

The joint load matrix $\mathbf{R}$ transforms into

$$\mathbf{Q}=\mathbf{T}^T\mathbf{R}=\left\{R_1+R_4+tR_5, R_3, R_6, R_2, \frac{1}{2}(R_4-R_1+tR_5), \frac{R_5}{c}\right\}$$

$\mathbf{K}_{11}$ coincides with the inextensional stiffness matrix found in Section 3, Example 4a. It is noted that only $\mathbf{K}_{22}$ depends on the member axial stiffnesses.

To discuss orders of magnitude, physical dimensions are made homogeneous by multiplying rotations and dividing moments by a representative length L. Connectivity coefficients are then of order of magnitude unity, or $O(1)$. Letting r be a representative radius of gyration, orders of magnitudes of $\mathbf{k}_b$ and $\mathbf{k}_a$ are such that

$$O(\mathbf{k}_b)=\varepsilon^2 O(\mathbf{k}_a) \tag{12–8}$$

where

$$\varepsilon=\frac{r}{L} \tag{12–9}$$

In the stiffness equations only $\mathbf{K}_{22}$ depends on the member axial stiffnesses. The stiffness coefficients are made all of comparable orders of magnitude by letting

$$\mathbf{K}'_{22}=\varepsilon^2\mathbf{K}_{22} \tag{12–10}$$

The stiffness equations take then the form

$$\begin{bmatrix} \mathbf{K}_{11} & \mathbf{K}_{12} \\ \varepsilon^2\mathbf{K}_{21} & \mathbf{K}'_{22} \end{bmatrix} \begin{Bmatrix} \mathbf{q}_1 \\ \mathbf{q}_2 \end{Bmatrix} = \begin{Bmatrix} \mathbf{Q}_1 \\ \varepsilon^2\mathbf{Q}_2 \end{Bmatrix} \tag{12–11}$$

For a loading that contains inextensional and extensional components of comparable magnitudes, Eq. (12–11) suggests that $O(\mathbf{q}_2) = \varepsilon^2 O(\mathbf{q}_1)$ and consequently that $\mathbf{q}_2$ may be neglected in the first set of equations. There results the inextensional approximation

$$\mathbf{K}_{11}\mathbf{q}_1 = \mathbf{Q}_1 \tag{12–12a}$$

$$\mathbf{K}'_{22}\mathbf{q}_2 = \varepsilon^2\mathbf{Q}_2 - \varepsilon^2\mathbf{K}_{21}\mathbf{q}_1 \tag{12–12b}$$

The solution of Eq. (12–12a) for $\mathbf{q}_1$ determines the bending deformations and forces. It is then substituted into the extensional stiffness equation (12–12b) in order to solve for $\mathbf{q}_2$ and determine the axial forces $\mathbf{V}_a$. However, these are usually statically determinate, in which case they may be determined more readily from the extensional equilibrium equations. In Example 1, $\boldsymbol{\alpha}_{a2} = \mathbf{I}$, and the second of Eq. (12–4) is the solution for $\mathbf{V}_a$ in terms of $\mathbf{Q}_2$ and the inextensional solution for $\mathbf{V}_b$.

For Eqs. (12–12) to be consistent with the underlying assumptions, the solution of Eq. (12–12b) for $\mathbf{q}_2$ must be such that $\mathbf{K}_{12}\mathbf{q}_2 \ll \mathbf{Q}_1$, or

$$\varepsilon^2\mathbf{K}_{12}\mathbf{K}'^{-1}_{22}\mathbf{K}_{21}\mathbf{K}^{-1}_{11}\mathbf{Q}_1 \ll \mathbf{Q}_1 \tag{12–13}$$

The solution of Eqs. (12–11) in the case of a purely extensional load $\mathbf{Q}_2$ is discussed in the next section.

Flexible Structures

Conditions that tend to invalidate the inextensional approximation are considered in what follows, together with the relative orders of magnitudes that describe structural behavior.

Elements of $\mathbf{K}_{11}, \mathbf{K}_{12}, \mathbf{K}_{21}$, and $\mathbf{K}'_{22}$ have a comparable order of magnitude. The matrices $(\mathbf{K}_{12}\mathbf{K}'^{-1}_{22})$ and $(\mathbf{K}_{21}\mathbf{K}^{-1}_{11})$ in Eq. (12–13) are thus nondimensional, but their orders of magnitude depend on the flexibilities of the structure. Consider for illustration an N-story rectangular frame. If the lateral stiffness of one story is $O(EI/L^3)$, then $\mathbf{K}^{-1}_{11}$ is of order $NO(L^3/EI)$. The term $\mathbf{K}_{21}\mathbf{K}^{-1}_{11}\mathbf{Q}_1$ represents axial forces. If vertical joint displacements are chosen as elements of $\mathbf{q}_2$, then corresponding elements of $\mathbf{K}_{21}\mathbf{K}^{-1}_{11}\mathbf{Q}_1$ are shear forces at the ends of the beams exerted along the columns. If column axial deformations are chosen as generalized displacements, elements of $\mathbf{K}_{21}\mathbf{K}^{-1}_{11}\mathbf{Q}_1$ are transformed into corresponding generalized forces. To the axial deformation of one column corresponds the sum of the shear forces just mentioned, above that column. The vertical joint displacements due to these forces reach the order of magnitude N relative to the axial deformations. Thus the term $\mathbf{K}_{12}\mathbf{K}'^{-1}_{22}$ also increases the order of magnitude of the term it multiplies. Letting, in general,

$$O(\mathbf{K}_{21}\mathbf{K}^{-1}_{11}) = F_1 \tag{12–14a}$$

$$O(\mathbf{K}_{12}\mathbf{K}'^{-1}_{22}) = F_2 \tag{12–14b}$$

the inextensional approximation is justified if

$$\varepsilon^2 F_1 F_2 \ll 1 \tag{12–15}$$

F_1 and F_2 may be considered as relative flexibilities.

For the multistory frame, $F_1 F_2 = N^2$. Assuming that vertical joint displacements, rather than axial deformations, are chosen as elements of $\mathbf{q}_2$, orders of magnitude of the solution may be expected to be

$$O(\mathbf{q}_1) = \frac{NL^3}{EI} O(\mathbf{Q}_1) \tag{12–16a}$$

$$O(\mathbf{q}_2) = \varepsilon^2 N^2 \frac{L^3}{EI} O(\mathbf{Q}_1) \tag{12–16b}$$

whereas member deformations are not affected by large displacements, so that

$$O(\mathbf{v}_b) = \frac{1}{N} O(\mathbf{q}_1) \tag{12–17a}$$

$$O(\mathbf{v}_a) = \frac{1}{N} O(\mathbf{q}_2) \tag{12–17b}$$

Axial forces, $\mathbf{V}_a = \mathbf{k}_a \mathbf{v}_a$, and bending forces (shears), $\mathbf{V}_b = \mathbf{k}_b \mathbf{v}_b$, are related by

$$O(\mathbf{V}_a) = NO(\mathbf{V}_b) \tag{12–18}$$

Illustration of the preceding in two cases in which the inextensional approximation is and is not justified, respectively, may be seen in Examples 2 and 3.

Example 2

The ten-story, three-bay rectangular frame shown with two loading conditions in Fig. 12–2*a* is analyzed, first, by assuming that only beams are inextensible and, then, by assuming all members to be inextensible.

Significant results are tabulated in (*c*). Their locations in the frame are identified by the notation shown in (*b*). Results of the first analysis are labeled "exact" and are tabulated above the corresponding results of the second analysis,

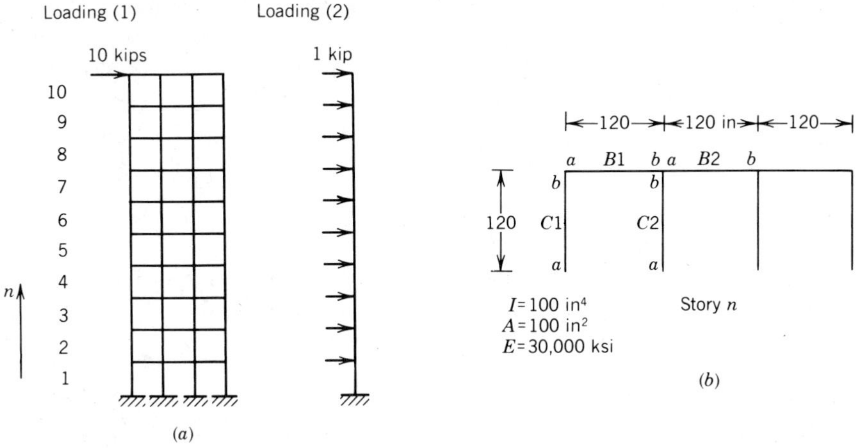

Maximum quantities	Loading (1) Location n	m	j	Loading (1) Exact / Inext	Loading (2) Location n	m	j	Loading (2) Exact / Inext
Moment, beams (k=in)	3	$B1$	a	223. / 225.	1	$B1$	a	194. / 194.
Moment, columns (k-in)	2	$C2$	a	192. / 191.	1	$C2$	a	193. / 193.
Axial force, columns (kip)	1	$C1$		32.9 / 33.4	1	$C1$		17.2 / 17.4
Floor displacement (in)	10			2.79 / 2.75	10			1.50 / 1.49
Joint rotation ×120 (in)	7	$B1$	a	0.206 / 0.200	1	$B1$	a	0.176 / 0.175

m = Member
j = Joint

(c)

FIG. 12–2. Exact and inextensional analysis, Example 2.

labeled "inext." For this frame, $\varepsilon = r/l = 1/120$, $\varepsilon^2 N^2 = 1/144 \ll 1$, and the inextensional approximation is thus justified.

To verify the order of magnitude relation (12–18), $O(\mathbf{V}_b)$ is taken as the maximum moment, which is made dimensionally homogeneous by dividing by the member length, 120 in. In both loading conditions the maximum moment has the order of magnitude of 200 k-in. Thus

$$O(N\mathbf{V}_b) = \frac{200}{120} \times 10 = 17$$

This is comparable with the maximum axial force of 32.9 k in loading (1) and 17.2 k in loading (2).

Equation (12–17a) is also verified with $O(\mathbf{q}_1)$ as the maximum floor displacement and $O(\mathbf{v}_b)$ representing bending deformations which are comparable to joint rotations.

Example 3

To treat an extreme case where the inextensional approximation is not valid, the frame of Fig. 12–2 is extended to become a 100-story frame. This is clearly only to test the mathematical behavior of the solution. In loading (1) the 10 k load is applied at the top floor, and loading (2) is made to consist of 0.1 k applied at

Maximum quantities	Loading (1)				Loading (2)			
	Location			Exact / Inext	Location			Exact / Inext
	n	m	j		n	m	j	
Moment, beams (k=in)	89	$B2$	a, b	238. / 175.	2	$B1$	a	217. / 219.
Moment, columns (k-in)	72	$C1$	a	201. / 188.	1	$C2$	a	196. / 195.
Axial force, columns (kip)	1	$C1$	a	309. / 352.	1	$C1$		158. / 176.
Floor displacement (in)	100			56.3 / 28.9	100			25.0 / 14.54
Joint rotation ×120 (in)	97	$B1$	a	0.620 / 0.200	31	$B1$	a	0.240 / 0.139

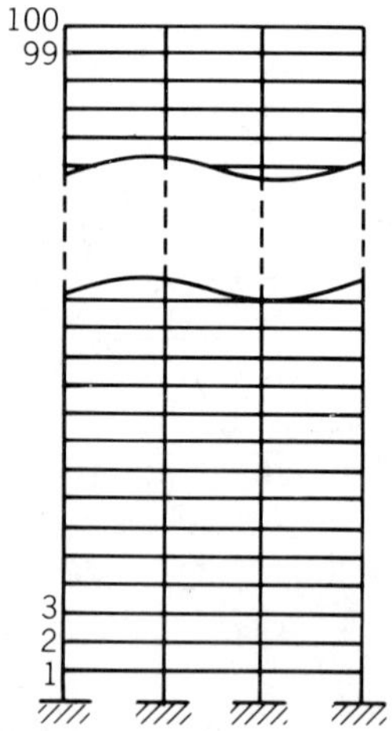

FIG. 12–3. One-hundred-story frame, Example 3.

each floor. Now $N = 100$ and $\varepsilon^2 N^2 = 1/1.44 = O(1)$. Maximum quantities obtained in the "exact" solution and the corresponding quantities obtained in the inextensional solution are tabulated in Fig. 12–3. The latter quantities are not necessarily maxima in their own solution. It is noted that unacceptable errors occur for the maximum displacements and that, although some moments are obtained accurately in the inextensional solution, the results as a whole are not accurate enough. The relative error in the maximum moment in the beams occurs with a comparable order of magnitude at several other locations and is exceeded at some. The axial forces in the interior columns of the first story as obtained through the inextensional solution have the wrong sense. The exact values in loadings (1) and (2) are 67.3 k and 24.3 k, respectively, and are tensile in column C_2. The results in the inextensional solution are 62.4 k and 31.3 k, respectively, and are compressive.

Although the inextensional solution is not applicable, the orders of magnitude relations remain valid because $\varepsilon^2 N^2 = O(1)$. According to Eq. (12–18), axial forces become large in comparison with the transverse forces. In loading (1), $O(\mathbf{V}_a) = 300$ k and $O(\mathbf{V}_b) = 238/120$ k, and in loading (2), $O(\mathbf{V}_a) = 158$ k, and $O(\mathbf{V}_b) = 217/120$ k. In both cases Eq. (12–18) is verified. It is also verified that the maximum floor displacement is $O(N)$ times the maximum bending deformation which in turn is comparable to the maximum joint rotation.

13. BEHAVIOR OF TRUSSED FRAMES

A rigid frame that does not have translational inextensional displacements is referred to here as a trussed frame. If the joints of such a frame are modified into pinned joints, the resulting structure would be kinematically stable. This modified structure will be referred as the substitute truss.

For a trussed frame, $\mathbf{q}_1$ consist of the joint rotations, and $\mathbf{q}_2$ of the joint translations. Elements of $\mathbf{Q}_1$ have the physical dimension of a moment and those of $\mathbf{Q}_2$ that of a force. It will be assumed, however, that elements of $\mathbf{q}$ and $\mathbf{Q}$ are made physically homogeneous, as done in the preceding section.

By contrast with frames having translational inextensional displacements, a trussed frame can only respond with extensional deformation modes to joint forces $\mathbf{Q}_2$. Behavior under joint moments $\mathbf{Q}_1$ is as described in the preceding section. If $\varepsilon^2 F_1 F_2 \ll 1$, the inextensional approximation may be used. The question to be considered here is the behavior due to joint forces $\mathbf{Q}_2$ applied alone. Equations (12–11) become

$$\begin{bmatrix} \mathbf{K}_{11} & \mathbf{K}_{12} \\ \varepsilon^2 \mathbf{K}_{21} & \mathbf{K}'_{22} \end{bmatrix} \begin{Bmatrix} \mathbf{q}_1 \\ \mathbf{q}_2 \end{Bmatrix} = \begin{Bmatrix} \mathbf{0} \\ \varepsilon^2 \mathbf{Q}_2 \end{Bmatrix} \tag{13–1}$$

Now the order of magnitude of $\mathbf{q}_2$ is governed by $\mathbf{K}'^{-1}_{22}$ and $\varepsilon^2 \mathbf{Q}_2$, and that of $\mathbf{q}_1$ is determined from the first of Eqs. (13–1). Since $\mathbf{K}_{11}$ and $\mathbf{K}_{12}$ are of comparable orders of magnitude, it is first assumed, subject to subsequent verification, that

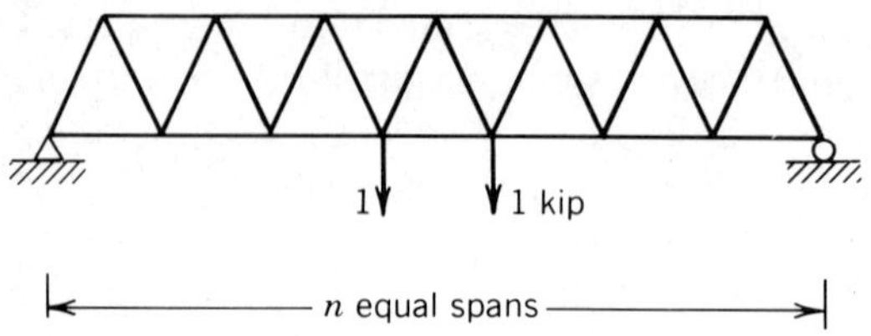

(a)

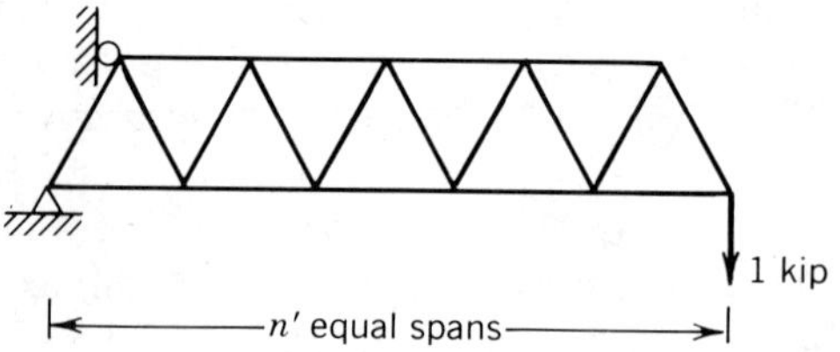

(b)

All members
$L = 120$ in
$A = 10$ in^2
$I = 100$ in^4
$E = 30{,}000$ ksi

	V_a kips		V_b kips	q_1 inch	q_2 inch
	Axial force		Bending moment/120	Rotation × 120	Translation
	Exact	Truss			
$n = 3$	1.15	1.15	0.00723	0.00193	0.00211
$n = 9$	4.59	4.61	0.0133	0.0114	0.0360
$n' = 9$	9.78	9.81	0.0205	0.0414	0.246
$n = 27$	14.94	15.01	0.0299	0.0974	0.883

(c)

	q_2/q_1	q_1/v_b	q_2/v_b	q_2/v_a	v_b/v_a	σ_b/σ_a
$n = 3$	1.1	0.5	0.5	4.5	9.1	0.24
$n = 9$	3.2	1.5	4.8	20.	4.2	0.11
$n' = 9$	5.9	3.3	20.	63.	3.0	0.08
$n = 27$	9.1	5.5	50.	147.	2.9	0.08

(d)

FIG. 13–1. Trussed frame behavior, Example 1.

$O(\mathbf{q}_1)$ does not exceed $O(\mathbf{q}_2)$. The term $\varepsilon^2\mathbf{K}_{21}\mathbf{q}_1$ is then neglected in the second of Eqs. (13–1), resulting in the approximation

$$\mathbf{K}'_{22}\mathbf{q}_2 = \varepsilon^2\mathbf{Q}_2 \tag{13–2a}$$

$$\mathbf{K}_{11}\mathbf{q}_1 = -\mathbf{K}_{12}\mathbf{q}_2 \tag{13–2b}$$

It is further noted that the contribution of $\mathbf{k}_b$ to $\mathbf{K}_{22}$ in Eq. (12–7c) is of order ε^2 relative to that of $\mathbf{k}_a$. $\mathbf{K}_{22}$ is thus approximated by

$$\mathbf{K}_{22} = \boldsymbol{\alpha}_{a2}^T\mathbf{k}_a\boldsymbol{\alpha}_{a2} \tag{13–3}$$

and becomes the stiffness matrix of the substitute pin-jointed truss. The approximation consists then in determining the joint translations $\mathbf{q}_2$ by analysis of the substitute truss. The joint rotations $\mathbf{q}_1$ are then determined by solving Eq. (13–2b). Bending stresses σ_b determined from $\mathbf{q}_1$ are called secondary stresses, but their order of magnitude may be comparable to the axial stresses σ_a in the substitute truss.

$\mathbf{K}_{11}$ is a stiffness associated with joint rotations, and $\mathbf{K}_{12}\mathbf{q}_2$ may be thought of as external restraining moments due to translations $\mathbf{q}_2$ without rotations. For a nonflexible structure, the relative flexibility associated with $\mathbf{K}_{11}^{-1}$ is expected to be $O(1)$, and the bending deformations $\mathbf{v}_b$ induced by $\mathbf{q}_2$ are expected to be $O(\mathbf{q}_2)$. The effect of increased flexibility of the substitute truss is expected to be more pronounced on the displacements $\mathbf{q}_2$ than on the deformations. Thus in both cases Eqs. (13–2) are justified.

Example 2

The preceding is illustrated using the trussed frames of Fig. 13–1. The frame of Fig. 13–1*a* was analyzed for spans $n = 3$, $n = 9$, and $n = 27$, and that of Fig. 13–1*b* was analyzed for $n' = 9$ spans. This latter case corresponds approximately to $n = 18$ in (*a*). Results for maximum quantities are tabulated in (*c*) and (*d*). In (*d*), $v_b = V_b L^3/EI$ and $v_a = V_a L/EA$, where L, A, E, and I are as shown in the figure. Further $\sigma_b/\sigma_a = \varepsilon v_b/v_a$, where $\varepsilon = r/L = 0.026$. Note that the actual bending stress is larger than the one computed by the ratio c/r, where c is the maximum distance to the neutral axis.

EXERCISES

Sections 1 to 3

1. For the frame of Fig. 3–1*a*, form the stiffness matrix by carrying out formally the product $\mathbf{a}^T\mathbf{k}_c\mathbf{a}$ in which $\mathbf{a}$ is partitioned by member ends and joints and $\mathbf{k}_c$ is partitioned by member ends. Ascertain that the result has the form of Fig. 3–1*d*.

2. Analyze the beams of Fig. P2 by the stiffness method. Draw in each case the bending moment diagram and the deformed shape.

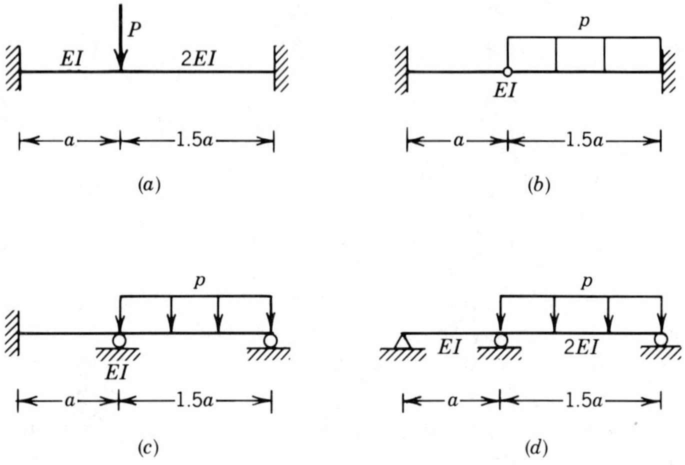

FIG. P2

3. Assemble the stiffness equations of the structures in Fig. P3 by the direct stiffness method. Assume inextensible members and a constant EI except where indicated, and use the least number of effective displacements. In

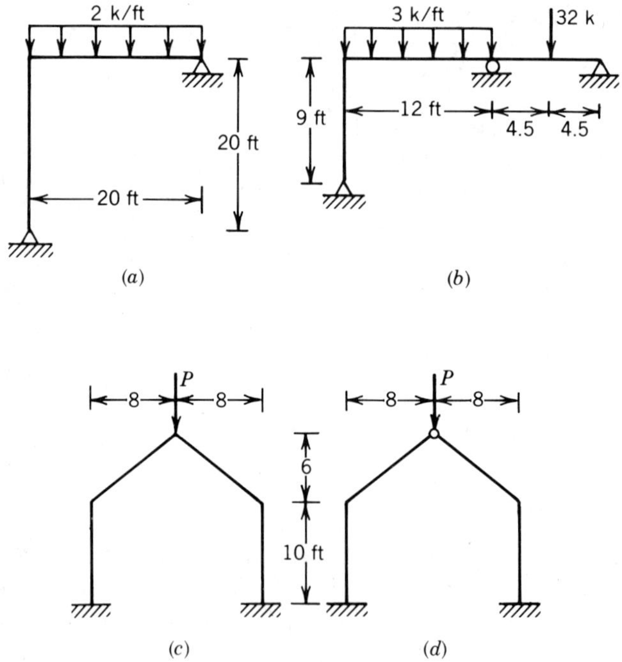

FIG. P3 (Continued on next page.)

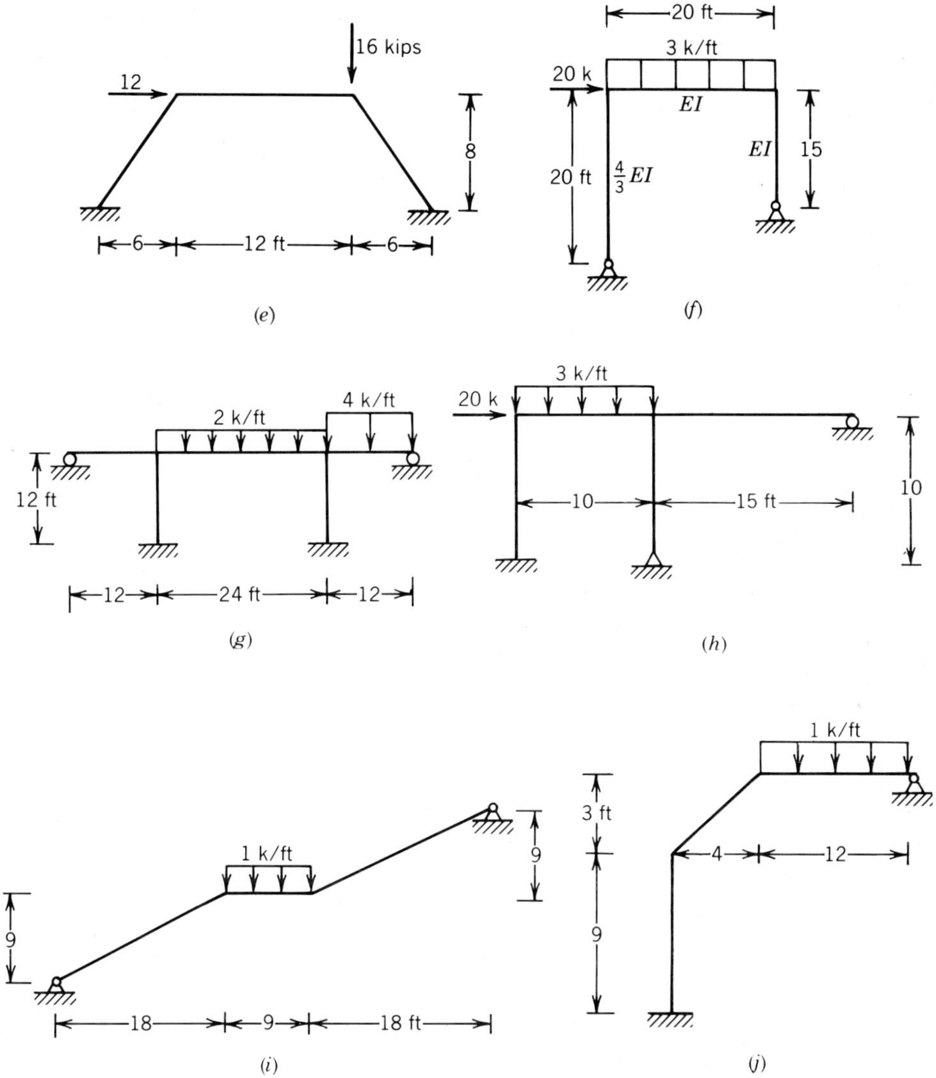

FIG. P3 *(continued)*

each case solve the equations, and draw the bending moment diagram and the deformed shape.

4. Do the preceding exercise for the frame of Fig. P3*d* by including the rotations at the pinned joint in the set of displacements.

5. Form the stiffness equations of the frame in Fig. 8/2.4–3*a*, assuming inextensible members and a constant *EI*.

6. Determine and sketch independent translational inextensional modes for the frames of Fig. P6.

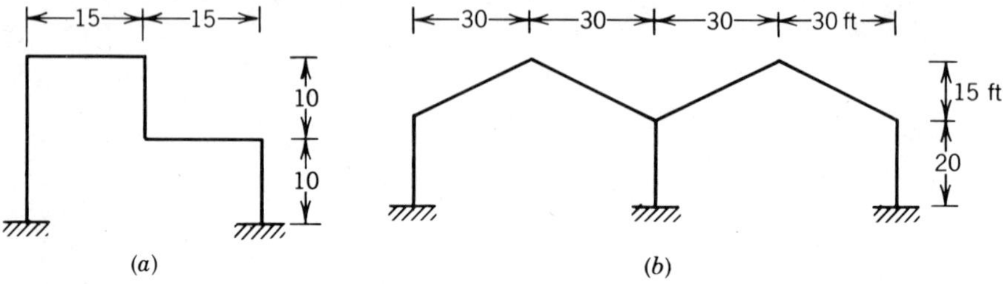

FIG. P6

7. For the frame of Fig. P6*b*, let all members have the properties $I = 3000\,\text{in}^4$, $A = 30\,\text{in}^2$, $E = 30{,}000$ ksi, $\alpha = 6.5(10^{-6})/°\text{F}$. Identify the joint displacements **r**, including member extensibility, and form the equivalent joint load for the following loading conditions:
   a. Uniform vertical load of 1 kip per horizontal foot.
   b. Uniform horizontal load of 1 kip per vertical foot applied on the first two members from the left.
   c. Uniform temperature drop of 70°F in the top members.
   d. Uniform temperature gradient of 35°F through the depth $d = 20$ in of the top members. Assume the outside temperature to be lower.
   e. A vertical settlement of 0.01 in at the middle support.

8. Do Exercise 7 for an inextensional analysis. Indicate the choice of displacements.

9. Using a computer, solve the stiffness equations for the structure and loadings of Exercises 7 and 8. Draw the bending moment diagrams.

10. Form the stiffness equations for the pin-jointed truss of Fig. P10. Obtain the solution with the help of a computer and determine the axial forces.

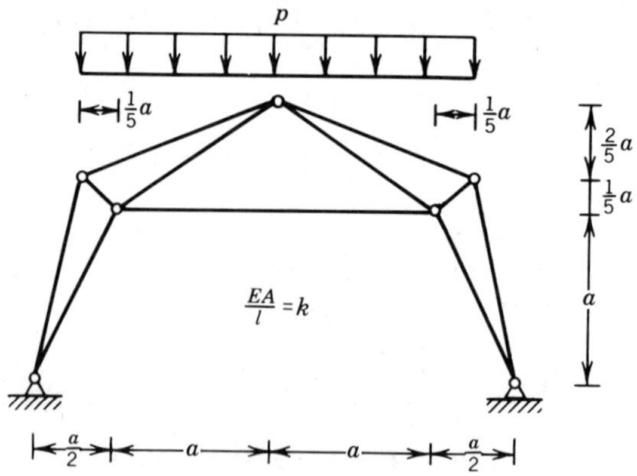

FIG. P10

Sections 4, 5

11. Interpret the elements of the stiffness matrix for the structure of Exercise 3, Fig. P3*j*, as done in Section 4.

12. Do Exercise 3 for the structures in Fig. P3*c*, *e*, *g*, taking advantage of structure symmetry.

13. Analyze the continuous beams in Fig. P13 by superposition of symmetric and antisymmetric loading conditions. Use the least number of effective displacements.

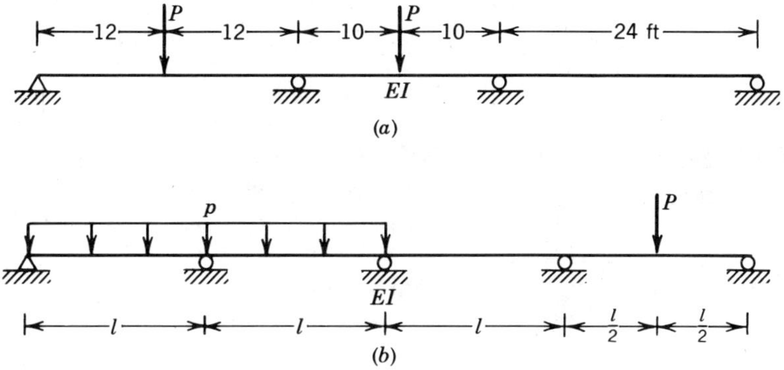

FIG. P13

Section 7

14. Do the static condensation of Section 7, Example 1, for a frame whose properties are EI_b for the beam and EI_c for the columns. Do a similar exercise for pinned supports.

15. Derive by static condensation of the degrees of freedom at joint 3 the reduced stiffness matrix of the alternate type for the two-element member of Fig. P15.

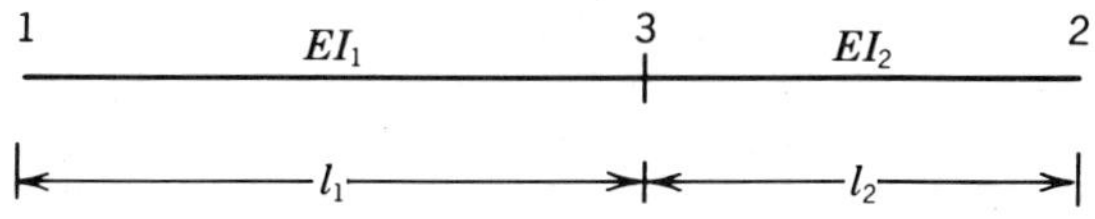

FIG. P15

16. Do Exercise 15 in the case where the two elements are pin connected at joint 3 and have the same *EI*.

Section 9

17. In the continuous beam of Fig. P17*a* the load $P = 1$ applied at x causes a bending moment M at A. A plot of M versus x, keeping A fixed, is called the influence line for moment. In Fig. P17*b* a rotational release is introduced at A, and a unit deformation $\Delta = 1$ is imposed at that release by means of some moment M', producing the deformed shape shown.

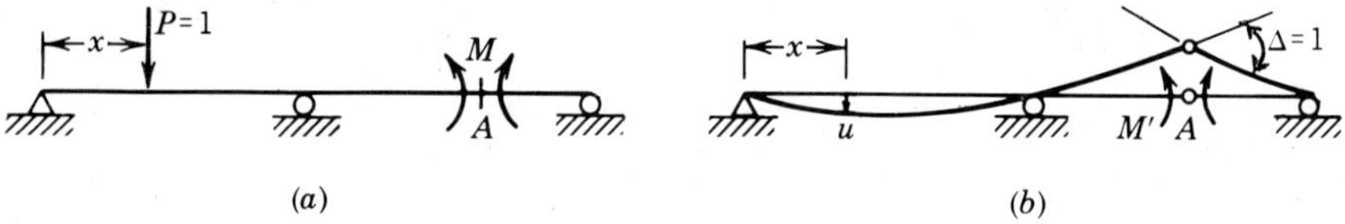

FIG. P17

a. Show by application of Betti's reciprocal theorem that $Pu + M\Delta = 0$, and deduce that the deformed shape in Fig. P17*b* is the influence line for M—that is, it is the plot of M at the fixed location A as x varies. (This is known as Müller–Breslau's principle.)
b. For a single simply supported span the release at A makes the beam geometrically unstable so that $M' = 0$. Draw Fig. P17*b* for that case.
c. Generalize the preceding to an arbitrary structure and to influence lines for any internal force.

Sections 12, 13

18. Determine the transformation $\mathbf{r} = \mathbf{Tq}$ to inextensional and extensional modes for the frame of Fig. 8/2.4–3*a*. Obtain the transformed load term of the stiffness equations.

19. Follow the order of magnitude analysis of Section 13 in the case of the frame of Fig. P19.

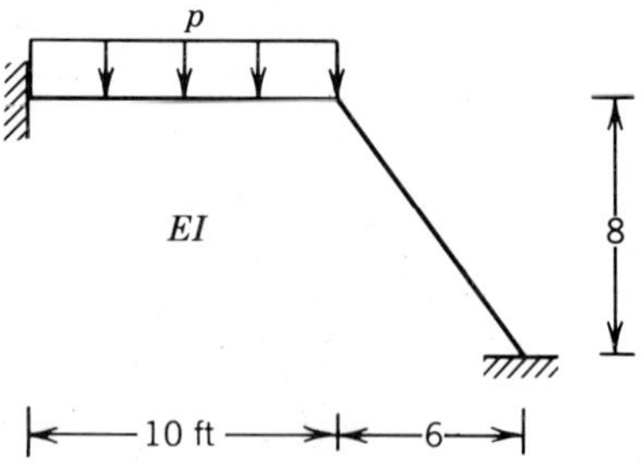

FIG. P19

11

FORCE (FLEXIBILITY) METHOD

Linear Theory

1. INTRODUCTION

The force method of analysis is a method for solving the set of governing equations (8/2.4–8) in which the force unknowns are determined first. These equations are

$$\mathbf{a}_v^T \mathbf{V} = \mathbf{R} - \mathbf{R}_p \tag{1–1a}$$

$$\mathbf{v} = \mathbf{a}_v \mathbf{r} \tag{1–1b}$$

$$\mathbf{v} = \mathbf{f}\mathbf{V} + \mathbf{v}_0 \tag{1–1c}$$

The member equilibrium solution

$$\mathbf{S} = \mathbf{B}\mathbf{V} + \mathbf{S}_p \tag{1–2}$$

completes the determination of the member end forces.

The force method will be presented first in the form of operations on Eqs. (1–1). To solve particular problems by this approach, a matrix programming language is a practical necessity, and what is needed is to form the connectivity matrix $\mathbf{a}_v$, the equivalent joint load $(\mathbf{R} - \mathbf{R}_p)$, and the initial deformations $\mathbf{v}_0$ (Chapter 8, Section 2.4).

The force method will also be presented in a form suitable for hand calculations, with the possible help of a calculator.

Both approaches are based on the static-kinematic properties of kinematically stable structures developed in Chapter 9, Section 2. Statically determinate structures are treated first because they are fundamental to a physical interpretation of the properties of statically indeterminate structures.

2. STATICALLY DETERMINATE STRUCTURES

2.1. General Solution

Statically determinate kinematically stable structures are characterized by a square nonsingular connectivity matrix $\mathbf{a}_v$. Letting

$$\mathbf{b} = (\mathbf{a}_v^T)^{-1} \tag{2.1–1}$$

Eqs. (1–1a, b) are inverted, and the governing equations are transformed into

$$\mathbf{V} = \mathbf{b}(\mathbf{R} - \mathbf{R}_p) \tag{2.1–2}$$

$$\mathbf{r} = \mathbf{b}^T\mathbf{v} \tag{2.1–3}$$

$$\mathbf{v} = \mathbf{f}\mathbf{V} + \mathbf{v}_0 \tag{2.1–4}$$

Having formed $\mathbf{a}_v$, $(\mathbf{R} - \mathbf{R}_p)$, and $\mathbf{v}_0$, the general solution proceeds by inverting $\mathbf{a}_v^T$ and then by evaluating member redundants through Eq. (2.1–2), member deformations through Eq. (2.1–4), and joint displacements through Eq. (2.1–3). If the device of support members is used, support reactions are determined as part of $\mathbf{V}$. Member end forces are determined through Eq. (1–2).

If there are support settlements, support members must be used for Eq. (2.1–3) to apply.

An individual joint displacement r_i is obtained from Eq. (2.1–3) in the form

$$r_i = \mathbf{b}_i^T\mathbf{v} \tag{2.1–5a}$$

where $\mathbf{b}_i$ is the ith column of $\mathbf{b}$. $\mathbf{b}_i$ is the set of member forces $\mathbf{V}$ in equilibrium with the joint load $R_i = 1$. Equation (2.1–5a) expresses the principle of virtual forces, using $R_i = 1$ and $\mathbf{b}_i$ as a virtual force system and r_i and $\mathbf{v}$ as the actual geometric system. r_i is then the external virtual work, and $\mathbf{b}_i^T\mathbf{v}$ the internal virtual work. For efficiency of calculations, $\mathbf{b}_i$ and $\mathbf{v}$ are partitioned by members so that Eq. (2.1–5a) takes the form

$$r_i = \sum \mathbf{b}_i^{mT} \mathbf{v}^m \tag{2.1–5b}$$

where the summation extends over the members.

Example 1

Consider the pin-jointed truss of Fig. 2.1–1*a*, with members and joints ordered as shown. The free joint displacements ordered by joints are

$$\begin{aligned} \mathbf{r} &= \{r_1^1 \ r_2^1 \ r_1^2 \ r_2^2 \ r_1^3 \ r_2^3 \ r_2^4\} \\ &= \{r_1 \ r_2 \ r_3 \ r_4 \ r_5 \ r_6 \ r_7\} \end{aligned}$$

where r_1^j and r_2^j are the displacements of joint j in the x and y directions, respectively. $\mathbf{a}_v$ is formed by rows from member matrices, $[\cos\theta \ \ \sin\theta]$, (Chapter 8, Section 2.4, Example 2). There comes

$\mathbf{a}_v =$

r_1^1	r_2^1	r_1^2	r_2^2	r_1^3	r_2^3	r_2^4
1		−1				
0.8	−0.6			−0.8	0.6	
			−1		1	
		1				
				0.8	0.6	
				1		
						1

For the loading shown in Fig. (2.1–1), $\mathbf{R}_p = 0$ and

$$\mathbf{R} = \{0 \ \ -15 \ \ 0 \ \ -30 \ \ 0 \ \ 0 \ \ 0\} \text{ kips}$$

The joint equilibrium equations, $\mathbf{a}_v^T \mathbf{V} = \mathbf{R}$, may now be solved by computer or by

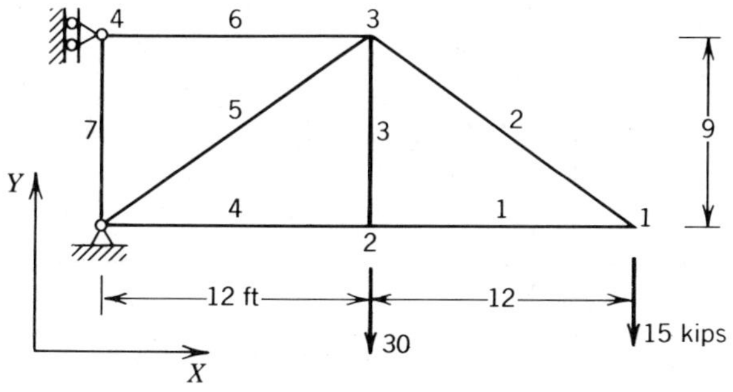

(*a*)

FIG. 2.1–1. Statically determinate truss, Examples 1, 2. (*Continued on following page.*)

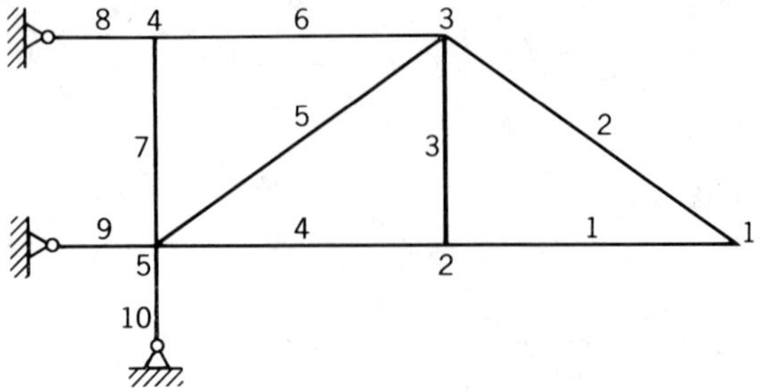

(b)

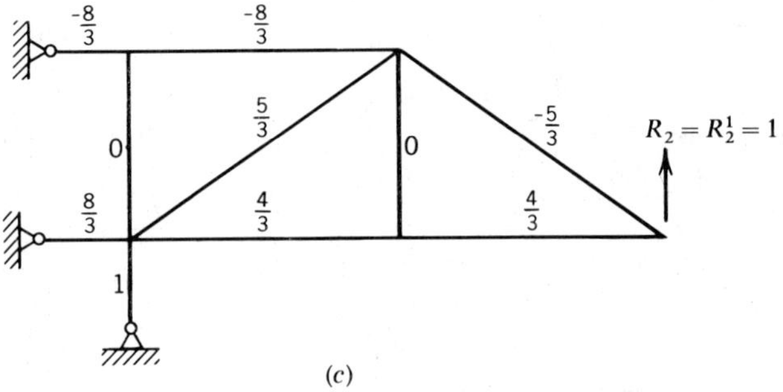

(c)

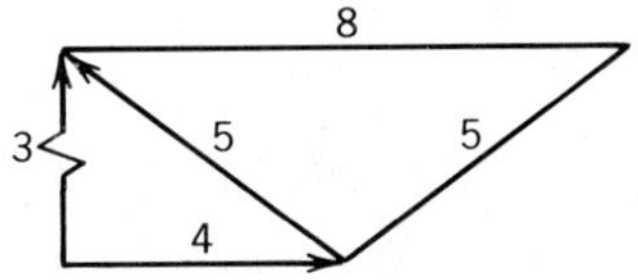

(d)

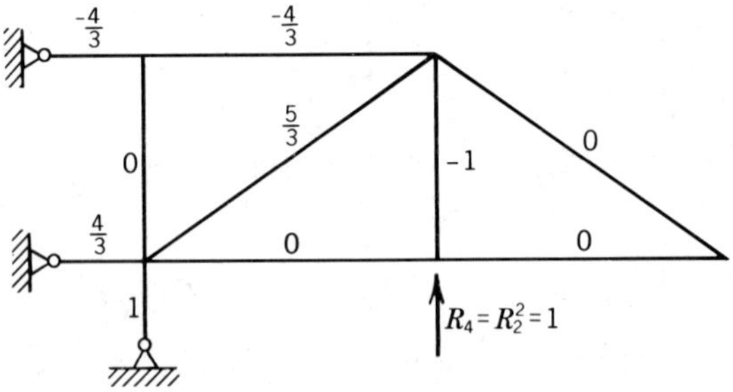

(e)

FIG. 2.1–1. *(continued)*

hand. The hand solution may be made to follow the method of joints. Instead of V_1^m, member axial forces will be denoted here by V_m. The first two equilibrium equations pertain to joint 1 and contain only V_1 and V_2 as unknowns. The coefficients of these equations are the first two columns of $\mathbf{a}_v$. Having V_1 and V_2, the next two equations pertain to joint 2 and contain only V_3 and V_4 as unknowns. Next V_5 and V_6 are determined. The last equation contains only V_7 which for the loading under consideration is zero.

The reactions at the supports are now found by solving the joint equilibrium equations in the directions of the prescribed displacements. The device of support members shown in Fig. 2.1–1*b* allows us to find the reactions as part of the member forces. $\mathbf{r}$ includes then the support displacements $\{r_1^4 \; r_1^5 \; r_2^5\}$ which are appended to the previous $\mathbf{r}$, and $\mathbf{V}$ includes the axial forces $\{V_8 \; V_9 \; V_{10}\}$ in the support members. $\mathbf{a}_v$ has thus three additional rows and columns which are the following:

$\mathbf{a}_v =$

	r_1^4	r_1^5	r_2^5
		-1	
		-0.8	-0.6
	-1		
			-1
- - - - - - - -	1		
		1	
			1

Inversion of $\mathbf{a}_v^T$ yields $\mathbf{b}$. To do this by hand, columns of $\mathbf{b}$ are determined one at a time. For example, column 1, or $\mathbf{b}_1$, is the set of member forces in equilibrium with $R_1 = R_1^1 = 1$ and is found readily to consist of $V_1 = V_4 = V_9 = 1$ and all other forces zero. Column 2, or $\mathbf{b}_2$, is the set of member forces in equilibrium with $R_2 = R_2^1 = 1$. The forces are conveniently determined by geometry, using the proportions shown in (*d*). The components of the force vector R_2^1 on the directions of members 1 and 2 are equal, respectively to V_1 and V_2. The geometry of this decomposition has the same proportions as the 3–4–5 triangle in (*d*). The components thus found are applied at the joint. A member is in tension if the corresponding component points out of the material and is in compression otherwise. The method of joints is thus applied geometrically, and the result is shown in (*c*) for $\mathbf{b}_2$ and in (*e*) for $\mathbf{b}_4$. Note that the sign of the axial force in a support member should be interpreted as that of an internal force in order to obtain the correct sense of the reaction. For example, member 8 in Fig. 2.1–1*c* has a compressive axial force $V_8 = -8/3$ which represents a support reaction acting in the positive X direction.

Determination of all columns of $\mathbf{b}$ by hand is a good statical, if tedious, exercise. For determining the member forces, $\mathbf{b}$ is usually not needed. Whether by hand or by computer, the joint equilibrium equations may be solved for several loading conditions without having to invert the coefficient matrix. However, if a large number of load combinations is to be considered, it may become convenient to evaluate **V** through the product **bR** for each load combination **R**. **b** is needed to determine the joint displacements and the flexibility matrix to be seen in Section 2.3.

For any one joint displacement, only one column of **b** is needed, as indicated in Eq. (2.1–5b). For example, the displacement r_2 is obtained through

$$r_2 = \mathbf{b}_2^T \mathbf{v}$$

where, from Fig. 2.1–1*c*,

$$\mathbf{b}_2 = \left\{ \frac{4}{3} \quad -\frac{5}{3} \quad 0 \quad \frac{4}{3} \quad \frac{5}{3} \quad -\frac{8}{3} \quad 0 \;\vdots\; -\frac{8}{3} \quad \frac{8}{3} \quad 1 \right\}$$

and **v** is the set of member deformations for the loading condition causing r_2. For the loading of Fig. 2.1–1*a*, using (*c*) and (*e*),

$$\mathbf{V} = \mathbf{b}_2 R_2 + \mathbf{b}_4 R_4 = -15\mathbf{b}_2 - 30\mathbf{b}_4$$

or

$$\mathbf{V} = \{-20 \quad 25 \quad 30 \quad -20 \quad -75 \quad 80 \quad 0 \;\vdots\; 80 \quad -80 \quad -45\} \text{ kips}$$

Then, if $f_m = l/EA$ is the axial flexibility of member m, with $f_m = 0$ for support members,

$$\mathbf{v} = \{-20f_1 \quad 25f_2 \quad 30f_3 \quad -20f_4 \quad -75f_5 \quad 80f_6 \quad 0 \;\vdots\; 0 \quad 0 \quad 0\}$$

and

$$r_2 = \mathbf{b}_2^T \mathbf{v} = \sum b_2^m v_m$$

$$= -\frac{1}{3}(80f_1 + 125f_2 + 80f_4 + 375f_5 + 640f_6)$$

For example, if all members have $l/A = 2\,\text{ft/in}^2$ and $E = 30{,}000$ ksi, then

$$f_m = l/EA = 2 \times 12/30000 = 1/1250 \text{ in/k}$$

and

$$r_2 = -\frac{1}{3 \times 1250}(80 + 125 + 80 + 375 + 640) = -0.347 \text{ in}$$

Example 2

As an example of a thermal loading condition, let $\Delta T = 50°\mathrm{F}$ be an increase in temperature of members 1 and 4, and let $\alpha = 6.5 \times 10^{-6}\ 1/°\mathrm{F}$ be the coefficient of thermal expansion. The free thermal elongation of members 1 and 4 is $v_I = \alpha \Delta T l = 0.0468$ in. Since $\mathbf{R} - \mathbf{R}_p = 0$, Eq. (2.1–2) yields $\mathbf{V} = 0$, and Eq. (2.1–4) yields

$$r_2 = \mathbf{b}_2^T \mathbf{v} = \frac{4}{3} \times 0.0468 + \frac{4}{3} \times 0.0468 = 0.1248 \text{ in}$$

A geometric error in member length is treated exactly as a thermal deformation. The same applies to a support displacement which may be treated as an initial deformation of a support member. For example, if joint 4 displaces 0.1 in in the X direction, member 8 in Fig. 2.1–1*b* has $v_I = 0.1$ so that

$$\mathbf{v} = \mathbf{v}_0 = \{0 \ 0 \ 0 \ 0 \ 0 \ 0 \ 0 \ 0.1 \ 0 \ 0\}$$

and

$$r_2 = \mathbf{b}_2^T \mathbf{v} = -\frac{8}{3} \times 0.1 \text{ in}$$

The displacement is that of a rigid body rotation of 0.1/108 rad about joint 5, which induces at joint 1 the Y displacement $-0.1 \times 24 \times 12/108$ as obtained here.

Example 3

Consider the frame of Fig. 2.1–2 with members and joints ordered as shown in (*a*). The members are oriented so that the ordering of member deformations $\mathbf{v}^m = \{v_1 \ v_2 \ v_3\}$ can be identified as shown in (*b*). $\mathbf{a}_v$ for this frame, before any support conditions are imposed, was formed in Fig. 8/2.4–2. For the present support conditions, the columns of $\mathbf{a}_v$ corresponding to the prescribed displacements are deleted. These are columns 1, 2, and 14. Elements of $\mathbf{r}$ are then $\mathbf{r} = \{r_3^1 \ \mathbf{r}^2 \ \mathbf{r}^3 \ \mathbf{r}^4 \ r_1^5 \ r_3^5\}$. The resulting $\mathbf{a}_v$ is 12×12 and will not be reproduced here.

The equivalent joint load $(-\mathbf{R}_p)$ and the initial deformations $\mathbf{v}_p$ are formed in Fig. 8/2.4–3 using the fundamental particular equilibrium state. In this state each member is simply supported and subjected to its external loading. From the results shown in Fig. 8/2.4–3,

$$-\mathbf{R}_p = \{0, \ -2.076, \ -6.107, \ 0, \ 4.515, \ -10.844, \ 0, \ -2.439, \ -3.049, \ 0, \ 0, \ 0\} \text{ kips}$$

$$\mathbf{v}_p = \left\{0, \ 0, \ 0, \ -\frac{40}{EA}, \ -\frac{106.25}{EI}, \ \frac{106.25}{EI}, \ \frac{40}{EA}, \ -\frac{80.04}{EI}, \ \frac{80.04}{EI}, \ 0, \ 0, \ 0\right\}$$

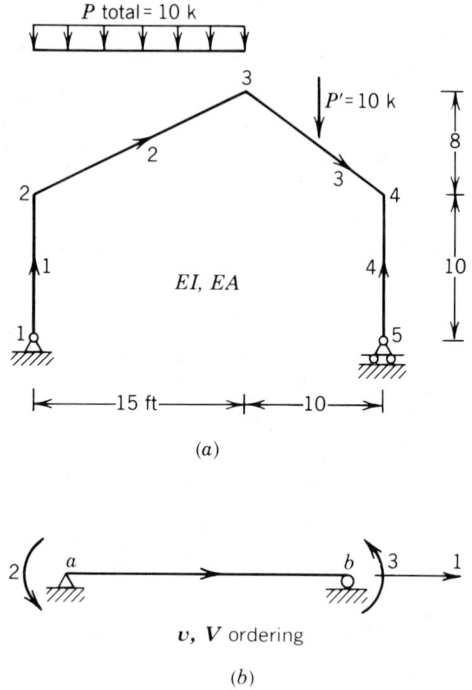

FIG. 2.1–2. Statically determinate frame, Examples 3, 4.

Numerical values in $\mathbf{v}_p$ are in units of kips and feet.

There remains to form $\mathbf{f} = \lceil \mathbf{f}^1 \ \mathbf{f}^2 \ \mathbf{f}^3 \ \mathbf{f}^4 \rfloor$. Member flexibility matrices $\mathbf{f}^m$ are as defined in Eq. (5/2–7), with $\beta = 0$ to neglect shear deformation. The matrices required to form the governing equations having been formed, the solution follows the general procedure. The solution for **V** yields for each member the axial force V_1 at end b and the end moments V_2 and V_3. The preceding illustrates a general method. The statical solution of the problem considered is readily obtained by hand (Section 2.4, Example 1).

Another convenient choice of a particular equilibrium state is one in which a member is pinned at both ends (Chapter 8, Section 2.4, Example 1b). From Fig. 8/2.4–4,

$$-\mathbf{R}_p = \{0,\ 0,\ -5,\ 0,\ 0,\ -10,\ 0,\ 0,\ -5,\ 0,\ 0,\ 0\}$$

$\mathbf{v}_p$ is then obtained from the preceding one by substituting zero for the axial deformations of members 2 and 3. The solution for **V** in the present formulation yields as before the member end moments, but V_1 for members 2 and 3 differs from the preceding case, because of the difference in axial forces in the two particular equilibrium states.

If the fixed joint state is used as a particular equilibrium state, $\mathbf{S}_p$ becomes the fixed end forces $\mathbf{S}_F$, and $\mathbf{v}_p$ becomes zero. This may be convenient if the fixed end forces are evaluated by available formulas in standard loading cases. It should be noted, however, that, fundamentally, the fixed end forces are determined by going through a statically determinate member analysis. It is also somewhat artificial to use stiffness properties in determining fixed end forces when the final statical solution represented by Eqs. (2.1–2) and (1–2) is independent of material properties.

Example 4

Two additional aspects of the preceding example that allow us to reduce the size of $\mathbf{a}_v$ are considered in what follows.

Member 1 in Fig. 2.1–2*a* has one rotational release at joint 1, and member 4 has two releases at joint 5. These releases allow us to form the effective force matrices $\mathbf{V}^1 = \{V_1^1 \; V_3^1\}$ and $\mathbf{V}^4 = \{V_1^4\}$ and corresponding deformation matrices $\mathbf{v}^1$ and $\mathbf{v}^4$. The released member displacements at the supports may now be deleted from $\mathbf{r}$. Corresponding elements of $(\mathbf{R} - \mathbf{R}_p)$ are also deleted. $\mathbf{a}_v$ is reduced to a 9×9 matrix by deleting the rows corresponding to the deleted deformations and the columns corresponding to the deleted joint displacements. The effective flexibility matrix of member 1 is $\mathbf{f} = \lceil l/EA \;\; l/3EI \rfloor$ and that of member 4 is $\mathbf{f} = [l/EA]$. The rest of the analysis is unchanged.

A further reduction in the size of the problem may be made by restricting $\mathbf{r}$ to inextensional displacements (Chapter 8, Section 4, Example 1a). There results five equilibrium equations having as unknowns five member end moments. There is no statical approximation in this but simply a derivation of generalized equilibrium equations that excludes the axial forces. If also inextensional deformations are assumed the effective $\mathbf{f}$ is obtained by deleting the l/EA terms. Member 4 is then not involved in the formulation. If a load were applied on member 4, its effect would appear in the equivalent joint load.

2.2. Properties of Statically Determinate Structures

Static-kinematic properties of statically determinate structures were established in Chapter 9, Section 2. The only additional consideration now is that member deformations are related to member forces by the member flexibility relations. These properties are summarized in what follows.

From Eq. (2.1–2) it is deduced that to any given $(\mathbf{R} - \mathbf{R}_p)$ corresponds a unique $\mathbf{V}$ satisfying joint equilibrium. In particular, if $\mathbf{R} - \mathbf{R}_p = 0$, then $\mathbf{V} = 0$. Thus the structure cannot have self-equilibriating internal forces.

For a given load the constitutive properties have no influence on the internal forces. They serve only to determine member deformations $\mathbf{v}$ which in turn determine the joint displacements $\mathbf{r}$ through Eq. (2.1–3).

Initial deformations $\mathbf{v}_I$ such as thermal deformations, geometric errors, and support settlements cannot cause internal forces. They cause joint displacements $\mathbf{r} = \mathbf{b}^T \mathbf{v}_I$.

Any set of deformations $\mathbf{v}$ is geometrically compatible with the joint displacements $\mathbf{r} = \mathbf{b}^T\mathbf{v}$. In particular, if $\mathbf{v} = 0$, then $\mathbf{r} = 0$. This last property is the expression of kinematic stability. Also the deformations $\mathbf{v}_p$ in the particular equilibrium state are compatible with joint displacements $\mathbf{r}_p = \mathbf{b}^T\mathbf{v}_p$.

2.3. Flexibility Matrix

The solution for $\mathbf{r}$, Eq. (2.1–3), expressed in terms of the loading by substituting for $\mathbf{v}$ from Eq. (2.1–4), and for $\mathbf{V}$ from Eq. (2.1–2), takes the form

$$\mathbf{r} = \mathbf{F}(\mathbf{R} - \mathbf{R}_p) + \mathbf{b}^T\mathbf{v}_0 \tag{2.3–1}$$

where

$$\mathbf{F} = \mathbf{b}^T\mathbf{f}\mathbf{b} \tag{2.3–2}$$

$\mathbf{F}$ is the structure flexibility matrix. Since $\mathbf{f}$ is symmetric, Eq. (2.3–2) shows that $\mathbf{F}$ is symmetric. Element F_{ij} of $\mathbf{F}$, in row i and column j, is evaluated through the formula

$$F_{ij} = \mathbf{b}_i^T\mathbf{f}\mathbf{b}_j \tag{2.3–3}$$

where $\mathbf{b}_i$ and $\mathbf{b}_j$ are the ith and jth columns of $\mathbf{b}$, respectively. F_{ij} is the displacement r_i caused by a unit joint load $R_j = 1$ acting alone. It is equal to F_{ji}, that is, to the displacement r_j caused by a unit joint load $R_i = 1$. This is an instance of Betti's reciprocal theorem.

Equations (2.3–2, 3) may be written in a form allowing us to evaluate $\mathbf{F}$ and F_{ij} by superposition of member contributions similarly to Eq. (2.1–5b). Thus

$$\mathbf{F} = \sum_m \mathbf{F}^m = \sum_m \mathbf{b}^{mT}\mathbf{f}^m\mathbf{b}^m \tag{2.3–4}$$

Forming $\mathbf{F}$ by superposition of $\mathbf{F}^m$ rather than through the product $\mathbf{b}^T\mathbf{f}\mathbf{b}$ avoids the wasted space and operations due to the sparsity of $\mathbf{f}$. It is noted, however, that contrary to forming of the stiffness matrix $\mathbf{K}$ by merging smaller size $\mathbf{K}^m$, $\mathbf{F}^m$ has the same size as $\mathbf{F}$. Except for particular structure connectivities, member forces $\mathbf{V}^m$ may be caused by all joint loads, so that $\mathbf{b}^m$ is not sparse, and consequently a systematic forming of a reduced $\mathbf{F}^m$ for merging into $\mathbf{F}$ is not possible.

Forming of $\mathbf{F}$ by superposition of member contributions may be applied to the individual element F_{ij} using the formula

$$F_{ij} = \sum_m F_{ij}^m \tag{2.3–5a}$$

where

$$F_{ij}^m = \mathbf{b}_i^{mT}\mathbf{f}^m\mathbf{b}_j^m \tag{2.3–5b}$$

and $\mathbf{b}_i^m$ and $\mathbf{b}_j^m$ are the ith and jth columns of $\mathbf{b}^m$, respectively. Equation (2.3–5) is convenient for manual calculations and is interpretable as a virtual work equation. $\mathbf{b}_i^m$ is the virtual $\mathbf{V}^m$ in equilibrium with the virtual load $R_i = 1$, and $\mathbf{f}^m\mathbf{b}_j^m$ is the actual $\mathbf{v}^m$ caused by the unit load $R_j = 1$. Equation (2.3–5) expresses that the external virtual work, F_{ij}, is equal to the internal virtual work, $\sum_m \mathbf{b}_i^{mT}\mathbf{v}^m$.

In particular problems the nonzero joint loads form a subset $(\mathbf{R}' - \mathbf{R}_p')$ of $(\mathbf{R} - \mathbf{R}_p)$. A corresponding effective flexibility matrix $\mathbf{F}'$ may be defined such that

$$\mathbf{r} = \mathbf{F}'(\mathbf{R}' - \mathbf{R}_p') + \mathbf{b}^T\mathbf{v}_0 \tag{2.3–6a}$$

where

$$\mathbf{F}' = \mathbf{b}^T\mathbf{f}\mathbf{b}' \tag{2.3–6b}$$

and $\mathbf{b}'$ is formed of the columns of $\mathbf{b}$ associated with the nonzero joint loads. Equations (2.3–6) may be further specialized to an arbitrary subset $\mathbf{r}''$ of $\mathbf{r}$ by deleting from $\mathbf{b}^T$ the rows associated with the deleted elements of $\mathbf{r}$. It may be applied, in particular, to the subset $\mathbf{r}'$ corresponding to $(\mathbf{R}' - \mathbf{R}_p')$ by replacing $\mathbf{b}^T$ with $\mathbf{b}'^T$.

We now consider the conditions under which the flexibility matrix is nonsingular. Since $\mathbf{b}^T = \mathbf{a}_v^{-1}$ is nonsingular, the condition for $\mathbf{b}^T\mathbf{f}\mathbf{b}$ to be nonsingular is that $\mathbf{f}$ be nonsingular. This in turn requires that there be no member having a singular flexibility matrix. Members having a singular flexibility matrix occur in the linear theory as support members, or more generally as members with constrained deformations, such as in inextensible frames. The joint displacements $\mathbf{r}$ cease then to be independent and a kinematic transformation to an independent set of displacements is required for defining a nonsingular flexibility matrix (see Section 7).

In the case of support members, deleting the rows and columns of $\mathbf{F}$ corresponding to the prescribed displacements $\mathbf{r}_s$ yields the effective nonsingular flexibility matrix.

Assuming then that $\mathbf{r}$ is the independent set of joint displacements, and noting that $\mathbf{b}^T = \mathbf{a}_v^{-1}$, Eq. (2.3–1) may be inverted into

$$\mathbf{K}\mathbf{r} = \mathbf{R} - \mathbf{R}_F \tag{2.3–7}$$

where

$$\mathbf{K} = \mathbf{F}^{-1} = \mathbf{a}_v^T\mathbf{k}\mathbf{a}_v \tag{2.3–8a}$$

$$\mathbf{R}_F = \mathbf{R}_p - \mathbf{K}\mathbf{b}^T\mathbf{v}_0 \tag{2.3–8b}$$

Equations (2.3–7) may be identified with the stiffness equations of the structure written for the unrestrained degrees of freedom. Equation (2.3–8b) also allows a direct interpretation of $\mathbf{R}_F$ as the joint required for equilibrium in the fixed joint state $\mathbf{r} = 0$.

2.4. Manual Methods

Manual methods may be defined as procedures suitable to solve a particular problem, by contrast with general purpose programmed methods which follow a preestablished procedure. In this general sense, manual methods may include the use of computers with various degrees of capability. These may vary from arithmetic functions of hand-held calculators to general matrix operations programmable with a language designed for that purpose. This latter capability, implemented preferably in an interactive way, is essential for allowing the student of matrix structural analysis to solve problems with relative ease.

Manual methods are discussed in this section in the narrow sense of what is practical for solving small problems with the possible help of a calculator. They may follow in part traditional methods which antedate matrix formulations. In certain instances traditional methods turn out to be efficient procedures for evaluating matrix expressions.

In the force method, statical solutions may be determined without formally forming the set of joint equilibrium equations and solving it as a system of simultaneous equations.

For trusses, the method of joints, when applicable, allows us to determine member axial forces by satisfying joint equilibrium one at a time. For the computation of a joint displacement Eq. (2.1–5b) takes the form $r_i = \sum_m b_i^m v^m$ in which v^m is obtained through the flexibility relation

$$v^m = f^m V^m + v_0^m \tag{2.4–1}$$

Thus

$$r_i = \sum_m f^m b_i^m V^m + \sum_m b_i^m v_0^m \tag{2.4–2}$$

To determine r_i, virtual axial forces $\mathbf{b}_i$ in equilibrium with the virtual joint load $R_i = 1$ are determined and are used in Eq. (2.4–2). This was illustrated in Section 2.1, Examples 1 and 2. A tabular organization of b_i^m, f^m, V^m, and v_0^m into columns whose rows correspond to the members is convenient for carrying out the calculations and for recording the basic elements of the analysis. For small problems member quantities may be simply read off graphs of the structure in performing the calculations of Eq. (2.4–2).

For statically determinate frames statical analysis by manual methods is usually made without forming explicitly an equivalent joint load. The bending moment diagram is drawn perpendicular to the members on a graph of the structure. A convention is chosen to draw the moment either on the tension side or on the compression side of the member. Axial and shear forces are determined separately, and diagrams are added if needed.

For any member the bending moment M has the form

$$M = M_h + M_p \tag{2.4–3}$$

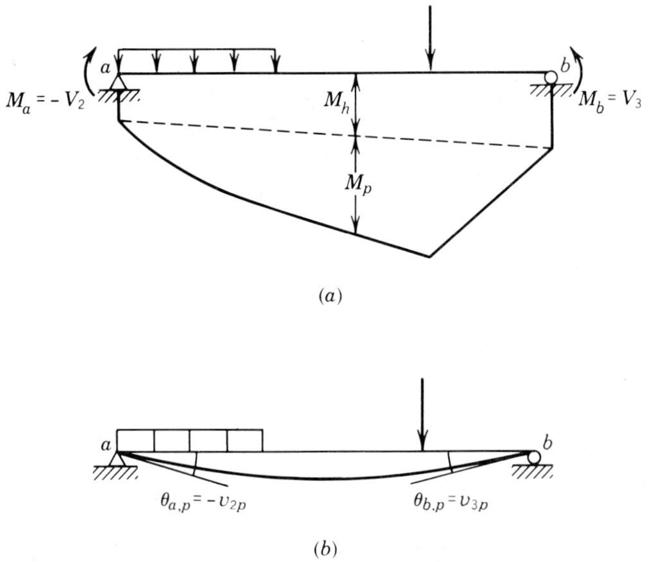

FIG. 2.4–1. General form of bending moment diagram (*a*); deformations in particular equilibrium state (*b*).

where M_h is due to the member statical redundants $\mathbf{V}^m$, and M_p is due to the member external load with $\mathbf{V}^m = 0$. This is shown in Fig. 2.4–1*a*, for $\mathbf{V}^m$ of the alternate type. The sign convention is such that

$$\mathbf{V}^m = \{V_2 \ \ V_3\} = \{-M_a \ \ M_b\} \tag{2.4–4}$$

M_h has the linear representation

$$M_h = \left(1 - \frac{x}{l}\right)M_a + \frac{x}{l}M_b \tag{2.4–5}$$

and M_p is the bending moment of the simply supported member subjected to its loading condition.

If M is first evaluated and plotted at member ends, the straight line diagram joining these plotted points and representing M_h is the bending moment caused by the equivalent joint load $(\mathbf{R} - \mathbf{R}_p)$. Portion M_p of the diagram is added for each member treated as simply supported and subjected to its own loading.

Example 1

For the frame of Fig. 2.4–2*a* the reactions are first determined through the moment equilibrium equation of the external force system taken at the support points. The bending moment M at any cross section may be evaluated as the moment at that cross section of the external forces on either side of it. It is plotted

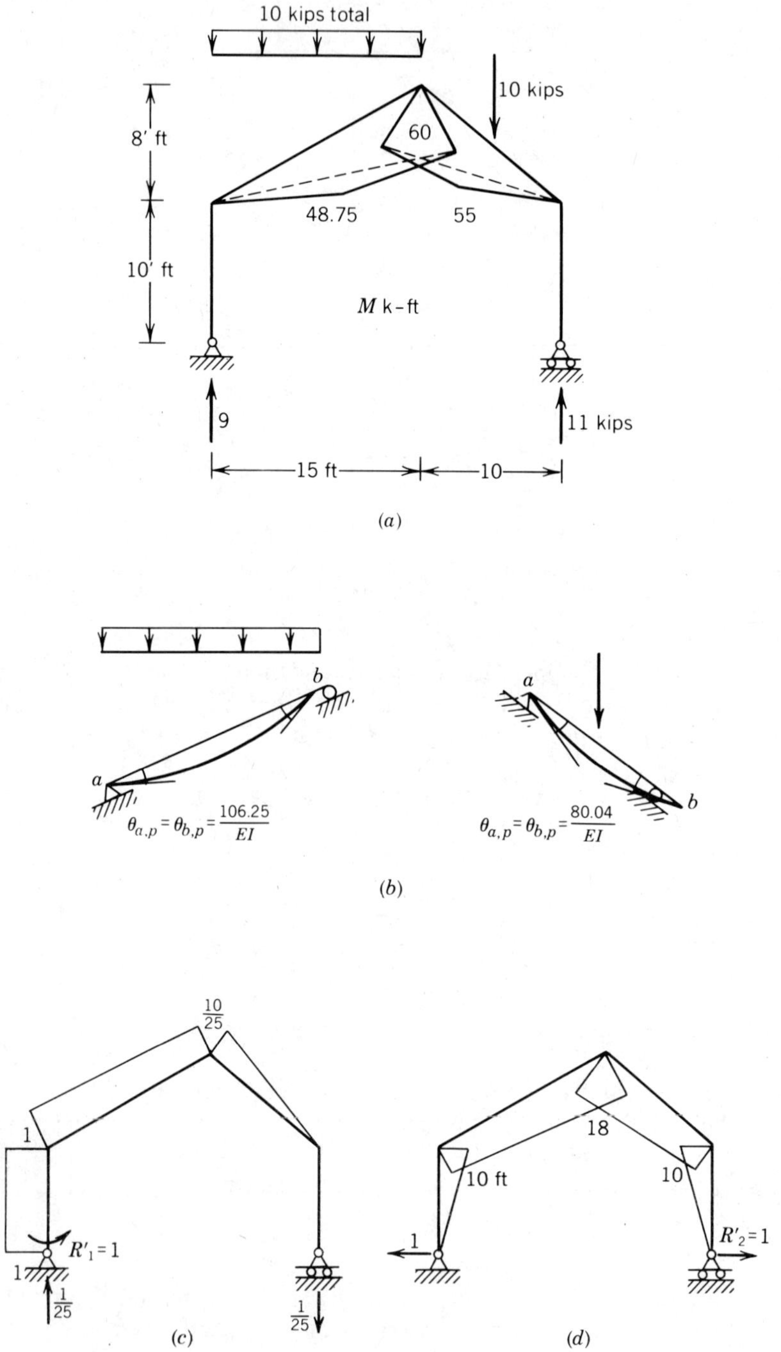

FIG. 2.4–2. Statically determinate frame, Examples 1, 2. (*Continued on next page.*)

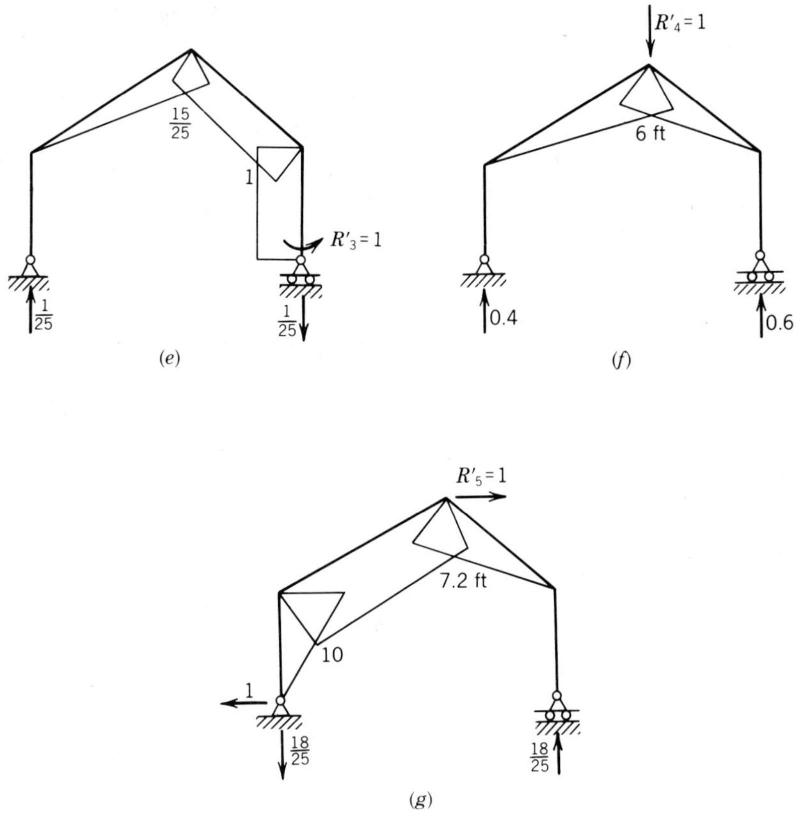

FIG. 2.4–2. *(continued)*

on the tension side. The dotted line joining the end points for a member separates M_h from M_p.

For computation of displacements axial and shear deformations are usually neglected. A displacement r_i is evaluated by application of the principle of virtual forces through the formula

$$r_i = \sum \int_0^l M' \chi \, dx \qquad (2.4\text{–}6a)$$

where

$$\chi = \frac{M}{EI} + \chi_I \qquad (2.4\text{–}6b)$$

and M' is the virtual bending moment due to the virtual load $R_i' = 1$. The member end values of M' are the elements of $\mathbf{b}_i$, that is, the ith column of $\mathbf{b}$ modified to conform to the sign convention of a bending moment and made effective by deleting the member axial forces. For member m,

$$\mathbf{b}_i^m = \{-M'_a \ M'_b\} \tag{2.4–7}$$

M' has an expression similar to Eq. (2.4–5) in terms of M'_a and M'_b.

A sign convention for deformations corresponding to that of bending moments is adopted, Fig. 2.4–1*b*, so that, similarly to Eq. (2.4–4),

$$\mathbf{v}^m = \{v_2 \ v_3\} = \{-\theta_a \ \theta_b\} \tag{2.4–8}$$

For a uniform member, neglecting shear deformation,

$$\mathbf{f}^m = \frac{l}{6EI}\begin{bmatrix} 2 & -1 \\ -1 & 2 \end{bmatrix} \tag{2.4–9}$$

The integrations in Eq. (2.4–6a) yield

$$\int_0^l \frac{M'M_h}{EI}dx = \mathbf{b}_i^{mT}\mathbf{f}^m\mathbf{V}^m = \frac{l}{3EI}\left(M_aM'_a + M_bM'_b + \frac{1}{2}M_aM'_b + \frac{1}{2}M_bM'_a\right) \tag{2.4–10a}$$

$$\int_0^l \frac{M'M_p}{EI}dx = \mathbf{b}_i^{mT}\mathbf{v}_p^m = M'_a\theta_{a,p} + M'_b\theta_{b,p} \tag{2.4–10b}$$

$$\int_0^l M'\chi_I\,dx = \mathbf{b}_i^{mT}\mathbf{v}_I^m = M'_a\theta_{a,I} + M'_b\theta_{b,I} \tag{2.4–10c}$$

The right-hand sides of these equations are evaluated for every member by reading the values M_a, M_b, M'_a, M'_b off the bending moment diagrams. For standard loadings formulas are available for $\theta_{a,p}$, $\theta_{b,p}$, $\theta_{a,I}$ and $\theta_{b,I}$. Also for a piecewise linear diagram of M_p, the integration formula in Eq. (2.4–10a) applies piecewise to the integral in Eq. (2.4–10b). Superposition of member contributions yields

$$r_i = \sum_m \mathbf{b}_i^{mT}\mathbf{f}^m\mathbf{V}^m + \sum_m \mathbf{b}_i^{mT}(\mathbf{v}_p^m + \mathbf{v}_I^m) \tag{2.4–11}$$

The first term in Eq. (2.4–11) is due to the equivalent joint load, and the second to the deformations $\mathbf{v}_0 = \mathbf{v}_p + \mathbf{v}_I$ in the particular equilibrium state.

Manual methods for evaluating displacements are seen to be efficient ways of evaluating the matrix products involved. To illustrate, the displacements at the bottom of the columns and the translational displacements at the top of the frame of Fig. 2.4–2*a* will be evaluated, assuming all members are uniform and have the same EI. To each desired displacement r_i, a virtual bending moment diagram due to $R'_i = 1$ is drawn as shown in (*c*), (*d*), (*e*), (*f*), and (*g*). r_i and R'_i correspond in ordering and sign convention, so that, the virtual work is $\mathbf{r}^T\mathbf{R}'$. Otherwise, the ordering and sign convention are arbitrary. Deformations of the simply supported members are shown in (*b*) and are obtained from previous calculations in Fig. 8/2.4–3. In evaluating the expression in Eq. (2.4–10a), moments that are plotted on the same side of the member have a positive product. Otherwise, they have a negative product. In Eq. (2.4–10b) a term such as $M'_a\theta_{a,p}$ represents virtual work. It is positive if M'_a and $\theta_{a,p}$ have the same sense and is negative otherwise. Equations (2.4–10) evaluate $r_iR'_1$ for $R'_i = 1$. The sense of R'_i is chosen arbitrarily. A positive r_i has the sense of R'_i, and a negative r_i has the opposite sense. Letting

$$r_{i,p} = \mathbf{b}_i^{mT}\mathbf{v}_p^m \tag{2.4–12a}$$

$$r_{i,R} = \mathbf{b}_i^{mT}\mathbf{f}^m\mathbf{V}^m \tag{2.4–12b}$$

typical calculations are

$$EIr_{1,p} = -(1)(106.25) - (\tfrac{10}{25})(106.25) - (\tfrac{10}{25})(80.04) = -180.766 \text{ k-ft}^2$$

$$EIr_{2,p} = (10)(106.25) + (18)(106.25) + (18)(80.04) + (10)(80.04) = 5216.12 \text{ k-ft}^3$$

and result in

$$EI\mathbf{r}_p = \{-180.766,\ 5216.12,\ 191.814,\ 1117.74,\ 2403.788\}$$

Similarly,

$$EIr_{1,R} = \frac{17}{6}\left[-2\left(\frac{10}{25}\right)(60) - (1)(60)\right] + \frac{12.806}{6}\left[-2\left(\frac{10}{25}\right)(60)\right]$$

$$= -408.448 \text{ k-ft}^2$$

$$EIr_{2,R} = \frac{17}{6}[2(18)(60) + (10)(60)] + \frac{12.806}{6}[2(18)(60) + (10)(60)]$$

$$= 13{,}710.76 \text{ k-ft}^3$$

$$EI\mathbf{r}_R = \{-408.448,\ 13710.76,\ 485.732,\ 3576.72,\ 5992.064\}$$

The total is

$$EI\mathbf{r} = \{-589.214,\ 18926.88,\ 677.546,\ 4694.46,\ 8395.852\}$$

Assuming $EI = 12(10^6)$ k-in^2, obtain in units of radians and inches,

$$EIr = \{-0.00707,\ 2.725,\ 0.00813,\ 0.676,\ 1.209\}$$

The deflected shape is drawn in Fig. 2.4 3.

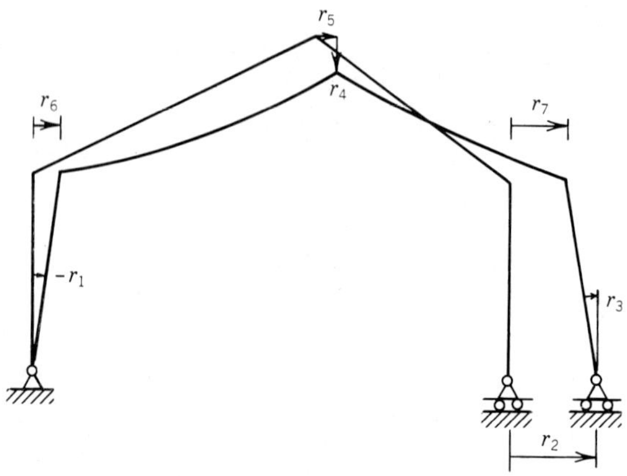

FIG. 2.4–3. Deflected shape for frame of Fig. 2.4–2*a*.

The frame being inextensible, the lateral displacements of the knees are determined by r_4 and r_5. The columns, having no bending moments, remain straight and rotate by r_1 and r_3, respectively. The rafters have curvatures corresponding to their bending moments.

The column rotations also allow us to evaluate the displacement of the knees and thus give a check on the numerical results. Using Fig. 2.4–3 and the geometric dimensions of Fig. 2.4–2*a*, there comes (in units of inches for column length)

$$r_6 = 120(-r_1) = 0.848 \text{ in}$$

$$r_7 = r_2 - 120r_3 = 1.749 \text{ in}$$

Inextensibility of the rafters requires that

$$15r_6 - 14r_5 + 8r_4 = 0$$

$$10r_7 - 10r_5 - 8r_4 = 0$$

It may be verified that the solution satisfies these equations.

Statics represented in Fig. 2.4–2*a* are seen to be slightly modified by the deformation. Geometrically nonlinear analysis (to be seen in Chapter 12) takes account of this. For the present problem the effect of geometric change on statics is negligible.

Example 2

Evaluation of flexibility coefficients for the frame of the preceding example, as defined in Eq. (2.3–5b), will now be illustrated. The method is the same as for any displacement. For example, F_{12} is the displacement r_1 caused by $R_2 = 1$. Thus diagram (*d*) in Fig. 2.4–2 represents M and diagram (*c*) represents M' in Eqs. (2.4–6). For F_{21} the roles of the two diagrams are reversed, and $F_{21} = F_{12}$. For F_{11} diagram (*c*) represents both M and M'. Applying the integration formula (2.4–10a), there comes, for example,

$$EIF_{11} = \frac{10}{3}(1+1+1) + \frac{17}{3}\left[1 + \left(\frac{10}{25}\right)^2 + \frac{10}{25}\right] + \frac{12.806}{3}\left(\frac{10}{25}\right)^2 = 19.523$$

$$EIF_{12} = -\frac{10}{6}(20+10) - \frac{17}{6}\left[20 + \left(\frac{20}{15}\right)(18) + 18 + \left(\frac{10}{25}\right)(10)\right]$$

$$-\frac{12.806}{6}\left[\left(\frac{20}{25}\right)(18) + \left(\frac{10}{25}\right)(10)\right] = 249.072$$

The flexibility matrix for the displacements $r_1 \ldots r_5$ is thus found as

$$\mathbf{EIF} = \begin{bmatrix} 19.523 & -249.072 & -4.938 & -40.845 & -167.014 \\ & 6667.608 & 268.212 & 1371.076 & 3055.291 \\ & & 20.407 & 48.573 & 75.288 \\ \text{symmetric} & & & 357.672 & 599.206 \\ & & & & 1823.048 \end{bmatrix} \qquad (2.4\text{–}13)$$

in units of kips and feet.

As a check, the displacements $EI\mathbf{r}_R$ of the preceding example may be obtained by multiplying $EI\mathbf{F}$ by the equivalent joint load. For the inextensible frame this load consist of $R_4 = 10$ kips. It is verified that $EI\mathbf{r}_R$ is equal to the fourth column of EIF multiplied by 10.

3. STATICALLY INDETERMINATE STRUCTURES: GENERAL SOLUTION AND PROPERTIES

The contents of Chapter 9, Sections 2 and 3, form a necessary background to the present developments.

The force method for statically indeterminate structures is developed from the transformed static-kinematic equations (9/2–14, 15) and the member flexibility relations. These equations are

$$\mathbf{V} = \mathbf{b}(\mathbf{R} - \mathbf{R}_p) + \mathbf{c}\mathbf{X} \qquad (3\text{–}1a)$$

$$\mathbf{r} = \mathbf{b}^T\mathbf{v} \qquad (3\text{–}1b)$$

$$\mathbf{c}^T\mathbf{v} = 0 \qquad (3\text{–}1c)$$

$$\mathbf{v} = \mathbf{f}\mathbf{V} + \mathbf{v}_p \qquad (3\text{–}1d)$$

The condition for compatibility of deformations (3–1c) is expressed in terms of the statical redundants $\mathbf{X}$ by substituting the flexibility relation for $\mathbf{v}$ and the equilibrium solution for $\mathbf{V}$. The result has the form

$$\mathbf{F}_{xx}\mathbf{X} + \mathbf{\Delta}_R + \mathbf{\Delta}_o = 0 \tag{3–2}$$

where

$$\mathbf{F}_{xx} = \mathbf{c}^T\mathbf{f}\mathbf{c} \tag{3–3}$$

$$\mathbf{\Delta}_R = \mathbf{c}^T\mathbf{f}\mathbf{b}(\mathbf{R} - \mathbf{R}_p) \tag{3–4a}$$

$$\mathbf{\Delta}_o = \mathbf{c}^T\mathbf{v}_o \tag{3–4b}$$

The solution proceeds by solving Eq. (3–2) for $\mathbf{X}$ and then evaluating $\mathbf{V}$ through Eq. (3–1a), $\mathbf{v}$ through Eq. (3–1d), and $\mathbf{r}$ through Eq. (3–1b).

$\mathbf{F}_{xx}$, $\mathbf{\Delta}_R$, and $\mathbf{\Delta}_o$ may be formed by superposition of member contributions. Row partitioning of Eqs. (3–1a) by members yields for member m

$$\mathbf{V}^m = \mathbf{V}_R^m + \mathbf{c}^m\mathbf{X} \tag{3–5a}$$

where

$$\mathbf{V}_R^m = \mathbf{b}^m(\mathbf{R} - \mathbf{R}_p) \tag{3–5b}$$

Equations (3–1b, c) may then be written in the form

$$\sum_m \mathbf{c}^{mT}\mathbf{v}^m = 0 \tag{3–6}$$

$$\mathbf{r} = \sum_m \mathbf{b}^{mT}\mathbf{v}^m \tag{3–7}$$

and Eqs. (3–3, 4) in the form

$$\mathbf{F}_{xx} = \sum_m \mathbf{c}^{mT}\mathbf{f}^m\mathbf{c}^m \tag{3–8}$$

$$\mathbf{\Delta}_R = \sum_m \mathbf{c}^{mT}\mathbf{f}^m\mathbf{V}_R^m \tag{3–9}$$

$$\mathbf{\Delta}_o = \sum_m \mathbf{c}^{mT}\mathbf{v}_0^m \tag{3–10}$$

The static-kinematic properties established in Chapter 9, Section 2, are now linked through the member flexibility relations. A summary presentation of these properties follows.

The internal forces in a statically indeterminate structure are governed by both statical and compatibility equations. Statical equations alone allow (NX) self-equilibriating and independent sets of member forces $\mathbf{V}$, which are, respectively, the (NX) columns of $\mathbf{c}$. The linear combination $\mathbf{cX}$, in which $\mathbf{X}$ is arbitrary, is the general solution for self-equilibriating member forces.

To the (NX) statical redundants correspond (NX) geometric compatibility equations to be satisfied by the member deformations $\mathbf{v}$, and thus also by the member forces through the flexibility properties.

By contrast with statically determinate structures, a loading condition consisting of initial deformations $\mathbf{v}_I$, such as support settlements, thermal variations, and geometric errors, causes self-equilibriating forces $\mathbf{V} = \mathbf{cX}$, unless the initial deformations happen to be compatible. The forces $\mathbf{cX}$ cause elastic deformations $\mathbf{fcX}$ that, when superimposed on $\mathbf{v}_I$, result in a compatible deformed state.

A physical interpretation of the force method by means of the primary structure is presented in Section 5.

4. CHOICE OF STATICAL REDUNDANTS IN MATRIX METHODS

Assuming $\mathbf{a}_v$, $(\mathbf{R} - \mathbf{R}_p)$, $\mathbf{v}_o$, and $\mathbf{f}$ have been formed, the main problem in applying the method is the selection of the statical redundants and the determination of $\mathbf{b}$ and $\mathbf{c}$.

An algorithm that operates on the connectivity matrix and that selects statical redundants and forms $\mathbf{b}$ and $\mathbf{c}$ is presented in Section 8. If statical redundants are selected manually, two cases may occur. In the first, $\mathbf{V}$ may be partitioned in the form $\{\mathbf{V}_1 \ \mathbf{V}_2\}$, and $\mathbf{V}_2 = \mathbf{X}$ is a valid choice of statical redundants. $\mathbf{b}$ and $\mathbf{c}$ are then formed by operations on $\mathbf{a}_v$ partitioned by rows, or on $\mathbf{a}_v^T$ partitioned by columns in the form $\mathbf{a}_v^T = [\mathbf{a}_1^T \, \mathbf{a}_2^T]$. These operations are defined in Eqs. (9/2–9) through (9/2–15). If the ordering of $\mathbf{V}$ does not allow us to choose as statical redundants the first or last (NX) elements, a reordering of $\mathbf{V}$, $\mathbf{v}$, and $\mathbf{f}$ may be made such that the preceding case is obtained. An alternate procedure that does not involve a reordering of the variables may be more convenient. $\mathbf{a}_2^T$ is formed by extracting the columns of $\mathbf{a}_v^T$ that correspond to the statical redundants $\mathbf{X}$, and $\mathbf{a}_1^T$ is formed of the remaining columns. Then $\mathbf{b}_1 = \mathbf{a}_1^{T-1}$, and $\mathbf{c}_1 = -\mathbf{b}_1\mathbf{a}_2^T$. $\mathbf{b}$ is formed by merging the rows of $\mathbf{b}_1$ into an $NV \times NR$ matrix in the same sequence that the columns of $\mathbf{a}_1^T$ occupy in $\mathbf{a}_v^T$. The remaining rows of $\mathbf{b}$ are zero. The same sequence is used to merge the rows of $\mathbf{c}_1$ into an $NV \times NX$ matrix. In the remaining NX rows an identity matrix is merged according to the sequence that the elements of $\mathbf{X}$ occupy in $\mathbf{V}$. The resulting matrix is $\mathbf{c}$. With $\mathbf{b}$ and $\mathbf{c}$ formed, the method follows the general equations.

In a general formulation support conditions may be represented by support members. Reactions at supports are thereby included in the set of member forces from which the statical redundants are to be chosen. However, by excluding prescribed displacements from **r** and thus not using support members to represent reactions, it is always possible to restrict the choice of structure statical redundants to be from member forces. Some of the redundants may happen to be equal by statics to reactions at supports. This occurs if a support joint has only one member end, and a member force at that end is chosen as redundant. If the structure is externally determinate, no support reaction may be chosen as redundant.

Example 1

Consider the frame of Fig. 4–1. To form $\mathbf{a}_v$, joint displacements are chosen as $\mathbf{r} = \{\mathbf{r}^2\,\mathbf{r}^3\,\mathbf{r}^4\}$, and member redundants are ordered by members into $\mathbf{V} = \{\mathbf{V}^1\,\mathbf{V}^2\,\mathbf{V}^3\,\mathbf{V}^4\}$. For this frame $NR = 9$, $NV = 12$, and $NX = 3$. The connectivity matrix $\mathbf{a}_v$ is formed in Fig. 8/2.4–2 with $\mathbf{r}_1$ and $\mathbf{r}_5$ included. For the present choice of **r** the first and last three columns are deleted. The choice of $\mathbf{V}^m$ is the alternate type. Some of the possible choices of statical redundants are shown with the corresponding primary structures in (*b*), (*c*), and (*d*). In case (*d*) submatrix $\mathbf{a}_2$ of $\mathbf{a}_v$ is formed of the first three rows and $\mathbf{a}_1$ of the remaining rows. In case (*b*) $\mathbf{X} = \{V_2^1\,V_3^2\,V_2^4\}$. $\mathbf{a}_2$ is then formed of rows 2, 6, and 11 of $\mathbf{a}_v$, and $\mathbf{a}_1$ of the remaining rows. The equivalent joint load $(\mathbf{R} - \mathbf{R}_p)$ and the initial deformations $\mathbf{v}_p$ are evaluated for a given loading, as done in Chapter 8, Example 1a, and shown in Fig. 8/2.4–3. A simpler choice of particular equilibrium state is made in Chapter 8, Example 1b, and shown in Fig. 8/2.4–4.

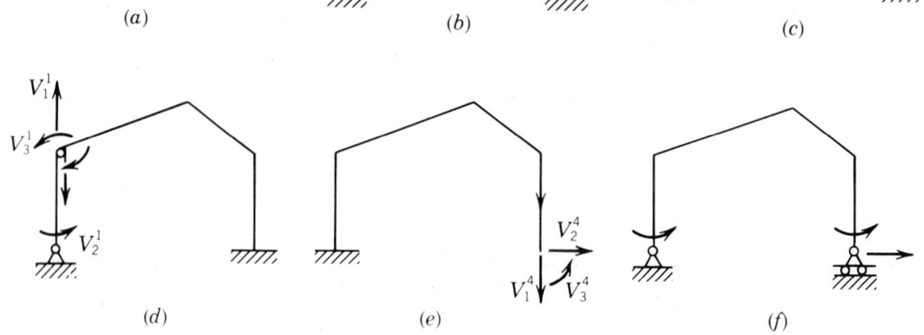

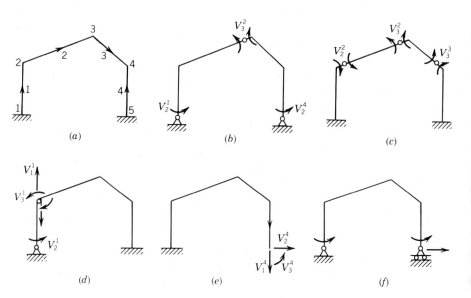

FIG. 4–1. Choice of statical redundants, Example 1.

If member redundants of the cantilever type are chosen for member 4, choices of **X** from the elements of **V** may be made as shown in (*e*) and (*f*).

Example 2

In Fig. 4–2*a* the bottom chord is continuous and is considered formed of three members. The remaining connections are assumed pinned. The count of effective member redundants NV is shown next to each member and totals 13. Joint displacements are shown in (*b*) and total 11. The structure is thus indeterminate to the second degree. Statical redundants must be chosen from the set of member forces because the structure is externally determinate. To form $\mathbf{a}_v$, member redundants of alternate type are chosen for members 1, 2, and 3. The members are oriented in order to identify elements of $\mathbf{V}^m = \{V_1^m \; V_2^m \; V_3^m\}$. Elements of $\mathbf{V}$ are identified as $\mathbf{V}^1 = \{V_1^1 \; V_3^1\}$, $\mathbf{V}^2 = \{V_1^2 \; V_2^2 \; V_3^2\}$, and $\mathbf{V}^3 = \{V_1^3 \; V_2^3\}$. For the remaining members $\mathbf{V}^m = \{V_1^m\}$. The effective flexibility matrices for members 1 and 3 are $f^m = \lceil l/EA \;\; l/3EI \rfloor$. For member 2, $\mathbf{f}^m$ is the standard 3×3 matrix, and for the remaining members $f^m = [l/EA]$. $\mathbf{a}_v$ is formed by one of the methods seen in Chapter 8, Section 2.4. Examples of choices of statical redundants are $\{V_1^6 \; V_1^9\}$, $\{V_2^2 \; V_3^2\}$, $\{V_3^1 \; V_2^3\}$. To the last two choices corresponds a pin-jointed truss as a primary structure.

Example 3

The truss of Fig. 5–1a is analyzed by hand in Section 5. The two statical redundants are chosen as one member axial force X_1 and one reaction represented as a support-member force X_2.

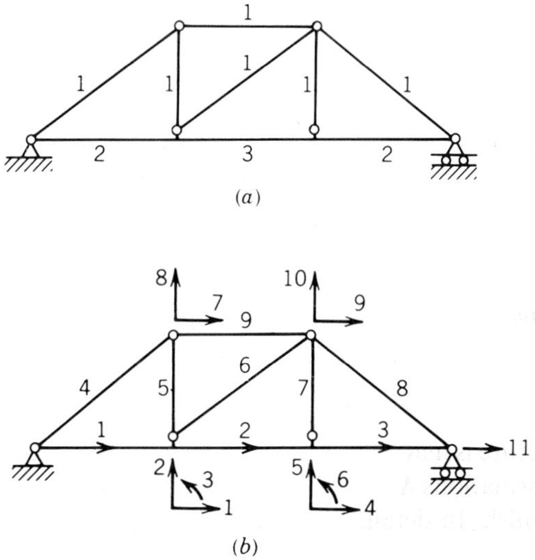

FIG. 4–2. Choice of statical redundants, Example 2.

Statical Redundants as Support Reactions

If a structure is externally statically indeterminate but is internally determinate, the statical redundants may be chosen as redundant reactions $\mathbf{R}_s$ at supports. The force method may then be formulated in a modified manner by treating the statical redundants as part of the joint load matrix $(\mathbf{R} - \mathbf{R}_p)$ rather than part of the member forces $\mathbf{V}$. With $\mathbf{R}_s$ included in $\mathbf{R}$, and $\mathbf{r}_s$ in $\mathbf{r}$, the connectivity matrix $\mathbf{a}_v$ is that of the primary structure obtained by releasing $\mathbf{r}_s$. The equilibrium solution, $\mathbf{V} = \mathbf{a}_v^{T-1}(\mathbf{R} - \mathbf{R}_p)$, is written in the form

$$\mathbf{V} = \mathbf{b}\mathbf{R}_f + \mathbf{c}\mathbf{X}$$

where $\mathbf{X}$ is the part of $(\mathbf{R} - \mathbf{R}_p)$ containing the reactions $\mathbf{R}_s$ and $\mathbf{R}_f$ is the remaining part. With $\mathbf{b}$ and $\mathbf{c}$ determined, the procedure follows the general equations. Note that the procedure of Example 1, in which prescribed displacements are deleted from $\mathbf{r}$, has an $\mathbf{a}_v$ matrix of smaller column size than in the present formulation.

5. INTERPRETATION THROUGH PRIMARY STRUCTURE: MANUAL METHODS

The concept of primary structure (Chapter 9, Section 5) is useful for a physical understanding of the equations of the force method and as a guide in manual methods of solution.

If $\mathbf{b}$ and $\mathbf{c}$ are partitioned by columns,

$$\mathbf{b} = [\mathbf{b}_1 \;\; \mathbf{b}_2 \ldots \mathbf{b}_{NR}]$$

$$\mathbf{c} = [\mathbf{c}_1 \;\; \mathbf{c}_2 \ldots \mathbf{c}_{NX}]$$

$\mathbf{b}_i$ is the set $\mathbf{V}$ in the primary structure in equilibrium with the load $R_i = 1$, and $\mathbf{c}_i$ is the set $\mathbf{V}$ in the primary structure in equilibrium with the load $X_i = 1$. X_i is applied externally as a pair of opposite forces on the two faces of release i.

An arbitrary set of member deformations $\mathbf{v}$ induces in the primary structure relative displacements at the releases, or deformations

$$\mathbf{\Delta} = \mathbf{c}^T\mathbf{v} \tag{5–1}$$

The primary structure is made to coincide in behavior with the given structure by requiring the deformations $\mathbf{\Delta}$ caused by the loading condition and by the statical redundants to vanish. In detail, the loading condition causes member forces

$$\mathbf{V}_R = \mathbf{b}(\mathbf{R} - \mathbf{R}_p),$$

member deformations $(\mathbf{v}_R + \mathbf{v}_0)$, where

$$\mathbf{v}_R = \mathbf{f}\,\mathbf{V}_R = \mathbf{f}\,\mathbf{b}(\mathbf{R} - \mathbf{R}_p)$$

$$\mathbf{v}_0 = \mathbf{v}_p + \mathbf{v}_I$$

and deformations at the releases $(\boldsymbol{\Delta}_R + \boldsymbol{\Delta}_0)$, where

$$\boldsymbol{\Delta}_R = \mathbf{c}^T\mathbf{v}_R = \mathbf{c}^T\mathbf{f}\,\mathbf{V}_R$$

$$\boldsymbol{\Delta}_0 = \mathbf{c}^T\mathbf{v}_0$$

The statical redundants **X** cause member forces

$$\mathbf{V}_x = \mathbf{c}\mathbf{X}$$

member deformations

$$\mathbf{v}_x = \mathbf{f}\,\mathbf{V}_x = \mathbf{f}\mathbf{c}\mathbf{X}$$

and deformations at the releases

$$\boldsymbol{\Delta}_x = \mathbf{c}^T\mathbf{v}_x = \mathbf{c}^T\mathbf{f}\mathbf{c}\mathbf{X}$$

The compatibility equation (3–2) expresses that

$$\boldsymbol{\Delta} = \boldsymbol{\Delta}_x + \boldsymbol{\Delta}_R + \boldsymbol{\Delta}_0 = 0$$

The compatibility equations have the ordering of the kinematic releases. The ith equation expresses that deformation Δ_i at release i vanishes. Letting δ_{ij} be the elements of $\mathbf{F}_{xx}$, the ith compatibility equation takes the form

$$\sum_{j=1}^{NX} \delta_{ij} X_j + \Delta_{i,R} + \Delta_{i,0} = 0 \qquad (5\text{–}2)$$

where

$$\delta_{ij} = \mathbf{c}_i^T\mathbf{f}\mathbf{c}_j = \sum_m \mathbf{c}_i^{mT}\mathbf{f}^m\mathbf{c}_j^m \qquad (5\text{–}3)$$

$$\Delta_{i,R} = \mathbf{c}_i^T\mathbf{f}\,\mathbf{V}_R = \sum_m \mathbf{c}_i^{mT}\mathbf{f}^m\mathbf{V}_R^m \qquad (5\text{–}4)$$

$$\Delta_{i,0} = \mathbf{c}_i^T\mathbf{v}_0 = \sum_m \mathbf{c}_i^{mT}\mathbf{v}_0^m \qquad (5\text{–}5)$$

In manual methods, forming the compatibility equations requires statical solutions to $(NX + 1)$ loading conditions of the primary structure. One solution

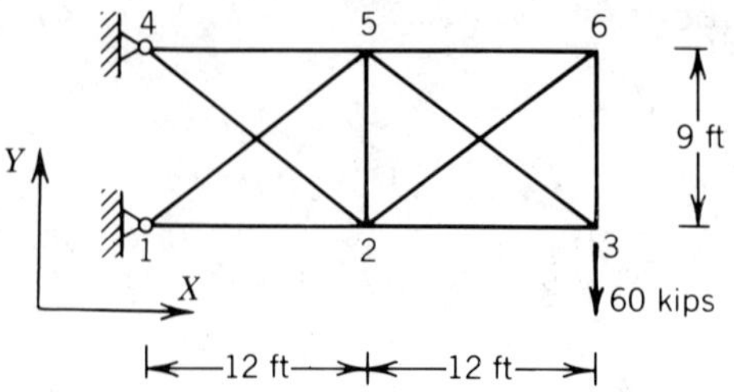

$$\frac{l}{A} = 1\ \text{ft/in}^2$$

$$E = 30{,}000\ \text{ksi}$$

$$\alpha = 6.5 \times 10^{-6}\ (\text{in/in})/^\circ\text{F}$$

(*a*)

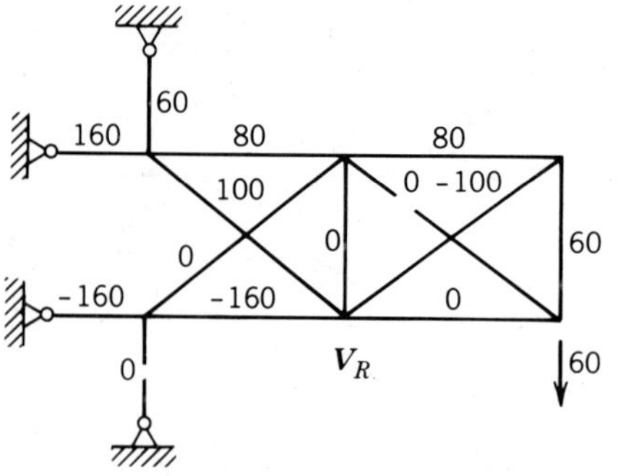

$$\Delta_{1,R} = \frac{l}{EA}[80(-4/5) - 100(1) - 60(3/5)]$$

$$= -200\frac{l}{EA}$$

$$\Delta_{2,R} = \frac{l}{EA}[80(-4/3) + 100(5/3) + 160(4/3)]$$

$$= \frac{820}{3}\frac{l}{EA}$$

(*b*)

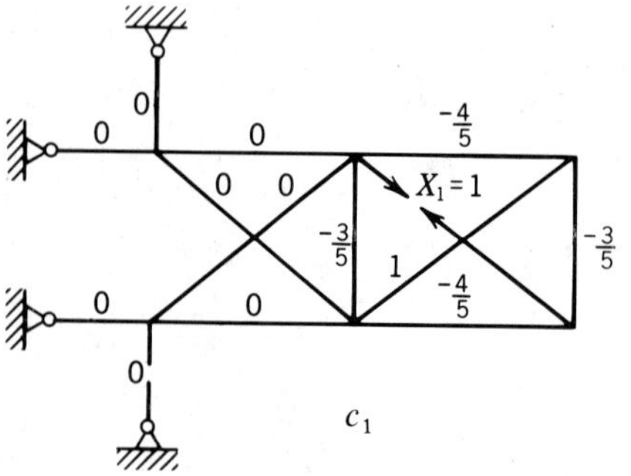

$$\delta_{11} = \frac{l}{EA}\left(\frac{9}{25} + 1 + 1 + \frac{16}{25} + \frac{16}{25} + \frac{9}{25}\right)$$

$$= 4\frac{l}{EA}$$

$$\delta_{12} = \delta_{21} = \frac{l}{EA}\left(\frac{3}{5}\right)$$

$$= 0.6\frac{l}{EA}$$

(*c*)

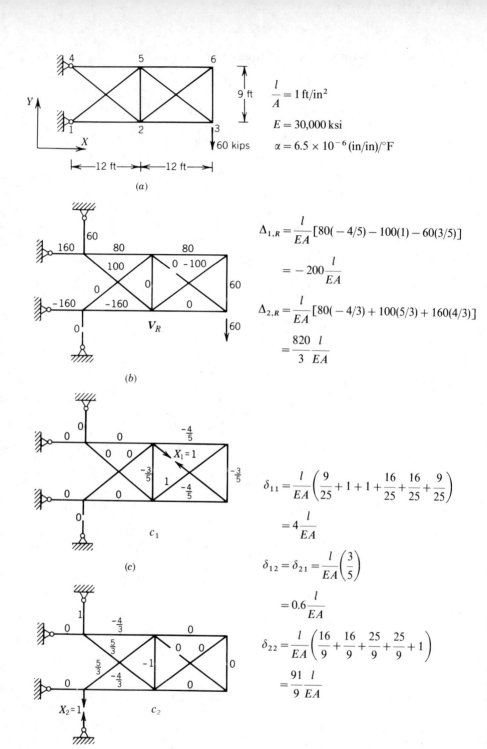

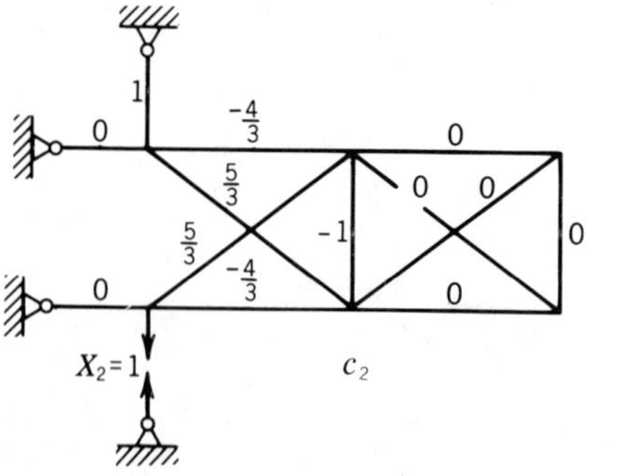

$$\delta_{22} = \frac{l}{EA}\left(\frac{16}{9} + \frac{16}{9} + \frac{25}{9} + \frac{25}{9} + 1\right)$$

$$= \frac{91}{9}\frac{l}{EA}$$

(*d*)

FIG. 5–1. Statically indeterminate truss, Examples 1, 2. (*Continued on next page.*)

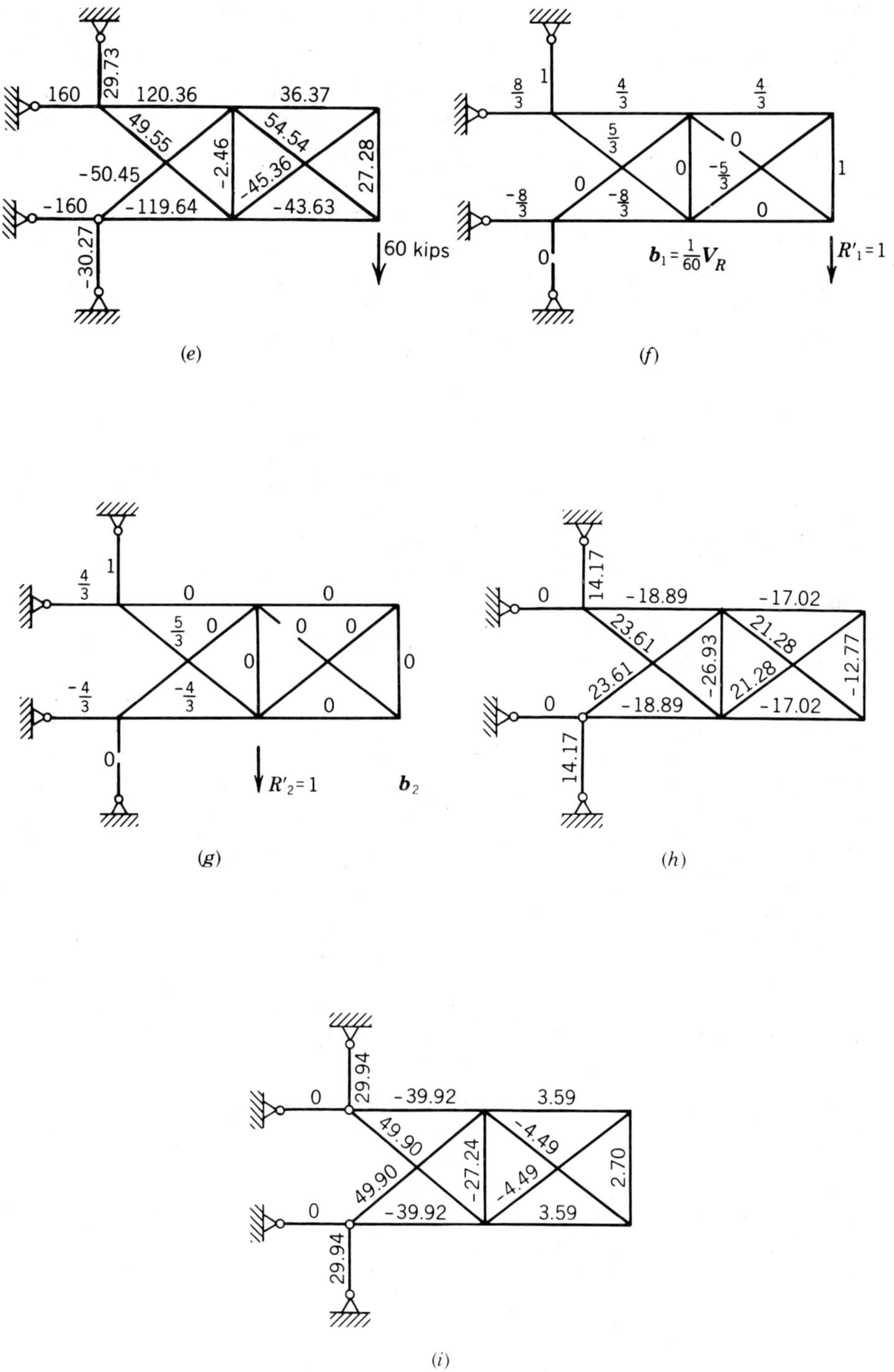

FIG. 5–1. (*continued*)

determines $\mathbf{V}_R$ due to the loading condition of the problem. The others determine the columns of $\mathbf{c}$ which are due, respectively, to the individual loadings $X_i = 1$, for $i = 1, 2, \ldots, NX$. The initial deformations $\mathbf{v}_0^m$ are determined separately for each member in its particular equilibrium state. $\mathbf{v}_p^m$ is caused by the member mechanical loading, and $\mathbf{v}_I^m$ consist of free thermal deformations or geometric errors. Evaluation of δ_{ij}, $\Delta_{i,R}$, and $\Delta_{i,0}$ is made by the procedure outlined for determinate structures. For frames expressions in Eqs. (5–3, 4) are conveniently evaluated from values read off the bending moment diagrams, with the possibility of adding axial and shear deformation terms if desired.

Example 1

Consider the truss of Fig. 5–1a with given loading, geometry and material data as shown. Let the object of the analysis be to determine the member axial forces, the reactions at supports, and the Y displacements of joints 2 and 3.

As an assemblage the truss is a triangulated system with one additional member, for example, member (3–5). Its supports have one restraint in addition to simple support conditions, for example, the Y restraint at joint 1. The truss is thus stable and statically indeterminate to the second degree. The statical redundants $\mathbf{X} = \{X_1 \ X_2\}$ that correspond to the preceding geometric analysis are the axial force in member (3–5) and the Y reaction at joint 1.

The general solution of the joint equilibrium equations is the superposition of $\mathbf{V}_R$ and $\mathbf{cX}$. $\mathbf{V}_R$ is due to the loading condition with $\mathbf{X} = 0$. $\mathbf{X}$ is made zero physically by releasing the axial continuity of member (3–5) and the Y restraint at joint 1. Member (3–5) is shown cut in (*b*). Also the device of support members is used, and release of the Y restraint at joint 1 is represented by cutting the corresponding support member. The resulting primary structure is subjected to the loading condition, and the axial forces are determined by the method of joints and are written next to the members in (*b*).

$\mathbf{c}$ is formed of two columns $\mathbf{c}_1$ and $\mathbf{c}_2$. $\mathbf{c}_1$ is the set of self-equilibriating forces in which $X_1 = 1$ and $X_2 = 0$. $\mathbf{c}_1$ is determined in the primary structure by loading released member (3–5) with a unit tensile axial force and determining the remaining axial forces by the method of joints. The results are shown in (*c*). $\mathbf{c}_2$ is determined similarly, and the results are shown in (*d*).

The preceding results are now processed to from the coefficient matrix $\mathbf{F}_{xx}$ and the right-hand side $-\mathbf{\Delta}_R$ of the compatibility equations. $\mathbf{\Delta}_0$ is zero in the present example because there is no member load.

$\mathbf{c}_i^m$ consists of one element c_i^m, and $\mathbf{f}^m$ of one element $f^m = l/EA$ which in the present problem is the same for all members. $\Delta_{i,R}$ and δ_{ij} are evaluated through the formulas

$$\Delta_{i,R} = \sum_m f^m c_i^m V_R^m$$

$$\delta_{ij} = \sum_m f^m c_i^m c_j^m$$

Values of c_i^m and V_R^m are read off the graphs, and calculations are shown in the figure. $\Delta_{1,R}$ is negative and represents a gap produced at release 1 by the loading in (*b*). $\Delta_{2,R}$ is positive and represents an overlap at release 2. δ_{11} is the overlap produced at release 1 by $X_1 = 1$. δ_{12} is the overlap at release 1 due to $X_2 = 1$ and is equal to the overlap δ_{21} at release 2 due to $X_1 = 1$. These results form the matrices

$$\mathbf{F}_{xx} = \frac{l}{EA}\begin{bmatrix} 4 & 0.6 \\ 0.6 & \dfrac{91}{9} \end{bmatrix}$$

$$\mathbf{\Delta}_R = \frac{l}{EA}\left\{ -200 \quad \frac{820}{3} \right\}$$

The compatibility equations are solved for **X**. Although there is no need to invert $\mathbf{F}_{xx}$, this will be done in view of other loading conditions to be considered subsequently. There comes

$$\mathbf{F}_{xx}^{-1} = \frac{EA}{l}\frac{9}{360.76}\begin{bmatrix} \dfrac{91}{9} & -0.6 \\ -0.6 & 4 \end{bmatrix}$$

and

$$\mathbf{X} = \mathbf{F}_{xx}^{-1}(-\mathbf{\Delta}_R) = \{54.540 \quad -30.269\} \text{ kips}$$

The axial forces are evaluated through

$$V^m = V_R^m + c_1^m X_1 + c_2^m X_2$$

and are shown in (*e*).

Evaluation of the displacements is similar to that in statically determinate structures. Displacement r_i is evaluated through the formula

$$r_i = \mathbf{b}_i^T \mathbf{v} = \sum_m b_i^m v^m$$

$\mathbf{b}_i$ is the set of (virtual) axial forces in the primary structure in equilibrium with the (virtual) joint load $R_i' = 1$. The desired displacements are labeled r_1 and r_2, respectively. The statics for $R_1' = 1$ and $R_2' = 1$ are carried out by the method of joints, and the results are shown in (*f*) and (*g*). For the loading under consideration $\mathbf{v}_p$ and $\mathbf{v}_I$ are zero. Thus $\mathbf{v} = \mathbf{fV}$, and

$$r_i = \sum_m f^m b_i^m V^m$$

V^m is read off (*e*), and b_1^m and b_2^m are read off (*f*) and (*g*), respectively. There comes

$$r_1 = 713.48 \frac{l}{EA} = 0.285 \text{ in}$$

$$r_2 = 242.10 \frac{l}{EA} = 0.097 \text{ in}$$

Example 2

Two separate additional loading conditions are now considered for the truss of Fig. 5–1*a*. A thermal loading condition in which $\Delta T = 50°\text{F}$ for members (4–5) and (5–6) and a support displacement down at joint 1 of 0.12 in.

The free thermal deformations of members (4–5) and (5–6) are equal to $v_I = \alpha \Delta T l = 6.5 \times 10^{-6} \times 50 \times 12 \times 12 = 0.0468$ in. For the present loading $\mathbf{v}_p = 0$. Thus

$$\Delta_{1,0} = \sum_m c_1^m v_I^m = -(4/5)(0.0468) = -0.03744 \text{ in}$$

$$\Delta_{2,0} = \sum_m c_2^m v_I^m = -(4/3)(0.0468) = -0.0624 \text{ in}$$

and

$$\mathbf{\Delta}_0 = \{-0.03744 \quad -0.0624\} \text{ in}$$

The solution for the statical redundants is

$$\mathbf{X} = \mathbf{F}_{xx}^{-1}(-\mathbf{\Delta}_0) = \frac{EA}{l} 10^{-3} \{8.510 \quad 5.666\}$$

With $EA/l = 30{,}000/12$ k/in, obtain

$$\mathbf{X} = \{21.275 \quad 14.166\} \text{ kips}$$

The axial forces are $\mathbf{c}_1 X_1 + \mathbf{c}_2 X_2$ and are shown in (*h*).

The settlement of 0.12 in down at joint 1 is in the support member an initial deformation $v_I^m = -0.12$ in. Then $\Delta_{i,0} = c_i^m v_I^m$, and form (c) and (*d*), $c_1^m = 0$ and $c_2^m = 1$. Thus $\Delta_{1,0} = 0$, and $\Delta_{2,0} = 1(-0.12) = -0.12$ in. The geometric meaning of this result is that the shortening of the support member causes in the primary structure a gap of 0.12 in the cut member, and no gap in the other redundant member. Then

$$\mathbf{X} = \mathbf{F}_{xx}^{-1}(-\mathbf{\Delta}_0) = \frac{EA}{l} 10^{-3} \{-1.796 \quad 11.974\}$$

$$= \{-4.490 \quad 29.937\} \text{ kips}$$

The axial forces are evaluated as for the thermal loading condition and are shown in (*i*).

Evaluations of displacements follows the same procedure as in Example 1. The axial forces to be used in the computations are those in (*h*) and (*i*), respectively.

A tabular presentation of the preceding analysis could be made as outlined in Section 2.4.

Example 3

An example of frame analysis is now treated. The frame is shown with its geometric and material properties and a loading condition in Fig. 5–2*a*. Shear deformation is neglected, and the members are treated as inextensible so that $\mathbf{V}^m$ excludes the member axial force. The alternate type of $\mathbf{V}^m$ is chosen because it may be represented by member end values in a bending moment diagram, as done in the case of determinate structures. A primary structure is obtained by releasing the support restraints at E and A. The corresponding statical redundants and the loading condition of the problem are shown acting on the primary structure in (*b*). Figures 5–2*c*, *d*, *e*, *f* show the bending moment diagrams in the primary structure due, respectively, to the loading condition and to $X_1 = 1, X_2 = 1,$ and $X_3 = 1$. The moments are drawn on the tension side. The member end values in these diagrams represent $\mathbf{V}_R, \mathbf{c}_1, \mathbf{c}_2,$ and $\mathbf{c}_3$, respectively. Deformations $\mathbf{v}_p$ of the loaded member in the particular equilibrium state are shown in (*g*). Note that, as seen in Section 2.4, the bending moment in (*c*) is formed of $M_h + M_p$. M_h is formed of the straight lines joining the member end values, and M_p is the bending moment in the particular equilibrium state. M_h is due to the equivalent joint load $(\mathbf{R} - \mathbf{R}_p)$ which need not be formed explicitly here. For the given loading, $\mathbf{v}_I = 0$ so that $\mathbf{v}_0 = \mathbf{v}_p$.

Now all the information needed to form $\mathbf{F}_{xx}, \mathbf{\Delta}_R,$ and $\mathbf{\Delta}_0$ is contained in (*c*), (*d*), (*e*), (*f*), and (*g*). The matrix products defining $\delta_{ij}, \Delta_{i,R},$ and $\Delta_{i,0}$ are of the same type as those evaluated for statically determinate structures, using Eqs. (2.4–10). δ_{ij} is evaluated through the formula in Eq. (2.4–10a), using for M_a, M'_a, M_b, M'_b the values from diagrams $\mathbf{c}_i$ and $\mathbf{c}_j$. For example,

$$EI\delta_{11} = \left(\frac{10}{3}\right)(25)^2(3) + \left(\frac{10}{3}\right)[(25)^2 + (15)^2 + (25)(15)] + \left(\frac{15}{3}\right)(15)^2$$

$$= 34,375/3 \text{ ft}^3$$

$$EI\delta_{12} = -\left(\frac{10}{3}\right)(25)(10)\left(\frac{3}{2}\right) - \left(\frac{10}{3}\right)(10)(25 + 15)\left(\frac{3}{2}\right) = -3.250 \text{ ft}^3$$

Similarly, using diagrams $\mathbf{c}_1$ and $\mathbf{V}_R$,

$$EI\Delta_{1,R} = -\left(\frac{10}{3}\right)(25)(350 + 150)\left(\frac{3}{2}\right) - \left(\frac{10}{3}\right)\left[25(150) + \left(\frac{1}{2}\right)(15)(150)\right]$$

$$= -78{,}750 \text{ k-ft}$$

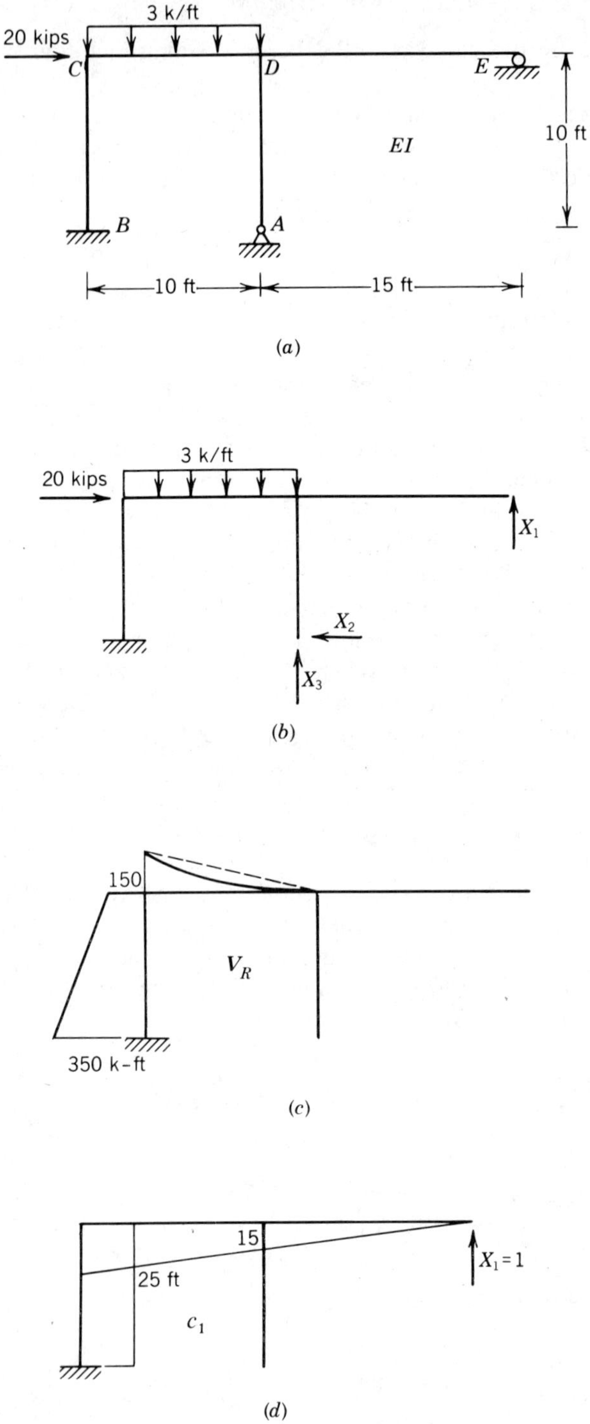

FIG. 5–2. Statically indeterminate frame, Examples 3, 4, 5. (*Continued on next page.*)

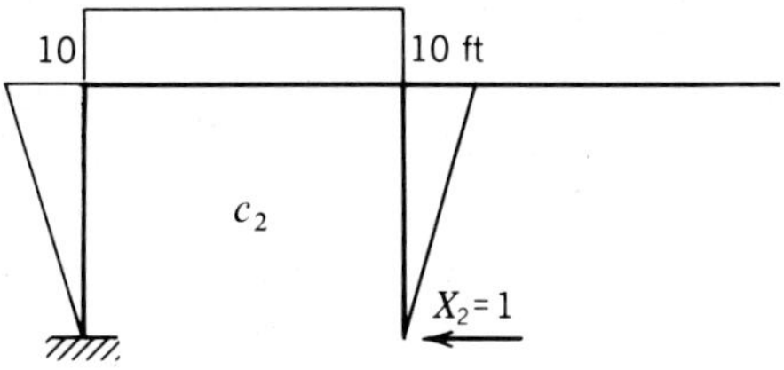

(e)

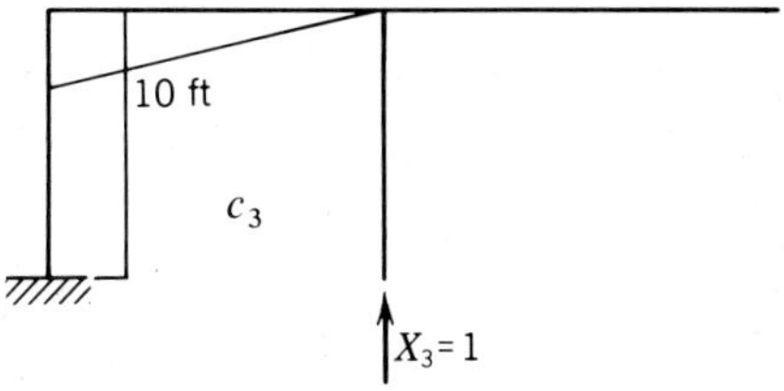

(f)

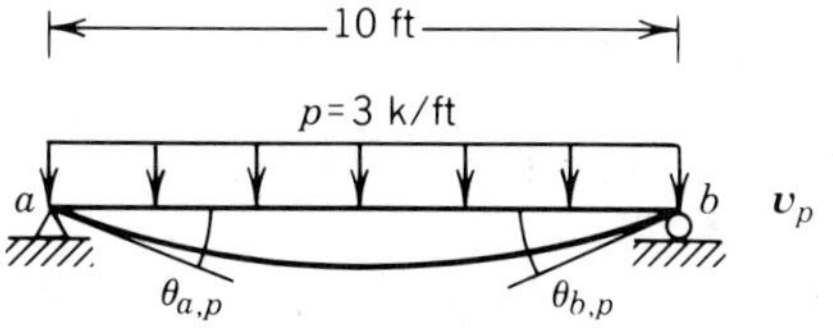

$$\theta_{a,p} = \theta_{b,p} = \frac{pl^3}{24EI} = \frac{1000}{8EI}$$

(g)

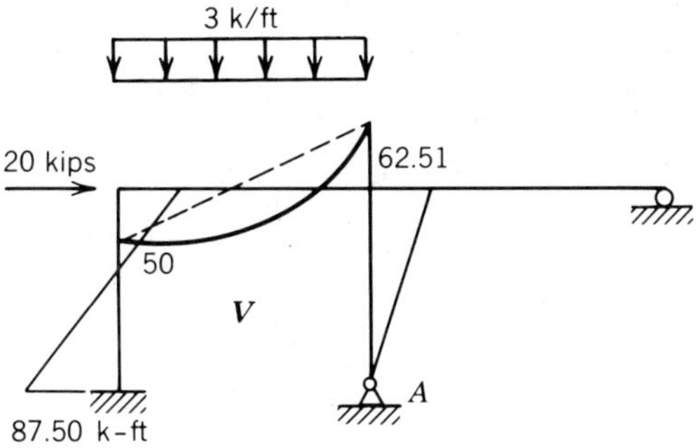

(h)

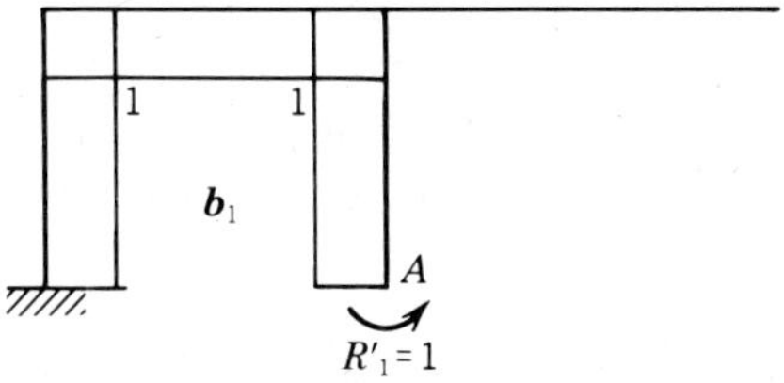

(i)

FIG. 5–2. (*continued*)

From diagram $\mathbf{c}_1$ and part (*g*) of Fig. 5–2

$$EI\Delta_{1,0} = (25 + 15)\frac{1000}{8} = 5{,}000 \text{ k-ft}^3$$

Completion of the calculations yields

$$EI\mathbf{F}_{xx} = \frac{1}{3}\begin{bmatrix} 34{,}375 & -9{,}750 & 10{,}750 \\ & 5{,}000 & -3{,}000 \\ \text{symmetric} & & 4{,}000 \end{bmatrix}$$

$$EI(\mathbf{\Delta}_R + \mathbf{\Delta}_0) = \left\{ -73{,}750 \quad \frac{47{,}500}{3} \quad -28{,}750 \right\}$$

The solution of the system of three compatibility equations $\mathbf{F}_{xx}\mathbf{X} = -(\mathbf{\Delta}_R + \mathbf{\Delta}_0)$ is found by Gauss elimination to be

$$\mathbf{X} = \{-0.00028 \quad 6.24972 \quad 26.25051\} \text{ kips}$$

X_1 is seen to be negligible for the loading considered. The bending moment diagram of the structure is now the result of superposition of diagram (*c*) and diagrams (*d*), (*e*), and (*f*) scaled, respectively, by X_1, X_2, and X_3. It is also obtained directly by statics of the primary structure loaded as in (*b*). It is shown in (*h*) and labeled **V**.

Evaluation of displacements is made through Eq. (3–7) in which $\mathbf{v} = \mathbf{fv} + \mathbf{v}_0$. Thus

$$\mathbf{r} = \mathbf{b}^T\mathbf{f}\,\mathbf{V} + \mathbf{b}^T\mathbf{v}_0 = \sum_m \mathbf{b}^{mT}\mathbf{f}^m\mathbf{V}^m + \sum_m \mathbf{b}^{mT}\mathbf{v}_0^m \qquad (5\text{–}6\text{a})$$

For displacement r_i,

$$\mathbf{r}_i = \mathbf{b}_i^T\mathbf{f}\,\mathbf{V} + \mathbf{b}_i^T\mathbf{v}_0 = \sum_m \mathbf{b}_i^{mT}\mathbf{f}^m\mathbf{V}^m + \sum_m \mathbf{b}_i^{mT}\mathbf{v}_0^m \qquad (5\text{–}6\text{b})$$

For example, let the rotation at A in Fig. 5–2a be called r_1. To evaluate r_1, a virtual load $R_1' = 1$ is applied to the primary structure as shown in (i). The bending moment diagram is labeled $\mathbf{b}_1$. Then from diagrams **V** and $\mathbf{b}_1$

$$EI\mathbf{b}_1^T\mathbf{f}\,\mathbf{V} = \frac{10}{3}(-87.50 + 50)\frac{3}{2} + \frac{10}{3}(50 - 62.51)\frac{3}{2} - \frac{10}{3}(62.51)\frac{3}{2}$$

$$= -762.55 \text{ k-ft}^2$$

and from diagram $\mathbf{b}_1$ and Fig. 5–2g

$$EI\mathbf{b}_1^T\mathbf{v}_0 = 2 \times \frac{1000}{8} = 250\,\text{k-ft}^2$$

thus

$$EIr_1 = -762.55 + 250 = -512.55\,\text{k-ft}^2$$

The negative result for r_1 indicates a sense opposite to the one chosen for R_1' in (i).

A geometric illustration of δ_{ij}, $\Delta_{i,R}$, and $\Delta_{i,0}$ is given in Fig. 5–3. The deformations are shown in the actual sense but not to scale. Note the physical meaning of the property $\delta_{ij} = \delta_{ji}$ for $i \neq j$. In (d), $\mathbf{\Delta}_R$ is caused by the equivalent joint load $(\mathbf{R} - \mathbf{R}_p)$. In ($e$) the deformed shape corresponds to the deformations of the particular equilibrium state. Note that only the loaded member is deformed.

Example 4

Analysis for a support settlement will now be illustrated. Let B in Fig. 5–2a undergo prescribed translations $v_{1,I}$ and $v_{2,I}$ and a rotation $v_{3,I}$. In the primary structure the displacements at B induce a rigid body displacement as shown in Fig. 5–4a. The deformations at the releases are obtained geometrically as

$$\Delta_{1,I} = -v_{2,I} + 25v_{3,I}\,\text{ft}$$

$$\Delta_{2,I} = -v_{1,I}$$

$$\Delta_{3,I} = -v_{2,I} + 10v_{3,I}\,\text{ft}$$

In the compatibility equations, $\mathbf{\Delta}_R = 0$, and

$$\mathbf{\Delta}_0 = \mathbf{\Delta}_I = \{\Delta_{1,I}\ \Delta_{2,I}\ \Delta_{3,I}\}$$

The rest of the analysis follows the general method.

In the matrix formulation a support settlement is treated as an initial deformation of a support member. A support member representing the fixed conditions at B is shown in Fig. 5–4b. The loading condition consists of initial deformations

$$\mathbf{v}_I^m = \{v_{1,I}\ v_{2,I}\ v_{3,I}\}$$

Then

$$\mathbf{\Delta}_0 = \mathbf{c}^{mT}\mathbf{v}_I^m$$

$\mathbf{c}^m$ is formed of three columns which consist of the forces of the support member

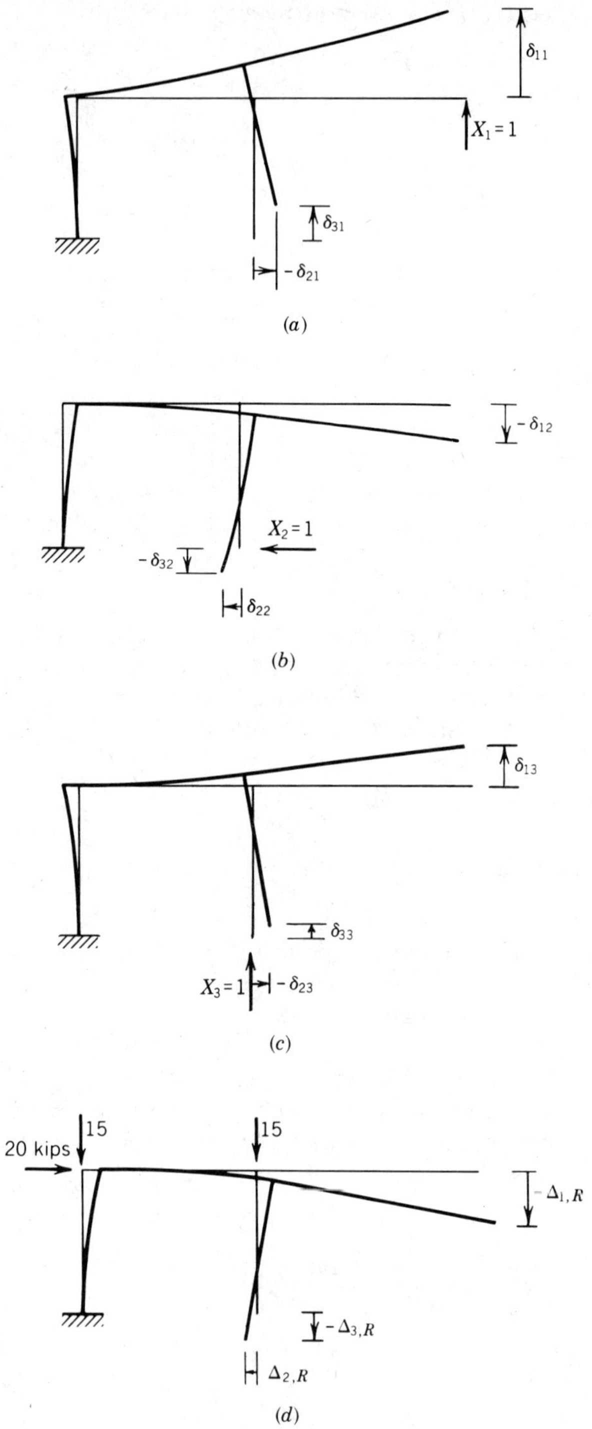

FIG. 5–3. Deformations in primary structure, Example 3. (*Continued on next page.*)

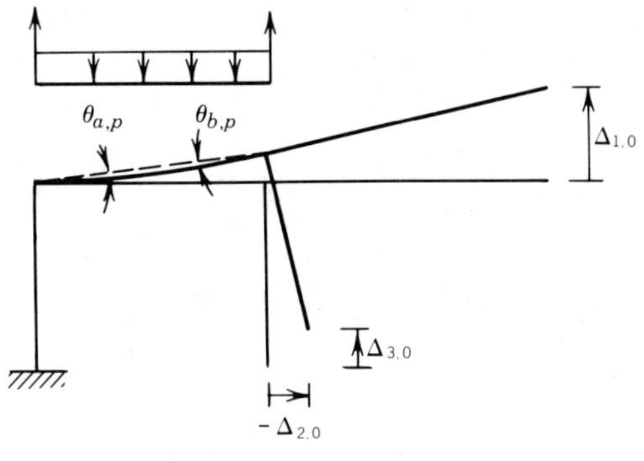

(*e*)

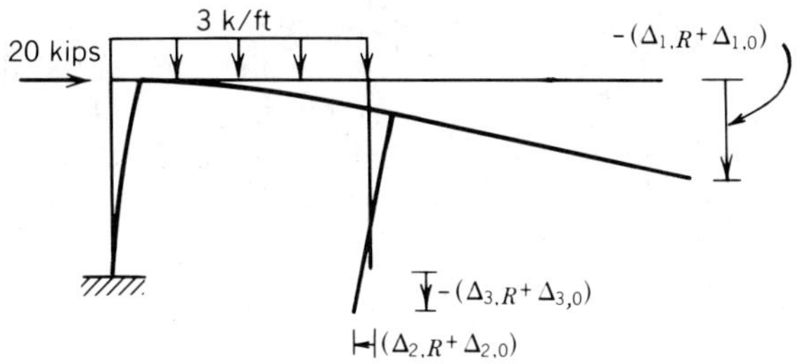

(*f*)

FIG. 5–3. (*continued*)

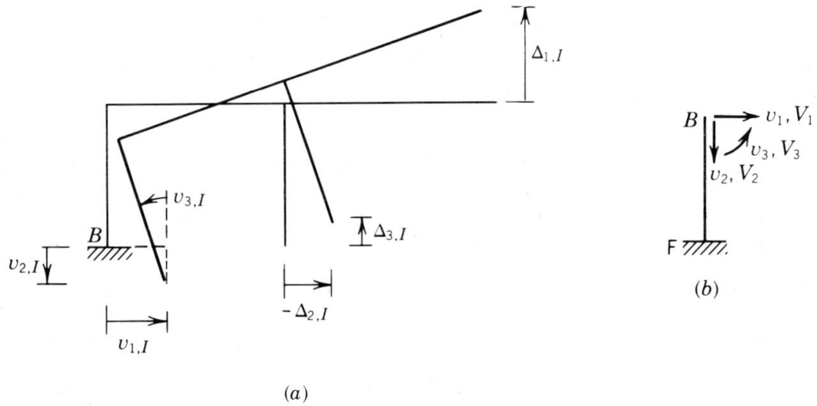

FIG. 5–4. Support settlement, Example 4.

caused, respectively, by unit values of the statical redundants. With the sign convention of Fig. 5–4*b*, there comes

$$\mathbf{c}^m = \begin{bmatrix} 0 & -1 & 0 \\ -1 & 0 & -1 \\ 25 & 0 & 10 \end{bmatrix}$$

It is verified that $\mathbf{c}^{mT}\mathbf{v}_I^m$ yields $\mathbf{\Delta}_I$ as found earlier.

Example 5

To illustrate the analysis for a thermal loading condition, let member *CD* in Fig. 5–2*a* be subjected to a thermal gradient through its depth that causes a free thermal curvature χ_I and free thermal deformations $\frac{1}{2}\chi_I l$, where $l = 10$ ft. The primary structure deforms freely similarly to Fig. 5–3*e* in which $\theta_{a,p}$ and $\theta_{b,p}$ are now $\theta_{a,I} = \theta_{b,I} = \frac{1}{2}\chi_I l$. $\mathbf{\Delta}_0$ is evaluated in the same way as in Example 3.

For a uniform temperature change of a member, the initial deformation is axial. In evaluating $\mathbf{\Delta}_0 = \mathbf{c}^T\mathbf{v}_I$, the effective terms in $\mathbf{c}$ are the (virtual) axial forces. After allowing for the axial initial deformations for evaluating $\mathbf{\Delta}_0$, the frame continues to be treated as inextensible, and $\mathbf{F}_{xx}$ remains unchanged in the compatibility equations.

Example 6. Elastic Center

Consider the frame of Fig. 5–5, and assume that the frame has a plane of geometric and material symmetry. A primary structure is conveniently chosen by cutting the structure by the plane of symmetry and choosing the statical redundants as shown in (*b*). The bending moment diagrams m_1 and m_3 due, respectively, to $X_1 = 1$ and $X_3 = 1$ are symmetric, and m_2 is antisymmetric. It follows that $\delta_{21} = \delta_{23} = 0$, and evaluation of $\delta_{11}, \delta_{22}, \delta_{33}$, and δ_{13} may deal only with half the structure. The compatibility equations are uncoupled into one equation for X_2 and a system of two equations for X_1 and X_3. Further in a symmetric loading, $\Delta_{2,R} = \Delta_{2,0} = 0$ and $X_2 = 0$, whereas for an antisymmetric loading $\Delta_{1,R} = \Delta_{1,0} = \Delta_{3,R} = \Delta_{3,0} = 0$ and $X_1 = X_3 = 0$. It was seen in Chapter 10, Section 5 how a general loading may be represented as the sum of a symmetric and an antisymmetric loading.

$\mathbf{F}_{xx}$ may be made into a diagonal matrix by a static transformation of $\mathbf{X}$ to a point *C* in the plane of symmetry. A physical model representing this consists of two rigid arms attached to each face of the cut and subjected to the new statical redundants at *C* as shown in (*c*). With axes *x*, *y* centered at *C*,

$$m_1 = yX_1 = y$$

$$m_2 = xX_2 = x$$

$$m_3 = X_3 = 1$$

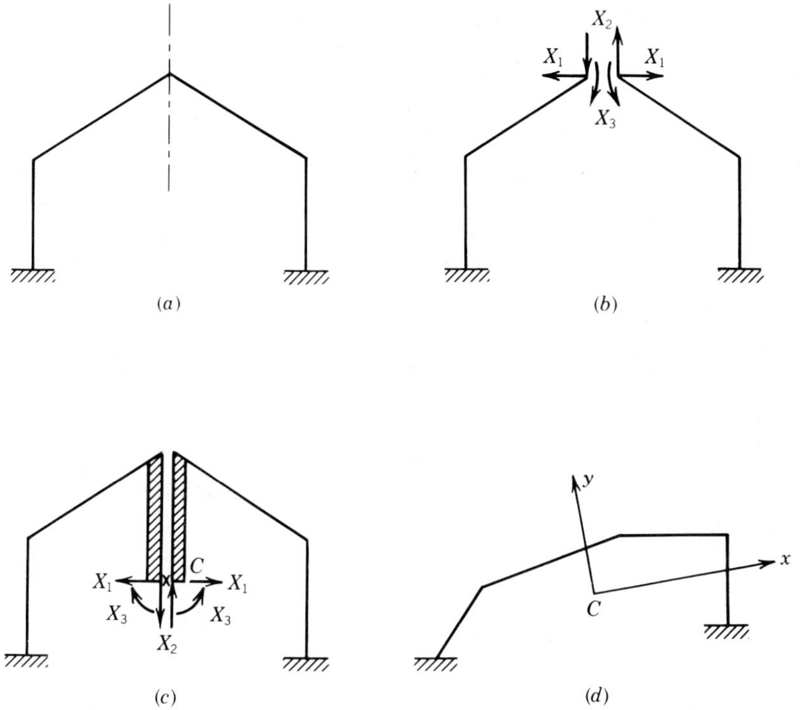

FIG. 5–5. Elastic center.

δ_{21} and δ_{23} continue to be zero, and δ_{13} is also zero if

$$\int y\,ds = 0$$

where the integral is taken along the member centerline of the frame. This integral is zero if C is the geometric centroid of the frame centerline. If EI is variable, C should be at the centroid of the centerline weighted with l/EI. C is called the elastic center.

The preceding may be generalized to an arbitrary frame as shown in (*d*). If the frame is cut at its intersection with the y axis, and rigid arms are connected to the faces of the cut and loaded at C with pairs of opposite forces as before, the conditions for $\mathbf{F}_{xx}$ to be a diagonal matrix are found to be

$$\int \frac{x}{EI}\,ds = \int \frac{y}{EI}\,ds = \int \frac{xy}{EI}\,ds = 0$$

These conditions are satisfied if C is placed at the centroid of the weighted centerline and axes x and y are chosen as its principal axes of inertia.

6. CONTINUOUS BEAMS

The formulation of compatibility equations for continuous beams is a particularly simple application of the general method.

Consider a general continuous beam as shown in Fig. 10/2–1*a*. A primary structure is obtained if the $(n-1)$ interior support restraints are released. Another one is obtained if the rotational continuity of the beam over the interior supports is released. This latter choice will be adopted. The statical redundants are then the bending moments over the interior supports. If an end of the beam is fixed against rotation, the bending moment at that end is an additional statical redundant. Let $X_1, X_2, \ldots, X_n$ be the bending moments at joints $1, 2, \ldots, n$. Figure 6–1*a* shows a general span *i* and the sign convention to be adopted. The geometric meaning of

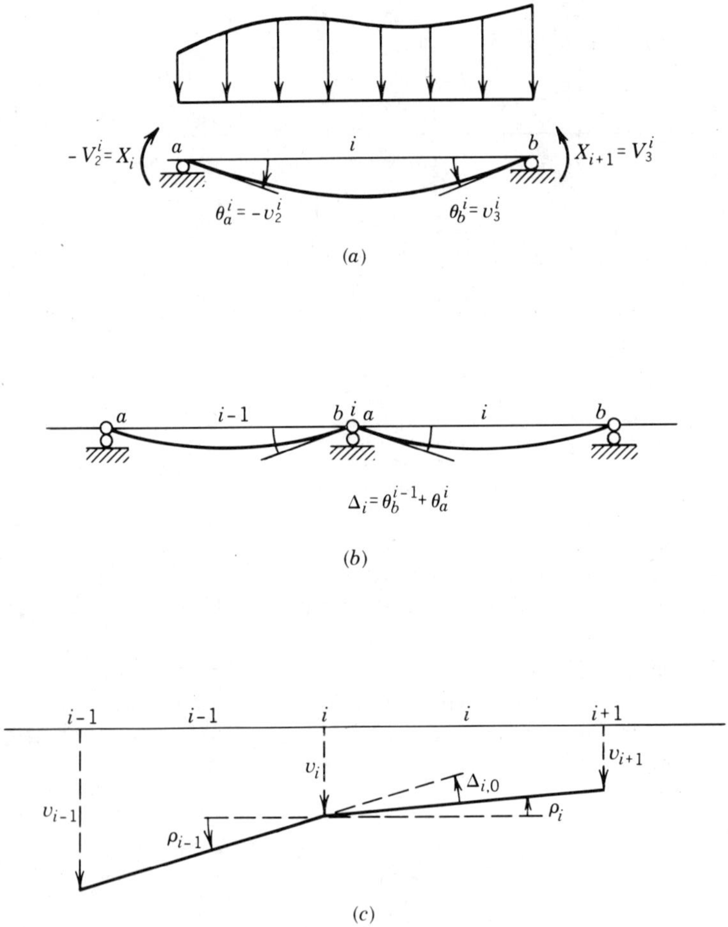

FIG. 6–1. Continuous beam.

Δ_i due to an arbitrary deformation of the primary structure is shown in (*b*). It is the angular discontinuity at release *i*,

$$\Delta_i = \theta_b^{i-1} + \theta_a^i \tag{6–1}$$

The flexibility equations for member *i* written in terms of θ_a^i, θ_b^i, X_i, and X_{i+1} are

$$\theta_a^i = f_{aa}^i X_i + f_{ab}^i X_{i+1} + \theta_{a,0}^i \tag{6–2a}$$

$$\theta_b^i = f_{ba}^i X_i + f_{bb}^i X_{i+1} + \theta_{b,0}^i \tag{6–2b}$$

For a uniform member without shear deformation

$$f_{aa} = f_{bb} = \frac{l}{3EI} \tag{6–3a}$$

$$f_{ab} = f_{ba} = \frac{l}{6EI} \tag{6–3b}$$

Substituting from Eq. (6–2) into the compatibility equation, $\Delta_i = 0$, yields

$$f_{ba}^{i-1} X_{i-1} + (f_{bb}^{i-1} + f_{aa}^i) X_i + f_{ab}^i X_{i+1} + \Delta_{i,0} = 0 \tag{6–4a}$$

where

$$\Delta_{i,0} = \theta_{b,0}^{i-1} + \theta_{a,0}^i \tag{6–4b}$$

This is known as the three-moment equation. If support 1 is fixed against rotation, the compatibility equation reduces to $\theta_a^1 = 0$. Equation (6–4) may be applied in that case, with $i = 1$ and zero flexibilities for a dummy span 0. Similarly, if support *n* is fixed against rotation, Eq. (6–4) may be applied, with $i = n$ and zero flexibilities for a dummy span *n*. If supports 1 or *n* are not rotationally restrained, then X_1 or X_n have prescribed values. In all cases the compatibility equations form a system in as many unknowns.

Note that the term $\mathbf{\Delta}_R$ in the general compatibility equation (3–4a) is zero in the present case. If a moment load is applied at an interior joint, it should be considered as a member load applied at either one of the two adjacent spans.

If members are uniform or if they are symmetric with respect to their respective midspans, then

$$f_{aa}^i = f_{bb}^i = f_i \tag{6–5a}$$

Let also

$$f_{ab}^i = f_{ba}^i = f_i' \tag{6–5b}$$

Equation (6–4a) becomes

$$f'_{i-1}X_{i-1}+(f_{i-1}+f_i)X_i+f'_iX_{i+1}+\Delta_{i,0}=0 \tag{6–6}$$

$\Delta_{i,0}$ is formed in general of two parts,

$$\Delta_{i,0}=\Delta_{i,p}+\Delta_{i,I} \tag{6–7}$$

$\Delta_{i,p}$ may be evaluated through the formulas for v_{2p} and v_{3p} derived in Chapter 5. Thermal loadings or support settlements cause $\Delta_{i,I}$. A general support settlement applied to the primary structure is shown in Fig. 6–1*c*. Rotation of span *i* is called ρ_i and is positive if counterclockwise. From the figure

$$\rho_{i-1}=\frac{v_{i-1}-v_i}{l_{i-1}} \tag{6–8a}$$

$$\rho_i=\frac{v_i-v_{i+1}}{l_i} \tag{6–8b}$$

and

$$\Delta_{i,0}=\rho_{i-1}-\rho_i \tag{6–8c}$$

7. FLEXIBILITY MATRICES

The generalized displacements of the primary structure corresponding, respectively, to $(\mathbf{R}-\mathbf{R}_p)$ and $\mathbf{X}$ are $\mathbf{r}$ and $\boldsymbol{\Delta}$. $\mathbf{r}=\mathbf{b}^T\mathbf{v}$ and $\boldsymbol{\Delta}=\mathbf{c}^T\mathbf{v}$ are expressed in terms of $(\mathbf{R}-\mathbf{R}_p)$ and $\mathbf{X}$, by substituting $\mathbf{v}=\mathbf{f}\mathbf{V}+\mathbf{v}_0$ and using Eq. (3–1a) to substitute for $\mathbf{V}$. There comes

$$\begin{Bmatrix}\mathbf{r}\\ \boldsymbol{\Delta}\end{Bmatrix}=\begin{bmatrix}\mathbf{F}_{RR} & \mathbf{F}_{Rx}\\ \mathbf{F}_{xR} & \mathbf{F}_{xx}\end{bmatrix}\begin{Bmatrix}\mathbf{R}-\mathbf{R}_p\\ \mathbf{X}\end{Bmatrix}+\begin{Bmatrix}\mathbf{b}^T\mathbf{v}_0\\ \boldsymbol{\Delta}_0\end{Bmatrix} \tag{7–1}$$

where

$$\mathbf{F}_{RR}=\mathbf{b}^T\mathbf{f}\mathbf{b} \tag{7–2a}$$

$$\mathbf{F}_{Rx}=\mathbf{F}_{xR}^T=\mathbf{b}^T\mathbf{f}\mathbf{c} \tag{7–2b}$$

$$\mathbf{F}_{xx}=\mathbf{c}^T\mathbf{f}\mathbf{c} \tag{7–2c}$$

The four matrices $\mathbf{F}_{RR}$, $\mathbf{F}_{Rx}=\mathbf{F}_{xR}^T$, and $\mathbf{F}_{xx}$ form a generalized flexibility matrix of the primary structure.

For the indeterminate structure, $\mathbf{\Delta} = 0$, so that

$$\mathbf{X} = -\mathbf{F}_{xx}^{-1}\mathbf{F}_{xR}(\mathbf{R} - \mathbf{R}_p) - \mathbf{F}_{xx}^{-1}\mathbf{\Delta}_0 \tag{7–3}$$

Substituting for $\mathbf{X}$ into the first part of Eq. (7–1), there comes

$$\mathbf{r} = (\mathbf{F}_{RR} - F_{Rx}\mathbf{F}_{xx}^{-1}F_{xR})(\mathbf{R} - \mathbf{R}_p) - \mathbf{F}_{Rx}\mathbf{F}_{xx}^{-1}\mathbf{\Delta}_0 + \mathbf{b}^T\mathbf{v}_0 \tag{7–4}$$

The flexibility matrix of the actual structure is the coefficient of $(\mathbf{R} - \mathbf{R}_p)$ in the preceding equation, or

$$\mathbf{F} = \mathbf{F}_{RR} - \mathbf{F}_{Rx}\mathbf{F}_{xx}^{-1}\mathbf{F}_{xR} \tag{7–5}$$

If the particular equilibrium state is chosen to be the fixed joint state $\mathbf{r} = 0$, then $\mathbf{R}_p$ becomes $\mathbf{R}_F$, and $\mathbf{v}_0$ and $\mathbf{\Delta}_0$ vanish. Equation (7–4) is thus equivalent to

$$\mathbf{r} = \mathbf{F}(\mathbf{R} - \mathbf{R}_F) \tag{7–6}$$

If the loading consists only of initial deformations $\mathbf{v}_I$, a convenient particular equilibrium state is the fundamental one in which $\mathbf{V} = 0$. In the absence of applied forces, $(\mathbf{R} - \mathbf{R}_p) = 0$, and $\mathbf{v}_0 = \mathbf{v}_I$. Equation (7–4) reduces then to

$$\mathbf{r} = (-\mathbf{F}_{Rx}\mathbf{F}_{xx}^{-1}\mathbf{c}^T + \mathbf{b}^T)\mathbf{v}_I \tag{7–7}$$

If support members are used in the formulation, Eq. (7–6) contains a subset equating $\mathbf{r}_s$ to the prescribed support displacements. This subset of equations is of no interest and may be deleted. The effective flexibility matrix applies then to the free joint displacements and corresponding joint loads. If there are no constraints on member deformations such as inextensibility conditions, elements of $\mathbf{r}$ are kinematically independent, and $\mathbf{F}$ is the inverse of the stiffness matrix, or

$$\mathbf{F} = \mathbf{K}^{-1} \tag{7–8}$$

A direct proof of Eq. (7–8) through matrix algebra is possible.

For inextensible frames and generally for structures with rigid inclusions, the forming of $\mathbf{F}_{xx}$ follows the general procedure and may in fact be simplified by zero flexibilities. As long as the primary structure is deformed by any nonzero $\mathbf{X}$, $\mathbf{F}_{xx}$ is nonsingular, and $\mathbf{F}$ may be formed by Eq. (7–5). However, if the elements of $\mathbf{r}$ are not kinematically independent, $\mathbf{F}$ is singular. In order to derive independent flexibility relations in this case, let the equations $\mathbf{r} = \mathbf{F}\mathbf{R}$ be partitioned in the form

$$\begin{Bmatrix} \mathbf{r}_1 \\ \mathbf{r}_2 \end{Bmatrix} = \begin{bmatrix} \mathbf{F}_{11} & \mathbf{F}_{12} \\ \mathbf{F}_{21} & \mathbf{F}_{22} \end{bmatrix} \begin{Bmatrix} \mathbf{R}_1 \\ \mathbf{R}_2 \end{Bmatrix}$$

and let $\mathbf{r}_2$ be the independent displacements. $\mathbf{r}_1$ is then expressed in terms of $\mathbf{r}_2$ in the form

$$\mathbf{r}_1 = \mathbf{T}_1 \mathbf{r}_2 \tag{7–9}$$

Thus

$$\mathbf{F}_{11} = \mathbf{T}_1 \mathbf{F}_{21} \tag{7–10a}$$

$$\mathbf{F}_{12} = \mathbf{T}_1 \mathbf{F}_{22} \tag{7–10b}$$

$$\mathbf{F}_{21} = \mathbf{F}_{12}^T = \mathbf{F}_{22} \mathbf{T}_1^T \tag{7–10c}$$

Substituting the last equation in the flexibility relation for $\mathbf{r}_2$, there comes

$$\mathbf{r}_2 = \mathbf{F}_{22}(\mathbf{R}_2 + \mathbf{T}_1^T \mathbf{R}_1) \tag{7–11}$$

The coefficient of $\mathbf{F}_{22}$ is the transformed generalized load. It is noted that it obeys the contragradient law associated with the kinematic transformation from $\mathbf{r}$ to $\mathbf{r}_2$.

8. PROGRAMMED SELECTION OF STATICAL REDUNDANTS

Consider a system of n simultaneous equations in n unknowns of the form

$$\mathbf{AX} = \mathbf{B} \tag{8–1}$$

where $\mathbf{A}$ is square nonsingular. The solution for $\mathbf{X}$ is

$$\mathbf{X} = \mathbf{A}^{-1}\mathbf{B} \tag{8–2}$$

If $\mathbf{B}$ is a rectangular matrix, $\mathbf{X}$ is a rectangular matrix of equal size. To each column of $\mathbf{B}$ corresponds a column of $\mathbf{X}$.

Methods for solving Eq. (8–1) do not necessarily form $\mathbf{A}^{-1}$. The Jordan–Gauss method transforms Eq. (8–1) into an equivalent system by a series of elementary operations each of which is a linear combination of two equations. At the ends of these steps $\mathbf{A}$ is transformed into an identity matrix, and $\mathbf{B}$ becomes a matrix $\mathbf{B}'$. The transformed system has the form $\mathbf{X} = \mathbf{B}'$. $\mathbf{B}'$ is the desired solution and is equal to $\mathbf{A}^{-1}\mathbf{B}$ but is obtained without forming $\mathbf{A}^{-1}$.

The Jordan–Gauss method may also be used to obtain $\mathbf{A}^{-1}$. It is only necessary for that purpose to start with $\mathbf{B} = \mathbf{I}$. The result will be $\mathbf{B}' = \mathbf{A}^{-1}\mathbf{B} = \mathbf{A}^{-1}$.

Before applying the Jordan–Gauss method to select statical redundants and to form the matrices $\mathbf{b}$ and $\mathbf{c}$, a description of its steps will be made. Refinements for numerical good conditioning are not presented, however. As the method proceeds, elements A_{ij} and B_{ij} of $\mathbf{A}$ and $\mathbf{B}$ have changing values, called current values at any one stage of the calculations. An elementary operation involves two equations, one of which is called a pivot equation and the other the transformed equation. The numerical operations are performed on the rows of **A** and **B** corresponding to these two equations. We define then the augmented matrix $\mathbf{A}' = [\mathbf{A} \quad \mathbf{B}]$ on which the operations are to be performed.

Let i be the pivot equation and j the transformed equation, and assume first that $A'_{ii} \neq 0$. Row i of $\mathbf{A}'$ is divided by A'_{ii}, resulting in a current pivot row in which $A'_{ii} = 1$. Row j is linearly combined with row i by substracting from it row i multiplied by A'_{ji}. Thus the following substitution is made: $A'_{jk} \leftarrow A'_{jk} - A'_{ji} A'_{ik}$, for k ranging over the columns of $\mathbf{A}'$. Current A'_{ji} is now zero. Keeping i fixed and varying j over all other rows of $\mathbf{A}'$, the current $\mathbf{A}'$ has $A'_{ii} = 1$ and zero in the remaining locations of the ith column.

At the start of the procedure $i = 1$. If $A'_{11} = 0$, a nonzero element in the first row of **A** is located, say in column c. This is always possible if **A** is nonsingular. Columns 1 and c of $\mathbf{A}'$ are interchanged, and the unknowns **X** are reordered by interchanging X_1 and X_c. Current A'_{11} is nonzero, and the method proceeds resulting in column 1 of $\mathbf{A}'$ having all zeros except $A'_{11} = 1$. In order to keep track of the correspondance between the current ordering of unknowns and the initial one, an array $L_1, L_2, \ldots, L_n$ is formed whose initial values are $L_j = j$, for $j = 1, 2, \ldots, n$. If columns 1 and c are interchanged when row 1 is the pivot row, values of L_1 and L_c are interchanged. At any stage of the calculations, the array $L_1, L_2, \ldots, L_n$ defines the current ordering of the unknowns in terms of the initial ordering. L_j is the initial sequential position of the unknown in current position j.

Pivot rows are taken sequentially for $i = 1, 2, \ldots, n$. Before the start of operations with pivot row i, all columns to the left of column i have zeros except on the main diagonal of **A** where there is 1.

If $A'_{ii} = 0$, a nonzero element to the right of A'_{ii} is located in the ith row of **A**, say in current column c. This is always possible if **A** is not singular. Column i and c and the values of L_i and L_c are, respectively, interchanged. Current A'_{ii} is nonzero, and the method proceeds resulting in column i of $\mathbf{A}'$ having all zeros except $A'_{ii} = 1$. At the end of the operations with the last pivot row, current **B** is the desired solution $\mathbf{B}'$, and current **A** is **I**.

In applying the Jordan–Gauss method to the joint equilibrium equations, the difference with the case previously considered is that $\mathbf{a}_v^T$ is rectangular. Whatever initial ordering of **V** is used, the joint equilibrium equations have the form

$$[\mathbf{a}_1^T \ \mathbf{a}_2^T]\begin{Bmatrix}\mathbf{V}_1\\ \mathbf{V}_2\end{Bmatrix} = \mathbf{R} - \mathbf{R}_p \tag{8–3}$$

where $\mathbf{a}_1^T$ is a square matrix. The Jordan–Gauss method is applied to the augmented matrix

$$\mathbf{A}' = [\mathbf{a}_1^T \ \mathbf{a}_2^T \ (\mathbf{R} - \mathbf{R}_p)] \tag{8–4}$$

but the interchange of columns may extend to all columns of $\mathbf{a}_v^T$, and the size of the array L_i is the number of elements (NV) in **V**. The procedure would fail only if the pivot equilibrium equation has all its coefficients zero, and this would occur only if the structure is kinematically unstable. Thus for a kinematically stable structure, the procedure is always successful. At the end of the operations current

$$\mathbf{A}' = [\mathbf{I} \ \ \mathbf{a}_2'^T \ \ \mathbf{R}'] \tag{8–5}$$

and the initial system of joint equilibrium equations is equivalent to

$$[\mathbf{I} \ \ \mathbf{a}_2'^T]\begin{Bmatrix}\mathbf{V}_1' \\ \mathbf{V}_2'\end{Bmatrix} = \mathbf{R}' \tag{8–6}$$

or

$$\mathbf{V}_1' = \mathbf{c}_1'\mathbf{V}_2' + \mathbf{V}_{1R}' \tag{8–7}$$

where $\mathbf{c}_1' = -\mathbf{a}_2'^T$ and $\mathbf{V}'_{1R} = \mathbf{R}'$. $\mathbf{V}_2'$ is the set of statical redundants selected by the method. To obtain the static matrix $\mathbf{b}_1'$ which expresses $\mathbf{V}'_{1R}$ in the form

$$\mathbf{V}'_{1R} = \mathbf{b}_1'(\mathbf{R} - \mathbf{R}_p) \tag{8–8}$$

the initial $\mathbf{A}'$ includes, instead of $(\mathbf{R} - \mathbf{R}_p)$, an identity matrix which is transformed by the Jordan–Gauss reduction into $\mathbf{b}_1'$.

Letting

$$\mathbf{V}_2' = \mathbf{X} \tag{8–9}$$

$$\mathbf{c}' = \begin{bmatrix}\mathbf{c}_1' \\ \mathbf{I}\end{bmatrix} \tag{8–10}$$

$$\mathbf{b}' = \begin{bmatrix}\mathbf{b}_1' \\ \mathbf{O}\end{bmatrix} \tag{8–11}$$

we can write

$$\mathbf{V}' = \mathbf{c}'\mathbf{X} + \mathbf{b}'(\mathbf{R} - \mathbf{R}_p) \tag{8–12}$$

To obtain the initial ordering of $\mathbf{V}$ such that

$$\mathbf{V} = \mathbf{c}\mathbf{X} + \mathbf{b}(\mathbf{R} - \mathbf{R}_p) \tag{8–13}$$

$\mathbf{c}$ and $\mathbf{b}$ are obtained by reordering the rows of $\mathbf{c}'$ and $\mathbf{b}'$ using the array L_i. The element in sequential position i in $\mathbf{V}'$ occupies sequential position L_i in $\mathbf{V}$. Thus rows i in $\mathbf{c}'$ and $\mathbf{b}'$ are placed in rows L_i of $\mathbf{c}$ and $\mathbf{b}$, respectively. If $\mathbf{c}'$ and $\mathbf{b}'$ are not reordered, it will be necessary to reorder the member flexibility matrix according to the ordering of $\mathbf{V}'$ and $\mathbf{v}'$. This ordering, however, does not usually correspond to the partitioning by members, and the computational efficiency that may be derived from this partitioning would be lost.

It is clear that the procedure for forming **b** and **c** with programmed selection of statical redundants is also applicable if any valid subset of statical redundants is chosen directly. The procedure is thus programmable with the provision for such a choice.

EXERCISES

Sections 1, 2

1. For the pin-jointed trusses of Fig. P1 do the following analyses:

a. Determine the member axial forces using Eq. (2.1–2) and forming the needed columns of **b** by statics.

b. Evaluate a typical joint displacement, assuming all members have the cross-sectional area $A = 10\,\text{in}^2$.

c. Evaluate two typical flexibility coefficients on and off the main diagonal, respectively.

d. Evaluate a typical joint displacement due to an initial deformation in one member of, say, 1 in or due to a support settlement.

2. Form the connectivity matrix $\mathbf{a}_v$ for any truss of Fig. P1, including support members, and use it as input to a computer program to obtain a complete solution for forces and joint displacements. Assume $A = 10\,\text{in}^2$ for all members, and obtain the flexibility matrix.

3. Write a program to form $\mathbf{a}_v$ for a pin-jointed truss by merging member matrices. The member input data consists then of the orientation angle and the joint numbers at the member ends. Order the joints and joint displacements sequentially, starting with 1. For a support member identify the supported end by the joint number 0.

4. The beams and frames of Fig. P4 are to be analyzed manually, neglecting axial and shear deformations. The beams in (*a*) and (*b*) are to be analyzed as two and three-member structures, respectively. For any of these structures obtain the following:

a. Draw the bending moment diagram due to the loading condition shown in the figure. In cases where there is a member load, outline the portion of the diagram that corresponds to the equivalent joint load $(\mathbf{R} - \mathbf{R}_p)$.

b. Evaluate a typical joint displacement, assuming all members are uniform and have the same EI, except where indicated.

c. Evaluate two typical flexibility coefficients.

d. Evaluate a typical joint displacement due to an initial curvature χ_I in one member. Do the same for a given initial axial deformation and for a given support settlement.

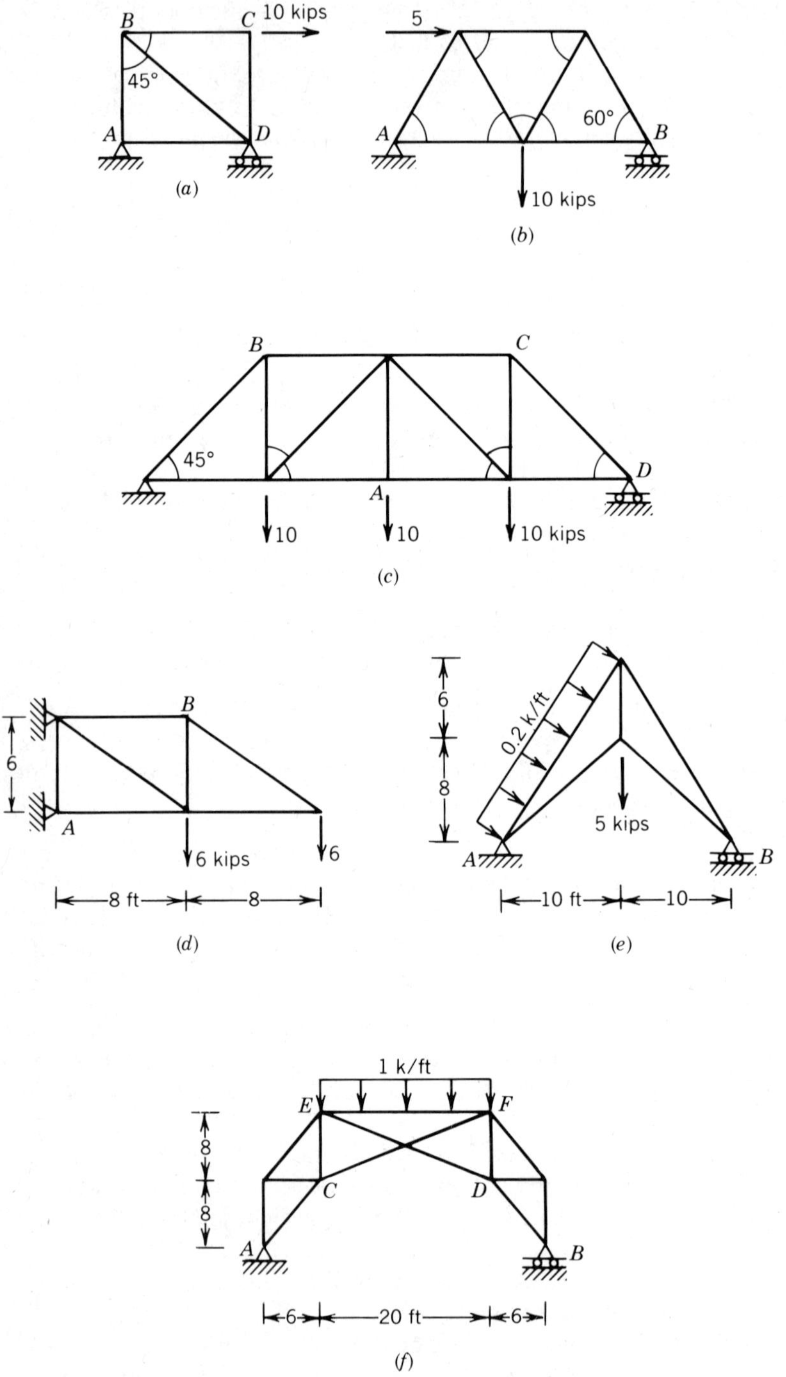
B
C
10 kips
45°
A
D
(a)
5
60°
A
B
10 kips
(b)
B
C
45°
D
A
10
10
10 kips
(c)
B
6
A
6 kips
6
8 ft
8
(d)
6
8
0.2 k/ft
5 kips
A
B
10 ft
10
(e)
1 k/ft
E
F
8
8
C
D
A
B
6
20 ft
6
(f)

FIG. P1

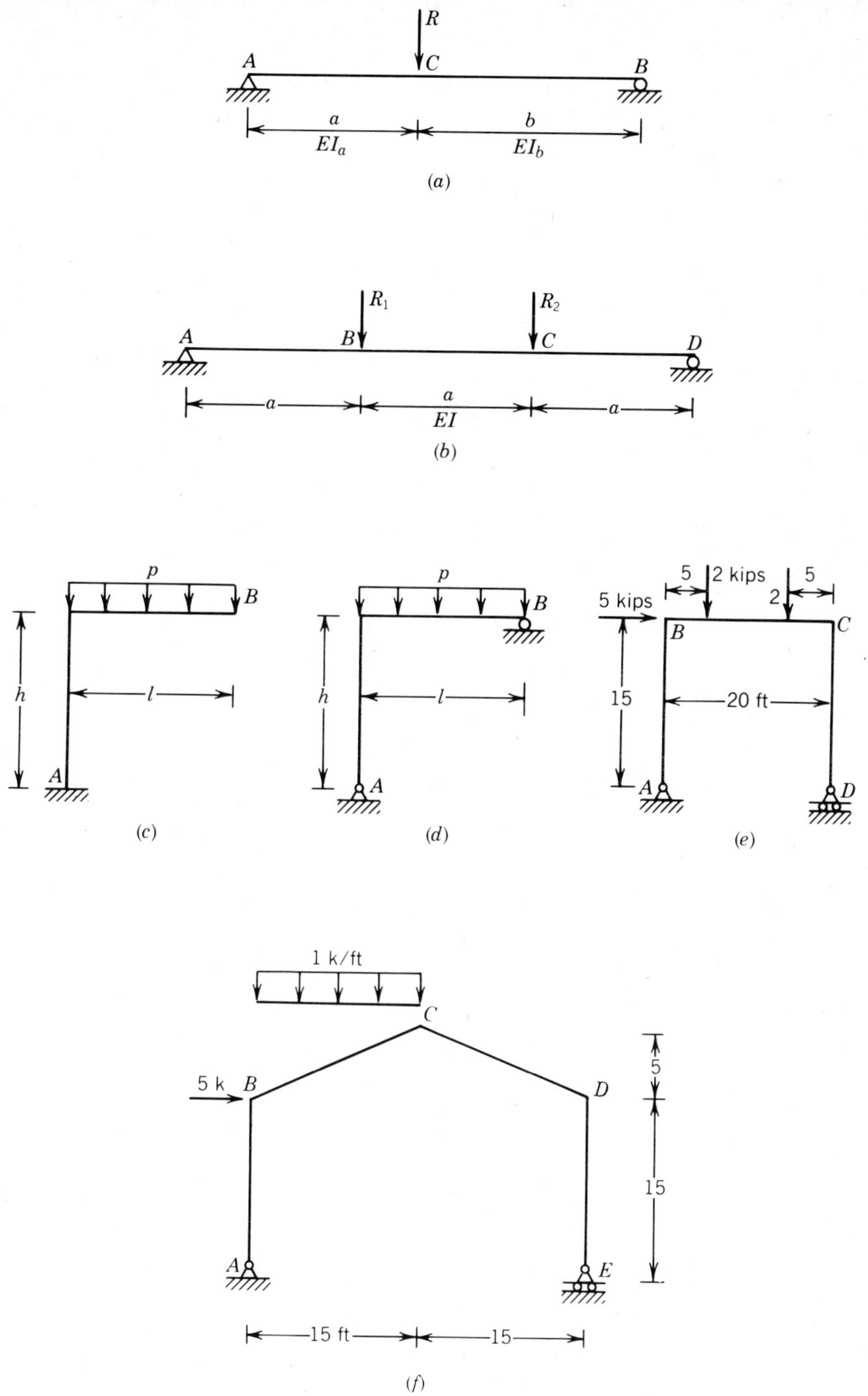

FIG. P4 (*Continued on following page.*)

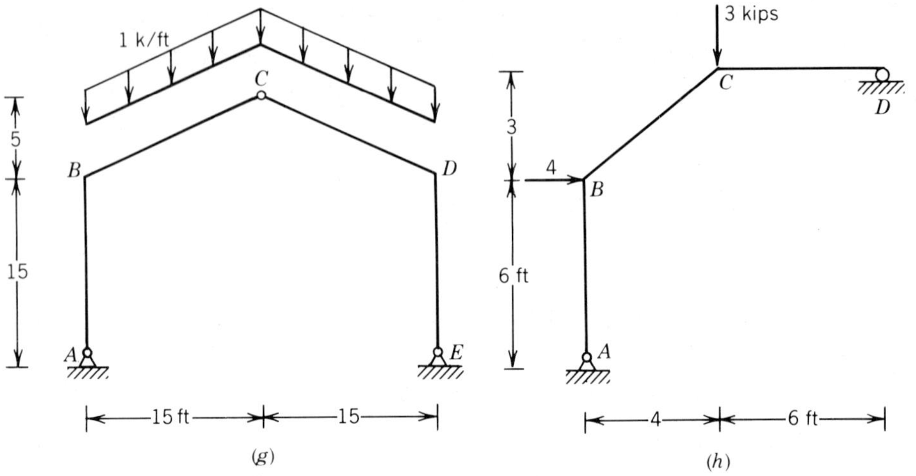

FIG. P4 *(continued)*

5. Write a computer program that assembles the connectivity matrix $\mathbf{a}_v$ for a rigid frame from member matrices $\bar{\mathbf{B}}^{mT}$. The input data consist of the member length, the orientation angle θ, and the joint numbers at the member ends. Include the option of choosing for any member the cantilever or the alternate type of deformations.

Sections 3 to 5

6. The pin-jointed trusses of Fig. P1 are made statically indeterminate as indicated here. For any of these trusses form **c** by statics, and determine the statical redundants and the member axial forces. Determine typical joint displacements and flexibility coefficients. Assume $A = 10\,\text{in}^2$ for all members.
   a. Add a member *AC* in Fig. P1*a*.
   b. Make *B* a pin support in Fig. P1*b*.
   c. Add a member *AB* in Fig. P1*d*.
   d. Add a member *AB*, and make *B* a pin support in Fig. P1*e*.
   e. Add a member *CD*, and make *B* a pin support in Fig. P1*f*.

7. Analyze the structure of Exercise 6 part (e) for a thermal loading producing in member *EF* a free thermal elongation of 0.1 in.

8. Repeat Exercise 6 by choosing in each case a different primary structure.

9. The structures of Exercise 4 are made statically indeterminate as indicated here. Analyze any of these structures manually by the flexibility method,

using bending moment diagrams. Show on sketches the deformations at the releases of the primary structure. Draw the resulting bending moment diagram for the loading condition shown in Fig. P4. Determine typical joint displacements and flexibility coefficients.

a. Fix the beam at *A* in Fig. P4*a*.
b. Fix the beam at *A* and *D* in Fig. P4*b*.
c. Introduce a pin support at *B* in Fig. P4*c*.
d. Make *B* a pin support in Fig. P4*d*.
e. Make *D* a pin support in Fig. P4*e*.
f. Make *E* a pin support in Fig. P4*f*.
g. Make *C* a rigid joint, and fix member ends at *A* and *E* in Fig. P4*g*.
h. Make *D* a pin support in Fig. P4*h*.

10. Repeat the analysis of any of the structures in Exercise 9 by choosing a different primary structure.

11. What is the result of the inextensional approximation if the frame in Fig. P4*h* is pinned at *D*?

12. Write a program to form **b** and **c** using as input $\mathbf{a}_v$ and a certain selection of statical redundants. Complete the program to carry out the analysis by the flexibility method.

Section 6

13. Analyze the continuous beam of Fig. P13 by the three-moment equation.

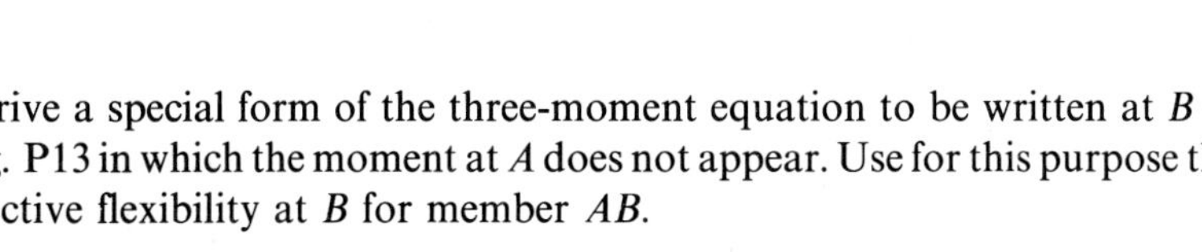

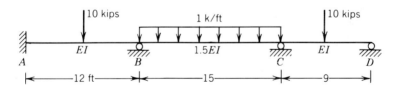

FIG. P13

14. Derive a special form of the three-moment equation to be written at *B* in Fig. P13 in which the moment at *A* does not appear. Use for this purpose the effective flexibility at *B* for member *AB*.

15. a. Consider a structure whose joints have only rotational displacements. Derive as a condition of rotation continuity a four-moment equation for points such as *A*, *B*, *C*, *D* in Fig. P15*a*.

b. Apply this equation to the analysis of the structure in Fig. P15*b*. Take advantage of symmetry and of effective column flexibility to reduce the number of simultaneous equations.

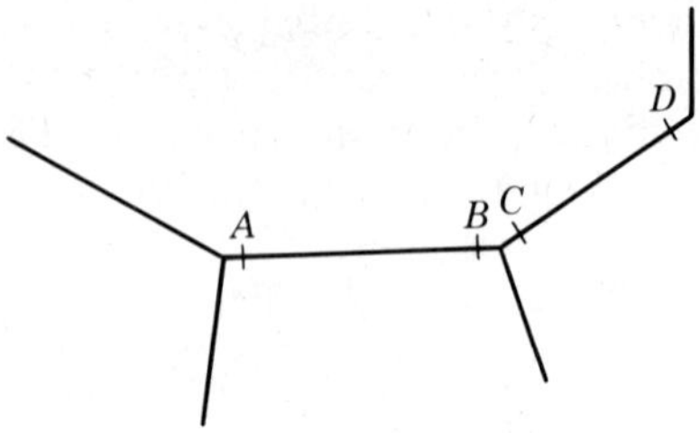

(a)

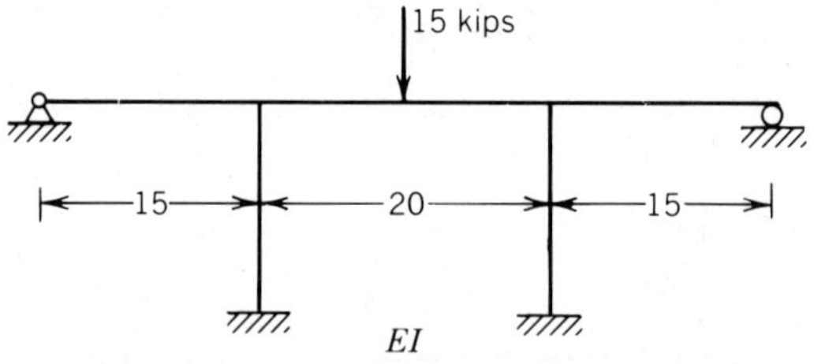

(b)

FIG. P15

16. Formulate the flexibility method for a structure having a plane of symmetry as done in the stiffness method in Chapter 10, Section 5. Apply the formulation to the analysis of a symmetric continuous beam.

12

NONLINEAR ANALYSIS

1. INTRODUCTION

This chapter outlines methods of solution of the governing equations of nonlinear theories established in Chapter 8 and discusses nonlinear behavior through examples. Some of these include material nonlinearity in the form of plastic hinges. Continuous beams are a relatively simple type of structure and are treated first, by both the displacement and the force method. The displacement method in second-order theory is then developed for general frames and is applied to particular types of structures. This is followed by a similar treatment of the force method. The general geometric nonlinearity due to large rotations of member-bound axes is treated by the stiffness method within the context of incremental and iterative methods in Sections 4.2 and 4.3, and by the force method in Section 12.

2. CONTINUOUS BEAMS: STIFFNESS METHOD

The stiffness equations of continuous beams, formulated in Chapter 10, Section 2, within the context of the linear theory, are applicable in the presence of axial forces if member stiffness matrices of beam-column theory are used. Frames without joint translations are treated similarly.

For uniform spans, each having a constant axial force, the general stiffness equation is Eq. (10/2–5), or

$$k'_{i-1}r_{i-1} + (k_{i-1} + k_i)r_i + k'_i r_{i+1} = R_{iT} \tag{2–1}$$

The stiffness coefficients for a general span, neglecting shear deformation, are

$$k = \frac{EI}{2l} \frac{\lambda(1 - \lambda \cot \lambda)}{\tan(\lambda/2) - (\lambda/2)} \tag{2–2a}$$

$$k' = \frac{EI}{2l} \frac{\lambda[(\lambda/\sin\lambda) - 1]}{\tan(\lambda/2) - (\lambda/2)} \tag{2–2b}$$

Subscripts in k and k' in Eq. (2–1) refer to spans, and subscripts in r and R refer to joints, Fig. 10/2–1. R_{iT} is the equivalent moment load at joint i. Formulas for fixed end moments needed to evaluate R_{iT} are derived in Chapter 6, Section 2.3.

Example 1

For the continuous beam of Fig. 2–1 we desire to evaluate the bending moments at the member ends and at the load point.

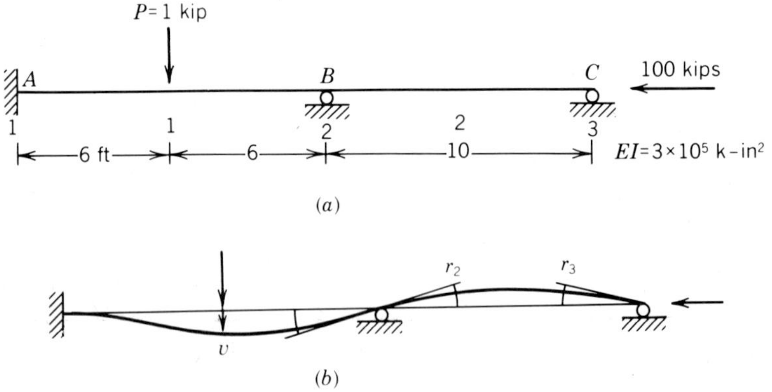

FIG. 2–1. Continuous beam, Section 2, Example 1; Section 3, Example 1. (*Continued on next page.*)

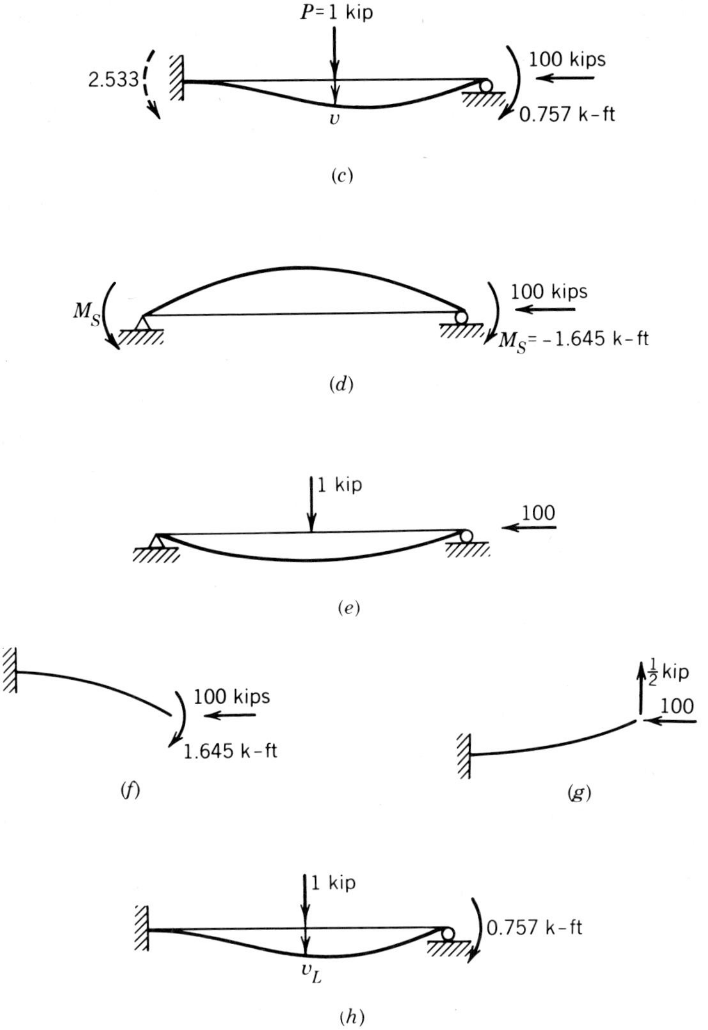

FIG. 2–1. *(continued)*

For member 1, or AB,

$$\lambda = \left(-\frac{V_1 l^2}{EI} \right)^{1/2} = \left(\frac{100(12 \times 12)^2}{3 \times 10^5} \right)^{1/2} = 2.629$$

the fixed end moments are

$$V_{2F} = -V_{3F} = \frac{Pl}{2\lambda} \tan\frac{\lambda}{4} = 1.761 \text{ k-ft}$$

and the stiffness coefficients, using Eqs. (2–2), are found to be

$$k_1 = 0.2484EI$$
$$k_1' = 0.1910EI$$

For member 2, or BC,

$$\lambda = 2.629 \times \frac{10}{12} = 2.191$$
$$k_1 = 0.3315EI$$
$$k_1' = 0.2187EI$$

From the fixed end moment of member 1 at joint 2, the equivalent joint load is found to be

$$R_{2T} = 1.761 \text{ k-ft}$$
$$R_{3T} = 0$$

Equation (2–1) is written at joint 2, with $i = 2$ and $r_1 = 0$. At joint 3, $i = 3$ and $k_3 = k_3' = 0$. There comes

$$0.5799r_2 + 0.2187r_3 = \frac{1.761}{EI}$$
$$0.2187r_2 + 0.3315r_3 = 0$$

The solution of these equations is

$$EIr_2 = 4.0425$$
$$EIr_3 = -2.6670$$

For member 1

$$\begin{Bmatrix} V_2 \\ V_3 \end{Bmatrix} = \begin{bmatrix} k_1 & k_1' \\ k_1' & k_1 \end{bmatrix} \begin{Bmatrix} 0 \\ r_2 \end{Bmatrix} + \begin{Bmatrix} V_{2F} \\ V_{3F} \end{Bmatrix} = \begin{Bmatrix} 2.533 \\ -0.757 \end{Bmatrix} \text{k-ft}$$

The stiffness equations for member 2 provide a check on the numerical calculations

$$\begin{Bmatrix} V_2 \\ V_3 \end{Bmatrix} = \begin{bmatrix} k_1 & k_1' \\ k_1' & k_1 \end{bmatrix} \begin{Bmatrix} r_2 \\ r_3 \end{Bmatrix} = \begin{Bmatrix} 0.757 \\ 0 \end{Bmatrix}$$

The deflected shape is shown in Fig. 2–1b, and member 1 is shown isolated in (c).

The bending moment at midspan of member 1 is obtained by adding $100v$ to the bending moment in the linear theory. This is

$$M = \frac{Pl}{4} + \frac{V_3 - V_2}{2} + 100v = 1.355 + 100v \text{ k-ft}$$

v will be determined by exact formulas before we illustrate an approximate evaluation using the magnification factor method.

The end moments of member 1 may be separated into a symmetrical part,

$$M_s = \frac{V_3 - V_2}{2} = -1.645 \text{ k-ft}$$

and an antisymmetric part that causes no deflection at midspan. M_s is shown in (*d*). The deflection v is obtained by superposition of the deflections in (*d*) and (*e*). These in turn may be evaluated using the flexibility matrix of the cantilever type for half the member, as shown in (*f*) and (*g*). For this cantilever

$$\lambda = 2.629 \times \tfrac{1}{2} = 1.3145$$

and the flexibility coefficients f_{22} and f_{23}, using Eq. (6/2.2–2) with $\alpha = 1$, are found to be

$$f_{22} = \frac{l^3}{EI}\frac{1}{\lambda^2}\left(\frac{\tan\lambda}{\lambda} - 1\right) = \frac{1}{EI}237.881$$

$$f_{23} = \frac{l^2}{EI}\frac{1 - \cos\lambda}{\lambda^2 \cos\lambda} = \frac{1}{EI}61.353$$

Then,

$$v = \left(\frac{1}{2}\right)f_{22} - 1.645 f_{23} = \frac{18.015}{EI} = 0.00865 \text{ ft}$$

and

$$M = 1.355 + 100v = 2.22 \text{ k-ft}$$

An approximate evaluation of v may be made using the formula

$$v = \frac{v_L}{1 - (V_1/V_{1,\text{cr}})}$$

in which v_L is the deflection in the linear theory, with the member as shown in (*h*), and $(V_1)_{\text{cr}}$ is the critical axial force of the proped cantilever. v_L is obtained through the formula,

$$v_L = \frac{7Pl^3}{768EI} - \frac{Ml^2}{32EI} = 0.005925\,\text{ft}$$

and $(V_1)_{cr}$ through Euler's formula with an effective length of $0.7l$, or

$$(V_1)_{cr} = \frac{\pi^2 EI}{(0.7l)^2} = 291.4$$

then

$$v = \frac{0.005925}{1-(100/291.4)} = 0.00902$$

and

$$M = 1.355 + 100v = 2.26\,\text{k-ft}$$

3. CONTINUOUS BEAMS: FLEXIBILITY METHOD

The three-moment equation derived in Chapter 11, Section 6, remains applicable in beam-column theory, provided the corresponding member flexibility equations are used. It applies to continuous beams and also to frames with or without joint translations, as will be seen through examples.

For a uniform member having an axial force V_1, the flexibility coefficients are obtained from Eqs. (6/2.1–8). For a general uniform member ab,

$$f_{aa} = f_{bb} = f = \frac{l}{EI\lambda^2}(1 - \lambda \cot \lambda) \tag{3–1a}$$

$$f_{ab} = f_{ba} = f' = \frac{l}{EI\lambda^2}\left(\frac{\lambda}{\sin \lambda} - 1\right) \tag{3–1b}$$

In these equations shear deformation is neglected by setting $\alpha = 1$, and the sign of f' is changed from that of f_{23} and f_{32} in order to conform to the sign convention adopted in Chapter 11, Section 6. The three-moment equation has the form

$$f'_{i-1}X_{i-1} + (f_{i-1} + f_i)X_i + f'_iX_{i+1} + \Delta_{i,0} = 0 \tag{3–2}$$

For a continuous beam without support settlement, $\Delta_{i,0}$ is defined in Eq. (11/6–4b). If support settlements take place causing member chord rotations ρ_{i-1} and ρ_i, as shown in Fig. 11/6–1c, then $\Delta_{i,0}$ includes the term $(\rho_{i-1} - \rho_i)$.

The derivation of the three-moment equation for a continuous beam remains

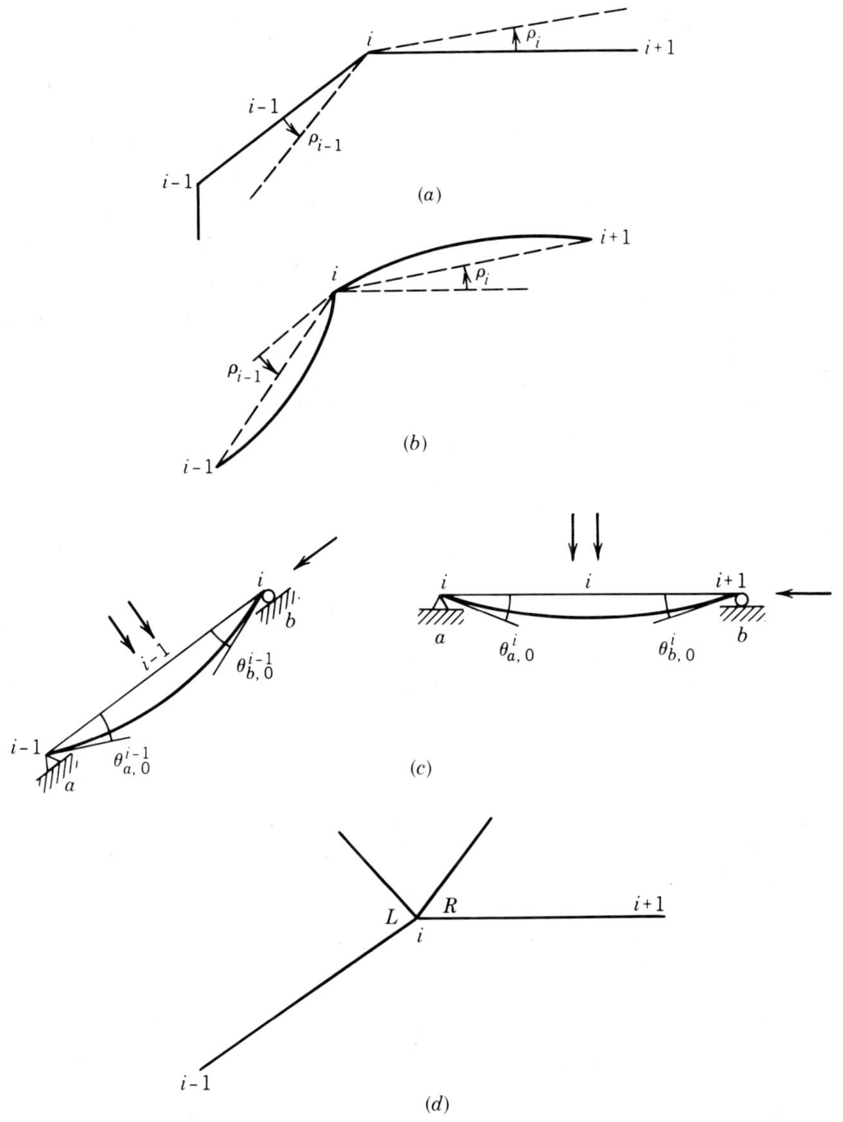

FIG. 3–1. Notation for three- and four-moment equations.

as is, for any pair of members connected rigidly at joint i as shown in Fig. 3–1a, and may be applied to inextensional frame analysis. If there are joint translations, as from (a) to (b), the general formula for $\Delta_{i,0}$ is

$$\Delta_{i,0} = \theta^{i-1}_{b,0} + \theta^{i}_{a,0} + \rho_{i-1} - \rho_i \tag{3–3}$$

$\theta_{a,0}$ and $\theta_{b,0}$, for a general member ab, are the rotations at ends a and b of the simply supported member caused by the member loading, including the axial force. A superscript identifies the member.

A further generalization deals with two members, as shown in Fig. 3–1*d*, where other members are connected at joint *i*. Letting X_i^L and X_i^R be the bending moments to the left and to the right of joint *i*, respectively, compatibility of rotation of the two members at joint *i* yields the four-moment equation.

$$f'_{i-1}X_{i-1} + f_{i-1}X_i^L + f_iX_i^R + f'_iX_{i+1} + \Delta_{i,0} = 0 \qquad (3\text{–}4)$$

Example 1

Consider the continuous beam problem treated by the stiffness method in Section 2, Example 1, and shown in Fig. 2–1. Calculations carried out there give, for member 1, $\lambda = 2.629$. Then from Eqs. 3–1, with $l = 12\,\text{ft}$,

$$EI\,f_1 = 9.84696$$

$$EI\,f'_1 = 7.57072$$

The 1 kip load at midspan of member 1 causes $\theta_{a,0}^1$ and $\theta_{b,0}^1$ which are obtained from Fig. 6/2.3–2, as

$$\theta_{a,0}^1 = \theta_{b,0}^1 = \frac{Pl^2}{2EI\lambda^2}\left[\frac{1}{\cos(\lambda/2)} - 1\right] \qquad (3\text{–}5)$$

With $P = 1\,\text{kip}$, and $l = 12\,\text{ft}$, obtain

$$EI\theta_{a,0}^1 = EI\theta_{b,0}^1 = 30.6764$$

For member 2, $\lambda = 2.191$. With $l = 10\,\text{ft}$, Eqs. (3–1) yield

$$EI\,f_2 = 5.34290$$

$$EI\,f'_2 = 3.52556$$

The three-moment equation is now applied at the fixed end and at the intermediate support. The fixed end condition at joint 1 may be represented by a dummy span, $i = 0$, having zero flexibilities, $f_0 = f'_0 = 0$. With $i = 1$, then $i = 2$, and with $X_3 = 0$, the three-moment equation yields

$$9.84696X_1 + 7.57072X_2 + 30.6764 = 0$$

$$7.57072X_1 + 15.18986X_2 + 30.6764 = 0$$

The solution is found by Gauss elimination to be

$$X_1 = -2.533\,\text{k-ft}$$

$$X_2 = -0.757\,\text{k-ft}$$

This coincides with the solution found by the stiffness method in Section 2.

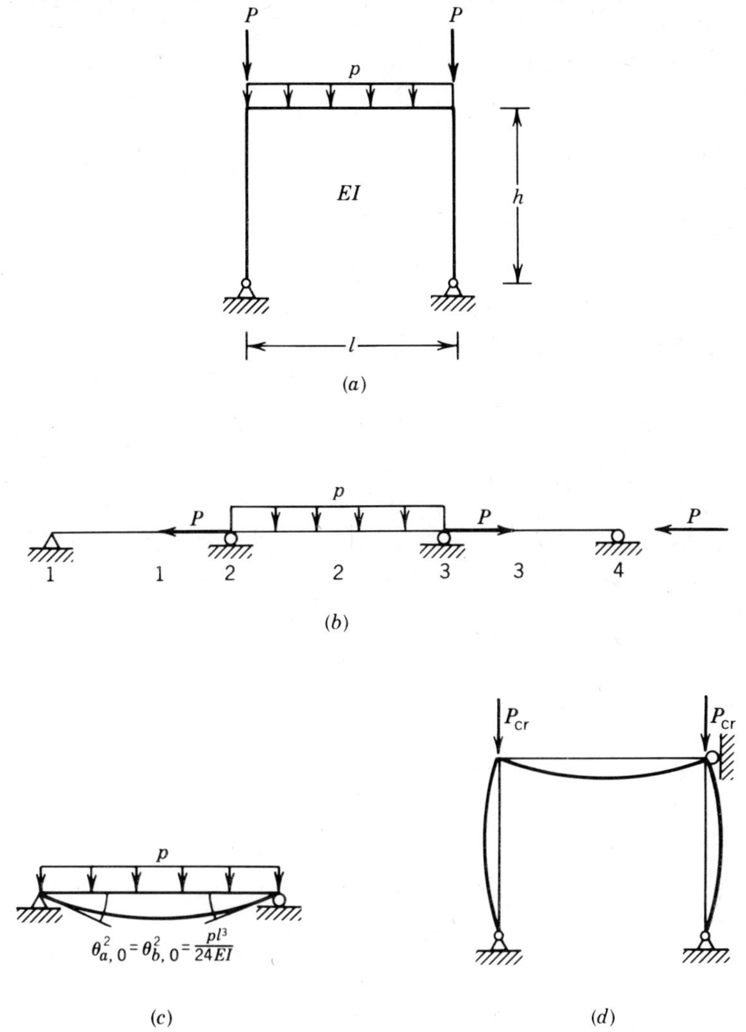

FIG. 3–2. Frame with sidesway, Example 2.

Example 2

The symmetric frame of Fig. 3–2*a* has no member chord rotation and may be analyzed as the continuous beam model shown in (*b*). For usual loadings the axial force in the beam of the frame has a negligible effect on the flexibility coefficients and is not represented in (*b*). Because of symmetry, $X_2 = X_3$. With $X_1 = 0$, the three-moment equation written at joint 2 yields

$$(f_1 + f_2 + f'_2)X_2 + \Delta_{2,0} = 0 \qquad (3\text{–}6)$$

where, from Fig. 3–2*c*,

$$\Delta_{2,0} = \frac{pl^3}{24EI}$$

f_1 is obtained through Eq. (3–1a) in which $l = h$, and

$$\lambda^2 = \frac{Ph^2}{EI} \tag{3–7}$$

For member 2 the linear theory applies, so that

$$f_2 + f'_2 = \frac{l}{2EI} \tag{3–8}$$

Equation (3–6) is readily solved for X_2.

Bifurcation buckling of the frame, as deduced from this analysis, would apply if lateral displacement is prevented as in (*d*). The condition is that the coefficient of X_2 in Eq. (3–6) vanishes, or

$$1 - \lambda \cot \lambda + \frac{\lambda^2}{2}\frac{l}{h} = 0 \tag{3–9}$$

For $l = h$, the lowest root is found to be

$$\lambda_{cr} = 3.591 \tag{3–10a}$$

The effective length of the column is

$$Kl = \frac{\pi}{\lambda} l = 0.875l \tag{3–10b}$$

Example 3

For the frame of Fig. 3–3, the load P is assumed to be large in comparison with R, so that the axial forces in the columns may be assumed equal to P in evaluating the flexibility coefficients. Assuming inextensional behavior, the deformation of the frame is antisymmetric, so that $X_2 = -X_3$, and $\rho_2 = 0$. In the three-moment equation written at joint 2, $X_1 = 0$, and

$$\Delta_{2,0} = \rho_1 = -\frac{r}{l} \tag{3–11}$$

$$(f_1 + f_2 - f'_2)X_2 - \frac{r}{h} = 0 \tag{3–12a}$$

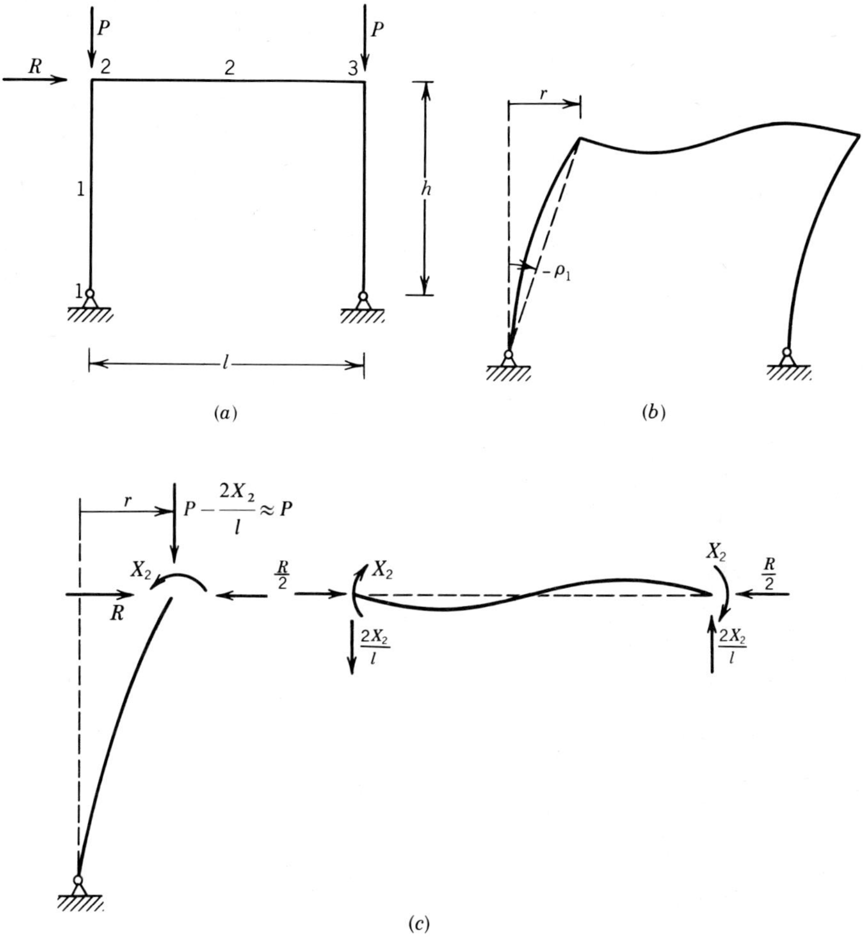

FIG. 3–3. Frame without sidesway, Example 3.

A second equation is provided by equilibrium of forces in (*c*). Moment equilibrium of the column about the pin support yields

$$X_2 - Pr - \frac{R}{2}h = 0 \tag{3–12b}$$

Solving Eqs. (3–12) for X_2 and r, there comes

$$X_2 = \frac{1}{2}\frac{Rh}{1 - Ph(f_1 + f_2 - f_2')} \tag{3–13a}$$

$$r = \frac{1}{2}\frac{Rh^2(f_1 + f_2 - f_2')}{1 - Ph(f_1 + f_2 - f_2')} \tag{3–13b}$$

In the preceding results,

$$f_1 + f_2 - f_2' = \frac{h}{EI\lambda^2}(1 - \lambda \cot \lambda) + \frac{l}{6EI} \tag{3–14}$$

The condition for bifurcation buckling under the load P, with $R = 0$, is that the denominator in the solution for r and X_2 vanishes. With $P = \lambda^2 EI/h^2$, there comes

$$\frac{\cot \lambda}{\lambda} - \frac{1}{6}\frac{l}{h} = 0 \tag{3–15}$$

For $l = h$, the smallest root is found by trial and error to be

$$\lambda_{cr} = 1.350 \tag{3–16a}$$

The effective column length is

$$Kl = \frac{\pi}{\lambda} l = 2.33l \tag{3–16b}$$

4. STIFFNESS METHOD

4.1. Finite Iterative Solution: Second-Order Theory

Governing equations based on member second-order theory have been formulated in Chapter 8, Sections 5.2 and 5.3. Equations (8/5.2–1) are in member local axes, and Eqs. (8/5.3–8) are in member-bound axes. Either set of equations may be used to derive the structure stiffness equations, by expressing the joint equilibrium equations in terms of the joint displacements. The first set yields

$$(\mathbf{K}_a + \mathbf{K}_b)\mathbf{r} + \mathbf{a}_1^T \mathbf{S}_{1N} = \mathbf{R} - \mathbf{R}_F \tag{4.1–1}$$

where

$$\mathbf{K}_a = \mathbf{a}_1^T \mathbf{k}_{11} \mathbf{a}_1 \tag{4.1–2a}$$

$$\mathbf{K}_b = \mathbf{a}^T \mathbf{k}_s \mathbf{a} \tag{4.1–2b}$$

$$\mathbf{R}_F = \mathbf{a}_1^T \mathbf{S}_{1F} + \mathbf{a}^T \mathbf{S}_F \tag{4.1–2c}$$

$$\mathbf{S}_{1N} = \{\mathbf{S}_{1N}^m\} = \{\tfrac{1}{2} k_{11}^m \mathbf{B}_1^m \mathbf{s}^{mT} \mathbf{g}_s^m \mathbf{s}^m\} \tag{4.1–2d}$$

$$\mathbf{B}_1^m = \{-1 \quad 1\} \tag{4.1–2e}$$

In these equations the notation separating member axial displacements $\mathbf{s}_1$ from transverse displacements $\mathbf{s}$ is used. $\mathbf{a}_1$ and $\mathbf{a}$ are the corresponding connectivity matrices and are the same as in the linear theory. $\mathbf{K}_a$ and $\mathbf{K}_b$ may thus be formed, respectively, by processing member axial stiffness matrices $\mathbf{k}_{11}^m$ and transverse stiffness matrices $\mathbf{k}_s^m$ by the direct stiffness method. The same applies to forming $\mathbf{a}_1^T\mathbf{S}_{1N}$ from $\mathbf{S}_{1N}^m$, and $\mathbf{R}_F$ from $\mathbf{S}_{1F}^m$ and $\mathbf{S}_F^m$. The member matrices involved are described in detail in Chapter 6, Sections 2.6 and 3.1. Nonlinearity in Eq. (4.1–1) appears in $\mathbf{S}_{1N}$ which is quadratic in $\mathbf{s}$, and in $\mathbf{K}_b$ which depends on the member axial forces. In exact relations, $\mathbf{g}_s$ and $\mathbf{R}_F$ also depend on the axial forces.

Equation (4.1–1) may be solved iteratively by placing the term $\mathbf{a}_1^T\mathbf{S}_{1N}$ on the right-hand side and starting with an approximate state $\mathbf{r}_0$ and approximate axial forces to evaluate $\mathbf{k}_s$, $\mathbf{S}_{1N}$, and $\mathbf{R}_F$. The solution for $\mathbf{r}$ is then used to reevaluate the right-hand side, and the process is repeated. Convergence tests may be performed on the displacements, or on the forces $(\mathbf{a}_1^T\mathbf{S}_N + \mathbf{R}_F)$, of two successive cycles. A convenient starting state is $\mathbf{r}_0 = 0$ whose stiffness equations, in the absence of initial axial forces, are those of the linear theory. Convergence usually occurs if the equilibrium state is not close to instability and is not on an unstable branch of the deformation path. The method described here is referred to as the finite iterative method. An incremental iterative method, known as the Newton–Raphson method, which has more reliable convergence properties is presented subsequently.

Example 1

The governing equations of the portal frame shown in Fig. 4.1–2*a* are now derived as an example. To illustrate the general equations, column extensibility is included, although it will be seen in Section 6 that an accurate simplified analysis is possible on the basis of inextensible members. The beam is assumed inextensible. The displacements $\mathbf{r} = \{r_1, r_2, \ldots, r_5\}$ are identified in (*b*). Member end displacements are ordered as shown in (*c*) and (*d*). The effect of the axial force on the beam stiffness matrix is neglected. It will be convenient to make all elements of $\mathbf{r}$ have the physical dimension of length, by multiplying the rotations by l, and to make all elements of $\mathbf{R}$ have the dimension of force by dividing monents by l. The beam stiffness matrix is then

$$\mathbf{k}_c^m = \frac{EI'}{l^3}\begin{bmatrix} 12 & & \text{symmetric} & \\ 6 & 4 & & \\ -12 & -6 & 12 & \\ 6 & 2 & -6 & 4 \end{bmatrix} \tag{4.1–3}$$

For a column the effective bending stiffness matrix $\mathbf{k}_s^m$ is also the reduced stiffness of the cantilever type, Eq. (6/2.2–3). With $\alpha = 1$, and adjustments for physical dimensions and sign convention, we obtain

$$\mathbf{k}_s^m = \frac{EI}{l^3}\,\frac{\lambda^2}{2[\tan(\lambda/2) - (\lambda/2)]}\begin{bmatrix} \lambda & \tan\dfrac{\lambda}{2} \\ \tan\dfrac{\lambda}{2} & \dfrac{\lambda}{\lambda}\,\dfrac{1}{\lambda} - \dfrac{1}{\tan\lambda} \end{bmatrix} \tag{4.1–4}$$

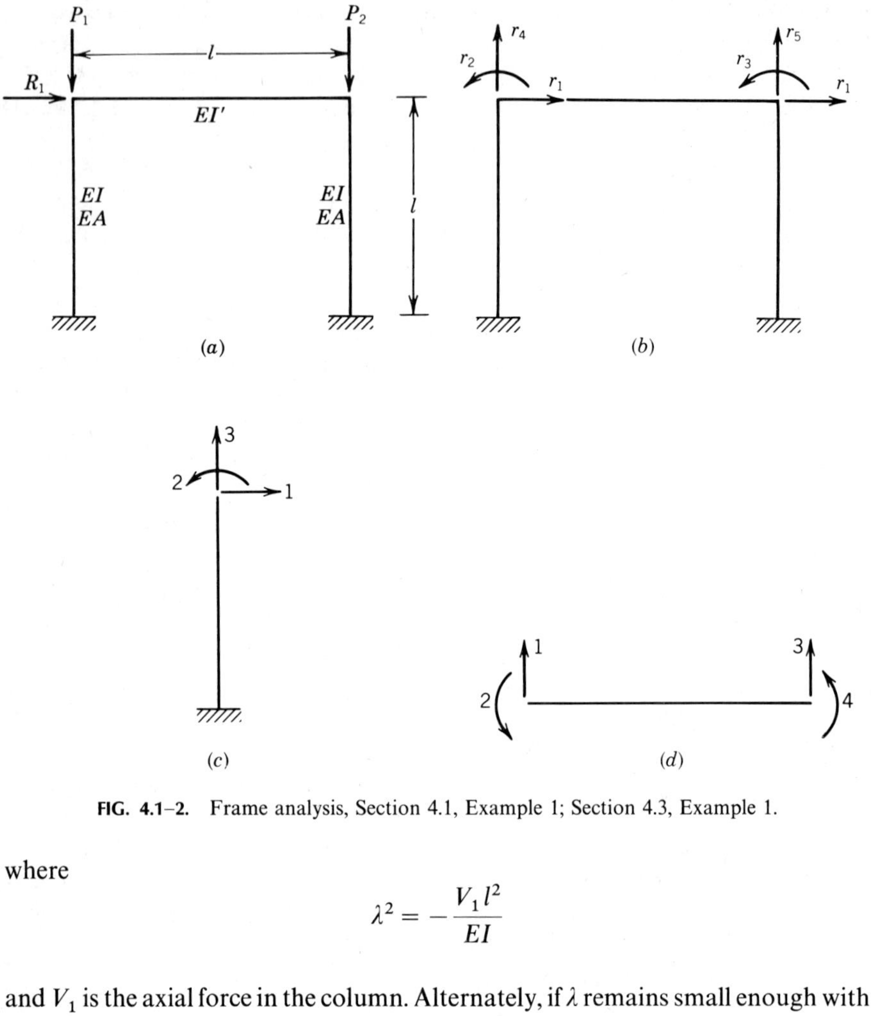

FIG. 4.1–2. Frame analysis, Section 4.1, Example 1; Section 4.3, Example 1.

where

$$\lambda^2 = -\frac{V_1 l^2}{EI}$$

and V_1 is the axial force in the column. Alternately, if λ remains small enough with respect to $\pi/2$, the approximate $\mathbf{k}_s$ of Eq. (6/3.1–28) may be used. This is

$$\mathbf{k}_s^m = \frac{EI}{l^3}\begin{bmatrix} 12 - \frac{6}{5}\lambda^2 & 6 - \frac{\lambda^2}{10} \\ 6 - \frac{\lambda^2}{10} & 4 - \frac{2\lambda^2}{15} \end{bmatrix} \qquad (4.1\text{–}5)$$

The terms in λ^2 in Eq. (4.1–5) are $V_1 \mathbf{g}_s^m$. $\mathbf{g}_s^m$ is thus extracted as

$$\mathbf{g}_s^m = \frac{1}{l}\begin{bmatrix} \frac{6}{5} & \frac{1}{10} \\ \frac{1}{10} & \frac{2}{15} \end{bmatrix} \qquad (4.1\text{–}6)$$

The assembled bending stiffness $\mathbf{K}_b$ is formed by merging the member stiffness matrices obtained here. The merging sequences for the beam, the left and right columns are, respectively,

$$4, 2, 5, 3$$
$$1, 2$$
$$1, 3$$

$\mathbf{K}_a$ is formed by merging EA/l on the main diagonal in positions 4 and 5.

For the loading shown in Fig. 4.1–2*a*,

$$\mathbf{R} = \{R_1 \;\; 0 \;\; 0 \;\; -P_1 \;\; -P_2\}$$

and $\mathbf{R}_F$ is zero. Equation (4.1–1) may be written in the form

$$\mathbf{Kr} = \mathbf{R} - \mathbf{R}_N \tag{4.1–7}$$

where

$$\mathbf{K} = \mathbf{K}_a + \mathbf{K}_b$$
$$\mathbf{R}_N = \mathbf{a}_1^T \mathbf{S}_{1N}$$

If the iteration procedure is started with zero displacements and zero axial forces, $\mathbf{k}_s^m$ is the stiffness matrix in the linear theory, and is obtained by letting $\lambda = 0$ in Eq. (4.1–5). In state $\mathbf{r} = 0$, $\mathbf{K}$ is obtained as

$$(K)_{r=0} = \frac{EI}{l^3}\begin{bmatrix} 24 & 6 & 6 & 0 & 0 \\ 6 & 4+4\frac{I'}{I} & 2\frac{I'}{I} & 6\frac{I'}{I} & -6\frac{I'}{I} \\ 6 & 2\frac{I'}{I} & 4+4\frac{I'}{I} & 6\frac{I'}{I} & -6\frac{I'}{I} \\ 0 & 6\frac{I'}{I} & 6\frac{I'}{I} & 12\frac{I'}{I}+\frac{Al^2}{I} & -12\frac{I'}{I} \\ 0 & -6\frac{I'}{I} & -6\frac{I'}{I} & -12\frac{I'}{I} & 12\frac{I'}{I}+\frac{Al^2}{I} \end{bmatrix}$$

and $\mathbf{S}_{1N}$ and $\mathbf{R}_N$ are zero. In the first cycle Eq. (4.1–7) reduces to the stiffness equation of the linear theory. Its solution for $\mathbf{r}$ is used to evaluate the column axial forces and $\mathbf{R}_N$, for use in the next iteration cycle. To do this, the effective displacements of the left column are $\mathbf{s}_1^L = \{r_4\}$ and $\mathbf{s}^L = \{r_1 \;\; r_2\}$. For the right column, $\mathbf{s}_1^R = \{r_5\}$, and $\mathbf{s}^R = \{r_1 \;\; r_3\}$. S_{1N} is evaluated using Eq. (4.1–2d) in which the effective $\mathbf{B}_1^m$ is $\{1\}$. Thus

$$S_{1N}^L = \frac{1}{2}\frac{EA}{l^2}[r_1 \ r_2]\begin{bmatrix} \frac{6}{5} & \frac{1}{10} \\ \frac{1}{10} & \frac{2}{15} \end{bmatrix}\begin{Bmatrix} r_1 \\ r_2 \end{Bmatrix}$$

$$S_{1N}^R = \frac{1}{2}\frac{EA}{l^2}[r_1 \ r_3]\begin{bmatrix} \frac{6}{5} & \frac{1}{10} \\ \frac{1}{10} & \frac{2}{15} \end{bmatrix}\begin{Bmatrix} r_1 \\ r_3 \end{Bmatrix}$$

The axial forces are evaluated through

$$V_1^L = \frac{EA}{l} r_4 + S_{1N}^L$$

$$V_1^R = \frac{EA}{l} r_5 + S_{1N}^R$$

$\mathbf{R}_N$ is formed by merging S_{1N}^L and S_{1N}^R in positions 4 and 5, respectively, so that

$$\mathbf{R}_N = \{0 \ 0 \ 0 \ S_{1N}^L \ S_{1N}^R\}$$

Member matrices $\mathbf{k}_s^m$ are evaluated using the current values of the axial forces and are used to assemble $\mathbf{K}$. Equation (4.1–7) is solved, and the process is repeated.

Numerical results obtained with a computer program by the finite iterative method and by the Newton–Raphson method (to be seen later) are shown in Fig. 4.1–3. Solutions are obtained for a fixed lateral load, $R_1 = 1$ kip, and variable vertical compressive loads P. Material nonlinearity is not considered in the solution.

Convergence to a relative error in r_1 of less than 10^{-3} was achieved in the finite iterative method with four cycles at $P = 100$, and five cycles at $P = 200$. At $P = 300$ eight cycles were required. In this and other problems it was found that the finite iterative method diverged well before the Newton–Raphson method.

Nonlinear behavior is illustrated in (*d*). If P were held constant and R_1 made variable, r_1 would vary essentially linearly with R_1 as long as the effect of R_1 on the member axial forces remains negligible in comparison with P. As $R_1 \to 0$, behavior tends to Euler-type buckling. Determination of the critical load P_{cr} is made by prescribing a fixed lateral displacement $r_1 = 0.1$ inch, and determining the required lateral load R_1 for variable P. At $P = P_{cr}$, $R_1 = 0$. Results are shown in Fig. 4.1–4*a*, *b*, *c*. It is found that points (R_1, P) on the graph (*c*) lie very nearly on a straight line. It is found by interpolation that $P_{cr} = 384$ kips.

The results shown in Fig. 4.1–3*b*, *c*, *d* are based on three significant digits for r_1.

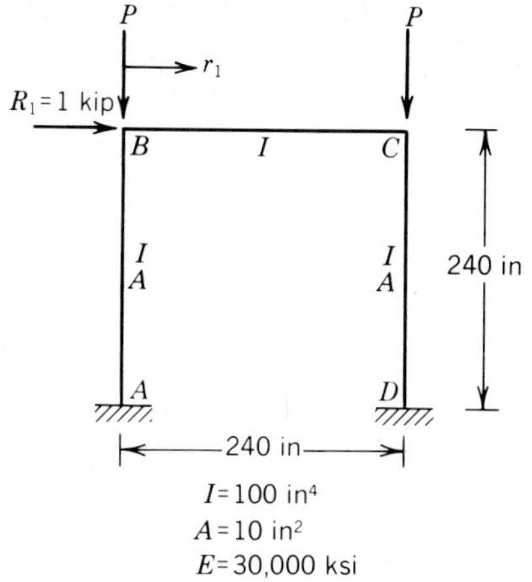

(a)

P	r_1	M_A	M_B	V_1^L	V_1^R	S_{1N}^L
0	0.275	68.6	51.4	0.428	−0.428	0.222
100	0.370	88.2	68.9	−99.4	−100.6	0.405
200	0.570	129.	105.	−199.1	−200.9	0.964
300	1.25	265.	229.	−298.1	−301.9	4.63
350	3.08	635.	564.	−345.3	−354.7	28.3
360	4.36	895.	799.	−353.3	−366.7	56.9
370	7.49	1532.	1372.	−358.6	−381.4	168.

Units: kip, inch

(b)

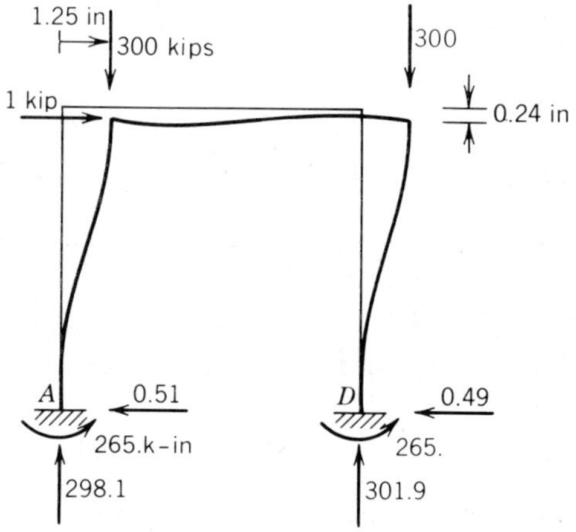

(c)

FIG. 4.1–3. Results of frame analysis, Section 4.1, Example 1; Section 4.3, Example 1. (*Continued on following page.*)

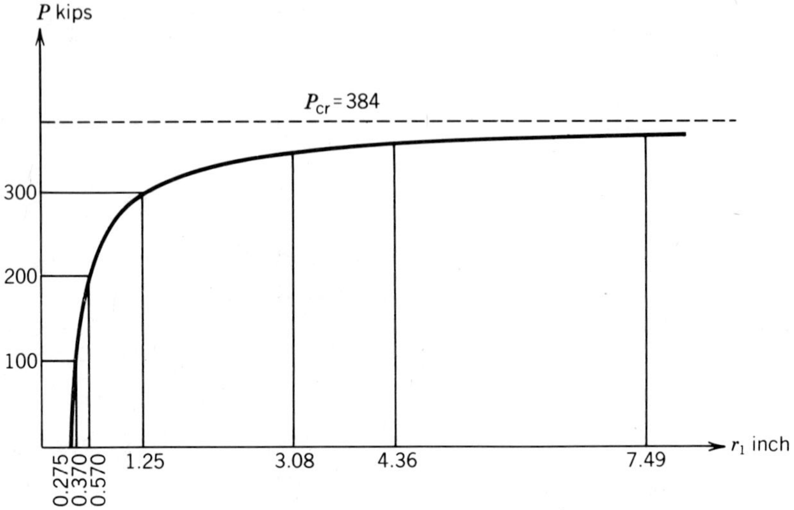

(*d*)

(*e*)

FIG. 4.1–3. (*continued*)

In (*b*), M_A and M_B are the bending moments at *A* and *B*, respectively. The moments at *D* and *C* are practically equal to those at *A* and *B*, respectively. The remaining symbols are as defined earlier in this section.

The equilibrium state for $P = 300$ kips is shown in (*c*). Note that the small variations from 300 kips of the axial forces in the columns are important in satisfying the moment equilibrium equation of the force system in (*c*). They form a couple equal to $1.9 \times 240 = 456$ k-in. In the case $P = 0$, this couple is only $0.43 \times 240 = 103$ k-in. Note that the increase in bending moment at *A* and *D* due to nonlinear behavior is less than $Pr_1 = 375$ k-in.

A free body diagram of the portion of a column between the support and the inflexion point is shown in Fig. 4.1–3*e*. The nonlinear effect on the bending moment is essentially $P\Delta$, where Δ is the lateral displacement at the inflexion

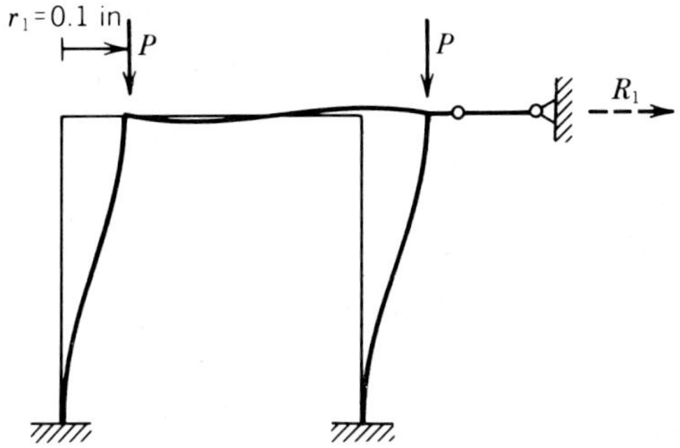

(a)

P(kip)	350	360	370	380	390
R_1(lb)	29.65	22.93	13.36	3.78	−5.81

(b)

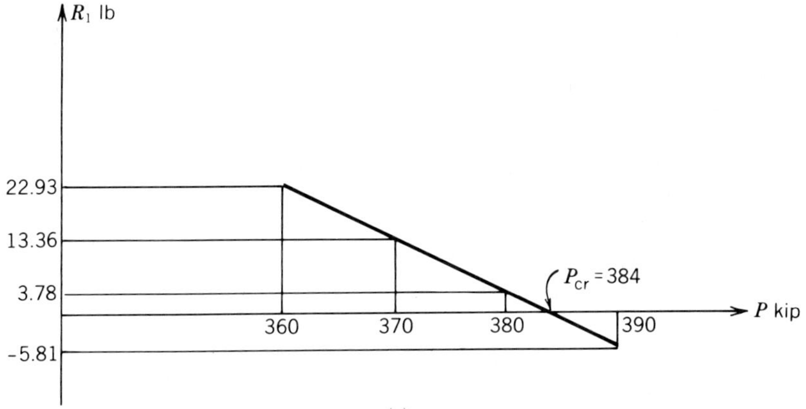

(c)

FIG. 4.1–4. Results of frame analysis, Section 4.1, Example 1; Section 4.3, Example 1.

point. Approximate procedures for determining the increase in bending moment due to nonlinear behavior may be seen in Exercises 4, 5, and 6.

4.2. Incremental Solution

Incremental sets of governing equations relate differential increments of member displacements, forces and joint displacements, to differential increments of loads. There is no need here to use a separate notation for axial and transverse components. The incremental equations in local axes are

$$d\mathbf{s} = \mathbf{a}d\mathbf{r} \tag{4.2–1a}$$

$$\mathbf{a}^T d\mathbf{S} = d\mathbf{R} \tag{4.2–1b}$$

$$d\mathbf{S} = \mathbf{k}_{\text{st}} d\mathbf{s} + d\mathbf{S}_0 \tag{4.2–1c}$$

The connectivity matrix $\mathbf{a}$ is constant and is the same as in the linear theory. $d\mathbf{R}$ is the joint load increment. The member incremental stiffness equation is obtained in Eq. (6/4–2) and, in a more general theory, in Chapter 7, Section 6.4.

Elimination of $d\mathbf{s}$ and $d\mathbf{S}$ from Eqs. (4.2–1) yields

$$\mathbf{K}_t d\mathbf{r} = d\mathbf{R}_T \tag{4.2–2}$$

where

$$\mathbf{K}_t = \mathbf{a}^T \mathbf{k}_{\text{st}} \mathbf{a} \tag{4.2–3a}$$

$$d\mathbf{R}_T = d\mathbf{R} - \mathbf{a}^T d\mathbf{S}_0 \tag{4.2–3b}$$

$\mathbf{K}_t$ is the assembled tangent stiffness matrix, and $\mathbf{a}^T d\mathbf{S}_0$ is an incremental equivalent joint load due to increments in member span loads. Assembling of $\mathbf{K}_t$ and of $\mathbf{a}^T d\mathbf{S}_0$ may be carried out by the direct stiffness method.

It is assumed in seeking an incremental solution that the loading condition of the problem is defined in terms of one parameter t, starting with zero load at $t = 0$, such that the structure is in equilibrium at any t. At $t = 0$ the structure is in the initial state $\mathbf{r} = 0$. For initially unstressed members all geometric matrices in $\mathbf{k}_{\text{st}}$ vanish, and $\mathbf{K}_t$ is the assembled stiffness matrix of the geometrically linear theory. If the process is started with initial axial forces, $\mathbf{k}_{\text{st}}$ includes member geometric stiffness matrices. The incremental solution proceeds by loading steps, in each of which a load increment is applied and the resulting equilibrium state determined. The solution is based on replacing differentials with finite differences and is thus approximate.

For the first load increment, $\mathbf{K}_t$ is formed in the state $\mathbf{r} = 0$, as described before. In general, before application of the ith load increment, the structure is in state $\mathbf{r} = \mathbf{r}_i$, with $\mathbf{s} = \mathbf{s}_i$ and $\mathbf{S} = \mathbf{S}_i$. $d\mathbf{R}_{T,i}$ is evaluated from the ith load increment, and member stiffnesses are evaluated in state $\mathbf{r}_i$ and used to assemble $\mathbf{K}_{t,i}$. Equation (4.2–2) is solved for $d\mathbf{r}_i$, and the deformation path is thus advanced to state $\mathbf{r}_{i+1} = \mathbf{r}_i + d\mathbf{r}_i$.

A graphical representation of the incremental solution is shown in Fig. 4.2–1 where R represents the load level and r some representative displacement. The solid curve represents the exact deformation path, and points $B_1, B_2, B_3, \ldots$ the approximate solutions corresponding to load levels $R_1, R_2, R_3, \ldots$, respectively. Point B_1 represents the solution by the geometrically linear theory for load level R_1.

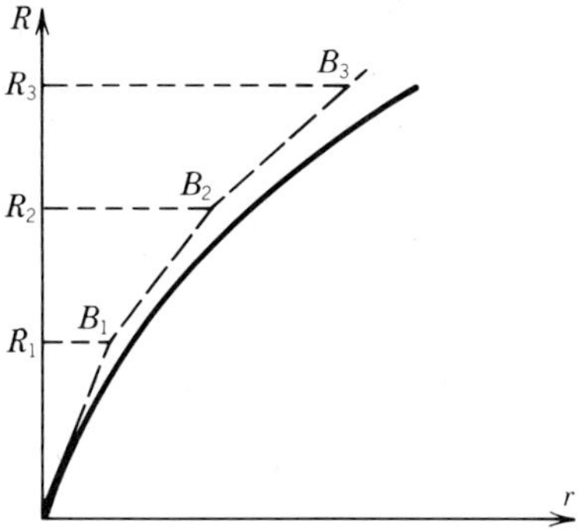

FIG. 4.2–1. Incremental solution.

The incremental procedure may be combined with the finite iterative method in order to obtain a more accurate solution after a few load increments. It is also possible to improve accuracy by iteration of the ith load step. Having obtained $d\mathbf{r}_i$ as described, $\mathbf{K}_{t,i}$ is reevaluated in the central state, $\mathbf{r}_{i+1/2} = \frac{1}{2}(\mathbf{r}_i + \mathbf{r}_{i+1})$, and a new solution is obtained for $d\mathbf{r}_i$. The Newton–Raphson method to be seen in the next section converges to the exact solution at a given load level and is more reliable. The incremental procedure is particularly useful, however, in cases where, because of material nonlinearity, member stiffness equations can be obtained only incrementally.

A state $\mathbf{r}$, at which $\mathbf{K}_t$ is singular, is a neutral equilibrium state. The equations $\mathbf{K}_t d\mathbf{r} = 0$ have then nontrivial solutions, and $\mathbf{r} + d\mathbf{r}$ is an alternate equilibrium state. Numerically large incremental displacements are obtained, as $\mathbf{K}_t$ becomes nearly singular.

Instability is represented on a deformation path, or load-displacement curve, in two basic ways. In one, the deformation path reaches a point with a horizontal tangent as shown in Fig. 4.2–2*a*. In the other, the deformation path reaches a point at which there is a discontinuity in slope, as shown in (*b*). In the first case, instability is reached through a gradually descreasing stiffness whose cause may be material, geometric, or a combination of both. An example of a geometric cause may be seen in Section 5, Example 4. In the second case, instability involves a discontinuity in incremental stiffness whose cause may be an abrupt change in material properties or an abrupt change in the geometry of deformation, as in Euler's column. If an abrupt change of geometry occurs, there is usually an unstable equilibrium path on a continuation of the initial deformation path, and a bifurcation of equilibrium, or Euler-type buckling, is said to occur. Instability due to material yielding, without appreciable geometric nonlinearity, is the special field of plastic analysis.

Analytically the critical load in bifurcation buckling also appears as an upper bound that is reached asymptotically. This occurs, for example, for an axially compressed column if the column is initially bent. The critical load is then a parameter that helps to describe nonlinear behavior at lower load levels, such as

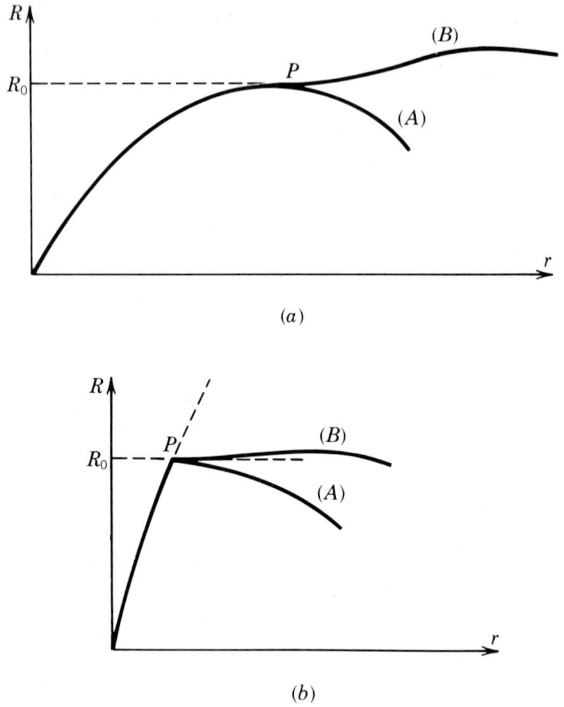

FIG. 4.2–2. Instability by decreasing stiffness (*a*); instability by bifurcation (*b*).

through the concept of magnification factor. A similar behavior may be seen for a frame in Fig. 4.1–3*d*.

Structural behavior beyond a point of instability is represented in Fig. 4.2–2. Descending curves (*A*) correspond to true or continuing instability, with R_0 representing the ultimate load that may be applied. Curves (*B*) in Fig. 4.2–2 have an ascending portion following the instability point. They represent an initial regain of stability followed eventually by renewed instability and failure.

4.3. Newton–Raphson Method

The Newton–Raphson method is an iterative procedure for solving a system of nonlinear equations that tends to an exact solution. The method may be applied with any constitutive properties that allow us to evaluate member force **S** from member displacements **s**. The equations to be solved are the joint equilibrium equations,

$$\mathbf{a}^T\mathbf{S} = \mathbf{R} \tag{4.3–1}$$

in which **S** is a function **S(s)** and $\mathbf{s} = \mathbf{ar}$. Assume that an approximate solution $(\mathbf{r}_0, \mathbf{s}_0, \mathbf{S}_0)$ has been determined, and let the exact solution be $(\mathbf{r}_0 + \delta\mathbf{r}, \mathbf{s}_0 +$

$\delta\mathbf{s}, \mathbf{S}_0 + \Delta\mathbf{S}$). Equation (4.3–1) becomes

$$\mathbf{a}^T \Delta\mathbf{S} = \mathbf{R} - \mathbf{a}^T \mathbf{S}_0 \tag{4.3–2}$$

If $\delta\mathbf{r}$ is small enough, $\Delta\mathbf{S}$ may be approximated by the first variation of the constitutive expression $\mathbf{S}(\mathbf{s})$, at a fixed value of the load, or

$$\delta\mathbf{S} = \mathbf{k}_{\text{st},0}\, \delta\mathbf{s} \tag{4.3–3}$$

$\mathbf{k}_{\text{st},0}$ is the member tangent stiffness matrix evaluated in state $(\mathbf{s}_0, \mathbf{S}_0)$, and $\delta\mathbf{s} = \mathbf{a}\delta\mathbf{r}$. In a theory with large rotations of member-bound axes, $\mathbf{k}_{st}$ is derived in Chapter 7, Section 6.4. Replacement of $\Delta\mathbf{S}$ with $\delta\mathbf{S}$ makes the system (4.3–2) linear in $\delta\mathbf{r}$ and leads to the assembled incremental stiffness equations

$$\mathbf{K}_{t,0}\, \delta\mathbf{r} = \Delta\mathbf{R}_0 \tag{4.3–4}$$

where

$$\mathbf{K}_{t,0} = \mathbf{a}^T \mathbf{k}_{st,0}\, \mathbf{a} \tag{4.3–5a}$$

$$\Delta\mathbf{R}_0 = \mathbf{R} - \mathbf{a}^T \mathbf{S}_0 \tag{4.3–5b}$$

Equation (4.3–4) is solved for $\delta\mathbf{r}$, and the process is repeated, starting with the new approximation, $\mathbf{r} = \mathbf{r}_0 + \delta\mathbf{r}$, $\mathbf{s} = \mathbf{a}\mathbf{r}$, and $\mathbf{S} = \mathbf{S}(\mathbf{s})$. The iterative process may be started with any approximate solution, in particular, with $\mathbf{r}_0 = 0$, $\mathbf{s}_0 = 0$, $\mathbf{S}_0 = 0$. In that case $\mathbf{k}_{st,0}$ and $\mathbf{K}_{t,0}$ are usually the matrices of the geometrically linear theory.

The process converges if $\delta\mathbf{r} \to 0$. Then $\Delta\mathbf{R}_0 \to 0$, and from Eq. (4.3–5b) the equilibrium equations are satisfied exactly in the limit. At any stage $\Delta\mathbf{R}_0$ is a residual joint load by which $\mathbf{S}_0$ violates the equilibrium equations. Convergence criteria are formulated in terms of norms $|\delta\mathbf{r}_0|$ or $|\Delta\mathbf{R}_0|$. Usual choices are the maximum absolute value and the "root mean square" of the elements of the matrix.

The tangent stiffness $\mathbf{K}_t$ is used in both the incremental method and the Newton–Raphson method, but in different ways. In the incremental solution the external load is varied by increments, and $\mathbf{K}_t$ determines the corresponding increments of displacements. In the Newton–Raphson method the load is constant, and at the end of each iteration cycle the solution obtained corresponds to a load that differs from the given load by a residual $\Delta\mathbf{R}_0$. The role of $\mathbf{K}_t$ is to determine small corrective displacements that at the limit make $\Delta\mathbf{R}_0 \to 0$. Thus, if the iterative process converges with some other matrix replacing $\mathbf{K}_t$, the solution at the limit still satisfies the governing equations and is an exact solution.

This property of the Newton–Raphson method allows a significant reduction of effort. For example, it allows $\mathbf{K}_t$ to remain unchanged for several iterations, and the decision to reevaluate $\mathbf{K}_t$ may be taken depending on the rate of convergence.

A graphic illustration of the Newton–Raphson method is shown in Fig. 4.3–1

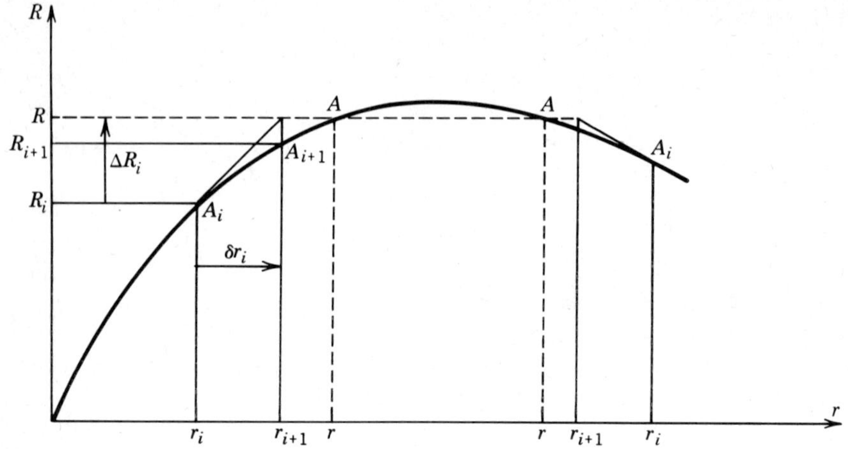

(a)

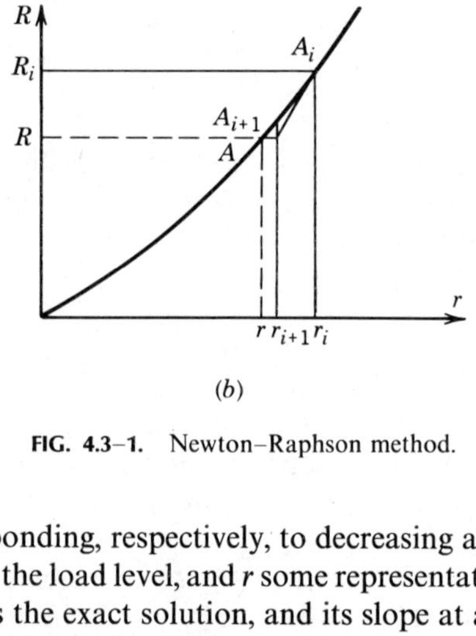

(b)

FIG. 4.3–1. Newton–Raphson method.

for two cases, corresponding, respectively, to decreasing and increasing tangent stiffness. R represents the load level, and r some representative displacement. The solid curve represents the exact solution, and its slope at any point is a tangent stiffness K_t, such that $dR = K_t dr$. Given a load level R, an approximate solution r_i corresponds to some load level $R_i \neq R$ and is represented by point A_i. Solving the equation

$$K_{t,i}\delta r_i = \Delta R_i$$

where $K_{t,i}$ is the slope at point A_i, and $\Delta R_i = R - R_i$, the next approximation is $\mathbf{r}_{i+1} = \mathbf{r}_i + \delta\mathbf{r}_i$ and is represented by point A_{i+1}. The Newton–Raphson method may be used in conjunction with the incremental solution. In Fig. 4.2–1 a Newton–Raphson iteration started at point B_3 would bring back the solution closer to the exact deformation path.

Convergence in the Newton–Raphson method occurs usually at an increasing rate and may be achieved at both stable and unstable portions of a deformation path, provided the starting approximation is close enough to the exact equilibrium state. The method diverges near a neutral equilibrium state as $\mathbf{K}_t$ becomes nearly singular.

Use of a fixed tangent stiffness in several cycles corresponds to using the slope of the tangent at point A_i in several successive cycles, before using again the tangent at the last approximate point reached. More on this topic may be seen in [8].

Example 1

The problem of Example 1, Section 4.1, was solved also by the Newton–Raphson method. Convergence to a relative error in r_1 of less than 10^{-3} was achieved in three cycles for $P = 100$, 200, and 300 kips, Fig. 4.1–3*d*. For $P = 350$, 360, and 370 five cycles or less were required. For $P = 380$ the process did not converge, and the determinant of the tangent stiffness matrix alternated in sign in successive cycles starting at the fifth cycle.

Example 2

The five-story, three-bay frame of Fig. 4.3–2*a* is analyzed for the loading shown, with variable *P*. Results are shown in (*b*). For $P = 100$ kips nonlinearity causes an increase in M_B from the value in the linear theory of 137 to 154 k-in, or

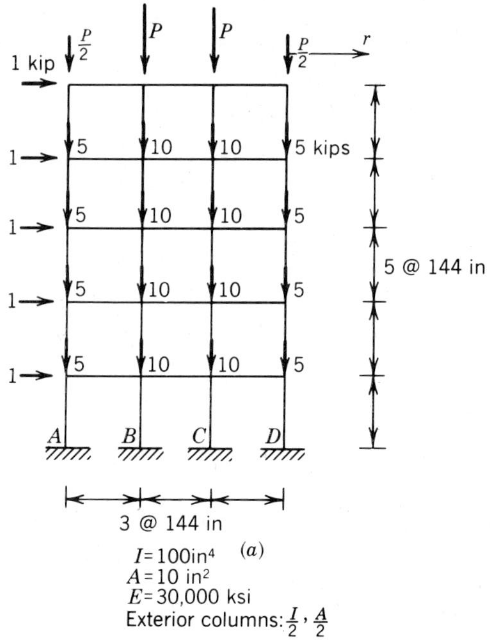

FIG. 4.3–2. Results for five-story frame, Section 4.3, Example 2; Chapter 13, Section 7, Example 6. (*Continued on following page.*)

P	r	M_A	M_B	V_A	V_B	V_C	V_D
0	0.763	68.3	137.	3.95	0.290	− 0.290	− 3.95
10	0.794	71.1	142.	− 20.9	− 49.7	− 50.3	− 29.1
20	0.804	71.7	144.	− 25.8	− 59.7	− 60.3	− 34.2
40	0.824	72.9	146.	− 35.7	− 79.7	− 80.3	− 44.3
80	0.867	75.5	151.	− 55.5	− 119.7	− 120.3	− 64.5
100	0.890	76.8	154.	− 65.4	− 139.7	− 140.3	− 74.6

units: kip, inch

(*b*)

M_A, M_B: bending moments
V_A, V_B, V_C, V_D: axial forces
Case $P = 0$: all vertical loads are zero

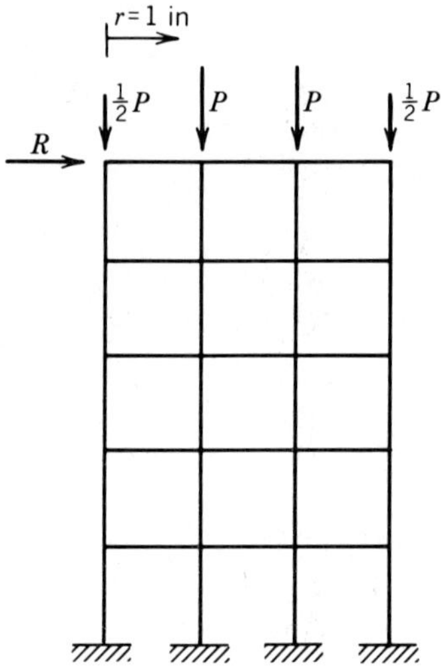

(*c*)

$r = 1$ in				
P(kip)	600	700	800	900
R(kip)	1.103	0.635	0.154	− 0.367

(*d*)

FIG. 4.3–2. (*continued*)

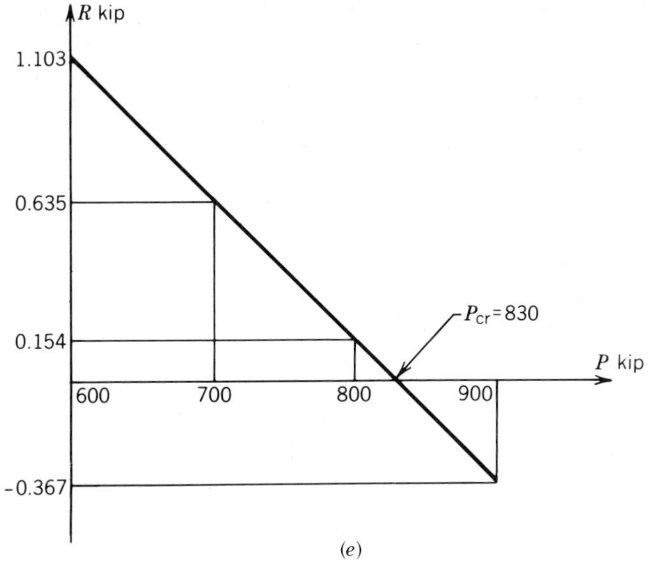

(e)

FIG. 4.3–2 *(continued)*

by 12.4%. The increase in r is from 0.763 to 0.890 in, or by 16.6%. The Newton–Raphson method converges for the load steps shown in (*b*) in three cycles or less, with a relative error on r of less than 10^{-3}.

To determine Euler's critical load, ignoring material yielding, a lateral displacement $r = 1$ in is prescribed, and the required lateral load R is determined for increasing values of P, as shown in (*c*). Results are tabulated in (*d*) and plotted in (*e*). A linear interpolation using the last two points yields

$$P_{cr} = 830 \text{ kips}$$

Actual instability of the frame of Fig. 4.3–2*a* would occur in the plastic range. The value of P_{cr} found on the basis of elastic behavior is useful however, for describing behavior in the elastic range and for determining an effective column length. See Exercises 7 and 8 at the end of this chapter.

5. PIN-JOINTED TRUSSES: STIFFNESS METHOD

Equation (4.1–1) is specialized in this section to the case of a pin-jointed truss. A general member *ab* subjected to a member load is shown in Fig. 5–1. The superscript *m* will be deleted from member matrices, for simplicity. Effective transverse variables are $\mathbf{s} = \{s_2^a \; s_2^b\}$ and $\mathbf{S} = \{S_2^a \; S_2^b\}$. Axial variables are $\mathbf{s}_1 = \{s_1^a \; s_1^b\}$ and $\mathbf{S}_1 = \{S_1^a \; S_1^b\}$.

The effective transverse stiffness relations are purely geometric. It will be

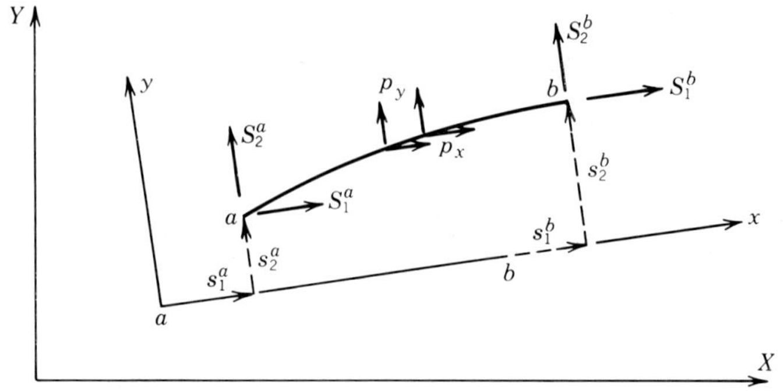

FIG. 5–1. Pin-ended member.

assumed that the geometric effect of the axial span load on the transverse stiffness equation, $\mathbf{S} = \mathbf{k}_s\mathbf{s} + \mathbf{S}_F$, is negligible. Then $\mathbf{k}_s$ is the string geometric stiffness,

$$\mathbf{k}_s = \mathbf{k}_G = V_1\mathbf{c}_1 = \frac{V_1}{l}\begin{bmatrix} 1 & -1 \\ -1 & 1 \end{bmatrix} \tag{5–1}$$

and $\mathbf{S}_F$ consists of the reactions of the pin-ended member due to the transverse span load. The axial stiffness relation is affected by the bent shape. It has the form

$$V_1 = k_{11}(s_1^b - s_1^a) + \frac{1}{2}\frac{k_{11}}{l}(s_2^b - s_2^a)^2 + V_{1F} + V_{1N} \tag{5–2}$$

The second term in Eq. (5–2) is due to the chord rotation and is $\frac{1}{2}k_{11}\mathbf{s}^T\mathbf{c}_1\mathbf{s}$. V_{1N} is the tie-rod force, caused by the transverse load with both member ends pin supported (see Chapter 6, Section 3.1, Example 5). The axial terms needed to form the stiffness equations are

$$\mathbf{S}_{1F} = (V_{1F} + V_{1N})\{-1 \quad 1\} + \left\{-\int_0^l p_x\,dx \quad 0\right\} \tag{5–3}$$

and

$$\mathbf{S}_{1N} = \frac{1}{2}k_{11}\mathbf{B}_1\mathbf{s}^T\mathbf{c}_1\mathbf{s} = \frac{1}{2}\frac{k_{11}}{l}(s_2^b - s_2^a)^2\{-1 \quad 1\} \tag{5–4}$$

The structure stiffness equation takes the form

$$(\mathbf{K}_a + \mathbf{K}_G)\mathbf{r} + \mathbf{a}_1^T\mathbf{S}_{1N} = \mathbf{R} - \mathbf{R}_F \tag{5–5}$$

$\mathbf{K}_a$ is the truss stiffness matrix in the linear theory, and $\mathbf{K}_G$ is the structure geometric stiffness matrix.

For forming the tangent stiffness matrix $\mathbf{K}_t$, the member tangent matrix (6/4–2) is specialized by letting $\mathbf{g}_s = \mathbf{c}_1$ and $\mathbf{k}_s = V_1 \mathbf{c}_1$. Thus

$$\mathbf{k}_{st} = \begin{bmatrix} \mathbf{k}_{11} & k_{11} \mathbf{B}_1 \mathbf{s}^T \mathbf{c}_1 \\ k_{11} \mathbf{c}_1 \mathbf{s} \mathbf{B}_1^T & V_1 \mathbf{c}_1 + k_{11} \mathbf{c}_1 \mathbf{s} \mathbf{s}^T \mathbf{c}_1 \end{bmatrix} \tag{5–6}$$

Example 1

Forming of the stiffness equations and their solution by the iterative method are now described for the truss of Fig. 5–2*a*, although such a truss is more conveniently analyzed by the force method. The two members are assumed to have the same axial stiffness $k_{11} = k$. The joint displacements are identified in (*b*). The effective member displacements in local coordinates, and the corresponding axial and geometric stiffness matrices and connectivity matrices are shown in (*c*). A superscript is used to refer to members.

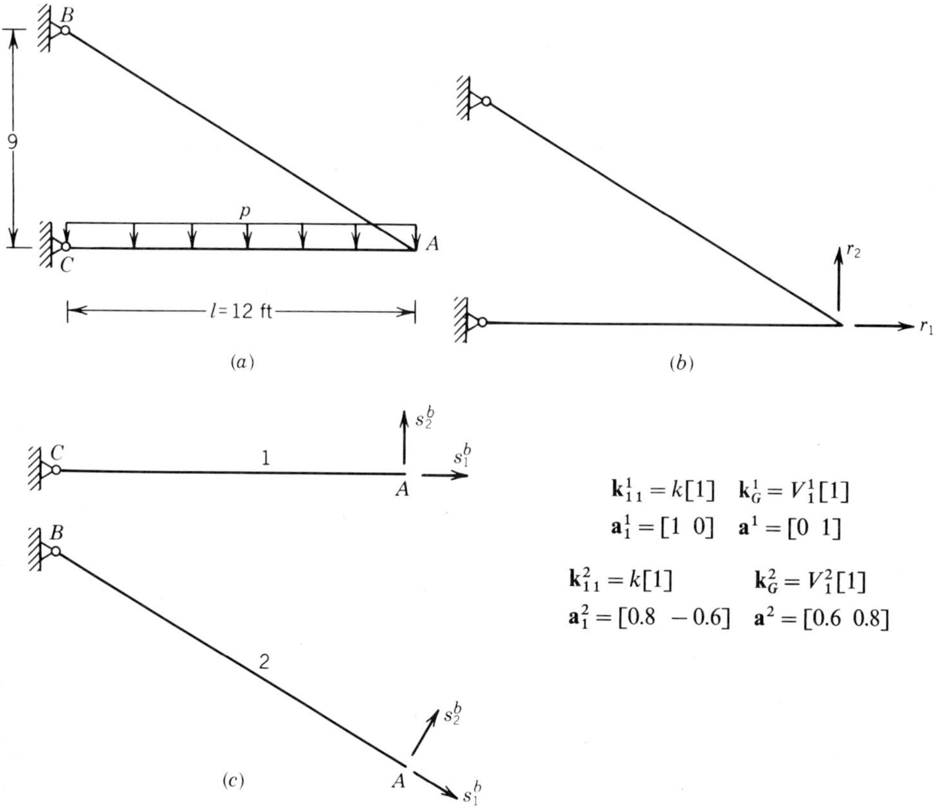

FIG. 5–2. Truss analysis, Example 1, 2.

Forming $\mathbf{K}_a$ and $\mathbf{K}_G$, there comes

$$\mathbf{K}_a = \sum_m \mathbf{a}_1^{mT} \mathbf{k}_{11}^m \mathbf{a}_1^m = k \begin{bmatrix} 1.64 & -0.48 \\ -0.48 & 0.36 \end{bmatrix}$$

$$\mathbf{K}_G = \sum_m \mathbf{a}^{mT} \mathbf{k}_G^m \mathbf{a}^m = \frac{V_1^1}{l} \begin{bmatrix} 0 & 0 \\ 0 & 1 \end{bmatrix} + \frac{V_1^2}{l} \frac{4}{5} \begin{bmatrix} 0.36 & 0.48 \\ 0.48 & 0.64 \end{bmatrix}$$

The effective member force matrices, $\mathbf{S}_1^m$ and $\mathbf{S}^m$, consist of one element each and will be designated by S_1^m and S^m, respectively. From Eq. (5–4)

$$S_{1N}^1 = \frac{1}{2} \frac{k}{l} (s_2^b)^2 = \frac{1}{2} \frac{k}{l} r_2^2$$

$$S_{1N}^2 = \frac{1}{2} \frac{k}{l} \frac{4}{5} (s_2^b)^2 = \frac{1}{2} \frac{k}{l} \frac{4}{5} \mathbf{r}^T \mathbf{a}^{2T} \mathbf{a}^2 \mathbf{r}$$

$$= \frac{1}{2} \frac{k}{l} \frac{4}{5} (0.36 r_1^2 + 0.96 r_1 r_2 + 0.64 r_2^2)$$

then

$$\mathbf{a}_1^T \mathbf{S}_{1N} = \{S_{1N}^1 + 0.8 S_{1N}^2, \ -0.6 S_{1N}^2\}$$

For the loading of Fig. 5–2*a*, $\mathbf{R} = 0$, and $\mathbf{R}_F$ is to be formed from the fixed end forces S_{1F}^1 and S_F^1 of member 1. S_{1F}^1 in the present case is the axial force in the tie rod, V_{1N}. S_F^1 is the transverse reaction at one end of the tie rod. Thus

$$S_F^1 = \frac{pl}{2}$$

and

$$\mathbf{R}_F = \begin{Bmatrix} S_{1F}^1 & \frac{pl}{2} \end{Bmatrix}$$

In the first cycle of the finite iterative method, the stiffness equations are

$$\mathbf{K}_a \mathbf{r} = -\mathbf{R}_F$$

whose solution, by inverting $\mathbf{K}_a$, is

$$\mathbf{r} = \frac{1}{k} \begin{bmatrix} 1 & \frac{4}{3} \\ \frac{4}{3} & \frac{41}{9} \end{bmatrix} \begin{Bmatrix} -S_{1F}^1 \\ -\frac{pl}{2} \end{Bmatrix} = \frac{pl}{2k} \begin{Bmatrix} -\frac{4}{3} \\ -\frac{41}{9} \end{Bmatrix} - \frac{S_{1F}^1}{k} \begin{Bmatrix} 1 \\ \frac{4}{3} \end{Bmatrix}$$

The axial forces have the expressions

$$V_1^1 = k\mathbf{a}_1^1\mathbf{r} + S_{1N}^1 + S_{1F}^1$$
$$V_1^2 = k\mathbf{a}_1^2\mathbf{r} + S_{1N}^2$$

Substituting the solution for $\mathbf{r}$ into these equations, there comes

$$V_1^1 = -\frac{4\,pl}{3\,2} + S_{1N}^1$$

$$V_1^2 = \frac{5\,pl}{3\,2} + S_{1N}^2$$

The terms in p are the axial forces in the linear theory. The contribution of S_{1F}^1 to these forces vanishes. This is to be expected because S_{1F}^1, which is the axial reaction in the state $\mathbf{r} = 0$, is applied in reverse sense as part of the equivalent joint load. A similar behavior would occur in any statically determinate truss. The nonlinear parts S_{1N}^m of the axial forces do depend, however, on S_{1F}^1.

For the second cycle $\mathbf{K}_G$ and $\mathbf{a}_1^T\mathbf{S}_{1N}$ are evaluated using the solution of the first cycle, and the process is iterated.

For the truss of the present example S_{1N}^1 and S_{1N}^2 are likely to be negligible. Member analysis, however, needs to consider the beam-column behavior of member 1.

Example 2

The tangent stiffness matrix $\mathbf{K}_t$ for the truss of Fig. 5–2 is formed in what follows. $\mathbf{K}_t$ is assembled from member tangent stiffness matrices $\mathbf{k}_{st}$ as defined in Eq. (5–6). In the present problem the effective matrices appearing in the expression of $\mathbf{k}_{st}$ are $\mathbf{k}_{11} = k[1]$, $\mathbf{B}_1 = [1]$, $\mathbf{s} = \{s_2^b\}$, and $\mathbf{c}_1 = l^{-1}[1]$. Thus for either member of the truss

$$\mathbf{k}_{\text{st}}^m = \begin{bmatrix} k & \dfrac{ks_2^b}{l} \\ \dfrac{ks_2^b}{l} & \dfrac{V_1}{l} + \dfrac{k(s_2^b)^2}{l} \end{bmatrix}$$

For member 1, $s_2^b = r_2$, and for member 2, $s_2^b = \mathbf{a}^2\mathbf{r} = 0.6r_1 + 0.8r_2$. $\mathbf{k}_{\text{st}}^m$ is now transformed to global coordinates, through the formula

$$\bar{\mathbf{k}}_{\text{st}}^m = \mathbf{a}_T^m\mathbf{k}_{\text{st}}^m\mathbf{a}_T^m$$

where $\mathbf{a}_T^m$ is the member connectivity matrix for both axial and transverse

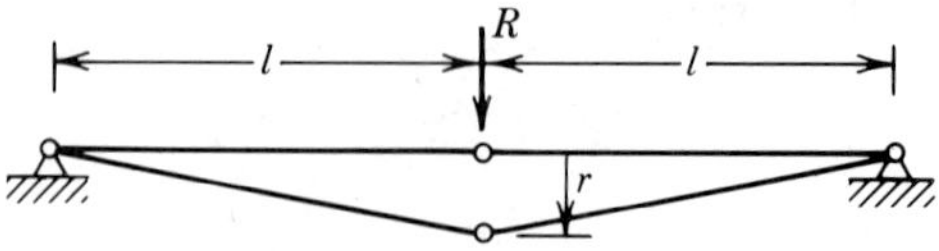

FIG. 5–3. Initially unstable structure, Example 3.

displacements,

$$\mathbf{a}_T^m = \begin{bmatrix} \mathbf{a}_1^m \\ \mathbf{a}^m \end{bmatrix}$$

For member 1, $\bar{\mathbf{k}}_{st}^1 = \mathbf{k}_{st}^1$. For member 2, from Fig. 5–2,

$$\mathbf{a}_T^2 = \begin{bmatrix} 0.8 & -0.6 \\ 0.6 & 0.8 \end{bmatrix}$$

Then

$$\mathbf{K}_t = \mathbf{k}_{st}^1 + \bar{\mathbf{k}}_{st}^2$$

Example 3

The two bar structure of Fig. 5–3 is unstable in the straight configuration but gains stability as it deflects and requires a nonlinear analysis. The two bars are identical and have the axial stiffness $k_{11} = k$. Because of symmetry, there is one joint degree of freedom, as shown, for which $\mathbf{K}_a$ is zero. The member geometric stiffness is

$$\mathbf{k}_G = \frac{V_1}{l}[1]$$

Since the two members have the same axial force, the assembled stiffness is

$$K_G = \mathbf{a}^T\mathbf{k}_G\mathbf{a} = 2\frac{V_1}{l}$$

Since the only displacement is transverse to the member local axis, there is no term $\mathbf{a}_1^T\mathbf{S}_{1N}$, and the stiffness equation reduces to

$$K_G r = R$$

whence

$$r = \frac{Rl}{2V_1}$$

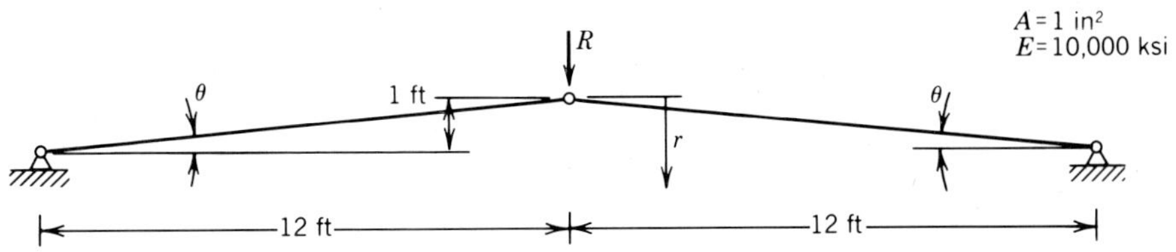

(a)

r	R	V_1	V_{1N}
1.	0.839	− 5.51	0.238
2.	1.461	− 10.54	0.951
3.	1.885	− 15.10	2.140
4.	2.131	− 19.18	3.805
5.	2.219	− 22.79	5.945
6.	2.168	− 25.92	8.561
7.	1.997	− 28.58	11.65
8.	1.726	− 30.76	15.22
9.	1.376	− 32.46	19.26
10.	0.965	− 33.69	23.78
11.	0.513	− 34.44	28.78
12.	0.040	− 34.72	34.24

Units: inch, kip
(b)

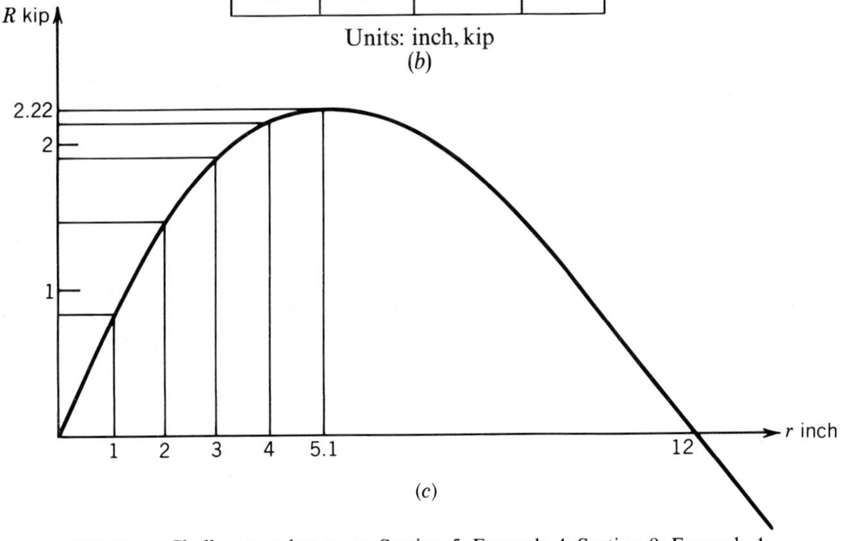

(c)

FIG. 5–4. Shallow two-bar truss, Section 5, Example 4; Section 9, Example 1.

From Eq. (5–2),

$$V_1 = \frac{kr^2}{2l}$$

The two preceding equations yield the cubic equation for r,

$$r^3 = \frac{Rl^2}{k}$$

and the cubic equation for V_1

$$V_1^3 = \frac{R^2kl}{8}$$

Example 4

The shallow two-bar truss of Fig. 5–4 is analyzed first by prescribing increasing values of r, then by prescribing increasing values of R and using the Newton–Raphson method.

The axial force V_1, its nonlinear part, $V_{1N} = k_{11}r^2/2l$, and the required load R are evaluated for r increasing from 1 to 12 in and tabulated in (*b*). A plot of R versus r is shown in (*c*). For $r = 12$ in the exact R is zero. The computed value is 0.04 kip and is in slight error because of the approximate character of the geometric stiffness matrix. The maximum value of $R = 2.22$ kips, corresponds to a snap-through condition. In that state the increase in member compressive forces due to further deflection is exactly counterbalanced by the geometric effect which decreases the resultant of these axial forces.

For analysis by the Newton–Raphson method R is given the successive values 1, 2, 2.1, 2.2, 2.21, 2.22, and 2.225. Satisfactory convergence is achieved in a few cycles for $R \leqslant 2.22$, and the results agree with those tabulated. For $R = 2.225$ no convergence is achieved.

In order to determine the descending part of the curve in (*c*), the Newton–Raphson iteration is started with a value of r on that side of the curve.

Effect of Initial Curvature

If a compression member has an initial curvature, the axial stiffness relation becomes nonlinear because of beam-column action. If the initial shape v_0 in Fig. 5–5*a* is assumed to be a half sine wave of amplitude e, the deformed shape v is shown in Eq. (6/2.5–3) to be a half sine wave of amplitude

$$a = \frac{e}{1 - \lambda^2/\pi^2} \tag{5–7}$$

where $\lambda^2 = -V_1l^2/EI$. Evaluating the chord shortening due to the deflection

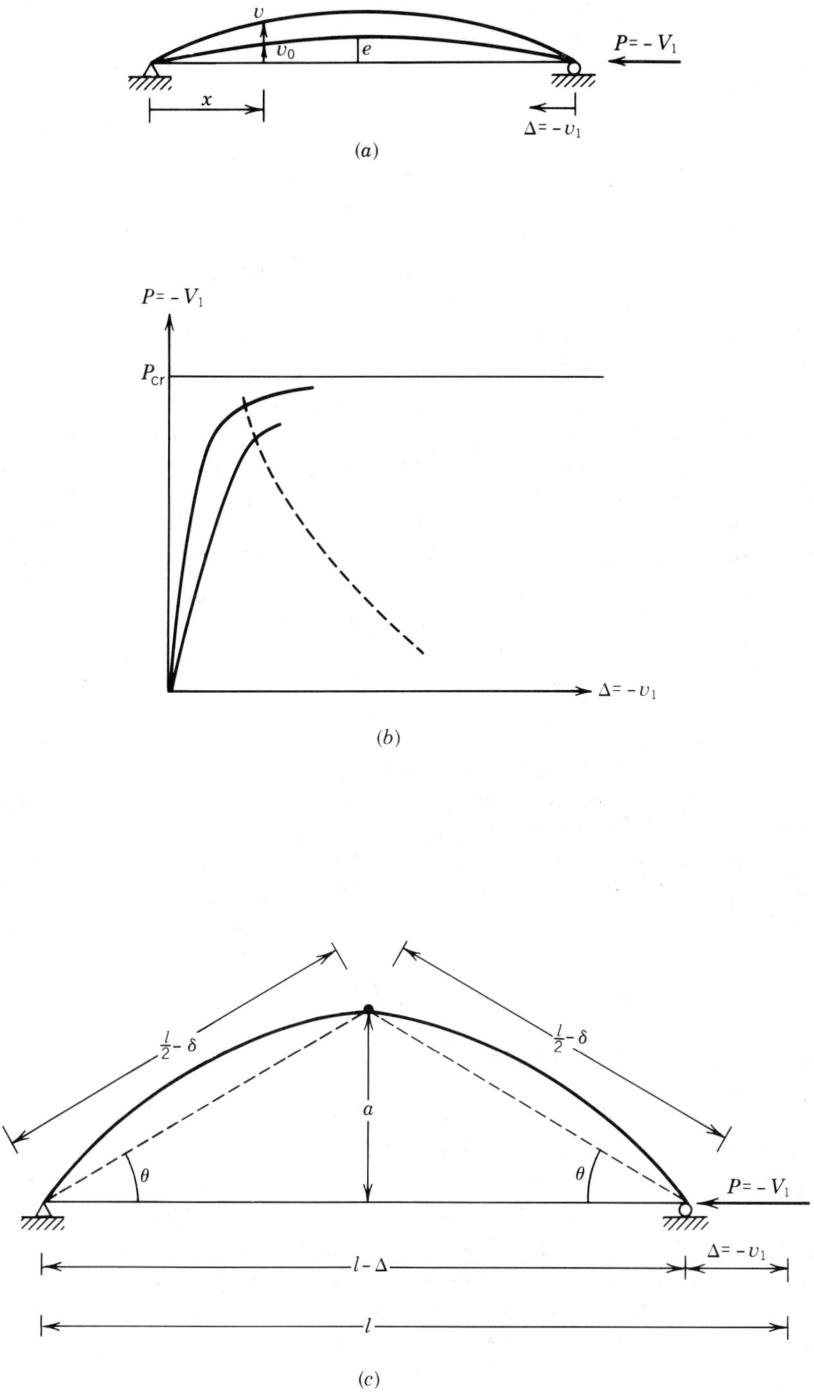

FIG. 5–5. Member with initial curvature, elastic and plastic behaviors.

$v - v_0$, the axial flexibility relation is found to be

$$v_1 = \frac{V_1 l}{EA} - \frac{\pi^2}{4l}(a^2 - e^2) \tag{5–8a}$$

or, with use of Eq. (5–7), and the notation $r^2 = I/A$,

$$v_1 = \frac{V_1 l}{EA}\left\{1 + \frac{e^2}{4r^2}\frac{2 - (\lambda^2/\pi^2)}{[1 - (\lambda^2/\pi^2)]^2}\right\} \tag{5–8b}$$

Equation (5–8b) is represented by the solid curves in Fig. 5–5*b*, for different values of e/r. To the larger e/r corresponds the lower stiffness. Equation (5–8b) defines an equivalent axial stiffness $k_{\text{eq}} = V_1/v_1$, function of V_1 for a given e. In an iterative process, k_{eq} is evaluated in a given cycle using the axial force obtained in the preceding cycle. Equation (5–8b) also allows to obtain by differentiation an incremental axial stiffness for use in an incremental method of solution.

Equation (5–8b) may also be applied as an approximation in the case where the member is subjected to a transverse load. The sinusoidal shape may then be considered as the first term of a Fourier series representation, and e as the deflection at midspan in the absence of the axial force.

Elastic–Plastic Behavior [9]

For an increasing compressive axial force Eq. (5–8b) ceases to be valid as the maximum compressive normal stress at midspan reaches the proportional limit of the material. Assuming the material is linearly elastic–perfectly plastic, a simplified way of treating the axial stiffness in the plastic range is to neglect the transition from first yield to the formation of a plastic hinge at midspan.

Fig. 5–5*c* shows the idealized member with a plastic hinge at midspan. The initial eccentricity is now ignored. By statics, the bending moment at midspan is

$$M'_P = Pa \tag{5–9a}$$

The plastic bending moment is a function of the axial force,

$$M'_P = f\left(\frac{P}{P_y}\right) \tag{5–9b}$$

where f is a function depending on the shape of the cross section and $P_y = A\sigma_y$ is the yield axial force. The member is now formed of two elastic members behaving each as a beam-column and connected at midspan by a plastic hinge. For a given P, Eqs. (5–9) determine M'_P and a. Elastic analysis of the two beam-columns allows to determine their chord shortening δ. Then, from Fig. 5–5*c*,

$$\Delta = l(1 - \cos\theta) + 2\delta\cos\theta \tag{5–10}$$

and

$$\sin\theta = \frac{a}{(l/2) - \delta} \tag{5–11}$$

δ is neglected with respect to $\frac{l}{2}$ in Eq. (5–11), but it is kept for further consideration in Eq. (5–10). Then, noting Eq. (5–9a), there comes

$$\sin\theta = \frac{2a}{l} = \frac{2M'_P}{Pl} \tag{5–12}$$

δ is determined using Eq. (6/2.1–11, 12) in which l is now $\frac{l}{2}$, $v_1 = -\delta$, $V_1 = -P\cos\theta$, $\mathbf{V} = \{M'_P \;\; 0\}$, and the stability parameter for the members of length $\frac{l}{2}$ is renamed λ'. There comes

$$\delta = \frac{Pl\cos\theta}{2EA} + \frac{(M'_P)^2}{lP^2\cos^2\theta}\left[\frac{\lambda'(2\lambda' + \sin 2\lambda')}{4\sin^2\lambda'} - 1\right] \tag{5–13a}$$

where

$$\lambda'^2 = \frac{Pl^2\cos\theta}{4EI} \tag{5–13b}$$

Substituting for δ into Eq. (5–10), and rearranging the trigonometric terms in λ', there comes

$$\Delta = l(1 - \cos\theta) + \frac{Pl}{EA}\cos^2\theta + \frac{l}{\cos\theta}\left(\frac{M'_P}{Pl}\right)^2\left(\frac{\lambda'^2}{\sin^2\lambda'} + \frac{\lambda'}{\tan\lambda'} - 2\right) \tag{5–14}$$

Eq. (5–14) is represented by the dotted curve in Fig. 5–5b.

For large θ, P is small, and the effect of δ on Δ is small. The $P - \Delta$ relation is then close to the case of two rigid bars connected by a plastic hinge. For small θ, such that $\theta^2 \ll 1$, the following simplified relations may be written

$$\theta = \frac{2M'_P}{Pl} \tag{5–15}$$

$$\lambda'^2 = \frac{Pl^2}{4EI} \tag{5–16}$$

$$\Delta = \frac{Pl}{EA} + l\left(\frac{M'_P}{Pl}\right)^2\left(\frac{\lambda'^2}{\sin^2\lambda'} + \frac{\lambda'}{\tan\lambda'}\right) \tag{5–17}$$

The dotted curve in Fig. 5–5*b* represents unstable equilibrium. An increase in P with increasing deformation would require by statics an increase in M'_P, and this is impossible because the plastic moment decreases as P increases. Thus P can only decrease such that $M'_P = Pa$ increases as deformation increases.

The effect on structural behavior of the loss of stability in one member depends on whether the structure is statically determinate. The removal of any one member causes a statically determinate structure to become a mechanism. The same effect results if a member reaches a state of zero incremental stiffness. The structure is then in a neutral equilibrium state, and mechanism motion is possible. An equilibrium state of the structure in which a member has a negative incremental stiffness, as on the dotted line of Fig. 5–5*b*, is unstable because the mechanism motion referred to earlier, and which now deforms the unstable member, requires negative external work. In the absence of external restraining forces the mechanism motion takes place dynamically and is the collapse mode of

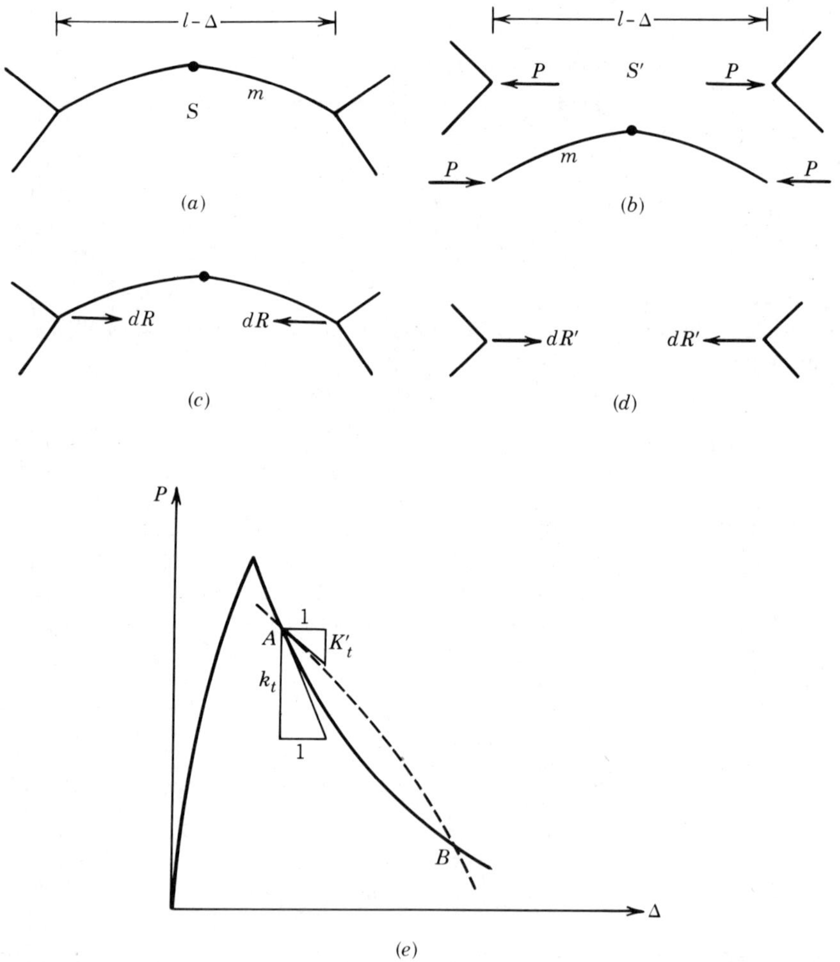

FIG. 5–6. Structure stability after yield in one member.

the structure. By contrast, if yield occurs in tension members, resulting in zero incremental stiffnesses, the structure may suffer from excessive deformations, but there is no dynamic collapse as long as none of the yielded members fractures.

In a statically indeterminate structure, removal of one member may leave a kinematically stable structure. Let m be the unstable compression member, S be the structure subjected to its loading condition, and S' be the structure obtained by removing member m and replacing it by its axial force P applied as a pair of opposite forces on the remaining structure, Fig. 5–6a, b. Let dR be a superimposed incremental load applied on S as shown in (c). The incremental stiffness of S corresponding to dR is

$$K_t = k_t + K'_t$$

where k_t is the incremental axial stiffness of member m, and K'_t is the incremental stiffness of S' associated with forces dR' as shown in (d). Since $k_t < 0$, the behavior of the structure depends on whether $k_t + K'_t$ is positive. In that case the equilibrium state is locally stable. If $k_t + K'_t < 0$, further chord shortening of member m cannot be prevented except by restraining external forces. In the absence of such forces, member m collapses without necessarily causing collapse of the total structure. That would depend on what happens to other members as member m collapses.

If the $P - \Delta$ relationships for m and S' are shown on the same graph, a point of intersection represents an equilibrium state of S. Since $dR' = -dP$, the slope of the curve for S' is $-K'_t$. In a case such as shown in Fig. 5–6e, $k_t + K'_t$ is negative at A and positive at B. Equilibrium at A is unstable. Snap-through to point B occurs and is accompanied by a redistribution of axial forces in S' and m.

6. INEXTENSIBLE FRAMES: STIFFNESS METHOD

Consider a structure that admits in the geometrically linear theory inextensional displacement modes, described by means of generalized displacements $\mathbf{q}_1$. The inextensional approximation in the linear theory is the result of a linear transformation, $\mathbf{r} = \mathbf{T}_1\mathbf{q}_1$, applied to the governing system of equations. If this transformation is now applied to the nonlinear governing equations, the order of magnitude analysis that justifies the inextensional approximation in the linear theory remains valid, under certain conditions. The displacement modes $\mathbf{q}_1$ are still referred to as inextensional, although only the linear portion of the axial deformation–displacement relation vanishes in these modes.

Consider first the complete linear transformation

$$\mathbf{r} = \mathbf{T}\mathbf{q} = [\mathbf{T}_1 \ \mathbf{T}_2]\{\mathbf{q}_1 \ \mathbf{q}_2\}$$

from $\mathbf{r}$ to an equal number of generalized displacements $\mathbf{q} = \{\mathbf{q}_1 \ \mathbf{q}_2\}$. The congruent and contragradient laws transform Eq. (4.1–1) into

$$\begin{bmatrix} \mathbf{K}_{11} & \mathbf{K}_{12} \\ \mathbf{K}_{21} & \mathbf{K}_{22}+\mathbf{K}'_a \end{bmatrix} \begin{Bmatrix} \mathbf{q}_1 \\ \mathbf{q}_2 \end{Bmatrix} + \begin{Bmatrix} \mathbf{O} \\ \mathbf{T}_2^T \mathbf{a}_1^T \mathbf{S}_{1N} \end{Bmatrix} = \begin{Bmatrix} \mathbf{Q}_1 \\ \mathbf{Q}_2 \end{Bmatrix} \tag{6–1}$$

where

$$\begin{bmatrix} \mathbf{K}_{11} & \mathbf{K}_{12} \\ \mathbf{K}_{21} & \mathbf{K}_{22} \end{bmatrix} = \begin{bmatrix} \mathbf{T}_1^T \\ \mathbf{T}_2^T \end{bmatrix} \mathbf{K}_b [\mathbf{T}_1 \ \ \mathbf{T}_2] \tag{6–2}$$

$$\mathbf{K}'_a = \mathbf{T}_2^T \mathbf{K}_a \mathbf{T}_2 \tag{6–3}$$

The first part of Eq. (6–1) expresses the inextensional equilibrium equations, that is, the equilibrium equations generated through the virtual inextensional modes $\delta\mathbf{q}_1$. $\mathbf{S}_{1N}$ contributes nothing to these equations because it is formed of pairs of equal and opposite forces whose virtual work in the inextensional modes is zero. As in the linear theory, the inextensional approximation may be shown to apply if $\mathbf{Q}_1$ is not small compared with $\mathbf{Q}_2$. It consists then in neglecting $\mathbf{q}_2$ in the first part of Eqs. (6–1). In the linear theory $\mathbf{K}_{22}$ is negligible compared with $\mathbf{K}'_a$, but the term $\mathbf{K}_{21}\mathbf{q}_1$ is not negligible. In the present theory $\mathbf{K}_{22}$ and $\mathbf{K}_{21}$ are modified by addition of geometric stiffness matrices that may or may not be negligible, depending on particular structural types and behavior. The second part of Eqs. (6–1) is thus kept without modification. The stiffness equations of the inextensional approximation take the form

$$\mathbf{K}_{11}\mathbf{q}_1 = \mathbf{Q}_1 \tag{6–4a}$$

$$(\mathbf{K}'_a + \mathbf{K}_{22})\mathbf{q}_2 = -\mathbf{K}_{21}\mathbf{q}_1 - \mathbf{T}_2^T \mathbf{a}_1^T \mathbf{S}_{1N} + \mathbf{Q}_2 \tag{6–4b}$$

Equations (6–4) may be solved iteratively. At the start of an iteration cycle the geometric matrices and $\mathbf{S}_{1N}$ are formed using the results from the preceding cycle. $\mathbf{K}'_a$ is the same as in the linear theory and is constant. The solution of Eq. (6–4a) for $\mathbf{q}_1$ is substituted into Eq. (6–4b) which is then solved for $\mathbf{q}_2$. The solution for $\mathbf{q}_1$ and $\mathbf{q}_2$ allows to evaluate $\mathbf{V}_1$ and $\mathbf{S}_{1N}$ for use in the next iteration cycle. However, iteration is not needed if the axial forces are not significantly affected by geometric nonlinearity. Their values may then be obtained through linear analysis and used to evaluate the geometric stiffness matrices needed to assemble $\mathbf{K}_{11}$.

Instability in the inextensional formulation occurs if, at a certain critical load level, compressive axial forces make $\mathbf{K}_{11}$ singular. This may occur through a bifurcation of equilibrium or in a continuous manner, with displacements tending to infinity as the point of instability is approached. In this latter case the governing equations cease in fact to be valid at a lower load (see Examples 3 and 4).

Example 1

The frame of Section 4.1, Example 1, shown in Fig. 4.1–3, is now analyzed according to the inextensional approximation. The inextensional displacements are $\{r_1 \ r_2 \ r_3\}$, where r_1 is the translation of member BC and r_2 and r_3 are the joint rotations. For evaluating the stiffness matrices of the columns, the axial forces may be considered with sufficient accuracy equal to P. Because of antisymmetry of the lateral load, $r_2 = r_3$, and only half the structure as shown in Fig. 6–1 need be analyzed. The stiffness coefficients are made dimensionally homogeneous, so that r_2 is the rotation multiplied by $l = 240$ in, and the member end moment is divided by l. Member AB has the stiffness matrix of Eq. (4.1–4) whose degrees of freedom coincide with r_1 and r_2, and member BB' has the effective stiffness $6EI/l \times 1/l^2$ associated with r_2. Merging of member stiffnesses yields the stiffness equations

$$\frac{EI}{l^3}\begin{bmatrix} c\lambda & c\tan\frac{\lambda}{2} \\ c\tan\frac{\lambda}{2} & c\left(\frac{1}{\lambda}-\frac{1}{\tan\lambda}\right)+6 \end{bmatrix}\begin{Bmatrix} r_1 \\ r_2 \end{Bmatrix} = \begin{Bmatrix} R_1 \\ 0 \end{Bmatrix} \tag{6–5}$$

where

$$c = \frac{\lambda^2}{2[\tan(\lambda/2) - (\lambda/2)]} \tag{6–6}$$

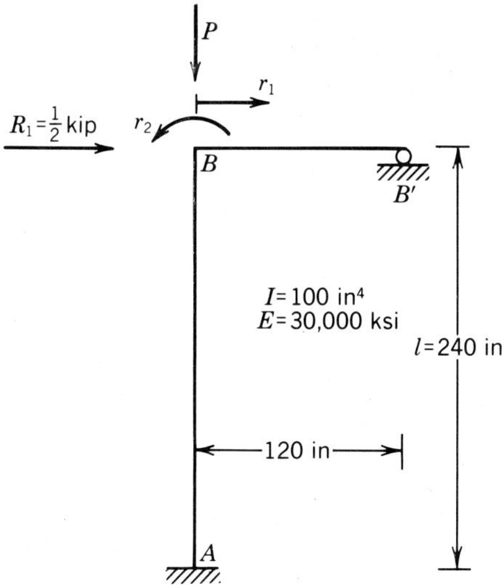

FIG. 6–1. Inextensional analysis, Example 1.

Elimination of r_2 yields

$$r_2 = -\frac{\tan(\lambda/2)}{1/\lambda - 1/\tan\lambda + (12/\lambda^2)[\tan(\lambda/2) - (\lambda/2)]} r_1 \qquad (6\text{–}7)$$

$$\frac{EI}{l^3}\frac{\lambda^3}{2[\tan(\lambda/2) - (\lambda/2)]}\left\{1 - \frac{\tan^2\lambda/2}{1 - \lambda/\tan\lambda + (12/\lambda)[\tan(\lambda/2) - (\lambda/2)]}\right\} r_1 = R_1 \qquad (6\text{–}8)$$

In the foregoing equation

$$\lambda^2 = \frac{Pl^2}{EI}$$

and $R_1 = 0.5$ kip. It is left as an exercise to verify that the solution for r_1 agrees with that recorded in Fig. 4.1–3*b*.

Having r_1 and r_2, member end forces are evaluated by means of the member stiffness equations.

Bifurcation buckling of the frame subjected to the load P, with $R_1 = 0$, occurs as the stiffness matrix $\mathbf{K}_{11}$, which is the coefficient matrix in Eq. (6–5), becomes singular. The condition $|\mathbf{K}_{11}| = 0$ is obtained by setting the coefficient of r_1 in Eq. (6–8) equal to zero, or

$$1 - \frac{\lambda}{\tan\lambda} + \frac{12}{\lambda}\left(\tan\frac{\lambda}{2} - \frac{\lambda}{2}\right) - \tan^2\frac{\lambda}{2} = 0$$

This equation may be simplified by expressing $(1 - \tan^2\lambda/2)$ as $2\tan(\lambda/2)/\tan\lambda$, and then factoring out $[\tan(\lambda/2) - (\lambda/2)]$. This latter factor as a higher root in λ than the other factor which, set to zero, yields

$$\frac{\lambda}{\tan\lambda} + 6 = 0 \qquad (6\text{–}9)$$

The lowest root of this equation is found by trial and error as

$$\lambda_{cr} = 2.716 \qquad (6\text{–}10)$$

whence

$$P_{cr} = \frac{\lambda^2 EI}{l^2} = 384 \text{ kips}$$

Example 2

Analysis of the frame of Fig. 6–2 is carried out in what follows using the inextensional approximation. As in the preceding example, the analysis is

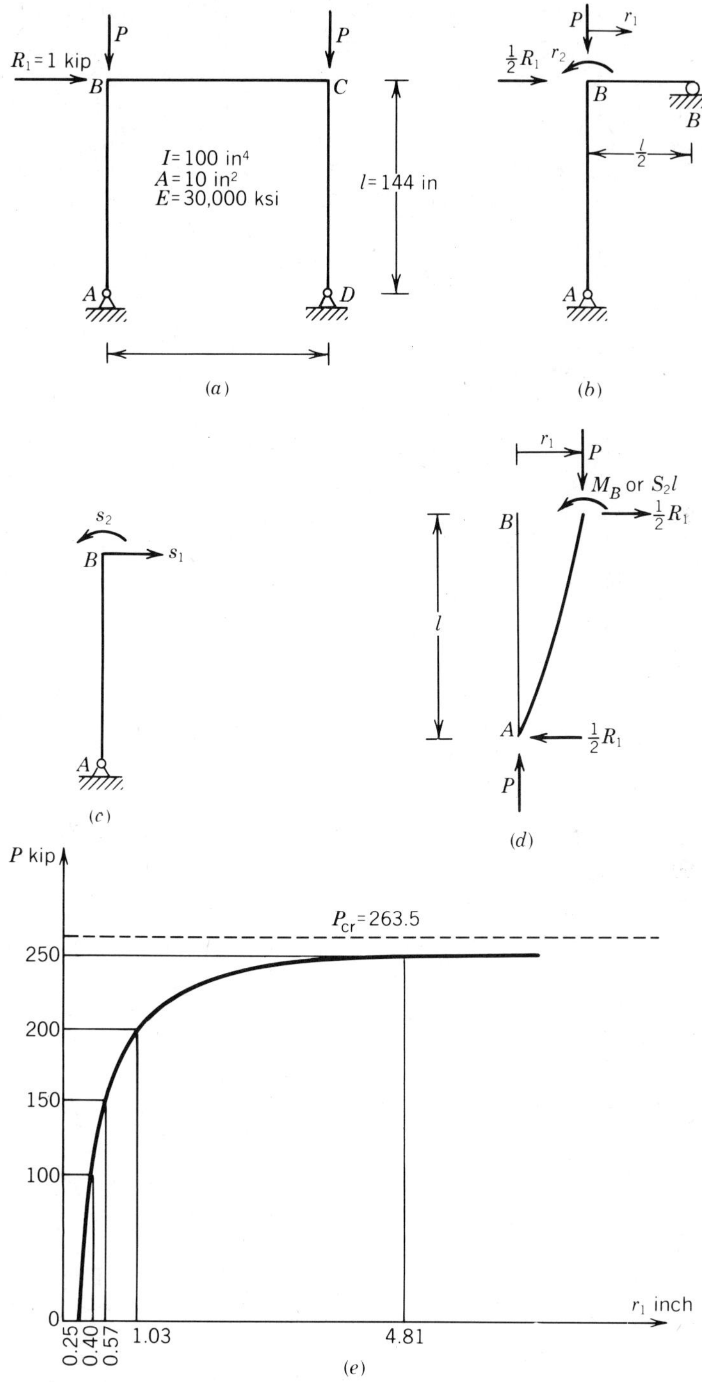

FIG. 6–2. Inextensional analysis, Example 2.

reduced to that of the structure shown in (b). For the columns the effective stiffness matrix $\mathbf{k}_s$ which takes into account the released rotation at the support is derived in Eq. (7/5–10b). This equation is made to apply to the choice of displacements shown in (c), by deleting the third row and column because of the support at A and by letting $\alpha = 1$ to neglect shear deformation. Using dimensionally homogeneous displacements, as done in the preceding example, there comes

$$\mathbf{k}_s = \frac{EI\lambda^2}{l^3(1-\lambda\cot\lambda)}\begin{bmatrix}\lambda\cot\lambda & 1\\ 1 & 1\end{bmatrix} \tag{6–11}$$

The effective rotational stiffness of member BB' is $6EI/l \times 1/l^2$. The assembled stiffness matrix is thus

$$\mathbf{K} = \frac{EI\lambda^2}{l^3(1-\lambda\cot\lambda)}\begin{bmatrix}\lambda\cot\lambda & 1\\ 1 & 1+\dfrac{6(1-\lambda\cot\lambda)}{\lambda^2}\end{bmatrix} \tag{6–12}$$

The loading term is

$$\mathbf{R} = \{\tfrac{1}{2}R_1 \quad 0\}$$

Noting that $P = \lambda^2 EI/l^2$, the solution of the stiffness equations is found to be

$$r_1 = \frac{1}{2}\frac{R_1 l}{P}\left[\frac{6}{\lambda(6\cot\lambda-\lambda)} - 1\right]$$

$$r_2 = -\frac{1}{2}\frac{R_1 l}{P}\frac{\lambda}{6\cot\lambda-\lambda}$$

The column forces at B, $\mathbf{S} = \{S_1 \quad S_2\}$, are evaluated through $\mathbf{S} = \mathbf{k}_s\mathbf{s}$, with $\mathbf{s} = \mathbf{r}$. This yields for S_1 the statically determinate value

$$S_1 = \frac{R_1}{2}$$

and for S_2

$$S_2 = \frac{R_1}{2}\frac{6}{\lambda(6\cot\lambda-1)}$$

A statics check on S_2 may be made through the force diagram of Fig. 6–2d, from which the bending moment at B is

$$S_2 l = Pr_1 + \frac{R_1 l}{2}$$

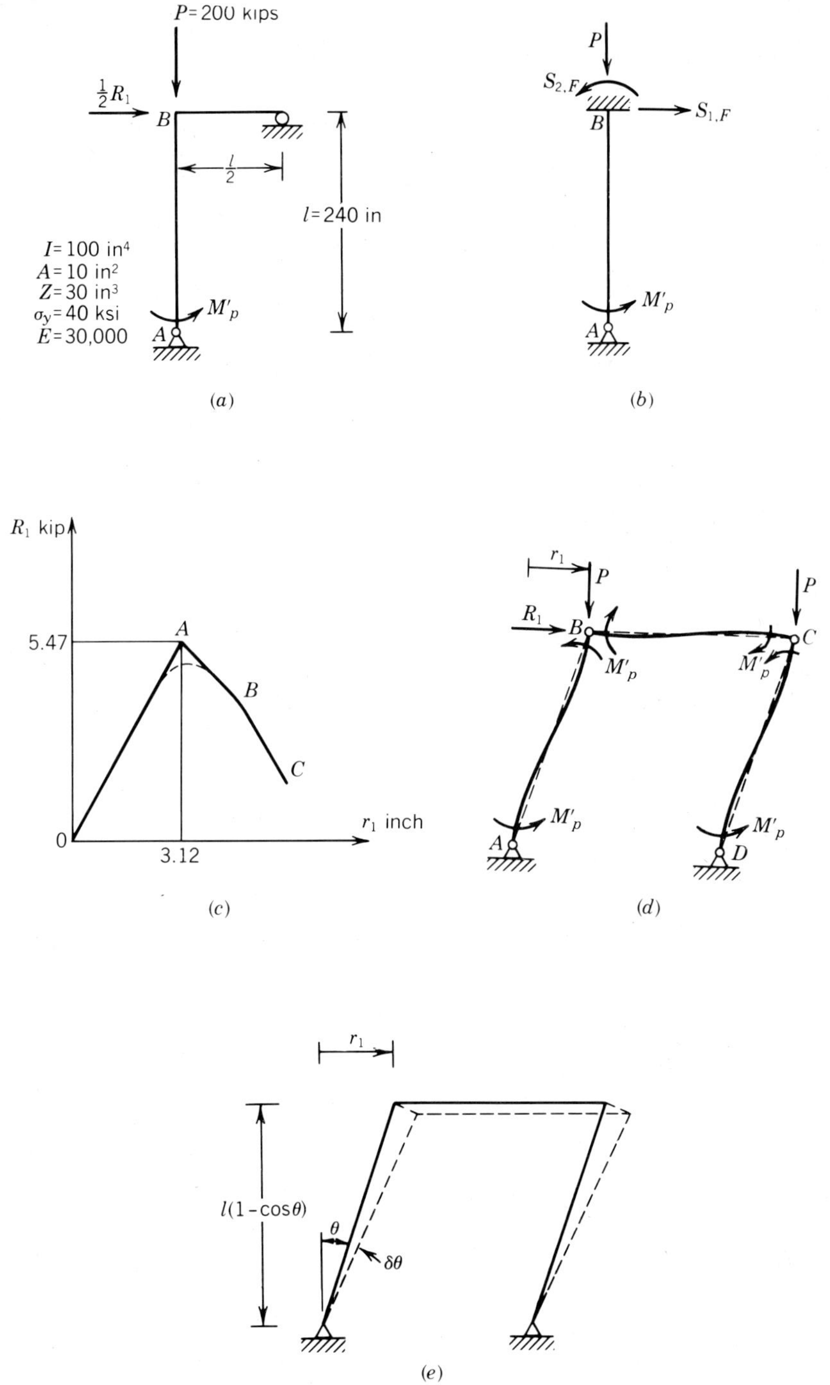

FIG. 6–3. Geometrically nonlinear elastic-plastic analysis, Example 3.

It is readily verified that the solution for r_1 and S_2 satisfies this last equation. A plot of P versus r_1, representing the solution just found, is shown in (*e*) for the data in (*a*). For example, for $P = 150$ kips, $r_1 = 0.575$ in. The part of the bending moment at B due to nonlinear behavior is $Pr_1 = 86$ k-in. It exceeds the linear part, $R_1 l/2 = 72$ k-in.

The critical load P_{cr} is obtained by equating the denominator in the expressions of r_1 and r_2 to zero. This yields

$$6 \cot \lambda - \lambda = 0 \tag{6–13}$$

The lowest root in λ is found by trial and error to be

$$\lambda_{cr} = 1.3496 \tag{6–14}$$

For the data of Fig. 6–2*a*,

$$P_{cr} = \frac{\lambda^2 EI}{l^2} = 263.5 \text{ kips}$$

The effective length of the columns is

$$Kl = \sqrt{\frac{P_E}{P_{cr}}}\, l = \frac{\pi}{\lambda_{cr}} l = 2.33l$$

Example 3

In this example nonlinear behavior in the plastic range is illustrated through a simplified analysis. Consider the frame of Fig. 4.1–3*a*, subjected to $P = 200$ kips and an increasing lateral load R_1. For P constant the variation in a column axial force due to variation in R_1 has a negligible effect on the column stiffness, so that behavior for variable R_1 is nearly linear. Using the tabulated value of $M_A = 129$ k-in for $R_1 = 1$ kip, and neglecting the small difference between M_A and M_D, we can write for a variable R_1,

$$M_A = M_D = 129 R_1 \text{ k-in}$$

Let M_P be the plastic moment capacity of the columns, and M'_P be the reduced plastic moment due to the presence of a compressive axial force. The reduction from M_P to M'_P may be assumed due to $P = 200$ kips for both columns. The maximum bending moment in the columns in the present case occurs at the fixed ends. Neglecting the transition from first yield to the formation of a plastic hinge, the preceding relation remains valid until R_1 reaches the value R_P such that

$$129 R_P = M'_P \text{ k-in}$$

Assuming an I cross section, the reduction in plastic moment is approximated by the interaction relation

$$\frac{M'_P}{M_P} = 1.176\left(1 - \frac{P}{P_y}\right), \qquad \frac{P}{P_y} > 0.15 \tag{6–15}$$

For a plastic modulus $Z = 30\,\text{in}^3$, and a yield stress $\sigma_y = 40\,\text{ksi}$, there comes $M_P = 30 \times 40 = 1200\,\text{k-in}$, $P_y = 10(40) = 400\,\text{kips}$, $M'_P = 1.176(1200)(1\text{–}200/400) = 705.6\,\text{k-in}$, and $R_P = 705.6/129 = 5.47\,\text{kips}$. As yield progresses at A and D, there is a reduction in incremental stiffness of the frame corresponding to increments in lateral load R_1. At the limit, when the plastic hinges are formed, there is no resistance to further deformation at A and D, so that the incremental stiffness of the frame becomes that for pin supports. To determine whether R_1 could in fact reach R_P, the load $P = 200\,\text{kips}$ is compared with the critical value of the pin-supported frame which, from Example 2 and Eq. (6–14), is

$$P_{\text{cr}} = (1.3496)^2 \frac{EI}{l^2} = 94.9\,\text{kips}$$

This result indicates that instability of the frame, as R_1 increases, occurs before the plastic hinges are completely formed. As yield starts at a value of R_1 slightly lower than R_P, the incremental stiffness of the frame decreases with a corresponding decrease of the critical axial load from the original value of 384 kips found in Example 1. The incremental stiffness becomes zero when the critical axial load reduces to the applied load of 200 kips. For a continually increasing lateral displacement, equilibrium becomes unstable and requires a decrease of the lateral load R_1.

To determine the unstable portion of the load-displacement relationship, before yield starts at B and C, the frame is analyzed elastically as pin supported, and subjected to R_1 and to the plastic moment M'_P at A and D. Only half the structure is considered as shown in Fig. 6–3*a*. The stiffness equations have the form

$$\mathbf{Kr} = \mathbf{R} - \mathbf{R}_F \tag{6–16}$$

where $\mathbf{K}$ and $\mathbf{R}$ are the same as in Example 2. $\mathbf{R}_F$ is determined from the fixed end forces at B caused by M'_P, as indicated in (*b*). These are obtained from the complete stiffness matrix $\mathbf{k}_s$, Eq. (6/2.6–6). The fourth column of this matrix, say $\{k_{14}\ k_{24}\ k_{34}\ k_{44}\}$, consists of member end forces caused by a unit rotation, $s_4 = 1$, which is identified here with a unit rotation at the pin support. Thus $M'_P = k_{44} s_4$, and, recalling that moments in the stiffness equations are divided by l, there comes

$$S_{1,F} = \frac{k_{14}}{k_{44}} M'_P = \frac{2\lambda \sin^2 \lambda/2}{\sin\lambda - \lambda\cos\lambda} \frac{M'_P}{l} \tag{6–17a}$$

$$S_{2,F} = \frac{k_{24}}{k_{44}} \frac{M'_P}{l} = \frac{\lambda - \sin\lambda}{\sin\lambda - \lambda\cos\lambda} \frac{M'_P}{l} \tag{6–17b}$$

The loading term in the stiffness equations is

$$\mathbf{R} - \mathbf{R}_F = \{\tfrac{1}{2}R_1 - S_{1,F}, -S_{2,F}\}$$

The solution is obtained by superposition of that due to **R** which was found in Example 2, to the solution due to $-\mathbf{R}_F$. There comes

$$r_1 = \frac{1}{2}\frac{R_1 l}{P}\left[\frac{6}{\lambda(6\cot\lambda - \lambda)} - 1\right] - \frac{M'_P}{P}\left[\frac{6}{6\cos\lambda - \lambda\sin\lambda} - 1\right] \tag{6–18a}$$

$$r_2 = -\frac{1}{2}\frac{R_1 l}{P}\frac{\lambda}{6\cot\lambda - \lambda} + \frac{M'_P}{P}\frac{\lambda^2}{6\cos\lambda - \lambda\sin\lambda} \tag{6–18b}$$

Using the numerical data of the problem, we obtain

$$r_1 = -1.02R_1 + 8.71$$

$$r_2 = 0.266R_1 - 3.31$$

For $R_1 = R_P = 5.47$, r_1 should be equal to the value obtained through the elastic analysis of the frame for fixed supports. The foregoing solution yields $r_1 = 3.13$ in. From Fig. 4.1–3*b*, $r_1 = 0.57R_P = 3.12$ in. The discrepancy may be due to two causes other than round off errors. These are that the solution in Fig. 4.1–3*b* includes the effect of column extensibility and that the assumed linearity of r_1 in terms of R_1, for a fixed P, is approximate. A complete agreement would occur if the analytic solution of Example 1 were used.

The elastic–perfectly plastic behavior is represented through lines *OAB* in Fig. 6–3*c*. In reality the sharp corner at *A* would be rounded to represent the transition from first yield to unstable behavior. Line *AB* remains valid until yield starts at joints *B* and *C*, followed by formation of plastic hinges. An equilibrium state with plastic hinges at *A*, *B*, *C*, and *D* is shown in (*d*). An equilibrium equation is derived by the principle of virtual displacements. A virtual displacement taking place from the equilibrium state is shown in (*e*). The internal virtual work is $4M'_P\delta\theta$, and the external virtual work is $R_1 l(1 - \cos\theta)\delta\theta + Pr_1\delta\theta \approx (R_1 l + 2P_1 r_1)\delta\theta$. Equating the two expressions, there comes

$$r_1 = -\frac{R_1}{2P}l + 2\frac{M'_P}{P} \tag{6–19}$$

Using the numerical data of the problem yields

$$r_1 = -0.6R_1 + 7.056$$

This is represented by line *BC* in Fig. 6–3*c*

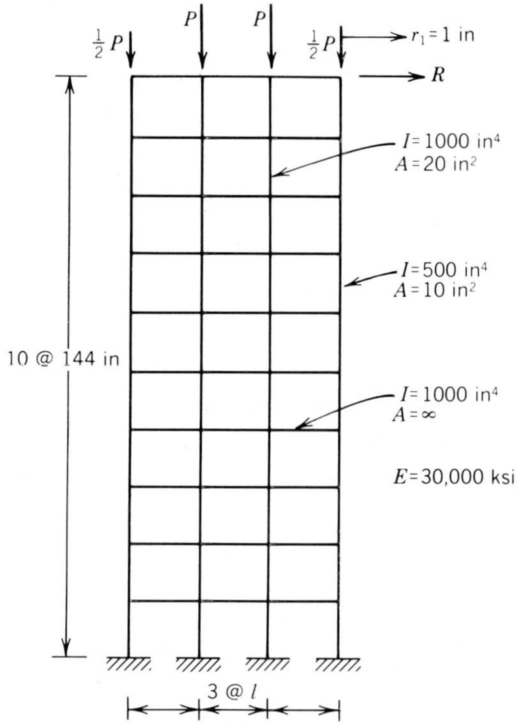

(a)

P(kip)	4800	5000	5200	5500
R(kip)	1.014	0.5066	– 0.0059	– 0.7906

"Exact," $P_{cr} = 5198$ kips

P(kip)	6600	8400	8600
R(kip)	3.997	– 0.3377	– 0.8004

Inextensional, $P_{cr} = 8254$ kips

(b) $l = 144$ in

P(kip)	4800	5000	5200	5500
R(kip)	1.648	1.175	0.6944	– 0.0496

"Exact," $P_{cr} = 5480$ kips

P(kip)	6000	6400	6600
R(kip)	1.007	0.0365	– 0.4834

Inextensional, $P_{cr} = 6414$ kips

(c) $l = 240$ in

FIG. 6–4. Exact and inextensional solution, Section 6, Example 4; Chapter 13, Section 7, Example 6.

Example 4

In this example the inextensional approximation is tested to find the critical load for a ten-story, three-bay rectangular frame and is found to be in significant error. The frame is shown in Fig. 6–4*a*. Typical member properties are indicated. Two cases of beam span are considered. The force R required to cause a lateral displacement of 1 in at the top floor is evaluated for various values of P. Results are shown in (*b*) and (*c*) for beam spans of 144 in and 240 in, respectively, and are labeled "exact." The inextensional approximation was simulated by assigning to the columns a cross-sectional area of 1000 in^2. P_{cr} is found by interpolation. It is seen that the inextensional approximation is in significant error. This example is discussed further in Chapter 13, Section 7, Example 6.

In Chapter 10, Section 12, it is seen that the inextensional approximation is justified if $\varepsilon^2 N^2 \ll 1$. In the present example the number of stories is $N = 10$, ε is the inverse of the slenderness ratio of the columns, $\varepsilon = \sqrt{50}/144$, and $\varepsilon^2 N^2 = 0.24$ is not small enough.

7. TRUSSED FRAMES: STIFFNESS METHOD

The transformation of the joint displacements into inextensional and extensional modes is also helpful to analyze the behavior of trussed frames, as seen in the linear theory, Chapter 10, Section 13. A trussed frame would remain kinematically stable if its joints were replaced with pinned joints. The truss thus obtained will be referred to as the substitute truss. The translational displacements of a trussed frame are all extensional and are $\mathbf{q}_2$. The inextensional displacements $\mathbf{q}_1$ are the joint rotations. The equivalent joint load consists of moments $\mathbf{Q}_1$ and forces $\mathbf{Q}_2$.

The behavior of a trussed frame could be examined separately for extensional and inextensional loading conditions, but superposition cannot be used because the separate loading conditions cause different axial forces. In a realistic loading condition joint moments do not cause axial forces large enough to bring about nonlinear behavior. However, the bending behavior due to joint moments is affected by the axial forces caused by the joint forces $\mathbf{Q}_2$. Although superposition does not apply, the final loading can be applied in an arbitrary sequence. Let $\mathbf{Q}_2$ be applied first, and assume for now that the axial forces caused by $\mathbf{Q}_2$ are far enough from making $\mathbf{K}_{11}$ singular. The order of magnitude considerations seen in the linear theory remain valid in the present theory. In the linear theory both $\mathbf{K}_{21}\mathbf{q}_1$ and $\mathbf{K}_{22}\mathbf{q}_2$ are negligible in Eq. (6–1). In the present theory the elastic stiffness terms remain negligible but not necessarily the geometric stiffness terms. Of these, those that are due to the string geometric stiffness $V_1\mathbf{c}_1$ are the same as for a pin-jointed truss. $V_1\mathbf{c}_1$ has no rotational degrees of freedom and thus contributes nothing to $\mathbf{K}_{21}$. The geometric stiffness term $V_1\mathbf{g}$, which is due to bending in the member-bound reference, does not alter the order of magnitude of the member bending stiffness. Its contribution to the extensional stiffness equations remains negligible as in the linear theory. The remaining geometric stiffness matrices, which are present only if the member axial force is variable, are assumed negligible. If they are not, they contribute terms similar to $V_1\mathbf{c}_1$ and do not alter this discussion. The result of these considerations is that $\mathbf{K}_{22}$ in Eq. (6–4b) may be formed only with the geometric stiffness $\mathbf{k}_G = V_1\mathbf{c}_1$, and $\mathbf{K}_{21}\mathbf{q}_1$ may be neglected. We thus obtain

$$(\mathbf{K}'_a + \mathbf{G}_{22})\mathbf{q}_2 = -\mathbf{T}_2^T\mathbf{a}_1^T\mathbf{S}_{1N} + \mathbf{Q}_2 \tag{7–1}$$

where

$$\mathbf{G}_{22} = \mathbf{T}_2^T(\mathbf{a}^T\mathbf{k}_G\mathbf{a})\mathbf{T}_2 = \mathbf{T}_2^T\mathbf{K}_G\mathbf{T}_2 \tag{7–2}$$

Allowing for difference in notation, Eq. (7–1) has the same stiffness matrix and the same loading term as the stiffness equation of the substitute truss, Eq. (5–5). The terms $\mathbf{S}_{1N}$ are, however, different in the two equations because of the difference in member-deformed shape between a pin-ended member in the substitute truss and a rigidly connected member in the trussed frame. This difference is not significant if the rotation of the member chord is the dominant cause of nonlinearity.

The preceding discussion was for a loading consisting of joint forces $\mathbf{Q}_2$. If now joint moments $\mathbf{Q}_1$ are also applied, the additional joint rotations may be obtained by means of the inextensional approximation in which the existing axial forces are used in forming the member bending stiffness matrices.

If these axial forces, which are caused by $\mathbf{Q}_2$, are large enough to make $\mathbf{K}_{11}$ singular, the structure becomes unstable in inextensional (rotational) modes. This topic is treated further in Chapter 13.

8. FORCE METHOD: SECOND-ORDER THEORY

The force method is developed starting with the governing equations in member-bound axes. These are Eqs. (8/5.3–8), except that the constitutive equation is taken in its flexibility form. It will be convenient to separate the axial deformation into a linear part v_{1L}, and a nonlinear part v_{1N}. Equations (8/5.3–8) take then the form

$$\begin{Bmatrix} \mathbf{v}_{1L} \\ \mathbf{v} \end{Bmatrix} = \begin{bmatrix} \mathbf{a}_{1v} \\ \mathbf{a}_v \end{bmatrix} \mathbf{r} \tag{8–1a}$$

$$\mathbf{a}_{1v}^T\mathbf{V}_1 + \mathbf{a}_v^T\mathbf{V}' + (\mathbf{K}_G + \mathbf{K}_{GF})\mathbf{r} = \mathbf{R} - \mathbf{a}_1^T\mathbf{S}_{1p} - \mathbf{a}^T\mathbf{S}_{pL} \tag{8–1b}$$

$$\begin{Bmatrix} \mathbf{v}_{1L} \\ \mathbf{v} \end{Bmatrix} = \begin{bmatrix} \mathbf{f}_{11} & \\ & \mathbf{f} \end{bmatrix} \begin{Bmatrix} \mathbf{V}_1 \\ \mathbf{V}' \end{Bmatrix} - \begin{Bmatrix} \mathbf{v}'_{1N} \\ 0 \end{Bmatrix} + \begin{Bmatrix} \mathbf{v}_{1,0} \\ \mathbf{v}_0 \end{Bmatrix} \tag{8–1c}$$

where

$$v'_{1N} = \tfrac{1}{2}\mathbf{s}^T\mathbf{c}_1\mathbf{s} + \tfrac{1}{2}\mathbf{v}^T\mathbf{g}\mathbf{v} = \tfrac{1}{2}\mathbf{s}^T\mathbf{g}_s\mathbf{s} \tag{8–2}$$

v'_{1N} is the change in length of the deformed member axis due to the transverse displacements.

The member flexibility relations are assumed obtained on the basis of an assumed shape. This includes the exact relations obtained for a uniform member and constant axial force.

The geometric stiffness appearing in the equilibrium equations is due to the rotation of the member-bound axes. $\mathbf{V}'$ is a modified set of member forces, obtained by removing from $\mathbf{V}$ geometric stiffness terms due to axial span loads and by placing these terms in $\mathbf{K}_{GF}$. If member axial forces are constant, then $\mathbf{V}' = \mathbf{V}$ and $\mathbf{K}_{GF} = 0$. The remaining terms in the equilibrium equation are the same as in the linear theory.

Equations (8–1) may be interpreted as pertaining to a modified linear problem as follows: the geometric stiffness term in Eq. (8–1b) is combined with the load term, the nonlinear term $\mathbf{v}'_{1N}$ is combined with the initial deformation term $\mathbf{v}_{1,0}$, and $\mathbf{f}$ is a modified bending flexibility due to axial forces. In an iterative method of solution, the modifications outlined here are made using current values of the variables, and the force method of the linear theory is used to solve the equations. The process is then iterated until a sufficient convergence is achieved.

In the general case of a statically indeterminate structure, the static and kinematic equations are solved in the form

$$\begin{Bmatrix} \mathbf{V}_1 \\ \mathbf{V}' \end{Bmatrix} = \begin{bmatrix} \mathbf{b}_1 \\ \mathbf{b} \end{bmatrix}(\mathbf{R}' - \mathbf{K}'_G\mathbf{r}) + \begin{bmatrix} \mathbf{c}_1 \\ \mathbf{c} \end{bmatrix}\mathbf{X} \tag{8–3a}$$

$$[\mathbf{c}_1^T \ \mathbf{c}^T]\begin{Bmatrix} \mathbf{v}_{1L} \\ \mathbf{v} \end{Bmatrix} = 0 \tag{8–3b}$$

$$\mathbf{r} = [\mathbf{b}_1^T \ \mathbf{b}^T]\begin{Bmatrix} \mathbf{v}_{1L} \\ \mathbf{v} \end{Bmatrix} \tag{8–3c}$$

where $\mathbf{K}'_G = \mathbf{K}_G + \mathbf{K}_{GF}$, and $\mathbf{R}'$ is the right-hand side of Eq. (8–1b). The coefficient matrices in Eqs. (8–3) are the same as in the linear theory, and $\mathbf{X}$ is a set of statical redundants. With use of the flexibility relations, Eqs. (8–3b, c) turn into

$$(\mathbf{I} + \mathbf{F}_{rr}\mathbf{K}'_G)\mathbf{r} - \mathbf{F}_{rx}\mathbf{X} = \mathbf{F}_{rr}\mathbf{R}' + \mathbf{b}_1^T(\mathbf{v}_{1,0} - \mathbf{v}'_{1N}) + \mathbf{b}^T\mathbf{v}_0 \tag{8–4a}$$

$$-\mathbf{F}_{xr}\mathbf{K}'_G\mathbf{r} + \mathbf{F}_{xx}\mathbf{X} = -\mathbf{F}_{xr}\mathbf{R}' - \mathbf{c}_1^T(\mathbf{v}_{1,0} - \mathbf{v}'_{1N}) - \mathbf{c}^T\mathbf{v}_0 \tag{8–4b}$$

where

$$\mathbf{F}_{rr} = \mathbf{b}_1^T\mathbf{f}_{11}\mathbf{b}_1 + \mathbf{b}^T\mathbf{f}\mathbf{b} \tag{8–5a}$$

$$\mathbf{F}_{rx} = \mathbf{F}_{xr}^T = \mathbf{b}_1^T\mathbf{f}_{11}\mathbf{c}_1 + \mathbf{b}^T\mathbf{f}\mathbf{c} \tag{8–5b}$$

$$\mathbf{F}_{xx} = \mathbf{c}_1^T\mathbf{f}_{11}\mathbf{c}_1 + \mathbf{c}^T\mathbf{f}\mathbf{c} \tag{8–5c}$$

Equations (8–4) may be solved iteratively in various ways. If solved in the form in which they are written, the solution for **r** and **X**, at the end of an iteration cycle, is used to evaluate $\mathbf{V}_1$ and $\mathbf{v}'_{1N}$ for use in the next cycle. In the procedure outlined earlier, the geometric stiffness terms are placed on the right-hand side, and fewer equations are solved simultaneously in each cycle. It is also possible to solve Eq. (8–4b) for **X** and to perform the iterations on **r**, within one cycle of the method. Equation (8.1–4a) takes then the form

$$\mathbf{r} = -\mathbf{F}\mathbf{K}'_G\mathbf{r} + \mathbf{r}_0 \tag{8–6}$$

where

$$\mathbf{F} = \mathbf{F}_{rr} - \mathbf{F}_{rx}\mathbf{F}_{xx}^{-1}\mathbf{F}_{xr} \tag{8–7}$$

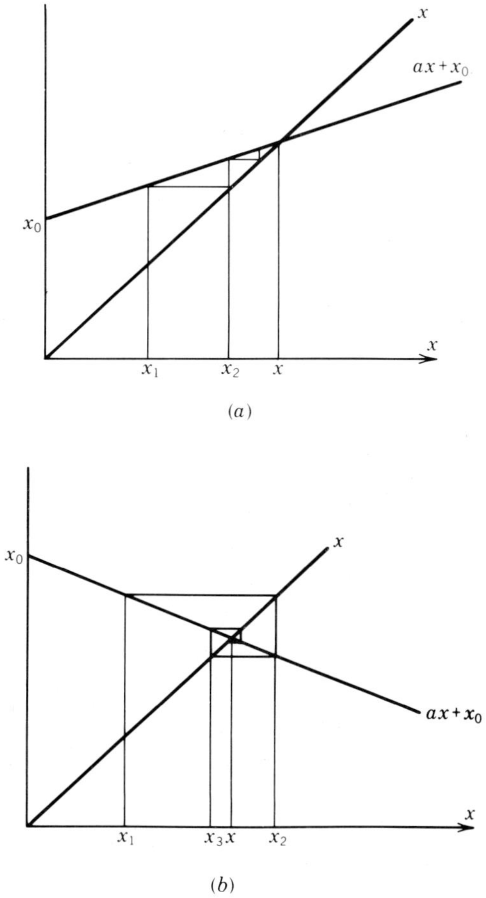

FIG. 8–1. Iterative solution of $x = ax + x_0$.

$\mathbf{F}$ is the flexibility matrix in the linear theory, with member bending flexibilities modified by the effect of axial forces.

Iterative evaluation of $\mathbf{r}$ may be shown to converge if the characteristic values of $-\mathbf{FK}'_G$ are less than unity. A representation of the iterative process is shown in Fig. 8–1 for an equation of the form

$$x = ax + x_0,$$

with $|a| < 1$. $x_1, x_2, \ldots$ represent the successive approximations to the solution x. If the line $ax + x_0$ tends to be parallel to the line x, the solution for x tends to infinity.

Near instability, the total coefficient matrix of $\mathbf{r}$ tends to be singular, one characteristic value tends toward unity, and the iterative process diverges. A criterion for instability is then

$$|\mathbf{I} + \mathbf{FK}'_G| = 0 \tag{8–8}$$

This is not the only criterion, however. In the case of continuous beams, it was seen in Section 3 that a criterion for instability is

$$|\mathbf{F}_{xx}| = 0 \tag{8–9}$$

A similar criterion holds for statically indeterminate structures having only joint rotations.

For statically determinate structures, $\mathbf{F} = \mathbf{F}_{rr}$. $\mathbf{F}_{rr}$ becomes infinite if a flexibility coefficient of any member becomes infinite. Individual member instability is thus also a criterion to be considered.

9. PIN-JOINTED TRUSSES: FLEXIBILITY METHOD

Equation (8–1) are specialized in what follows to the case of pin-jointed trusses. The effective joint displacements are the joint translations. Effective member variables are $\mathbf{s}^m = \{s_2^a \;\; s_2^b\}$ and $\mathbf{S}^m = \{S_2^a \;\; S_2^b\}$. Effective deformations and member statical redundants are the axial ones, v_1 and V_1.

Possible bending of the member will be kept in the formulation. With zero member end moments, the bending deformations caused by a transverse load are $\mathbf{v} = \mathbf{v}_0$. Then, from Eq. (8–2), with $\mathbf{c}_1$ from Eq. (5–1),

$$v'_{1N} = \frac{1}{2l}(s_2^b - s_2^a)^2 + \frac{1}{2}\mathbf{v}_0^T \mathbf{g} \mathbf{v}_0 \tag{9–1}$$

Geometric stiffness terms due to axial span loads are assumed negligible for simplicity. Then $\mathbf{K}'_G = \mathbf{K}_G$, and the governing equations are

$$\mathbf{v}_{1L} = \mathbf{a}_{1v}\mathbf{r} \tag{9–2a}$$

$$\mathbf{a}_{1v}^T\mathbf{V}_1 + \mathbf{K}_G\mathbf{r} = \mathbf{R}' \tag{9–2b}$$

$$\mathbf{v}_{1L} = \mathbf{f}_{11}\mathbf{V}_1 - \mathbf{v}'_{1N} + \mathbf{v}_{1,0} \tag{9–2c}$$

The static and kinematic equations are solved in the form

$$\mathbf{V}_1 = \mathbf{b}_1(\mathbf{R}' - \mathbf{K}_G\mathbf{r}) + \mathbf{c}_1\mathbf{X} \tag{9–3a}$$

$$\mathbf{r} = \mathbf{b}_1^T\mathbf{v}_{1L} \tag{9–3b}$$

$$\mathbf{c}_1^T\mathbf{v}_{1L} = 0 \tag{9–3c}$$

The compatibility equations expressed in terms of $\mathbf{r}$ and $\mathbf{X}$ are obtained from the general ones, Eqs. (8–4, 5), by deleting the terms in $\mathbf{b}$ and $\mathbf{c}$. Since $\mathbf{f}_{11}$ is independent of the axial forces, the flexibility matrices $\mathbf{F}_{rr}$, $\mathbf{F}_{rx}$, and $\mathbf{F}_{xx}$ are the same as in the geometrically linear theory. In the iterative solution, $\mathbf{X}$ is eliminated once from Eqs. (8–4), and the iterations are performed only on Eq. (8–6).

Example 1

Consider the two-bar truss of Fig. 5–4*a*. Because of symmetry $\mathbf{V}_1$ may be taken to consist of one member axial force V_1, and $\mathbf{r}$ consists of one displacement r. Equation (9–3a, b) are specialized in the form

$$V_1 = b_1(R - K_G r)$$

$$r = 2b_1 v_{1L}$$

b_1 is obtained by joint equilibrium in the undeformed geometry. Noting the sign convention for R in Fig. 5–4*a*, there comes

$$b_1 = \frac{-1}{2\sin\theta}$$

From Eq. (9–2c), with $v_{1,0} = 0$ and v'_{1N} from Eq. (9–1),

$$v_{1L} = \frac{V_1 l}{EA} - \frac{1}{2l}(r\cos\theta)^2$$

K_G is obtained by transforming $k_G = V_1/l$ to global coordinates, and multiplying the result by 2, to account for the two members. There comes

$$K_G = 2\frac{V_1}{l}\cos^2\theta$$

and

$$V_1 = \frac{-1}{2\sin\theta}\left(R - 2\frac{V_1 r}{l}\cos^2\theta\right) \tag{9–4}$$

$$r = \frac{1}{\sin\theta}\left(-\frac{V_1 l}{EA} + \frac{r^2\cos^2\theta}{2l}\right) \tag{9–5}$$

Letting

$$x = \frac{r\cos^2\theta}{l\sin\theta} \tag{9–6}$$

Equation (9–4) is solved for V_1 in terms of R and x, and Eq. (9–5) is then used to determine R in terms of x. This yields

$$V_1 = -\frac{R}{2(1-x)\sin\theta} \tag{9–7a}$$

$$R = 2EA\frac{\sin^3\theta}{\cos^2\theta}x(1-x)\left(1-\frac{x}{2}\right) \tag{9–7b}$$

For a given R, Eq. (9–7b) is a cubic equation in x whose solution determines r and V_1 by means of Eqs. (9–6) and (9–7), respectively. If Eq. (9–7b) is used to plot R versus r, the load-deflection curve of Fig. 5–4*c* is obtained.

To obtain the snap-through condition, Eq. (9–7b) is differentiated with respect to R and x, and a nontrivial $dx \neq 0$, with $dR = 0$, is sought. This gives $(1-x)^2 = \frac{1}{3}$, or

$$x = 1 - \frac{1}{\sqrt{3}} \tag{9–8}$$

The load at snap-through is obtained from Eq. (9–7b) as

$$R_{cr} = \frac{2EA}{3\sqrt{3}}\frac{\sin^2\theta}{\cos^2\theta} \tag{9–9a}$$

For small θ the preceding result simplifies to

$$R_{cr} = \frac{2EA}{3\sqrt{3}}\theta^3 \tag{9–9b}$$

10. INEXTENSIBLE FRAMES: FLEXIBILITY METHOD

In the flexibility method the inextensibility assumption may be implemented in the general procedure simply by neglecting the terms involving the axial flexibility $\mathbf{f}_{11}$. The flexibility matrices in Eqs. (8–5) are then formed using only the matrices $\mathbf{b}$ and $\mathbf{c}$ and the 2×2 member flexibility matrices $\mathbf{f}$.

For a formulation of governing equations in terms of inextensional displacements, the transformation $\mathbf{r} = \mathbf{Tq}$ of Section 6 is applied to the connectivity equations, and the equilibrium equations are transformed according to the contragradient law. We thus obtain from Eqs. (8–1a, b)

$$\begin{Bmatrix} \mathbf{v}_{1L} \\ \mathbf{v} \end{Bmatrix} = \begin{bmatrix} \mathbf{a}_{1L} \\ \mathbf{a}_v \end{bmatrix} [\mathbf{T}_1 \ \ \mathbf{T}_2] \begin{Bmatrix} \mathbf{q}_1 \\ \mathbf{q}_2 \end{Bmatrix} = \begin{bmatrix} 0 & \boldsymbol{\alpha}_a \\ \boldsymbol{\alpha}_1 & \boldsymbol{\alpha}_2 \end{bmatrix} \begin{Bmatrix} \mathbf{q}_1 \\ \mathbf{q}_2 \end{Bmatrix}$$

$$\begin{bmatrix} \mathbf{T}_1^T \\ \mathbf{T}_2^T \end{bmatrix} [\mathbf{a}_{1L}^T \mathbf{V}_1 + \mathbf{a}_v^T \mathbf{V}' + \mathbf{K}_G'(\mathbf{T}_1\mathbf{q}_1 + \mathbf{T}_2\mathbf{q}_2)] = \begin{bmatrix} \mathbf{T}_1^T \\ \mathbf{T}_2^T \end{bmatrix} \mathbf{R}'$$

These equations are separated into two parts corresponding, respectively, to extensional and inextensional degrees of freedom. The extensional part is

$$\mathbf{v}_{1L} = \boldsymbol{\alpha}_a \mathbf{q}_2 \tag{10–1a}$$

$$\boldsymbol{\alpha}_a^T \mathbf{V}_1 + \boldsymbol{\alpha}_2^T \mathbf{V}' + \mathbf{G}_{21}\mathbf{q}_1 + \mathbf{G}_{22}\mathbf{q}_2 = \mathbf{Q}_2 \tag{10–1b}$$

and the inextensional part is

$$\mathbf{v} = \boldsymbol{\alpha}_1 \mathbf{q}_1 + \boldsymbol{\alpha}_2 \mathbf{q}_2 \tag{10–2a}$$

$$\boldsymbol{\alpha}_1^T \mathbf{V}' + \mathbf{G}_{11}\mathbf{q}_1 + \mathbf{G}_{12}\mathbf{q}_2 = \mathbf{Q}_1 \tag{10–2b}$$

where

$$\begin{bmatrix} \mathbf{G}_{11} & \mathbf{G}_{12} \\ \mathbf{G}_{21} & \mathbf{G}_{22} \end{bmatrix} = \begin{bmatrix} \mathbf{T}_1^T \\ \mathbf{T}_2^T \end{bmatrix} \mathbf{K}_G' [\mathbf{T}_1 \ \ \mathbf{T}_2] \tag{10–3a}$$

$$\begin{Bmatrix} \mathbf{Q}_1 \\ \mathbf{Q}_2 \end{Bmatrix} = \begin{bmatrix} \mathbf{T}_1^T \\ \mathbf{T}_2^T \end{bmatrix} \mathbf{R}' \tag{10–3b}$$

In the inextensional approximation the terms in $\mathbf{q}_2$ are neglected in Eqs. (10–2). The governing equations are then

$$\mathbf{v} = \boldsymbol{\alpha}_1 \mathbf{q}_1 \tag{10–4a}$$

$$\boldsymbol{\alpha}_1^T \mathbf{V}' + \mathbf{G}_{11}\mathbf{q}_1 = \mathbf{Q}_1 \tag{10–4b}$$

$$\mathbf{v} = \mathbf{f}\mathbf{V}' + \mathbf{v}_0 \tag{10–4c}$$

The force method for solving Eqs. (10–4) is similar to that of the general case. Equation (10–4a, b) are solved in the form

$$\mathbf{V}' = \mathbf{b}(\mathbf{Q}_1 - \mathbf{G}_{11}\mathbf{q}_1) + \mathbf{cX} \tag{10–5a}$$

$$\mathbf{q}_1 = \mathbf{b}^T \mathbf{v} \tag{10–5b}$$

$$\mathbf{c}^T \mathbf{v} = 0 \tag{10–5c}$$

The last two equations expressed in terms of $\mathbf{q}_1$ and $\mathbf{X}$ take the form

$$(\mathbf{I} + \mathbf{F}_{rr}\mathbf{G}_{11})\mathbf{q}_1 - \mathbf{F}_{rx}\mathbf{X} = \mathbf{F}_{rr}\mathbf{Q}_1 + \mathbf{b}^T\mathbf{v}_0 \qquad (10\text{–}6a)$$

$$-\mathbf{F}_{xr}\mathbf{G}_{11}\mathbf{q}_1 + \mathbf{F}_{xx}\mathbf{X} = -\mathbf{F}_{xr}\mathbf{Q}_1 - \mathbf{c}^T\mathbf{v}_0 \qquad (10\text{–}6b)$$

where

$$\mathbf{F}_{rr} = \mathbf{b}^T\mathbf{f}\mathbf{b} \qquad (10\text{–}7a)$$

$$\mathbf{F}_{rx} = \mathbf{F}_{xr}^T = \mathbf{b}^T\mathbf{f}\mathbf{c} \qquad (10\text{–}7b)$$

$$\mathbf{F}_{xx} = \mathbf{c}^T\mathbf{f}\mathbf{c} \qquad (10\text{–}7c)$$

The coefficient matrices in Eqs. (10–6) depend on the member axial forces which remain governed by the extensional set of equations. The solution, in general, is iterative. An iteration cycle consists in solving the inextensional equations for $\mathbf{X}$ and $\mathbf{q}_1$, then solving the extensional set for $\mathbf{V}_1$ and $\mathbf{q}_2$. However, iteration is not needed if the compressive axial forces are not significantly affected by geometric nonlinearity.

Example 1

For the frame of Fig. 10–1*a*, the load P is assumed to be large in comparison with R_1, so that the axial forces in the columns may be assumed equal to P in evaluating the flexibility coefficients. Assuming inextensional behavior, the deformation of the frame is antisymmetric, and the problem is reduced to the one shown in (b).

Equations (10–4) are now written for the frame in (*b*). Inextensional displacements, $\mathbf{q}_1 = \{r_1 \;\; r_2\}$, and effective deformations, $\mathbf{v} = \{v_1 \;\; v_2 \;\; v_3\}$, are shown in (*c*). For convenience the general sign convention is here replaced by the one shown in (*c*). The connectivity equation is the same as in the linear theory and is found by geometry as

$$\begin{Bmatrix} v_1 \\ v_2 \\ v_3 \end{Bmatrix} = \begin{bmatrix} \dfrac{1}{h} & 0 \\ \dfrac{1}{h} & -1 \\ 0 & 1 \end{bmatrix} \begin{Bmatrix} r_1 \\ r_2 \end{Bmatrix}$$

The geometric stiffness matrix $\mathbf{G}_{11}$ in Eq. (10–3a) is obtained from the string geometric matrix of the column. With $V_1 = -P$,

$$\mathbf{G}_{11} = -\frac{P}{h}\begin{bmatrix} 1 & 0 \\ 0 & 0 \end{bmatrix}$$

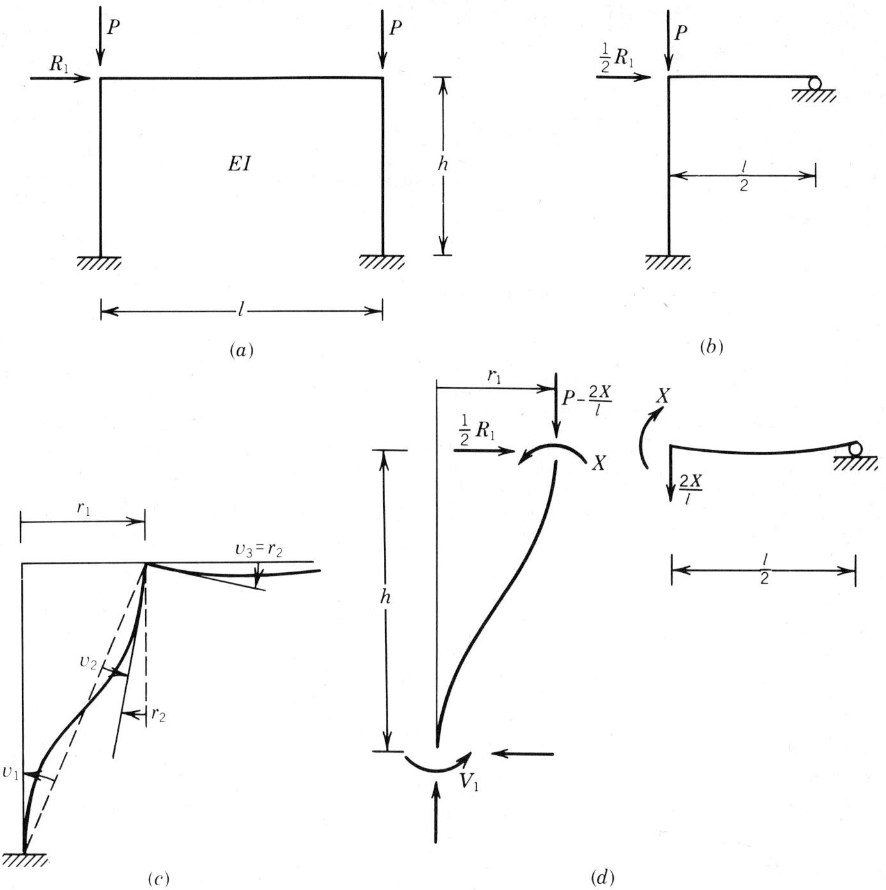

FIG. 10–1. Frame analysis, Example 1.

Here $\mathbf{V}' = \mathbf{V}$, and Eq. (10–4b) takes the form

$$\begin{bmatrix} \dfrac{1}{h} & \dfrac{1}{h} & 0 \\ 0 & -1 & 1 \end{bmatrix} \begin{Bmatrix} V_1 \\ V_2 \\ V_3 \end{Bmatrix} - \begin{Bmatrix} \dfrac{Pr_1}{h} \\ 0 \end{Bmatrix} = \begin{Bmatrix} \dfrac{1}{2}R_1 \\ 0 \end{Bmatrix}$$

The member forces $\{V_1 \; V_2 \; V_3\}$ correspond to the deformations shown in (*c*). Choosing as statical redundant $X = V_3$, the equilibrium equations are solved in the form

$$\begin{Bmatrix} V_1 \\ V_2 \\ V_3 \end{Bmatrix} = \begin{bmatrix} h & 1 \\ 0 & -1 \\ 0 & 0 \end{bmatrix} \begin{Bmatrix} \dfrac{1}{2}R_1 + \dfrac{Pr_1}{h} \\ \\ 0 \end{Bmatrix} + \begin{Bmatrix} -1 \\ 1 \\ 1 \end{Bmatrix} X \tag{10–8}$$

The member flexibility matrix **f** is

$$\mathbf{f} = \begin{bmatrix} f & -f' & 0 \\ -f' & f & 0 \\ 0 & 0 & \dfrac{l}{6EI} \end{bmatrix}$$

The upper 2 × 2 submatrix is the flexibility matrix of the alternate type for the column. From Eq. (6/2.1–8), with $\alpha = 1$,

$$f = \frac{h}{EI\lambda^2}(1 - \lambda \cot \lambda) \tag{10–9a}$$

$$f' = \frac{h}{EI\lambda^2}\left(\frac{\lambda}{\sin \lambda} - 1\right) \tag{10–9b}$$

where

$$\lambda^2 = \frac{Ph^2}{EI}$$

Matrices **b** and **c** are the coefficient matrices on the right-hand side of Eq. (10–8). We thus obtain for the flexibility matrices in Eqs. (10–7),

$$\mathbf{F}_{rr} = \begin{bmatrix} h^2 f & h(f + f') \\ h(f + f') & 2(f + f') \end{bmatrix}$$

$$\mathbf{F}_{xr} = -h(f + f') - 2(f + f')$$

$$\mathbf{F}_{xx} = 2(f + f') + \frac{l}{6EI}$$

then

$$\mathbf{F}_{rr}\mathbf{G}_{11} = -P\begin{bmatrix} hf & 0 \\ (f + f') & 0 \end{bmatrix}$$

$$\mathbf{F}_{xr}\mathbf{G}_{11} = P[(f + f') \;\; 0]$$

Equations (10–6) yield the three simultaneous equations

$$(1 - Phf)r_1 + h(f + f')X = \tfrac{1}{2}h^2 f R_1$$

$$-P(f + f')r_1 + r_2 + 2(f + f')X = \tfrac{1}{2}h(f + f')R_1$$

$$-P(f + f')r_1 + \left[2(f + f') + \frac{l}{6EI}\right]X = \tfrac{1}{2}h(f + f')R_1$$

Elimination of r_1 from the last two equations yields

$$r_2 = \frac{lX}{6EI}$$

The solution for X is found to be

$$X = \frac{1}{2} \frac{hR_1(f + f')}{(l/6EI)(1 - Phf) + 2(f + f') - Ph(f^2 - f'^2)}$$

With f and f' from Eqs. (10–9), and with $\lambda^2 = Ph^2/EI$, X may be put in the form

$$X = \frac{1}{2} hR_1 \frac{\tan \lambda \tan (\lambda/2)}{\lambda[\tan \lambda + (\lambda l/h)/6]} \qquad (10\text{–}10)$$

r_1 is then found as

$$r_1 = \frac{R_1 h^3}{EI} \frac{\tan (\lambda/2) - (\lambda/2) + (l/12h)\lambda(1 - \lambda \cot \lambda)}{\lambda^3[1 + (l/6h)\lambda \cot \lambda]} \qquad (10\text{–}11)$$

Using the data of the frame in Fig. 4.1–3a, with $P = 300$ kips, we obtain

$$\lambda^2 = \frac{Ph^2}{EI} = \frac{300(240)^2}{30{,}000(100)} = 5.76$$

and

$$X = 228 \text{ k-in}$$

$$r_1 = 1.240 \text{ in}$$

The forces at the top end of the column are shown in Fig. 10–1*d*. The determined axial force $(P - 2X/l)$ differs negligibly from the assumed P and no iteration is required. Equilibrium of forces in (*d*) yields for the moment V_1 at the fixed end

$$V_1 = \frac{hR_1}{2} + \left(P - \frac{2X}{l}\right) r_1 - X = 262 \text{ k-in}$$

If the first of Eqs. (10–8) is used, $(P - 2X/l)$ is replaced by P because P is the assumed axial force in the geometric stiffness matrix.

Bifurcation buckling of the frame, subjected to P, with $R_1 = 0$, occurs if the denominator of Eq. (10–11) is zero. The condition is

$$\frac{\lambda}{\tan \lambda} + \frac{6h}{l} = 0 \qquad (10\text{–}12)$$

This is in agreement with the result found for $h = l$ by the stiffness method, Eq. (6–9).

11. TRUSSED FRAMES: FLEXIBILITY METHOD

The inextensional displacements for a trussed frame are the joint rotations, and the extensional displacements are the joint translations. Bound axes are chosen along the member chords so that $\mathbf{K}'_G$ corresponds to translational degrees of freedom only. The transformation (10–3a) yields then

$$\mathbf{G}_{11} = \mathbf{G}_{12} = \mathbf{G}_{21} = 0 \tag{11–1}$$

Equations (10–1) and (10–2) simplify, respectively, to

$$\mathbf{v}_{1L} = \boldsymbol{\alpha}_a \mathbf{q}_2 \tag{11–2a}$$

$$\boldsymbol{\alpha}_a^T \mathbf{V}_1 + \boldsymbol{\alpha}_2^T \mathbf{V} + \mathbf{G}_{22} \mathbf{q}_2 = \mathbf{Q}_2 \tag{11–2b}$$

and

$$\mathbf{v} = \boldsymbol{\alpha}_1 \mathbf{q}_1 + \boldsymbol{\alpha}_2 \mathbf{q}_2 \tag{11–3a}$$

$$\boldsymbol{\alpha}_1^T \mathbf{V} = \mathbf{Q}_1 \tag{11–3b}$$

The last set of equations is the same as in the linear theory. It is to be complemented with member bending flexibility relations that depend on V_1.

The extensional behavior of a trussed frame is governed by Eqs. (11–2) in which $\mathbf{V}$ is neglected and by the axial flexibility relations. These equations are the same as for the substitute truss, except for the term $\mathbf{v}'_{1N}$ which depends on the deformed shape of the member as well as on the rotation of the member chord. Assuming the effect of bending on v'_{1N} may be neglected, analysis of the substitute truss yields the axial forces $\mathbf{V}_1$. These are then used to determine the bending flexibility matrices, and an inextensional analysis is carried out for bending behavior. With $\mathbf{G}_{11} = 0$, Eqs. (10–6) specialize to

$$\mathbf{q}_1 = \mathbf{F}_{rx} \mathbf{X} + \mathbf{F}_{rr} \mathbf{Q}_1 + \mathbf{b}^T \mathbf{v}_0 \tag{11–4a}$$

$$\mathbf{F}_{xx} \mathbf{X} = -\mathbf{F}_{xr} \mathbf{Q}_1 - \mathbf{c}^T \mathbf{v}_0 \tag{11–4b}$$

Geometric nonlinearity affects Eqs. (11–4) only in the dependence of bending flexibilities on the axial forces. The analysis is otherwise linear. Examples have been treated in Section 3, for a continuous beam and for a frame.

Instability of the frame in joint-rotation modes is detected by a nearly singular $\mathbf{F}_{xx}$ and large joint rotations $\mathbf{q}_1$. The corresponding instability criterion is

$$|\mathbf{F}_{xx}| = 0 \tag{11–5}$$

12. LARGE ROTATION THEORY, FLEXIBILITY METHOD: INCREMENTAL AND NEWTON–RAPHSON METHODS

Finite governing equations, with large rotations of member-bound axes, have the form

$$\mathbf{a}_v^T \mathbf{V} - \mathbf{R} + \mathbf{R}_p = 0 \tag{12–1a}$$

$$\mathbf{v} - \mathbf{v}(\mathbf{r}) = 0 \tag{12–1b}$$

$$\mathbf{v} - \mathbf{v}(\mathbf{V}) = 0 \tag{12–1c}$$

Equation (12–1a) is the joint equilibrium equation, and Eqs. (12–1b) and (12–1c) represent symbolically the expressions of finite member deformations in terms of joints displacement, and finite constitutive properties.

Incremental equations are derived in Chapter 8, Section 6, in the form

$$\mathbf{a}_v^T d\mathbf{V}' + (\mathbf{K}_G + \mathbf{K}_{GF}) d\mathbf{r} = d\mathbf{R} - \mathbf{a}^T d\mathbf{S}_{p0} \tag{12–2a}$$

$$d\mathbf{v} = \mathbf{a}_v d\mathbf{r} \tag{12–2b}$$

$$d\mathbf{v} = \mathbf{f}_t d\mathbf{V}' + d\mathbf{v}_0 \tag{12–2c}$$

Matrices $\mathbf{a}_v$, $\mathbf{K}_G$, $\mathbf{K}_{GF}$, $d\mathbf{S}_{p0}$, $\mathbf{f}_t$, and $d\mathbf{v}_0$ are all displacement dependent. Details may be found in Chapter 8, Section 6, and in Chapter 7, Section 6.4.

Equations (12–2) are solved by the force method following the general procedure outlined in the second-order theory. Incremental solutions involve no iterations. They determine approximately a deformation path due to a prescribed loading sequence.

In the Newton–Raphson method the incremental variables are corrections to an approximate solution $(\mathbf{r}_0, \mathbf{v}_0, \mathbf{V}_0)$ at a constant load level. Equations (12–1b, c) may be satisfied exactly in each cycle, and their residuals are thus zero. The equations are then

$$\mathbf{a}_v^T \delta\mathbf{V}' + (\mathbf{K}_G + \mathbf{K}_{GF}) \delta\mathbf{r} = \mathbf{R} - \mathbf{R}_p - \mathbf{a}_v^T \mathbf{V}_0 \tag{12–3a}$$

$$\delta\mathbf{v} - \mathbf{a}_v \delta\mathbf{r} = 0 \tag{12–3b}$$

$$\delta\mathbf{v} - \mathbf{f}_t \delta\mathbf{V}' = 0 \tag{12–3c}$$

In applying the preceding equations to the second-order theory, $\mathbf{a}_v$ is different from the connectivity matrix considered in Section 8 and subsequent sections. First, the notation here does not separate axial from bending variables. More important, however, the axial deformations are not separated here into linear and nonlinear parts. The corresponding terms in $\mathbf{a}_v$ are displacement dependent, whereas in preceding sections $\mathbf{a}_{1v}$ is only the constant part of the axial connectivity relation. Thus, if $\mathbf{V}_1$ is varied by $\delta\mathbf{V}_1$ in Eq. (8–1b), the term $(d\mathbf{K}_G)\mathbf{r}$ which is linear in $\delta\mathbf{V}_1$ should be combined with the term $\mathbf{a}_{1v}^T \delta\mathbf{V}_1$. The result is contained in the term $\mathbf{a}_v^T \delta\mathbf{V}'$ in Eq. (12–3a).

EXERCISES

Sections 2, 3

1. In the continuous beam of Fig. 2–1 replace A with a pin support, and let the transverse load be 0.1 k/ft over the two spans. Analyze the beam by the stiffness method, determine the bending moment at B, and estimate the maximum positive moments within the spans. Formulate a procedure for determining exactly the maximum positive moments.

2. Do the analysis of Exercise 1 by the flexibility method.

3. Do the analysis of Section 3, Example 2, by the stiffness method. Use effective column stiffnesses.

Sections 4 to 7

4. Refer to the structure of Fig. 4.1–2*a* and Example 1, Section 4.1. Form the equivalent joint load of the stiffness equations for a distributed lateral load p on the left column.

5. Perform an equilibrium check on the force system of Fig. 4.1–3*c*.

6. Refer to the tabulated solution for r_1 in Fig. 4.1–3*b*. Verify that a good approximation for r_1 is obtained by using the magnification factor $(1-P/P_{cr})^{-1}$ on the value of r_1 for $P=0$. Do the same for the frame of Fig. 6–2.

7. a. Refer to the tabulated solution for r_1, M_A, and M_B in Fig. 4.1–3*b*. Assume that the inflexion point in column AB may be approximately located on the basis of a linear bending moment diagram and that the lateral displacement of the inflexion point may be approximately evaluated on the basis of a linear variation of displacement with distance from the fixed end A. Show that $P\Delta + (M_A)_{P=0}$ is a good approximation to the tabulated M_A.
   b. Verify that the inflexion point in column AB shifts toward the middle of the column as P increases.
   c. Verify that a satisfactory approximation to M_A is obtained as outlined in (a) if the inflexion point is assumed to be at midcolumn height.

8. On the basis of Exercises 6 and 7, devise an approximate procedure for evaluating the effect of geometric nonlinearity on bending moments using results of a linear analysis and assuming that P_{cr} has been determined. Apply this procedure in Exercise 10.

9. Refer to the frame of Fig. 4.3–2. Verify that the magnification factor $(1-P/P_{cr})^{-1}$ provides an approximation for the lateral displacement as done in Exercise 6. Investigate using average column values instead of P and P_{cr}.

10. Refer to the frame of Fig. 4.3–2. Apply the procedure of Exercise 8 to approximate the bending moments at A and B, assuming that the lateral displacement at the first floor is one-fifth that at the fifth floor.

11. Form the geometric stiffness matrix of the truss of Fig. P11, assuming that the axial forces are determined with sufficient accuracy through linear analysis.

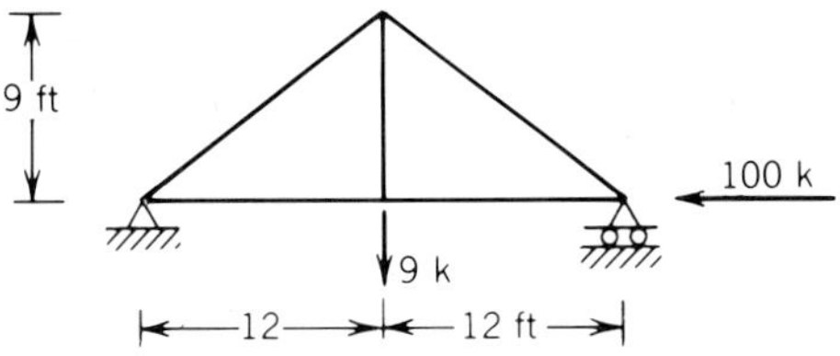

FIG. P11

12. a. For the truss of Fig. 5–2*a*, establish a formula for the maximum normal stress at midspan of member CA. Assume axial forces are accurately determined through linear analysis.

b. Assuming a linearly elastic–perfectly plastic material with a yield stress $\sigma_y = 36$ ksi, determine the load p_y that causes first yield in member CA, using the following data for an I cross section $A = 3.83\ \text{in}^2$, $d =$ depth $= 4.16$ in, $I = 11.3\ \text{in}^4$, $E = 30{,}000$ ksi.

c. Let $p = \frac{1}{2}p_y$. What concentrated vertical force P applied at A would cause first yield in member CA?

13. Do part (c) of Exercise 12 in the case where P acts alone and member AC has an initial eccentricity of 0.1 in at midspan.

14. Formulate Eq. (5–14) for the nondimensional variables Δ/l and P/P_{cr}, and in terms of nondimensional parameters Z/S, l/r, σ_y/E. Draw the dotted curve in Fig. 5–5*b* for $l/r = 80$, $\sigma_y/E = 1.2 \times 10^{-3}$, and two cross-sectional shapes that follow:

a. I cross section having $Z/S = 1.15$. Assume Eq. (6–15) applies.

b. Rectangular cross section. Use Eq. (4/8–17, 18).

15. Verify that the inextensional analysis resulting in Eq. (6–8) is in agreement with the results that include column extensibility recorded in Fig. 4.1–3.

16. Apply the Newton–Raphson procedure to find the lowest root of Eq. (6–9).

17. Derive the stiffness equations for Example 1, Section 6, using the approximate member stiffness matrix established in Chapter 6, Section 3.1. Obtain P_{cr}.

18. Repeat Exercise 17 for the frame of Fig. 6–2, using effective column stiffness matrices. Repeat the exercise by changing the support at D to a roller support.

19. For the frame of Fig. P19 determine the bending moment at the ends and at midspan of member AB.

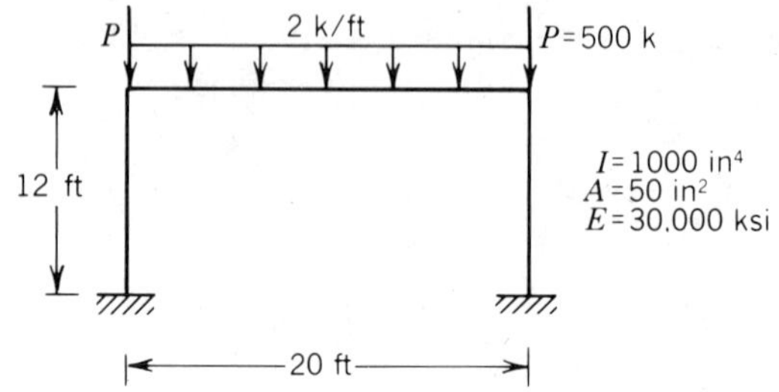

FIG. P19

20. Repeat Exercise 19 assuming that the columns are pinned at the supports.

21. Discuss the use of approximate member stiffness matrices in Exercises 19 and 20.

22. Refer to the frame of Fig. 6–2. Assuming P_{cr} is given, use a linear analysis and the magnification factor $(1 - P/P_{\text{cr}})^{-1}$ to determine the bending moment at B and C.

23. Refer to Example 3, Section 6. Outline how to prepare a plot of P versus r_1 for a fixed R_1.

24. Refer to Example 3, Section 6, and to Eq. (6–19). In a geometrically linear theory, r_1 is ignored in applying the principle of virtual displacements to the mechanism of Fig. 6–3d. Compare the ultimate value of R_1 predicted by the geometrically linear theory to the one found in Example 3.

25. Figure P25 shows two rigid bars connected with a rotational spring of constant stiffness k.
    a. Establish the stiffness equation relating R to r without assuming θ to be small.
    b. Show that $P_{\text{cr}} = k/a$.
    c. Plot P/P_{cr} versus r/r_0 where r_0 is the value of r for a given R and $P = 0$.
    d. Plot P/P_{cr} versus r/a for $R = 0$. How is this plot modified if k varies

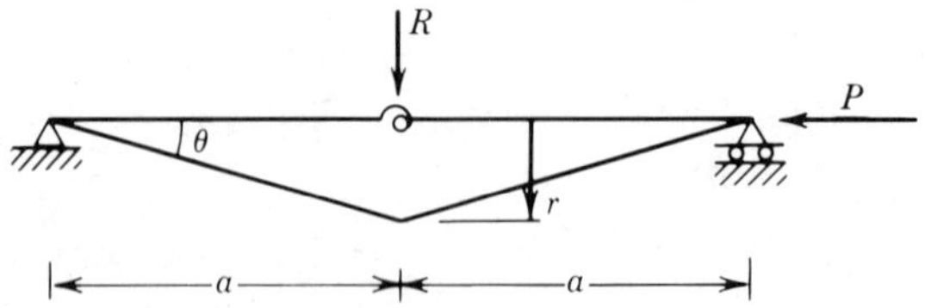

FIG. P25

linearly with r according to the relation

$$k = k_0(1 + \alpha r)$$

where k_0 and α are constant?

e. Discuss the possible use of part (d) as a model for the start of inelastic postbuckling behavior of a straight member.

Sections 8 to 11

26. a. Write the governing equations (8–1) for the truss of Fig. P11. Note that for a structure subjected only to joint loads, $\mathbf{v}_{1,0}$, $\mathbf{v}_0$, $\mathbf{S}_{1p}$, and $\mathbf{S}_p$ are zero and that $\mathbf{V}'$ and $\mathbf{K}'_G$ reduce, respectively, to $\mathbf{V}$ and $\mathbf{K}_G$. Further $\mathbf{V}$ and $\mathbf{v}$ are zero in the present problem.

b. Interpret Eqs. (8–1) as pertaining to a modified linear problem in which the joint load $-\mathbf{K}_G\mathbf{r}$ is applied in addition to the actual joint load. Show on sketches how the elements of $-\mathbf{K}_G\mathbf{r}$ are formed in terms of member axial forces and joint displacements.

c. As a first approximation, neglect $\mathbf{v}'_{1N}$, and determine the transverse displacement at midspan, assuming that the axial forces are those of the linear theory and a constant l/EA. Examine whether an iteration is necessary.

27. Analyze the frame of Fig. 6–2 by the flexibility method, as done in Section 10, Example 1. Compare the results with those of Section 3, Example 3.

28. Analyze the frame of Fig. 10–1 by the three-moment equation. Compare the results to those of the example treated in Section 10.

29. Analyze the frames of the two preceding exercises by superimposing the joint load $-\mathbf{K}_G\mathbf{r}$ and performing a linear analysis.

13

LINEARIZED STABILITY ANALYSIS

1. INTRODUCTION

The concept of stability of equilibrium, and the work and potential-energy criterion of stability, presented for a member in Chapter 6, Section 5, remain applicable to a structure. In Chapter 12 stability problems were encountered within the context of nonlinear analysis.

This chapter deals specifically with the problem of determining critical loads for which a neutral equilibrium state exists. A neutral state admits a neighboring equilibrium state that is governed by the incremental governing equations of Chapter 12, with no load increment. It will be assumed that the change in geometry prior to instability may be neglected. The problem becomes then one of bifurcation of equilibrium from state $\mathbf{r} = 0$, and the only difference between the incremental equations and the equations of the linear theory is due to the effect of member axial forces. This effect appears analytically in two ways. In a stiffness formulation in fixed axes, the effect of axial forces is all represented in the member stiffness equations. In a formulation in bound axes, the effect of axial forces appears separately in the joint equilibrium equations and in the member stiffness or flexibility equations. The governing equations are thus linear in the displacements, but they depend on the axial forces as parameters. The theory is called linearized stability analysis and takes the form of a characteristic-value problem.

Since the equations of linearized stability analysis are incremental, the theory includes elastic as well as inelastic behavior. The application to inelastic buckling of straight columns, treated in Chapter 6, Section 5.2, may be generalized to a structure. The special subject of plastic analysis, in which instability is due to the formation of plastic hinges, requires specific developments for determining the possible mechanism modes and will not be treated here. Problems including both plastic behavior and geometric nonlinearity were treated, however, in examples in Chapter 12.

Linearized stability analysis appears as a buckling problem, but it pertains also to problems in which bifurcation of equilibrium does not actually occur. The critical axial force determined for a straight column is also a parameter by which the nonlinear behavior of an initially bent or transversely loaded column is described. A similar behavior occurs for frames.

Critical values of member axial forces determine member effective lengths and are important parameters in defining safety in structural design.

Linearized stability analysis is treated by the stiffness method in Sections 2 to 7 and by the flexibility method in Sections 8 to 13.

2. STIFFNESS METHOD FOR LINEARIZED STABILITY ANALYSIS: GENERAL PROCEDURE

For a linearized stability analysis by the stiffness method, the member tangent stiffness matrix $\mathbf{k}_{st}$ is obtained by setting $\mathbf{s} = 0$ in Eq. (6/4–2), thereby obtaining

$$\mathbf{k}_{st} = \begin{bmatrix} \mathbf{k}_{11} & \\ & \mathbf{k}_s \end{bmatrix} \tag{2–1}$$

$\mathbf{k}_s$ is to be evaluated for the member axial force obtained through a geometrically linear analysis.

For a linearly elastic material $\mathbf{k}_s$ is the 4×4 bending stiffness matrix of the second-order theory, and $\mathbf{k}_{11}$ is the 2×2 axial stiffness matrix of the linear theory. In the present context these matrices are incremental. Their defining formulas are applicable to nonlinear material behavior if tangent material rigidities are used.

The structure tangent stiffness matrix $\mathbf{K}_t$ is assembled from member matrices $\mathbf{k}_{st}$. It coincides with the matrix $(\mathbf{K}_a + \mathbf{K}_b)$ of the finite stiffness equations

(12/4.1–1). An alternate equilibrium state **r** exists if the homogeneous equations

$$\mathbf{K}_t \mathbf{r} = 0 \tag{2–2}$$

have a nontrivial solution. The instability criterion is then

$$|\mathbf{K}_t| = 0 \tag{2–3}$$

Linearized stability analysis is concerned with finding lowest load levels at which $|\mathbf{K}_t| = 0$. The loading of a structure, in general, may be expressed linearly in terms of several load parameters that allow us to investigate different loading sequences. In a given analysis, only one load parameter, γ, is made to vary. Equation (2–2) represents then a characteristic-value problem, and Eq. (2–3) is the characteristic equation. The lowest root γ_{cr} of Eq. (2–3) defines the critical load, and the solution of Eq. (2–2) the corresponding buckling shape.

The mathematical type of the characteristic equation depends on the manner in which $\mathbf{k}_s$ depends on the axial force. The exact $\mathbf{k}_s$, derived in Eq. (6/2.6–6) for a uniform member with a constant axial force, leads to a transcendental characteristic equation. The approximate $\mathbf{k}_s$, derived on the basis of an assumed shape by Castigliano's first theorem, Eq. (6/3.1–20, 21), has the form

$$\mathbf{k}_s = \mathbf{k}_{sO} + \mathbf{k}_{sG} \tag{2–4}$$

in which $\mathbf{k}_{sG}$ is the total geometric stiffness matrix and is proportional to the axial force. Assuming for simplicity that there is only one load parameter, $\mathbf{k}_{sG}$ is proportional to γ, and the assembled stiffness equations take the form

$$(\mathbf{K}_0 + \gamma \mathbf{K}_{GT})\mathbf{r} = 0 \tag{2–5}$$

$\mathbf{K}_0$ is the stiffness matrix in the geometrically linear theory, and $\gamma\mathbf{K}_{GT}$ is the total geometric stiffness matrix, both based on an assumed shape.

Equation (2–5) represents a linear characteristic-value problem, with symmetric matrices, for which solution techniques are well developed [10].

In using an approximate $\mathbf{k}_s$, it should be ascertained that it is accurate enough at the critical load level. Accuracy of $\mathbf{k}_s$ may be assessed from that of the stiffness matrix in a member-bound reference, which was examined in Chapter 6, Section 3.1. It is found that for the cantilever type, the stiffness coefficients are accurate enough for a stability parameter λ reaching π, that is, for an axial force equal to Euler's critical load. For the alternate type it is found that a relative error of the order of 5% occurs in the stiffness coefficients as λ reaches 0.9π, that is, for an axial force equal to 80% of Euler's critical load.

An improved member stiffness matrix may also be used. It involves a larger number of degrees of freedom than the member end displacements. If the degrees of freedom are condensed, the resulting matrix depends nonlinearly on the axial force.

A nonlinear characteristic equation could be solved by trial and error, or iteratively as by the Newton–Raphson method.

In general, the accuracy of an approximate stiffness matrix improves, at a fixed axial force, with decreasing member slenderness. Thus a member could be treated as formed of several members for which an accurate stiffness matrix is more readily formed.

For nonlinear material behavior the tangent rigidities required to form the member stiffness are usually unknown at the critical load. An iterative process is then followed by which the tangent rigidities are assumed at the start of an iteration cycle and are adjusted in the following cycle. A simple illustration of this procedure is the determination of the inelastic buckling load of an axially compressed column by the tangent modulus theory.

3. BUCKLING OF PIN-JOINTED TRUSSES: STIFFNESS METHOD

For pin-jointed trusses the tangent stiffness matrix in linearized stability analysis is the coefficient of $\mathbf{r}$ in Eq. (12/5–5). This is $\mathbf{K}_t = \mathbf{K}_a + \mathbf{K}_G$, where $\mathbf{K}_a$ is the stiffness matrix of the geometrically linear theory and $\mathbf{K}_G$ is the geometric stiffness matrix, which is assembled from the string geometric stiffness

$$\mathbf{k}_G = V_1 \mathbf{c}_1 = \frac{V_1}{l}\begin{bmatrix} 1 & -1 \\ -1 & 1 \end{bmatrix} \tag{3–1}$$

The displacements associated with $\mathbf{k}_G$ are the transverse member end displacements $\mathbf{s} = \{s_2^a \; s_2^b\}$. $\mathbf{K}_a$ is assembled from the axial stiffness matrix

$$\mathbf{k}_{11} = k\begin{bmatrix} 1 & -1 \\ -1 & 1 \end{bmatrix} \tag{3–2}$$

where k is the member axial stiffness, which is denoted generally as k_{11}. The displacements associated with $\mathbf{k}_{11}$ are the axial ones, $\mathbf{s}_1 = \{s_1^a \; s_1^b\}$.

The possibility of individual member buckling as an Euler column is investigated independently. Its effect on the behavior of the structure is discussed in Chapter 12.

Example 1

For the truss of Fig. 3–1*a* members 1 and 2 have axial stiffnesses k_1 and k_2, respectively. Letting $c = \cos\theta$ and $s = \sin\theta$, the stiffness matrix $\mathbf{K}_a$ is formed as in the linear theory and is found to be

$$\mathbf{K}_a = \begin{bmatrix} c^2 k_1 & csk_1 \\ csk_1 & s^2 k_1 + k_2 \end{bmatrix}$$

The effective $\mathbf{k}_G$ for member 2 is

$$\mathbf{k}_G = -\frac{P}{l}[1]$$

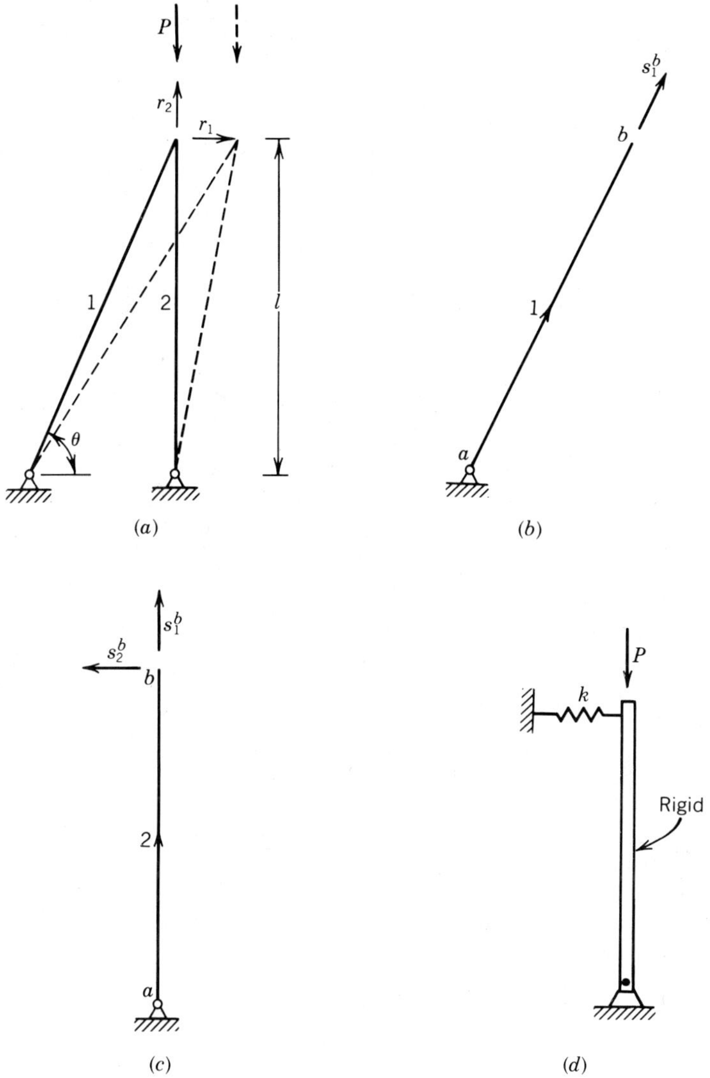

FIG. 3–1. Buckling of two-bar truss, Section 3, Example 1; Section 9, Example 1.

and the transverse connectivity relation is

$$s_2^b = \mathbf{ar} = [-1 \;\; 0]\{r_1 \;\; r_2\}$$

Member 1 has zero axial force in the prebuckling state. Thus

$$\mathbf{K}_G = \mathbf{a}^T\mathbf{k}_G\mathbf{a} = -\frac{P}{l}\begin{bmatrix} 1 & 0 \\ 0 & 0 \end{bmatrix}$$

and

$$\mathbf{K}_t = \mathbf{K}_a + \mathbf{K}_G = \begin{bmatrix} c^2k_1 - \dfrac{P}{l} & csk_1 \\ csk_1 & s^2k_1 + k_2 \end{bmatrix}$$

The characteristic equation is

$$|\mathbf{K}_t| = \left(c^2k_1 - \frac{P}{l}\right)(s^2k_1 + k_2) - c^2s^2k_1 = 0$$

The root may be put in the form

$$P_{\mathrm{cr}} = k_1 l \frac{c^2}{1 + s^2k_1/k_2} \tag{3–3}$$

The alternate equilibrium state, shown in dotted lines in Fig. 3–1*a*, is the characteristic shape, solution of $\mathbf{K}_t\mathbf{r} = 0$, for $P = P_{\mathrm{cr}}$. It is indeterminate to within a multiplicative constant.

The preceding solution is valid, provided P_{cr} is less than Euler's load for the column, or $P_{\mathrm{cr}} < \pi^2 EI/l^2$. Euler buckling of the vertical member could occur first, unless $c^2 \ll 1$ or k_1 is small enough.

If $k_1/k_2 \ll 1$, the vertical member is effectively infinitely rigid. Equation (3–3) reduces then to $P_{\mathrm{cr}} = k_1 lc^2$. For a direct formulation of the buckling problem, treating the vertical member as infinitely rigid, $\mathbf{r}$ consists of r_1, and $\mathbf{K}_a$ is due only to the inclined member. $\mathbf{K}_a$ and $\mathbf{K}_G$ reduce each to the term in position (1, 1) in the previously determined matrices. Then $\mathbf{K}_t = c^2k_1 - P/l$, and $K_t = 0$ yields the solution found earlier.

The system treated here has the representation shown in Fig. 3–1*d* in which the spring stiffness is $k = k_1c^2$.

Example 2

The three pin-connected bars in Fig. 3–2*a* are infinitely rigid. The displacements $\mathbf{r} = \{r_1 \;\; r_2\}$ are as shown. $\mathbf{K}_a$ is due to the springs and is $\mathbf{K}_a = \lceil k_1 \;\; k_2 \rfloor$. The effective $\mathbf{k}_G^m$ for the three members are

$$\mathbf{k}_G^1 = \mathbf{k}_G^3 = -\frac{P}{a}[1]$$

$$\mathbf{k}_G^2 = -\frac{P}{a}\begin{bmatrix} 1 & -1 \\ -1 & 1 \end{bmatrix}$$

$\mathbf{k}_G^2$ is merged into the $2 \times 2\mathbf{K}_G$, $\mathbf{k}_G^1$ is merged in location (1, 1), and $\mathbf{k}_G^3$ in location

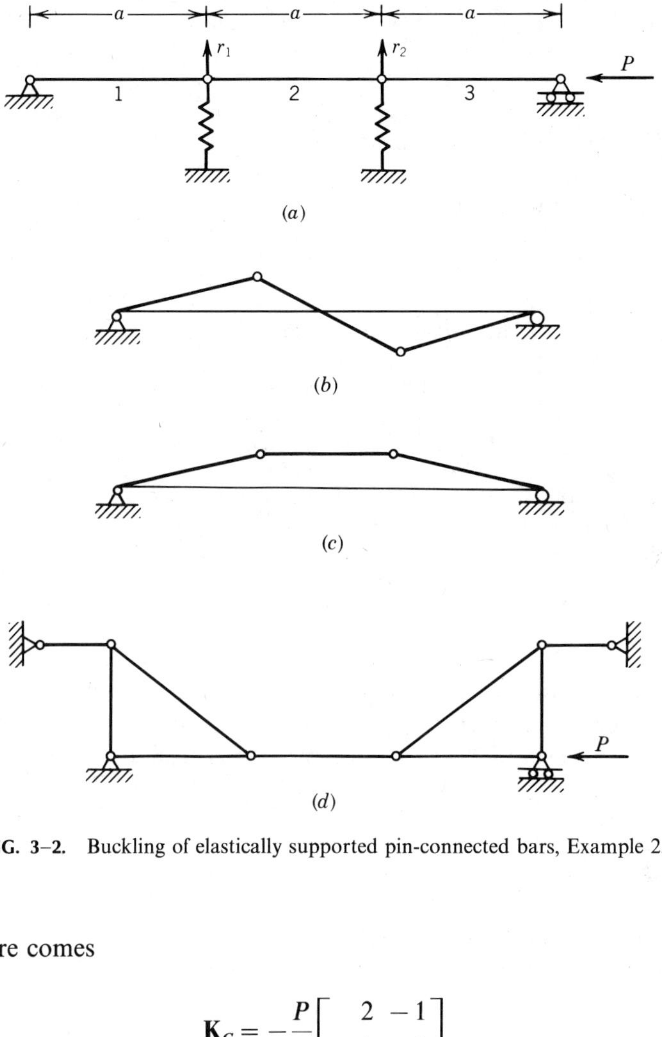

FIG. 3–2. Buckling of elastically supported pin-connected bars, Example 2.

(2, 2). There comes

$$\mathbf{K}_G = -\frac{P}{a}\begin{bmatrix} 2 & -1 \\ -1 & 2 \end{bmatrix}$$

The tangent stiffness matrix is

$$\mathbf{K}_t = \mathbf{K}_a + \mathbf{K}_G = \begin{bmatrix} k_1 - \dfrac{2P}{a} & \dfrac{P}{a} \\ \dfrac{P}{a} & k_2 - \dfrac{2P}{a} \end{bmatrix}$$

The characteristic equation $|\mathbf{K}_t| = 0$ may be put in the form

$$3P^2 - 2a(k_1 + k_2)P + a^2 k_1 k_2 = 0$$

For the case $k_1 = k_2 = k$, the roots are

$$P_1 = \tfrac{1}{3}ka$$
$$P_2 = ka$$

The critical load is the lowest root

$$P_{\text{cr}} = \tfrac{1}{3}ka \tag{3–4}$$

The buckled shape is the solution of $\mathbf{K}_t\mathbf{r} = 0$ for $P = \frac{1}{3}ka$, or

$$\begin{bmatrix} \frac{1}{3}k & \frac{1}{3}k \\ \frac{1}{3}k & \frac{1}{3}k \end{bmatrix} \begin{Bmatrix} r_1 \\ r_2 \end{Bmatrix} = 0$$

Thus $r_1 = -r_2$, and the shape is as shown in Fig. 3–2*b*.

To the second root, $P_2 = ka$, corresponds the equation

$$\begin{bmatrix} -k & k \\ k & -k \end{bmatrix} \begin{Bmatrix} r_1 \\ r_2 \end{Bmatrix} = 0$$

and the characteristic shape $r_1 = r_2$, as shown in (*c*). This is a possible but unstable alternate equilibrium state. If member 2 is prevented from rotating without restraining its translation, then P_2 becomes the critical load.

The system shown in Fig. 3–2*a* may be considered as a model representing the truss of Fig. 3–2(*d*).

Example 3

Consider the truss of Fig. 3–3*a* with displacements $\mathbf{r} = \{r_1 \ldots r_7\}$ as shown. $\mathbf{K}_a$ and $\mathbf{K}_G$ are 7 × 7 matrices. $\mathbf{K}_G$ is assembled from $\mathbf{k}_G^m$ for members 1 and 2. For both members

$$\mathbf{k}_G = -\frac{P}{a}[1]$$

$\mathbf{K}_G$ is formed by merging $\mathbf{k}_G$ twice in location (1, 1), whence

$$\mathbf{K}_G = -\frac{P}{a}\left[\begin{array}{c|c} 2 & \mathbf{O} \\ \hline \mathbf{O} & \mathbf{O} \end{array}\right]_{7\times 7}$$

$\mathbf{K}_a$ may be partitioned in the form

$$\mathbf{K}_a = \begin{bmatrix} K_{11} & \mathbf{K}_{12} \\ \mathbf{K}_{21} & \mathbf{K}_{22} \end{bmatrix}_{7\times 7}$$

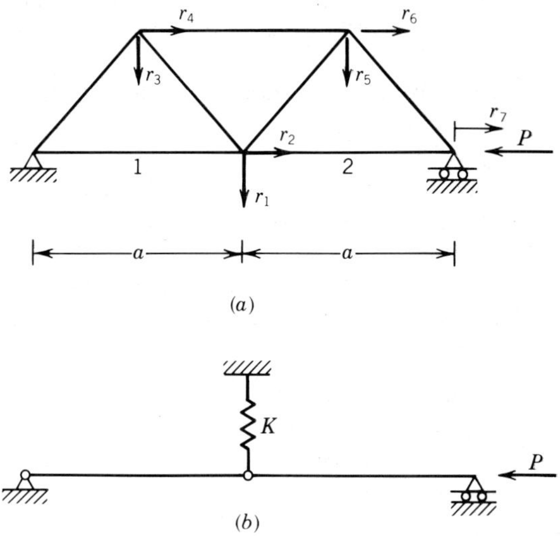

FIG. 3–3. Buckling of pin-jointed truss, Section 3, Example 3; Section 9, Example 2.

The homogeneous stiffness equations are then

$$\mathbf{K}_t\mathbf{r} = \begin{bmatrix} K_{11} - \dfrac{2P}{a} & \mathbf{K}_{12} \\ \mathbf{K}_{21} & \mathbf{K}_{22} \end{bmatrix} \begin{Bmatrix} r_1 \\ \mathbf{r}_2 \end{Bmatrix} = 0$$

where

$$\mathbf{r}_2 = \{r_2 \dots r_7\}$$

Now $\mathbf{r}_2$ is condensed by solving the second part of this equations for $\mathbf{r}_2$ and substituting into the first part. There comes

$$\mathbf{r}_2 = -\mathbf{K}_{22}^{-1}\mathbf{K}_{21}r_1$$

and

$$\left(K - \frac{2P}{a}\right)r_1 = 0$$

where

$$K = K_{11} - \mathbf{K}_{12}\mathbf{K}_{22}^{-1}\mathbf{K}_{21}$$

Then

$$P_{\text{cr}} = \frac{Ka}{2}$$

The truss is equivalent to the bar-spring system shown in (*b*). The spring stiffness K, which was determined here by static condensation, is more efficiently determined by the flexibility method as $K = F^{-1}$, where F is the displacement r_1 caused by a unit force $R_1 = 1$. See Example 2 of Section 9.

4. BUCKLING OF CONTINUOUS BEAMS: STIFFNESS METHOD

The stiffness equations for continuous beams are presented in Chapter 12, Section 2. For linearized stability analysis the loading terms on the right-hand sides of these equations are set to zero. Using, for simplicity, the notation adopted for uniform spans, the homogeneous stiffness equation at joint i, Eq. (12/2–1), is

$$k'_{i-1} r_{i-1} + (k_{i-1} + k_i) r_i + k'_i r_{i+1} = 0 \qquad (4\text{–}1)$$

The indexes refer to spans or to joints as shown in Fig. 10/2–1.

If the exact formulas for the member stiffness coefficients, Eqs. (12/2–2), are used, the characteristic value problem is transcendental and is solved by trial and error.

The approximate stiffness coefficients based on the deformed shape of the linear theory and a constant axial force are

$$k = \frac{EI}{l}\left(4 - \frac{2\lambda^2}{15}\right) \qquad (4\text{–}2a)$$

$$k' = \frac{EI}{l}\left(2 + \frac{\lambda^2}{30}\right) \qquad (4\text{–}2b)$$

where

$$\lambda^2 = -\frac{V_1 l^2}{EI} \qquad (4\text{–}3)$$

Use of these stiffness coefficients leads to a linear characteristic-value problem.

Example 1

For the continuous beam shown in Fig. 4–1, Eq. (4–1) is written at joint 2, and the end conditions $r_1 = r_3 = 0$ are enforced. There comes $(k_1 + k_2) r_2 = 0$.

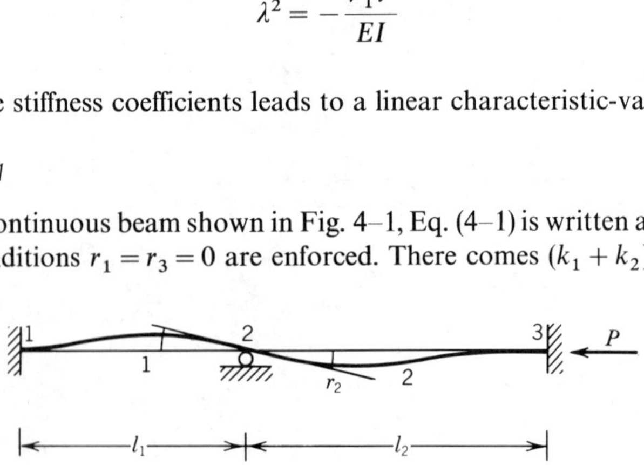

FIG. 4–1. Buckling of continuous beam, Section 4, Example 1; Section 10, Example 2.

Consider first the case where $k_1 = k_2 = k$. The condition $k = 0$ is the condition for buckling of the proped cantilever for which $\lambda_{cr} = \pi/0.7$. Because the two members have the same rotational stiffness k, they buckle without interaction when $k = 0$.

Consider now the case where the two members have properties I_1, l_1 and I_2, l_2, respectively. Since both have the same axial force, Eq. (4–3) yields the relation

$$\frac{\lambda_1}{\lambda_2} = \frac{l_1}{l_2}\sqrt{\frac{I_2}{I_1}} \tag{4–4a}$$

The stiffness coefficients are obtained from Eq. (12/2–2a). The characteristic equation, expressed as $k_1/k_2 + 1 = 0$, takes the form

$$\frac{1 - \lambda_1 \cot \lambda_1}{1 - \lambda_2 \cot \lambda_2} \, \frac{\tan(\lambda_2/2) - (\lambda_2/2)}{\tan(\lambda_1/2) - (\lambda_1/2)} + \sqrt{\frac{I_2}{I_1}} = 0 \tag{4–4b}$$

To find the lowest root of this equation by trial, it is helpful to have lower or upper bounds.

Assume $\lambda_1 < \lambda_2$. If the members are considered independently as proped cantilevers, they would buckle each at $\lambda = \pi/0.7$. Since $\lambda_2 > \lambda_1$, λ_2 reaches $\pi/0.7$ first, and member 2 would buckle first. Thus for the continuous beam, member 1 will restrain member 2, so that $\lambda_{2,cr} > \pi/0.7$. Conversely, member 1 is helped by member 2 to buckle below the critical load of the proped cantilever, so that $\lambda_{1,cr} < \pi/0.7$. Taking into account the relation between λ_1 and λ_2, we obtain the bounds

$$\frac{\pi}{0.7}\frac{l_1}{l_2}\sqrt{\frac{I_2}{I_1}} < \lambda_{1,cr} < \frac{\pi}{0.7} \tag{4–4c}$$

For example, let $I_1 = I_2$, and $l_2 = 1.5 l_1$. Then $\lambda_2 = 1.5\lambda_1$. The root λ_1 is found by trial with three significant digits as $\lambda_{1,cr} = 3.47$. The effective length of member 1 is thus $\pi l_1/3.27 = 0.91 l_1$, and that of member 2 is $\pi l_2/(1.5 \times 3.47) = 0.60 l_2$.

Use of the approximate stiffness coefficients in this example would be in significant error because they cease to be accurate at lower values than the critical values found.

Example 2

Consider the two-span continuous beam shown in Fig. 4–2*a*. Equation (4–1) is written at joints 2 and 3, with $i = 2$ and $i = 3$, respectively. The condition at joint 1 is $r_1 = 0$, and at joint 3, Eq. (4–1) reduces to the stiffness equation of member 2 at joint 3. Formally, $k_3 = k'_3 = 0$ in Eq. (4–1). There comes

$$(k_1 + k_2) r_2 + k'_2 r_3 = 0$$

$$k'_2 r_2 + k_2 r_3 = 0$$

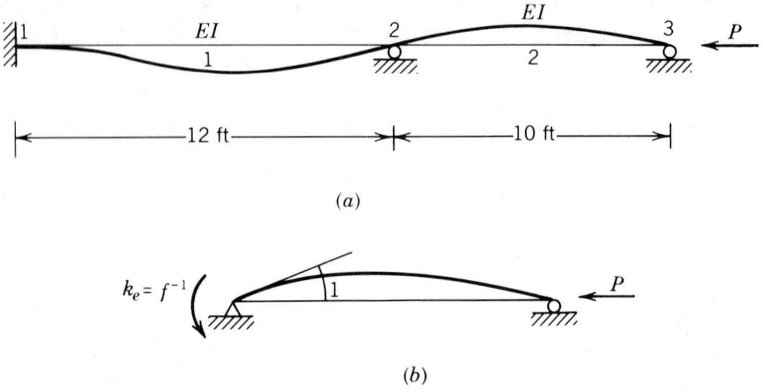

FIG. 4–2. Buckling of continuous beam, Section 4, Example 2; Section 10, Example 1.

Elimination of r_3 yields

$$(k_1 + k_{2,e})r_2 = 0$$

where

$$k_{2,e} = k_2 - \frac{(k_2')^2}{k_2}$$

$k_{2,e}$ is the effective stiffness of member 2, defined at one end with the other end unrestrained against rotation, Fig. 4–2*b*. It is equal to the inverse of the flexibility coefficient f in Eq. (12/3–1a). Deleting the subscript 2, the result is written for a general member as

$$k_e = f^{-1} = \frac{EI\lambda^2}{l} \frac{1}{1 - \lambda \cot \lambda} \tag{4–5}$$

Using subscripts 1 and 2 to refer to members 1 and 2, the characteristic equation is found as

$$k_1 + k_{2,e} = \frac{EI}{2l_1} \frac{\lambda_1(1 - \lambda_1 \cot \lambda_1)}{\tan(\lambda_1/2) - (\lambda_1/2)} + \frac{EI\lambda_2^2}{l_2} \frac{1}{1 - \lambda_2 \cot \lambda_2} = 0 \tag{4–6}$$

Since both members have the same axial force and the same EI, λ_1, and λ_2 have the ratio

$$\frac{\lambda_1}{\lambda_2} = \frac{l_1}{l_2} \tag{4–7a}$$

The characteristic equation may be put in the form

$$(1-\lambda_1 \cot\lambda_1)(1-\lambda_2 \cot\lambda_2)+2\lambda_2\left(\tan\frac{\lambda_1}{2}-\frac{\lambda_1}{2}\right)=0 \qquad (4\text{–}7b)$$

To seek the appropriate solution of this equation, consider the two members separated at joint 2. The member having the lowest critical load will be stiffened by the other member as the two members are joined. For member 2 alone, the critical λ_2 is π, and for member 1 alone, the critical λ_1 is $\pi/0.7 = 1.43\pi$. Since $l_1 = 1.2l_2$, and $\lambda_1 = 1.2\lambda_2, \lambda_2$ reaches the value π before λ_1 reaches the value 1.43π. Thus in the continuous beam member 2 is stiffened by member 1, and $\lambda_{2,\mathrm{cr}} > \pi$. Conversely, $\lambda_{1,\mathrm{cr}} < 1.43\pi$. We thus obtain the bounds

$$\pi < \lambda_{2,\mathrm{cr}} < 1.192\pi$$

or, equivalently

$$1.2\pi < \lambda_{1,\mathrm{cr}} < 1.43\pi$$

The root of the characteristic equation is found to three significant digits to be

$$\lambda_{2,\mathrm{cr}} = 3.38$$

The effective length of member 2 is $\pi l_2/3.38 = 0.93l_2$. For member 1, $\lambda_{1,\mathrm{cr}} = 1.2\lambda_{2,\mathrm{cr}} = 4.05$, and the effective length is $\pi l_1/4.05 = 0.78l_1$.

5. BUCKLING OF FRAMES: STIFFNESS METHOD, INEXTENSIONAL AND EXTENSIONAL MODES

The description of rigid frame behavior into inextensional and extensional displacement modes may be applied to the buckling displacements, and results in a simplification of the instability problem, from both the point of view of the mathematical solution and the physical behavior involved.

The transformed homogeneous incremental equations of linearized stability are obtained from Eq. (12/6–1) in the form

$$\begin{bmatrix} \mathbf{K}_{11} & \mathbf{K}_{12} \\ \mathbf{K}_{21} & \mathbf{K}_a + \mathbf{K}_{22} \end{bmatrix} \begin{Bmatrix} \mathbf{q}_1 \\ \mathbf{q}_2 \end{Bmatrix} = 0 \qquad (5\text{–}1)$$

The behavior of frames in their inextensional and extensional modes was discussed in Chapter 12, Sections 6 and 7.

First consider the case where, at instability, $(\mathbf{K}_a + \mathbf{K}_{22})$ is not nearly singular. The inextensional approximation then applies, and the term in $\mathbf{q}_2$ may be neglected in the inextensional stiffness equations which reduce to

$$\mathbf{K}_{11}\mathbf{q}_1 = 0 \tag{5–2}$$

The assumption that $(\mathbf{K}_a + \mathbf{K}_{22})$ is not nearly singular means that the structure is not near instability in its extensional displacement modes. Frames having translational inextensional modes, and frames having only rotational inextensional modes, are both covered by this assumption. Typical frames of the former type are unbraced rectangular frames. Their buckling modes involve horizontal floor displacements and joint rotations. Instability of such frames in extensional modes corresponds to loadings and stiffness distributions that do not occur in practice. A typical inextensional buckling mode is shown in Fig. 5–1*a*. In (*b*) buckling is also inextensional. However, a rigid bar supported on flexible springs would buckle in an extensional mode as in (*c*).

Frames whose inextensional modes are purely rotational, such as trussed frames, could become unstable in both their rotational and translational modes, depending on the configuration of the structure and its loading. The buckling mode of a trussed frame is rotational if the frame is not near instability in an extensional mode. This latter type of instability may in turn be approximated by that of the substitute truss, as seen in Chapter 12, Section 7.

An example is provided by the truss of Fig. 3–1*a* considered now as a trussed frame with a rigid joint at the load point. In inextensional buckling, the vertical member buckles as a compressed column elastically restrained against translations and against rotation at one end. This would occur if the critical load is not near that determined in Section 3 for the extensional buckling of the substitute truss.

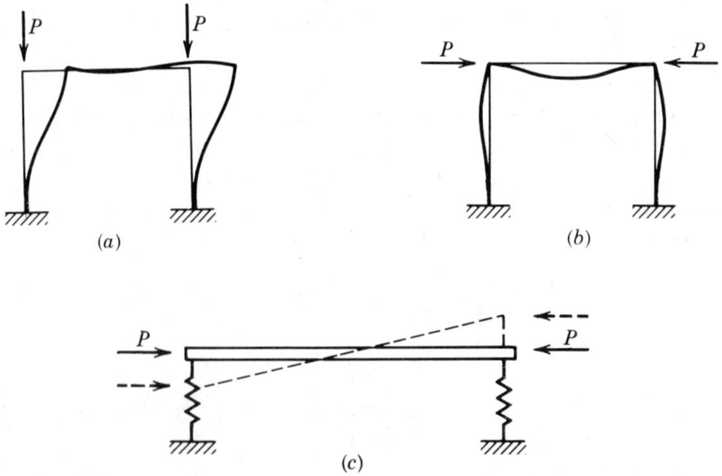

FIG. 5–1. Buckling modes.

In the inextensional buckling of trussed frames, $\mathbf{K}_{11}$ is assembled from the 2×2 reduced stiffness matrix of the alternate type.

Extensional buckling of a trussed frame is approximated by that of the substitute truss, as formulated in the preceding section.

6. BUCKLING OF FRAMES WITHOUT JOINT TRANSLATIONS: STIFFNESS METHOD

Linearized stability analysis of inextensional frames whose joint degrees of freedom are only rotational is a generalization of the analysis of continuous beams seen in a preceding section. Such frames include trussed frames that are not near instability in an extensional mode.

The effective member stiffness matrix is the 2×2 reduced stiffness matrix $\mathbf{k}$ of the alternate type. For uniform members, with a constant compressive axial force, the member stiffness matrix has the form

$$\mathbf{k} = \begin{bmatrix} k & k' \\ k' & k \end{bmatrix} \tag{6–1}$$

where k and k' are as defined in Eqs. (12/2 – 2). The approximate formulas (4–2) can rarely be used because without joint translations the axial force of a given member, at instability of the structure, is usually larger than Euler's critical axial force. Members whose axial forces are small have the stiffness coefficients of the linear theory.

A typical homogeneous stiffness equation at a rigid joint i, such as shown in Fig. 6–1, has the form

$$\left(\sum_m k_m\right) r_i + \sum_m k'_m r_j = 0 \tag{6–2}$$

where subscript m refers to the members framing into joint i, r_j is the rotation at the far end j of member m, and r_i is the rotation of joint i.

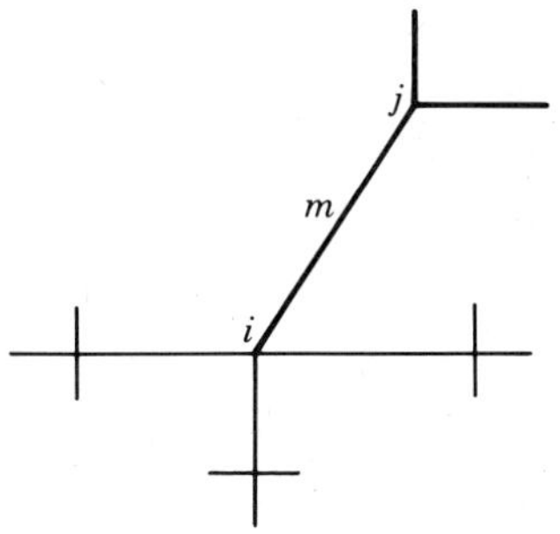

FIG. 6–1. Typical member in a frame.

Equation (6–2) may be generalized to the case where some members are pin connected at i or at j. If a member is pin connected at i, it has no stiffness due to the rotation of joint i, and there is no carry-over moment to joint i due to the rotation at j. Thus for such a member k_m and k'_m are deleted from the stiffness equation at joint i. If a member is rigidly connected at i and pin connected at j, its effective stiffness at joint i is k_e, as defined in Eq. (4–5). In Eq. (6–2) k_m is replaced with $k_{m,e}$, and k'_m is set equal to zero.

Example 1. Portal Frame without Sidesway, Fixed Base

The frame of Fig. 6–2 is symmetric and is assumed restrained against sidesway displacement. This restraint could in fact be due to a pin-connected diagonal brace that makes sidesway buckling extensional, and at a higher load than the inextensional one.

The stiffness equation at joint 2 is

$$k'_1 r_1 + (k_1 + k_2) r_2 + k'_2 r_3 = 0$$

in which $r_1 = 0$, and because of symmetry, $r_3 = -r_2$. The characteristic equation is thus

$$k_1 + k_2 - k'_2 = 0$$

Member 2 has the stiffness coefficients of the linear theory, $k_2 = 4EI_b/l$, and $k'_2 = 2EI_b/l$. For the columns k_1 is as given in Eq. (12/2–2a). We thus obtain the characteristic equation

$$\frac{\lambda(1 - \lambda \cot \lambda)}{\tan(\lambda/2) - (\lambda/2)} + 4\frac{I_b/l}{I_c/h} = 0 \qquad (6\text{–}3a)$$

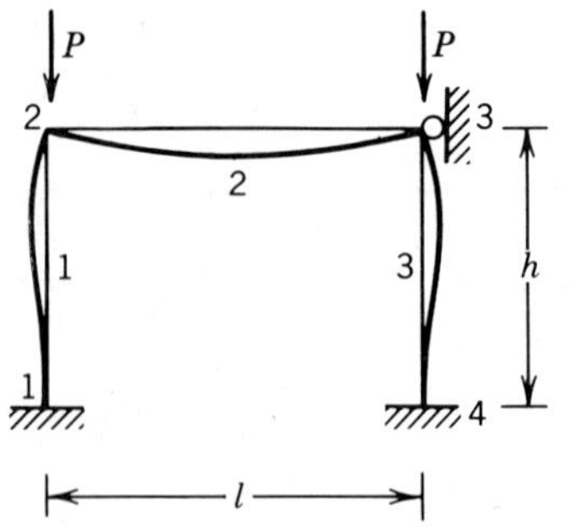

FIG. 6–2. Buckling of frame without sidesway, Example 1.

where

$$\lambda^2 = \frac{Ph^2}{EI_c} \tag{6–3b}$$

The beam stiffens the columns, considered as proped cantilevers. A lower bound for λ_{cr} is the value for the proped cantilever, $\pi/0.7$, and an upper bound is the value for the column restrained against rotation at both ends, 2π. Thus

$$1.43\pi < \lambda_{cr} < 2\pi \tag{6–3c}$$

Example 2. Portal Frame without Sidesway, Pinned Base

The frame of Fig. 6–3 differs from that in Fig. 6–2 only in that the columns are pinned to the supports. In the stiffness equation at joint 2, k_1' is set to zero, and k_1 becomes the effective stiffness $k_{1,e}$ for a pinned far end. Because of symmetry $r_3 = -r_2$. The characteristic equation is $k_{1,e} + k_2 - k_2' = 0$, and $k_{1,e}$ is obtained from Eq. (4–5). We thus obtain

$$\frac{\lambda^2}{1 - \lambda \cot \lambda} + 2\frac{I_b/l}{I_c/h} = 0 \tag{6–4a}$$

The limits for λ_{cr} are

$$\pi < \lambda_{cr} < 1.43\pi \tag{6–4b}$$

The lower limit corresponds to a beam with negligible bending stiffness, for which the columns buckle as Euler columns. The upper limit corresponds to an infinitely stiff beam, for which the columns buckle as proped cantilevers.

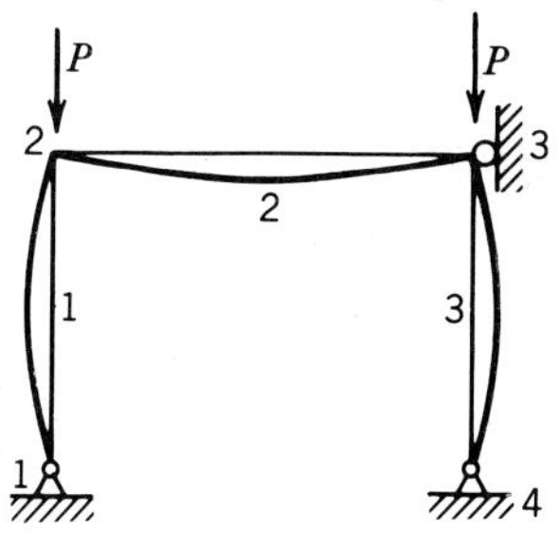

FIG. 6–3. Buckling of frame without sidesway, Section 6, Example 2; Section 12, Example 2.

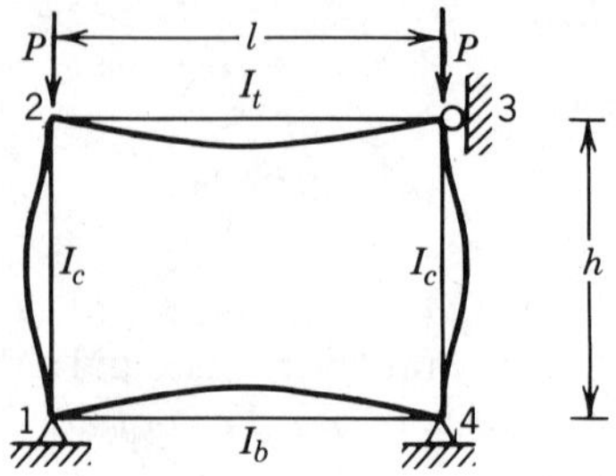

FIG. 6–4. Buckling of frame without sidesway, Example 3.

Example 3. Building Frame without Sidesway, Type 1

For the frame of Fig. (6–4), the stiffness equations at joints 1 and 2, are

$$k_c' r_2 + (k_c + k_b) r_1 + k_b' r_4 = 0$$
$$k_c' r_1 + (k_c + k_t) r_2 + k_t' r_3 = 0$$

where subscript c refers to the column and subscripts b and t refer to the bottom and top beams, respectively. The frame and its buckled shape have a vertical plane of symmetry. Thus $r_4 = -r_1$, and $r_3 = -r_2$. The stiffness equations become

$$(k_c + k_b - k_b') r_1 + k_c' r_2 = 0$$
$$k_c' r_1 + (k_c + k_t - k_t') r_2 = 0$$

For the beams

$$k_b - k_b' = \frac{2EI_b}{l}$$

$$k_t - k_t' = \frac{2EI_t}{l}$$

Setting the determinant of the stiffness matrix to zero, there comes

$$\left(k_c + \frac{2EI_b}{l}\right)\left(k_c + \frac{2EI_t}{l}\right) - k_c'^2 = 0 \qquad (6\text{–}5)$$

k_c is obtained by applying Eq. (12/2–2a) to the column. Letting

$$\kappa = \frac{h}{EI_c} k_c = \frac{\lambda(1 - \lambda \cot \lambda)}{2 \tan(\lambda/2) - \lambda} \qquad (6\text{–}6a)$$

$$\kappa' = \frac{h}{EI_c} k'_c = \frac{\lambda(\lambda/\sin\lambda - 1)}{2\tan(\lambda/2) - \lambda} \tag{6–6b}$$

$$G_b = \frac{I_c/h}{I_b/l} \tag{6–6c}$$

$$G_t = \frac{I_c/h}{I_t/l} \tag{6–6d}$$

Eq. (6–5) takes the form

$$\left(\kappa + \frac{2}{G_b}\right)\left(\kappa + \frac{2}{G_t}\right) - \kappa'^2 = 0 \tag{6–7a}$$

By a reasoning similar to that of preceding examples, the limits of λ_{cr} are

$$\pi < \lambda_{cr} < 2\pi \tag{6–7b}$$

The solution of Eq. (6–7a) is represented by means of an alignment chart or a nomograph, Fig. 6–6. For given values of $G_b = G_A$ and $G_t = G_B$, the straight line joining the corresponding points on the G_A and G_B axes intersects the K axis at the ordinate, $K = \pi/\lambda_{cr}$, which is the effective length factor of the column.

The nomograph is used to estimate column effective length in frame design by replacing G_b and G_t by expressions of the form

$$G = \frac{\sum(I_c/h)}{\sum(I/l)} \tag{6–8}$$

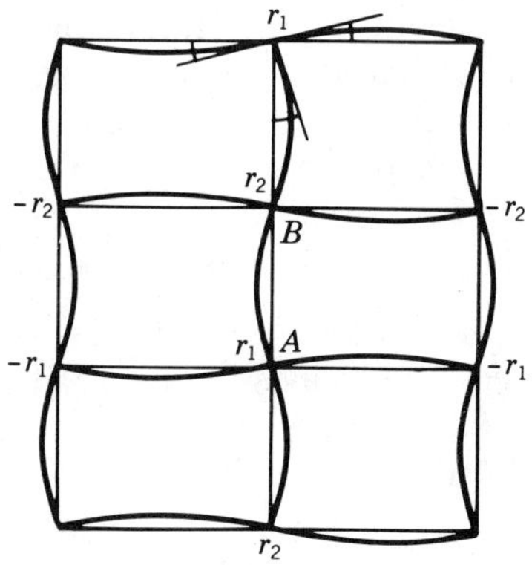

FIG. 6–5. Assumed buckling shape, Example 3.

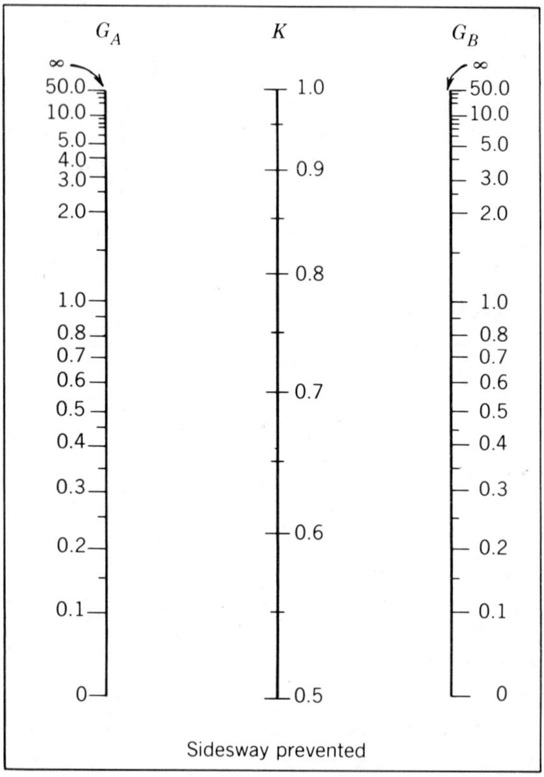

FIG. 6–6. Alignment chart for buckling of building frame without sidesway, Example 3. (by permission of UNITED ENGINEERS INC., Boston, Mass.)

where the summation sign extends over the members framing into a column end. A justification of this generally approximate procedure may be made by considering Fig. 6–5 which shows a portion of a frame surrounding column AB and consisting of three floors and two bays. If columns are identical, the beams at each floor level are identical, and the buckled shape is assumed as shown, the characteristic equation is found to coincide with Eq. (6–7a), and the same G_b and G_t are obtained by Eqs. (6–6c, d) and (6–8).

Example 4. Building Frame without Sidesway, Type 2

The frame of Fig. 6–7 has a horizontal plane of symmetry, and all exterior member ends are assumed unrestrained against rotation. The interior column buckles symmetrically, so that $r_2 = -r_1$. The effective stiffness for this symmetric deformation is $k_s = k - k'$, which, using Eqs. (12/2–2), is found to be

$$k_s = k - k' = \frac{EI}{l} \lambda \cot \frac{\lambda}{2} \tag{6–9}$$

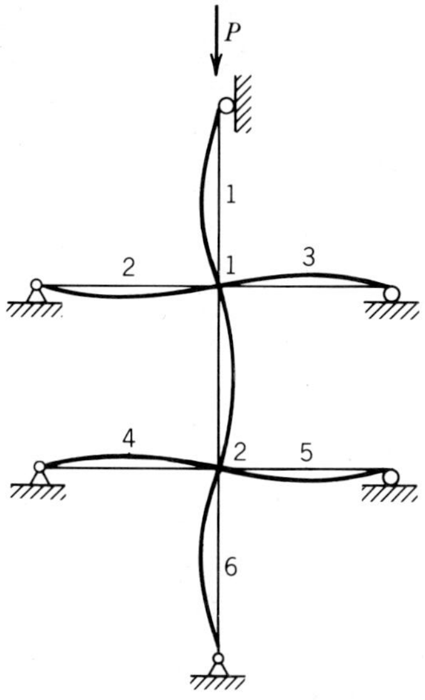

FIG. 6–7. Buckling of frame without sidesway, Example 4.

Using the effective stiffnesses of the members, the stiffness equation for the symmetric buckling mode is

$$(k_s + k_{1,e} + k_{2,e} + k_{3,e})r_1 = 0$$

where $k_{1,e}$ is given in Eq. (4–5), and $k_{2,e}$, and $k_{3,e}$ are equal to $3EI/l$ (in each case the EI and l of the member in question).

Assuming the columns are identical, with parameters I_c and h, and the beams to have all the same ratio I_b/l, the characteristic equation takes the form

$$\frac{\lambda^2}{1 - \lambda \cot \lambda} + \lambda \cot \frac{\lambda}{2} + 6\frac{I_b/l}{I_c/h} = 0 \tag{6–10}$$

The middle column is stiffer than Euler's column because of the restraint provided by the other members, and the two remaining columns are less stiff than the proped cantilever because the restraint at the interior joints is not infinite. Since all columns have the same λ, it follows that

$$\pi < \lambda_{cr} < 1.43\pi \tag{6–11}$$

If there is no plane of symmetry, the stiffness equations at joints 1 and 2 are, respectively,

$$(k + k_{1,e} + k_{2,e} + k_{3,e})r_1 + k'r_2 = 0$$

$$k'r_1 + (k + k_{4,e} + k_{5,e} + k_{6,e})r_2 = 0$$

k and k' are the stiffness coefficients of the middle column, and all other stiffnesses are effective for a pinned far end, as defined earlier. The characteristic equation is obtained by setting the determinant to zero. We thus obtain

$$(k + k_{1,e} + k_{2,e} + k_{3,e})(k + k_{4,e} + k_{5,e} + k_{6,e}) - k'^2 = 0 \qquad (6\text{–}12)$$

Bounds on the critical load may be established by considerations similar to those made in the preceding case.

Equation (6–12) is readily generalized to the case of a member connected rigidly at each of its ends to any number of members which are pinned at their far ends.

Example 5. Compression Member of a Trussed Frame

If the truss of Fig. 6–8*a* is built with rigid joints, it becomes a trussed frame whose instability may be investigated by neglecting joint translations. The axial forces in the members are obtained through a linear analysis of the substitute truss.

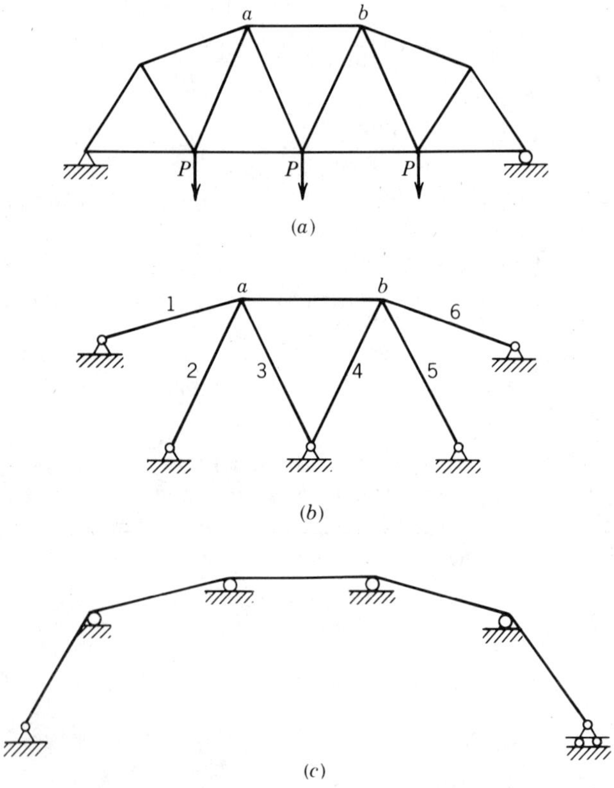

FIG. 6–8. Buckling of compression member of trussed frame, Examples 5, 6.

A simplified stability analysis deals only with one compression member and the members framing into its ends. The far ends of these members are assumed free to rotate. Neglecting the rotational restraints at the far ends places the simplified analysis on the safe side by lowering the critical load.

For investigating the stability of member *ab* in Fig. 6–8*a*, the simplified analysis deals with the frame shown in (*b*). The characteristic equation has the form of Eq. (6–12). For each compression member the λ to be used in evaluating its stiffness is related through Eq. (4–3) to its axial force, as determined by linear truss analysis. For the loading shown in Fig. 6–8*a*, the axial forces are linear in P which is the load parameter of the characteristic-value problem. Tension members may be given the approximate effective stiffness $3EI/l$.

Example 6. Continuous Compression Chord of a Truss

Consider the truss of Fig. 6–8*a*, and assume that the top chord is continuous but that the web members are pin connected to it. Stability of the top chord is investigated as for the continuous beam shown in (*c*). The axial forces in the various members are obtained through linear truss analysis of the structure in (*a*).

7. BUCKLING OF FRAMES WITH JOINT TRANSLATIONS: STIFFNESS METHOD

This section deals with frames having translational degrees of freedom, and whose buckling modes are inextensional, as discussed in Section 5. The tangent stiffness matrix for the inextensional degrees of freedom, denoted as $\mathbf{K}_{11}$ in Eqs. (5–2), will now be denoted simply as $\mathbf{K}$, and the general notation $\mathbf{r}$ will be used instead of $\mathbf{q}_1$. $\mathbf{K}$ is assembled from the 4×4 member matrices $\mathbf{k}_s$. Use of the exact $\mathbf{k}_s$, Eq. (6/2.6–6), leads to a transcendental characteristic equation, whereas the approximate $\mathbf{k}_s$ of Eq. (6/3.1–35) leads to a linear characteristic-value problem.

Example 1. Nonuniform Member with Fixed Ends

Figure 7–1 shows a fixed ended member formed of two uniform members, a and b, of properties EI_a and EI_b, respectively. The joint displacements $\mathbf{r} = \{r_1 \ r_2\}$ are as shown. Member connectivity relations are $\mathbf{s} = \mathbf{r}$. The stiffness matrix for

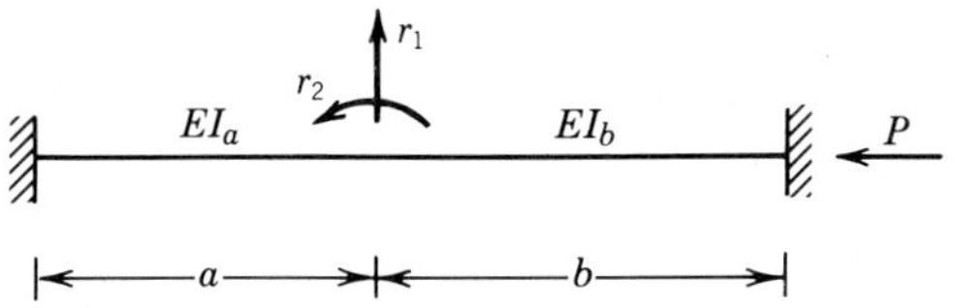

FIG. 7–1. Nonuniform member, Example 1.

member a is obtained by deleting from the complete $\mathbf{k}_s$ rows and columns 1 and 2, and for member b, by deleting rows and columns 3 and 4. These matrices are also the cantilever-type matrix, Eq. (6/2.2–3). Rewriting this matrix, with $\alpha = 1$ to neglect shear deformation, there comes

$$\mathbf{k} = \frac{EI\lambda^2}{2l^2[\tan(\lambda/2) - (\lambda/2)]}\begin{bmatrix} \dfrac{\lambda}{l} & -\tan\dfrac{\lambda}{2} \\ -\tan\dfrac{\lambda}{2} & \dfrac{l}{\lambda}\left(1 - \dfrac{\lambda}{\tan\lambda}\right) \end{bmatrix} \tag{7–1}$$

Equation (7–1) applies to member a with $I = I_a$, $l = a$, and $\lambda = \lambda_a$. For member b the off-diagonal terms should be changed in sign, and $I = I_b$, $l = b$, and $\lambda = \lambda_b$. Since the axial force is the same in both members,

$$\frac{EI_a\lambda_a^2}{a^2} = \frac{EI_b\lambda_b^2}{b^2} = P \tag{7–2}$$

The assembled $\mathbf{K}$ is here the sum of the member matrices, or

$$\mathbf{K} = \mathbf{k}_a + \mathbf{k}_b$$

The characteristic equation, $|\mathbf{K}| = 0$, may be put after trigonometric transformations in the form

$$\frac{\tan(\lambda_b/2)}{\tan(\lambda_a/2)} + \frac{\tan(\lambda_a/2)}{\tan(\lambda_b/2)} - \frac{l}{b}\frac{\lambda_b}{\tan\lambda_a} - \frac{l}{a}\frac{\lambda_a}{\tan\lambda_b} + \sqrt{\frac{I_b}{I_a}} + \sqrt{\frac{I_a}{I_b}} = 0 \tag{7–3}$$

in which $l = a + b$, and from Eq. (7–2),

$$\frac{\lambda_a}{\lambda_b} = \frac{a}{b}\sqrt{\frac{I_b}{I_a}}$$

Equation (7–3) is solved by trials. To obtain bounds on P_{cr}, assume that

$$I_b > I_a$$

then P_{cr} is less than the critical load for a uniform member having $I = I_b$ and larger than the critical load for a uniform member having $I = I_a$. Thus

$$\frac{4\pi^2 EI_a}{l^2} < P_{cr} < \frac{4\pi^2 EI_b}{l^2} \tag{7–4a}$$

Bounds on $\lambda_{b,cr}$ are obtained, using Eq. (7–2), as

$$4\pi^2\frac{b^2}{l^2}\frac{I_a}{I_b} < \lambda_{b,cr}^2 < 4\pi^2\frac{b^2}{l^2} \tag{7–4b}$$

For example, for $a = b = \frac{l}{2}$, and $I_b = 2I_a$, $\lambda_{b,\text{cr}}$ is found by trials to be

$$\lambda_{b,\text{cr}} = 2.540$$

The effective length of the member, based on I_b is $\pi l/2\lambda_b = 0.618l$.

Use of the approximate member stiffness matrix in the present problem yields accurate results because the approximate cantilever-type matrix, Eq. (6/3.1–28), is accurate enough for values of λ reaching π. The stiffness matrix takes then the form

$$\mathbf{K} = \begin{bmatrix} \dfrac{12EI_a}{a^3} + \dfrac{12EI_b}{b^3} - \dfrac{12}{10}\left(\dfrac{1}{a} + \dfrac{1}{b}\right)P & -\dfrac{6EI_a}{a^2} + \dfrac{6EI_b}{b^2} \\ -\dfrac{6EI_a}{a^2} + \dfrac{6EI_b}{b^2} & \dfrac{4EI_a}{a} + \dfrac{4EI_b}{b} - \dfrac{2}{15}(a+b)P \end{bmatrix} \tag{7–5}$$

For the data used here, the smaller root of the characteristic equation $|\mathbf{K}| = 0$ yields for λ_b the value 2.57 which is a good approximation to the axact value 2.54 found earlier.

Example 2. Nonuniform Member, Simply Supported

Figure 7–2 shows a simply supported member, made of two uniform members a and b, of properties EI_a and EI_b, respectively. The effective joint displacements are chosen as shown. The problem is similar to that of Example 1, the difference being in the effective member matrices. For a rotational release at the right end of a member, the effective $\mathbf{k}_s$ is derived in Eq. (7/5–10b). It applies to member b of the present problem, after deleting the third row and column which correspond to the zero transverse displacement at the right end. Thus, with $\alpha = 1$ to neglect shear deformation,

$$\mathbf{k}_s = \frac{EI\lambda^2}{l^3(\sin\lambda - \lambda\cos\lambda)} \begin{bmatrix} \lambda\cos\lambda & l\sin\lambda \\ l\sin\lambda & l^2\sin\lambda \end{bmatrix} \tag{7–6a}$$

Another form of Eq. (7–6a) will prove convenient in what follows. This is

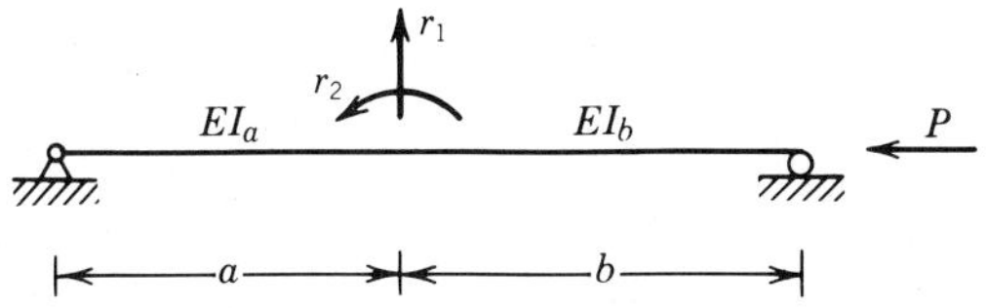

FIG. 7–2. Nonuniform member, Section 7, Example 2; Section 13, Example 1.

Eq. (7/5–10a) in which the geometric stiffness $V_1 \mathbf{c}_1$ is written separately, and the rest is expressed in terms of the effective stiffness coefficient k_e. We thus have, equivalently to Eq. (7–6a),

$$\mathbf{k}_s = \frac{1}{l^2}\begin{bmatrix} k_e + V_1 l & k_e l \\ k_e l & k_e l^2 \end{bmatrix} \tag{7–6b}$$

where k_e is defined in Eq. (4–5). Equation (7–6b) applies to member b, with $V_1 = -P$, $k_e = k_{b,e}$, $l = b$, $I = I_b$, and $\lambda = \lambda_b$. It applies to member a by changing the sign of the off-diagonal terms, and with $k_e = k_{a,e}$, $l = a$, $I = I_a$, and $\lambda = \lambda_a$. Since the axial force is the same for members a and b, λ_a and λ_b are related by Eq. (7–2). The assembled stiffness matrix is

$$\mathbf{K} = \begin{bmatrix} \dfrac{k_{a,e}}{a^2} + \dfrac{k_{b,e}}{b^2} - P\left(\dfrac{1}{a} + \dfrac{1}{b}\right) & \dfrac{k_{b,e}}{b} - \dfrac{k_{a,e}}{a} \\ \dfrac{k_{b,e}}{b} - \dfrac{k_{a,e}}{a} & k_{a,e} + k_{b,e} \end{bmatrix} \tag{7–7}$$

The characteristic equation takes the form

$$\frac{Pab}{a+b}\left(\frac{1}{k_{a,e}} + \frac{1}{k_{b,e}}\right) - 1 = 0 \tag{7–8a}$$

and reduces to

$$\frac{\tan \lambda_a}{\tan \lambda_b} + \sqrt{\frac{I_b}{I_a}} = 0 \tag{7–8b}$$

Bounds on λ_a or λ_b are readily established, as in the previous example, by comparison to the case of a uniform member. For example, let $a = b = l/2$ and $I_b = 2I_a$. From Eq. (7–2), $\lambda_a = \sqrt{2}\lambda_b$. Equation (7–8b), expressed in terms of λ_b, becomes

$$\frac{\tan(\sqrt{2}\lambda_b)}{\sqrt{2}\tan \lambda_b} + 1 = 0$$

Bounds on $\lambda_{b,\text{cr}}$ are

$$\frac{\pi}{2\sqrt{2}} < \lambda_{b,\text{cr}} < \frac{\pi}{2}$$

Solution by trials yields $\lambda_{b,\text{cr}} = 1.266$ and $P_{\text{cr}} = \lambda^2_{b,\text{cr}}(EI_b/b^2) = 6.41\ (EI_b/l^2)$. The effective member length based on I_b is $\pi l/\sqrt{6.41} = 1.24l$.

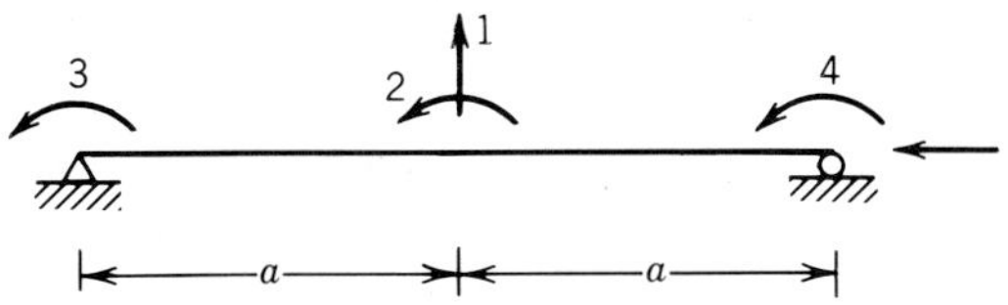

FIG. 7–3. Two-element representation of one member, Example 3.

Example 3. Use of Approximate Member Stiffness Matrix

To illustrate further the use of the approximate stiffness matrix, a simply supported uniform member will be treated as a two-member structure. Assume first that the effective member stiffness, for a released end rotation, is not readily available, so that the degrees of freedom are chosen as shown in Fig. 7–3. The approximate $\mathbf{k}_s$ is given in Eq. (6/3.1–35) in which $\mu^2 = 0$. For the right half of the member the third row and column are deleted because of the support condition. The merging sequence for the resulting matrix, into the 4×4 assembled stiffness matrix, is 1, 2, 4. For the left half of the member the first row and column are deleted from $\mathbf{k}_s$, and the merging sequence is 3, 1, 2. The assembled stiffness matrix is thus found as

$$\mathbf{K} = \frac{EI}{a^3}\begin{bmatrix} 24 - \frac{12}{5}\lambda^2 & & & \text{symmetric} \\ 0 & 8 - \frac{4\lambda^2}{15} & & \\ -a\left(6 - \frac{\lambda^2}{10}\right) & a^2\left(2 + \frac{\lambda^2}{30}\right) & a^2\left(4 - \frac{2\lambda^2}{15}\right) & \\ a\left(6 - \frac{\lambda^2}{10}\right) & a^2\left(2 + \frac{\lambda^2}{30}\right) & 0 & a^2\left(4 - \frac{2\lambda^2}{15}\right) \end{bmatrix} \tag{7–9}$$

The stiffness equations, $\mathbf{Kr} = 0$, form a linear characteristic-value problem. Because of symmetry the solution of interest is such that $r_2 = 0$ and $r_3 = -r_4$. These conditions satisfy identically the second stiffness equation and make the third and fourth into equivalent equations. The two independent stiffness equations reduce to

$$\begin{bmatrix} 24 - \frac{12}{5}\lambda^2 & -a\left(12 - \frac{\lambda^2}{5}\right) \\ -a\left(12 - \frac{\lambda^2}{5}\right) & a^2\left(8 - \frac{4\lambda^2}{15}\right) \end{bmatrix} \begin{Bmatrix} r_1 \\ r_2 \end{Bmatrix} = 0$$

Setting the determinant equal to zero, the equation obtained may be put in the form

$$3\lambda^4 - 104\lambda^2 + 240 = 0$$

The lowest root is $\lambda^2 = 2.486$ for which $P = \lambda^2 EI/a^2 = 9.94EI/l^2$. The relative error is only $9.94/\pi^2 - 1 = 0.007$. The second root corresponds to a higher mode of symmetric buckling which is not of interest in this problem.

The matrix of Eq. (7–9) could also be used to find the critical load of a fixed ended member. The conditions $r_3 = r_4 = 0$ are enforced by deleting the third and fourth column. The approximate solution is $\lambda^2 = 10$, and the exact one is $\lambda^2 = \pi^2 = 9.87$.

Consider again the member of Fig. 7–3, but assume that the two parts of the member are not identical. The characteristic-value problem, which is formulated with four degrees of freedom, becomes numerically more complicated. Use of effective degrees of freedom from the outset, as done in Example 2 and shown in Fig. 7–2, would be advantageous. To do this, the approximate $\mathbf{k}_s$ for a member with its right end pinned is obtained through Eq. (7–6b) in which k_e is the approximate effective rotational stiffness. To determine k_e, we start with the 2 × 2 member stiffness of the alternate type,

$$\mathbf{k} = \frac{EI}{l}\begin{bmatrix} 4 - \dfrac{2\lambda^2}{15} & 2 + \dfrac{\lambda^2}{30} \\ 2 + \dfrac{\lambda^2}{30} & 4 - \dfrac{2\lambda^2}{15} \end{bmatrix} \tag{7–10}$$

Using the general notation for deformations $\{v_2 \ \ v_3\}$ and member end moments $\{V_2 \ \ V_3\}$, the rotational release at the right end causes $V_3 = 0$. Then from the second stiffness equation

$$v_3 = -\frac{2 + \lambda^2/30}{4 - 2\lambda^2/15} v_2$$

Substituting for v_3 into the first stiffness equation, the result has the form

$$V_2 = k_e v_2$$

where

$$k_e = \frac{EI}{l} \frac{(6 - \lambda^2/10)(2 - \lambda^2/6)}{4 - 2\lambda^2/15} \tag{7–11}$$

Equation (7–8a) may now be applied to the problem of Fig. 7–2, with $k_{a,e}$ and $k_{b,e}$ determined through Eq. (7–11). For the particular case of two identical member segments, $k_{a,e} = k_{b,e}$ is obtained from Eq. (7–11) in which l is replaced with $l/2$, and

$\lambda^2 = Pl^2/4EI$. Equation (7–8a) becomes

$$\frac{Pl^2}{4EI}\frac{4-2\lambda^2/15}{(6-\lambda^2/10)(2-\lambda^2/6)}-1=0$$

Noting that $Pl^2/4EI = \lambda^2$, the characteristic equation may be seen to reduce to the one found earlier.

Example 4. Cantilever with Rigid End Portion

In Fig. 7–4 the end portion of length a of the cantilever is assumed to be infinitely rigid. The deformable portion is governed by the stiffness equations of the cantilever type, Eq. (6/2.2–3), in which the load term is $\{0 \ \text{P}ar_2\}$ and $\mathbf{v} = \{r_1 \ r_2\}$. Noting that $EI\lambda^2/l^2 = P$, the member stiffness equations, divided through by P, are

$$\frac{1}{2[\tan(\lambda/2)-(\lambda/2)]}\begin{bmatrix}\dfrac{\lambda}{l} & -\tan\dfrac{\lambda}{2}\\ -\tan\dfrac{\lambda}{2} & \dfrac{l}{\lambda}\left(1-\dfrac{\lambda}{\tan\lambda}\right)\end{bmatrix}\begin{Bmatrix}r_1\\ r_2\end{Bmatrix}-\begin{Bmatrix}0\\ ar_2\end{Bmatrix}=0$$

The determinant of this homogeneous system of equations may be shown to contain in the numerator the factor $\tan(\lambda/2)-(\lambda/2)$. The characteristic equation reduces to

$$\lambda \tan \lambda = \frac{l}{a} \tag{7–12}$$

A lower bound on P_{cr} is the critical load of a uniform cantilever of length $(l+a)$ and of bending rigidity EI. An upper bound is the critical load of a uniform cantilever of length l.

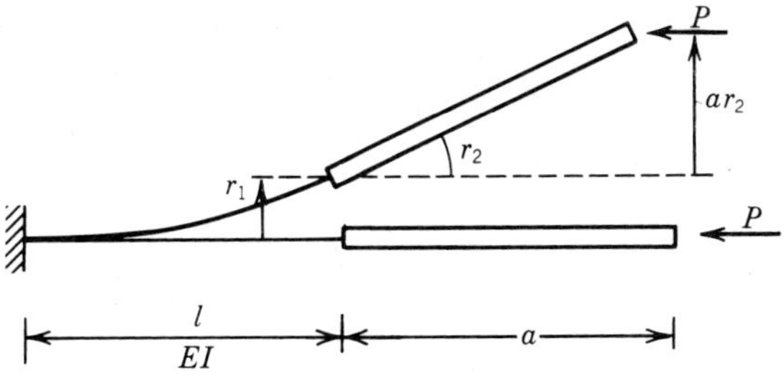

FIG. 7–4. Cantilever with rigid part, Example 4.

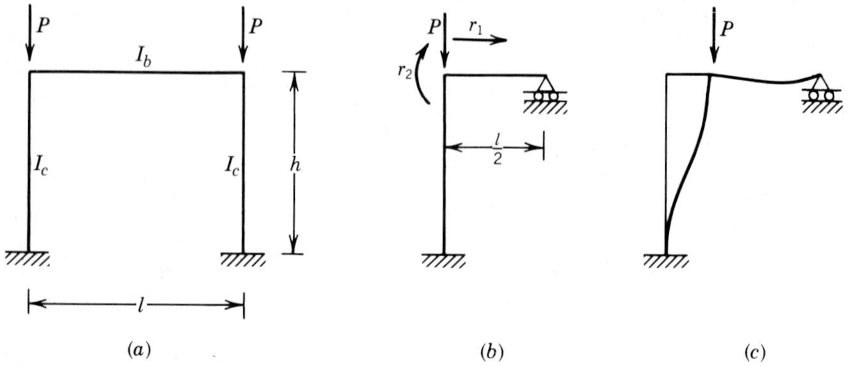

FIG. 7–5. Frame with sidesway, Example 5.

Example 5. Portal Frame with Sidesway, Fixed Base

For the frame of Fig. 7–5*a* treated as inextensible, the buckled state is antisymmetric, and only half the structure, as shown in (*b*), need be considered. The stiffness matrix of the column is as given in Eq. (7–1). For the beam the effective rotational stiffness is $6EI_b/l$. The determinant of the assembled stiffness matrix may be simplified as in the preceding example, and the characteristic equation takes the form

$$\frac{\tan\lambda}{\lambda} + \frac{1}{6}\frac{I_c/h}{I_b/l} = 0 \tag{7–13a}$$

Bounds on λ_{cr} are obtained by varying the beam stiffness from 0, for which $\lambda_{cr} = \pi/2$, to ∞ for which $\lambda_{cr} = \pi$. Thus

$$\frac{\pi}{2} < \lambda_{cr} < \pi \tag{7–13b}$$

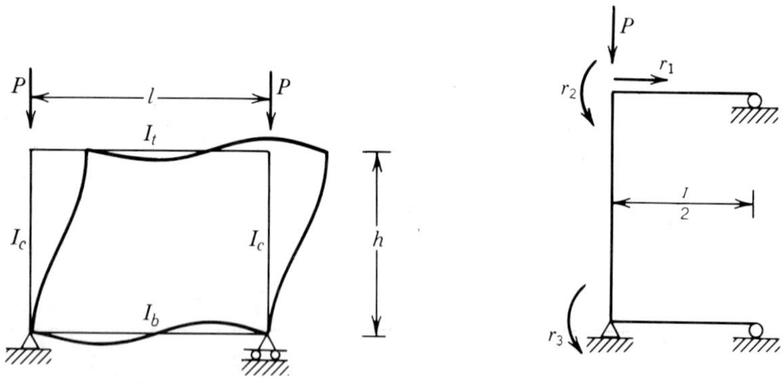

FIG. 7–6. Frame with sidesway, Section 7, Example 6; Section 13, Example 5.

Example 6. Building Frame with Sidesway

The rigid frame of Fig. 7–6*a* has a vertical plane of symmetry and buckles in an antisymmetric mode. Only half the structure, as shown in (*b*), need be considered in the analysis. The effective column stiffness matrix for the degrees of freedom shown in (*b*) is the first 3 × 3 submatrix in Eq. (6/2.6–6). For the beams the effective rotational stiffnesses are $6EI_t/l$ and $6EI_b/l$, which correspond to displacements r_2 and r_3, respectively. The determinant of the stiffness equations may be shown to contain the factor $(\tan \lambda/2 - \lambda/2)^2$, and the characteristic equation may be reduced to the form

$$\frac{1}{\lambda \tan \lambda} = \frac{G_t G_b}{6(G_t + G_b)} - \frac{6}{\lambda^2 (G_t + G_b)} \tag{7–14}$$

where

$$G_t = \frac{I_c/h}{I_t/l} \tag{7–15a}$$

$$G_b = \frac{I_c/h}{I_b/l} \tag{7–15b}$$

Bounds on λ_{cr} are obtained by varying the beam stiffnesses from 0, for which $\lambda_{cr} = 0$, to ∞ for which $\lambda_{cr} = \pi$. Thus

$$0 < \lambda_{cr} < \pi \tag{7–16}$$

The effective column length factor is thus larger than one and tends to infinity as the stiffnesses of the beams tend to zero.

As an example, for $G_t = G_b = 1$, a solution by trials yields $\lambda_{cr} = 2.385$, and an effective column length factor of $\pi/\lambda_{cr} = 1.32$.

The solution of Eq. (7–14) for λ_{cr} is represented on a nomograph, Fig. 7–7, similarly to that of the frame without sidesway considered in Section 6, Example 3. It is used to estimate effective column lengths in frame design by replacing Eqs. (7–15) with

$$G_t = \frac{\sum (I_c/h)}{\sum (I_t/l)} \tag{7–17a}$$

$$G_b = \frac{\sum (I_c/h)}{\sum (I_b/l)} \tag{7–17b}$$

In Fig. 7–7, G_t and G_b are named G_A and G_B.

This approximate procedure is now tested in two cases for which the critical load was obtained in Chapter 12 through an exact formulation.

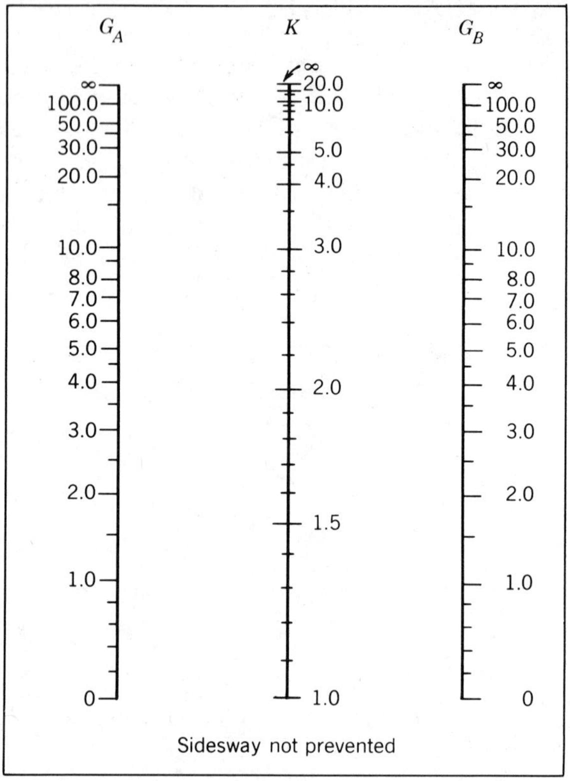

Fig. 7–7. Alignment chart for buckling of building frame with sidesway, Example 6. (by permission of UNITED ENGINEERS INC., Boston, Mass.)

For the five-story frame of Fig. 12/4.3–2 loaded as in (*c*), it was found that $P_{\text{cr}} = 830$ kips. The effective column length for interior columns, whose axial force is P, is thus

$$K = \sqrt{\frac{P_E}{P_{\text{cr}}}} = \left[\frac{\pi^2(30{,}000)(100)}{(144)^2(830)}\right]^{1/2} = 1.31$$

If Eq. (7–14) is used, G_t and G_b depend on the particular column. For an interior column of the first floor, Eqs. (7–17) yield $G_t = 1$ and $G_b = 0$. For an interior column of the top floor, $G_t = \frac{1}{2}$ and $G_b = 1$. For the remaining interior columns, $G_t = 1$ and $G_b = 1$. The effective length factor, $K = \pi/\lambda_{\text{cr}}$, for these three cases—as obtained by solving Eq. (7–14) or, equivalently, through the nomograph of Fig. 7–7—are found to be 1.24, 1.32, and 1.16, respectively.

Consider now the ten-story frame of Fig. 12/6–4, for which results indicate that the inextensional approximation for predicting P_{cr} is in significant error. Since Eq. (7–14) is itself based on an inextensional formulation, it should be expected to be also in error and to yield results comparable to those of the inextensional approximation in Fig. 12/6–4. The lowest critical load predicted by Eq. (7–14) is for the columns of the intermediate floors, that is, floors 2 through 9 including in this example the exterior columns which have a second moment of area half that of the interior columns. For the case where the beam span is $l = 144$ in, the lowest λ_{cr} is obtained for $G_t = G_b = 1$, and the corresponding effective length factor is $K = 1.32$. The same value for K is obtained through the inextensional approximation in Fig. 12/6–4 in which $P_{cr} = 8254$ kips. The actual critical load is 5198 kips, and the corresponding effective column length factor is $K = \sqrt{P_E/P_{cr}} = 1.66$.

For the case where the beam span is $l = 240$ in, the lowest λ_{cr} corresponds to $G_t = G_b = 1.67$. The effective column length factor is found through Eq. (7–14), as well as through the inextensional approximation in Fig. 12/6–4, to be nearly 1.5, whereas the correct value corresponds to $P_{cr} = 5480$ kips, or $K = 1.61$.

In addition to the error in predicting effective column lengths, the inextensional approximation erroneously predicts a lower critical load for the wider frame. This is an inherent property of Eq. (7–14) in which G_t and G_b increase as the I/l ratio of the beams decreases. Results in the present example indicate that the critical load may first increase with increasing beam span, and that this is associated with column extensibility. It is clear that with a continuously increasing beam span, a point is reached where the lower stiffness of the beams causes a decrease in the critical load.

Example 7. Member with Intermediate Axial Load, Out of Plane Buckling of a Truss Member

Consider the simply supported member shown in Fig. 7–8*a*, subjected to the compressive axial loads P and P_b. The buckled shape and the choice of displacements are shown in (*b*). Forming of the stiffness matrix follows the procedure of Example 2, with the difference here that members a and b have different axial forces. In member a the compressive axial force is

$$P_a = P + P_b$$

Assuming members a and b are uniform, their stability parameters are

$$\lambda_a^2 = \frac{P_a a^2}{EI_a}$$

$$\lambda_b^2 = \frac{P_b b^2}{EI_b}$$

The member stiffness matrices are obtained by application of Eq. (7–6b). For member b, $l = b$, $k_e = k_{b,e}$, and $V_1 = -P_b$. For member a the sign of the off-diagonal terms is changed, and $l = a$, $k_e = k_{a,e}$, and $V_1 = -P_a$. The assembled

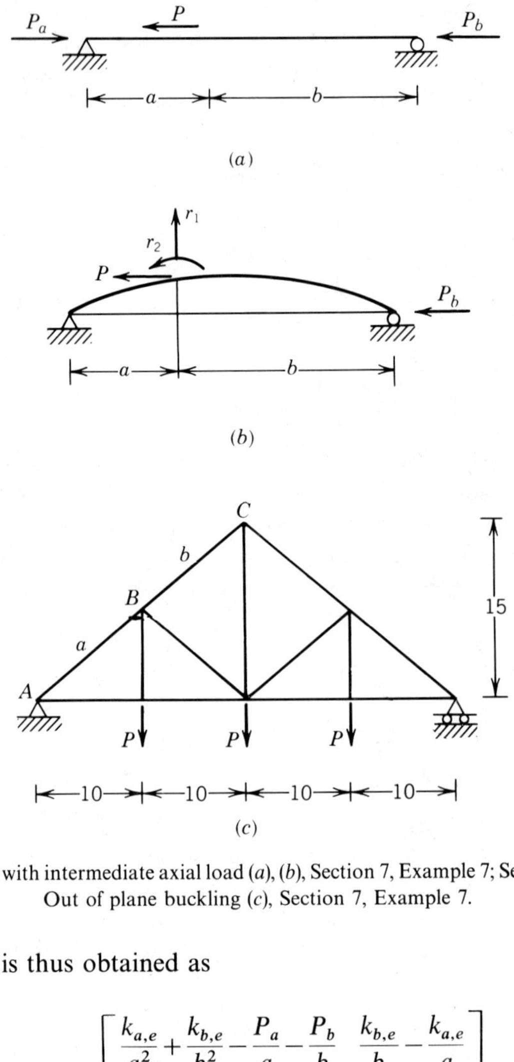

FIG. 7–8. Member with intermediate axial load (*a*), (*b*), Section 7, Example 7; Section 13, Example 2. Out of plane buckling (*c*), Section 7, Example 7.

stiffness matrix is thus obtained as

$$\mathbf{K} = \begin{bmatrix} \dfrac{k_{a,e}}{a^2} + \dfrac{k_{b,e}}{b^2} - \dfrac{P_a}{a} - \dfrac{P_b}{b} & \dfrac{k_{b,e}}{b} - \dfrac{k_{a,e}}{a} \\ \dfrac{k_{b,e}}{b} - \dfrac{k_{a,e}}{a} & k_{a,e} + k_{b,e} \end{bmatrix} \tag{7–18}$$

where $k_{a,e}$ and $k_{b,e}$ are obtained by applying Eq. (4–5) to members a and b, respectively. This yields

$$k_{a,e} = \frac{aP_a}{1 - \lambda_a \cot \lambda_a}$$

$$k_{b,e} = \frac{bP_b}{1 - \lambda_b \cot \lambda_b}$$

The characteristic equation takes the form

$$\frac{ab}{(a+b)^2}(bP_a + aP_b)\left(\frac{1}{k_{a,e}} + \frac{1}{k_{b,e}}\right) - 1 = 0 \qquad (7\text{–}19a)$$

and reduces, if $P_aP_b \neq 0$, to

$$\left(\frac{b}{a} + \frac{P_b}{P_a}\right)\lambda_a \cot\lambda_a + \left(\frac{a}{b} + \frac{P_a}{P_b}\right)\lambda_b \cot\lambda_b - \frac{P_a}{P_b} - \frac{P_b}{P_a} + 2 = 0 \qquad (7\text{–}19b)$$

λ_a and λ_b have the ratio

$$\frac{\lambda_a}{\lambda_b} = \frac{a}{b}\sqrt{\frac{P_aI_b}{P_bI_a}} \qquad (7\text{–}20)$$

To bound the critical loads, $P_{a,\mathrm{cr}}$ and $P_{b,\mathrm{cr}}$, let P_E be the critical load of the member when in uniform compression. Then

$$P_{b,\mathrm{cr}} < P_E < P_{a,\mathrm{cr}} \qquad (7\text{–}21)$$

As an application, consider the truss of Fig. 7–8*c*, and assume that members *AB* and *BC* form a continuous member that may buckle out of the plane of the truss. In such buckling there are restraints against rotations at *A* and *C* and against translation at *B*. If such restraints are small and are neglected, the error will be on the safe side for predicting the buckling load. Buckling of member *AC* is then governed by Eq. (7–19). P_a and P_b are determined by linear analysis of the substitute truss as

$$P_a = \frac{5}{2}P$$

$$P_b = \frac{5}{3}P$$

With $a/b = 1$, and $P_a/P_b = 3/2$, Eq. (7–19b) yields

$$\lambda_a \cot\lambda_a + \frac{3}{2}\lambda_b \cot\lambda_b - \frac{1}{10} = 0$$

Assuming $I_a = I_b$, Eq. (7–20) yields $\lambda_b/\lambda_a = \sqrt{2/3}$, and from Eq. (7–21),

$$\lambda_{b,\mathrm{cr}} < \frac{\pi}{2} < \lambda_{a,\mathrm{cr}}$$

The root for λ_a is found by trials as $\lambda_{a,\mathrm{cr}} = 1.719$. The effective length factor of

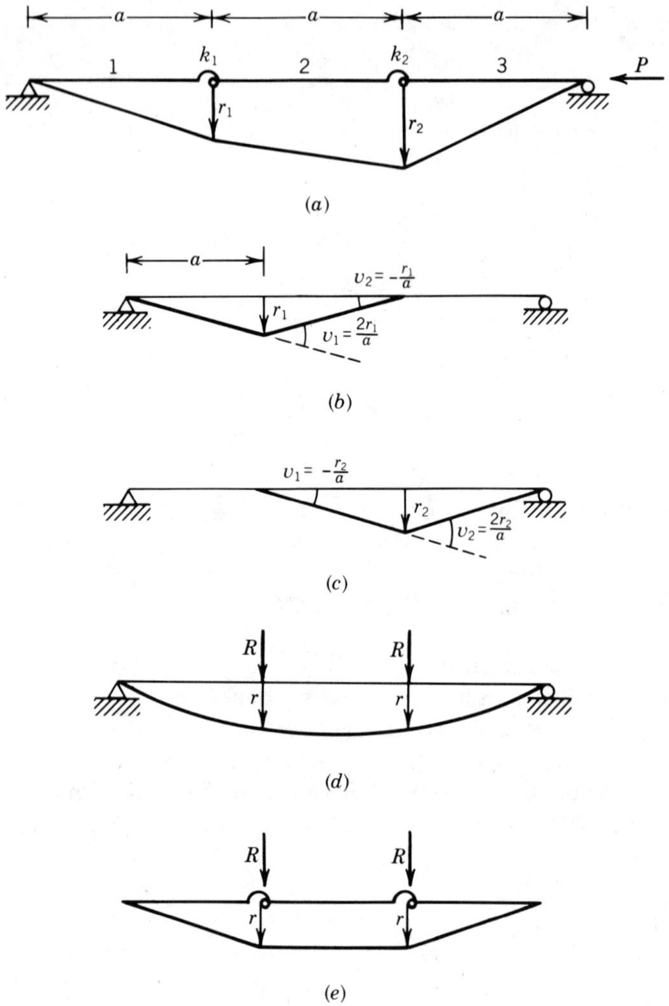

FIG. 7–9. Rigid bar system, Example 8.

member AC based on P_a is then

$$K = \frac{\pi}{2\lambda_{a,\mathrm{cr}}} = 0.914$$

As another application, consider the case where $P_b = 0$. Then $P_a = P$, and $k_{b,e} = 3EI_b/b$. Equation (7–19a) yields

$$\lambda_a \cot \lambda_a - \frac{1}{3}\frac{b}{a}\frac{I_a}{I_b}\lambda_a^2 + \frac{l^2}{b^2} - 1 = 0 \tag{7–22}$$

where $l = a + b$. A lower bound on P_{cr} is the critical axial force of the uniformly compressed member. For example, for $a = b$ and $I_a = I_b$, we obtain $2\lambda_{a,cr} > \pi$. The root of Eq. (7–22) is found by trials as $\lambda_{a,cr} = 2.1602$. The critical load is

$$P_{cr} = (2\lambda_{a,cr})^2 EI/l^2 = 18.67 EI/l^2 \tag{7–23}$$

Example 8. Rigid Bar System

In Fig. 7–9*a* is shown a system formed of three rigid bars connected with rotational springs of stiffnesses k_1 and k_2, respectively. The deformation modes are shown in (*b*) and (*c*). The connectivity equations are

$$\begin{Bmatrix} v_1 \\ v_2 \end{Bmatrix} = \frac{1}{a}\begin{bmatrix} 2 & -1 \\ -1 & 2 \end{bmatrix}\begin{Bmatrix} r_1 \\ r_2 \end{Bmatrix}$$

The stiffness matrix of the linear theory is

$$\mathbf{K}_0 = \mathbf{a}^T\mathbf{k}_0\mathbf{a} = \frac{1}{a^2}\begin{bmatrix} 2 & -1 \\ -1 & 2 \end{bmatrix}\begin{bmatrix} k_1 & \\ & k_2 \end{bmatrix}\begin{bmatrix} 2 & -1 \\ -1 & 2 \end{bmatrix}$$

$$= \frac{1}{a^2}\begin{bmatrix} 4k_1 + k_2 & -2(k_1 + k_2) \\ -2(k_1 + k_2) & k_1 + 4k_2 \end{bmatrix} \tag{7–24}$$

The geometric stiffness matrix $\mathbf{K}_G$ is assembled from $\mathbf{k}_G$ for the three members. For members 1 and 3,

$$\mathbf{k}_G = -\frac{P}{a}[1]$$

For member 2,

$$\mathbf{k}_G = -\frac{P}{a}\begin{bmatrix} 1 & -1 \\ -1 & 1 \end{bmatrix}$$

Merging of the member matrices yields

$$\mathbf{K}_G = -\frac{P}{a}\begin{bmatrix} 2 & -1 \\ -1 & 2 \end{bmatrix} \tag{7–25}$$

The characteristic equation for buckling is

$$|\mathbf{K}_0 + \mathbf{K}_G| = 0$$

Consider the case

$$k_1 = k_2 = k$$

Then

$$\mathbf{K} = \mathbf{K}_0 + \mathbf{K}_G = \frac{1}{a^2}\begin{bmatrix} 5k - 2Pa & -4k + Pa \\ -4k + Pa & 5k - 2Pa \end{bmatrix}$$

The buckling mode is symmetric, with $r_1 = r_2$. Thus

$$5k - 2Pa - 4k + Pa = 0$$

or

$$P_{\text{cr}} = \frac{k}{a} \tag{7–26}$$

The system considered here may be chosen as a model for approximating a beam of span $3a$. The approximation is to be made by equating model and beam deflections for the loading shown in (d) and (e). For the model, the stiffness equations, using $\mathbf{K}_0$ and $r_1 = r_2 = r$, yield

$$R = \frac{k}{a^2} r$$

The spring stiffness is thus defined through

$$k = a^2 \frac{R}{r}$$

To test the approximation, consider a uniform beam for which it may be established that

$$\frac{R}{r} = \frac{6EI}{5a^3}$$

Equation (7–26) yields $P_{\text{cr}} = aR/r = 6EI/5a^2$, or $P_{\text{cr}} = 10.8EI/l^2$. The error is $10.8/\pi^2 - 1 = 0.094$. For a member that may not be as easily analyzed as the uniform member, but for which the stiffness R/r may be evaluated, an adequate approximation should be obtained through Eq. (7–26), corrected by the factor $\pi^2/10.8$.

8. FLEXIBILITY METHOD FOR LINEARIZED STABILITY ANALYSIS: GENERAL PROCEDURE

Linearized stability analysis by the flexibility method consists in seeking a nontrivial solution to the homogeneous incremental equations (12/8–4), or

$$(\mathbf{I} + \mathbf{F}_{rr}\mathbf{K}'_G)\mathbf{r} - \mathbf{F}_{rx}\mathbf{X} = 0 \tag{8–1a}$$

$$-\mathbf{F}_{xr}\mathbf{K}'_G\mathbf{r} + \mathbf{F}_{xx}\mathbf{X} = 0 \tag{8–1b}$$

In Eqs. (8–1), $\mathbf{F}_{rr}$, $\mathbf{F}_{rx}$, $\mathbf{F}_{xr}$, and $\mathbf{F}_{xx}$ are flexibility matrices of the primary structure and are defined in Eqs. (12/8–5). They are formed by the same methods as in the geometrically linear theory, except that member bending flexibilities are those of beam-column theory. The axial forces required to form these flexibilities are determined through a geometrically linear analysis.

For a uniform linearly elastic member and a constant axial force, the flexibility matrix of the alternate type has the form

$$\mathbf{f} = \begin{bmatrix} f & -f' \\ -f' & f \end{bmatrix}$$

where, from Eq. (6/2.1–8), with $\alpha = 1$ to neglect shear deformation,

$$f = \frac{l}{EI\lambda^2}(1 - \lambda \cot \lambda) = -\frac{1}{V_1 l}(1 - \lambda \cot \lambda) \tag{8–2a}$$

$$f' = \frac{l}{EI\lambda^2}\left(\frac{\lambda}{\sin \lambda} - 1\right) = -\frac{1}{V_1 l}\left(\frac{\lambda}{\sin \lambda} - 1\right) \tag{8–2b}$$

$\mathbf{K}'_G$ is the geometric stiffness due to the rotation of the member-bound reference. It is assembled from member matrices $\mathbf{k}_G$ and $\mathbf{k}_{GF}$.

The alternate type of bound reference will be usually chosen in applications. $\mathbf{k}_G$ is then the string geometric stiffness:

$$\mathbf{k}_G = \frac{V_1}{l}\begin{bmatrix} 1 & -1 \\ -1 & 1 \end{bmatrix} \tag{8–3}$$

The displacements associated with $\mathbf{k}_G$ are the transverse translations $\mathbf{s} = \{s_2^a \ \ s_2^b\}$. $\mathbf{k}_{GF}$ is due to the axial span load and is derived in Chapter 6, Section 3.3. For simplicity $\mathbf{K}'_G$ is replaced by $\mathbf{K}_G$ in what follows.

Equation (8–1b) is the condition for compatibility of deformations in a statically indeterminate structure. Equation (8–1a) is the condition for an alternate equilibrium state $\mathbf{r}$ to exist. This alternate equilibrium state may be investigated by methods of the geometrically linear theory provided modified member bending flexibilities are used. This is done by defining the "geometric" joint load

$$\mathbf{R}_G = -\mathbf{K}_G \mathbf{r} \tag{8–4}$$

and rewriting Eqs. (8–1) in the form

$$\mathbf{r} = \mathbf{F}_{rx}\mathbf{X} + \mathbf{f}_{rr}\mathbf{R}_G \tag{8–5a}$$

$$\mathbf{F}_{xR}\mathbf{R}_G + \mathbf{F}_{xx}\mathbf{X} = 0 \tag{8–5b}$$

Examples of stability problems by the force method have been treated within the context of nonlinear analysis in Chapter 12, Section 3, and instability criteria have been formulated in Chapter 12, Section 8. Applications to particular types of structures are developed in this chapter after discussing some aspects of the characteristic-value problem represented by Eqs. (8–1).

The exact flexibility relations (8–2) lead to a transcendental characteristic-value problem. The characteristic equation may be solved by trials, as done in the stiffness method. This equation remains in general nonlinear if an approximate linearized expression of **f** is derived because of the product $\mathbf{F}_{rr}\mathbf{K}_G$ in Eq. (8–1a). Exceptions occur, however, in which instability does not involve $\mathbf{K}_G$.

The nonlinear characteristic equation may also be solved iteratively and by the Newton-Raphson method. In the latter case the problem is linearized by expanding member flexibility coefficients in Taylor's series in the neighborhood of a certain load level.

An approximate **f** may also be used by expanding Eqs. (8–2) in powers of λ^2. This yields

$$f = \frac{l}{3EI}\left(1 + \frac{\lambda^2}{15} + \frac{\lambda^4}{280} + \ldots\right) \tag{8–6a}$$

$$f' = \frac{l}{6EI}\left(1 + \frac{7}{60}\lambda^2 + \frac{31}{2520}\lambda^4 + \ldots\right) \tag{8–6b}$$

If terms of order λ^4 and higher are neglected, the terms in λ^4 provide an estimate of the error. For example, for $\lambda = \pi/2$, $\lambda^4/280 = 0.02$, and $31\lambda^4/2520 = 0.07$. Thus a relative error of the order of 7% occurs if the axial force reaches one-fourth of Euler's critical load. Use of the approximate formulas is thus more limited than in the stiffness method.

9. BUCKLING OF PIN-JOINTED TRUSSES: FLEXIBILITY METHOD

Compression members of a pin-jointed truss may become individually unstable as Euler columns. In a statically determinate truss the instability of one member causes instability of the truss as a whole. In a statically indeterminate truss the effect depends on the stiffness of the structure if the unstable member is removed. Behavior of a truss in such a condition was discussed in Chapter 12, Section 5. In linearized stability analysis the member axial flexibilities are assumed constant, so individual instability of truss members must be considered separately. The theory deals then only with stability of the structure as a whole.

Elimination of **X** from Eqs. (8–1) yields the homogeneous equation in **r**

$$(\mathbf{I} + \mathbf{F}\mathbf{K}_G)\mathbf{r} = 0 \tag{9–1}$$

where

$$\mathbf{F} = \mathbf{F}_{rr} - \mathbf{F}_{rx}\mathbf{F}_{xx}^{-1}\mathbf{F}_{xr} \tag{9–2}$$

Since axial flexibilities are constant, $\mathbf{F}$ is the flexibility matrix of the truss in the geometrically linear theory. Using the concept of the "geometric" joint load $\mathbf{R}_G = -\mathbf{K}_G\mathbf{r}$, Eq. (9–1) becomes the flexibility relation

$$\mathbf{r} = \mathbf{F}\mathbf{R}_G \tag{9–3}$$

If Eq. (9–1) is premultiplied by $\mathbf{F}^{-1}$, a stiffness formulation of the problem is obtained. There is, however, a practical difference between the two formulations because the choice of effective degrees of freedom in the force method may be different and simpler than in the stiffness method.

Example 1

Consider the truss of Fig. 3–1*a* which was analyzed for stability by the stiffness method in Section 3, Example 1. Choosing first the same displacements as in the stiffness method, $\mathbf{K}_G$ was found to be

$$\mathbf{K}_G = -\frac{P}{l}\begin{bmatrix} 1 & 0 \\ 0 & 0 \end{bmatrix}$$

To form the flexibility matrix, we use the formula for statically determinate structures

$$\mathbf{F} = \mathbf{b}^T\mathbf{f}\mathbf{b}$$

$\mathbf{b}$ is determined by statics. Column i of $\mathbf{b}$ consists of the member axial forces in equilibrium with the joint load $R_i = 1$. $\mathbf{b}$ is thus found to be

$$\mathbf{b} = \begin{bmatrix} \dfrac{1}{\cos\theta} & 0 \\ -\tan\theta & 1 \end{bmatrix}$$

Letting $\mathbf{f} = \lceil f_1 \ \ f_2 \rfloor$ be the member flexibility, we obtain

$$\mathbf{F} = \begin{bmatrix} \dfrac{f_1}{\cos^2\theta} + f_2\tan^2\theta & -f_2\tan\theta \\ -f_2\tan\theta & f_2 \end{bmatrix}$$

The governing equation (9–1) takes the form

$$(\mathbf{I} + \mathbf{F}\mathbf{K}_G)\mathbf{r} = \begin{bmatrix} 1 - \dfrac{P}{l}\left(\dfrac{f_1}{\cos^2\theta} + f_2\tan^2\theta\right) & 0 \\ \dfrac{P}{l}f_2\tan\theta & 1 \end{bmatrix}\begin{Bmatrix} r_1 \\ r_2 \end{Bmatrix} = 0$$

The determinant of the coefficient matrix vanishes if the first term vanishes. Thus

$$P_{cr} = \frac{l}{f_1} \frac{\cos^2\theta}{1 + (f_2/f_1)\sin^2\theta} \tag{9–4}$$

This result concides with that found by the stiffness method in Eq. (3–3) in which k_1 and k_2 are f_1^{-1} and f_2^{-1}, respectively.

To illustrate the interpretation of $-\mathbf{K}_G\mathbf{r}$ as a joint load, Fig. 9–1*a* shows the effect of r_1 on the orientation of the member axial force $-V_1 = P$ in the vertical member. There is an increment in the horizontal component equal to $-Pr_1/l$, whereas the change in the vertical component is of higher order in r_1 and is ignored. This increment represents $\mathbf{K}_G\mathbf{r}$ in the present problem. The geometric joint load $\mathbf{R}_G$ is $-\mathbf{K}_G\mathbf{r}$ as shown in (*b*). Setting the horizontal displacement caused by this joint load, as determined by linear analysis, equal to r_1 yields the characteristic equation of the problem.

The matrix formulation of this problem could be done with only r_1 as the effective displacement. The reason for this appears on considering the product

$$\mathbf{F}\mathbf{R}_G = \begin{bmatrix} F_{11} & F_{12} \\ F_{21} & F_{22} \end{bmatrix} \begin{Bmatrix} R_{1,G} \\ R_{2,G} \end{Bmatrix}$$

Since $R_{2,G} = 0$, only F_{11} is needed to evaluate the displacement r_1 caused by $\mathbf{R}_G$.

Effective Degrees of Freedom

The observation made here may be generalized to the case where the effective degrees of freedom, for which there is a geometric stiffness, form a subset $\mathbf{r}_1$ of $\mathbf{r}$.

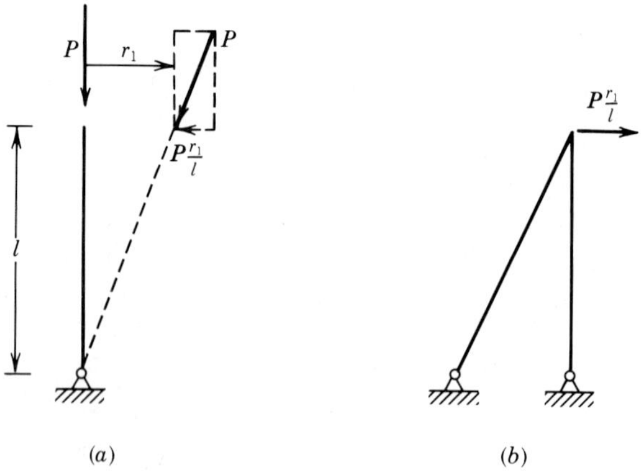

FIG. 9–1. Geometric joint load, Example 1.

Letting $\mathbf{r} = \{\mathbf{r}_1 \ \mathbf{r}_2\}$, $\mathbf{K}_G$ is partitioned in the form,

$$\mathbf{K}_G = \begin{bmatrix} \mathbf{K}_{11,G} & \mathbf{O} \\ \mathbf{O} & \mathbf{O} \end{bmatrix} \tag{9–5}$$

and similarly,

$$\mathbf{F} = \begin{bmatrix} \mathbf{F}_{11} & \mathbf{F}_{12} \\ \mathbf{F}_{21} & \mathbf{F}_{22} \end{bmatrix}$$

Equation (9–1) is then partitioned into

$$(\mathbf{I} + \mathbf{F}_{11}\mathbf{K}_{11,G})\mathbf{r}_1 = 0 \tag{9–6a}$$

$$\mathbf{r}_2 + \mathbf{F}_{21}\mathbf{K}_{11,G}\mathbf{r}_1 = 0 \tag{9–6b}$$

The first of Eqs. (9–6) represents the characteristic-value problem to be solved. The second serves only to define $\mathbf{r}_2$ in terms of $\mathbf{r}_1$.

In treating directly a problem where $\mathbf{K}_G$ is known to have the form of Eq. (9–5), it is advantageous to deal only with the effective displacements $\mathbf{r}_1$ and associated effective matrices $\mathbf{F}_{11}$ and $\mathbf{K}_{11,G}$. These quantities are then denoted by unindexed symbols.

Example 2

Consider the truss of Fig. 9–2*a*. $\mathbf{K}_G$ is formed from contributions of the stressed members, which are only members 1 and 2. Since the effective displacements in $\mathbf{k}_G$ are the transverse member end displacements, there is one effective displacement r, as shown. Accordingly the scalar notation k_G, K_G and F is used. For k_G we have for both members 1 and 2,

$$k_G = -\frac{P}{l}$$

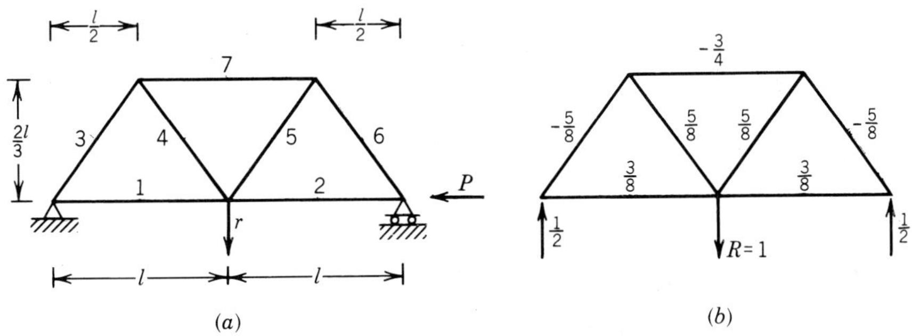

FIG. 9–2. Buckling of truss, Example 2.

Assembly of K_G yields

$$K_G = -\frac{2P}{l}$$

The flexibility F is the displacement r caused by a load $R = 1$. The axial forces caused by $R = 1$ are found by statics and are shown in (b). They form the matrix

$$\mathbf{b} = \tfrac{1}{8}\{3 \;\; 3 \;\; -5 \;\; 5 \;\; 5 \;\; -5 \;\; -6\}$$

Assuming all members have the same axial flexibility f, F is obtained through

$$F = \mathbf{b}^T\mathbf{f}\mathbf{b} = f\mathbf{b}^T\mathbf{b} = \frac{77}{32}f$$

The governing flexibility equation of the buckled state is

$$(1 + FK_G)r = 0$$

For a nontrivial solution

$$1 + FK_G = 1 - \frac{77}{16}f\frac{P}{l} = 0$$

or

$$P_{\text{cr}} = \frac{16l}{77f}$$

Treatment of the preceding problem by the stiffness method was only outlined in Section 3, Example 3, because reduction of the degrees of freedom required a static condensation of large size.

Example 3

Consider the rigid bar system of Fig. 9–3. The bars are assumed rigid, and the springs have flexibilities f_1 and f_2. The geometric stiffness matrix of the bottom bar is $-P/l[1]$, and the assembled geometric stiffness is

$$\mathbf{K}_G = -\frac{P}{l}\begin{bmatrix} 2 & -1 \\ -1 & 1 \end{bmatrix}$$

The flexibility matrix in the present case if $\mathbf{F} = \mathbf{f} = \lceil f_1 \;\; f_2 \rfloor$. We thus obtain

$$(\mathbf{I} + \mathbf{F}\mathbf{K}_G)\mathbf{r} = \begin{bmatrix} 1 - \dfrac{2f_1 P}{l} & f_1\dfrac{P}{l} \\ f_2\dfrac{P}{l} & 1 - f_2\dfrac{P}{l} \end{bmatrix}\begin{Bmatrix} r_1 \\ r_2 \end{Bmatrix} = 0 \qquad (9\text{–}7)$$

The characteristic equation is obtained by setting the determinant of the matrix equal to zero. There comes

$$P^2 - \left(\frac{2l}{f_2} + \frac{l}{f_1}\right)P + \frac{l^2}{f_1 f_2} = 0 \tag{9–8}$$

Assume now that

$$f_1 = f_2 = f$$

the roots of Eq. (9–8) are then

$$P_{cr} = P_1 = \frac{l}{f}\frac{3-\sqrt{5}}{2}$$

$$P_2 = \frac{l}{f}\frac{3+\sqrt{5}}{2}$$

The buckled shape is determined by solving Eq. (9–7) in which $P = P_{cr}$. From the first equation,

$$\frac{r_1}{r_2} = \frac{\sqrt{5}-3}{2(\sqrt{5}-2)} = -\frac{1+\sqrt{5}}{2} = -1.618$$

The second equation should give the same ratio for r_1/r_2 as a check. The buckled shape is shown in Fig. 9 – 3*b*.

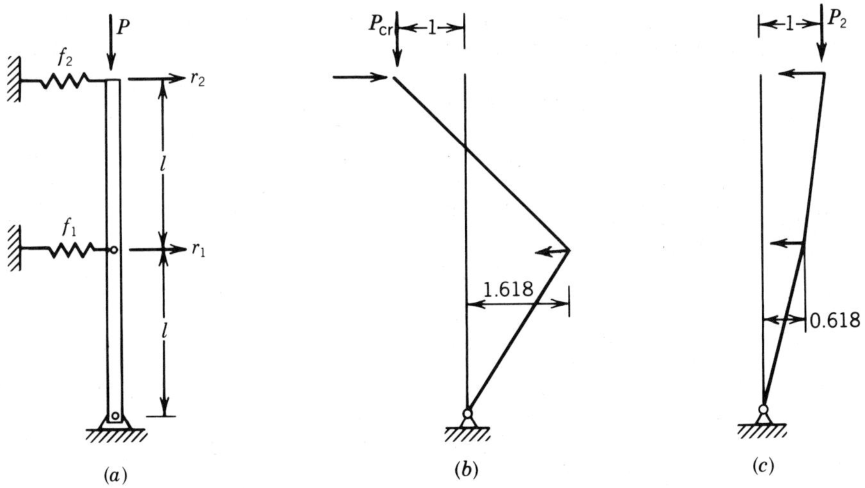

FIG. 9–3. Buckling of rigid bar system, Example 3.

The characteristic shape corresponding to the root P_2 is also a possible equilibrium shape but is unstable. For $P = P_2$, Eqs. (9–7) yield

$$\frac{r_1}{r_2} = \frac{\sqrt{5}+3}{2(\sqrt{5}+2)} = \frac{\sqrt{5}-1}{2} = 0.618$$

The corresponding shape is shown in Fig. 9–3*c*.

10. BUCKLING OF CONTINUOUS BEAMS: FLEXIBILITY METHOD

For continuous beams **r** consists of joint rotations. Since $\mathbf{K}_G$ corresponds only to translational degrees of freedom, Eq. (8.1–b) reduces to

$$\mathbf{F}_{xx}\mathbf{X} = 0 \tag{10–1}$$

X consists of the bending moments over supports. The general individual equation in Eq. (10–1) is the homogeneous three-moment equation (12/3–2), or

$$f'_{i-1}X_{i-1} + (f_{i-1} + f_i)X_i + f'_iX_{i+1} = 0 \tag{10–2}$$

Equation (10–2) applies to two adjacent spans $i-1$ and i. For a uniform member f and f' are as given in Eqs. (8–2). The notation and sign convention for the statical redundants is defined in Chapter 11, Section 6.

Example 1

For the continuous beam of Fig. 4–2a, $X_3 = 0$, and $\mathbf{X} = \{X_1 \;\; X_2\}$. Equation (10–2), written at joints 1 and 2, yields

$$f_1X_1 + f'_1X_2 = 0$$
$$f'_1X_1 + (f_1 + f_2)X_2 = 0$$

Elimination of X_1 yields

$$(f_2 + f_{1,e})X_2 = 0$$

where

$$f_{1,e} = f_1 - \frac{f'^2_1}{f_1} \tag{10–3a}$$

$f_{1,e}$ is the effective flexibility of member 1, defined at one end with the other end fixed against rotation. It is the inverse of the stiffness coefficient k in Eq. (12/2–2a). Deleting the subscript, the formula for a general member is

$$f_e = \frac{2l}{EI} \frac{\tan(\lambda/2) - (\lambda/2)}{\lambda(1 - \cot \lambda)} \qquad (10\text{–}3b)$$

The characteristic equation is

$$f_2 + f_{1,e} = 0 \qquad (10\text{–}4)$$

where f_2 is obtained by applying Eq. (8–2a) to member 2, and $f_{1,e}$ by applying Eq. (10–4) to member 1. The characteristic equation for the same problem, by the stiffness method, is Eq. (4–6). With $k_1 = f_{1,e}^{-1}$ and $k_{2,e} = f_2^{-1}$, it is equivalent to Eq. (10–4).

Example 2

For the continuous beam of Fig. 4–1, use of the three-moment equation results in a system of three homogeneous equations in three unknowns X_1, X_2, X_3. A more efficient procedure is to write one compatibility equation at joint 2, using effective flexibility coefficients for fixed far ends. The equation is then

$$(f_{1,e} + f_{2,e})X_2 = 0$$

The characteristic equation is equivalent to the one derived by the stiffness method in Section 4, Example 1. The correspondance between the two methods is $f_{1,e} = k_1^{-1}$ and $f_{2,e} = k_2^{-1}$.

Example 3

For the continuous beam of Fig. 10–1 the three-moment equation is written at joints 2 and 3, with the end conditions $X_1 = X_4 = 0$. There comes

$$(f_1 + f_2)X_2 + f'_2 X_3 = 0$$

$$f'_2 X_2 + (f_2 + f_3)X_3 = 0$$

The characteristic equation, obtained by setting the determinant equal to zero, is

$$(f_1 + f_2)(f_2 + f_3) - f_2'^2 = 0 \qquad (10\text{–}5)$$

The three spans have the same axial force P, so that their stability parameters are

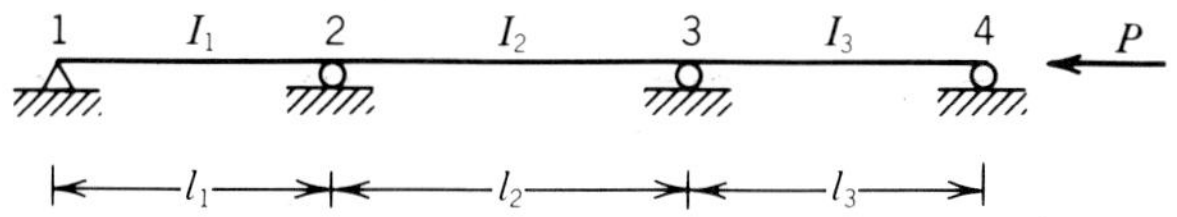

FIG. 10–1. Buckling of continuous beam, Example 3.

related through

$$\frac{\lambda_1^2 EI_1}{l_1^2} = \frac{\lambda_2^2 EI_2}{l_2^2} = \frac{\lambda_3^2 EI_3}{l_3^2} = P$$

To find the proper root of Eq. (10–5), lower or upper bounds on P_{cr} are helpful. Considerations for determining such bounds were developed in the examples treated previously by the stiffness method. For example, a lower bound on P_{cr} is the lowest of Euler's loads for the three spans, considered separately, because the weakest span can only be strengthened by continuity with one or two other spans. If such a span is the middle span, it lowers the buckling loads of the two other spans, considered each as an Euler column.

Consider now the case where Fig. 10–1 has a plane of symmetry. Then $f_1 = f_3$, and $X_2 = X_3$. The left-hand side of Eq. (10–5) may be written as the product of two factors. The factor that must be chosen for the characteristic equation is the one obtained by setting $X_2 = X_3$ in either one of the compatibility equations. The other factor corresponds to an unstable antisymmetric characteristic shape and a higher axial force. The characteristic equation is then

$$f_1 + f_2 + f_2' = 0$$

The quantity

$$f_s = f + f' = \frac{l}{EI\lambda} \tan\frac{\lambda}{2} \tag{10–6}$$

is a member flexibility for symmetric deformation, which could be used to write directly the compatibility equation at either joint 2 or joint 3 and thus to obtain

$$f_1 + f_{2,s} = 0$$

Using Eq. (8–2a) for f_1, and (10–6) for $f_{2,s}$, there comes

$$\frac{l_1}{EI_1\lambda_1^2}(1 - \lambda_1 \cot\lambda_1) + \frac{l_2}{EI_2\lambda_2} \tan\frac{\lambda_2}{2} = 0$$

The stability parameters are related through

$$\frac{\lambda_1^2 EI_1}{l_1^2} = \frac{\lambda_2^2 EI_2}{l_2^2} = P$$

The two foregoing equations may be written in the form

$$1 - \lambda_1 \cot\lambda_1 + \frac{l_1}{l_2}\lambda_2 \tan\frac{\lambda_2}{2} = 0 \tag{10–7a}$$

$$\frac{\lambda_1}{\lambda_2} = \frac{l_1}{l_2}\sqrt{\frac{I_2}{I_1}} \tag{10–7b}$$

Consider the case $I_2 = 2I_1$, $l_1 = l_2$. Then $\lambda_1/\lambda_2 = \sqrt{2}$. P_{cr} is larger than Euler's load for the exterior spans and smaller than that for the interior span. Thus $\lambda_1 > \pi$, and $\lambda_2 < \pi$. The bounds on $\lambda_{1,cr}$ are then

$$\pi < \lambda_{1,cr} < \pi\sqrt{2}$$

In terms of λ_1, Eq. (10–7a) takes the form

$$1 - \lambda_1 \cot \lambda_1 + \frac{\lambda_1}{\sqrt{2}} \tan \frac{\lambda_1}{2\sqrt{2}} = 0$$

The solution is found by trials as

$$\lambda_{1,cr} = 3.535$$

The effective length factor of the exterior spans is $\pi/\lambda_1 = 0.89$, and that of the interior span is $\pi/\lambda_2 = \sqrt{2}\pi/\lambda_1 = 1.26$.

In the case of three identical spans, each span buckles as an Euler column without rotational restraint at the interior joint, and thus with $X_2 = X_3 = 0$. It is of interest to examine how this appears analytically. The compatibility equations presume the existence of nontrivial statical redundants in the buckled state. If these are zero, the existence of nonzero deformations requires infinite member flexibilities, as occurs in the critical state of Euler's column. For $\lambda_1 = \lambda_2 = \lambda$, the flexibilities f_1 and $f_{2,s}$ tend to ∞ with the same sign, as $\lambda \to \pi$. If the case $\lambda_1 = \lambda_2$ is approached as a limit of cases where, say, $I_1 < I_2$, then $\lambda_2 < \pi < \lambda_1$. As $I_1 \to I_2$, λ_1 tends to π from above, f_1 tends to $-\infty$, λ_2 tends to π from below, and $f_{2,s}$ tends to $+\infty$, whereas for any given I_1 and I_2, the characteristic equation, $f_1 + f_{2,s} = 0$, is satisfied.

In the stiffness method the characteristic equation takes the form $f_1^{-1} + f_{2,s}^{-1} = 0$ and is equivalent to the present equation.

11. BUCKLING OF FRAMES: FLEXIBILITY METHOD, INEXTENSIONAL AND EXTENSIONAL MODES

Separation of frame behavior into inextensional and extensional modes was discussed in the context of nonlinear analysis by the force method in Chapter 12, Sections 10 and 11. Examples involving instability were also treated. Further applications of instability in inextensional modes are treated in the following section. Frames without joint translations are treated first. Their theory is the same as that of trussed frames seen in Chapter 12, and it covers continuous beams as a particular case. Extensional buckling of trussed frames is approximated by that of the substitute truss and will not be dealt with further. Frames with translational inextensional modes are then considered.

12. BUCKLING OF FRAMES WITHOUT JOINT TRANSLATIONS: FLEXIBILITY METHOD

For frames without joint translations the geometric stiffness terms due to $\mathbf{K}_G$ vanish. Such frames are usually statically indeterminate. A buckling mode involving nonzero statical redundants $\mathbf{X}$ satisfies the homogeneous equations

$$\mathbf{F}_{xx}\mathbf{X} = 0 \tag{12–1}$$

The possibility of buckling with $\mathbf{X} = 0$ requires that buckling be the same as for the primary structure. However, since there are no joint translations and thus no geometric effect on the joint equilibrium equations, there can be no self-equilibriating member end moments in the primary structure. The only possibility for buckling under those conditions is one of individual members as Euler columns. Analytically $\mathbf{K}_G\mathbf{r} = 0$, but $\mathbf{F}_{rr}$ and $\mathbf{F}_{xx}$ have unbounded elements. An example of such an occurrence may be seen in Section 10, Example 3.

Equation (12–1) is formed by the same method as in the geometrically linear theory, using, however, the bending flexibilities $\mathbf{f}$ of beam-column theory. $\mathbf{F}_{xx}$ is evaluated through the formula

$$\mathbf{F}_{xx} = \mathbf{c}^T\mathbf{f}\mathbf{c} = \sum_m \mathbf{c}^{mT}\mathbf{f}^m\mathbf{c}^m \tag{12–2}$$

Letting $\mathbf{c}_i$ be the ith column of $\mathbf{c}$, element (i, j) of $\mathbf{F}_{xx}$ is denoted δ_{ij} and is evaluated through

$$\delta_{ij} = \mathbf{c}_i^T\mathbf{f}\mathbf{c}_j = \sum_m \mathbf{c}_i^{mT}\mathbf{f}^m\mathbf{c}_j^m \tag{12–3}$$

$\mathbf{c}_i^m$ consists of the moments at the ends of member m, due to the load $X_i = 1$ applied on the primary structure. A convenient formula is established for evaluating $\mathbf{c}_i^{mT}\mathbf{f}^m c_j^m$ for a uniform member. c_i^m and c_j^m have the form

$$\mathbf{c}_i^m = \{-m_a \quad m_b\}$$
$$\mathbf{c}_j^m = \{-m_a' \quad m_b'\}$$

where subscripts a and b refer to the ends of member m, and bending moments m_a, m_b, m_a' and m_b' are considered positive if they produce tension on a given side of the member. The formula to be established is

$$\mathbf{c}_i^{mT}\mathbf{f}^m\mathbf{c}_j^m = [-m_a \quad m_b]\begin{bmatrix} f & -f' \\ -f' & f \end{bmatrix}\begin{Bmatrix} -m_a' \\ m_b' \end{Bmatrix}$$

or

$$\mathbf{c}_i^{mT}\mathbf{f}^m\mathbf{c}_j^m = f(m_am_a' + m_bm_b') + f'(m_am_b' + m_bm_a') \tag{12–4}$$

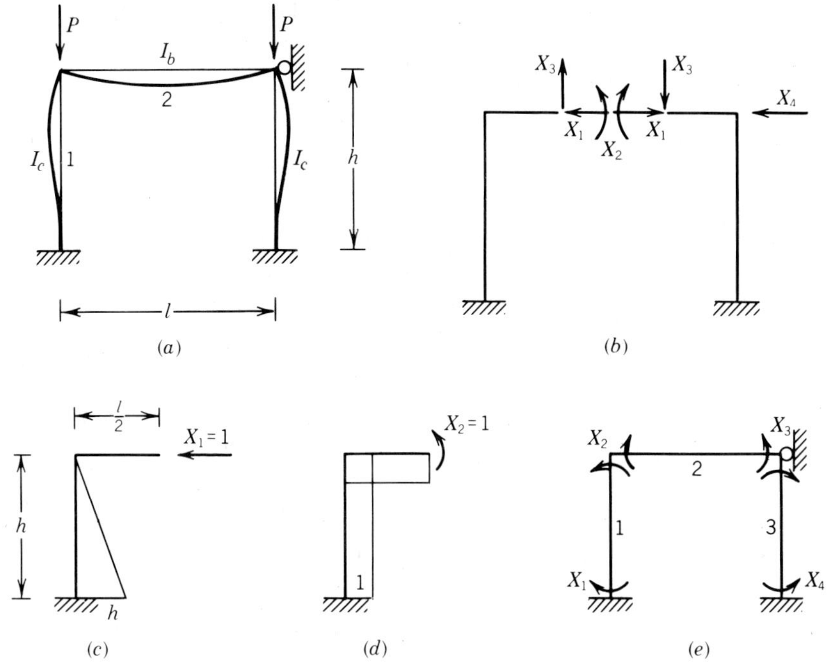

FIG. 12–1. Frame without sidesway, Example 1.

f and f' are evaluated for a uniform member with a constant compressive axial force through Eqs. (8–2). If the member is not in compression,

$$f = \frac{l}{3EI} \tag{12–5a}$$

$$f' = \frac{l}{6EI} \tag{12–5b}$$

For manual solutions the three- and four-moment equations provide an equivalent approach to forming the compatibility equations. If the moments in these equations are all expressed by statics in terms of the statical redundants $\mathbf{X}$, the coefficient matrix is $\mathbf{F}_{xx}$.

Example 1. Portal Frame without Sidesway, Fixed Base

The frame of Fig. 12–1*a* is symmetric and is assumed restrained against sidesway displacement. It buckles in a symmetric deformation mode, as shown. The frame is indeterminate to the fourth degree. A choice of statical redundants is shown in (*b*). Because of symmetry, $X_3 = X_4 = 0$, and only the compatibility equations associated with X_1 and X_2 need be written. The bending moment diagrams due to $X_1 = 1$ and $X_2 = 1$ are shown in (*c*) and (*d*), respectively. The member end values define $\mathbf{c}_1$ and $\mathbf{c}_2$. Using formula (12–4), there comes

$$\delta_{11} = h^2 f_1$$
$$\delta_{12} = \delta_{21} = h(f_1 + f'_1)$$
$$\delta_{22} = 2(f_1 + f'_1) + 2(f_2 + f'_2)$$

The compatibility equations are thus

$$\begin{bmatrix} h^2 f_1 & h(f_1 + f'_1) \\ h(f_1 + f'_1) & 2(f_1 + f'_1) + 2(f_2 + f'_2) \end{bmatrix} \begin{Bmatrix} X_1 \\ X_2 \end{Bmatrix} = 0 \tag{12–6}$$

For the column f_1 and f'_1 are as defined in Eqs. (8–2), and for the beam of length $\frac{l}{2}$, $f_2 = \frac{1}{2}(l/3EI_b)$ and $f'_2 = \frac{1}{2}(l/6EI_b)$. Equating the determinant to zero, the result may be put in the form

$$f_{1,e} + \frac{l}{2EI_b} = 0 \tag{12–7}$$

where $f_{1,e}$ is the effective flexibility defined in Eq. (10–3). Applying this equation to the column, Eq. (12–7) is found to be equivalent to Eq. (6–3a) which was found by the stiffness method.

The problem is more directly solved using the three-moment equation because this equation incorporates the intermediate calculations of the elements of $\mathbf{F}_{xx}$. Using the notation of Fig. 12–1*e*, the three-moment equation, written at the ends of column 1 and with $X_3 = X_2$, yields

$$\begin{bmatrix} f_1 & f'_1 \\ f'_1 & f_1 + f_2 + f'_2 \end{bmatrix} \begin{Bmatrix} X_1 \\ X_2 \end{Bmatrix} = 0 \tag{12–8}$$

The characteristic equation is readily seen to coincide with Eq. (12–7).

Still a more direct way is to write the three-moment equation once at the top of either column, using the effective flexibility f_e for a fixed far end and the symmetry property $X_2 = X_3$. The coefficient of such an equation is the left-hand side of Eq. (12–7).

Example 2. Portal Frame without Sidesway, Pinned Base

The frame to be considered differs from that of Fig. 12–1*a* only in having the columns pinned at the base and is shown with a symmetric buckled shape in Fig. 6–3. The three-moment equation written at the top of the left column is the second of Eqs.(12–8) in which $X_1 = 0$. The characteristic equation is thus found to be

$$f_1 + f_2 + f'_2 = 0 \tag{12–9}$$

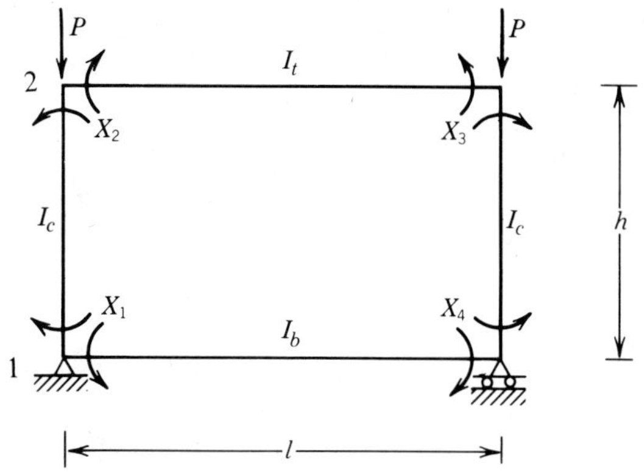

FIG. 12–2. Frame without sidesway, Example 3.

With f_1 from Eq. (8–2a) for the column, and $f_2 + f'_2 = l/2EI_b$, Eq. (12–9) coincides with Eq. (6–4a) found by the stiffness method.

Example 3. Building Frame without Sidesway, Type 1

The three-moment equation is written for the frame of Fig. 12–2 at joints 1 and 2. There comes, with indexes b, c, and t to refer to the members,

$$f'_b X_4 + (f_b + f_c)X_1 + f'_c X_2 = 0 \tag{12–10a}$$

$$f'_c X_1 + (f_c + f_t)X_2 + f'_t X_3 = 0 \tag{12–10b}$$

The frame has a vertical plane of symmetry and buckles in a symmetric mode, as shown in Fig. 6–4. Thus $X_4 = X_1$, and $X_3 = X_2$. Equations (12–10) become

$$(f_c + f_b + f'_b)X_1 + f'_c X_2 = 0$$

$$f'_c X_1 + (f_c + f_t + f'_t)X_2 = 0$$

For the beams $f + f' = l/2EI$. The characteristic equation takes the form

$$\left(f_c + \frac{l}{2EI_b}\right)\left(f_c + \frac{l}{2EI_t}\right) - f_c'^2 = 0 \tag{12–11}$$

Equation (12–11) should be equivalent to Eq. (6–5) found by the stiffness method. To show this, we have for the column $\mathbf{f} = \mathbf{k}^{-1}$. Thus, with $\mathbf{k}$ as defined in Eq. (6–1),

$$f = \frac{k}{k^2 - k'^2} \tag{12–12a}$$

$$f' = \frac{k'}{k^2 - k'^2} \tag{12–12b}$$

and consequently

$$f^2 - f'^2 = \frac{1}{k^2 - k'^2} \tag{12–13}$$

Using these relations in Eq. (12–11), and multiplying it by $4(EI_b/l)\,(EI_t/l)$, the result coincides with the characteristic equation found by the stiffness method.

Example 4

The frame of Fig. 12–3*a* is statically indeterminate to the fourth degree. A choice of statical redundants is shown in (*b*). The moment diagrams due to $X_1 = 1$ and $X_2 = 1$ are shown in (*c*) and (*d*). Using formula (12–4), there comes

$$\delta_{11} = f_1 + f_3$$
$$\delta_{22} = f_1 + f_4$$
$$\delta_{12} = -f_1$$

The moment diagrams due to $X_3 = 1$ and $X_4 = 1$ are similar to the preceding ones and are not shown. The remaining elements of $\mathbf{F}_{xx}$ are obtained as

$$\delta_{13} = \delta_{14} = 0$$
$$\delta_{23} = f'_4$$
$$\delta_{24} = 0$$
$$\delta_{33} = f_2 + f_4$$
$$\delta_{44} = f_2 + f_5$$
$$\delta_{34} = -f_2$$

Thus

$$\mathbf{F}_{xx} = \begin{bmatrix} f_1 + f_3 & -f_1 & 0 & 0 \\ -f_1 & f_1 + f_4 & f'_4 & 0 \\ 0 & f'_4 & f_2 + f_4 & -f_2 \\ 0 & 0 & -f_2 & f_2 + f_5 \end{bmatrix} \tag{12–14}$$

The determinant $|\mathbf{F}_{xx}|$, divided by $f_1 f_2 f_3 f_5$ and set equal to zero, may be put in the form

$$\left(\frac{1}{f_1} + \frac{1}{f_3}\right)\left(\frac{1}{f_2} + \frac{1}{f_5}\right)(f_4^2 - f_4'^2) + f_4\left(\frac{1}{f_1} + \frac{1}{f_3} + \frac{1}{f_2} + \frac{1}{f_5}\right) + 1 = 0 \tag{12–15}$$

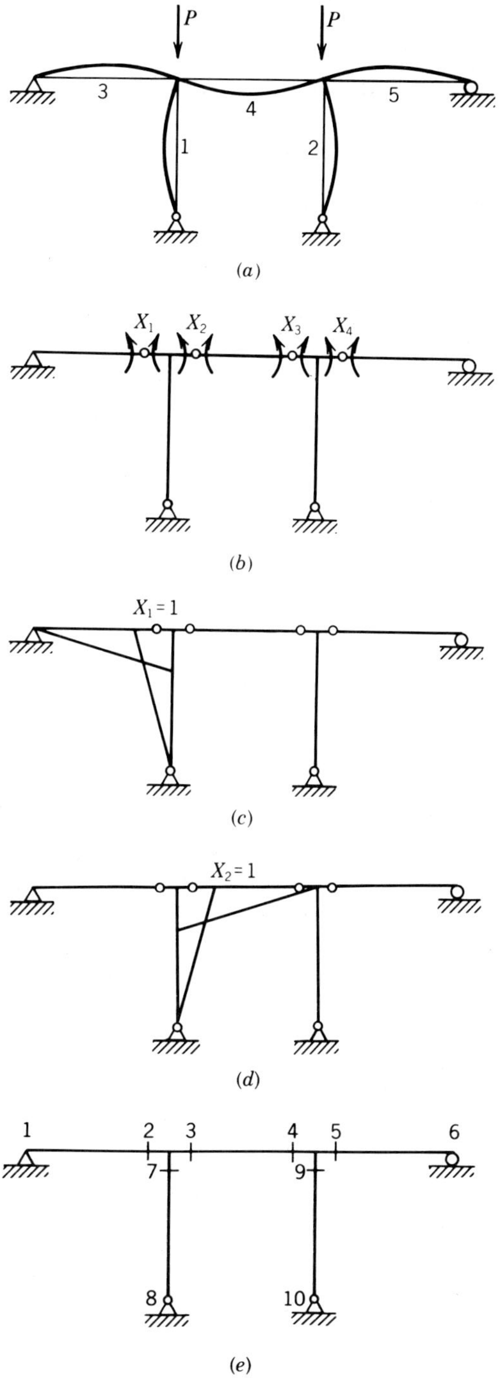

FIG. 12–3. Frame of Example 4.

For the present problem the stiffness method is simpler because the structure has only two degrees of freedom. Equation (6–12) may be applied to the present structure by using the appropriate notation. In Eq. (6–12) k and k' pertain to member 4. The other stiffnesses are effective for a pinned far end. We thus obtain

$$\left(k_4+\frac{1}{f_1}+\frac{1}{f_3}\right)\left(k_4+\frac{1}{f_2}+\frac{1}{f_5}\right)-k_4'^2=0 \tag{12–16}$$

If Eq. (12–16) is divided through by $(k_4^2-k_4'^2)$, and Eqs. (12–12, 13) are used for member 4, the result coincides with Eq. (12–15).

To illustrate use of the four-moment equation, the homogeneous Eq. (12/3–4) is applied to the following paths in Fig. 12–3*e*

1	2	7	8
4	3	7	8
3	4	9	10
6	5	9	10

With $M_1=M_8=M_{10}=M_6=0$, $M_7=X_2-X_1$, and $M_9=X_4-X_3$, the coefficient matrix of the system of four equations in $\{X_1\ X_2\ X_3\ X_4\}$ coincides with $\mathbf{F}_{xx}$ in Eq. (12–14).

13. BUCKLING OF FRAMES WITH JOINT TRANSLATIONS: FLEXIBILITY METHOD

Inextensional buckling of frames with joint translations is governed by Eqs. (8–1) in which $\mathbf{r}$ represents the inextensional degrees of freedom. $\mathbf{F}_{xx}$ may be assumed nonsingular as long as buckling without joint translations occurs at a different load level than with joint translations. Equation (8–1) may then be reduced by elimination of $\mathbf{X}$ to Eq. (9–1), as was done for trusses.

For statically determinate structures Eq. (8–1b) vanishes, and Eq. (8–1a) has the form of the general equation (9–1).

In some of the examples to follow, the characteristic equation is derived by both the general matrix formulation and the three-moment equation, including joint translations.

Example 1. Nonuniform Member, Simply Supported

The nonuniform member of Fig. (7–2) is formed of two uniform members a and b. The effective displacement is the one for which there is a geometric stiffness, that is, r_1. It will be denoted r and the scalar notation K_G and F will be used. K_G is readily formed as

$$K_G=-P\left(\frac{1}{a}+\frac{1}{b}\right)$$

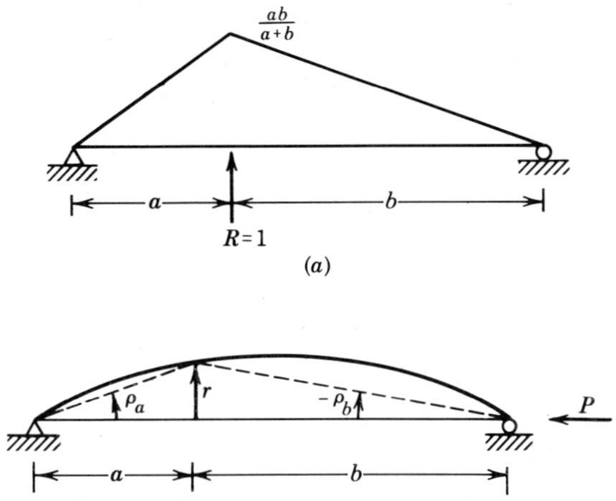

FIG. 13–1. Buckling of nonuniform member, Example 1.

The member is statically determinate, so

$$F = \mathbf{b}^T \mathbf{f} \mathbf{b}$$

$\mathbf{b}$ is defined by means of the bending moment diagram due to $R = 1$, shown in Fig. 13–1*a*. The formula established for $\mathbf{c}_i^T \mathbf{f} \mathbf{c}_j$ in Eq. (12–4) is applicable to the evaluation of F and yields

$$F = \left(\frac{ab}{a+b}\right)^2 (f_a + f_b) \tag{13–1}$$

The characteristic equation $1 + FK_G = 0$ yields

$$P = \left(\frac{1}{a} + \frac{1}{b}\right) \frac{1}{f_a + f_b} \tag{13–2}$$

Equation (13–2) agrees with the one found by the stiffness method in Eq. (7–8a) and may be solved for P_{cr} in a similar manner.

The preceding problem may be solved using the three-moment equation with joint translations, Eq. (12/3–2, 3), applied to the buckled state, Fig. 13–1*b*. Letting M be the bending moment at the joint of members a and b, there comes

$$(f_a + f_b)M + \rho_a - \rho_b = 0 \tag{13–3}$$

where $\rho_a = r/a$, $\rho_b = -r/b$, and by statics, $M = -Pr$. Substituting these relations

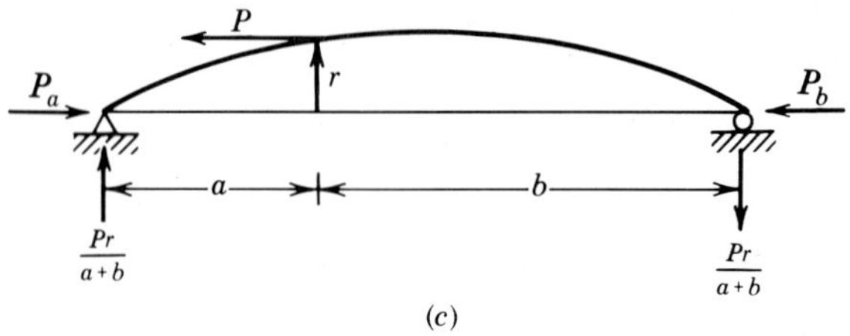

FIG. 13–2. Member with intermediate axial force, Example 2.

into Eq. (13–3), and setting the coefficient of r to zero, yields the characteristic equation for P found earlier.

Example 2. Member with Intermediate Axial Load, Simply Supported

For the member of Fig. 7–8a the solution procedure is similar to that of Example 1, and the effective displacement is the same. The geometric stiffness is $K_G = -P_a/a - P_b/b$, and F is as obtained in Eq. (13–1). The characteristic equation $1 + FK_G = 0$ is thus

$$1 - \left(\frac{P_a}{a} + \frac{P_b}{b}\right)\left(\frac{ab}{a+b}\right)^2 (f_a + f_b) = 0 \tag{13–4}$$

and is equivalent to Eq. (7–19a) which was found by the stiffness method.

To solve the problem by the three-moment equation, the only change to Eq. (13–3) is the expression of M. From Fig. 13–2,

$$M = -\left(P_b r + Pr\frac{b}{a+b}\right) = -\frac{P_b a + P_a b}{a+b} r$$

Equation (13–3) yields then Eq. (13–4)

Example 3. Nonuniform Cantilever with Intermediate Axial Load

Figure 13–3a shows the buckled shape of a cantilever subjected to the axial loads P_1 and P_2. Spans l_1 and l_2 are assumed each to be uniform. $\mathbf{K}_G$ is assembled from member matrices $\mathbf{k}_G$ as

$$\mathbf{K}_G = -\begin{bmatrix} \dfrac{P_1 + P_2}{l_1} + \dfrac{P_2}{l_2} & -\dfrac{P_2}{l_2} \\ -\dfrac{P_2}{l_2} & \dfrac{P_2}{l_2} \end{bmatrix}$$

$\mathbf{F}$ is evaluated through, $\mathbf{F} = \mathbf{b}^T\mathbf{f}\mathbf{b}$, where $\mathbf{b}$ is defined by means of the bending

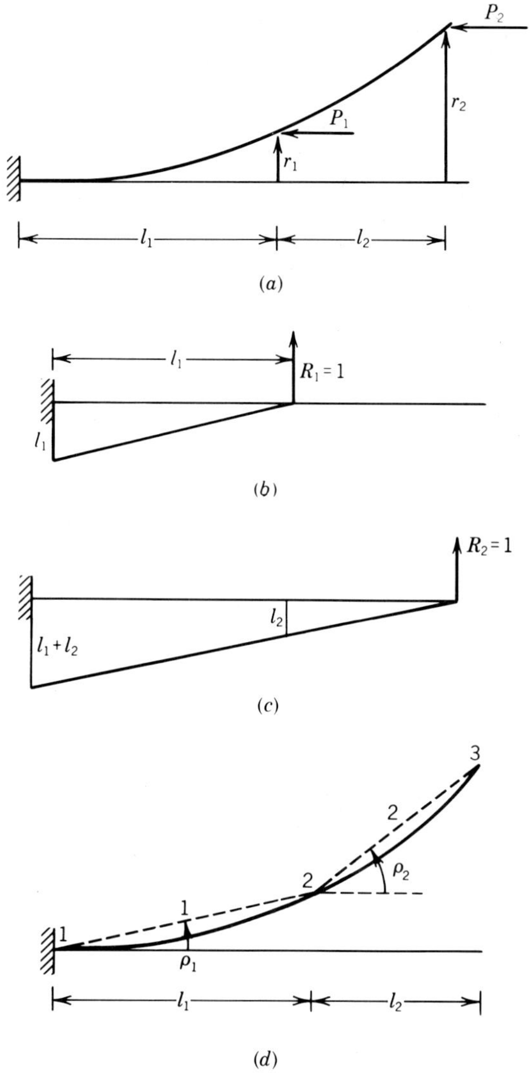

FIG. 13–3. Cantilever with intermediate axial force, Example 3.

moment diagrams due to $R_1 = 1$ and $R_2 = 1$, respectively, and shown in Fig. 13–3*b*, *c*.

Using formula (12–4) to evaluate the elements of **F**, there comes,

$$F_{11} = l_1^2 f_1$$

$$F_{22} = [(l_1 + l_2)^2 + l_2^2] f_1 + 2l_2(l_1 + l_2) f'_1 + l_2^2 f_2$$

$$F_{12} = l_1(l_1 + l_2) f_1 + l_1 l_2 f'_1$$

The characteristic equation, $|\mathbf{I}+\mathbf{FK}_G|=0$, takes the form

$$\begin{aligned}1-P_2l_2(2f_1+2f'_2+f_2)-(P_1+P_2)l_1f_1\\ +P_2(P_1+P_2)l_1l_2[f_1^2-f_1'^2+f_1f_2]=0\end{aligned} \tag{13–5}$$

The problem may also be solved by the three-moment equations. With the notation of Fig. 13–3*d*, the equation is written at joints 1 and 2 and yields

$$f_1M_1+f'_1M_2-\rho_1=0$$

$$f'_1M_1+(f_1+f_2)M_2+\rho_1-\rho_2=0$$

where

$$\rho_1=\frac{r_1}{l_1}$$

$$\rho_2=\frac{r_2-r_1}{l_2}$$

and by statics,

$$M_1=P_2r_2+P_1r_1$$

$$M_2=P_2(r_2-r_1)$$

Substituting for ρ_1, ρ_2, M_1, and M_2 into the three-moment equations, the result is a homogeneous linear system of two equations in r_1 and r_2. Setting the determinant equal to zero yields Eq. (13–5).

As an application, let

$$P_1=P_2=P$$

$$l_1=l_2=l$$

Then

$$P=\frac{\lambda_1^2EI_1}{2l^2}=\frac{\lambda_2^2EI_2}{l^2}$$

Equation (13–5) simplified to

$$\tan\lambda_1\tan\lambda_2-\frac{2\lambda_2}{\lambda_1}=0 \tag{13–6}$$

Consider further the case of a uniform cantilever; then

$$\lambda_1=\lambda_2\sqrt{2}$$

$2P_{cr}$ is larger than the critical load of the cantilever subjected to a constant axial force and is lower than the critical load of a cantilever of length l. Thus

$$\frac{\pi^2 EI}{16l^2} < 2P_{cr} < \frac{\pi^2 EI}{4l^2}$$

or, in terms of λ_1,

$$\frac{\pi}{4} < \lambda_{1,cr} < \frac{\pi}{2}$$

The root of Eq. (13–6) for λ_1 is found to be

$$\lambda_{1,cr} = 1.017 \qquad (13\text{–}7)$$

The effective length factor of the cantilever based on the load $2P$ is

$$K = \frac{\pi}{2\lambda_{1,cr}} = 1.54$$

Example 4. Portal Frame with Sidesway, Pinned Base

The portal frame of Fig. 13–4 is symmetric and buckles in an antisymmetric mode as shown. The three-moment equation written at joint 2 is

$$(f_c + f_b)M_2 + f'_b M_3 + \rho_c = 0$$

where ρ_c is the chord rotation of the column. In a consistent sign convention, ρ_c is negative if clockwise, and M_2 and M_3 positive if they produce tension on the

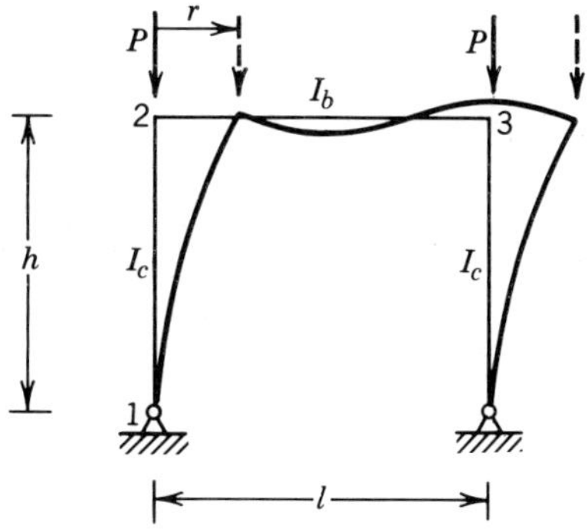

FIG. 13–4. Frame pinned at base, Example 4.

inside of the frame. Thus

$$\rho_c = -\frac{r}{h}$$

and by antisymmetry,

$$M_3 = -M_2$$

Although the frame is statically indeterminate to the first degree, antisymmetry makes it determinate. The horizontal reactions at the supports must be equal vectors, and their sum must vanish because there is no horizontal force applied. They are thus zero. The change in the vertical reactions due to buckling affects the bending moments with second-order terms in r and may be ignored. Thus

$$M_2 = Pr$$

Expressing the three-moment equation in terms of r yields the characteristic equation

$$f_c + f_b - f'_b - \frac{1}{Ph} = 0 \tag{13–8}$$

where

$$f_b - f'_b = \frac{l}{6EI_b}$$

and f_c is obtained by applying Eq. (8–2a) to the column. With

$$P = \frac{\lambda^2 EI_c}{h^2}$$

Eq. (13–8) takes the form

$$\frac{\cot\lambda}{\lambda} = \frac{1}{6}\frac{I_c/h}{I_b/l} \tag{13–9}$$

Bounds on λ are obtained by varying the beam stiffness from 0 to ∞. At the lower limit $\lambda_{cr} = 0$, and at the upper limit the columns buckle as cantilevers fixed at the top. Thus

$$0 < \lambda_{cr} < \frac{\pi}{2} \tag{13–10}$$

The effective length factor is larger than 2 and may reach ∞.

Example 5. Building Frame with Sidesway, Type 1

The frame of Fig. 7–6*a* is reduced for analysis to that shown in (*b*). The effective degree of freedom is the lateral displacement r_1 which will be denoted here as r.

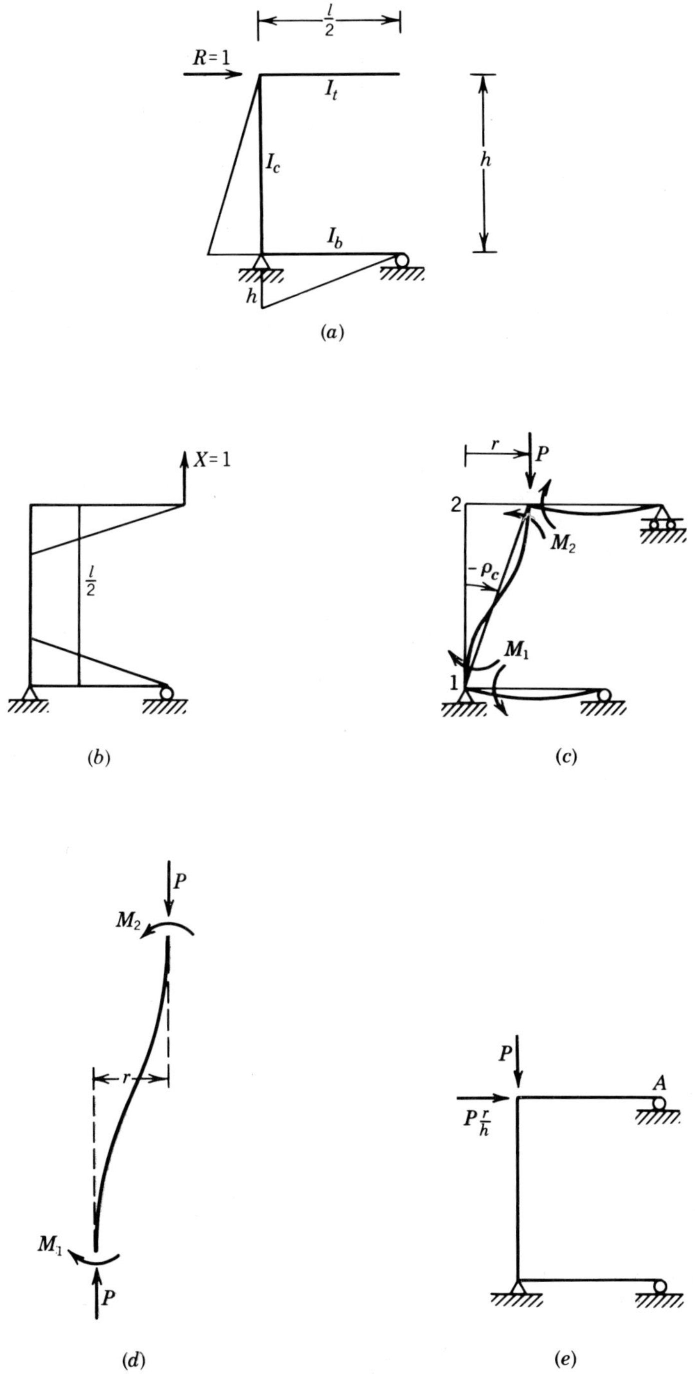

FIG. 13–5. Frame with sidesway, Example 5.

We have

$$K_G = -\frac{P}{h}$$

The characteristic equation is then

$$1 - \frac{P}{h} F = 0 \tag{13–11}$$

where F is the flexibility of the structure. The structure is statically indeterminate to the first degree, and F is evaluated using Eq. (9–2), or

$$F = F_{rr} - F_{rx} F_{xx}^{-1} F_{xr}$$

The matrices **b** and **c**, for evaluating the flexibilities in this equation, are defined by means of the bending moment diagrams in Figs. 13–5*a* and 13–5*b*, respectively. Using formula (12–4), there comes

$$F_{rr} = \mathbf{b}^T \mathbf{f} \mathbf{b} = h^2(f_c + f_b)$$

$$F_{rx} = \mathbf{b}^T \mathbf{f} \mathbf{c} = -\frac{hl}{2}(f_c + f'_c + f_b)$$

$$F_{xr} = F_{rx}$$

$$F_{xx} = \mathbf{c}^T \mathbf{f} \mathbf{c} = \frac{l^2}{2}\left(f_c + f'_c + \frac{1}{2} f_t + \frac{1}{2} f_b \right)$$

whence

$$F = h^2(f_c + f_b) - \frac{h^2}{2} \frac{(f_c + f'_c + f_b)^2}{f_c + f'_c + \frac{1}{2} f_t + \frac{1}{2} f_b} \tag{13–12}$$

Substituting for F into Eq. (13–11), the result may be written in the form

$$[2 - Ph(f_c - f'_c)](f_c + f'_c) + (1 - Phf_c)(f_t + f_b) - Phf_t f_b = 0 \tag{13–13}$$

In this equation f_t and f_b are defined for beams of length $l/2$. Thus

$$f_t = \frac{l}{6EI_t}$$

$$f_b = \frac{l}{6EI_b}$$

f_c and f'_c are obtained from Eqs. (8–2) applied to the column. With

$$P = \frac{\lambda^2 EI_c}{h^2}$$

there comes

$$1 - Phf_c = \lambda \cot \lambda$$

$$2 - Ph(f_c - f'_c) = \lambda \cot \frac{\lambda}{2}$$

$$f_c + f'_c = \frac{h}{EI_c \lambda} \tan \frac{\lambda}{2}$$

Using these results, Eq. (13–13) may be put in the form of Eq. (7–14) which was derived by the stiffness method.

To solve this problem by the three-moment equation, we obtain with the notation of Fig. 13–5c

$$(f_c + f_b)M_1 + f'_c M_2 - \rho_c = 0$$

$$f'_c M_1 + (f_c + f_t)M_2 + \rho_c = 0$$

where ρ_c is the chord rotation of the column and is negative in the figure. Thus

$$\rho_c = -\frac{r}{h}$$

Now it is needed to express r by statics, in terms of M_1 and M_2. A direct way of doing this is to consider the column in the buckled state, Fig. 13–5*d*. The horizontal components of the column end forces must be zero because the beams cannot provide a horizontal restraint. The shear forces in the beams are infinitesimals of order r, which do not affect the vertical member end forces. These are then equal to P, as shown. Equilibrium of forces in (*d*) gives

$$Pr = M_2 - M_1 \tag{13–14}$$

thus

$$\rho_c = -\frac{r}{h} = \frac{M_1 - M_2}{Ph}$$

Substituting ρ_c into the three-moment equations, there comes

$$\begin{bmatrix} f_c + f_b - \dfrac{1}{Ph} & f'_c + \dfrac{1}{Ph} \\ f'_c + \dfrac{1}{Ph} & f_c + f_t - \dfrac{1}{Ph} \end{bmatrix} \begin{Bmatrix} M_1 \\ M_2 \end{Bmatrix} = 0 \tag{13–15}$$

Setting the determinant equal to zero, the characteristic equation may be put in the form found earlier.

A comment on the derivation of Eq. (13–14) is that the statics could be performed in the undeformed geometry by subjecting the structure to the geometric joint load

$$R_G = -K_G r = \frac{Pr}{h}$$

as shown in Fig. 13–5*e*. If X is the reaction at A, there comes

$$M_2 = \frac{l}{2} X$$

$$M_1 = \frac{l}{2} X - Pr$$

Elimination of X yields the equilibrium equation found earlier.

EXERCISES

Sections 1 to 6

1. Determine P_{cr} for the rigid bar of Fig. 5–1*c*. Let the springs have the stiffness k and the bar be of length l. Do the problem by an elementary method and then as an application of the matrix formulation of the stiffness method.

2. Determine P_{cr} and the buckled shape for the rigid bar system of Fig. 3–2*a* assuming the spring stiffnesses are k and $2k$, respectively. Determine also the second characteristic value of P and the corresponding shape.

3. Determine P_{cr} for the truss of Fig. P3. Assume all bars have the axial stiffness k.

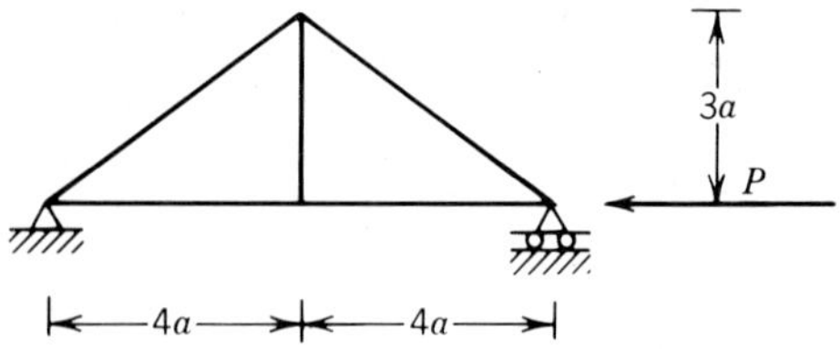

FIG. P3

4. Determine P_{cr} for the continuous beam of Fig. 4–1 assuming $l_1 = l_2$ and $I_1 = 2I_2$.

5. Do Exercise 4 assuming ends 1 and 3 are simply supported.

6. Determine the effective column length for a portal frame without sidesway by solving Eq. (6–3a) for the three cases $I_c l/I_b h = 0.5$, 1, 2. Check the solution with the nomograph of Fig. 6–6.

7. Determine the effective column lengths in Fig. 6–3 for a span to height ratio of 2 and for the values $\frac{1}{2}$, 1, and 2 of I_c/I_b.

8. Do Exercise 7 for the frame of Fig. 6–4. Assume $I_t = I_b$.

9. Determine the effective length of the middle column in Fig. 6–7 assuming all columns have the same I/l ratio and all beams have the ratio $2I/l$.

10. Determine P_{cr} for the braced frame of Fig. P10 assuming the brace prevents sidesway buckling.

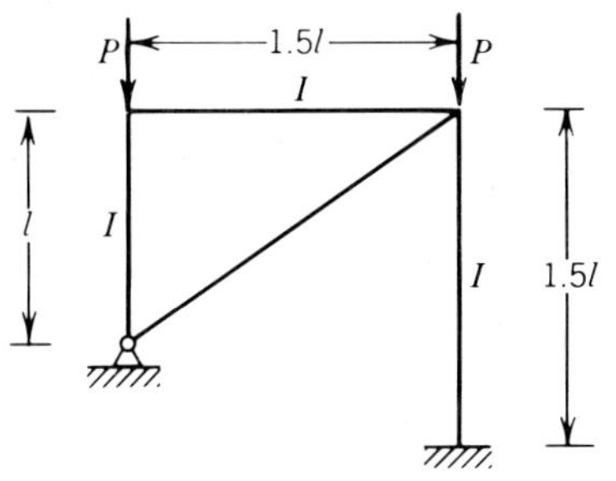

FIG. P10

11. Determine P_{cr} in Fig. P11 by the stiffness method.

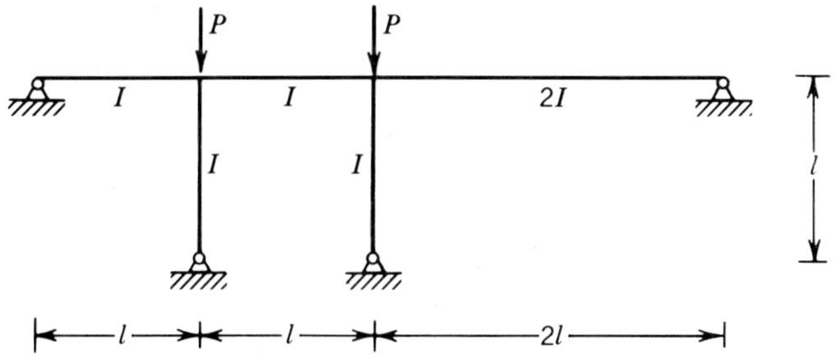

FIG. P11

12. Derive the characteristic equation for the beam shown in Fig. P12. Indicate how Fig. P12 represents a model for the problem of Eq. (6–12).

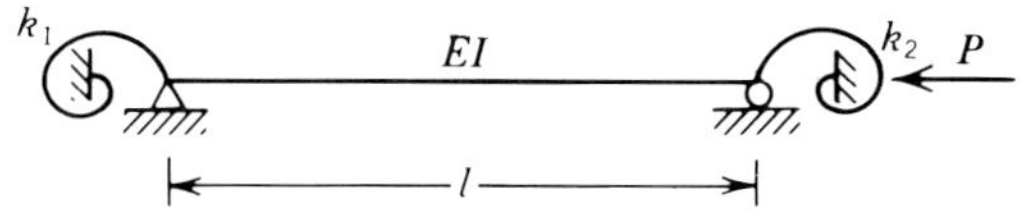

FIG. P12

13. Determine the effective length of member ab in the trussed frame of Fig. P13. Assume all members have properties l and I. Use the simplified approach of Section 6, Example 5.

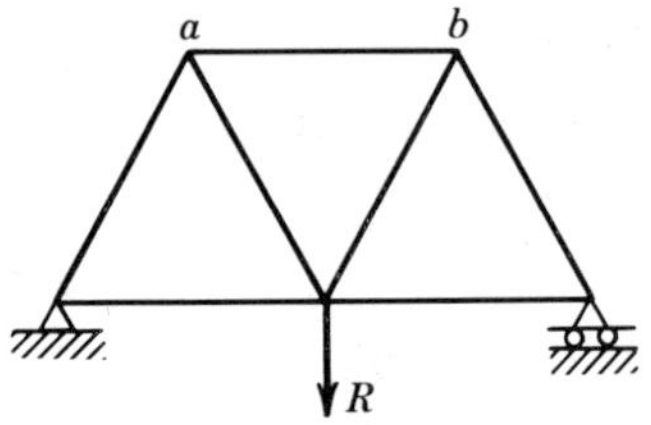

FIG. P13

14. Determine the effective length of member ab of the continuous compression chord in Fig. P14. Use the approach of Section 6, Example 6, and assume a uniform chord.

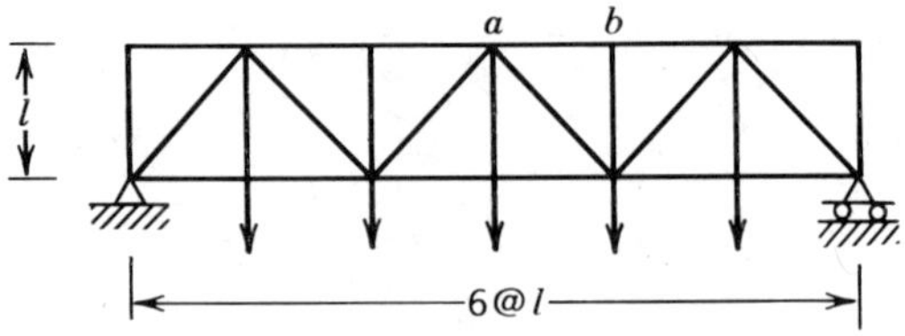

FIG. P14

15. Determine P_{cr} in Fig. P15.

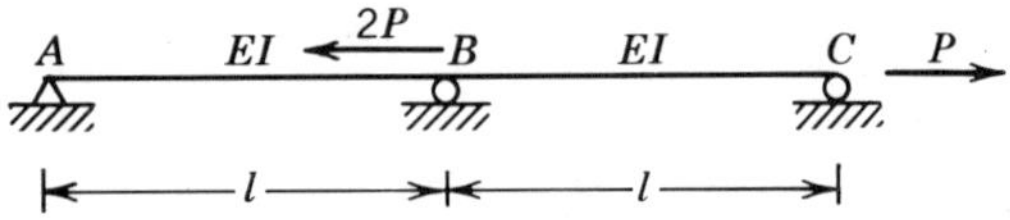

FIG. P15

Section 7

16. Determine P_{cr} in Figs. 7–1 and 7–2, if $a=2b$, and for the two cases $I_a=2I_b$ and $I_a=\frac{1}{2}I_b$.

17. Determine P_{cr} in Fig. P17 assuming the end segments are infinitely rigid.

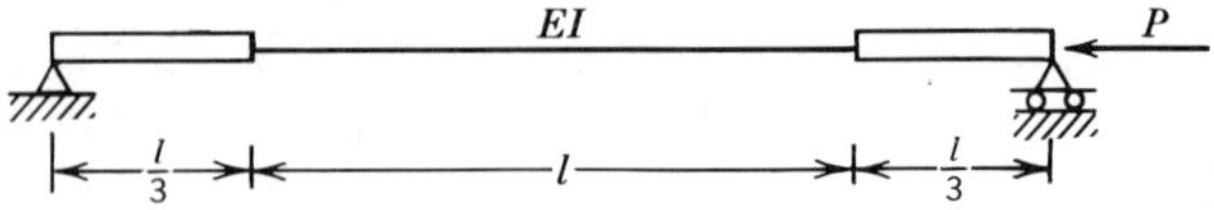

FIG. P17

18. Do Exercise 17 using an approximate stiffness matrix such as established in Chapter 6, Section 3. Do the same for Example 4, Section 7, with $a=l/2$. Compare exact and approximate solutions.

19. Derive Eqs. (7–13a) and (7–14).

20. Determine P_{cr} in Fig. 7–5*a* by solving Eq. (7–13a), and compare the result to that deduced from the nomograph of Fig. 7–7 for the following cases:

(a) $I_b = I_c, l = h$.

(b) $I_b = I_c, l = 2h$.

21. Derive the characteristic equation for a frame similar to that in Fig. 7–5*a* in which one column has the length 1.5*h*.

22. Derive the characteristic equation for the problem of Section 7, Example 6, using the approximate stiffness matrix for the columns.

23. Consider a rectangular frame of constant story height *h* and constant beam span *l*. If all columns have the same I_c and all beams the same I_b, classify the columns according to the pair of values (G_t, G_b) in Eq. (7–17). Do the same if all exterior columns have $\frac{1}{2}I_c$.

24. Determine P_{cr} in Fig. 7–8*a*, for $a = b$ and $P_b = P$.

25. In Fig. P14 assume that the continuous top chord is restrained against out-of-plane displacement at every other joint. Estimate the critical load for out-of-plane buckling. State the approximations involved.

26. Use the rigid bar system treated in Example 8, Section 7, as a model for determining P_{cr} in Fig. P26. Try the correction factor suggested, and compare the result to that of an exact analysis.

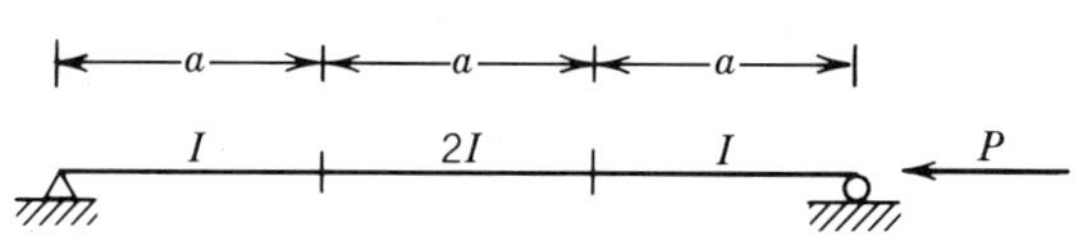

FIG. P26

Sections 8 to 12

27. Do any of these preceding exercises by the flexibility method: Exercises 1, 2, 3, 4, 5, 10, 11, 14, and 15. Use the three-moment equation where applicable.

28. Do Exercises 11 and 13 by the four-moment equation.

Section 13

29. Do Exercises 16 and 24 by the three-moment equation.

30. Do Exercises 17, 21, and 24 by the flexibility method.

31. Formulate by the flexibility method the characteristic equation for the rigid bar system of Example 8, Section 7.

32. Derive by the flexibility method the characteristic equation for a frame similar to the one of Fig. 13–4 in which one column has the length $1.5h$.

33. Do Exercise 32 by the three-moment equation.

14

SPACE STRUCTURES

1. INTRODUCTION

A space structure is an assemblage of members whose axes are not coplanar in the undeformed geometry or cease to be coplanar because of the loading. A single member is included in this definition if it is not in a state of plane deformation.

Space structures may be classified according to their configuration or their structural action. For example, the structure may be formed of several interconnected plane frames as is the case in a usual building frame, or the structure may be a surfacelike framework covering a space. The structural action depends on the configuration as well as on the types of member to joint connections and on the loading.

The basic concepts and approach for developing governing equations of space structures are the same as for plane structures. However, a general nonlinear theory of the structural member similar to the one developed in Chapter 2, but applicable to space structures, is mathematically more complex than a direct generalization. The complexity is associated mainly with finite rigid body rotations, as will be seen in Chapter 15. A limited nonlinear theory is presented in Sections 6 and 7.

The matrix notation and the matrix equations developed for planar structures remain generally the same for spatial structures but with an increase in degrees of freedom and corresponding forces. Attention will be concentrated on what is specific to spatial structures in the governing equations.

2. COORDINATE TRANSFORMATION

In a spatial structure the orientation of the member axis, or local x axis, is not sufficient to define the local reference frame. The orientation of the y and z axis must also be defined by means of a rotation angle about the x axis, from some origin fixed in the global reference. For usual members the x axis is the centroidal line, and the y and z axes are chosen as the principal axes of inertia. Let

$$\vec{\mathbf{e}}_0 = \{\vec{e}_{1,0} \;\; \vec{e}_{2,0} \;\; \vec{e}_{3,0}\}$$

be the base vectors of the global reference, and let

$$\vec{\mathbf{e}} = \{\vec{e}_1 \;\; \vec{e}_2 \;\; \vec{e}_3\}$$

be the base vectors of the local reference. Each base consists of unit orthogonal vectors forming a right-handed system. $\vec{\mathbf{e}}$ has on $\vec{\mathbf{e}}_0$ a component representation of the form

$$\begin{Bmatrix} \vec{e}_1 \\ \vec{e}_2 \\ \vec{e}_3 \end{Bmatrix} = \begin{bmatrix} \lambda_{11} & \lambda_{12} & \lambda_{13} \\ \lambda_{21} & \lambda_{22} & \lambda_{23} \\ \lambda_{31} & \lambda_{32} & \lambda_{33} \end{bmatrix} \begin{Bmatrix} \vec{e}_{1,0} \\ \vec{e}_{2,0} \\ \vec{e}_{3,0} \end{Bmatrix} \tag{2–1a}$$

or

$$\vec{\mathbf{e}} = \boldsymbol{\lambda} \vec{\mathbf{e}}_0 \tag{2–1b}$$

Since $\vec{\mathbf{e}}$ and $\vec{\mathbf{e}}_0$ are orthonormal,

$$\lambda_{ij} = \vec{e}_i \cdot \vec{e}_{j,0} \tag{2–2}$$

and $\vec{\mathbf{e}}_0$ has on $\vec{\mathbf{e}}$ the component representation

$$\vec{\mathbf{e}}_0 = \boldsymbol{\lambda}^T \vec{\mathbf{e}} \tag{2–3}$$

$\boldsymbol{\lambda}$ satisfies the orthogonality property

$$\boldsymbol{\lambda}^T = \boldsymbol{\lambda}^{-1} \tag{2–4}$$

Let $\vec{u}$ be a vector of components $\mathbf{u} = \{u_1 \ u_2 \ u_3\}$ on $\vec{\mathbf{e}}$, and $\bar{\mathbf{u}} = \{\bar{u}_1 \ \bar{u}_2 \ \bar{u}_3\}$ on $\vec{\mathbf{e}}_0$. Then

$$\vec{u} = \mathbf{u}^T \vec{\mathbf{e}} = \bar{\mathbf{u}}^T \vec{\mathbf{e}}_0 \tag{2–5}$$

If $\vec{\mathbf{e}}_0 = \boldsymbol{\lambda}^T \vec{\mathbf{e}}$ is substituted into Eq. (2–5), there comes the transformation

$$\mathbf{u} = \boldsymbol{\lambda} \bar{\mathbf{u}} \tag{2–6}$$

The nine elements of $\boldsymbol{\lambda}$ are determined for each member from the initial geometry. Let $\mathbf{X}^a = \{X_1^a \ X_2^a \ X_3^a\}$ be the global coordinates of member end a. $\mathbf{X}^b$ is defined similarly at end b. The member length is

$$l = [(\mathbf{X}^b - \mathbf{X}^a)^T (\mathbf{X}^b - \mathbf{X}^a)]^{1/2} \tag{2–7}$$

As $\vec{e}_1$ is the unit vector oriented from a to b, its components are

$$\lambda_{1i} = \frac{X_i^b - X_i^a}{l}, \qquad i = 1, 2, 3 \tag{2–8}$$

The orientation of $\vec{e}_2$ may be defined by means of an angle of rotation β about $\vec{e}_1$ whose origin is an axis $\vec{e}'_2$ which is defined as follows [11]. If the member is not parallel to the global Yaxis, $\vec{e}'_2$ is in the plane $(\vec{e}_1, \vec{e}_{2,0})$, is orthogonal to $\vec{e}_1$, and is oriented so that $\vec{e}'_2 \cdot \vec{e}_{2,0} > 0$. This is shown in Fig. 2–1a. β is given the sign convention of the right-hand rule about $\vec{e}_1$. If member ab is parallel to $\vec{e}_{2,0}$, $\vec{e}'_3$ is chosen equal to $\vec{e}_{3,0}$, and $\vec{e}'_2$ is oriented so that $\vec{\mathbf{e}}' = \{\vec{e}_1 \ \vec{e}'_2 \ \vec{e}'_3\}$ is a right-handed system, Fig. 2.1b. Let $\vec{\mathbf{e}} = \mathbf{T}_1 \vec{\mathbf{e}}'$ and $\vec{\mathbf{e}}' = \mathbf{T}_2 \vec{\mathbf{e}}_0$; then $\vec{\mathbf{e}} = \mathbf{T}_1 \mathbf{T}_2 \vec{\mathbf{e}}_0$, and consequently

$$\boldsymbol{\lambda} = \mathbf{T}_1 \mathbf{T}_2 \tag{2–9}$$

$\mathbf{T}_1$ is readily defined in terms of β as

$$\mathbf{T}_1 = \begin{bmatrix} 1 & 0 & 0 \\ 0 & \cos\beta & \sin\beta \\ 0 & -\sin\beta & \cos\beta \end{bmatrix} \tag{2–10}$$

To form $\mathbf{T}_2$, we have

$$\vec{e}'_3 = \frac{\vec{e}_1 \times \vec{e}_{2,0}}{|\vec{e}_1 \times \vec{e}_{2,0}|}$$

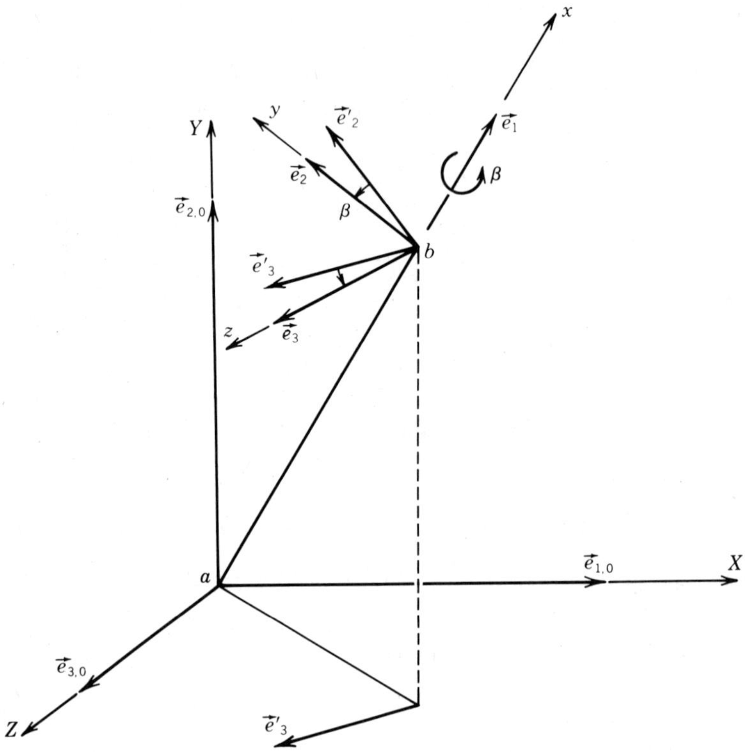

(*a*)

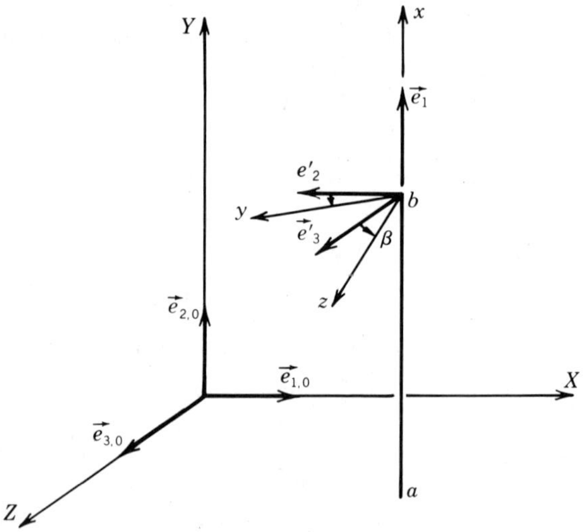

(*b*)

FIG. 2–1. Coordinate transformation.

$$\vec{e}'_2 = \vec{e}'_3 \times \vec{e}_1 = \left[-\frac{\lambda_{11}\lambda_{12}}{\alpha} \quad \alpha \quad -\frac{\lambda_{13}\lambda_{12}}{\alpha} \right]$$

where

$$\alpha = (\lambda_{13}^2 + \lambda_{11}^2)^{1/2} = (1 - \lambda_{12}^2)^{1/2} \tag{2–11}$$

$\mathbf{T}_2$ is formed from the components of $\vec{e}_1$, $\vec{e}'_2$, and $\vec{e}'_3$ found before, or

$$\mathbf{T}_2 = \begin{bmatrix} \lambda_{11} & \lambda_{12} & \lambda_{13} \\ -\dfrac{\lambda_{11}\lambda_{12}}{\alpha} & \alpha & -\dfrac{\lambda_{13}\lambda_{12}}{\alpha} \\ -\dfrac{\lambda_{13}}{\alpha} & 0 & \dfrac{\lambda_{11}}{\alpha} \end{bmatrix} \tag{2–12}$$

$\boldsymbol{\lambda}$ may now be formed by executing the product $\mathbf{T}_1\mathbf{T}_2$.

If ab is parallel to $\vec{e}_{2,0}$, then $\vec{e}_1 = \pm\vec{e}_{2,0}$, $\lambda_{12} = \pm 1$, and $\alpha = 0$. The sign of λ_{12} depends on the naming of the member ends. $\vec{e}'_3$ is chosen in either case as $\vec{e}'_3 = \vec{e}_{3,0}$, and $\vec{e}'_2 = \vec{e}'_3 \times \vec{e}_1$. Thus

$$\mathbf{T}_2 = \begin{bmatrix} 0 & \lambda_{12} & 0 \\ -\lambda_{12} & 0 & 0 \\ 0 & 0 & 1 \end{bmatrix} \tag{2–13}$$

3. LINEAR THEORY OF THE STRUCTURAL MEMBER

3.1. Static Analysis

The stress resultants acting on a positive cross section at x, Fig. 3.1–1, are denoted

$$\mathbf{S}^x = \{\mathbf{N}\ \mathbf{M}\} = \{N_1\ N_2\ N_3\ M_1\ M_2\ M_3\} \tag{3.1–1}$$

where $\mathbf{N} = \{N_1\ N_2\ N_3\}$ are the components of the resultant force $\vec{N}$ and $\mathbf{M} = \{M_1\ M_2\ M_3\}$ are the components of the resultant moment $\vec{M}$ on the local reference. $\vec{M}$ is defined at the reference point of the cross section which is at the intersection with the local x axis. The resultants acting at ends a and b are

$$\mathbf{S}^a = -(\mathbf{S}^x)_{x=0} = \{-\mathbf{N}^a\ -\mathbf{M}^a\} \tag{3.1–2a}$$

$$\mathbf{S}^b = (\mathbf{S}^x)_{x=l} = \{\mathbf{N}^b\ \mathbf{M}^b\} \tag{3.1–2b}$$

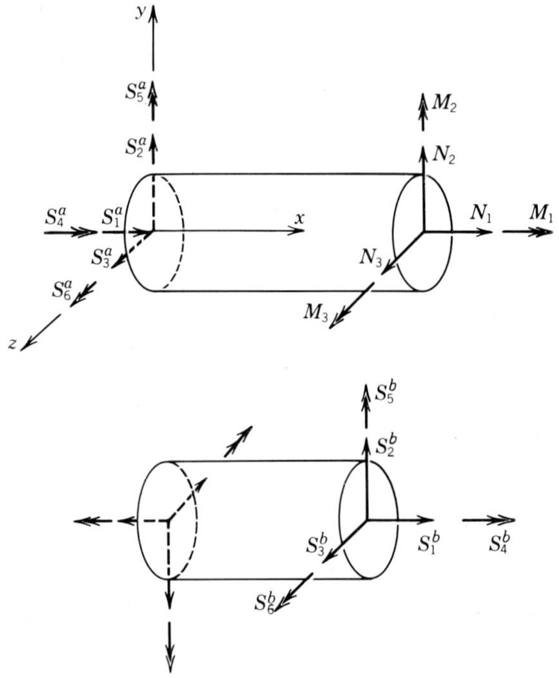

FIG. 3.1–1. Stress resultants and end forces.

The member span load intensity is denoted

$$\mathbf{p} = \{\mathbf{q}\ \ \mathbf{m}\} \tag{3.1–3}$$

where $\mathbf{q}$ are force components and $\mathbf{m}$ are moment components on the local axes.

In the geometrically linear theory, forces are considered acting in the undeformed geometry. N_1 is the axial force, N_2 and N_3 are shear forces, M_1 is the torsional moment, and M_2 and M_3 are bending moments.

The force and moment rigid body equilibrium equations of the member have the form

$$\mathbf{S}^a + \mathbf{T}_b^a \mathbf{S}^b + \mathbf{P}^a = 0 \tag{3.1–4}$$

where $\mathbf{T}_b^a$ is the force transformation matrix giving $\mathbf{T}_b^a\mathbf{S}^b$ as the statical equivalents of $\mathbf{S}^b$ at material point a, or

$$\mathbf{T}_b^a = \left[\begin{array}{ccc|c} & \mathbf{I} & & \mathbf{0} \\ \hline 0 & 0 & 0 & \\ 0 & 0 & -l & \mathbf{I} \\ 0 & l & 0 & \end{array}\right]_{6\times 6} \tag{3.1–5}$$

and $\mathbf{P}^a$ is formed of the resultants at a of the member span load,

$$\mathbf{P}^a = \int_0^l \mathbf{T}_x^a \mathbf{p}\, dx \tag{3.1–6}$$

The force transformation matrix from point x to point a is

$$\mathbf{T}_x^a = \left[\begin{array}{ccc|c} & \mathbf{I} & & 0 \\ \hline 0 & 0 & 0 & \\ 0 & 0 & -x & \mathbf{I} \\ 0 & x & 0 & \end{array}\right]_{6\times 6} \tag{3.1–7}$$

Equations (3.1–4) form a system of 6 equations in 12 unknowns

$$\mathbf{S} = \{\mathbf{S}^a \ \mathbf{S}^b\} \tag{3.1–8}$$

Choosing $\mathbf{S}^b$ as member statical redundants $\mathbf{V}$, Eqs. (3.1–4) are solved in the form

$$\mathbf{S} = \mathbf{B}\mathbf{V} + \mathbf{S}_p \tag{3.1–9}$$

where

$$\mathbf{B} = \begin{bmatrix} \mathbf{B}^a \\ \mathbf{B}^b \end{bmatrix} = \begin{bmatrix} -\mathbf{T}_b^a \\ \mathbf{I} \end{bmatrix} \tag{3.1–10}$$

and

$$\mathbf{S}_p = \begin{Bmatrix} \mathbf{S}_p^a \\ \mathbf{S}_p^b \end{Bmatrix} = \begin{Bmatrix} -\mathbf{P}^a \\ \mathbf{0} \end{Bmatrix} \tag{3.1–11}$$

Evaluation of $\mathbf{S}^x$ in terms of $\mathbf{V}$ and $\mathbf{p}$ is made through an equation similar to Eq. (3.1–4) by replacing point a with point x and $\mathbf{S}^a$ with $-\mathbf{S}^x$. There comes

$$\mathbf{S}^x = \mathbf{b}\mathbf{V} + \mathbf{S}_p^x \tag{3.1–12}$$

where

$$\mathbf{b} = \mathbf{T}_b^x \tag{3.1–13a}$$

$$\mathbf{S}_p^x = \int_x^l \mathbf{T}_\xi^x \mathbf{p}\, d\xi \tag{3.1–13b}$$

3.2. Kinematic Analysis

In engineering beam theory a cross section has a rigid body displacement $(\vec{u}, \vec{\varphi})$ characterized by the translation $\vec{u}$ of a reference point in the cross section and by a free rotation vector $\vec{\varphi}$. The components of these vectors on the local reference are denoted $\mathbf{u}$ and $\boldsymbol{\varphi}$, respectively. At ends a and b the notation is

$$\mathbf{s}^a = \{\mathbf{u}^a \;\; \boldsymbol{\varphi}^a\} = \{u_1^a \;\; u_2^a \;\; u_3^a \;\; \varphi_1^a \;\; \varphi_2^a \;\; \varphi_3^a\} \tag{3.2–1a}$$

$$\mathbf{s}^b = \{\mathbf{u}^b \;\; \boldsymbol{\varphi}^b\} = \{u_1^b \;\; u_2^b \;\; u_3^b \;\; \varphi_1^b \;\; \varphi_2^b \;\; \varphi_3^b\} \tag{3.2–1b}$$

and for the member

$$\mathbf{s} = \{\mathbf{s}^a \;\; \mathbf{s}^b\} \tag{3.2–2}$$

As in the planar theory, member deformations $\mathbf{v}$ of the cantilever type, corresponding to the member statical redundants $\mathbf{V}$, are defined as the displacements at b, $\mathbf{v} = \{\mathbf{u}_r \;\; \boldsymbol{\varphi}_r\}$, relative to the member-bound reference attached to the cross section at a. A consequence of the definition is that $\mathbf{V}^T\delta\mathbf{v} = \mathbf{S}^T\delta\mathbf{s}$ for any set of self-equilibriating forces $\mathbf{S} = \mathbf{BV}$. Thus $\delta\mathbf{v} = \mathbf{B}^T\delta\mathbf{s}$, and since $\mathbf{B}$ is constant in the present linear theory,

$$\mathbf{v} = \mathbf{B}^T\mathbf{s} \tag{3.2–3a}$$

In more detail

$$\mathbf{v} = -\mathbf{T}_b^{aT}\mathbf{s}^a + \mathbf{s}^b \tag{3.2–3b}$$

$$\mathbf{v} = \{\mathbf{u}_r \;\; \boldsymbol{\varphi}_r\} \tag{3.2–4}$$

$$\mathbf{u}_r = -\mathbf{u}^a + \mathbf{u}^b + l\{0 \;\; -\varphi_3^a \;\; \varphi_2^a\} \tag{3.2–5a}$$

$$\boldsymbol{\varphi}_r = \boldsymbol{\varphi}^b - \boldsymbol{\varphi}^a \tag{3.2–5b}$$

For a direct kinematic derivation the deformation vectors $(\vec{u}_r, \vec{\varphi}_r)$ are defined as characterizing the rigid body displacement at b relative to the member-bound reference. Assuming rotations squared are negligible compared with unity, rigid body rotations may be treated as infinitesimal. From basic rigid body kinematics the displacements at b relative to the bound reference at a are

$$\vec{\varphi}_r = \vec{\varphi}^b - \vec{\varphi}^a \tag{3.2–6a}$$

$$\vec{u}_r = \vec{u}^b - \vec{u}^a - \vec{\varphi}^a \times l\vec{e}_1 \tag{3.2–6b}$$

The term $\vec{u}^a + \vec{\varphi}^a \times l\vec{e}_1$ is the displacement of a point fixed in the bound reference and coinciding with point b in the initial state.

The scalar deformations are components of $\vec{u}_r$ and $\vec{\varphi}_r$ on the bound axes. In the geometrically linear theory, however, these components reduce to components on the local axes and coincide with those established earlier.

Strain-Displacement Relations

The kinematic analysis may be pursued, as in Chapter 2, to derive strains $\{\boldsymbol{\varepsilon} \;\; \boldsymbol{\chi}\}$ by identifying deformation for a beam slice dx with $\{\boldsymbol{\varepsilon} \;\; \boldsymbol{\chi}\}\, dx$. The notation here is modified from that of Chapter 2. Here $\boldsymbol{\varepsilon}$ corresponds to the force resultants **N**, and $\boldsymbol{\chi}$ to the moment resultants **M**. Applying Eqs. (3.2–5) to a segment dx, and letting $\mathbf{u}_r = \boldsymbol{\varepsilon}\, dx$ and $\boldsymbol{\varphi}_r = \boldsymbol{\chi}\, dx$, the strain components are obtained as

$$\varepsilon_1 = u_{1,x} \tag{3.2–7a}$$

$$\varepsilon_2 = u_{2,x} - \varphi_3 \tag{3.2–7b}$$

$$\varepsilon_3 = u_{3,x} + \varphi_2 \tag{3.2–7c}$$

$$\chi_1 = \varphi_{1,x} \tag{3.2–7d}$$

$$\chi_2 = \varphi_{2,x} \tag{3.2–7e}$$

$$\chi_3 = \varphi_{3,x} \tag{3.2–7f}$$

ε_1 is the axial strain, ε_2 and ε_3 are transverse shear strains, χ_1 is the rate of twist, and χ_2 and χ_3 are curvature strains.

Having followed the approach of Chapter 2, the preceding derivation ensures the applicability of the principles of virtual displacements and of virtual forces.

The strain distribution over a cross section is obtained by changing the reference point for displacements. Letting $(\mathbf{u}, \boldsymbol{\varphi})$ be defined at the origin of axes (y, z) in the cross section, and $(\mathbf{u}', \boldsymbol{\varphi}')$ be defined at point (y, z), the rigid body kinematic transformation is

$$u'_1 = u_1 - y\varphi_3 + z\varphi_2 \tag{3.2–8a}$$

$$u'_2 = u_2 - z\varphi_1 \tag{3.2–8b}$$

$$u'_3 = u_3 + y\varphi_1 \tag{3.2–8c}$$

and

$$\varphi'_i = \varphi_i, \qquad i = 1, 2, 3 \tag{3.2–8d}$$

The strains $(\boldsymbol{\varepsilon}', \boldsymbol{\chi}')$ at the reference point (y, z) are related to $(\mathbf{u}', \boldsymbol{\varphi}')$ through the same relations as Eqs. (3.2–7). We thus obtain

$$\varepsilon'_1 = \varepsilon_1 - y\chi_3 + z\chi_2 \tag{3.2–9a}$$

$$\varepsilon'_2 = \varepsilon_2 - z\chi_1 \tag{3.2–9b}$$

$$\varepsilon'_3 = \varepsilon_3 + y\chi_1 \tag{3.2–9c}$$

$$\chi'_i = \chi_i, \qquad i = 1, 2, 3 \tag{3.2–9d}$$

3.3. Constitutive Equations

Constitutive equations for axial and flexural deformations are derived by the same method as that followed in Chapter 4. For y and z axes along the principal axes of inertia, the bending constitutive properties are found to uncouple into those of planar behavior in the two principal planes of inertia.

In torsion, the kinematic model of deformation adopted so far is incomplete except for circular cross sections and tubes. It is refined by considering in addition to the rigid body displacement of a cross section an axial warping displacement. The subject is covered in texts on strengths of materials and on elasticity [2, p. 255] [12]. Only a qualitative description and some results will be presented.

In pure torsion and with warping unrestrained, the torsional constitutive equation has the form

$$M_1 = GJ\chi_1 \tag{3.3–1}$$

where G is the shear modulus and J is a geometric constant depending on the shape and dimensions of the cross section, called the torsional constant.

In combined bending and torsion, with warping unrestrained, Eq. (3.3–1) remains valid provided the torsional moment is defined at a point S of the cross section which is called the shear center. If the cross section has axes of symmetry, S lies on such axes. Thus if there are two axes of symmetry, S coincides with the centroid C. In general, the shear center has coordinates (x_s, y_s) with respect to centroidal axes. The torsional moment M'_1 at S is related to the moment M_1 at the centroid through

$$M'_1 = M_1 - y_s N_3 + z_s N_2 \tag{3.3–2}$$

and Eq. (3.3–1) is replaced by

$$M'_1 = GJ\chi_1 \tag{3.3–3}$$

The transformation from M_1 to M'_1 transforms the work correspondance between $(\mathbf{N}, \mathbf{M})$ and (ε, χ). If $M_1 = M'_1 + y_s N_3 - z_s N_2$ is substituted into the work expression $\mathbf{N}^T\delta\varepsilon + \mathbf{M}^T\delta\chi$, the coefficients of N_2 and N_3 become, respectively, the incremental shear strains $\delta\varepsilon'_2$ and $\delta\varepsilon'_3$ defined at the shear center. ε'_2 and ε'_3 are obtained from Eqs. (3.2–9b, c) in which $z = z_s$ and $y = y_s$, or

$$\varepsilon'_2 = \varepsilon_2 - z_s\chi_1 = u'_{2,x} - \varphi_3 \tag{3.3–4a}$$

$$\varepsilon'_3 = \varepsilon_3 + y_s\chi_1 = u'_{3,x} + \varphi_2 \tag{3.3–4b}$$

Using principal axes of inertia, the strain energy density takes the form

$$U = \tfrac{1}{2}EA\varepsilon_1^2 + \tfrac{1}{2}GA_2\varepsilon_2'^2 + \tfrac{1}{2}GA_3\varepsilon_3'^2 + \tfrac{1}{2}GJ\chi_1^2 + \tfrac{1}{2}EI_2\chi_2^2 + \tfrac{1}{2}EI_3\chi_3^2 \tag{3.3–5a}$$

Partial derivatives of U with respect to $\varepsilon_1, \varepsilon'_2, \varepsilon'_3, \chi_1, \chi_2$, and χ_3 yield the constitutive relations for N_1, N_2, N_3, M'_1, M_2, and M_3, respectively.

For neglecting shear deformation, ε'_2 and ε'_3 are set to zero. Then

$$\varphi_2 = -u'_{3,x} \tag{3.3–5a}$$

$$\varphi_3 = u'_{2,x} \tag{3.3–5b}$$

and

$$\chi_2 = -u'_{3,xx} \tag{3.3–6a}$$

$$\chi_3 = u'_{2,xx} \tag{3.3–6b}$$

The independent displacements are now u_1, u'_2, u'_3, and φ_1. Note that the centroid is the reference point for defining ε_1, M_2, and M_3, and the shear center is the reference point for defining M'_1, χ_2, and χ_3.

The preceding theory is known as St-Venant's theory and is based on the assumption of unrestrained warping at the end cross sections of the member. For a structural member warping is usually restrained because of geometric continuity at the member's ends. Restrained warping causes an increase in the torsional stiffness of the member whose importance depends on the shape of the cross section. It is significant for thin-walled open cross sections such as the I and U shapes whose plate elements bend in their own respective planes as a result of warping restraint. It is negligible for full cross sections, such as for a rectangle and for thin-walled cross sections whose plate elements intersect at one point, such as in an angle. In this latter case the point common to the plate elements is the shear center, and rotation about this point causes no inplane bending of the plate elements.

Warping restraint causes the warping displacement to vary longitudinally and thus to induce axial strains. The theory by which the warping displacement is determined in terms of the angle of twist if known as Vlassov's theory and is based on neglecting shear deformation along the middle line of the thin-walled cross section. The distribution of axial strain over a cross section is found to be proportional to $\chi_{1,x}$, and the associated strain energy density takes the form

$$U_w = \tfrac{1}{2}EK(\chi_{1,x})^2 \tag{3.3–7}$$

where K is a geometric constant called the warping constant. Equation (3.3–7) leads to defining a statical quantity B called bimoment, as

$$B = \frac{\partial U_w}{\partial(\chi_{1,x})} = EK\chi_{1,x} \tag{3.3–8}$$

The total strain energy due to torsion is

$$\bar{U}_t = \int_0^l [\tfrac{1}{2}GJ\chi_1^2 + \tfrac{1}{2}EK(\chi_{1,x})^2]\,dx \tag{3.3–9}$$

An assumed representation of φ_1 in terms of generalized displacements and use of Castigliano's first theorem allow us to derive a stiffness relation for torsion. The condition for restrained warping at a member end is $\chi_1 = \varphi_{1,x} = 0$. The condition for free warping is $\chi_{1,x} = \varphi_{1,xx} = 0$. In choosing shape functions for use in the energy method, the geometric conditions on φ_1 and $\varphi_{1,x}$ must be satisfied. The condition expressing free warping is statical and need not be satisfied by the shape functions, although an improved accuracy may be expected if it is.

The axial stresses caused by restrained warping have zero resultant force and zero resultant moments. These stresses are accompanied by shear stresses having a zero resultant force but a nonzero torque about the shear center. Evaluation of this torque for a uniform cross section yields the constitutive equation

$$M'_1 = GJ\chi_1 - EK\chi_{1,xx} \tag{3.3–10}$$

Equation (3.3–10) provides an alternate procedure for deriving the torsional stiffness relation for a uniform member (see Example 1).

To complete this presentation, Eq. (3.3–10) is shown to be consistent with Eq. (3.3–9). $\bar{U}_t$ may be evaluated as $\bar{U}_t = \int_0^l \frac{1}{2} M'_1 \chi_1 \, dx$, using the expression of M'_1 from Eq. (3.3–10). Integration by parts of the warping term, $-\frac{1}{2}EK\chi_1\chi_{1,xx}$ transforms the integral into that of Eq. (3.3–9) and yields boundary terms $-\frac{1}{2}EK\chi_1\chi_{1,x}|_0^l$. These terms are zero whether warping is free or restrained.

Example 1

The torsional stiffness of a uniform cantilever subjected to a torque M_1 at the free end is determined in what follows by integrating Eq. (3.3–10). In the present case there are no shear forces, and $M_1 = M'_1$. Letting

$$\lambda_1^2 = \frac{GJl^2}{EK} \tag{3.3–11}$$

the general solution of Eq. (3.3–10) is

$$\chi_1 = \frac{M_1}{GJ} + a\cosh\frac{\lambda_1 x}{l} + b\sinh\frac{\lambda_1 x}{l}$$

Integration of $\varphi_{1,x} = \chi_1$ yields

$$\varphi_1 = \frac{M_1 x}{GJ} + \frac{al}{\lambda_1}\sinh\frac{\lambda_1 x}{l} + \frac{bl}{\lambda_1}\cosh\frac{\lambda_1 x}{l} + c$$

For a cantilever fixed at $x = 0$, one end condition is $(\varphi_1)_{x=0} = 0$. The two other end conditions depend on whether warping is restrained. For application to a structural member rigidly connected at both ends, the conditions to be used are those of restrained warping, or $\chi_1 = 0$ at $x = 0$ and $x = l$. The three preceding conditions determine the integration constants. After simplification of the hyperbolic functions involved, the twist at $x = l$ is found to be

$$(\varphi_1)_{x=l} = \frac{M_1 l}{GJ}\left(1 - \frac{2}{\lambda_1}\tanh\frac{\lambda_1}{2}\right) \tag{3.3–12}$$

As an application, consider an I cross section having depth $d = 6.25$ in, flange width $b = 4$ in, and a uniform thickness $t = 0.25$ in. The warping constant is given through the formula

$$K = I_f \frac{h^2}{2} \tag{3.3–13}$$

where I_f is the moment of inertia of one flange about its strong axis, and h is the distance between the middle lines of the flanges, or $h = 6$ in. Thus

$$K = 0.25 \frac{(4)^3}{12} \frac{(6)^2}{2} = 24 \text{ in}^6$$

The torsional constant is

$$J = \tfrac{1}{3} h t^3 + 2(\tfrac{1}{3}) b t^3 = 0.07292 \text{ in}^4$$

With $G = E/2(1 + \nu)$ and $\nu = 0.3$, and assuming $l = 8$ ft, obtain

$$\lambda_1 = \left[\frac{1}{2(1.3)} \frac{J l^2}{K} \right]^{1/2} = 3.282$$

and

$$1 - \frac{2}{\lambda_1} \tanh \frac{\lambda_1}{2} = 0.435$$

The flexibility coefficient in Eq. (3.3–12) may be put in the form l_e/GJ, or equivalently in the form l/GJ_e, where l_e and J_e are effective quantities. It is seen that warping restraint at both ends of the member increases the torsional stiffness of the member by a factor of $1/0.435 = 2.30$.

If warping is restrained only at $x = 0$, the condition of free warping at $x = l$ is $\chi_{1,x} = 0$. In that case the solution takes the form

$$(\varphi_1)_{x=l} = \frac{M_1 l}{GJ} \left(1 - \frac{1}{\lambda_1} \tanh \lambda_1 \right) \tag{3.3–14}$$

With λ_1 as determined here, the torsional stiffness is found to increase by a factor of 1.44.

Effective Torsional Stiffness

For implementing general methods of structural analysis, member stiffness and flexibility matrices are needed that include the effect of warping restraint in torsion and that could be specialized to various conditions at the member ends.

To derive torsional stiffness relations by Castigliano's theorem, the generalized displacements are chosen as

$$\mathbf{q} = \{\varphi_1^a \ \chi_1^a \ \varphi_1^b \ \chi_1^b\}$$

where a and b refer to the member ends. The shape functions are the third-degree polynomials used earlier in the planar theory, or

$$\begin{aligned}\varphi_1 &= [3(1-t)^2 - 2(1-t)^3]\varphi_1^a + l[(1-t)^2 - (1-t)^3]\chi_1^a \\ &\quad + (3t^2 - 2t^3)\varphi_1^b + l(t^3 - t^2)\chi_1^b\end{aligned} \tag{3.3–15}$$

where $t = x/l$. Evaluation of $\bar{U}_t$ in Eq. (3.3–9) yields $\bar{U}_t = \frac{1}{2}\mathbf{q}^T\mathbf{k}_T\mathbf{q}$ and

$$\mathbf{k}_T = \begin{bmatrix} \frac{6GJ}{5l} + \frac{12EK}{l^3} & \frac{GJ}{10} + \frac{6EK}{l^2} & -\frac{6GJ}{5l} - \frac{12EK}{l^3} & \frac{GJ}{10} + \frac{6EK}{l^2} \\ & \frac{2}{15}GJl + \frac{4EK}{l} & -\frac{GJ}{10} - \frac{6EK}{l^2} & -\frac{GJl}{30} + \frac{2EK}{l} \\ & & \frac{6GJ}{5l} + \frac{12EK}{l^3} & -\frac{GJ}{10} - \frac{6EK}{l^2} \\ \text{symmetric} & & & \frac{2}{15}GJl + \frac{4EK}{l} \end{bmatrix} \tag{3.3–16}$$

The torsional stiffness equations are $\mathbf{Q} = \mathbf{k}_T\mathbf{q}$ where $\mathbf{Q}$ is the set of generalized external forces due to the end actions and to the torsional span load if any. Considering only end actions,

$$\mathbf{Q} = \{-M_1^a \ B^a \ M_1^b \ B^b\}$$

where B is the bimoment and M_1 is the torsional moment.

The presence of the two generalized displacements χ_1^a and χ_1^b requires a discussion on how to use $\mathbf{k}_T$ in the direct stiffness method. Consider first a member that needs to be discretized into several elements, each having a torsional stiffness matrix similar to $\mathbf{k}_T$. The member stiffness is assembled from element stiffnesses by the direct stiffness method which is based on the geometric continuity of $\mathbf{q}$ at the internal joints or nodes. An external torsional loading is represented by equivalent torsional moments based on virtual work equivalence, but it does not contribute bimoments. If the generalized displacements χ_1 at the internal nodes are statically condensed, the result is an effective member stiffness having $\mathbf{q}$ as effective displacements. There remains to examine the conditions at the member ends. The idealization of a structural joint as a rigid body suggests the absence of any warping at the member joint interface. This constraint could be removed without difficulty at the internal nodes of the member considered here, but to do so at a structural joint connecting several member ends having arbitrary orientations would require a deformable joint having warping stiffness properties associated, respectively, with the member end connections. We will consider only two limiting idealizations corresponding to free and restrained warping, respectively, and illustrate the method on the matrix of Eq. (3.3–16) applied to the member as a whole. In all cases χ_1^a and χ_1^b will be eliminated, resulting in an effective torsional stiffness matrix of the form

$$\mathbf{k}_T = k\begin{bmatrix} 1 & -1 \\ -1 & 1 \end{bmatrix} \tag{3.3–17}$$

k is now determined for the idealized cases mentioned before.

Free Warping at Both Ends

When free warping occurs at both ends, $B^a = B^b = 0$, and Eq. (3.3–17) is obtained by static condensation of χ_1^a and χ_1^b. The result is as expected

$$k = \frac{GJ}{l} \tag{3.3–18}$$

Restrained Warping at Both Ends

In the case of restrained warping at both ends, $\chi_1^a = \chi_1^b = 0$. The effective stiffness is obtained by deleting the rows and columns associated with χ_1^a and χ_1^b in Eq. (3.3–16). There comes

$$k = \frac{6}{5}\frac{GJ}{l} + 12\frac{EK}{l^3} \tag{3.3–19}$$

For the cross section of Example 1, Eq. (3.3–19) yields $k = 2.31\,GJ/l$. The exact numerical factor was found to be 2.30.

Restrained Warping at One End

Assuming warping is restrained at end a and free at end b, we have $\chi_1^a = 0$ and $B^b = 0$. The effective $\mathbf{k}_T$ is formed by first deleting the row and column associated with χ_1^a and then condensing χ_1^b. The result has the form of Eq. (3.3–17) where

$$k = \frac{GJ}{l}\,\frac{1.125 + 39(EK/GJl^2) + 90(EK/GJl^2)^2}{1 + 30(EK/GJl^2)} \tag{3.3–20}$$

For the cross section of Example 1, Eq. (3.3–20) yields $k = 1.46\,GJ/l$. The exact numerical factor was found to be 1.44.

Note that Eqs. (3.3–19) and (3.3–20) do not tend to Eq. (3.3–18) as $EK \to 0$.

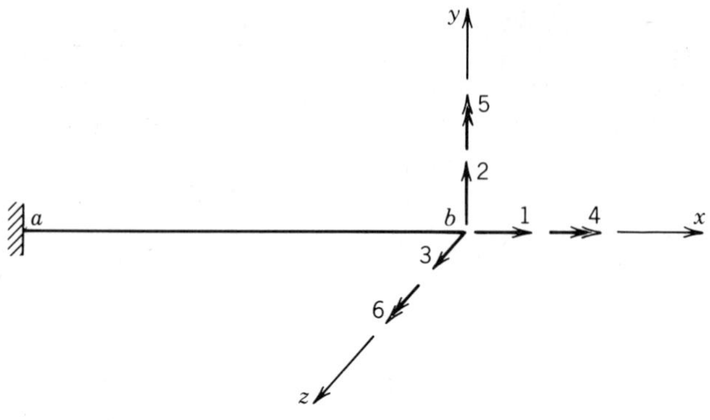

FIG. 3.4–1. Ordering of **v** and **V**, cantilever type.

3.4. Stiffness and Flexibility Equations

Member deformations **v** and statical redundants **V** are ordered as indicated in Fig. 3.4–1. Because the torsional moment is defined at the shear center, the transformation, Eq. (3.3–2), is used to define V_4' as

$$V_4' = V_4 - y_s V_3 + z_s V_2 \tag{3.4–1a}$$

Equation (3.4–1a) and the identity relations

$$V_i' = V_i, \quad i \neq 4 \tag{3.4–1b}$$

define a linear static transformation on **V**. The contragradient transformation is

$$v_2 = v_2' + z_s v_4' \tag{3.4–2a}$$

$$v_3 = v_3' - y_s v_4' \tag{3.4–2b}$$

and

$$v_i = v_i', \qquad i = 1, 4, 5, 6 \tag{3.4–2c}$$

Equations (3.4–2) are recognizable as a rigid body rotation transformation.

For the shear forces V_2 and V_3, applied at the shear center S, to produce bending without twisting, it is necessary that V_2 and V_3 produce no torsional moment M_1' at any cross section, that is, that the line of shear centers be parallel to the x axis. In such a case the torsional and bending flexibility relations for $\mathbf{v}'$ and $\mathbf{V}'$ are uncoupled. The flexibility relations have the form

$$\mathbf{v}' = \mathbf{f}'\mathbf{V}' + \mathbf{v}_0' \tag{3.4–3}$$

For a uniform member, neglecting shear deformation in bending,

$$\mathbf{f}' = \begin{bmatrix} \frac{l}{EA} & & & & & \\ & \frac{l^3}{3EI_3} & & & & \frac{l^2}{2EI_3} \\ & & \frac{l^3}{3EI_2} & & -\frac{l^2}{2EI_2} & \\ & & & \frac{l}{GJ_e} & & \\ & & -\frac{l^2}{2EI_2} & & \frac{l}{EI_2} & \\ & \frac{l^2}{2EI_3} & & & & \frac{l}{EI_3} \end{bmatrix} \tag{3.4–4}$$

In Eq. (3.4–4) I_2 and I_3 are the second moments of area of the cross section about the principal axes of inertia, y and z, respectively. G is the shear modulus, and J_e is an effective torsional constant that takes into account warping restraint.

The static-kinematic transformation from $(\mathbf{V}', \mathbf{v}')$ to $(\mathbf{V}, \mathbf{v})$ is of the form

$$\mathbf{V}' = \mathbf{L}\mathbf{V} \tag{3.4–5a}$$

$$\mathbf{v} = \mathbf{L}^T\mathbf{v}' \tag{3.4–5b}$$

The flexibility relation transforms into

$$\mathbf{v} = \mathbf{f}\mathbf{V} + \mathbf{v}_0 \tag{3.4–6}$$

where

$$\mathbf{f} = \mathbf{L}^T\mathbf{f}'\mathbf{L} \tag{3.4–7}$$

$$\mathbf{v}_0 = \mathbf{L}^T\mathbf{v}_0' \tag{3.4–8}$$

From Eqs. (3.4–1)

$$\mathbf{L} = \begin{bmatrix} 1 & 0 & 0 & 0 & 0 & 0 \\ 0 & 1 & 0 & 0 & 0 & 0 \\ 0 & 0 & 1 & 0 & 0 & 0 \\ 0 & z_s & y_s & 1 & 0 & 0 \\ 0 & 0 & 0 & 0 & 1 & 0 \\ 0 & 0 & 0 & 0 & 0 & 1 \end{bmatrix} \tag{3.4–9}$$

The stiffness relations for $\mathbf{V}'$ and $\mathbf{v}'$ are the inverse of Eqs. (3.4–3), or

$$\mathbf{V}' = \mathbf{k}'\mathbf{v}' + \mathbf{V}'_F \tag{3.4–10}$$

where

$$\mathbf{k}' = \mathbf{f}'^{-1} \tag{3.4–11a}$$

$$\mathbf{V}'_F = -\mathbf{k}'\mathbf{v}'_0 \tag{3.4–11b}$$

For a uniform member, neglecting shear deformation in bending,

$$\mathbf{k}' = \begin{bmatrix} \frac{EA}{l} & & & & & \\ & \frac{12EI_3}{l^3} & & & & -\frac{6EI_3}{l^2} \\ & & \frac{12EI_2}{l^3} & & \frac{6EI_2}{l^2} & \\ & & & \frac{GJ_e}{l} & & \\ & & \frac{6EI_2}{l^2} & & \frac{4EI_2}{l} & \\ & -\frac{6EI_2}{l^2} & & & & \frac{4EI_3}{l} \end{bmatrix} \tag{3.4–12}$$

For transformation of $\mathbf{k}'$ to the centroidal axes (x, y, z), we have the kinematic transformation

$$\mathbf{v}' = \mathbf{T}\mathbf{v} \tag{3.4–13}$$

where

$$\mathbf{T} = (\mathbf{L}^T)^{-1} \tag{3.4–14}$$

The transformation formulas are

$$\mathbf{V} = \mathbf{T}^T\mathbf{V}' \tag{3.4–15a}$$

$$\mathbf{k} = \mathbf{T}^T\mathbf{k}'\mathbf{T} \tag{3.4–15b}$$

$\mathbf{T}$ is obtained by solving Eqs. (3.4–2) for $\mathbf{v}'$. There comes

$$\mathbf{T} = \begin{bmatrix} 1 & & & & & \\ & 1 & & -z_s & & \\ & & 1 & y_s & & \\ & & & 1 & & \\ & & & & 1 & \\ & & & & & 1 \end{bmatrix} \tag{3.4–16}$$

Complete Stiffness Relations

The complete stiffness relations in local coordinates and in global coordinates are obtained as described in Chapter 5, Section 3, and Chapter 7, Section 2.3. The transformation matrices required are **B** and $\boldsymbol{\lambda}^a$ and $\boldsymbol{\lambda}^b$. **B** is derived in Eq. (3.1–10), and $\boldsymbol{\lambda}^a = \boldsymbol{\lambda}^b$ is derived in Section 2.

4. GOVERNING EQUATIONS AND METHODS OF ANALYSIS

4.1. General Structures

The mathematical formulation of the assemblage of members into a structure, which lead to the governing equations presented in Chapter 8, is the same as a procedure for planar or for space structures.

The definition of joints, effective degrees of freedom and joint displacements, the derivation of connectivity and of joint equilibrium equations, and the procedure for forming the connectivity matrices need only to be adjusted for size to apply to spatial structures. The governing equations of planar structures may now be considered as resulting from the general equations through the choice of effective displacement and force matrices.

The mathematical formulation of the static-kinematic properties of Chapter 9 applies also to space structures. The geometric study of kinematic stability becomes, however, more complex, as will be seen in Section 5. Also unchanged are the matrix formulations of the stiffness and of the flexibility methods, the direct stiffness method, the programmed selection of statical redundants, and the general description of inextensional and extensional behaviors. Manual methods become, however, impractical except in simple cases. From the point of view of behavior there are similarities between a shell-like structure made out of beam elements and a continuous shell and between a transversely loaded grid and a continuous slab.

4.2. Grid Structures

Consider a planar assemblage of members in the (X, Y) plane, and let the six displacements of a rigid joint be ordered into translations $\{r_1\ r_2\ r_3\}$ and rotations $\{r_4\ r_5\ r_6\}$ with respect to the global axes (X, Y, Z). It is assumed that each member has a principal plane of inertia containing its own x axis and the Z axis. If the external loading consists of Z forces and X and Y moments, each member will be in a combined state of torsion and of bending in the (x, Z) plane. Such a structure and behavior will be referred to as a grid. The effective joint displacements in a grid structure, Fig. 4.2–1, are

$$\mathbf{r} = \{r_3\ r_4\ r_5\} \tag{4.2–1}$$

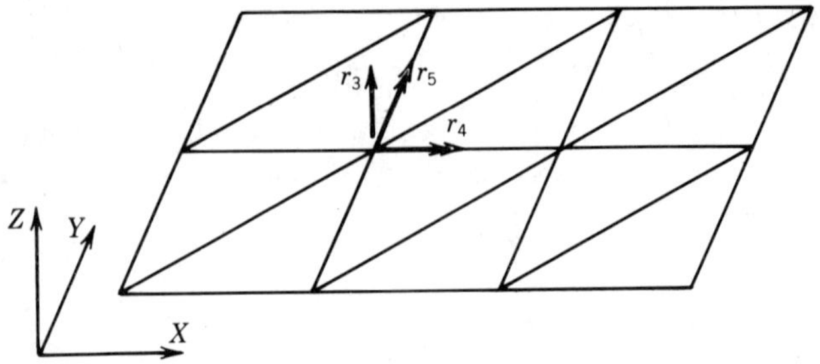

FIG. 4.2–1. Degrees of freedom for grid structure.

Effective joint forces, member end displacements, and member end forces correspond to the joint degrees of freedom defined here. The effective set of governing equations is a corresponding subset of the complete set of equations. The member deformations and statical redundants are

$$\mathbf{v} = \{v_3 \ \ v_4 \ \ v_5\} \tag{4.2–2a}$$

$$\mathbf{V} = \{V_3 \ \ V_4 \ \ V_5\} \tag{4.2–2b}$$

and the member end displacements and forces are

$$\mathbf{s} = \{\mathbf{s}^a \ \ \mathbf{s}^b\} = \{s_3^a \ \ s_4^a \ \ s_5^a \ \ s_3^b \ \ s_4^b \ \ s_5^b\} \tag{4.2–3a}$$

$$\mathbf{S} = \{\mathbf{S}^a \ \ \mathbf{S}^b\} = \{S_3^a \ \ S_4^a \ \ S_5^a \ \ S_3^b \ \ S_4^b \ \ S_5^b\} \tag{4.2–3b}$$

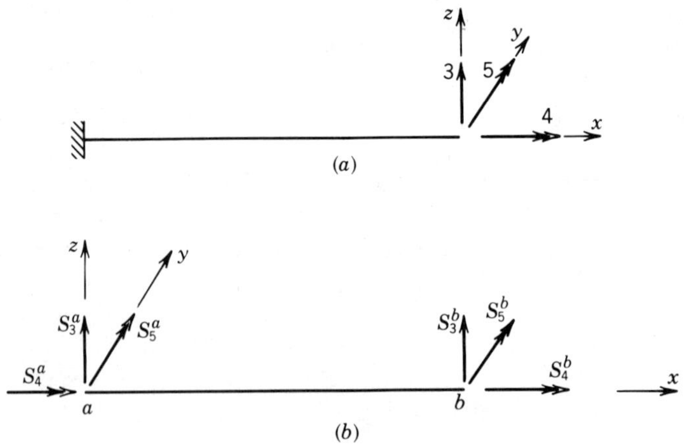

FIG. 4.2–2. Member variables in grid structure.

as shown in Fig. 4.2–2. The static transformation matrix from end b to end a is

$$\mathbf{T}_b^a = \begin{bmatrix} 1 & 0 & 0 \\ 0 & 1 & 0 \\ -l & 0 & 1 \end{bmatrix} \tag{4.2–4}$$

The member static-kinematic matrix $\mathbf{B}$ is as obtained in Eq. (3.1–10), or

$$\mathbf{B} = \begin{bmatrix} \mathbf{B}^a \\ \mathbf{B}^b \end{bmatrix} = \begin{bmatrix} -\mathbf{T}_b^a \\ \mathbf{I} \end{bmatrix} \tag{4.2–5}$$

where $\mathbf{I}$ is 3×3.

The flexibility matrix $\mathbf{f}'$, defined with the shear center as reference point for torsional moment V'_4 and for translational deformation v'_3, is extracted from Eq. (3.4–4) as

$$\mathbf{f}' = \begin{bmatrix} \dfrac{l^3}{3EI} & 0 & -\dfrac{l^2}{2EI} \\ 0 & \dfrac{l}{GJ_e} & 0 \\ -\dfrac{l^2}{2EI} & 0 & \dfrac{l}{EI} \end{bmatrix} \tag{4.2–6}$$

where I is the principal moment of inertia about the y axis.

If the shear center is not in the plane (x, Z), transformation to $\mathbf{V}$ and $\mathbf{v}$ obeys Eq. (3.4–7) in which $\mathbf{L}$ is extracted from Eq. (3.4–9) as

$$\mathbf{L} = \begin{bmatrix} 1 & 0 & 0 \\ -y_s & 1 & 0 \\ 0 & 0 & 1 \end{bmatrix} \tag{4.2–7}$$

Similarly, $\mathbf{k}'$ and $\mathbf{T} = (\mathbf{L}^T)^{-1}$ are obtained as

$$\mathbf{k}' = \begin{bmatrix} \dfrac{12EI}{l^3} & 0 & \dfrac{6EI}{l^2} \\ 0 & \dfrac{GJ_e}{l} & 0 \\ \dfrac{6EI}{l^2} & 0 & \dfrac{4EI}{l} \end{bmatrix} \tag{4.2–8}$$

and

$$\mathbf{T} = \begin{bmatrix} 1 & y_s & 0 \\ 0 & 1 & 0 \\ 0 & 0 & 1 \end{bmatrix} \tag{4.2–9}$$

Transformation of $\mathbf{k}'$ to centroidal axes obeys Eq. (3.4–15b).

The complete stiffness matrix in local coordinates is

$$\mathbf{k}_c = \mathbf{BkB}^T \tag{4.2–10}$$

and may be put in the form

$$\mathbf{k}_c = \begin{bmatrix} \mathbf{T}_b^a\mathbf{k}\mathbf{T}_b^{aT} & -\mathbf{T}_b^a\mathbf{k} \\ -\mathbf{k}\mathbf{T}_b^{aT} & \mathbf{k} \end{bmatrix} \tag{4.2–11}$$

For transformation to global coordinates the only information needed is the orientation angle θ of the member x axis with respect to the X axes. The member displacement transformation is

$$\mathbf{s} = \boldsymbol{\lambda}\bar{\mathbf{s}} \tag{4.2–12}$$

where

$$\boldsymbol{\lambda} = \begin{bmatrix} \boldsymbol{\lambda}^a & \\ & \boldsymbol{\lambda}^b \end{bmatrix} \tag{4.2–13}$$

and

$$\boldsymbol{\lambda}^a = \boldsymbol{\lambda}^b = \begin{bmatrix} 1 & 0 & 0 \\ 0 & \cos\theta & \sin\theta \\ 0 & -\sin\theta & \cos\theta \end{bmatrix} \tag{4.2–14}$$

Then

$$\bar{\mathbf{S}} = \boldsymbol{\lambda}^T\mathbf{S} \tag{4.2–15}$$

and

$$\bar{\mathbf{k}}_c = \boldsymbol{\lambda}^T\mathbf{k}_c\boldsymbol{\lambda} \tag{4.2–16}$$

Other transformation formulas are

$$\bar{\mathbf{B}} = \boldsymbol{\lambda}^T\mathbf{B} \tag{4.2–17}$$

and

$$\bar{\mathbf{k}}_c = \bar{\mathbf{B}}\mathbf{k}\bar{\mathbf{B}}^T \tag{4.2–18}$$

4.3. Truss Structures

In the formulation to follow, it is assumed that rotational continuity is completely released at the member joint connections. This may be an approximation to actual conditions whose justification is similar to the use of the substitute truss in planar trussed frame analysis.

Member loads are assumed converted into equivalent joint loads. Bending is thus treated separately, and the general member is treated in what follows as a two-force member. The effective member matrices are

$$\mathbf{v} = \{v_1\} \tag{4.3–1a}$$

$$\mathbf{s} = \{\mathbf{s}^a \ \mathbf{s}^b\} = \{s_1^a \ s_1^b\} \tag{4.3–1b}$$

$$\bar{\mathbf{s}} = \{\bar{\mathbf{s}}^a \ \bar{\mathbf{s}}^b\} = \{\bar{s}_1^a \ \bar{s}_2^a \ \bar{s}_3^a \ \bar{s}_1^b \ \bar{s}_2^b \ \bar{s}_3^b\} \tag{4.3–1c}$$

$$\mathbf{B}^T = [-1 \ \ 1] \tag{4.3–1d}$$

$$\boldsymbol{\lambda} = \begin{bmatrix} \boldsymbol{\lambda}^a & \\ & \boldsymbol{\lambda}^b \end{bmatrix} \tag{4.3–1e}$$

$$\boldsymbol{\lambda}^a = \boldsymbol{\lambda}^b = [\lambda_{11} \ \lambda_{12} \ \lambda_{13}] = [c_1 \ c_2 \ c_3] \tag{4.3–1f}$$

where $\lambda_{11}, \lambda_{12}, \lambda_{13}$ are the direction cosines of the x axis as defined in Eq. (2–8) and are renamed for convenience c_1, c_2, and c_3.

Static matrices **V**, **S**, $\bar{\mathbf{S}}$ correspond to **v**, **s**, and $\bar{\mathbf{s}}$, respectively.

The axial stiffness k is transformed into the stiffness matrix $\mathbf{k}_c$ in local coordinates the same as for planar structures; that is,

$$\mathbf{k}_c = \mathbf{B}k\mathbf{B}^T = k\begin{bmatrix} 1 & -1 \\ -1 & 1 \end{bmatrix} \tag{4.3–2}$$

Transformation to global coordinates yields

$$\bar{\mathbf{k}}_c = \bar{\mathbf{B}}k\bar{\mathbf{B}}^T \tag{4.3–3}$$

where

$$\bar{\mathbf{B}}^T = \mathbf{B}^T\boldsymbol{\lambda} = [-\boldsymbol{\lambda}^a \ \ \boldsymbol{\lambda}^b] = [-c_1 \ \ -c_2 \ \ -c_3 \ \ c_1 \ \ c_2 \ \ c_3] \tag{4.3–4}$$

Thus

$$\bar{\mathbf{k}}_c = k\left|\begin{array}{ccc|c} c_1^2 & c_1c_2 & c_1c_3 & \\ c_2c_1 & c_2^2 & c_2c_3 & (-) \\ c_3c_1 & c_3c_2 & c_3^2 & \\ \hline & (-) & & (+) \end{array}\right| \tag{4.3–5}$$

Forming of the structure stiffness matrix **K** by the direct stiffness method follows the procedure outlined for planar structures.

In the force method $\mathbf{a}_v$ is formed by merging member matrices $\bar{\mathbf{B}}^{mT}$ which are defined without the superscript m in Eq. (4.3–4). The formulation of the force method in terms of $\mathbf{a}_v$ is the same as for planar trusses. It can be made as systematic as the displacement method and is a more efficient procedure for statically determinate trusses or for a low degree of statical indeterminacy.

The method of joints for solving joint equilibrium equations requires that there be not more than three unknown member forces at the joint whose equilibrium equations are being solved. The method is applicable, in particular, to simple assemblages, as will be seen in Section 5. The basic operation of the method of joints for spatial trusses is to decompose an applied force vector $\vec{R}$ into three components F_1, F_2, F_3 on three generally nonorthogonal member directions. Letting $\vec{t}_1, \vec{t}_2, \vec{t}_3$ be unit vectors along the members, it is required to determine F_1, F_2, F_3 such that

$$F_1\vec{t}_1 + F_2\vec{t}_2 + F_3\vec{t}_3 = \vec{R} \tag{4.3–6}$$

Letting

$$\vec{t}'_1 = \vec{t}_2 \times \vec{t}_3 \tag{4.3–7a}$$

$$\vec{t}'_2 = \vec{t}_3 \times \vec{t}_1 \tag{4.3–7b}$$

$$\vec{t}'_3 = \vec{t}_1 \times \vec{t}_2 \tag{4.3–7c}$$

F_1 is determined by taking the scalar product of Eq. (4.3–6) by $\vec{t}'_1$ and noting that $\vec{t}'_1 \cdot \vec{t}_2 = \vec{t}'_1 \cdot \vec{t}_3 = 0$. F_2 and F_3 are determined similarly. Letting

$$\Delta = \vec{t}_1 \cdot (\vec{t}_2 x \vec{t}_3) = \vec{t}_2 \cdot (\vec{t}_3 x \vec{t}_1) = \vec{t}_3 \cdot (\vec{t}_1 x \vec{t}_2) \tag{4.3–8}$$

the solution is

$$F_i = \frac{\vec{t}'_i \cdot \vec{R}}{\Delta}, \qquad i = 1, 2, 3 \tag{4.3–9}$$

For calculations, the components on the global axes of a unit vector along a member are the direction cosines defined in Eq. (2–8) and renamed $\{c_1 \ c_2 \ c_3\}$ in Eq. (4.3–1f). Components of $\vec{t}'_1$, $\vec{t}'_2$, and $\vec{t}'_3$ on the global axes are obtained from Eqs. (4.3–7) by applying the formulas for evaluating vector product components. The scalar product in Eq. (4.3–9) is then evaluated in terms of these components and of the given component of $\vec{R}$ on the global axes.

Some particular statical properties may be deduced from Eq. (4.3–6). If $\vec{R} = 0$ at a joint of three noncoplanar bars, the bar forces are zero. If $\vec{R}$ is oriented along one of the three bars, the axial forces in the other two bars are zero. If $\vec{R}$ is in the plane of two bars, the axial force in the third bar is zero. If any number of members

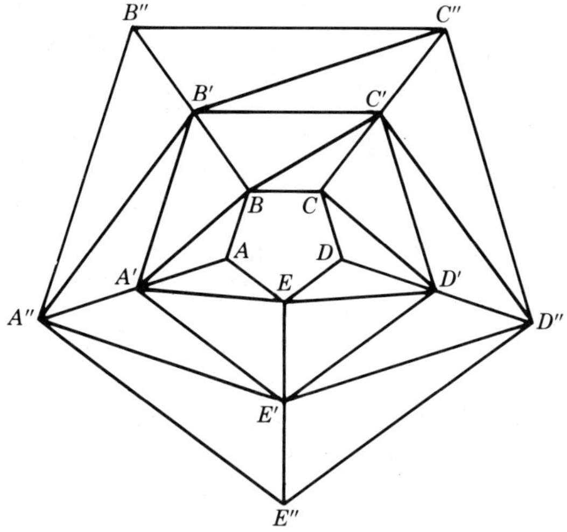

(a)

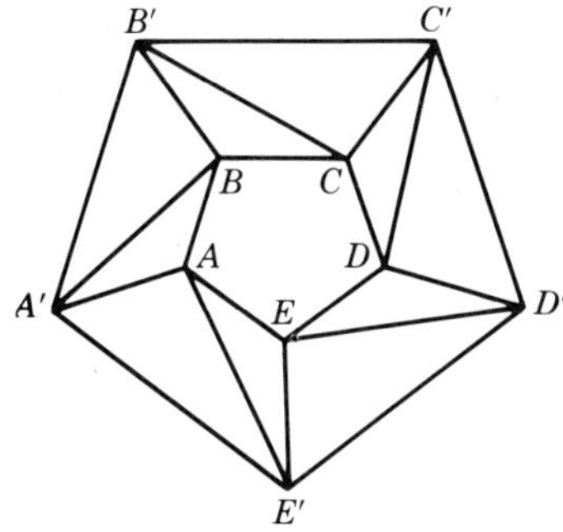

(b)

FIG. 4.3–1. Schwedler dome.

at a joint are coplanar except one, the axial force in that member is determined by projecting the vector equilibrium equation on the normal to the coplanar members. This procedure applied in turn to each member of a set of three noncoplanar members results in Eq. (4.3–9).

As an application of the method of joints consider the domelike structure shown in plan in Fig. 4.3–1. It is of a type called Schwedler dome [13]. The structure is formed of three pentagonal rings $ABCDE$, $A'B'C'D'E'$, and $A''B''C''D''E''$ lying, respectively, in different horizontal planes and such that AB, $A'B'$, and $A''B''$ are parallel, with similar arrangements for the other sides. The three rings are connected by meridional bars such as AA' and $A'A''$ and by diagonals such as BA' and BC'. Support conditions at the base $A''B''C''D''E''$ and other arrangements of the diagonals are discussed later.

Given any applied joint load, the method of joints may start at either joint A or joint D at each of which there are only three members that also are not coplanar. The joint equilibrium equation may be solved at the top ring in the sequence A, D, E, C, B. The axial forces in the members of the top ring and in all members connecting it to the lower ring are thus determined. A similar sequence of joint processing may be followed at the next ring, that is, A', D', E', C', B'.

At each step of these sequences the joint being processed has only three unknown and noncoplanar member forces. The remaining unknowns are the forces in the members of the base ring and in the support links. Before discussing support conditions, the alternate arrangement of the diagonals shown in Fig. 4.3–1b is considered. At each joint of the top ring there are four members, three of which, however, are coplanar. The bar force in the noncoplanar member is determined as seen earlier by projection of the vector equilibrium equation on the normal to the plane of the three coplanar members. Starting, for example, at A, the force in AB is determined. The method of joints is then applied in the sequence B, C, D, E. At each step there are only three unknown noncoplanar member forces. Equilibrium of joint A in the plane $AEE'A'$ determines the forces in the diagonals AA' and AE'.

Consider now support conditions for the base ring in Fig. 4.3–1a. To prescribe such conditions, one way that allows a continuation of the method of joints is to provide a vertical link, say, at A'', two links at each of joints B'', C'', D'', and three links equivalent to a hinge at E''. Further the two links at B'' are not coplanar with $B''C''$, those at C'' are not coplanar with $C''D''$, and those at D'' are not coplanar with $D''E''$. With such support conditions joint A'' is processed first, thereby determining the link force and the forces in members $A''B''$ and $A''E''$. Next B'', C'',

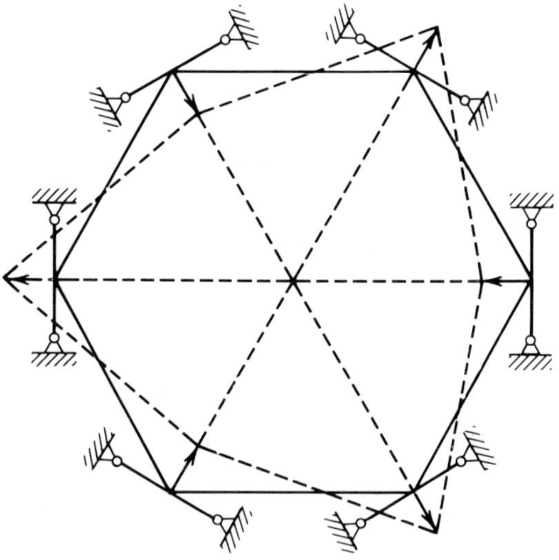

FIG. 4.3–2. Unstable support conditions.

and D'' are processed. The remaining unknowns are now the three reactions provided by the hinge at E'' and are determined by the equilibrium equations of that joint.

The feasibility of the procedure under an arbitrary joint load indicates that the structure is kinematically stable.

If the support conditions are changed so that two links are provided at A'' and E'', the method of joints ceases to be applicable at the base ring. The joint equilibrium equations may then be solved simultaneously, provided the base ring and support links form a kinematically stable system.

The preceding study may be readily generalized to domes of general polygonal rings.

An unstable support system consisting of two links at each joint of the base ring is shown in Fig. 4.3–2. The hexagonal ring is regular, and the two links at each joint are in a vertical plane perpendicular to the radius. These links are shown in horizontal projection in the figure. A possible mechanism motion is shown in dotted lines. The joints can move horizontally by a constant displacement in the radial direction, alternately in the outward and inward directions. Such a support arrangement would be unstable for any even-sided regular polygon.

5. KINEMATIC STABILITY OF SPATIAL STRUCTURES

5.1. Kinematic Stability of a Rigid Body

Support of a body against rigid body motion may be represented as in the case of planar structures by means of rigid support links. Referring to Fig. 5.1–1, a typical support link prevents displacement of point A along the link but allows freely an arbitrary displacement at A in the plane perpendicular to the link. Accordingly the interaction at A is a force in the direction of the link. For discussing kinematic stability, only infinitesimal displacements are considered, and a support link may equivalently be considered hinged at both ends.

Support of a rigid body against translational and rotational motions is

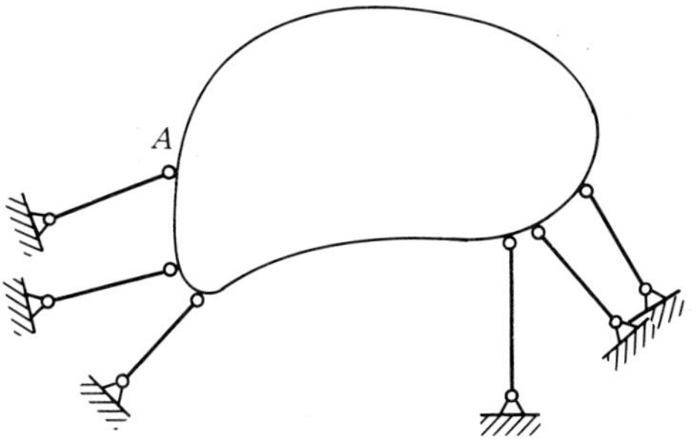

FIG. 5.1–1. Six-link support.

considered separately in what follows before considering support against general rigid body motion.

For a translational displacement of a supported body to be possible, it must be orthogonal to all support links. Thus any purely translational motion is prevented if support links are not orthogonal to a single direction, or equivalently, if they are not parallel to a single plane. This is achieved through a minimum of three links whose orientations are not coplanar but otherwise arbitrary. In particular, links may or may not intersect. A hinge (ball and socket) support corresponds to three noncoplanar support links intersecting at one point.

For a rotational displacement of a supported body to be possible, every support link must intersect or be parallel to the axis of rotation; otherwise, the displacement vector at the supported point would not be orthogonal to the link. Thus any purely rotational motion is prevented if the support links do not intersect a single axis.

The most general rigid body motion is a superposition of a translation $\delta\vec{u}$ and a rotation $\delta\vec{\omega}$. At point A of the body the displacement is

$$\delta\vec{u}_A = \delta\vec{u} + \delta\vec{\omega} \times \vec{r}_A \qquad (5.1\text{–}1)$$

where $\vec{r}_A = OA$, and O is the reference point at which $\delta\vec{u}$ is defined. If a support link at point A is oriented by a unit vector $\vec{t}_A$ the support condition is

$$\vec{t}_A \cdot \delta\vec{u}_A = \vec{t}_A \cdot \delta\vec{u} + \delta\vec{\omega} \cdot (\vec{r}_A \times \vec{t}_A) = 0 \qquad (5.1\text{–}2)$$

This condition is a linear homogeneous equation in the components of $\delta\vec{u}$ and $\delta\vec{\omega}$. For a stable support condition the set of such equations corresponding to the set of support links should have only the null solution $\delta\vec{u} = 0$, $\delta\vec{\omega} = 0$. Thus a minimum of six links is required whose disposition is such as not to make the coefficient matrix of the set of support conditions singular. Necessary conditions to achieve this are those found earlier to prevent purely translational and purely rotational motions, that is, that the six links not intersect a single axis nor be parallel to a single plane. This last condition is in fact a particular case of the first whereby links parallel to a single plane intersect the line at infinity in that plane, and the translational motion they allow is the limit of a rotation about an axis receding to infinity in the plane parallel to the links.

A stable support condition by means of six equivalent links is termed simple. Two examples are shown in Fig. 5.1–2. In (*a*) the three links at A are equivalent to a hinge support. The three other links prevent any rotation about A because there is no axis passing through A and intersecting them. The support condition in (*a*) remains stable if any link is moved to another location by simply sliding it along its axis. If link BC is oriented along BD, line AD becomes an axis of rotation. If link BC is made parallel to DF, the support condition remains stable. If link DF is placed at B without changing its orientation, AG becomes an axis of rotation.

In Fig. 5.1–2*b*, A, B, and C are parallel links. Motion in the plane orthogonal to

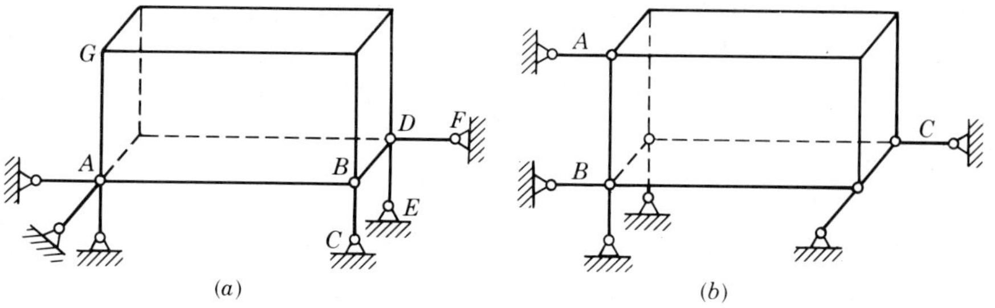

FIG. 5.1–2. Stable support conditions.

them is prevented by the three other links. The same point of view may be applied to Fig. 5.1–2*a*.

The necessary conditions for separate kinematic stability against translation and rotation are not sufficient against general rigid body motion. This may be seen geometrically. Given $\delta\vec{u}$ at a point O, and $\delta\vec{\omega}$, it is always possible to find a point O' at which $\delta\vec{u}$ is colinear with $\delta\vec{\omega}$. The axis passing through O' and oriented by $\delta\vec{\omega}$ is called the screw axis. The motion is one of translation along and rotation about the screw axis. The displacement of a point A is shown in Fig. 5.1–3*a*. $\vec{k}$ is a unit vector along the screw axis, and $\vec{t}_r$ and $\vec{t}_\theta$ are radial and circumferential unit vectors at A, respectively, in the plane perpendicular to $\vec{k}$. Any link orthogonal to $\delta\vec{u}_A$ does not restrain the motion. In particular, a link orthogonal to $\delta\vec{u}_A$ in the plane $(\vec{k}, \vec{t}_\theta)$ is such that

$$\vec{k} \cdot \vec{t}_A = \cos\alpha \tag{5.1–3}$$

and from Fig. 5.1–3*a*

$$\tan\alpha = \frac{\delta u}{r\,\delta\omega} \tag{5.1–4}$$

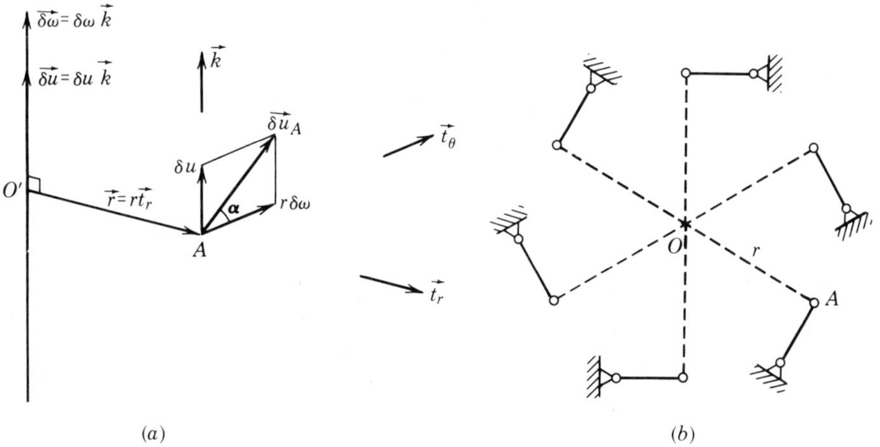

FIG. 5.1–3. Screw motion.

A disposition of six such links is shown in projection on a plane perpendicular to the screw axis in Fig. 5.1–3*b*. All links have the same *r* and the same inclination α to the screw axis. The six links do not intersect a single axis, nor are they parallel to a single plane, and are not sufficient to restrain the motion $(\delta\vec{u}, \delta\vec{\omega})$ defined in Eq. (5.1–4).

Consider now a body supported by six links. Let $\vec{R}$ and $\vec{M}$ be the resultant force and resultant moment at a point *O*, of an applied force system. The equations of equilibrium of the body are

$$\sum F_A \vec{t}_A = -\vec{R} \tag{5.1–5a}$$

$$\sum \vec{r}_A \times F_A \vec{t}_A = -\vec{M} \tag{5.1–5b}$$

where the sum extends over the support links and $F_A \vec{t}_A$ is the reaction force exerted by the link at point *A*. Equations (5.1–5) form a linear system of six scalar equations in the six reactions F_A. It is shown in what follows that the condition for the system of equilibrium equations to be linearly independent, and thus to have a unique solution for arbitrarily given loads, is that the support links prevent any rigid body displacement. In a virtual rigid body displacement the virtual work of the reaction forces $F_A \vec{t}_A$ is, using Eq. (5.1–1),

$$\sum F_A \vec{t}_A \cdot \delta\vec{u}_A = \left(\sum F_A \vec{t}_A\right) \cdot \delta\vec{u} + \left(\sum \vec{r}_A \times F_A \vec{t}_A\right) \cdot \delta\vec{\omega} \tag{5.1–6}$$

Similarly the virtual work of the applied load is

$$\vec{R} \cdot \delta\vec{u} + \vec{M} \cdot \delta\vec{\omega} \tag{5.1–7}$$

First, it is noted, in view of Eqs. (5.1–5), that the total external virtual work of a force system in equilibrium vanishes for arbitrary $\delta\vec{u}$ and $\delta\vec{\omega}$. This is the necessary part of the theorem of virtual displacements.

The expression of virtual work in Eq. (5.1–6) is an arbitrary linear combination of the left-hand sides of the equilibrium equations. The condition for linear independence is that there be no virtual displacement for which $\vec{t}_A \cdot \delta\vec{u}_A$ vanishes at all support points *A*. This is also the condition seen earlier for the support links to prevent any rigid body displacement.

5.2. Rigid Spatial Assemblages

The rigid spatial assemblage that corresponds in type to the planar triangulated assemblage is based on the tetrahedron as the basic rigid assemblage. Joints are assumed hinged so that member rotations at a joint are independent.

Starting with a triangle as a base, a tetrahedron is assembled by connecting three bars, each to one vertex of the triangle and together at the remaining ends to form a joint not in the plane of the base triangle. The process may be repeated indefinitely, starting with a triangular face of the existing assemblage as a base for

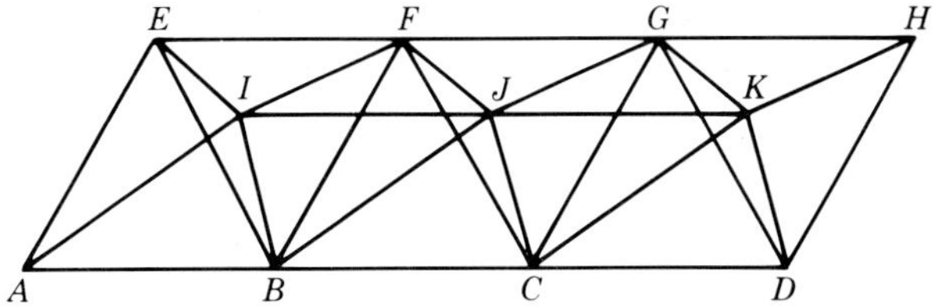

FIG. 5.2–1. Tetrahedral assemblage.

a new tetrahedron. An example is shown in Fig. 5.2–1. Starting with triangle EAB as a base, a tetrahedron is formed by joining members IE, IA, and IB at I. The next base triangle is IEB, and the new joint is F. At this point the assemblage is a pyramid with a diagonal EB in its base. Next joint J is formed with FIB as a base, joint C with JFB as a base, and joint G with CJF as a base. At this point a second pyramid similar to the first has been formed. The two pyramids have bar BF in common, and their summits are connected by bar IJ. The rest of the figure is a continuation of the preceding construction.

If a tetrahedral assemblage is simply supported against rigid body displacement, there results a kinematically stable truss structure that is also statically determinate. The statical determinacy is established by counting three equilibrium equations per joint and finding the result equal to the number of bar forces including the six forces in six support links.

A simple assemblage is a generalization of the tetrahedral one in which three new bars are connected to any three noncolinear points of the existing rigid assemblage and to a new joint without being coplanar.

A compound assemblage is formed by connecting a rigid assemblage to another rigid assemblage with six links whose disposition prevents relative rigid body motion. The assemblage of Fig. 5.2–1 may be viewed as formed of separate tetrahedra connected by six links. For example, $IEAB$ and $JFCG$ are two tetrahedra connected by the six bars FE, FI, FB, JI, JB, and CB. It is also possible to consider the assemblage of Fig. 5.2–1 as formed of pyramidal assemblages. For example, pyramids $IABFE$ and $JBCGF$ have bar BF in common, and their summits are connected by bar IJ.

This modular construction which proceeds in the longitudinal direction $ABCD$ may be extended in the transverse direction EA so as to form an L shape, and that results also in a tetrahedral assemblage. However, if a rectangular pattern is made, the assemblage will have more members than needed to be tetrahedral. If simply supported, it would be statically indeterminate.

A nontetrahedral assemblage is shown in Fig. 5.2–2. It consists of two plane rectangular grids, named top and bottom grids, respectively, connected by diagonals [14]. The assemblage may be considered formed of pyramidal assemblages having their bases in the top grid and their summits in the bottom grid.

The pyramids differ from those considered in Fig. 5.2–1 in that their bases have

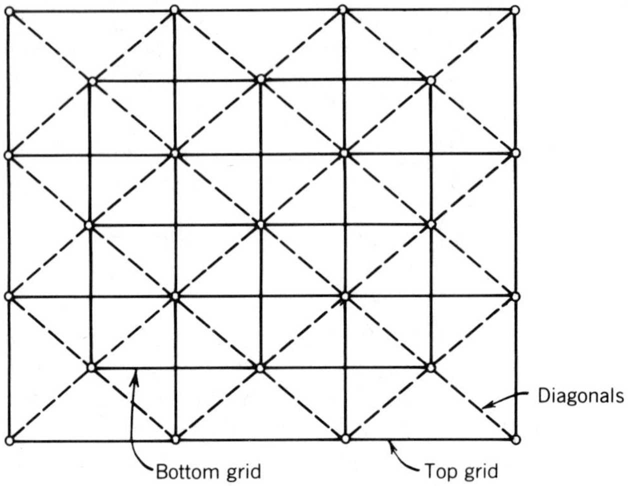

FIG. 5.2–2. Space assemblage.

no diagonals. If only one row of pyramids is considered, the assemblage is not rigid because of the missing diagonals in the bases of the pyramids. If two rows are considered, the count of joints yields $NJ = 18$, and the count of members yields $NM = 48$. With six support links there are 54 member forces and $3 \times NJ = 54$ joint equilibrium equations. It is difficult to show geometrically that the assemblage is r:gid. However, this seems to be the case, and if simply supported, the assemblage would be statically determinate. In the case of Fig. 5.2–2 the assemblage is found statically indeterminate.

In actual constructions the joints are not hinged and are thus capable of transmitting internal moments. Nonrigid assemblages have translational inextensional modes that may involve significant bending, whereas the concept of the substitute truss seen in planar structures applies to rigid assemblages.

A type of assemblage similar to that of Fig. 5.2–2 but in which the two grids are inclined one relative to the other is shown in Fig. 5.2–3. The assemblage is formed of four pyramids whose summits are in the bottom grid and their bases in the top grid. A typical pyramid has A as summit and $BCDE$ as a base. A count of joints and members yields $NJ = 16$ and $NM = 36$, respectively. Since $3 \times NJ > NM + 6$, the module is not a rigid assemblage. The assemblage of Fig. 5.2–3 has shorter members in the top grid, which is an advantage for compression members. It can itself be used as a module in a larger assemblage.

The use of modular construction allows us to use in the analysis the substructure technique presented in Chapter 10, which treats the module as the basic structural element.

Surfacelike assemblages referred to as Schwedler domes were examined in Section 4.3. Other types known as geodesic domes have found many applications, ranging from a single-family dwelling to extremely large exhibition halls [15]. A basic solid from which various geodesic domes may be generated is the regular

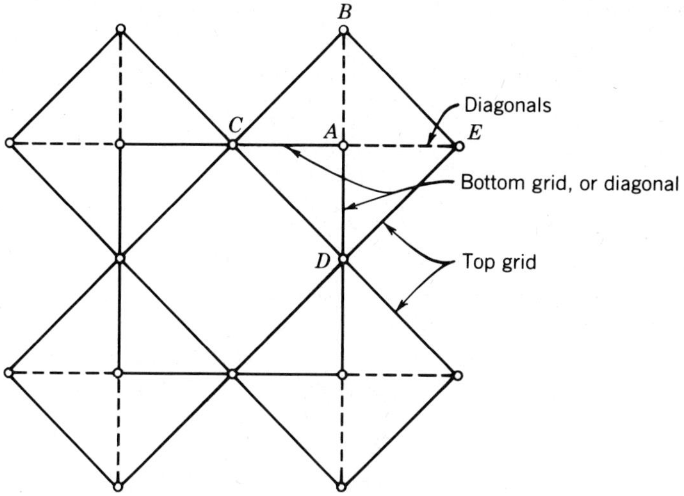

FIG. 5.2–3. Space assemblage, nonrigid.

icosahedron. This is a solid bounded by 20 equal equilateral triangles, having 30 sides and 12 vertexes lying on a sphere, Fig. 5.2–4*a*. The 12 vertexes may be grouped into pairs lying at the extremities of 6 diameters. If any pair of opposite vertexes A and B is set apart, the remaining 10 vertexes form two regular pentagons lying in parallel planes orthogonal to AB. An assemblage of 30 bars along the sides of the icosahedron is rigid. If supported against rigid body displacements by means of 6 links, the number of force unknowns including reactions is 36, and this is equal to the number of joint equilibrium equations. The truss structure thus defined is stable and statically determinate.

In order to follow more closely a spherical shape, a triangular face may be subdivided into triangles as shown in (*b*), and the nodes of the subdivision are projected on the sphere of the icosahedron. It is clear that the subdivision shown

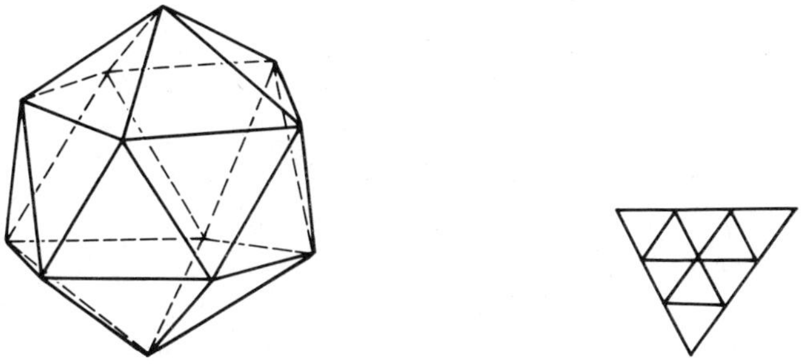

(*a*) (*b*)

FIG. 5.2–4. Icosahedron and possible subdivision of triangular face.

in (*b*) may be generalized to any number of subdivisions per side. The resulting assemblage remains rigid, and a count of members and joints would show that it is internally statically determinate. A complete assemblage may be truncated in various ways and supported to obtain a final structural shape.

6. NONLINEAR THEORY OF SPACE TRUSSES

A geometrically nonlinear analysis of space trusses is developed in what follows as a generalization of the formulation for plane trusses. The fundamental kinematic relation in the formulation is the relation defining the axial deformation v_1 in terms of member end displacements. v_1 is the change in member chord length whose exact expression is

$$v_1 = [(l + s_1^b - s_1^a)^2 + (s_2^b - s_2^a)^2 + (s_3^b - s_3^a)^2]^{1/2} - l \tag{6–1}$$

Equation (6–1) is now simplified, assuming that the square of the member rotation is comparable in order of magnitude with v_1/l. Accordingly the series development of the square root term is truncated keeping the second-order terms in $\mathbf{s}$. Further the term $(s_1^b - s_1^a)^2$ is neglected as in the planar theory. There comes

$$v_1 = s_1^b - s_1^a + \frac{1}{2l}[(s_2^b - s_2^a)^2 + (s_3^b - s_3^a)^2] \tag{6–2a}$$

or, in matrix form

$$v_1 = \mathbf{B}_1^T \mathbf{s}_1 + \frac{1}{2}\mathbf{s}^T \mathbf{c}_1 \mathbf{s} \tag{6–2b}$$

where

$$\mathbf{s}_1 = \{s_1^a \;\; s_1^b\} \tag{6–3a}$$

$$\mathbf{s} = \{s_2^a \;\; s_3^a \;\; s_2^b \;\; s_3^b\} \tag{6–3b}$$

$$\mathbf{B}_1^T = [-1 \;\; 1] \tag{6–3c}$$

$$\mathbf{c}_1 = \frac{\partial^2 v_1}{\partial \mathbf{s} \partial \mathbf{s}^T} = \frac{1}{l}\begin{bmatrix} 1 & 0 & -1 & 0 \\ 0 & 1 & 0 & -1 \\ -1 & 0 & 1 & 0 \\ 0 & -1 & 0 & 1 \end{bmatrix} \tag{6–3d}$$

The incremental deformation δv_1 is

$$\delta v_1 = [\mathbf{B}_1^T \;\; \mathbf{B}_N^T]\begin{Bmatrix} \delta \mathbf{s}_1 \\ \delta \mathbf{s} \end{Bmatrix} \tag{6–4}$$

where

$$\mathbf{B}_N = \mathbf{c}_1 \mathbf{s} \tag{6–5}$$

The static relation forming a contragradient pair with Eq. (6–4) is

$$\begin{Bmatrix} \mathbf{S}_1 \\ \mathbf{S} \end{Bmatrix} = \begin{bmatrix} \mathbf{B}_1 \\ \mathbf{B}_N \end{bmatrix} V_1 \tag{6–6}$$

Incrementing Eq. (6–6), there comes

$$d\mathbf{S}_1 = \mathbf{B}_1 \, dV_1 \tag{6–7a}$$

$$d\mathbf{S} = \mathbf{B}_N d\mathbf{V}_1 + \mathbf{k}_G d\mathbf{s} \tag{6–7b}$$

where

$$\mathbf{k}_G = V_1 \mathbf{c}_1 \tag{6–8}$$

The equations established here may be seen to be the same in matrix form as in the case of planar structures. Having the geometric stiffness matrix $\mathbf{k}_G$, formulation of nonlinear governing equations, methods of analysis such as the stiffness and flexibility methods, forming of the tangent stiffness matrix, and instability analysis all follow the procedure outlined for planar structures.

7. GEOMETRIC NONLINEARITY IN SPACE FRAMES

The general nonlinear behavior of a structural member is not a simple generalization of planar behavior. The presence of torsion causes a coupling of bending, torsion, and axial behaviors. These aspects will appear in the general theory to be developed in Chapter 15. A simplified theory with a limited degree of geometric nonlinearity is based on neglecting the coupling between bending and torsion. The result is a juxtaposition of bending in two principal planes, of torsion and of axial compression or tension. The axial strain is generalized from the planar theory to

$$\varepsilon = \frac{du}{dx} + \frac{1}{2}\left(\frac{dv}{dx}\right)^2 + \frac{1}{2}\left(\frac{dw}{dx}\right)^2 \tag{7–1}$$

where u is the axial displacement and v and w are orthogonal displacement components in the plane of the cross section.

Member flexibility and stiffness matrices of the cantilever type are assembled from axial, bending, and torsion terms similarly to Eqs. (3.4–4) and (3.4–12). For a uniform member and a constant compressive axial force, the bending flexibility coefficients are obtained by applying Eq. (6/2.2–2) in the two principal planes. For bending in the plane (x, z), the stability parameter, omitting shear deformation for simplicity, is

$$\lambda_2^2 = \frac{-V_1 l^2}{EI_2} \tag{7.2–a}$$

and for bending in the (x, y) plane,

$$\lambda_3^2 = \frac{-V_1 l^2}{EI_3} \tag{7.2–b}$$

The axial flexibility relation is obtained similarly to Eq. (6/2.1–11) in which the nonlinear term becomes the sum of two such terms corresponding to the two principal planes of bending.

The member reduced stiffness matrix is formed similarly to the flexibility matrix and is its inverse. For the cantilever type the bending stiffness coefficients are defined in Eq. (6/2.2–3).

The formulas derived on the basis of an assumed shape in planar behavior may also be applied in the present case. Thus in Eqs. (6/3.1–22) the geometric matrix $\mathbf{g}$ is now a 4×4 matrix of the form

$$\mathbf{g} = \begin{bmatrix} \mathbf{g}_v & \\ & \mathbf{g}_w \end{bmatrix} \tag{7–3}$$

If the same shape functions are used in the two principal planes, then $\mathbf{g}_v = \mathbf{g}_w$.

The preceding considerations apply as well to flexibility and stiffness matrices of the alternate type. The corresponding member-bound reference may be defined to be attached to end a and to have as axis $x_{1,r}$ the member chord. A second axis $x_{2,r}$ may be defined to be perpendicular to the chord and in the plane containing the chord and the principal y axis of the cross section. The possible displacements of the member ends with respect to the bound reference are an axial translation and an axial rotation at end b, and two bending rotations at each member end. These relative displacements form the deformations of the alternate type, Fig. 7–1b, c.

For formulating the governing equations of the structure as done in Chapter 8, the member static-kinematic matrix $\mathbf{B}$ is needed. It is recalled that $\mathbf{B}$ expresses the member end forces $\mathbf{S} = \{\mathbf{S}^a \ \mathbf{S}^b\}$ in the local reference in the form

$$\mathbf{S} = \mathbf{BV} + \mathbf{S}_p \tag{7–4}$$

and it also expresses incremental deformations in the form

$$\delta\mathbf{v} = \mathbf{B}^T \delta\mathbf{s} \tag{7–5}$$

Simplifications in the expression of $\mathbf{B}$ depend on assumptions concerning the rotation of the member-bound reference. Simplifications based on the planar beam-colum theory of Chapter 3 and on uncoupling torsion from bending are presented in what follows. For deformations and member statical redundants of the cantilever type, and the orderings shown in Fig. 7–1a, we obtain by generalizing Eqs. (3/2.1–1) and appending a relation for torsional deformations

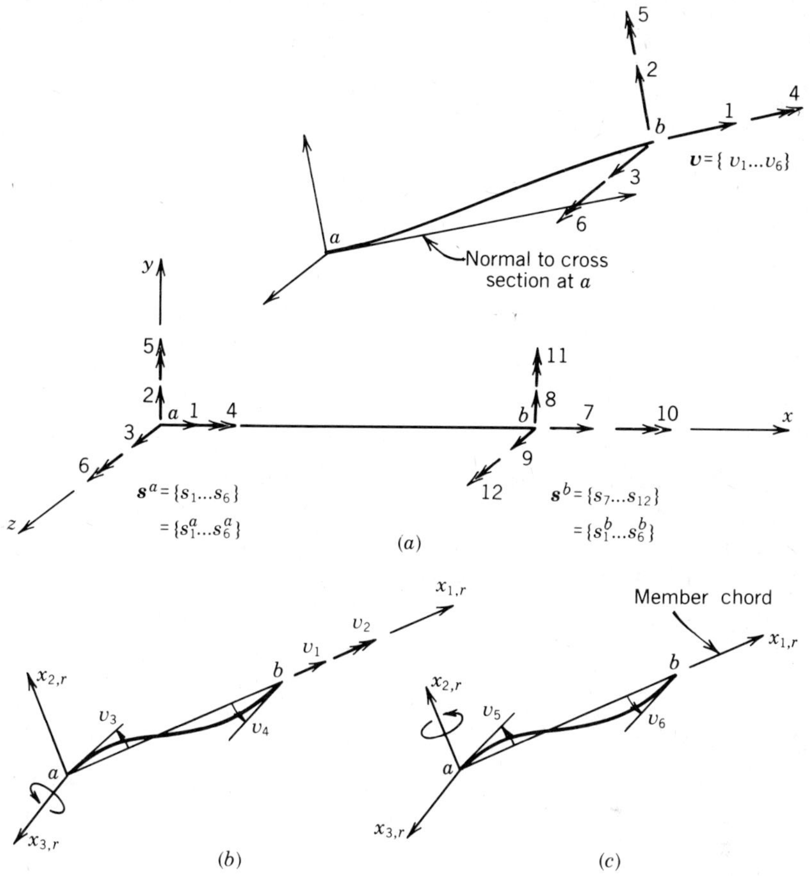

FIG. 7–1. Member variables.

$$v_1 = s_1^b - s_1^a + s_6^a(s_2^b - s_2^a - \tfrac{1}{2}ls_6^a) - s_5^a(s_3^b - s_3^a + \tfrac{1}{2}ls_5^a) \tag{7–6a}$$

$$v_2 = s_2^b - s_2^a - ls_6^a \tag{7–6b}$$

$$v_3 = s_3^b - s_3^a + ls_5^a \tag{7–6c}$$

$$v_4 = s_4^b - s_4^a \tag{7–6d}$$

$$v_5 = s_5^b - s_5^a \tag{7–6e}$$

$$v_6 = s_6^b - s_6^a \tag{7–6f}$$

We then have

$$\mathbf{B}^T = \frac{\partial \mathbf{v}}{\partial \mathbf{s}^T} = \left[\frac{\partial \mathbf{v}}{\partial \mathbf{s}^{aT}} \quad \frac{\partial \mathbf{v}}{\partial \mathbf{s}^{bT}} \right] = [\mathbf{B}^{aT} \ \ \mathbf{B}^{bT}] \tag{7–7}$$

$$\mathbf{B}^{aT} = \frac{\partial \mathbf{v}}{\partial \mathbf{s}^{aT}} = \begin{bmatrix} -1 & -s_6^a & s_5^a & 0 & -(s_3^b - s_3^a + l s_5^a) & s_2^b - s_2^a - l s_6^a \\ 0 & -1 & 0 & 0 & 0 & -l \\ 0 & 0 & -1 & 0 & l & 0 \\ 0 & 0 & 0 & -1 & 0 & 0 \\ 0 & 0 & 0 & 0 & -1 & 0 \\ 0 & 0 & 0 & 0 & 0 & -1 \end{bmatrix}$$

$$\mathbf{B}^{bT} = \frac{\partial \mathbf{v}}{\partial \mathbf{s}^{bT}} = \begin{bmatrix} 1 & s_6^a & -s_5^a & 0 & 0 & 0 \\ 0 & 1 & 0 & 0 & 0 & 0 \\ 0 & 0 & 1 & 0 & 0 & 0 \\ 0 & 0 & 0 & 1 & 0 & 0 \\ 0 & 0 & 0 & 0 & 1 & 0 \\ 0 & 0 & 0 & 0 & 0 & 1 \end{bmatrix}$$

For deformations of the alternate type ordered as shown in Fig. 7–1*b*, *c*, the axial deformation v_1 is defined as in Eq. (6–2), and the torsional deformation v_2 has the same expression as in the preceding case. The bending deformations are obtained similarly to Eqs. (3/2.1–3b, c). We thus obtain

$$v_1 = s_1^b - s_1^a + \frac{1}{2l}(s_2^b - s_2^a)^2 + \frac{1}{2l}(s_3^b - s_3^a)^2 \tag{7–8a}$$

$$v_2 = s_4^b - s_4^a \tag{7–8b}$$

$$v_3 = s_6^a - \frac{1}{l}(s_2^b - s_2^a) \tag{7–8c}$$

$$v_4 = s_6^b - \frac{1}{l}(s_2^b - s_2^a) \tag{7–8d}$$

$$v_5 = s_5^a + \frac{1}{l}(s_3^b - s_3^a) \tag{7–8e}$$

$$v_6 = s_5^b + \frac{1}{l}(s_3^b - s_3^a) \tag{7–8f}$$

and

$$\mathbf{B}^{aT} = \frac{\partial \mathbf{v}}{\partial \mathbf{s}^{aT}} = \begin{bmatrix} -1 & -(s_2^b - s_2^a)/l & -(s_3^b - s_3^a)/l & 0 & 0 & 0 \\ 0 & 0 & 0 & -1 & 0 & 0 \\ 0 & 1/l & 0 & 0 & 0 & 1 \\ 0 & 1/l & 0 & 0 & 0 & 0 \\ 0 & 0 & -1/l & 0 & 1 & 0 \\ 0 & 0 & -1/l & 0 & 0 & 0 \end{bmatrix} \tag{7–9a}$$

$$
\mathbf{B}^{bT} = \frac{\partial \mathbf{v}}{\partial \mathbf{s}^{bT}} = \begin{bmatrix} 1 & (s_2^b - s_2^a)/l & (s_3^b - s_3^a)/l & 0 & 0 & 0 \\ 0 & 0 & 0 & 1 & 0 & 0 \\ 0 & -1/l & 0 & 0 & 0 & 0 \\ 0 & -1/l & 0 & 0 & 0 & 1 \\ 0 & 0 & 1/l & 0 & 0 & 0 \\ 0 & 0 & 1/l & 0 & 1 & 0 \end{bmatrix} \tag{7–9b}
$$

The geometric stiffness matrix $\mathbf{k}_G$ associated with the motion of the member bound reference is

$$
\mathbf{k}_G = V_1 \mathbf{c}_1 = V_1 \frac{\partial^2 v_1}{\partial \mathbf{s} \partial \mathbf{s}^T} \tag{7–10}
$$

For the cantilever type

$$
\mathbf{c}_1 = \begin{array}{c} \\ \end{array}
\begin{array}{cccccccccccc} s_1^a & s_2^a & s_3^a & s_4^a & s_5^a & s_6^a & s_1^b & s_2^b & s_3^b & s_4^b & s_5^b & s_6^b \end{array}
$$

$$
\mathbf{c}_1 = \left[\begin{array}{c|c|c|c|c|c|c|c|c|c|c|c}
 & & & & & & & & & & & \\ \hline
 & & & & & -1 & & & & & & \\ \hline
 & & & & 1 & & & & & & & \\ \hline
 & & & & & & & & & & & \\ \hline
 & & 1 & & -l & & & & -1 & & & \\ \hline
 & -1 & & & & -l & & 1 & & & & \\ \hline
 & & & & & & & & & & & \\ \hline
 & & & & & 1 & & & & & & \\ \hline
 & & & & -1 & & & & & & & \\ \hline
 & & & & & & & & & & & \\ \hline
 & & & & & & & & & & & \\ \hline
 & & & & & & & & & & &
\end{array}\right] \tag{7–11}
$$

and for the alternate type,

$c_1 = \frac{1}{l}$

s_1^a	s_2^a	s_3^a	s_4^a	s_5^a	s_6^a	s_1^b	s_2^b	s_3^b	s_4^b	s_5^b	s_6^b
	1						−1				
		1						−1			
	−1						1				
		−1						1			

(7–12)

The three alternate formulations of governing equations seen in Chapter 8 are directly applicable to space structures and so are the methods of analysis developed in Chapter 12 for the stiffness and flexibility methods, and the linearized stability analysis developed in Chapter 13. It should be kept in mind, however, that the present theory only takes into account the effect of the axial force in investigating alternate equilibrium configurations. In Chapter 15 geometric nonlinearity is formulated for large rotations of the member-bound reference.

EXERCISES

Section 2

1. The structure in Fig. P1 is shown in projection on the (X, Z) and (X, Y) planes, respectively. Assume all members have a principal plane of inertia

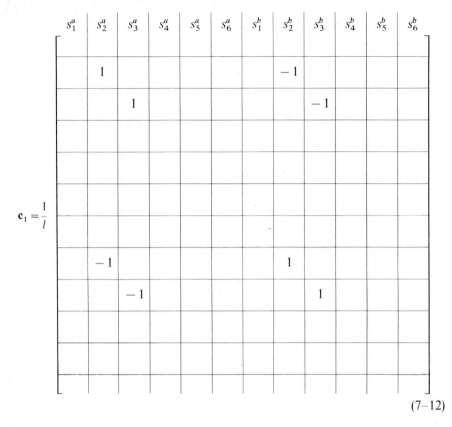

containing the member axis and the Y axis, so that $\beta = 0$. Determine the coordinate transformation matrix $\boldsymbol{\lambda}$ for each member.

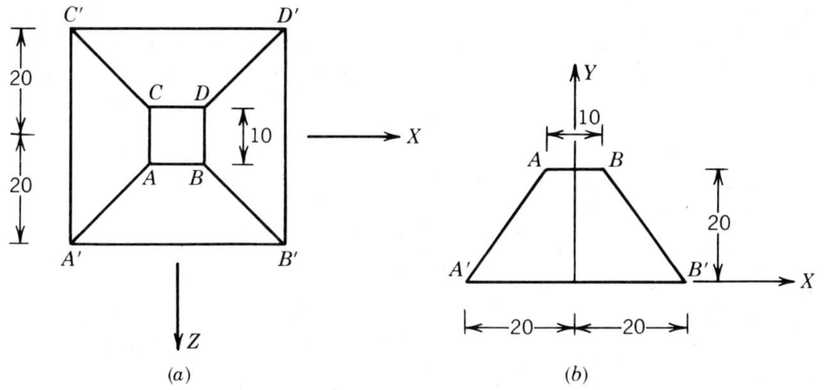

FIG. P1

2. Write a program to form the coordinate transformation matrix $\boldsymbol{\lambda}$ given the angle β and the position coordinates of the member ends.

3. Reformulate the procedure for determining the coordinate transformation matrix $\boldsymbol{\lambda}$ presented in Section 2 if the Z axis instead of the Y axis is used to define the angle β; keep the global axes a right-handed system.

Sections 3 to 5

4. Determine the member end forces in the local reference frame for the members in Figs. P4a and P4b.

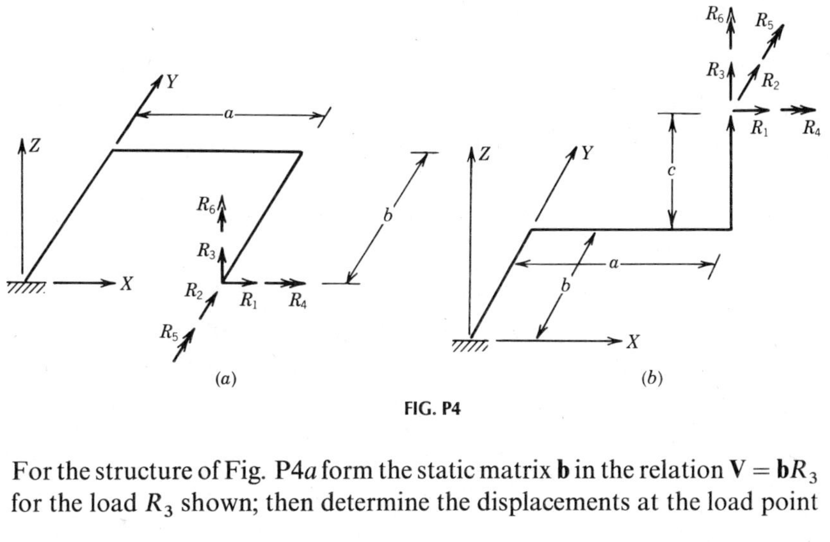

FIG. P4

5. For the structure of Fig. P4a form the static matrix $\mathbf{b}$ in the relation $\mathbf{V} = \mathbf{b}R_3$ for the load R_3 shown; then determine the displacements at the load point

due to R_3. Assume all members are uniform and inextensible with properties EI_2, EI_3, and GJ where EI_3 refers to bending in the plane of the frame.

6. Do an exercise similar to Exercise 5 for the structure in Fig. P4*b* and for the load R_2. Assume EI_3 refers to bending in the plane (X, Y) for the members lying in that plane and for bending in the plane (X, Z) for the remaining member.

7. Let y and z be the principal axes of inertia of the cross section in Fig. P7. Let $\mathbf{M} = \{M_y \ M_z\}, \boldsymbol{\chi} = \{\chi_y \ \chi_z\}, \mathbf{M} = \mathscr{k}\boldsymbol{\chi}$, and $\mathscr{k} = \lceil EI_y \ EI_z \rfloor$. Treating $(\mathbf{M}, \boldsymbol{\chi})$ as a static-kinematic pair (Chapter 7), transform $\mathbf{M}$, $\boldsymbol{\chi}$, and $\mathscr{k}$ to axes y', z'.

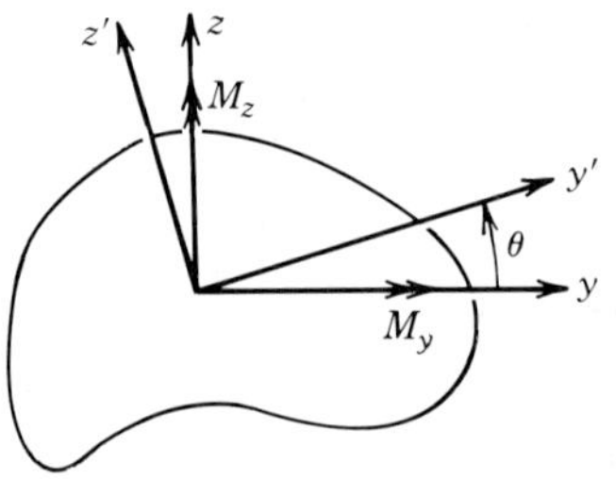

FIG. P7

8. Obtain the detailed expressions of the flexibility and stiffness coefficients in Eqs. (3.4–7) and (3.4–15b).

9. Write a program to form the complete stiffness matrix in global coordinates given the data of Exercise 2 and the cross section properties EA, EI_2, EI_3, GJ_e, y_s, and z_s.

10. Form effective flexibility and reduced stiffness matrices for a member having one rotational release at one end. Consider in turn releases about the directions of the three local axes.

11. Do the preceding exercise for a complete rotational release.

12. For the members in the two preceding exercises, identify effective variables and the corresponding static-kinematic matrix $\mathbf{B}$ and coordinate transformation matrix $\boldsymbol{\lambda}$. Define the matrix operations for obtaining the effective stiffness matrix in global coordinates, starting with the flexibility matrix of the member.

13. Figure P13 shows a rectangular grid (Section 4.2). Establish the three stiffness equations associated with the degrees of freedom of joint 1. Assume all members have properties EI and GJ.

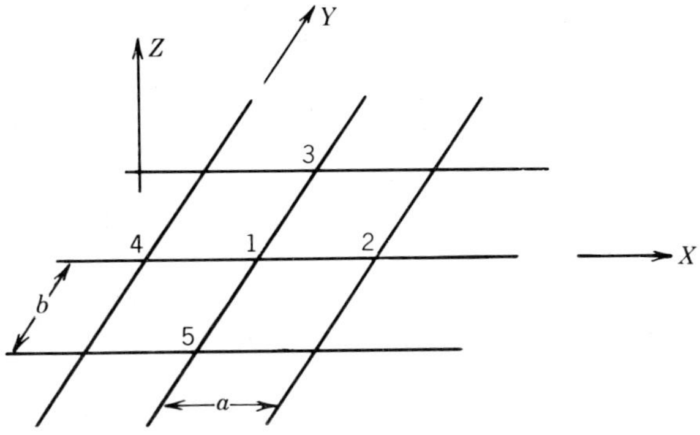

FIG. P13

14. Make the assemblage of Fig. P1 into a Schwedler dome, and devise a stable statically determinate support system. Determine the member axial forces for a load P at A acting in the negative Y direction. Do the same for equal loads P applied at A, B, C, and D.

15. Write a program to form the connectivity matrix $\mathbf{a}_v$ for a space truss by assembling member matrices $\bar{\mathbf{B}}^{mT}$. Use as input data the member joint connectivity table and the joint coordinates.

16. Write a program for analysis of a statically determinate space truss by the force method, using the program of Exercise 15 as a subroutine.

17. The structure of Fig. P17 is formed of four identical columns fixed at the base and rigidly connected to an infinitely rigid platform. Assuming the principal planes of the columns to be parallel to the (X, Z) and (Y, Z) planes, respectively, derive the stiffness equations of the structure using as generalized displacements the rigid body displacements of the platform. Outline how the load term is formed from an arbitrarily prescribed loading.

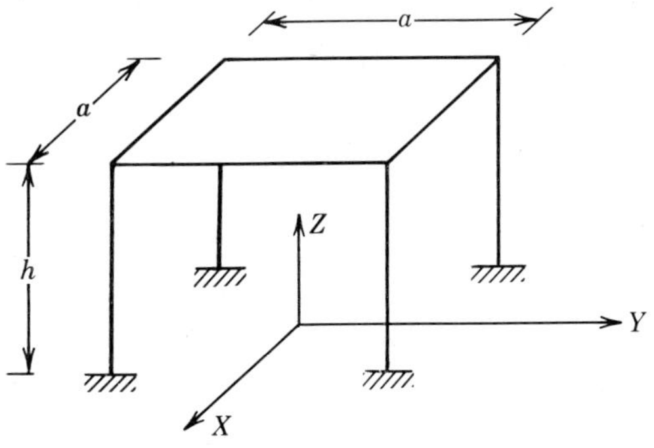

FIG. P17

18. Determine whether the support systems in Fig. P18 are stable.

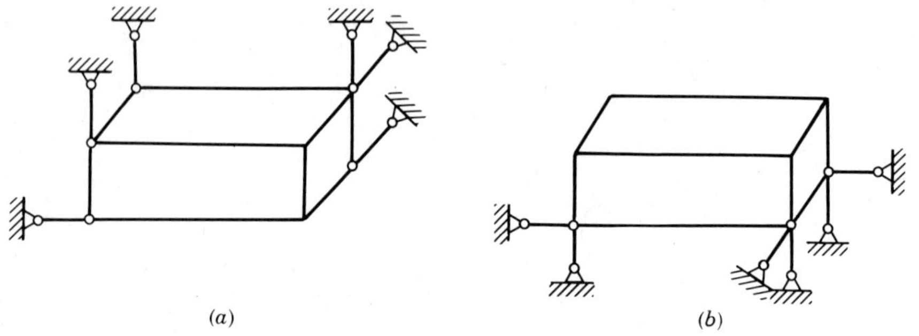

FIG. P18

19. Show that if a six-link support system is unstable, it is also statically indeterminate.

20. Determine the forces in the support links of Fig. P20.

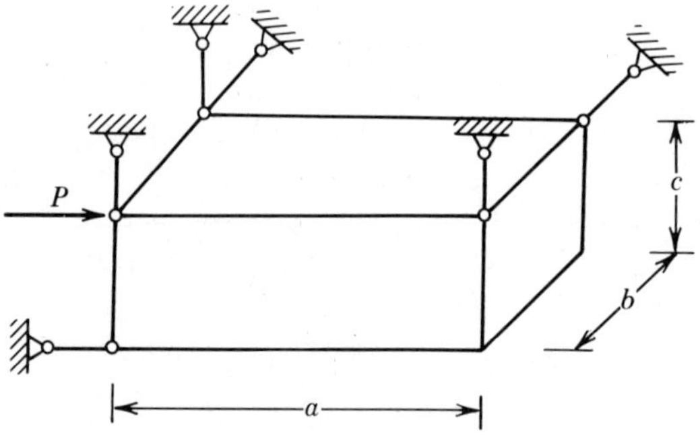

FIG. P20

21. Let the structure of Exercise 17 be modified so that all joints become universal joints, and let the four vertical faces of the structure be braced by taught diagonal cables. Discuss the stability of the structure by a geometric approach.

Section 6

22. In the truss of Fig. P22 members DB and DC are infinitely rigid, and member DA has an axial stiffness k. Form the stiffness equations of the structure in the second-order theory, and determine the critical load in terms of the given data.

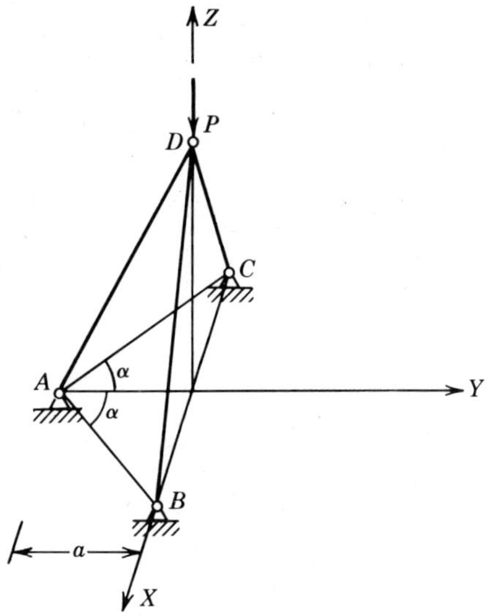

FIG. P22

15

GENERAL NONLINEAR THEORY OF THE STRUCTURAL MEMBER

1. INTRODUCTION

A geometrically nonlinear theory for large deformations and displacements is developed in this chapter for the structural member of a space structure. The approach is similar to that followed in Chapter 2 for planar behavior. Because finite rigid body rotation becomes mathematically more complex in spatial behavior, it is treated as a separate subject in Sections 2 to 4. It is then applied in Section 5 to formulate an exact deformation theory based on the extended Euler-Bernoulli assumption. Sections 6, 7, and 8 complete the exact static-kinematic theory. A simplified second-order theory consistent with the principle of virtual displacements is derived in Sections 9 and 10. Governing differential equations are established in Section 11 and are applied to stability analysis in Section 12. A variational formulation is developed in Section 13 and is applied to stability analysis in Sections 14 and 15. The method of virtual forces in linearized stability analysis is developed in Section 16.

2. FINITE ROTATION

Consider a finite rotation of angle ψ about an axis oriented by a unit vector $\vec{k}$. A vector $\vec{e}_0$ is rotated into vector $\vec{e}$, as shown in Fig. 2–1. To a positive ψ corresponds a rotation about $\vec{k}$ according to the right-hand rule. $\psi\vec{k}$ is called Rodriguez' rotation vector.

For expressing $\vec{e}$ in terms of $\vec{e}_0$, ψ, and $\vec{k}$, it will be convenient to define the modified rotation vector [16]

$$\vec{\varphi} = 2\tan\frac{\psi}{2}\vec{k} \tag{2–1}$$

and to determine $\overrightarrow{OB}$. From Fig. 2–1a, $\overrightarrow{OB} = \vec{e}_0 + \overrightarrow{A_0B} = \vec{e} + \overrightarrow{AB}$. The geometry of the vector product allows to write $\overrightarrow{A_0B} = \frac{1}{2}\vec{\varphi}\times\vec{e}_0$, and $\overrightarrow{AB} = -\frac{1}{2}\vec{\varphi}\times\vec{e}$. Thus

$$\overrightarrow{OB} = \vec{e}_0 + \tfrac{1}{2}\vec{\varphi}\times\vec{e}_0 = \vec{e} - \tfrac{1}{2}\vec{\varphi}\times\vec{e} \tag{2–2}$$

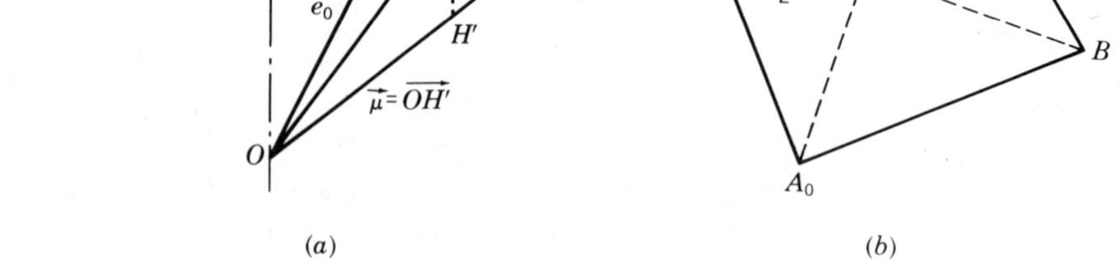

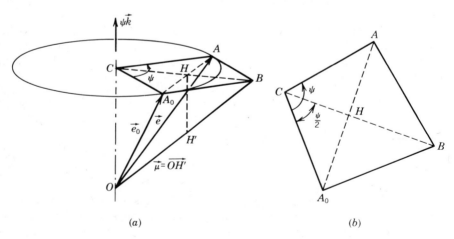

FIG. 2–1. Finite rotation.

We can write $\vec{e} - \vec{e}_0 = \overrightarrow{A_0 A} = \vec{\varphi} \times \overrightarrow{CH} = \cos^2(\psi/2)\vec{\varphi} \times \overrightarrow{CB} = \cos^2(\psi/2)\vec{\varphi} \times \overrightarrow{OB}$. With $\overrightarrow{OB}$ as determined before, we obtain

$$\vec{e} = \vec{e}_0 + \cos^2\frac{\psi}{2}\vec{\varphi} \times \left(\vec{e}_0 + \frac{1}{2}\vec{\varphi} \times \vec{e}_0\right) \tag{2–3}$$

Equation (2–3) may be written in the form

$$\vec{e} = \vec{e}_0 - \vec{\mu} \times \vec{\varphi} \tag{2–4}$$

where

$$\vec{\mu} = \cos^2\frac{\psi}{2}\overrightarrow{OB} = \cos^2\frac{\psi}{2}\left(\vec{e}_0 + \frac{1}{2}\vec{\varphi} \times \vec{e}_0\right) = \cos^2\frac{\psi}{2}\left(\vec{e} - \frac{1}{2}\vec{\varphi} \times \vec{e}\right) \tag{2–5}$$

$\vec{\mu}$ is shown in Fig. 2–1.

The rotation $\vec{\varphi}$ is now applied to a triad of vectors $\vec{\mathbf{e}}_0 = \{\vec{e}_{1,0}\ \vec{e}_{2,0}\ \vec{e}_{3,0}\}$. The rotated triad $\vec{\mathbf{e}} = \{\vec{e}_1\ \vec{e}_2\ \vec{e}_3\}$ is obtained by writing Eq. (2–4) for each vector. In matrix form

$$\vec{\mathbf{e}} = \vec{\mathbf{e}}_0 - \vec{\boldsymbol{\mu}} \times \vec{\varphi} \tag{2–6}$$

where

$$\vec{\boldsymbol{\mu}} = \{\vec{\mu}_1\ \vec{\mu}_2\ \vec{\mu}_3\} \quad \text{and}$$

$$\vec{\boldsymbol{\mu}} = \cos^2\frac{\psi}{2}\left(\vec{\mathbf{e}}_0 - \frac{1}{2}\vec{\mathbf{e}}_0 \times \vec{\varphi}\right) = \cos^2\frac{\psi}{2}\left(\vec{\mathbf{e}} + \frac{1}{2}\vec{\mathbf{e}} \times \vec{\varphi}\right) \tag{2–7}$$

The triad $\vec{\mathbf{e}}_0$ is assumed orthonormal and right handed. Thus the rotated triad $\vec{\mathbf{e}}$ as well as $\vec{\mathbf{e}}_0$ satisfy the relations

$$\vec{\mathbf{e}} \cdot \vec{\mathbf{e}}^T = \mathbf{I} \tag{2–8a}$$

$$\vec{\mathbf{e}} \times \vec{\mathbf{e}}^T = \begin{bmatrix} 0 & \vec{e}_3 & -\vec{e}_2 \\ -\vec{e}_3 & 0 & \vec{e}_1 \\ \vec{e}_2 & -\vec{e}_1 & 0 \end{bmatrix} \tag{2–8b}$$

The component representation of $\vec{\mathbf{e}}$ on $\vec{\mathbf{e}}_0$ and some results associated with the choice of $\vec{\mathbf{e}}_0$ and $\vec{\mathbf{e}}$ as base vectors are derived in what follows after adopting an appropriate notation for the component representation of the vector product.

Component Representation

For any vector $\vec{a} = \mathbf{a}^T\vec{\mathbf{e}}_0 = [a_1\ a_2\ a_3]\vec{\mathbf{e}}_0$, we introduce the notation

$$\vec{\mathbf{e}}_0 \times \vec{a} = -\underset{\sim}{\mathbf{a}}\vec{\mathbf{e}}_0 \tag{2–9}$$

where

$$\underset{\sim}{\mathbf{a}} = -\underset{\sim}{\mathbf{a}}^T = \begin{bmatrix} 0 & a_3 & -a_2 \\ -a_3 & 0 & a_1 \\ a_2 & -a_1 & 0 \end{bmatrix} \tag{2–10}$$

The vector product $\vec{a} \times \vec{b}$ is evaluated as $\mathbf{a}^T\vec{\mathbf{e}}_0 \times \vec{b}$, or as $-\mathbf{b}^T\vec{\mathbf{e}}_0 \times \vec{a}$, and takes the form

$$\vec{a} \times \vec{b} = (\underset{\sim}{\mathbf{b}}\mathbf{a})^T\vec{\mathbf{e}}_0 = -(\underset{\sim}{\mathbf{a}}\mathbf{b})^T\vec{\mathbf{e}}_0 \tag{2–11}$$

Useful relations are obtained by expressing the components of the vector expressions $(\vec{\mathbf{e}}_0 \times \vec{a}) \times \vec{b}$ and $\vec{\mathbf{e}}_0 \times (\vec{a} \times \vec{b})$, and those of the vector relation

$$(\vec{a} \times \vec{b}) \times \vec{c} = (\vec{a} \times \vec{c}) \times \vec{b} - (\vec{b} \times \vec{c}) \times \vec{a} \tag{2–12}$$

We thus obtain

$$\underset{\sim}{\mathbf{a}}\underset{\sim}{\mathbf{b}} = \mathbf{b}\mathbf{a}^T - (\mathbf{a}^T\mathbf{b})\mathbf{I} \tag{2–13a}$$

$$\underset{\sim}{\mathbf{ab}} = \mathbf{b}\mathbf{a}^T - \mathbf{a}\mathbf{b}^T \tag{2–13b}$$

$$\mathbf{a}^T\underset{\sim}{\mathbf{b}}\underset{\sim}{\mathbf{c}} = \mathbf{a}^T\underset{\sim}{\mathbf{c}}\underset{\sim}{\mathbf{b}} - \mathbf{b}^T\underset{\sim}{\mathbf{c}}\underset{\sim}{\mathbf{a}} \tag{2–13c}$$

We now obtain the component representations of $\vec{\varphi}$ and $\vec{\boldsymbol{\mu}}$ on $\vec{\mathbf{e}}_0$ and $\vec{\mathbf{e}}$, and the representation of $\vec{\mathbf{e}}$ on $\vec{\mathbf{e}}_0$. Being along the axis of rotation, $\vec{\varphi}$ has the same components $\boldsymbol{\varphi} = \{\varphi_1 \;\; \varphi_2 \;\; \varphi_3\}$ on $\vec{\mathbf{e}}_0$ and on $\vec{\mathbf{e}}$. Thus

$$\vec{\varphi} = \boldsymbol{\varphi}^T\vec{\mathbf{e}}_0 = \boldsymbol{\varphi}^T\vec{\mathbf{e}} \tag{2–14}$$

Substituting $\vec{\mathbf{e}}_0 \times \vec{\varphi} = -\underset{\sim}{\boldsymbol{\varphi}}\vec{\mathbf{e}}_0$ and $\vec{\mathbf{e}} \times \vec{\varphi} = -\underset{\sim}{\boldsymbol{\varphi}}\vec{\mathbf{e}}$ in Eq. (2–7), $\vec{\boldsymbol{\mu}}$ takes the representations

$$\vec{\boldsymbol{\mu}} = \boldsymbol{\mu}\vec{\mathbf{e}}_0 = \boldsymbol{\mu}^T\vec{\mathbf{e}} \tag{2–15}$$

where

$$\boldsymbol{\mu} = \cos^2\frac{\psi}{2}\left(\mathbf{I} + \frac{1}{2}\underset{\sim}{\boldsymbol{\varphi}}\right) \tag{2–16}$$

A result that will be useful in subsequent applications is

$$\boldsymbol{\mu}^{-1} = \mathbf{I} - \tfrac{1}{2}\underset{\sim}{\boldsymbol{\varphi}} + \tfrac{1}{4}\boldsymbol{\varphi}\boldsymbol{\varphi}^T \tag{2–17}$$

The product $\vec{\boldsymbol{\mu}} \times \vec{\varphi}$ may be evaluated in two ways using, respectively, the component representations on $\vec{\mathbf{e}}_0$ and on $\vec{\mathbf{e}}$. There comes

$$\vec{\boldsymbol{\mu}} \times \vec{\varphi} = -\boldsymbol{\mu}\underset{\sim}{\boldsymbol{\varphi}}\vec{\mathbf{e}}_0 = -\boldsymbol{\mu}^T\underset{\sim}{\boldsymbol{\varphi}}\vec{\mathbf{e}} \tag{2–18}$$

Equation (2–6) takes the form

$$\vec{\mathbf{e}} = \boldsymbol{\lambda}\vec{\mathbf{e}}_0 \tag{2–19}$$

where

$$\boldsymbol{\lambda} = \mathbf{I} + \boldsymbol{\mu}\underset{\sim}{\boldsymbol{\varphi}} = \mathbf{I} + \underset{\sim}{\boldsymbol{\varphi}}\boldsymbol{\mu} = \mathbf{I} + \cos^2\frac{\psi}{2}\left(\underset{\sim}{\boldsymbol{\varphi}} + \frac{1}{2}\underset{\sim}{\boldsymbol{\varphi}}\underset{\sim}{\boldsymbol{\varphi}}\right) \tag{2–20}$$

The inverse of Eq. (2–19) is obtainable geometrically by reversing the rotation vector and thus changing $\underset{\sim}{\boldsymbol{\varphi}}$ into $-\underset{\sim}{\boldsymbol{\varphi}}$. The result is consistent with the orthogonality property and is

$$\boldsymbol{\lambda}^{-1} = \boldsymbol{\lambda}^T = \mathbf{I} + \cos^2\frac{\psi}{2}\left(-\underset{\sim}{\boldsymbol{\varphi}} + \frac{1}{2}\underset{\sim}{\boldsymbol{\varphi}}\underset{\sim}{\boldsymbol{\varphi}}\right) \tag{2–21}$$

The term $\underset{\sim}{\boldsymbol{\varphi}}\underset{\sim}{\boldsymbol{\varphi}}$ is expressed using Eq. (2–13a) in the form

$$\underset{\sim}{\boldsymbol{\varphi}}\underset{\sim}{\boldsymbol{\varphi}} = \boldsymbol{\varphi}\boldsymbol{\varphi}^T - \varphi^2\mathbf{I} \tag{2–22a}$$

where

$$\varphi^2 = \boldsymbol{\varphi}^T\boldsymbol{\varphi} = \varphi_1^2 + \varphi_2^2 + \varphi_3^2 = 4\tan^2\frac{\psi}{2} \tag{2–22b}$$

Noting that

$$\cos^2\frac{\psi}{2} = \left(1 + \tan^2\frac{\psi}{2}\right)^{-1} = \frac{4}{4+\varphi^2} \tag{2–23}$$

Eq. (2–20) becomes

$$\boldsymbol{\lambda} = \mathbf{I} + \frac{2}{4+\varphi^2}\begin{bmatrix} -(\varphi_2^2+\varphi_3^2) & 2\varphi_3 + \varphi_1\varphi_2 & -2\varphi_2 + \varphi_1\varphi_3 \\ -2\varphi_3 + \varphi_1\varphi_2 & -(\varphi_3^2+\varphi_1^2) & 2\varphi_1 + \varphi_2\varphi_3 \\ 2\varphi_2 + \varphi_1\varphi_3 & -2\varphi_1 + \varphi_2\varphi_3 & -(\varphi_1^2+\varphi_2^2) \end{bmatrix} \tag{2–24}$$

The component representation of a vector $\vec{a}$ on $\vec{\mathbf{e}}_0$ and $\vec{\mathbf{e}}$ is defined as

$$\vec{a} = \mathbf{a}^T\vec{\mathbf{e}}_0 = \mathbf{a}'^T\vec{\mathbf{e}} \tag{2–25}$$

The transformation law between $\mathbf{a}$ and $\mathbf{a}'$ is obtained by substituting $\vec{\mathbf{e}} = \boldsymbol{\lambda}\vec{\mathbf{e}}_0$ in the preceding relation. There comes

$$\mathbf{a}' = \boldsymbol{\lambda}\mathbf{a} \tag{2-26}$$

$\boldsymbol{\varphi}$ and $\boldsymbol{\mu}$ satisfy particular relations that are obtained from Eqs. (2–14) and (2–15) by substituting in turn $\vec{\mathbf{e}} = \boldsymbol{\lambda}\vec{\mathbf{e}}_0$ and $\vec{\mathbf{e}}_0 = \boldsymbol{\lambda}^T\vec{\mathbf{e}}$. These relations are

$$\boldsymbol{\varphi} = \boldsymbol{\lambda}^T\boldsymbol{\varphi} = \boldsymbol{\lambda}\boldsymbol{\varphi} \tag{2-27}$$

$$\boldsymbol{\mu} = \boldsymbol{\mu}^T\boldsymbol{\lambda} = \boldsymbol{\lambda}\boldsymbol{\mu}^T \tag{2-28}$$

3. INCREMENTAL ROTATION

An incremental rotation is an infinitesimal rotation superimposed on a finite rotation. Let $\vec{\varphi}$ be a rotation vector rotating $\vec{\mathbf{e}}_0$ into $\vec{\mathbf{e}}$, and let $\delta\vec{\varphi}$ be an infinitesimal variation of $\vec{\varphi}$. The infinitesimal rigid body motion $\delta\vec{\mathbf{e}}$ of $\vec{\mathbf{e}}$ has an infinitesimal rotation vector $\delta\vec{\omega}$ with the property

$$\delta\vec{\mathbf{e}} = -\vec{\mathbf{e}} \times \delta\vec{\omega} \tag{3-1}$$

The expression of $\delta\vec{\omega}$ in terms of $\delta\vec{\varphi}$ and $\vec{\varphi}$ is derived in what follows and is obtained in Eq. (3–5).

To express $\delta\vec{\mathbf{e}}$, we proceed from Eq. (2–2) applied to the triads $\vec{\mathbf{e}}_0$ and $\vec{\mathbf{e}}$. From Eq. (2–2),

$$\vec{\mathbf{e}} - \vec{\mathbf{e}}_0 = -\tfrac{1}{2}(\vec{\mathbf{e}} + \vec{\mathbf{e}}_0) \times \vec{\varphi} \tag{3-2}$$

Variation of both sides of Eq. (3–2) yields

$$\delta\vec{\mathbf{e}} = -\tfrac{1}{2}\delta\vec{\mathbf{e}} \times \vec{\varphi} - \tfrac{1}{2}(\vec{\mathbf{e}} + \vec{\mathbf{e}}_0) \times \delta\vec{\varphi} \tag{3-3}$$

The term $\delta\vec{\mathbf{e}} \times \vec{\varphi}$ in this equation is evaluated using the expression of $\delta\vec{\mathbf{e}}$, given by that same equation, and developing the double vector product $(\delta\vec{\mathbf{e}} \times \vec{\varphi}) \times \vec{\varphi}$. There comes

$$-\tfrac{1}{2}\delta\vec{\mathbf{e}} \times \vec{\varphi} = \tfrac{1}{4}(\delta\vec{\mathbf{e}}\cdot\vec{\varphi})\vec{\varphi} - \tfrac{1}{4}\varphi^2\delta\vec{\mathbf{e}} + \tfrac{1}{4}[(\vec{\mathbf{e}} + \vec{\mathbf{e}}_0) \times \delta\vec{\varphi}] \times \vec{\varphi}$$

To evaluate $(\delta\vec{\mathbf{e}}\cdot\vec{\varphi})\vec{\varphi}$, we have $(\vec{\mathbf{e}} - \vec{\mathbf{e}}_0)\cdot\vec{\varphi} = 0$ and thus $\delta\vec{\mathbf{e}}\cdot\vec{\varphi} = -(\vec{\mathbf{e}} - \vec{\mathbf{e}}_0)\cdot\delta\vec{\varphi}$. Then

$$(\delta\vec{\mathbf{e}}\cdot\vec{\varphi})\vec{\varphi} = -[(\vec{\mathbf{e}} - \vec{\mathbf{e}}_0)\cdot\delta\vec{\varphi}]\vec{\varphi} = -(\vec{\mathbf{e}} - \vec{\mathbf{e}}_0) \times (\vec{\varphi} \times \delta\vec{\varphi})$$

Using these two results, Eq. (3–3) yields

$$(1 + \tfrac{1}{4}\varphi^2)\delta\vec{\mathbf{e}} = -\tfrac{1}{4}(\vec{\mathbf{e}} - \vec{\mathbf{e}}_0) \times (\vec{\varphi} \times \delta\vec{\varphi}) - \tfrac{1}{2}(\vec{\mathbf{e}} + \vec{\mathbf{e}}_0) \times \delta\vec{\varphi}$$
$$+ \tfrac{1}{4}[(\vec{\mathbf{e}} + \vec{\mathbf{e}}_0) \times \delta\vec{\varphi}] \times \vec{\varphi}$$

The last term in this equation is transformed using Eq. (2–12). The result contains the term $(\vec{\mathbf{e}} + \vec{\mathbf{e}}_0) \times \vec{\varphi}$ which is transformed using Eq. (3–2). We thus obtain after simplification

$$(1 + \tfrac{1}{4}\varphi^2)\delta\vec{\mathbf{e}} = -\vec{\mathbf{e}} \times (\delta\vec{\varphi} + \tfrac{1}{2}\vec{\varphi} \times \delta\vec{\varphi}) \tag{3–4}$$

Equation (3–4) has the form of Eq. (3–1) and yields

$$\delta\vec{\omega} = \cos^2\frac{\psi}{2}\left(\delta\vec{\varphi} + \frac{1}{2}\vec{\varphi} \times \delta\vec{\varphi}\right) \tag{3–5}$$

For a component representation of $\delta\vec{\omega}$, substitute $\delta\vec{\varphi} = \delta\boldsymbol{\varphi}^T\vec{\mathbf{e}}_0$ in Eq. (3–5), and obtain, after factoring out $\delta\boldsymbol{\varphi}^T$,

$$\delta\vec{\omega} = \cos^2\frac{\psi}{2}\delta\boldsymbol{\varphi}^T\left(\vec{\mathbf{e}}_0 - \frac{1}{2}\vec{\mathbf{e}}_0 \times \vec{\varphi}\right) \tag{3–6a}$$

Noting Eq. (2–7) the preceding result is

$$\delta\vec{\omega} = \delta\boldsymbol{\varphi}^T\vec{\boldsymbol{\mu}} \tag{3–6b}$$

Equation (3–6) represents the remarkable geometric property that $\delta\boldsymbol{\varphi}$ are the components of $\delta\vec{\omega}$ on the triad $\vec{\boldsymbol{\mu}}$. Letting $\delta\vec{\omega} = \boldsymbol{\delta\omega}^T\vec{\mathbf{e}}_0 = \boldsymbol{\delta\omega}'^T\vec{\mathbf{e}}$, and using the representation of $\vec{\boldsymbol{\mu}}$ in Eqs. (2–15) and (2–16), we obtain

$$\boldsymbol{\delta\omega} = \boldsymbol{\mu}^T\delta\boldsymbol{\varphi} = \cos^2\frac{\psi}{2}\left(\mathbf{I} - \frac{1}{2}\underset{\sim}{\boldsymbol{\varphi}}\right)\delta\boldsymbol{\varphi} \tag{3–7a}$$

$$\boldsymbol{\delta\omega}' = \boldsymbol{\mu}\delta\boldsymbol{\varphi} = \cos^2\frac{\psi}{2}\left(\mathbf{I} + \frac{1}{2}\underset{\sim}{\boldsymbol{\varphi}}\right)\delta\boldsymbol{\varphi} \tag{3–7b}$$

$\boldsymbol{\delta\omega}'$ and $\boldsymbol{\delta\omega}$ are related through the vector transformation formula

$$\boldsymbol{\delta\omega}' = \boldsymbol{\lambda\delta\omega} \tag{3–8}$$

It should be noted that whereas $\delta\boldsymbol{\varphi}$ is the differential of $\boldsymbol{\varphi}$, $\boldsymbol{\delta\omega}$ is not, in general, an exact differential. An exception occurs in plane motion.

The incremental rotation has also the representation $\delta\vec{\mathbf{e}} = \delta\boldsymbol{\lambda}\vec{\mathbf{e}}_0$. To relate $\delta\boldsymbol{\lambda}$ to $\boldsymbol{\delta\omega}$ and $\boldsymbol{\delta\omega}'$, the vector product $\delta\vec{\mathbf{e}} = -\vec{\mathbf{e}} \times \delta\vec{\omega}$ is evaluated using the component representations on $\vec{\mathbf{e}}_0$ and on $\vec{\mathbf{e}}$. There comes

$$\delta\vec{\mathbf{e}} = \boldsymbol{\lambda}\underset{\sim}{\boldsymbol{\delta\omega}}\vec{\mathbf{e}}_0 = \underset{\sim}{\boldsymbol{\delta\omega}}'\vec{\mathbf{e}} \tag{3–9}$$

whence

$$\delta\boldsymbol{\lambda} = \boldsymbol{\lambda}\underset{\sim}{\boldsymbol{\delta\omega}} = \underset{\sim}{\boldsymbol{\delta\omega}}'\boldsymbol{\lambda} \tag{3–10}$$

Consider finally a vector $\vec{u} = \mathbf{u}^T\vec{\mathbf{e}}_0 = \mathbf{u}'^T\vec{\mathbf{e}}$ that varies by $\delta\vec{u}$ as $\vec{\mathbf{e}}$ varies by $\delta\vec{\mathbf{e}}$. We have $\mathbf{u} = \boldsymbol{\lambda}^T\mathbf{u}'$, and $\delta\mathbf{u} = \boldsymbol{\lambda}^T\delta\mathbf{u}' + \delta\boldsymbol{\lambda}^T\mathbf{u}'$. Using Eq. (3–10) for $\delta\boldsymbol{\lambda}$, the transformation for incremental components is obtained as

$$\delta\mathbf{u} = \boldsymbol{\lambda}^T\delta\mathbf{u}' - \delta\underset{\sim}{\boldsymbol{\omega}}\mathbf{u} = \boldsymbol{\lambda}^T\delta\mathbf{u}' + \underset{\sim}{\mathbf{u}}\delta\boldsymbol{\omega} \tag{3–11}$$

4. RELATIVE ROTATION

Given rotation vectors $\vec{\varphi}$ and $\vec{\varphi}'$ rotating $\vec{\mathbf{e}}_0$ into $\vec{\mathbf{e}}$ and $\vec{\mathbf{e}}'$, respectively, it is proposed to determine the rotation vector $\vec{\varphi}_r$ that rotates $\vec{\mathbf{e}}$ into $\vec{\mathbf{e}}'$. From Eq. (3–2)

$$\vec{\mathbf{e}} - \vec{\mathbf{e}}_0 = -\tfrac{1}{2}(\vec{\mathbf{e}} + \vec{\mathbf{e}}_0) \times \vec{\varphi} \tag{4–1}$$

$$\vec{\mathbf{e}}' - \vec{\mathbf{e}}_0 = -\tfrac{1}{2}(\vec{\mathbf{e}}' + \vec{\mathbf{e}}_0) \times \vec{\varphi}' \tag{4–2}$$

The problem is to determine $\vec{\varphi}_r$ such that

$$\vec{\mathbf{e}}' - \vec{\mathbf{e}} = -\tfrac{1}{2}(\vec{\mathbf{e}}' + \vec{\mathbf{e}}) \times \vec{\varphi}_r \tag{4–3}$$

Forming $\vec{\mathbf{e}}' - \vec{\mathbf{e}}$ using Eqs. (4–1) and (4–2), the result may be rearranged in the form

$$\vec{\mathbf{e}}' - \vec{\mathbf{e}} = -\tfrac{1}{2}(\vec{\mathbf{e}}' + \vec{\mathbf{e}}) \times (\vec{\varphi}' - \vec{\varphi}) - \tfrac{1}{2}(\vec{\mathbf{e}}' - \vec{\mathbf{e}}_0) \times \vec{\varphi} + \tfrac{1}{2}(\vec{\mathbf{e}} - \vec{\mathbf{e}}_0) \times \vec{\varphi}'$$

The last two terms in the equation are transformed using Eqs. (4–1) and (4–2) and the transformation formula (2–12) for the double vector product. We thus obtain

$$\begin{aligned}\vec{\mathbf{e}}' - \vec{\mathbf{e}} = &-\tfrac{1}{2}(\vec{\mathbf{e}}' + \vec{\mathbf{e}}) \times (\vec{\varphi}' - \vec{\varphi} - \tfrac{1}{2}\vec{\varphi}' \times \vec{\varphi}) \\ &+ \tfrac{1}{4}[(\vec{\mathbf{e}}' - \vec{\mathbf{e}}_0) \times \vec{\varphi}] \times \vec{\varphi}' - \tfrac{1}{4}[(\vec{\mathbf{e}} - \vec{\mathbf{e}}_0) \times \vec{\varphi}'] \times \vec{\varphi}\end{aligned}$$

Developing the double vector products and noting that $(\vec{\mathbf{e}}' - \vec{\mathbf{e}}_0)\cdot\vec{\varphi}' = (\vec{\mathbf{e}} - \vec{\mathbf{e}}_0)\cdot\vec{\varphi} = 0$, the preceding equation takes the form

$$(1 + \tfrac{1}{4}\vec{\varphi}'\cdot\vec{\varphi})(\vec{\mathbf{e}}' - \vec{\mathbf{e}}) = -\tfrac{1}{2}(\vec{\mathbf{e}}' + \vec{\mathbf{e}}) \times (\vec{\varphi}' - \vec{\varphi} - \tfrac{1}{2}\vec{\varphi}' \times \vec{\varphi})$$

and determines $\vec{\varphi}_r$ as

$$\vec{\varphi}_r = \frac{1}{1 + \tfrac{1}{4}\vec{\varphi}'\cdot\vec{\varphi}}(\vec{\varphi}' - \vec{\varphi} - \tfrac{1}{2}\vec{\varphi}' \times \vec{\varphi}) \tag{4–4}$$

The case of an incremental rotation seen in the preceding section is obtained from this derivation by setting $\vec{\varphi}' = \vec{\varphi} + \delta\vec{\varphi}$. $\vec{\varphi}_r$ becomes an infinitesimal whose principal part is $\delta\vec{\omega}$ as obtained in Eq. (3–5).

Components of $\vec{\varphi}_r$ on $\vec{\mathbf{e}}_0$ are readily obtained in terms of $\boldsymbol{\varphi}$ and $\boldsymbol{\varphi}'$. Of interest is the component representation on $\vec{\mathbf{e}}$ and $\vec{\mathbf{e}}'$,

$$\vec{\varphi}_r = \boldsymbol{\varphi}_r^T \vec{\mathbf{e}} = \boldsymbol{\varphi}_r^T \vec{\mathbf{e}}' \tag{4–5}$$

We have $\vec{\varphi} = \boldsymbol{\varphi}^T \vec{\mathbf{e}}$ and $\vec{\varphi}' = \boldsymbol{\varphi}'^T \vec{\mathbf{e}}_0 = \boldsymbol{\varphi}'^T \boldsymbol{\lambda}^T \vec{\mathbf{e}}$. Thus

$$\vec{\varphi}' \times \vec{\varphi} = -\,\boldsymbol{\varphi}'^T \boldsymbol{\lambda}^T \underset{\sim}{\boldsymbol{\varphi}} \vec{\mathbf{e}}$$

The components of Eq. (4–4) on $\vec{\mathbf{e}}$ are found to be

$$\boldsymbol{\varphi}_r = \frac{1}{1 + \frac{1}{4}\boldsymbol{\varphi}^T \boldsymbol{\varphi}'}\left(\boldsymbol{\lambda}\boldsymbol{\varphi}' - \boldsymbol{\varphi} - \tfrac{1}{2}\underset{\sim}{\boldsymbol{\varphi}}\boldsymbol{\lambda}\boldsymbol{\varphi}'\right) \tag{4–6}$$

This equation contains the term $(\mathbf{I} - \frac{1}{2}\underset{\sim}{\boldsymbol{\varphi}})\boldsymbol{\lambda}\boldsymbol{\varphi}'$ in which the coefficient of $\boldsymbol{\varphi}'$ is proportional to $\boldsymbol{\mu}^T \boldsymbol{\lambda}$, which is shown in Eq. (2–28) to be equal to $\boldsymbol{\mu}$. Equation (4–6) simplifies to

$$\boldsymbol{\varphi}_r = \frac{1}{1 + \frac{1}{4}\boldsymbol{\varphi}^T \boldsymbol{\varphi}'}\left(\boldsymbol{\varphi}' - \boldsymbol{\varphi} + \tfrac{1}{2}\underset{\sim}{\boldsymbol{\varphi}}\boldsymbol{\varphi}'\right) \tag{4–7}$$

The transformation matrix $\boldsymbol{\lambda}_r$ in the relation

$$\vec{\mathbf{e}}' = \boldsymbol{\lambda}_r \vec{\mathbf{e}} \tag{4–8}$$

is expressed in terms of $\boldsymbol{\varphi}_r$ by applying Eq. (2–24), or for a more compact form, Eq. (2–20). Thus

$$\boldsymbol{\lambda}_r = \mathbf{I} + \cos^2\frac{\psi_r}{2}\left(\underset{\sim}{\boldsymbol{\varphi}}_r + \frac{1}{2}\underset{\sim}{\boldsymbol{\varphi}}_r \underset{\sim}{\boldsymbol{\varphi}}_r\right) \tag{4–9}$$

We also have $\vec{\mathbf{e}}' = \boldsymbol{\lambda}' \vec{\mathbf{e}}_0 = \boldsymbol{\lambda}' \boldsymbol{\lambda}^T \vec{\mathbf{e}}$ so that

$$\boldsymbol{\lambda}_r = \boldsymbol{\lambda}' \boldsymbol{\lambda}^T \tag{4–10}$$

The angle of rotation ψ_r is related to φ_r through Eq. (2–22b), or equivalently, Eq. (2–23). To determine $\varphi_r^2 = \vec{\varphi}_r \cdot \vec{\varphi}_r$, we have from Eq. (4–4),

$$\varphi_r^2 = \frac{1}{(1 + \frac{1}{4}\vec{\varphi}' \cdot \vec{\varphi})^2}\left(\varphi'^2 + \varphi^2 - 2\vec{\varphi}' \cdot \vec{\varphi} + \tfrac{1}{4}|\vec{\varphi}' \times \vec{\varphi}|^2\right)$$

where

$$|\vec{\varphi}' \times \vec{\varphi}|^2 = \varphi'^2 \varphi^2 - (\vec{\varphi}' \cdot \vec{\varphi})^2$$

From these results the following relation is readily established

$$1+\frac{\varphi_r^2}{4}=\frac{(1+\varphi^2/4)(1+\varphi'^2/4)}{(1+\frac{1}{4}\vec{\varphi}'\cdot\vec{\varphi})^2} \tag{4–11}$$

or noting Eq. (2–23),

$$\cos^2\frac{\psi_r}{2}=\cos^2\frac{\psi}{2}\cos^2\frac{\psi'}{2}\left(1+\tan\frac{\psi}{2}\tan\frac{\psi'}{2}\cos\theta\right)^2 \tag{4–12}$$

where $\cos\theta=\vec{k}\cdot\vec{k}'$. From Eq. (4–12)

$$\cos\frac{\psi_r}{2}=\pm\left(\cos\frac{\psi}{2}\cos\frac{\psi'}{2}+\sin\frac{\psi}{2}\sin\frac{\psi'}{2}\cos\theta\right) \tag{4–13a}$$

The preceding may also be written in the form

$$\cos\frac{\psi_r}{2}=\pm\left(\cos^2\frac{\theta}{2}\cos\frac{\psi'-\psi}{2}+\sin^2\frac{\theta}{2}\cos\frac{\psi'+\psi}{2}\right) \tag{4–13b}$$

For $-\pi<\psi_r\leqslant\pi$, the sign to be chosen in Eqs. (4–13) is the one for which $\cos\psi_r/2>0$.

Consider now infinitesimal rotations of $\vec{\mathbf{e}}$ and $\vec{\mathbf{e}}'$ whose infinitesimal rotation vectors are $\delta\vec{\omega}$ and $\delta\vec{\omega}'$, respectively. Thus $\delta\vec{\mathbf{e}}=-\vec{\mathbf{e}}\times\delta\vec{\omega}$ and $\delta\vec{\mathbf{e}}'=-\vec{\mathbf{e}}'\times\delta\vec{\omega}'$. The motion of $\vec{\mathbf{e}}'$ relative to $\vec{\mathbf{e}}$ is the difference between the absolute motion $\delta\vec{\mathbf{e}}'$ and the carrying motion of $\vec{\mathbf{e}}$ which is $-\vec{\mathbf{e}}'\times\delta\vec{\omega}$. The result is $\delta\vec{\mathbf{e}}'+\vec{\mathbf{e}}'\times\delta\vec{\omega}=-\vec{\mathbf{e}}'\times(\delta\vec{\omega}'-\delta\vec{\omega})$. Thus the infinitesimal rotation vector of the motion of $\vec{\mathbf{e}}'$ relative to $\vec{\mathbf{e}}$ is

$$\delta\vec{\omega}_r=\delta\vec{\omega}'-\delta\vec{\omega} \tag{4–14}$$

Note that $\delta\vec{\omega}_r$ and $\delta\vec{\varphi}_r$ are not related by Eq. (3–5). In order to apply Eq. (3–5), $\delta\vec{\varphi}_r$ must be interpreted as a variation of $\vec{\varphi}_r$ relative to $\vec{\mathbf{e}}$, whereas δ indicates the absolute variation, that is, the variation in the fixed reference $\vec{\mathbf{e}}_0$. The relative variation of $\vec{\varphi}_r$ is $\delta\boldsymbol{\varphi}_r^T\vec{\mathbf{e}}$. Letting $\boldsymbol{\delta\omega}_r$ be the components of $\delta\vec{\omega}_r$ on $\vec{\mathbf{e}}$, we can apply Eq. (3–7a), or

$$\boldsymbol{\delta\omega}_r=\boldsymbol{\mu}_r^T\,\delta\boldsymbol{\varphi}_r=\cos^2\frac{\psi_r}{2}\left(\mathbf{I}-\frac{1}{2}\underset{\sim}{\boldsymbol{\varphi}_r}\right)\delta\boldsymbol{\varphi}_r \tag{4–15}$$

5. DEFORMATION THEORY

5.1. Displacements

Consider a member with a fixed local reference frame defined by the orthonormal triad $\vec{\mathbf{e}}_0$. In the initial state $\vec{e}_{1,0}$ is along the member x axis, and $\vec{e}_{2,0}$ and $\vec{e}_{3,0}$

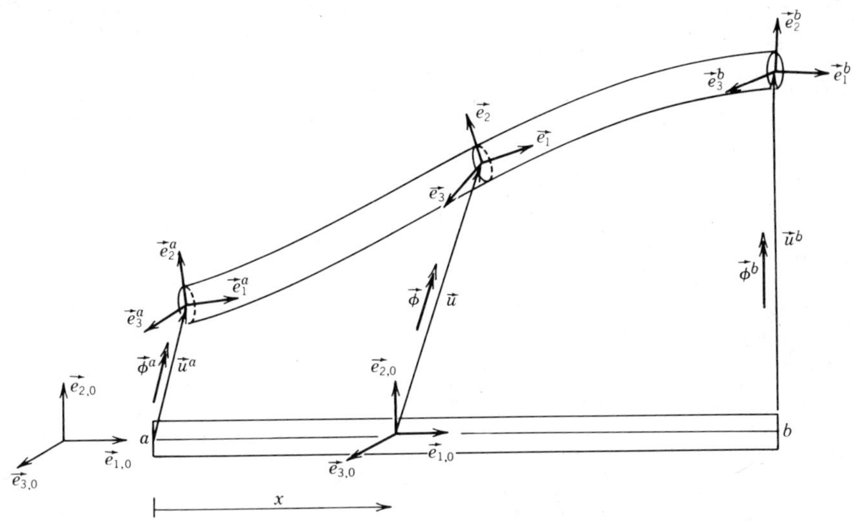

FIG. 5.1–1. Displacements: translation $\vec{u} = \mathbf{u}^T\vec{\mathbf{e}}_0$; rotation $\vec{\varphi} = \boldsymbol{\varphi}^T\vec{\mathbf{e}}_0 = \boldsymbol{\varphi}^T\vec{\mathbf{e}}$.

have some definite orientation in the plane of a cross section. The deformation theory to be developed is based on assuming that the general cross section at x displaces as a rigid body. The displaced position of a cross section is thus defined by means of a translation vector $\vec{u}$ and a rotation vector $\vec{\varphi}$, Fig. 5.1–1. $\vec{u}$ pertains to the material point initially on the x axis, which is called the reference point, and $\vec{\varphi}$ is the modified rotation vector introduced in Eq. (2–1). The deformed state of the member is defined by $\vec{u}$ and $\vec{\varphi}$ as functions of x. For any cross section a triad $\vec{\mathbf{e}}_0$ centered at its reference point is considered. A triad that is attached to the cross section and coincides with $\vec{\mathbf{e}}_0$ in the initial state displaces into $\vec{\mathbf{e}}$. The determination of the orientation of $\vec{\mathbf{e}}$ relative to $\vec{\mathbf{e}}_0$ has been the subject of Section 2. The scalar displacements at x are defined as components on $\vec{\mathbf{e}}_0$, or

$$\vec{u}, \vec{\varphi} = \mathbf{u}^T\vec{\mathbf{e}}_0, \boldsymbol{\varphi}^T\vec{\mathbf{e}}_0 \tag{5.1–1}$$

As seen in Section 2, $\vec{\varphi}$ has the same components on $\vec{\mathbf{e}}_0$ and on $\vec{\mathbf{e}}$.

A superscript is used to refer to displacements at ends a and b of the member. Member end displacements are defined as $\{\mathbf{u}^a \;\; \boldsymbol{\varphi}^a \;\; \mathbf{u}^b \;\; \boldsymbol{\varphi}^b\}$. The triads attached at a and b are $\vec{\mathbf{e}}^a$ and $\vec{\mathbf{e}}^b$, respectively. The magnitudes of $\vec{\varphi}^a$ and $\vec{\varphi}^b$ are denoted by indexes as φ_a and φ_b and their squares as φ_a^2 and φ_b^2. We have the coordinate transformation formulas

$$\vec{\mathbf{e}}^a, \vec{\mathbf{e}}^b = \boldsymbol{\lambda}^a\vec{\mathbf{e}}_0, \boldsymbol{\lambda}^b\vec{\mathbf{e}}_0 \tag{5.1–2}$$

where $\boldsymbol{\lambda}^a$ and $\boldsymbol{\lambda}^b$ are obtained by application of Eq. (2–20) or (2–24) for $\boldsymbol{\varphi}^a$ and $\boldsymbol{\varphi}^b$, respectively.

5.2. Deformations

Deformations of the cantilever type are defined as displacements relative to the triad $\vec{\mathbf{e}}^a$ attached to the cross section at a. The deformations thus defined at end b will be called member deformations.

The rigid body displacement of end b relative to $\vec{\mathbf{e}}^a$ is described by the relative translation $\vec{u}_r$ and the relative rotation $\vec{\varphi}_r$. $\vec{\varphi}_r$ is determined by applying Eq. (4–4) with $\varphi = \vec{\varphi}^a$ and $\vec{\varphi}' = \vec{\varphi}^b$. $\vec{u}_r$ is equal to $\vec{u}^b - \vec{u}'^b$, where $\vec{u}'^b$ is the carrying displacement. This is the displacement of point b' fixed in $\vec{\mathbf{e}}^a$ and coinciding with point b in the initial state. The position vector of point b' in the initial state is $l\vec{e}_{1,0}$. It is translated through $\vec{u}^a$ and rotated into $l\vec{e}_1^a$, Fig. 5.2–1. We thus have

$$\vec{u}_r = \vec{u}^b - \vec{u}^a - l(\vec{e}_1^a - \vec{e}_{1,0}) \tag{5.2–1a}$$

$$\vec{\varphi}_r = \frac{1}{1 + \frac{1}{4}\vec{\varphi}^a \cdot \vec{\varphi}^b}\left(\vec{\varphi}^b - \vec{\varphi}^a - \frac{1}{2}\vec{\varphi}^b \times \vec{\varphi}^a\right) \tag{5.2–1b}$$

Scalar deformations are defined as the components of $\vec{u}_r$ and $\vec{\varphi}_r$ on $\vec{\mathbf{e}}^a$. Since $\vec{\varphi}_r$ is the rotation vector rotating $\vec{\mathbf{e}}^a$ into $\vec{\mathbf{e}}^b$, it has the same components on $\vec{\mathbf{e}}^a$ and $\vec{\mathbf{e}}^b$. The notation is

$$\vec{u}_r = \mathbf{u}_r^T \vec{\mathbf{e}}^a \tag{5.2–2a}$$

$$\vec{\varphi}_r = \boldsymbol{\varphi}_r^T \vec{\mathbf{e}}^a = \boldsymbol{\varphi}_r^T \vec{\mathbf{e}}^b \tag{5.2–2b}$$

$\boldsymbol{\varphi}_r$ is obtained by applying Eq. (4–7) in which $\boldsymbol{\varphi}'$ is $\boldsymbol{\varphi}^b$ and $\boldsymbol{\varphi}$ is $\boldsymbol{\varphi}^a$. In taking the components of Eq. (5.2–1a) on $\vec{\mathbf{e}}^a$, displacement components $\mathbf{u}$, which are defined

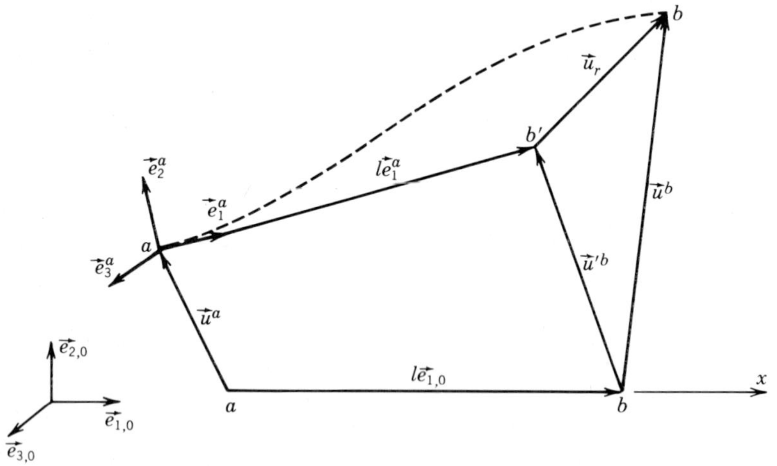

FIG. 5.2–1. Translational deformation $\vec{u}_r = \mathbf{u}_r^T \vec{\mathbf{e}}^a$.

on $\vec{\mathbf{e}}_0$, transform into $\boldsymbol{\lambda}^a\mathbf{u}$. This is applied also to $\vec{e}_{1,0}$ whose components on $\vec{\mathbf{e}}_0$ are

$$\mathbf{I}_1 = \{1 \;\; 0 \;\; 0\} \tag{5.2–3}$$

$\vec{e}_1^{\,a}$ has also $\mathbf{I}_1$ as components on $\vec{\mathbf{e}}^a$. We thus obtain

$$\mathbf{u}_r = \boldsymbol{\lambda}^a(\mathbf{u}^b - \mathbf{u}^a + l\mathbf{I}_1) - l\mathbf{I}_1 \tag{5.2–4a}$$

$$\boldsymbol{\varphi}_r = \frac{1}{1 + \frac{1}{4}\boldsymbol{\varphi}^{aT}\boldsymbol{\varphi}^b}\left(\boldsymbol{\varphi}^b - \boldsymbol{\varphi}^a + \frac{1}{2}\underset{\sim}{\boldsymbol{\varphi}}^a\boldsymbol{\varphi}^b\right) \tag{5.2–4b}$$

5.3. Strains

To define strains at cross section x, a beam slice of width dx is considered as a member ab in the treatment of the preceding section. End a is now at x and end b at $x + dx$, and $l = dx$. Superscript a is deleted, and displacements at b are obtained by incrementing displacements at a as functions of x. Thus $\vec{u}^a = \vec{u}$, $\vec{u}^b = \vec{u} + d\vec{u}$, $\vec{\varphi}^a = \vec{\varphi}$, $\vec{\varphi}^b = \vec{\varphi} + d\vec{\varphi}$, $\vec{\mathbf{e}}^a = \vec{\mathbf{e}}$, and $\boldsymbol{\lambda}^a = \boldsymbol{\lambda}$. $\vec{u}_r$ and $\vec{\varphi}_r$ in Eqs. (5.2–1) become infinitesimals whose principal parts are $\vec{\varepsilon}\,dx$ and $\vec{\chi}\,dx$, respectively. We thus obtain the strain vectors

$$\vec{\varepsilon} = \vec{u}_{,x} + \vec{e}_{1,0} - \vec{e}_1 \tag{5.3–1a}$$

$$\vec{\chi} = \frac{1}{1 + \frac{1}{4}\varphi^2}\left(\vec{\varphi}_{,x} + \frac{1}{2}\vec{\varphi} \times \vec{\varphi}_{,x}\right) \tag{5.3–1b}$$

where $(\;\;)_{,x}$ represents the derivative with respect to x. Scalar strain components are defined as components of the strain vectors on $\vec{\mathbf{e}}$,

$$\vec{\varepsilon}, \vec{\chi} = \boldsymbol{\varepsilon}^T\vec{\mathbf{e}}, \boldsymbol{\chi}^T\vec{\mathbf{e}} \tag{5.3–2}$$

There comes

$$\boldsymbol{\varepsilon} = \boldsymbol{\lambda}(\mathbf{u}_{,x} + \mathbf{I}_1) - \mathbf{I}_1 \tag{5.3–3a}$$

$$\boldsymbol{\chi} = \frac{1}{1 + \frac{1}{4}\varphi^2}\left(\boldsymbol{\varphi}_{,x} + \frac{1}{2}\underset{\sim}{\boldsymbol{\varphi}}\boldsymbol{\varphi}_{,x}\right) = \boldsymbol{\mu}\boldsymbol{\varphi}_{,x} \tag{5.3–3b}$$

The kinematic interpretation of $\vec{\chi}$ is that $\vec{\chi}dx$ is the infinitesimal rotation vector of the motion of $\vec{\mathbf{e}}$ along the deformed centerline from x to $x + dx$. Thus $d\vec{\mathbf{e}} = -\vec{\mathbf{e}} \times \vec{\chi}dx$, and

$$\vec{\mathbf{e}}_{,x} = -\vec{\mathbf{e}} \times \vec{\chi} = \underset{\sim}{\boldsymbol{\chi}}\vec{\mathbf{e}} \tag{5.3–4}$$

Letting $\vec{\mathbf{e}} = \boldsymbol{\lambda}\vec{\mathbf{e}}_0$ in this equation yields the differentiation formula for $\boldsymbol{\lambda}$,

$$\boldsymbol{\lambda}_{,x} = \underset{\sim}{\boldsymbol{\chi}}\boldsymbol{\lambda} \tag{5.3–5}$$

As obtained here, $\boldsymbol{\chi}$ coincides with that obtained by the Cosserats [17].

5.4. Incremental Deformations

Incremental deformations are differential increments or first variations of $\mathbf{u}_r$ and $\boldsymbol{\varphi}_r$, respectively, corresponding to displacement variations $\delta\mathbf{u}^a$, $\delta\boldsymbol{\varphi}^a$, $\delta\mathbf{u}^b$, $\delta\boldsymbol{\varphi}^b$. Variation of the expression of $\mathbf{u}_r$ in Eq. (5.2–4a), and use of the properties $\delta\boldsymbol{\lambda} = \boldsymbol{\lambda}\delta\underset{\sim}{\boldsymbol{\omega}}$ and $\delta\underset{\sim}{\boldsymbol{\omega}}\mathbf{u} = -\underset{\sim}{\mathbf{u}}\delta\boldsymbol{\omega}$ established in Eqs. (3–10) and (2–11), yields

$$\delta\mathbf{u}_r = \boldsymbol{\lambda}^a(\delta\mathbf{u}^b - \delta\mathbf{u}^a) - \boldsymbol{\lambda}^a(\underset{\sim}{\mathbf{u}}^b - \underset{\sim}{\mathbf{u}}^a + l\underset{\sim}{\mathbf{I}}_1)\delta\boldsymbol{\omega}^a \tag{5.4–1a}$$

This expression may also be put in the form

$$\delta\mathbf{u}_r = \boldsymbol{\lambda}^a(\delta\mathbf{u}^b - \delta\mathbf{u}^a) - (\underset{\sim}{\mathbf{u}}_r + l\underset{\sim}{\mathbf{I}}_1)\delta\boldsymbol{\omega}'^a \tag{5.4–1b}$$

where $\delta\boldsymbol{\omega}'^a$ are the components of $\delta\vec{\omega}^a$ on $\vec{\mathbf{e}}^a$. These two relations are expressed in terms of $\delta\boldsymbol{\varphi}^a$ by means of Eqs. (3–7). They become, respectively,

$$\delta\mathbf{u}_r = \boldsymbol{\lambda}^a(\delta\mathbf{u}^b - \delta\mathbf{u}^a) - \boldsymbol{\lambda}^a(\underset{\sim}{\mathbf{u}}^b - \underset{\sim}{\mathbf{u}}^a + l\underset{\sim}{\mathbf{I}}_1)\boldsymbol{\mu}^{aT}\delta\boldsymbol{\varphi}^a \tag{5.4–2a}$$

$$\delta\mathbf{u}_r = \boldsymbol{\lambda}^a(\delta\mathbf{u}^b - \delta\mathbf{u}^a) - (\underset{\sim}{\mathbf{u}}_r + l\underset{\sim}{\mathbf{I}}_1)\boldsymbol{\mu}^a\delta\boldsymbol{\varphi}^a \tag{5.4–2b}$$

$\delta\boldsymbol{\varphi}_r$ may be formed by varying the expression of $\boldsymbol{\varphi}_r$ in Eq. (5.2–4b). The result, however, is not readily put in compact matrix form. We start rather with the infinitesimal rotation vector $\delta\vec{\omega}_r$ of the incremental motion of $\vec{\mathbf{e}}^b$ relative to $\vec{\mathbf{e}}^a$. Using Eq. (4–14), the components of $\delta\vec{\omega}_r$ on $\vec{\mathbf{e}}^a$ are

$$\delta\boldsymbol{\omega}_r = \boldsymbol{\lambda}^a(\delta\boldsymbol{\omega}^b - \delta\boldsymbol{\omega}^a) = \boldsymbol{\lambda}_r^T\delta\boldsymbol{\omega}'^b - \delta\boldsymbol{\omega}'^a \tag{5.4–3}$$

where

$$\boldsymbol{\lambda}_r = \boldsymbol{\lambda}^b\boldsymbol{\lambda}^{aT} \tag{5.4–4}$$

$\delta\boldsymbol{\omega}_r$ is expressed in terms of $\delta\boldsymbol{\varphi}^a$ and $\delta\boldsymbol{\varphi}^b$ by means of Eqs. (3–7), and $\delta\boldsymbol{\varphi}_r$ is obtained from Eq. (4–15) as $\delta\boldsymbol{\varphi}_r = \boldsymbol{\mu}_r^{T-1}\delta\boldsymbol{\omega}_r$. Using the property $\boldsymbol{\lambda}\boldsymbol{\mu}^T = \boldsymbol{\mu}$, we obtain

$$\delta\boldsymbol{\omega}_r = \boldsymbol{\lambda}^a\boldsymbol{\mu}^{bT}\delta\boldsymbol{\varphi}^b - \boldsymbol{\mu}^a\delta\boldsymbol{\varphi}^a = \boldsymbol{\lambda}_r^T\boldsymbol{\mu}^b\delta\boldsymbol{\varphi}^b - \boldsymbol{\mu}^a\delta\boldsymbol{\varphi}^a \tag{5.4–5}$$

$$\delta\boldsymbol{\varphi}_r = \boldsymbol{\mu}_r^{-1}\boldsymbol{\mu}^b\delta\boldsymbol{\varphi}^b - \boldsymbol{\mu}_r^{-1T}\boldsymbol{\mu}^a\delta\boldsymbol{\varphi}^a \tag{5.4–6}$$

$\boldsymbol{\mu}_r^{-1}$ is expressed in terms of $\boldsymbol{\varphi}_r$ through Eq. (2–17).

5.5. Incremental Strains

Incremental strains $\delta\boldsymbol{\varepsilon}$ and $\delta\boldsymbol{\chi}$ are defined similarly to incremental deformations $\delta\mathbf{u}_r$ and $\delta\boldsymbol{\varphi}_r$. Their expressions in terms of $\delta\mathbf{u}$, $\delta\boldsymbol{\varphi}$, and $\boldsymbol{\delta\omega}$ may be obtained by specializing $\delta\mathbf{u}_r$ and $\delta\boldsymbol{\varphi}_r$ for a beam slice dx, or directly by variation of the strain-displacement relations. We thus obtain the alternate expressions

$$\delta\boldsymbol{\varepsilon} = \boldsymbol{\lambda}\delta\mathbf{u}_{,x} - \boldsymbol{\lambda}(\underset{\sim}{\mathbf{u}}_{,x} + \underset{\sim}{\mathbf{I}}_1)\boldsymbol{\delta\omega} \tag{5.5–1a}$$

$$\delta\boldsymbol{\varepsilon} = \boldsymbol{\lambda}\delta\mathbf{u}_{,x} - (\underset{\sim}{\boldsymbol{\varepsilon}} + \underset{\sim}{\mathbf{I}}_1)\boldsymbol{\delta\omega}' \tag{5.5–1b}$$

$\delta\boldsymbol{\varepsilon}$ is expressed in terms of $\delta\boldsymbol{\varphi}$ by means of Eqs. (3.7).

To determine $\delta\boldsymbol{\chi}$, we consider a beam slice dx for which $\boldsymbol{\varphi}_r = \boldsymbol{\chi}dx$ and thus $\delta\boldsymbol{\varphi}_r = \delta\boldsymbol{\chi}dx$. In applying Eq. (5.4–6) to such a beam slice, we have $\boldsymbol{\mu}_r = \mathbf{I} + \frac{1}{2}\underset{\sim}{\boldsymbol{\varphi}}_r = \mathbf{I} + \frac{1}{2}\underset{\sim}{\boldsymbol{\chi}}dx$, $\boldsymbol{\mu}_r^{-1} = \mathbf{I} - \frac{1}{2}\underset{\sim}{\boldsymbol{\chi}}dx$, $\boldsymbol{\mu}^a\delta\boldsymbol{\varphi}^a = \boldsymbol{\mu}\delta\boldsymbol{\varphi}$, and $\boldsymbol{\mu}^b\delta\boldsymbol{\varphi}^b = \boldsymbol{\mu}\delta\boldsymbol{\varphi} + (\boldsymbol{\mu}\delta\boldsymbol{\varphi})_{,x}dx$. The principal part of $\delta\boldsymbol{\varphi}_r$ is thus obtained as

$$\delta\boldsymbol{\chi} = (\boldsymbol{\mu}\delta\boldsymbol{\varphi})_{,x} - \underset{\sim}{\boldsymbol{\chi}}\boldsymbol{\mu}\delta\boldsymbol{\varphi} \tag{5.5–2a}$$

or

$$\delta\boldsymbol{\chi} = (\boldsymbol{\delta\omega}')_{,x} - \underset{\sim}{\boldsymbol{\chi}}\boldsymbol{\delta\omega}' \tag{5.5–2b}$$

Equation (5.5–2b) expresses the components on $\vec{\mathbf{e}}$ of the vector equation

$$\delta\boldsymbol{\chi}^T\vec{\mathbf{e}} = (\delta\vec{\omega})_{,x} \tag{5.5–3}$$

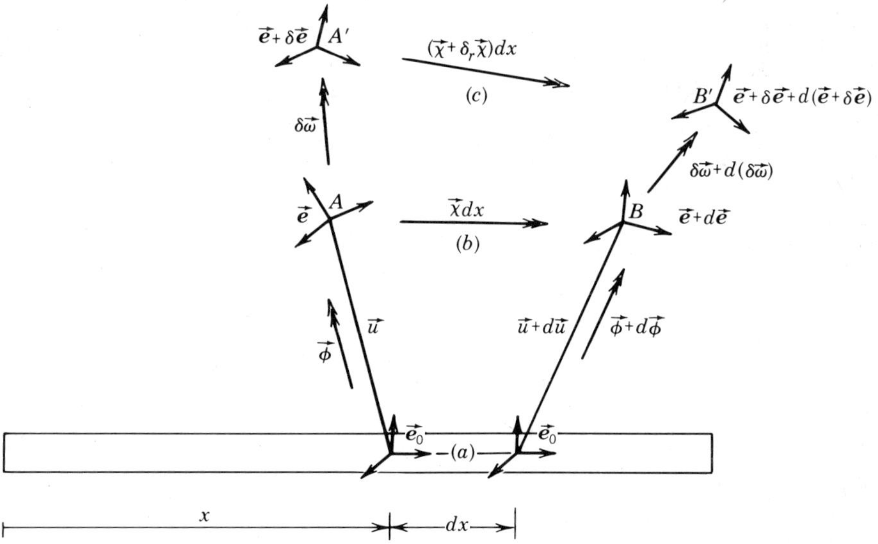

FIG. 5.5–1. Kinematic interpretation of curvature strain $\vec{\chi}$ and of incremental curvature strain $\delta_r\vec{\chi}$.

A kinematic interpretation of this equation may be seen through Fig. 5.5–1. A beam slice dx is shown in the initial state (a), in the deformed state (b), and in the incrementally deformed state (c). Differential increments as x changes by dx are indicated by the symbol d. Variations at a fixed x are indicated by the symbol δ. δ is also used to indicate that $\delta\vec{\omega}$ is infinitesimal, but it does not have in this instance the meaning of a variation. Four infinitesimal motions of the triad $\vec{\mathbf{e}}$ are identified in Fig. 5.5–1 by their respective infinitesimal rotation vectors: $\delta\vec{\omega}$ from A to A', $\delta\vec{\omega} + d(\delta\vec{\omega})$ from B to B', $\vec{\chi}\,dx$ from A to B, and $(\vec{\chi} + \delta_r\vec{\chi})dx$ from A' to B'. $\delta_r\vec{\chi}$ is seen geometrically as the variation of $\vec{\chi}$ relative to $\vec{\mathbf{e}}$ and is thus equal to the left-hand side of Eq. (5.5–3). Using the property of the rotation vector of a relative motion, Eq. (4–14), we can write $(\vec{\chi} + \delta_r\vec{\chi})dx = -\delta\vec{\omega} + \vec{\chi}\,dx + \delta\vec{\omega} + d(\delta\vec{\omega})$, and this yields $\delta_r\vec{\chi} = (\delta\vec{\omega})_{,x}$, as found in Eq. (5.5–3).

5.6. Strain Distribution over the Cross Section

The strain distribution over the cross section is needed in deriving constitutive equations for a beam slice from three-dimensional stress-strain relations. It is also needed for changing the reference point at which beam strains are defined. Limitations of the deformation theory with respect to warping restraint in torsion have been discussed in Chapter 14 and are dealt with separately.

Consider in a cross section two material points O and A. Let O be the reference point at which $\vec{u}$ is defined, and let the position vector of A be $\overrightarrow{OA} = \mathbf{r}^T\vec{\mathbf{e}}_0 = \{0 \;\; y \;\; z\}^T\vec{\mathbf{e}}_0$. The rotation of the cross section rotates $\mathbf{r}^T\vec{\mathbf{e}}_0$ into $\mathbf{r}^T\vec{\mathbf{e}}$. The displacement $\vec{u}^A$ of point A is the sum of the translation $\vec{u}$ and of the rotational displacement $\mathbf{r}^T\vec{\mathbf{e}} - \mathbf{r}^T\vec{\mathbf{e}}_0$. Components of $\vec{u}^A$ on $\vec{\mathbf{e}}_0$ are thus

$$\mathbf{u}^A = \mathbf{u} + (\boldsymbol{\lambda}^T - \mathbf{I})\mathbf{r} \tag{5.6–1}$$

This equation applies to any pair of material points in a cross section. If point A is chosen as reference point, strains $\boldsymbol{\varepsilon}'$ are obtained by applying Eq. (5.3–3a) with $\mathbf{u}^A$ replacing $\mathbf{u}$. Curvature strains depend only on $\boldsymbol{\varphi}$ and are unchanged. Taking into account Eq. (5.3–5), we obtain

$$\boldsymbol{\varepsilon}' = \boldsymbol{\varepsilon} - \underset{\sim}{\boldsymbol{\chi}}\mathbf{r} = \boldsymbol{\varepsilon} + \underset{\sim}{\mathbf{r}}\boldsymbol{\chi} \tag{5.6–2}$$

or

$$\varepsilon_1' = \varepsilon_1 + z\chi_2 - y\chi_3 \tag{5.6–2a}$$

$$\varepsilon_2' = \varepsilon_2 - z\chi_1 \tag{5.6–2b}$$

$$\varepsilon_3' = \varepsilon_3 + y\chi_1 \tag{5.6–2c}$$

The normal strain ε_1 is generally distinct from the axial strain ε. An axial material fiber $\vec{e}_{1,0}dx$ along the reference axis deforms into $(\vec{e}_1 + \vec{\varepsilon})dx$ of length $(1 + \varepsilon)dx$.

Thus

$$(1+\varepsilon)^2 = (\vec{e}_1 + \vec{\varepsilon})^2 = 1 + 2\varepsilon_1 + \varepsilon_1^2 + \varepsilon_2^2 + \varepsilon_3^2 \tag{5.6–3a}$$

or

$$\varepsilon = (1 + 2\varepsilon_1 + \varepsilon_1^2 + \varepsilon_2^2 + \varepsilon_3^2)^{1/2} - 1 \tag{5.6–3b}$$

Equations (5.6–3) are written similarly for ε' at a general point.

6. MEMBER EQUILIBRIUM

The force and moment resultants at a positive cross section are denoted $\vec{N}$ and $\vec{M}$, respectively, Fig. 6–1. The resultants acting at end b are $\vec{N}^b$ and $\vec{M}^b$, and those acting at end a are $-\vec{N}^a$ and $-\vec{M}^a$. Stress resultants $\mathbf{N}$ and $\mathbf{M}$ are defined, respectively, as components of $\vec{N}$ and $\vec{M}$ on $\vec{\mathbf{e}}$, or

$$\vec{N}, \vec{M} = \mathbf{N}^T\vec{\mathbf{e}}, \mathbf{M}^T\vec{\mathbf{e}}$$

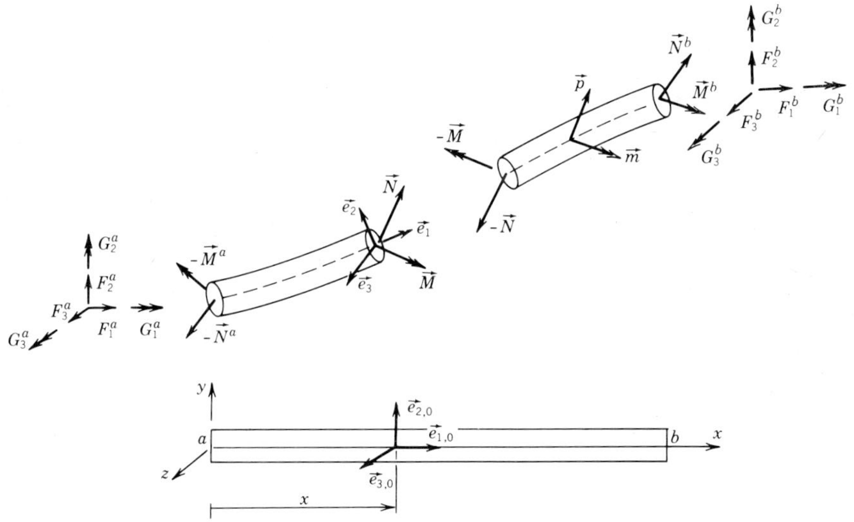

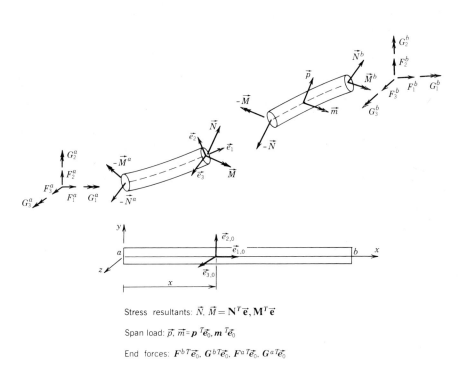

Stress resultants: $\vec{N}, \vec{M} = \mathbf{N}^T\vec{\mathbf{e}}, \mathbf{M}^T\vec{\mathbf{e}}$

Span load: $\vec{p}, \vec{m} = \boldsymbol{p}\,^T\vec{\mathbf{e}}_0, \boldsymbol{m}\,^T\vec{\mathbf{e}}_0$

End forces: $\boldsymbol{F}^{bT}\vec{\mathbf{e}}_0, \boldsymbol{G}^{bT}\vec{\mathbf{e}}_0, \boldsymbol{F}^{aT}\vec{\mathbf{e}}_0, \boldsymbol{G}^{aT}\vec{\mathbf{e}}_0$

FIG. 6–1 Stress resultants $\vec{N}, \vec{M} = \mathbf{N}^T\vec{\mathbf{e}}, \mathbf{M}^T\vec{\mathbf{e}}$
Span load intensities $\vec{p}, \vec{m} = \mathbf{p}^T\vec{\mathbf{e}}_0, \mathbf{m}^T\vec{\mathbf{e}}_0$; end forces $\mathbf{F}^{bT}\vec{\mathbf{e}}_0, \mathbf{G}^{bT}\vec{\mathbf{e}}_0, \vec{\mathbf{F}}^{aT}\vec{\mathbf{e}}_0, \vec{\mathbf{G}}^{aT}\vec{\mathbf{e}}_0$.

Member end forces in local fixed coordinates are defined as components on $\vec{\mathbf{e}}_0$, or

$$\vec{N}^b, \vec{M}^b = \mathbf{F}^{bT}\vec{\mathbf{e}}_0, \mathbf{G}^{bT}\vec{\mathbf{e}}_0$$

$$-\vec{N}^a, -\vec{M}^a = \mathbf{F}^{aT}\vec{\mathbf{e}}_0, \mathbf{G}^{aT}\vec{\mathbf{e}}_0$$

From Fig. 5.2–1 the chord vector ab in the deformed state is

$$\overrightarrow{ab} = \vec{u}_r + l\vec{e}_1^{\,a} = \vec{u}^b - \vec{u}^a + l\vec{e}_{1,0} \tag{6–1a}$$

or

$$\overrightarrow{ab} = (\mathbf{u}_r + l\mathbf{I}_1)^T\vec{\mathbf{e}}^a = (\mathbf{u}^b - \mathbf{u}^a + l\mathbf{I}_1)^T\vec{\mathbf{e}}_0 \tag{6–1b}$$

Force and moment equilibrium equations are then

$$\vec{N}^b - \vec{N}^a + \vec{P} = 0 \tag{6–2a}$$

$$\vec{M}^b - \vec{M}^a + \overrightarrow{ab} \times \vec{N}^b + \vec{Q}^a = 0 \tag{6–2b}$$

where $\vec{P}$ and $\vec{Q}^a$ are the resultants of the span load at point a. Using the second expression of $\overrightarrow{ab}$ in Eq. (6–1b), the components of the vector equilibrium equations on $\vec{\mathbf{e}}_0$ are

$$\mathbf{F}^b + \mathbf{F}^a + \mathbf{P} = 0 \tag{6–3a}$$

$$\mathbf{G}^b + \mathbf{G}^a - (\underset{\sim}{\mathbf{u}}^b - \underset{\sim}{\mathbf{u}}^a + l\underset{\sim}{\mathbf{I}}_1)\mathbf{F}^b + \mathbf{Q}^a = 0 \tag{6–3b}$$

Following the same approach as in the planar theory, Eqs. (6–3) are solved in terms of the force and moment resultants at end b, which are then the reduced forces $\mathbf{V}$ of the cantilever type. Elements of $\mathbf{V}$ are defined, however, as components $\{\mathbf{F}_r\ \mathbf{G}_r\}$ on the bound reference $\vec{\mathbf{e}}^a$, Fig. 6–2. We thus obtain

$$\begin{Bmatrix} \mathbf{F}^a \\ \mathbf{G}^a \\ \mathbf{F}^b \\ \mathbf{G}^b \end{Bmatrix} = \left[\begin{array}{c|c} -\boldsymbol{\lambda}^{aT} & \mathbf{0} \\ (\underset{\sim}{\mathbf{u}}^b - \underset{\sim}{\mathbf{u}}^a + l\underset{\sim}{\mathbf{I}}_1)\boldsymbol{\lambda}^{aT} & -\boldsymbol{\lambda}^{aT} \\ \boldsymbol{\lambda}^{aT} & \mathbf{0} \\ \mathbf{0} & \boldsymbol{\lambda}^{aT} \end{array}\right] \begin{Bmatrix} \mathbf{F}_r \\ \mathbf{G}_r \end{Bmatrix} + \begin{Bmatrix} -\mathbf{P} \\ -\mathbf{Q}^a \\ \mathbf{0} \\ \mathbf{0} \end{Bmatrix} \tag{6–4a}$$

or, in a more compact form

$$\mathbf{S} = \mathbf{B}\mathbf{V} + \mathbf{S}_p \tag{6–4b}$$

The stress resultants are determined in terms of $\mathbf{V}$ by solving equilibrium equations similar to Eqs. (6–3a), written for the portion of the beam between the

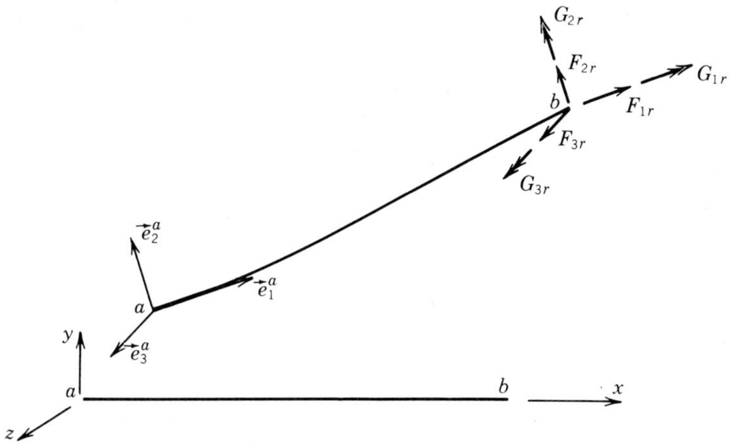

FIG. 6–2. Member reduced forces.

cross section at x and end b, and transforming force and moment components to the $\vec{\mathbf{e}}$ reference. There comes

$$\mathbf{N} = \boldsymbol{\lambda}\mathbf{F}^b + \mathbf{N}_p \tag{6–5a}$$

$$\mathbf{M} = \boldsymbol{\lambda}\mathbf{G}^b - \boldsymbol{\lambda}[\underset{\sim}{\mathbf{u}}^b - \underset{\sim}{\mathbf{u}} + (l - x)\underset{\sim}{\mathbf{I}}_1]\mathbf{F}^b + \mathbf{M}_p \tag{6–5b}$$

or, in terms of $\mathbf{V}$,

$$\mathbf{N} = \boldsymbol{\lambda}\boldsymbol{\lambda}^{aT}\mathbf{F}_r + \mathbf{N}_p \tag{6–6a}$$

$$\mathbf{M} = \boldsymbol{\lambda}\boldsymbol{\lambda}^{aT}\mathbf{G}_r - \boldsymbol{\lambda}[\underset{\sim}{\mathbf{u}}^b - \underset{\sim}{\mathbf{u}} + (l - x)\underset{\sim}{\mathbf{I}}_1]\boldsymbol{\lambda}^{aT}\mathbf{F}_r + \mathbf{M}_p \tag{6–6b}$$

$\mathbf{N}_p$ and $\mathbf{M}_p$ in these equations are the resultants of the member span load over the portion of the member between the cross section at x and end b.

Differential equations of equilibrium are derived by specializing Eqs. (6–2) to a beam slice dx. The result is

$$\vec{N}_{,x} + \vec{p} = 0 \tag{6–7a}$$

$$\vec{M}_{,x} + (\vec{\varepsilon} + \vec{e}_1) \times \vec{N} + \vec{m} = 0 \tag{6–7b}$$

where $\vec{p} = \mathbf{p}^T\vec{\mathbf{e}}_0$ and $\vec{m} = \mathbf{m}^T\vec{\mathbf{e}}_0$ are the force and moment load intensities per unit dx.

To obtain scalar equilibrium equations, we make use of Eq. (5.3–4) to write the differentiation formula

$$(\mathbf{N}^T\vec{\mathbf{e}})_{,x} = (\mathbf{N}^T_{,x} + \mathbf{N}^T\underset{\sim}{\boldsymbol{\chi}})\vec{\mathbf{e}} \tag{6–8}$$

Components on $\vec{e}$ of Eqs. (6–7) are then

$$\mathbf{N}_{,x} - \underset{\sim}{\chi}\mathbf{N} + \lambda\mathbf{p} = 0 \tag{6–9a}$$

$$\mathbf{M}_{,x} - \underset{\sim}{\chi}\mathbf{M} - (\underset{\sim}{\varepsilon} + \underset{\sim}{\mathbf{I}}_1)\mathbf{N} + \lambda\mathbf{m} = 0 \tag{6–9b}$$

7. VIRTUAL WORK: STIFFNESS PROPERTIES, STRUCTURE GOVERNING EQUATIONS

Basic ideas and methods presented in Chapter 2, Section 4, are applied here to the spatial member.

Consider an arbitrary infinitesimal variation of the deformed state $(\mathbf{u}, \boldsymbol{\varphi})$, defined by the arbitrary variations $(\delta\mathbf{u}, \delta\boldsymbol{\varphi})$. With $\delta\boldsymbol{\varphi}$ is associated the infinitesimal rotation vector $\delta\vec{\omega}$. From Eq. (3–7a),

$$\boldsymbol{\delta\omega} = \boldsymbol{\mu}^T\delta\boldsymbol{\varphi} = \frac{1}{1+\frac{1}{4}\varphi^2}\left(\mathbf{I} - \frac{1}{2}\underset{\sim}{\boldsymbol{\varphi}}\right)\delta\boldsymbol{\varphi} \tag{7–1}$$

The virtual work of the member end forces is $\mathbf{S}^T\boldsymbol{\delta}\mathbf{s}_\omega$, where

$$\boldsymbol{\delta}\mathbf{s}_\omega = \{\delta\mathbf{u}^a \;\; \boldsymbol{\delta\omega}^a \;\; \delta\mathbf{u}^b \;\; \boldsymbol{\delta\omega}^b\} = \{\delta\mathbf{u}^a \;\; \boldsymbol{\mu}^{aT}\delta\boldsymbol{\varphi}^a \;\; \delta\mathbf{u}^b \;\; \boldsymbol{\mu}^{bT}\delta\boldsymbol{\varphi}^b\} \tag{7–2}$$

For a member in equilibrium $\mathbf{S} = \mathbf{BV} + \mathbf{S}_p$. The virtual work of $\mathbf{BV}$ is $(\mathbf{BV})^T\boldsymbol{\delta}\mathbf{s}_\omega = \mathbf{V}^T\mathbf{B}^T\boldsymbol{\delta}\mathbf{s}_\omega$. As was done in Chapter 2, Section 4.1, elements of $\mathbf{B}^T\boldsymbol{\delta}\mathbf{s}_\omega$ are interpreted as components of infinitesimal translation and rotation vectors at end b relative to the member-bound reference $\vec{e}^a$, that is,

$$\boldsymbol{\delta}\mathbf{v}_\omega = \{\delta\mathbf{u}_r \;\; \boldsymbol{\delta\omega}_r\} = \mathbf{B}^T\boldsymbol{\delta}\mathbf{s}_\omega \tag{7–3}$$

With $\mathbf{B}$ as defined in Eq. (6–4), it may be verified that $\mathbf{B}^T\boldsymbol{\delta}\mathbf{s}_\omega$ is indeed formed of the expressions of $\delta\mathbf{u}_r$ and $\boldsymbol{\delta\omega}_r$ found earlier in Eqs. (5.4–1a) and (5.4–3). The virtual work of the member end forces is then

$$\mathbf{S}^T\boldsymbol{\delta}\mathbf{s}_\omega = \mathbf{V}^T\boldsymbol{\delta}\mathbf{v}_\omega + \mathbf{S}_p^T\boldsymbol{\delta}\mathbf{s}_\omega \tag{7–4}$$

Note that the infinitesimal rotations in $\boldsymbol{\delta}\mathbf{s}_\omega$ and $\boldsymbol{\delta}\mathbf{v}_\omega$ are not total differentials.

The method followed in Chapter 2, Section 4.2, by which the member is subdivided into elements and Eq. (7–4) is applied to the general element, leads to the equation of virtual work:

$$\int_0^l (\mathbf{N}^T\delta\boldsymbol{\varepsilon} + \mathbf{M}^T\delta\boldsymbol{\chi})\,dx = \mathbf{S}^T\boldsymbol{\delta}\mathbf{s}_\omega + \int_0^l (\mathbf{p}^T\delta\mathbf{u} + \mathbf{m}^T\boldsymbol{\delta\omega})\,dx \tag{7–5a}$$

in which $\boldsymbol{\delta\omega} = \boldsymbol{\mu}^T\delta\boldsymbol{\varphi}$, and similarly, $\boldsymbol{\delta}\mathbf{s}_\omega$ is expressed in terms of displacement variations by means of Eq. (7–2). Equation (7–5a) takes the form

$$\int_0^l (\mathbf{N}^T \delta\boldsymbol{\varepsilon} + \mathbf{M}^T \delta\boldsymbol{\chi})\, dx = \mathbf{S}_\varphi^T \delta\mathbf{s} + \int_0^l [\mathbf{p}^T \delta\mathbf{u} + (\boldsymbol{\mu}\mathbf{m})^T \delta\boldsymbol{\varphi}]\, dx \qquad (7\text{–}5b)$$

where

$$\mathbf{S}_\varphi = \{\mathbf{F}^a \;\; \boldsymbol{\mu}^a\mathbf{G}^a \;\; \mathbf{F}^b \;\; \boldsymbol{\mu}^b\mathbf{G}^b\} \qquad (7\text{–}6a)$$

$$\mathbf{s} = \{\mathbf{u}^a \;\; \boldsymbol{\varphi}^a \;\; \mathbf{u}^b \;\; \boldsymbol{\varphi}^b\} \qquad (7\text{–}6b)$$

As in Chapter 2, Section 4.3, Eq. (7–5) may be established analytically as a necessary and sufficient condition for equilibrium, if $\delta\boldsymbol{\varepsilon}$ and $\delta\boldsymbol{\chi}$ are kinematically compatible with arbitrary virtual displacements $(\delta\mathbf{u}, \delta\boldsymbol{\varphi})$. Also Eq. (7–5) may be established as a necessary and sufficient condition for kinematic compatibility of $\delta\boldsymbol{\varepsilon}$ and $\delta\boldsymbol{\chi}$ with $\delta\mathbf{u}$, $\delta\boldsymbol{\varphi}$, $\mathbf{u}$, and $\boldsymbol{\varphi}$ if the force system is arbitrary but statically admissible in the deformed state $(\mathbf{u}, \boldsymbol{\varphi})$. An incremental compatibility equation for $\{\delta\mathbf{u}_r \;\; \delta\boldsymbol{\omega}_r\}$ is derived by using Eq. (7–4) and a virtual self-equilibriating force system in Eq. (7–5). The incremental rotations $\delta\boldsymbol{\varphi}_r$ are then expressed in terms of $\delta\boldsymbol{\omega}_r$ by the inverse of Eq. (4–15), or

$$\delta\boldsymbol{\varphi}_r = \boldsymbol{\mu}_r^{-1T}\delta\boldsymbol{\omega}_r = (\mathbf{I} + \tfrac{1}{2}\underset{\sim}{\boldsymbol{\varphi}_r} + \tfrac{1}{4}\boldsymbol{\varphi}_r\boldsymbol{\varphi}_r^T)\delta\boldsymbol{\omega}_r \qquad (7\text{–}7)$$

Generalized Moments

The virtual work expression of the member end moments defines the moment terms $\mathbf{G}_\varphi^a = \boldsymbol{\mu}^a\mathbf{G}^a$ and $\mathbf{G}_\varphi^b = \boldsymbol{\mu}^b\mathbf{G}^b$ occurring in Eq. (7–6a) as generalized moments in Lagrange's sense of a generalized force. In general, to an external moment vector $\vec{G} = \mathbf{G}^T\vec{\mathbf{e}}_0$ applied at a cross section where the rotation vector is $\vec{\varphi} = \boldsymbol{\varphi}^T\vec{\mathbf{e}}_0$ corresponds the generalized moment vector $\vec{G}_\varphi = \mathbf{G}_\varphi^T\vec{\mathbf{e}}_0$ such that

$$\mathbf{G}_\varphi = \boldsymbol{\mu}\mathbf{G} = \cos^2\frac{\psi}{2}\left(\mathbf{I} + \frac{1}{2}\underset{\sim}{\boldsymbol{\varphi}}\right)\mathbf{G} \qquad (7\text{–}8a)$$

$$\vec{G}_\varphi = \cos^2\frac{\psi}{2}\left(\vec{G} - \frac{1}{2}\vec{\varphi} \times \vec{G}\right) \qquad (7\text{–}8b)$$

Because generalized moments are the natural quantities to work with in virtual work methods, the member equilibrium relation (6–4b) is now transformed accordingly. The virtual work expression of moment components $\mathbf{G}_r$ defined on the member-bound axes through virtual rotations relative to that reference defines

$$\mathbf{G}_{\varphi r} = \boldsymbol{\mu}_r\mathbf{G}_r = \cos^2\frac{\psi_r}{2}\left(\mathbf{I} + \frac{1}{2}\underset{\sim}{\boldsymbol{\varphi}_r}\right)\mathbf{G}_r \qquad (7\text{–}8c)$$

The virtual work of self-equilibriating member end forces $\mathbf{V}^T\boldsymbol{\delta}\mathbf{v}_\omega$ takes the form $\mathbf{V}_\varphi^T\boldsymbol{\delta}\mathbf{v}$, where $\mathbf{v}$ are the deformations

$$\mathbf{v} = \{\mathbf{u}_r \;\; \boldsymbol{\varphi}_r\} \tag{7–9a}$$

and $\mathbf{V}_\varphi$ are corresponding generalized forces

$$\mathbf{V}_\varphi = \{\mathbf{F}_r \;\; \mathbf{G}_{\varphi r}\} \tag{7–9b}$$

The equilibrium relation between $\mathbf{S}_\varphi$ and $\mathbf{V}_\varphi$ is obtained using Eqs. (7–6a), (6–4b), and the inverse of Eq. (7–8c). It takes the form

$$\mathbf{S}_\varphi = \mathbf{B}_\varphi \mathbf{V}_\varphi + \mathbf{S}_{\varphi p} \tag{7–10}$$

$\mathbf{B}_\varphi$ is a generalized static-kinematic matrix whose kinematic property is obtained from equivalence of the virtual work expressions $(\mathbf{B}_\varphi \mathbf{V}_\varphi)^T\delta\mathbf{s} = \mathbf{V}_\varphi^T\delta\mathbf{v}$, as

$$\delta\mathbf{v} = \mathbf{B}_\varphi^T\delta\mathbf{s} \tag{7–11a}$$

Equation (7–11a) is written in detail in Eqs. (5.4–2) and (5.4–6) from which

$$\mathbf{B}_\varphi^T = \frac{\partial\mathbf{v}}{\partial\mathbf{s}^T} = \begin{bmatrix} -\boldsymbol{\lambda}^a & -\boldsymbol{\lambda}^a\underset{\sim}{\boldsymbol{\Delta}}\boldsymbol{\mu}^{aT} & \boldsymbol{\lambda}^a & \mathbf{0} \\ \mathbf{0} & -\boldsymbol{\mu}_r^{T-1}\boldsymbol{\mu}^a & \mathbf{0} & \boldsymbol{\mu}_r^{-1}\boldsymbol{\mu}^b \end{bmatrix} \tag{7–11b}$$

where $\underset{\sim}{\boldsymbol{\Delta}}$ and some useful relations are given as

$$\boldsymbol{\Delta} = \mathbf{u}^b - \mathbf{u}^a + l\mathbf{I}_1 = \boldsymbol{\lambda}^{aT}(\mathbf{u}_r + l\mathbf{I}_1) \tag{7–11c}$$

$$\underset{\sim}{\boldsymbol{\Delta}} = \underset{\sim}{\mathbf{u}}^b - \underset{\sim}{\mathbf{u}}^a + l\underset{\sim}{\mathbf{I}}_1 = \boldsymbol{\lambda}^{aT}(\underset{\sim}{\mathbf{u}}_r + l\mathbf{I}_1)\boldsymbol{\lambda}^a \tag{7–11d}$$

$$\boldsymbol{\lambda}^a\underset{\sim}{\boldsymbol{\Delta}}\boldsymbol{\mu}^{aT} = (\underset{\sim}{\mathbf{u}}_r + l\underset{\sim}{\mathbf{I}}_1)\boldsymbol{\mu}^a \tag{7–11e}$$

Equations (7–10) and (7–11a) are generalizations of similar equations established in the planar theory, and $\mathbf{B}_\varphi$ plays a similar important role in the present theory.

Stiffness Properties by Castigliano's First Theorem

Considerations of actual work done along a deformation path and specialization to the case of an elastic material follow the procedure of Chapter 2, Section 5. The existence of a strain energy density function U and the principle of virtual displacements lead then to Castigliano's first theorem. Thus, if the strain energy $\bar{U}$ is expressed in terms of $\mathbf{v} = \{\mathbf{u}_r \;\; \boldsymbol{\varphi}_r\}$ as generalized displacements, stiffness relations are obtained in the form

$$\mathbf{V}_\varphi = \frac{\partial\bar{U}}{\partial\mathbf{v}} - \mathbf{Q}_{\varphi p} \tag{7–12}$$

where $\mathbf{Q}_{\varphi p}$ are the generalized forces due to the member span load.

The strain energy density function needed to apply Eq. (7–12) is determined in

Section 10. Explicit derivation of member stiffness properties for use in structural analysis is carried out in Chapter 16. As a preparation for this topic we consider how structure governing equations with spatial geometric nonlinearity are formulated.

Structure Governing Equations

Consider an assemblage of members, and assume for simplicity of presentation that all joints are rigid and are point joints. Displacements $\mathbf{r}^j$ at joint j are the components on the global axes of a translation vector $\vec{u}^j$ and a rotation vector $\vec{\varphi}^j$. Connectivity equations for the set of members have the form

$$\mathbf{s} = \mathbf{a}\mathbf{r} \tag{7–13a}$$

and the connectivity matrix $\mathbf{a}$ is the same as in the linear theory as it involves the same vector component transformation from global to local axes. The principle of virtual displacements applied to the set of joints isolated as free bodies yields the joint equilibrium equations

$$\mathbf{a}^T\mathbf{S}_\varphi = \mathbf{R}_\varphi \tag{7–13b}$$

where the moment terms in $\mathbf{R}_\varphi$ are generalized moments defined similarly to the elements of $\mathbf{S}_\varphi$. Assuming that member stiffness equations have been established by application of Castigliano's theorem, stiffness relations for $\mathbf{S}_\varphi$ are obtained by means of Eqs. (7–12) and (7–10) as

$$\mathbf{S}_\varphi = \mathbf{B}_\varphi \frac{\partial \bar{U}}{\partial \mathbf{v}} - \mathbf{B}_\varphi \mathbf{Q}_{\varphi p} + \mathbf{S}_{\varphi p} \tag{7–13c}$$

The formulation of governing equations in terms of generalized moments is preferable to a formulation in terms of the actual moments because, as will be seen subsequently, they have symmetric tangent stiffness and geometric stiffness matrices, whereas the actual moments do not have this property. Further joint moment equilibrium is expressed in terms of generalized moment vectors in the same way as in terms of actual moment vectors. This property may be deduced from Eq. (7–8b) which shows that if several moments $\vec{G}$ acting on a rigid joint have a zero sum, then the corresponding $\vec{G}_\varphi$ also have a zero sum.

Solution methods of the governing equations are usually incremental, or iterative, as seen in the planar theory. Incremental equations are formulated in the next section.

8. INCREMENTAL GOVERNING EQUATIONS: GEOMETRIC STIFFNESS

Member Tangent Stiffness in Bound Reference

Member incremental stiffness equations are essential for implementing solution procedures of nonlinear governing equations and of stability equations. It is advantageous to establish stiffness relations in the bound reference because simplifying assumptions may be made that leave arbitrary the rigid body motion of that reference. Incremental relations in the bound reference may be expressed by differentiating Eq. (7–12). We thus obtain

$$d\mathbf{V}_{\varphi} = \mathbf{k}_{\varphi} d\mathbf{v} - d\mathbf{Q}_{\varphi p} \tag{8–1}$$

where

$$\mathbf{k}_{\varphi} = \frac{\partial^2 \bar{U}}{\partial \mathbf{v} \partial \mathbf{v}^T} \tag{8–2}$$

Equation (8–2) shows that $\mathbf{k}_{\varphi}$ is symmetric. Note that the moment terms in $d\mathbf{V}_{\varphi}$ consist of generalized moments $d\mathbf{G}_{\varphi r}$. To show the difference with $d\mathbf{G}_r$, we have, by inverting Eq. (7–8c) and noting Eq. (2–17),

$$\mathbf{G}_r = \boldsymbol{\mu}_r^{-1} \mathbf{G}_{\varphi r} = (\mathbf{I} - \tfrac{1}{2}\underset{\sim}{\boldsymbol{\varphi}}_r + \tfrac{1}{4}\boldsymbol{\varphi}_r \boldsymbol{\varphi}_r^T)\mathbf{G}_{\varphi r} \tag{8–3a}$$

Differentiation of Eq. (8–3a) and rearrangement of terms yield

$$d\mathbf{G}_r = \boldsymbol{\mu}_r^{-1} d\mathbf{G}_{\varphi r} + [\tfrac{1}{2}\underset{\sim}{\mathbf{G}}_{\varphi r} + \tfrac{1}{4}\boldsymbol{\varphi}_r \mathbf{G}_{\varphi r}^T + \tfrac{1}{4}(\boldsymbol{\varphi}_r^T \mathbf{G}_{\varphi r})\mathbf{I}]d\boldsymbol{\varphi}_r \tag{8–3b}$$

It is seen that the transformation from generalized moments $d\mathbf{G}_{\varphi r}$ to moments $d\mathbf{G}_r$ results in a nonsymmetric tangent stiffness matrix.

The term $d\mathbf{Q}_{\varphi p}$ in Eq. (8–1) depends on the span load and contributes a stiffness matrix if $\mathbf{Q}_{\varphi p}$ is displacement dependent. Examples are treated in Section 14.

Member Tangent Stiffness in Fixed Reference

The problem to be addressed now is the transformation of Eq. (8–1) into complete incremental stiffness relations in the local (fixed) reference. We have, by incrementing Eq. (7–10a), $d\mathbf{S}_{\varphi} = \mathbf{B}_{\varphi} d\mathbf{V}_{\varphi} + d\mathbf{B}_{\varphi} \mathbf{V}_{\varphi} + d\mathbf{S}_{\varphi p}$. The first term is expressed using Eqs. (8–1) and (7–11a). The second term is linear in $d\mathbf{s}$ and has thus the form

$$d\mathbf{B}_{\varphi} \mathbf{V}_{\varphi} = \mathbf{k}_{G\varphi} d\mathbf{s} \tag{8–4}$$

where $\mathbf{k}_{G\varphi}$ is a geometric stiffness matrix. Noting the property $\delta\mathbf{v} = \mathbf{B}_{\varphi}^T \delta\mathbf{s}$ and following the same procedure as in the planar theory, we obtain similarly to Eq. (7/6.4–12),

$$\mathbf{k}_{G\varphi} = \sum V_{i\varphi} \mathbf{c}_i \tag{8–5a}$$

$$\mathbf{c}_i = \frac{\partial^2 v_i}{\partial \mathbf{s} \partial \mathbf{s}^T} \tag{8–5b}$$

The resulting stiffness equations take the form

$$d\mathbf{S}_{\varphi} = \mathbf{k}_{c\varphi} d\mathbf{s} + d\mathbf{S}_{\varphi F} \tag{8–6a}$$

where

$$\mathbf{k}_{c\varphi} = \mathbf{B}_\varphi \mathbf{k}_\varphi \mathbf{B}_\varphi^T + \mathbf{k}_{G\varphi} \tag{8–6b}$$

$$d\mathbf{S}_{\varphi F} = d\mathbf{S}_{\varphi p} - \mathbf{B}_\varphi d\mathbf{Q}_{\varphi p} \tag{8–6c}$$

It is noted that $\mathbf{k}_{c\varphi}$ is symmetric because $\mathbf{k}_\varphi$ and $\mathbf{k}_{G\varphi}$ are.

Structure Incremental Governing Equations

Incremental governing equations for a structure are obtained by incrementing Eqs. (7–13) and using Eqs. (8–6) for the set of members. We thus obtain

$$d\mathbf{s} = \mathbf{a}d\mathbf{r} \tag{8–7a}$$

$$\mathbf{a}^T d\mathbf{S}_\varphi = d\mathbf{R}_\varphi \tag{8–7b}$$

$$d\mathbf{S}_\varphi = \mathbf{k}_{c\varphi} d\mathbf{s} + d\mathbf{S}_{\varphi F} \tag{8–7c}$$

Equations (8–7) are a generalization of similar equations established in the planar nonlinear theory. $d\mathbf{R}_\varphi$ and $d\mathbf{S}_{\varphi F}$ may contribute stiffness matrices which depend, respectively, on the joint and member span loads. We let then

$$d\mathbf{R}_\varphi = -\mathbf{K}_R d\mathbf{r} + d\mathbf{R}_0 \tag{8–8a}$$

$$d\mathbf{S}_{\varphi F} = -\mathbf{k}_{cF} d\mathbf{s} + d\mathbf{S}_0 \tag{8–8b}$$

The assembled incremental stiffness equations take the form

$$\mathbf{K}d\mathbf{r} = d\mathbf{R} \tag{8–9}$$

where

$$\mathbf{K} = \mathbf{a}^T(\mathbf{k}_{c\varphi} + \mathbf{k}_{cF})\mathbf{a} + \mathbf{K}_R \tag{8–10a}$$

$$d\mathbf{R} = d\mathbf{R}_0 + \mathbf{a}^T d\mathbf{S}_0 \tag{8–10b}$$

Equation (8–10a) shows that $\mathbf{K}$ is formed of two parts, depending respectively on member properties and on the external load. This separation is an important practical aspect of computer applications. The part that depends only on member properties, that is, $\mathbf{a}^T\mathbf{k}_{c\varphi}\mathbf{a}$, is assembled by the direct stiffness method seen in the linear theory. Further the matrices from which $\mathbf{k}_{c\varphi}$ is formed in Eq. (8–6b) are separable in two parts. $\mathbf{k}_\varphi$ is the only matrix that depends on material properties, whereas $\mathbf{B}_\varphi$ and $\mathbf{k}_{G\varphi}$ are static-kinematic and are determined without approximations in Eqs. (7–11) and (8–5). An explicit determination of $\mathbf{k}_{G\varphi}$ suitable for numerical evaluation is made after the following comment.

To apply numerically the incremental governing equations, there remains to determine an element stiffness $\mathbf{k}_\varphi$ and to establish methods for determining the tangent stiffness of external loads. The rest of this chapter will serve this end, but it is mostly devoted to the analysis of a single member. Structural analysis by the direct stiffness method is taken up again in Chapter 16.

Determination of $\mathbf{k}_{G\varphi}$

An explicit expression of $\mathbf{k}_{G\varphi}$ obtained from Eqs. (8–5) would be too cumbersome and would involve six 12 × 12 matrices corresponding to the six deformations. A direct derivation using Eq. (8–4) leads to a compact expression of $\mathbf{k}_{G\varphi}$ and is outlined in what follows. The relation $\mathbf{S}_\varphi = \mathbf{B}_\varphi \mathbf{V}_\varphi$ yields the expressions of $\mathbf{F}^a$, $\mathbf{G}^a_\varphi$, $\mathbf{F}^b$, and $\mathbf{G}^b_\varphi$. Differentials of these expressions as functions of $\mathbf{s}$, with $\mathbf{F}_r$ and $\mathbf{G}_{\varphi r}$ fixed, are formed making use of the mathematical results of Sections 2 and 3 and, in particular, of the expressions of $\boldsymbol{\mu}$ and $\boldsymbol{\mu}^{-1}$ and of the relations $\delta\boldsymbol{\lambda} = \boldsymbol{\lambda}\delta\underset{\sim}{\boldsymbol{\omega}}$ and $\delta\boldsymbol{\omega} = \boldsymbol{\mu}^T\delta\boldsymbol{\varphi}$. Details are omitted for brevity. The result has the form

$$\begin{Bmatrix} d\mathbf{F}^a \\ d\mathbf{G}^a_\varphi \\ d\mathbf{F}^b \\ d\mathbf{G}^b_\varphi \end{Bmatrix} = \begin{bmatrix} & \mathbf{k}^{aa}_{u\varphi} & & \\ \mathbf{k}^{aa}_{\varphi u} & \mathbf{k}^{aa}_{\varphi\varphi} & \mathbf{k}^{ab}_{\varphi u} & \mathbf{k}^{ab}_{\varphi\varphi} \\ & \mathbf{k}^{ba}_{u\varphi} & & \\ & \mathbf{k}^{ba}_{\varphi\varphi} & & \mathbf{k}^{bb}_{\varphi\varphi} \end{bmatrix} \begin{Bmatrix} d\mathbf{u}^a \\ d\boldsymbol{\varphi}^a \\ d\mathbf{u}^b \\ d\boldsymbol{\varphi}^b \end{Bmatrix} \tag{8–11}$$

where

$$\mathbf{k}^{aa}_{u\varphi} = -\underset{\sim}{\mathbf{F}}^b\boldsymbol{\mu}^{aT} = -\boldsymbol{\lambda}^{aT}\underset{\sim}{\mathbf{F}}_r\boldsymbol{\mu}^a \tag{8–12a}$$

$$\mathbf{k}^{aa}_{\varphi u} = (\mathbf{k}^{aa}_{u\varphi})^T = -\mathbf{k}^{ab}_{\varphi u} = -(\mathbf{k}^{ba}_{u\varphi})^T \tag{8–12b}$$

$$\mathbf{k}^{ba}_{\varphi\varphi} = -\tfrac{1}{2}\boldsymbol{\mu}^{bT}[-\underset{\sim}{\mathbf{G}}_{\varphi r} + \tfrac{1}{2}\boldsymbol{\varphi}_r\mathbf{G}^T_{\varphi r} + \tfrac{1}{2}(\boldsymbol{\varphi}_r^T\mathbf{G}_{\varphi r})\mathbf{I}]\boldsymbol{\mu}_r^{T-1}\boldsymbol{\mu}^a \tag{8–12c}$$

$$\mathbf{k}^{ab}_{\varphi\varphi} = -\tfrac{1}{2}\boldsymbol{\mu}^{aT}[\underset{\sim}{\mathbf{G}}_{\varphi r} + \tfrac{1}{2}\boldsymbol{\varphi}_r\mathbf{G}^T_{\varphi r} + \tfrac{1}{2}(\boldsymbol{\varphi}_r^T\mathbf{G}_{\varphi r})\mathbf{I}]\boldsymbol{\mu}_r^{-1}\boldsymbol{\mu}^b \tag{8–12d}$$

$$\begin{aligned} \mathbf{k}^{bb}_{\varphi\varphi} = \tfrac{1}{2}&\left(1 + \frac{\varphi_b^2}{4}\right)^{-1} \\ &\times [\underset{\sim}{\mathbf{G}}_{\varphi r} - \tfrac{1}{2}\boldsymbol{\varphi}_r\mathbf{G}^T_{\varphi r} + \tfrac{1}{2}\mathbf{G}_{\varphi r}\boldsymbol{\varphi}_r^T + \tfrac{1}{4}(\boldsymbol{\varphi}_r^T\mathbf{G}_{\varphi r})\underset{\sim}{\boldsymbol{\varphi}}_r - \boldsymbol{\mu}^{bT}\boldsymbol{\mu}_r^{T-1}\mathbf{G}_{\varphi r}\boldsymbol{\varphi}^{bT}] \\ &+ \tfrac{1}{2}\boldsymbol{\mu}^{bT}[-\underset{\sim}{\mathbf{G}}_{\varphi r} + \tfrac{1}{2}\boldsymbol{\varphi}_r\mathbf{G}^T_{\varphi r} + \tfrac{1}{2}(\boldsymbol{\varphi}_r^T\mathbf{G}_{\varphi r})\mathbf{I}]\boldsymbol{\mu}_r^{-1}\boldsymbol{\mu}^b \end{aligned} \tag{8–12e}$$

$$\begin{aligned} \mathbf{k}^{aa}_{\varphi\varphi} = \boldsymbol{\mu}^a&(\mathbf{F}^b\boldsymbol{\Delta}^T - (\boldsymbol{\Delta}^T\mathbf{F}^b)\mathbf{I})\boldsymbol{\mu}^{aT} + \tfrac{1}{2}(1 + \tfrac{1}{4}\varphi_a^2)^{-1}[\boldsymbol{\Delta}\mathbf{F}^{bT} - \mathbf{F}^b\boldsymbol{\Delta}^T - \boldsymbol{\mu}^a\underset{\sim}{\boldsymbol{\Delta}}\mathbf{F}^b\boldsymbol{\varphi}^{aT}] \\ &- \tfrac{1}{2}\left(1 + \frac{\varphi_a^2}{4}\right)^{-1} \\ &\times [\underset{\sim}{\mathbf{G}}_{\varphi r} + \tfrac{1}{2}\boldsymbol{\varphi}_r\mathbf{G}^T_{\varphi r} - \tfrac{1}{2}\mathbf{G}_{\varphi r}\boldsymbol{\varphi}_r^T + \tfrac{1}{4}(\boldsymbol{\varphi}_r^T\mathbf{G}_{\varphi r})\underset{\sim}{\boldsymbol{\varphi}}_r - \boldsymbol{\mu}^{aT}\boldsymbol{\mu}_r^{-1}\mathbf{G}_{\varphi r}\boldsymbol{\varphi}^{aT}] \\ &+ \tfrac{1}{2}\boldsymbol{\mu}^{aT}[\underset{\sim}{\mathbf{G}}_{\varphi r} + \tfrac{1}{2}\boldsymbol{\varphi}_r\mathbf{G}^T_{\varphi r} + \tfrac{1}{2}(\boldsymbol{\varphi}_r^T\mathbf{G}_{\varphi r})\mathbf{I}]\boldsymbol{\mu}_r^{T-1}\boldsymbol{\mu}^a \end{aligned} \tag{8–12f}$$

In the last equation $\boldsymbol{\Delta}$ is defined in Eq. (7–11c) and $\mathbf{F}^b$ may be replaced with $\boldsymbol{\lambda}^{aT}\mathbf{F}_r$ if all terms are desired in terms of $\mathbf{F}_r$ and $\mathbf{G}_{\varphi r}$. Symmetry of $\mathbf{k}_{G\varphi}$ is apparent in Eqs. (8–12a, b) but is hidden in the remaining expressions. A direct verification of symmetry is possible but is long and tedious.

9. SECOND-ORDER THEORY

Kinematic Analysis

A simplification of the general static-kinematic formulation presented in the preceding sections is now carried out. In so doing, we neglect the axial strain and squares and products of rotations with respect to unity. The simplifying assumptions are applied in what follows to the absolute displacements in the fixed local reference $\vec{\mathbf{e}}_0$. If they are to be restricted to displacements relative to the bound reference, all is needed is to set $\mathbf{u}^a = 0$ and $\boldsymbol{\varphi}^a = 0$.

The simplified theory to be developed will be referred to as a second-order theory. The criterion for consistency of approximations is that the principle of virtual displacements remain satisfied.

For $\psi^2 \ll 1$ we can write $\psi = 2\tan\psi/2$. It follows that there is no distinction between the rotation vector $\psi\vec{k}$ and the modified vector $\vec{\varphi}$ defined in Eq. (2–1).

Implementation of the simplifying assumption in Eqs. (2–16) and (2–20) yields, after neglecting nonlinear terms in $\boldsymbol{\varphi}$ with respect to $\mathbf{I}$,

$$\boldsymbol{\mu} = \mathbf{I} + \tfrac{1}{2}\underset{\sim}{\boldsymbol{\varphi}} \tag{9–1}$$

$$\boldsymbol{\lambda} = \mathbf{I} + \boldsymbol{\mu}\underset{\sim}{\boldsymbol{\varphi}} = \mathbf{I} + \underset{\sim}{\boldsymbol{\varphi}} \tag{9–2}$$

However, in approximating $(\boldsymbol{\lambda} - \mathbf{I})$, second-order terms in $\boldsymbol{\varphi}$ are not negligible in comparison with the first-order terms. Thus

$$\boldsymbol{\lambda} - \mathbf{I} = \boldsymbol{\mu}\underset{\sim}{\boldsymbol{\varphi}} = (\mathbf{I} + \tfrac{1}{2}\underset{\sim}{\boldsymbol{\varphi}})\underset{\sim}{\boldsymbol{\varphi}} \tag{9–3}$$

Then, to within errors of order φ^2 with respect to unity,

$$\boldsymbol{\mu}^T = \boldsymbol{\mu}^{-1} = \mathbf{I} - \tfrac{1}{2}\underset{\sim}{\boldsymbol{\varphi}} \tag{9–4}$$

$$\boldsymbol{\lambda}^T = \boldsymbol{\lambda}^{-1} = \mathbf{I} - \underset{\sim}{\boldsymbol{\varphi}} \tag{9–5}$$

$$\boldsymbol{\delta\omega} = \boldsymbol{\mu}^T\delta\boldsymbol{\varphi} = (\mathbf{I} - \tfrac{1}{2}\underset{\sim}{\boldsymbol{\varphi}})\delta\boldsymbol{\varphi} \tag{9–6}$$

In the preceding relations

$$\underset{\sim}{\boldsymbol{\varphi}} = \begin{bmatrix} 0 & \varphi_3 & -\varphi_2 \\ -\varphi_3 & 0 & \varphi_1 \\ \varphi_2 & -\varphi_1 & 0 \end{bmatrix} \tag{9–7}$$

In simplifying the deformation-displacement relations, Eqs. (5.2–4), $\mathbf{u}_r$ has the expression $\boldsymbol{\lambda}^a(\mathbf{u}^b - \mathbf{u}^a) + l(\boldsymbol{\lambda}^a - \mathbf{I})\mathbf{I}_1$. $\boldsymbol{\lambda}^a$ and $(\boldsymbol{\lambda}^a - \mathbf{I})$ are approximated differently, as seen in Eqs. (9–2) and (9–3). The resulting expression is further simplified by the order of magnitude assumption

$$\frac{u_1^b - u_1^a}{l} \ll 1 \tag{9–8}$$

Neglecting $u_1^b - u_1^a$ with respect to l is represented by introducing the notation

$$\underset{\sim}{\boldsymbol{\varphi}}^a(\mathbf{u}^b - \mathbf{u}^a + l\mathbf{I}_1) = \underset{\sim}{\boldsymbol{\varphi}}_0^a(\mathbf{u}^b - \mathbf{u}^a) + l\underset{\sim}{\boldsymbol{\varphi}}^a\mathbf{I}_1 \tag{9–9a}$$

or, equivalently,

$$\underset{\sim}{\boldsymbol{\varphi}}^a(\mathbf{u}^b - \mathbf{u}^a + l\mathbf{I}_1) = \underset{\sim}{\boldsymbol{\varphi}}^a(\mathbf{u}_0^b - \mathbf{u}_0^a) + l\underset{\sim}{\boldsymbol{\varphi}}^a\mathbf{I}_1 \tag{9–9b}$$

where

$$\underset{\sim}{\boldsymbol{\varphi}}_0 = \begin{bmatrix} 0 & \varphi_3 & -\varphi_2 \\ 0 & 0 & \varphi_1 \\ 0 & -\varphi_1 & 0 \end{bmatrix} \tag{9–10}$$

and

$$\mathbf{u}_0 = \{0 \ u_2 \ u_3\} \tag{9–11}$$

The simplified deformation-displacement relations take the form

$$\mathbf{u}_r = \boldsymbol{\lambda}_0^a(\mathbf{u}^b - \mathbf{u}^a) + l\boldsymbol{\mu}^a\underset{\sim}{\boldsymbol{\varphi}}^a\mathbf{I}_1 \tag{9–12a}$$

$$\boldsymbol{\varphi}_r = \boldsymbol{\varphi}^b - \boldsymbol{\varphi}^a + \tfrac{1}{2}\underset{\sim}{\boldsymbol{\varphi}}^a\boldsymbol{\varphi}^b \tag{9–12b}$$

where

$$\boldsymbol{\lambda}_0 = \mathbf{I} + \underset{\sim}{\boldsymbol{\varphi}}_0 = \begin{bmatrix} 1 & \varphi_3 & -\varphi_2 \\ 0 & 1 & \varphi_1 \\ 0 & -\varphi_1 & 1 \end{bmatrix} \tag{9–13a}$$

$$\boldsymbol{\mu}\underset{\sim}{\boldsymbol{\varphi}} = \underset{\sim}{\boldsymbol{\varphi}}\boldsymbol{\mu} = \underset{\sim}{\boldsymbol{\varphi}} + \tfrac{1}{2}\underset{\sim}{\boldsymbol{\varphi}}\underset{\sim}{\boldsymbol{\varphi}} \tag{9–13b}$$

Writing out Eqs. (9–12) in detail, there comes

$$u_{1r} = u_1^b - u_1^a + \varphi_3^a\left(u_2^b - u_2^a - \frac{l}{2}\varphi_3^a\right) - \varphi_2^a\left(u_3^b - u_3^a + \frac{l}{2}\varphi_2^a\right) \tag{9–14a}$$

$$u_{2r} = u_2^b - u_2^a - l\varphi_3^a + \varphi_1^a\left(u_3^b - u_3^a + \frac{l}{2}\varphi_2^a\right) \tag{9–14b}$$

$$u_{3r} = u_3^b - u_3^a + l\varphi_2^a - \varphi_1^a\left(u_2^b - u_2^a - \frac{l}{2}\varphi_3^a\right) \tag{9–14c}$$

$$\varphi_{1r} = \varphi_1^b - \varphi_1^a + \tfrac{1}{2}(\varphi_3^a\varphi_2^b - \varphi_2^a\varphi_3^b) \tag{9–15a}$$

$$\varphi_{2r} = \varphi_2^b - \varphi_2^a + \tfrac{1}{2}(\varphi_1^a\varphi_3^b - \varphi_3^a\varphi_1^b) \tag{9–15b}$$

$$\varphi_{3r} = \varphi_3^b - \varphi_3^a + \tfrac{1}{2}(\varphi_2^a\varphi_1^b - \varphi_1^a\varphi_2^b) \tag{9–15c}$$

The strain-displacement relations, Eq. (5.3–3), are simplified similarly with the assumption $u_{1,x} \ll 1$. There comes

$$\boldsymbol{\varepsilon} = \boldsymbol{\lambda}_0 \mathbf{u}_{,x} + \boldsymbol{\mu} \underset{\sim}{\boldsymbol{\varphi}} \mathbf{I}_1 \tag{9–16a}$$

$$\boldsymbol{\chi} = \boldsymbol{\varphi}_{,x} + \tfrac{1}{2} \underset{\sim}{\boldsymbol{\varphi}} \boldsymbol{\varphi}_{,x} \tag{9–16b}$$

or, in detail,

$$\varepsilon_1 = u_{1,x} + \varphi_3(u_{2,x} - \tfrac{1}{2}\varphi_3) - \varphi_2(u_{3,x} + \tfrac{1}{2}\varphi_2) \tag{9–17a}$$

$$\varepsilon_2 = u_{2,x} - \varphi_3 + \varphi_1(u_{3,x} + \tfrac{1}{2}\varphi_2) \tag{9–17b}$$

$$\varepsilon_3 = u_{3,x} + \varphi_2 - \varphi_1(u_{2,x} - \tfrac{1}{2}\varphi_3) \tag{9–17c}$$

$$\chi_1 = \varphi_{1,x} + \tfrac{1}{2}(\varphi_3 \varphi_{2,x} - \varphi_2 \varphi_{3,x}) \tag{9–18a}$$

$$\chi_2 = \varphi_{2,x} + \tfrac{1}{2}(\varphi_1 \varphi_{3,x} - \varphi_3 \varphi_{1,x}) \tag{9–18b}$$

$$\chi_3 = \varphi_{3,x} + \tfrac{1}{2}(\varphi_2 \varphi_{1,x} - \varphi_1 \varphi_{2,x}) \tag{9–18c}$$

Equations (9–16) are consistent with Eqs. (9–12) in that they may be derived by specializing Eqs. (9–12) to a beam slice dx.

Static Analysis

Simplified expressions of $\delta\mathbf{u}_r$ and $\boldsymbol{\delta\omega}_r$ in terms of $\delta\mathbf{u}^a$, $\boldsymbol{\delta\omega}^a$, $\delta\mathbf{u}^b$, $\boldsymbol{\delta\omega}^b$ will ensure consistency of simplifications in the equilibrium equations. $\delta\mathbf{u}_r$ is conveniently obtained by simplifying the original expression (5.4–1a). $\boldsymbol{\lambda}^a$ is replaced with $\boldsymbol{\lambda}_0^a$ in the first term, and $(u_1^b - u_1^a)$ is deleted from $(\mathbf{u}^b - \mathbf{u}^a)$ in the second term. We thus obtain

$$\delta\mathbf{u}_r = \boldsymbol{\lambda}_0^a(\delta\mathbf{u}^b - \delta\mathbf{u}^a) - \boldsymbol{\lambda}^a(\underset{\sim}{\mathbf{u}}_0^b - \underset{\sim}{\mathbf{u}}_0^a + l\underset{\sim}{\mathbf{I}}_1)\boldsymbol{\delta\omega}^a \tag{9–19a}$$

$$\boldsymbol{\delta\omega}_r = \boldsymbol{\lambda}^a(\boldsymbol{\delta\omega}^b - \boldsymbol{\delta\omega}^a) \tag{9–19b}$$

These equations represent the matrix equation $\boldsymbol{\delta}\mathbf{v}_\omega = \mathbf{B}^T \boldsymbol{\delta}\mathbf{s}_\omega$. The consistent equilibrium equations $\mathbf{S} = \mathbf{B}\mathbf{V} + \mathbf{S}_p$ are therefore

$$\begin{Bmatrix} \mathbf{F}^a \\ \mathbf{G}^a \\ \mathbf{F}^b \\ \mathbf{G}^b \end{Bmatrix} = \left[\begin{array}{c|c} -\boldsymbol{\lambda}_0^{aT} & \mathbf{0} \\ (\underset{\sim}{\mathbf{u}}_0^b - \underset{\sim}{\mathbf{u}}_0^a + l\underset{\sim}{\mathbf{I}}_1)\boldsymbol{\lambda}^{aT} & -\boldsymbol{\lambda}^{aT} \\ \boldsymbol{\lambda}_0^{aT} & \mathbf{0} \\ \mathbf{0} & \boldsymbol{\lambda}^{aT} \end{array}\right] \begin{Bmatrix} \mathbf{F}_r \\ \mathbf{G}_r \end{Bmatrix} + \begin{Bmatrix} -\mathbf{P} \\ -\mathbf{Q}^a \\ \mathbf{0} \\ \mathbf{0} \end{Bmatrix} \tag{9–20}$$

Because $\boldsymbol{\lambda}_0 \neq \boldsymbol{\lambda}$, vector component transformations from the $\vec{\mathbf{e}}$ reference to the $\vec{\mathbf{e}}_0$ reference are made differently for forces and for moments. Further the orthogonality property $\boldsymbol{\lambda}^T = \boldsymbol{\lambda}^{-1}$ is not satisfied by $\boldsymbol{\lambda}_0$. Thus vector transformation laws of force and displacement components no longer coincide, but they remain governed by the contragradient law,

$$\mathbf{u}' = \boldsymbol{\lambda}_0 \mathbf{u} \tag{9–21a}$$

$$\mathbf{F} = \boldsymbol{\lambda}_0^T \mathbf{N} \tag{9–21b}$$

where $\mathbf{u}$ and $\mathbf{u}'$ are the components of a displacement vector on $\vec{\mathbf{e}}_0$ and $\vec{\mathbf{e}}$, respectively, and $\mathbf{F}$ and $\mathbf{N}$ are the components of a force vector on $\vec{\mathbf{e}}_0$ and $\vec{\mathbf{e}}$, respectively. The inverse transformations of Eqs. (9–21) require $\boldsymbol{\lambda}_0^{-1}$ which is obtained from Eq. (9–13a), after neglecting squares and product of rotations, as

$$\boldsymbol{\lambda}_0^{-1} = \mathbf{I} - \underset{\sim}{\boldsymbol{\varphi}}_0 = \begin{bmatrix} 1 & -\varphi_3 & \varphi_2 \\ 0 & 1 & -\varphi_1 \\ 0 & \varphi_1 & 1 \end{bmatrix} \tag{9–22}$$

The equilibrium solution, Eq. (6–5), becomes

$$\mathbf{N} = \boldsymbol{\lambda}_0^{-1T} \mathbf{F}^b + \mathbf{N}_p \tag{9–23a}$$

$$\mathbf{M} = \boldsymbol{\lambda} \mathbf{G}^b - \boldsymbol{\lambda}[\underset{\sim}{\mathbf{u}}_0^b - \underset{\sim}{\mathbf{u}}_0 + (l - x)\underset{\sim}{\mathbf{I}}_1]\mathbf{F}^b + \mathbf{M}_p \tag{9–23b}$$

In detail, Eqs. (9–23) have the form

$$\begin{Bmatrix} N_1 \\ N_2 \\ N_3 \end{Bmatrix} = \begin{bmatrix} 1 & 0 & 0 \\ -\varphi_3 & 1 & \varphi_1 \\ \varphi_2 & -\varphi_1 & 1 \end{bmatrix} \begin{Bmatrix} F_1^b \\ F_2^b \\ F_3^b \end{Bmatrix} + \begin{Bmatrix} N_{1p} \\ N_{2p} \\ N_{3p} \end{Bmatrix} \tag{9–24a}$$

$$\begin{aligned} \begin{Bmatrix} M_1 \\ M_2 \\ M_3 \end{Bmatrix} = &\begin{bmatrix} 1 & \varphi_3 & -\varphi_2 \\ -\varphi_3 & 1 & \varphi_1 \\ \varphi_2 & -\varphi_1 & 1 \end{bmatrix} \begin{Bmatrix} G_1^b \\ G_2^b \\ G_3^b \end{Bmatrix} - \begin{bmatrix} 1 & \varphi_3 & -\varphi_2 \\ -\varphi_3 & 1 & \varphi_1 \\ \varphi_2 & -\varphi_1 & 1 \end{bmatrix} \\ &\times \begin{bmatrix} 0 & (u_3^b - u_3) & -(u_2^b - u_2) \\ -(u_3^b - u_3) & 0 & (l - x) \\ (u_2^b - u_2) & -(l - x) & 0 \end{bmatrix} \begin{Bmatrix} F_1^b \\ F_2^b \\ F_3^b \end{Bmatrix} + \begin{Bmatrix} M_{1p} \\ M_{2p} \\ M_{3p} \end{Bmatrix} \end{aligned} \tag{9–24b}$$

It is noted that, contrary to planar second-order theory, all deformation-displacement relations and all strain-displacement relations in the present theory are nonlinear. If further simplifications are made by neglecting nonlinear terms except in the axial deformation-displacement relations, the consistently derived equilibrium equations would contain the nonlinear effect of the axial force on bending, as in planar beam-column theory, but would loose any coupling between bending and torsion. This is the basis for the limited nonlinear member theory in Section 7, Chapter 14. Coupling between bending and torsion is essential in the study of the lateral stability of the member.

Geometric Stiffness Matrix of Second-Order Theory

The geometric stiffness $\mathbf{k}_{G\varphi}$ associated with generalized moments may be obtained by applying Eq. (8–5) in which v_i are as defined in Eqs. (9–14, 15). Since $\mathbf{v}$ is at most of second order in $\mathbf{s}$, $\mathbf{k}_{G\varphi}$ is independent of $\mathbf{s}$ and is linear in $\mathbf{V}_\varphi$. Further $\mathbf{v}$ is linear in u_1^a and u_1^b. $\mathbf{k}_{G\varphi}$ is thus also obtained from the exact formulas (8–12) expressed in terms of $\mathbf{V}_\varphi$ by deleting the rows and columns corresponding to du_1^a and du_1^b and by setting the displacements to zero. The result has the form

$$\mathbf{k}_{G\varphi} = \begin{bmatrix} & \underset{\sim}{\mathbf{F}}^T_{r,0} & & \\ \underset{\sim}{\mathbf{F}}_{r,0} & \mathbf{k}^{aa}_{\varphi\varphi} & -\underset{\sim}{\mathbf{F}}_{r,0} & -\tfrac{1}{2}\underset{\sim}{\mathbf{G}}_{\varphi r} \\ & -\underset{\sim}{\mathbf{F}}^T_{r,0} & & \\ & \tfrac{1}{2}\underset{\sim}{\mathbf{G}}_{\varphi r} & & \end{bmatrix} \tag{9–25a}$$

where $\underset{\sim}{\mathbf{F}}_{r,0}$ is obtained by making zero the first column of $\underset{\sim}{\mathbf{F}}_r$ and

$$\mathbf{k}^{aa}_{\varphi\varphi} = \frac{l}{2}(\mathbf{F}_r\mathbf{I}_1^T + \mathbf{I}_1\mathbf{F}_r^T) - lF_{1r}\mathbf{I} = \begin{bmatrix} 0 & \frac{l}{2}F_{2r} & \frac{l}{2}F_{3r} \\ \frac{l}{2}F_{2r} & -lF_{1r} & 0 \\ \frac{l}{2}F_{3r} & 0 & -lF_{1r} \end{bmatrix} \tag{9–25b}$$

10. THEORY WITHOUT SHEAR DEFORMATION: CONSTITUTIVE EQUATIONS

In order to complete the second-order theory, constitutive equations are derived for a linear material, with flexural shear deformation neglected. For generality the shear center S is assumed to be distinct from the centroid C.

Strain-Displacement Relations

In order to neglect shear deformation, the shear strains ε_2 and ε_3 at the shear center are set equal to zero. It will be convenient then to choose the shear center as reference point for defining the translational displacements $\mathbf{u} = \{u_1\ u_2\ u_3\}$. Displacements at the centroid will be designated $\mathbf{u}_c = \{u_{1c}\ u_{2c}\ u_{3c}\}$. The reference line for translational displacements in the strain-displacement relations (9–17) is now considered to be the line of shear centers, and the underlying assumption $u_{1,x} \ll 1$ is thus also assumed to hold along this line. The conditions $\varepsilon_2 = \varepsilon_3 = 0$ form two simultaneous equations for φ_2 and φ_3 whose solution, with $\varphi_1^2 \ll 1$, is

$$\varphi_2 = -u_{3,x} + \tfrac{1}{2}\varphi_1 u_{2,x} \tag{10–1a}$$

$$\varphi_3 = u_{2,x} + \tfrac{1}{2}\varphi_1 u_{3,x} \tag{10–1b}$$

The remaining strain-displacement relations, with φ_1^2, $(u_{2,x})^2$, and $(u_{3,x})^2 \ll 1$, become

$$\varepsilon_1 = u_{1,x} + \tfrac{1}{2}(u_{2,x})^2 + \tfrac{1}{2}(u_{3,x})^2 \tag{10–2}$$

$$\chi_1 = \varphi_{1,x} - \tfrac{1}{2}u_{2,x}u_{3,xx} + \tfrac{1}{2}u_{3,x}u_{2,xx} \tag{10–3a}$$

$$\chi_2 = -u_{3,xx} + \varphi_1 u_{2,xx} \tag{10–3b}$$

$$\chi_3 = u_{2,xx} + \varphi_1 u_{3,xx} \tag{10–3c}$$

Constitutive Equations

Constitutive equations for the normal force and bending moments are based on the stress-strain relation $\sigma = E\varepsilon'$ for axial material fibers. The strain distribution over the cross section is obtained by applying Eq. (5.6–2), taking the shear center as reference point for $\boldsymbol{\varepsilon}$. The origin of axes (y, z) is kept at the centroid, however, and the coordinates of the shear center are denoted (y_s, z_s). We thus obtain

$$\varepsilon_1' = \varepsilon_1 + (z - z_s)\chi_2 - (y - y_s)\chi_3 \tag{10–4a}$$

$$\varepsilon_2' = -(z - z_s)\chi_1 \tag{10–4b}$$

$$\varepsilon_3' = (y - y_s)\chi_1 \tag{10–4c}$$

The axial strain ε' is obtained from Eq. (5.6–3b) by a series development truncated to second-order terms in the strains and simplified further by the assumption $(\varepsilon_1')^2 \ll \varepsilon_1'$. There comes

$$\varepsilon' = \varepsilon_1' + \tfrac{1}{2}(\varepsilon_2')^2 + \tfrac{1}{2}(\varepsilon_3')^2 \tag{10–5}$$

and, using Eqs. (10–4),

$$\varepsilon' = \varepsilon_{1c} + z\chi_2 - y\chi_3 + \tfrac{1}{2}r'^2\chi_1^2 \tag{10–6}$$

where

$$\varepsilon_{1c} = \varepsilon_1 - z_s\chi_2 + y_s\chi_3 \tag{10–7a}$$

$$r'^2 = (y - y_s)^2 + (z - z_s)^2 \tag{10–7b}$$

ε_{1c} is the normal strain at the centroid.

Constitutive equations for the normal force N_1 and bending moments M_2 and M_3 about the centroidal axes are obtained by integration over the cross section of the normal stress and its moments. The axial stress-strain relation $\sigma = E\varepsilon'$ may be applied equivalently to the normal stress because squares of rotations are negligible compared with unity. Assuming the y and z axes are the principal axes, and using Eq. (10–6) for ε', the result is shown in Eqs. (10–8a, b, c).

The twisting deformation causes an axial material fiber at distance r' from the shear center axis to take the inclination $\chi_1 r'$ to the normal to the cross section. The axial stress $E\varepsilon'$ has then in the plane of the cross section the component $E\varepsilon'\chi_1 r'$ whose moment about the shear center is $M_{1\sigma} = \int E\varepsilon'\chi_1 r'^2 dA$. It is assumed that the complete constitutive equation for torsion is obtained by adding $M_{1\sigma}$ to the expression of the linear theory, Eq. (14/3.3–10). The result is given in Eqs. (10–8d, e). An alternate expression of $M_{1\sigma}$ which is useful in stability analysis is given in Eq. (10–8f) and is obtained by solving Eqs. (10–8a, b, c) for ε_{1c}, χ_1, and χ_2 and substituting into Eq. (10–8e). The constitutive equations are thus found in the form

$$N_1 = EA(\varepsilon_{1c} + \tfrac{1}{2}\rho^2\chi_1^2) \qquad (10\text{–}8a)$$

$$M_2 = EI_2(\chi_2 + \tfrac{1}{2}\beta_2\chi_1^2) \qquad (10\text{–}8b)$$

$$M_3 = EI_3(\chi_3 - \tfrac{1}{2}\beta_3\chi_1^2) \qquad (10\text{–}8c)$$

$$M_{1s} = GJ\chi_1 - EK\chi_{1,xx} + M_{1\sigma} \qquad (10\text{–}8d)$$

where $M_{1\sigma}$ has the two equivalent expressions

$$M_{1\sigma} = (EA\rho^2\varepsilon_{1c} + EI_2\beta_2\chi_2 - EI_3\beta_3\chi_3)\chi_1 + EH\chi_1^3 \qquad (10\text{–}8e)$$

$$M_{1\sigma} = (\rho^2 N_1 + \beta_2 M_2 - \beta_3 M_3)\chi_1 + E\bar{H}\chi_1^3 \qquad (10\text{–}8f)$$

ρ^2, β_2, β_3, H, and $\bar{H}$ are geometric constants,

$$\rho^2 = \frac{I_s}{A} = \frac{1}{A}\int (r')^2 dA = \frac{1}{A}\int (y^2 + z^2) dA + (y_s^2 + z_s^2) \qquad (10\text{–}8g)$$

$$\beta_2 = \frac{1}{I_2}\int z r'^2 dA = \frac{1}{I_2}\int z(y^2 + z^2) dA - 2z_s \qquad (10\text{–}8h)$$

$$\beta_3 = \frac{1}{I_3}\int y r'^2 dA = \frac{1}{I_3}\int y(y^2 + z^2) dA - 2y_s \qquad (10\text{–}8i)$$

$$H = \tfrac{1}{2}\int (r')^4 dA \qquad (10\text{–}8j)$$

$$\bar{H} = H - \tfrac{1}{2}A\rho^4 - \tfrac{1}{2}I_2\beta_2^2 - \tfrac{1}{2}I_3\beta_3^2 \qquad (10\text{–}8k)$$

Strain Energy: Energy Theorems

The strain energy density associated with the axial strain is the integral of $\tfrac{1}{2}E(\varepsilon')^2$ over the cross section. Adding the torsion strain energy, Eq. (14/3.3–9), we obtain

$$U = \tfrac{1}{2}EA\varepsilon_{1c}^2 + \tfrac{1}{2}EI_2\chi_2^2 + \tfrac{1}{2}EI_3\chi_3^2 + \tfrac{1}{2}GJ\chi_1^2 + \tfrac{1}{2}EK(\chi_{1,x})^2 + U_N \qquad (10\text{–}9a)$$

where

$$U_N = \tfrac{1}{2}(\rho^2 EA\varepsilon_{1c} + \beta_2 EI_2\chi_2 - \beta_3 EI_3\chi_3)\chi_1^2 + \tfrac{1}{4}EH\chi_1^4 \qquad (10\text{–}9b)$$

For consistency of energy theorems with the principle of virtual displacements, it should be verified that the first variation of the strain energy is equal to the internal virtual work. Considering U as a function of $\varepsilon_{1,c}, \chi_1, \chi_{1,x}, \chi_2$, and χ_3, it is readily verified that the constitutive expressions of N_1, M_2, and M_3 are the partial derivatives of U with respect to $\varepsilon_{1,c}$, χ_2, and χ_3, respectively. For the torsional terms let

$$M_{1\chi} = \frac{\partial U}{\partial \chi_1} = GJ\chi_1 + M_{1\sigma} \qquad (10\text{–}10a)$$

$$B = \frac{\partial U}{\partial(\chi_{1,x})} = EK\chi_{1,x} \qquad (10\text{–}10b)$$

the constitutive equation for M_{1s} is then

$$M_{1s} = M_{1\chi} - B_{,x} \qquad (10\text{–}10c)$$

and

$$\delta U = N_1\delta\varepsilon_{1c} + M_{1\chi}\delta\chi_1 + B\delta\chi_{1,x} + M_2\delta\chi_2 + M_3\delta\chi_3 \qquad (10\text{–}11)$$

In evaluating $\int_0^l \delta U\,dx$, the term $B\delta\chi_{1,x}$ is integrated by parts, and at a member end, $B = 0$ if warping is free and $\delta\chi_1 = 0$ if warping is restrained. We thus obtain

$$\int_0^l \delta U\,dx = \int_0^l (N_1\delta\varepsilon_{1c} + M_{1s}\delta\chi_1 + M_2\delta\chi_2 + M_3\delta\chi_3)\,dx \qquad (10\text{–}12)$$

There remains to verify that the virtual work expression $\mathbf{N}^T\delta\boldsymbol{\varepsilon} + \mathbf{M}^T\delta\boldsymbol{\chi}$, in which $\delta\boldsymbol{\varepsilon}$ and $\mathbf{M}$ are defined at the same reference line, transforms into the integrand in Eq. (10–12). With the line of shear centers as the reference, the expression is $N_1\delta\varepsilon_1 + M_{1s}\delta\chi_1 + M_{2s}\delta\chi_2 + M_{3s}\delta\chi_3$. Transformation of M_{2s} and M_{3s} to the centroid yields $M_{2s} = M_2 - z_s N_1$ and $M_{3s} = M_3 + y_s N_1$, and the virtual work expression becomes $N_1(\delta\varepsilon_1 - z_s\delta\chi_2 + y_s\delta\chi_3) + M_{1s}\delta\chi_1 + M_2\delta\chi_2 + M_3\delta\chi_3$. The coefficient of N_1 is seen to be $\delta\varepsilon_{1c}$, and the virtual work expression is the integrand in Eq. (10–12).

Usefulness of nonlinear terms in the constitutive equations is illustrated in the two following examples.

Example 1. Torsional Buckling

Consider a straight member in pure compression whose cross section has two axes of symmetry. Apart from flexural buckling, the member may buckle in a torsional mode with its axis remaining straight. To investigate this possibility, it is required kinematically that $\chi_1 \neq 0$ and $\chi_2 = \chi_3 = 0$. Statically, it is required that $M_{1,x} = 0$, $M_2 = M_3 = 0$, and that there be no change in the axial force. This last condition determines by means of Eq. (10–8a) the change in normal strain ε_{1c} in terms of χ_1. The constants β_2 and β_3 are zero because of the assumed symmetry of the cross section, and the bending constitutive equations (10–8b, c) are satisfied by $M_2 = M_3 = 0$ and $\chi_2 = \chi_3 = 0$. In the torsional equation (10–8d, f) the term in χ_1^3 is neglected because χ_1 is incremental, and M_{1s} coincides with the centroidal moment M_1. Noting that $\chi_1 = \varphi_{1,x}$, the condition $M_{1,x} = 0$ yields

$$EK\varphi_{1,xxxx} - (GJ + \rho^2 N_1)\varphi_{1,xx} = 0 \tag{10–13}$$

If $K = 0$, a nontrivial solution of Eq. (10–13) for $\varphi_{1,xx}$ exists, provided $GJ + \rho^2 N_1 = 0$. The critical axial force is then

$$N_{1,\text{cr}} = -\frac{GJ}{\rho^2} \tag{10–14}$$

Various end conditions on φ_1, $\varphi_{1,x}$, or $\varphi_{1,xx}$ are possible with this solution. It is noted that $N_{1,\text{cr}}$ is independent of member length.

In the general case where $K \neq 0$, let

$$\lambda^2 = -\frac{(GJ + \rho^2 N_1)l^2}{EK} \tag{10–15}$$

and assume that the compressive axial force is large enough for λ^2 to be positive. Equation (10–13) is analogous to that of planar flexural buckling, with correspondance between the stability parameters λ^2 and between types of boundary conditions. If warping and rotations are restrained at both ends, the conditions are $\varphi_{1,x} = 0$ and $\varphi_1 = 0$, at $x = 0$ and $x = l$. The problem is analogous to the flexural buckling of a compressed member with both ends fixed against transverse displacements and rotations. The critical value for λ^2 is $4\pi^2$, and from Eq. (10–15),

$$N_{1,\text{cr}} = -\frac{GJ}{\rho^2}\left(1 + \frac{\lambda_{\text{cr}}^2 EK}{GJl^2}\right) \tag{10–16}$$

If warping is free and rotations are retrained at both ends, the conditions are $\varphi_{1,xx} = 0$ and $\varphi_1 = 0$, at $x = 0$ and $x = l$. The problem is analogous to Euler's column, and $\lambda_{\text{cr}}^2 = \pi^2$ in Eq. (10–16).

If the preceding end conditions are applied each at one end, the analogous problem is that of buckling of the proped cantilever for which $\lambda_{\text{cr}}^2 = (\pi/0.7)^2$. Finally, if one end is completely restrained and the other is free to warp and to rotate, the problem is analogous to that of the cantilever, and $\lambda_{\text{cr}}^2 = \pi^2/4$.

Example 2. Nonlinear Torsion [2, p. 286]

To illustrate a case where the nonlinear terms in χ_1 in the constitutive equations may be important, consider a thin rectangular cross section of thickness t and width b, and let the member be subjected to a torsional moment M_1. For the rectangular cross section $J = bt^3/3$, and $K = 0$. Also $\beta_2 = \beta_3 = 0$, and $y_s = z_s = 0$. Assuming $t \ll b$, Eqs. (10–8g, j, k) yield $I_s = tb^3/12$, $\rho^2 = b^2/12$, $H = tb^5/160$, and $\bar{H} = tb^5/360$. From Eq. (10–8a) and $N_1 = 0$, obtain $\varepsilon_{1c} = -\frac{1}{2}\rho^2\chi_1^2$. Then from Eq. (10–8d, f),

$$M_1 = G\frac{bt^3}{3}\chi_1\left(1 + \frac{E}{G}\frac{b^4\chi_1^2}{120t^2}\right) \tag{10–17}$$

The normal stress $\sigma = E\varepsilon'$ is obtained using Eq. (10–6) in which $\chi_2 = \chi_3 = 0$ and $r'^2 = r^2 = y^2 + z^2 \approx z^2$. Thus

$$\sigma = E\varepsilon' = \frac{E}{2}\left(z^2 - \frac{b^2}{12}\right)\chi_1^2$$

σ varies parabolically along the width of the cross section. It is maximum and tensile at the ends, $z = \pm b/2$. At the center it is compressive and takes half its maximum value of $\sigma_{\max} = Eb^2\chi_1^2/12$.

To compare orders of magnitude, the maximum shear strain $\gamma_{\max}$ in St-Venant's torsion is $\chi_1 t$, and the corresponding stress is $\tau_{\max} = G\chi_1 t$. The ratio $\sigma_{\max}/\tau_{\max}$ is proportional to b^2/t^2 and to $\gamma_{\max}$ and may become significant if b/t is sufficiently large. Similarly the nonlinear term in Eq. (10–17) may contribute a nonnegligible amount to the torsional rigidity.

11. GOVERNING DIFFERENTIAL EQUATIONS

Governing equations for the displacements $\{u_1\ u_2\ u_3\ \varphi_1\}$ are obtained by substituting in the constitutive equations (10–8a, b, c, d) the strain-displacement relations and the equilibrium solution (9–24). This solution must be transformed, however, so that the bending moments are defined at the centroid and the torsional moment and translational displacements are defined at the shear center.

The reference point for displacements and moments in Eqs. (9–24) is considered to be the shear center. Accordingly $\mathbf{M}$ and $\mathbf{G}^b$ in these equations are renamed $\mathbf{M}_s$ and $\mathbf{G}_s^b$, respectively, and $\mathbf{M}$ and $\mathbf{G}^b$ will be the notation at the centroid. Transformation formulas for bending moments are

$$M_2 = M_{2s} + z_s N_1 \tag{11–1a}$$

$$M_3 = M_{3s} - y_s N_1 \tag{11–1b}$$

where y_s and z_s are the coordinates of the shear center with respect to centroidal axes. We thus obtain

$$N_1 = F_1^b + N_{1p} \tag{11–2a}$$

$$\begin{aligned} M_{1s} &= G_{1s}^b + G_{2s}^b\varphi_3 - G_{3s}^b\varphi_2 + F_1^b[\varphi_2(u_2^b - u_2) + \varphi_3(u_3^b - u_3)] \\ &\quad + F_2^b[-u_3^b + u_3 - (l-x)\varphi_2] + F_3^b[u_2^b - u_2 - (l-x)\varphi_3] + M_{1sp} \end{aligned} \tag{11–2b}$$

$$\begin{aligned} M_2 &= G_{2s}^b + G_{3s}^b\varphi_1 - G_{1s}^b\varphi_3 + F_1^b[z_s + u_3^b - u_3 - \varphi_1(u_2^b - u_2)] \\ &\quad + F_2^b[\varphi_3(u_3^b - u_3) + (l-x)\varphi_1] + F_3^b[-\varphi_3(u_2^b - u_2) - l + x] \\ &\quad + M_{2p} \end{aligned} \tag{11–2c}$$

$$\begin{aligned} M_3 &= G_{3s}^b - G_{2s}^b\varphi_1 + G_{1s}^b\varphi_2 - F_1^b[y_s + u_2^b - u_2 + \varphi_1(u_3^b - u_3)] \\ &\quad + F_2^b[l - x - \varphi_2(u_3^b - u_3)] + F_3^b[(l-x)\varphi_1 + \varphi_2(u_2^b - u_2)] + M_{3p} \end{aligned} \tag{11–2d}$$

Moments $\mathbf{G}_s^b$ may be transformed to the centroid by the vector relation

$$\vec{G}_s^b = \vec{G}^b - \mathbf{r}^T\boldsymbol{\lambda}^b\vec{\mathbf{e}}_0 \times \vec{F}^b \tag{11–3}$$

where $\mathbf{r} = \{0 \;\; y_s \;\; z_s\}$. The components on $\vec{\mathbf{e}}_0$ of this equation are

$$G_{1s}^b = G_1^b + F_2^b(z_s + y_s\varphi_1^b) + F_3^b(-y_s + z_s\varphi_1^b) \tag{11–3a}$$

$$G_{2s}^b = G_2^b + F_1^b(-z_s - y_s\varphi_1^b) + F_3^b(z_s\varphi_2^b - y_s\varphi_3^b) \tag{11–3b}$$

$$G_{3s}^b = G_3^b + F_1^b(y_s - z_s\varphi_1^b) + F_2^b(-z_s\varphi_2^b + y_s\varphi_3^b) \tag{11–3c}$$

For a member subjected only to end forces, N_{1p}, M_{1sp}, M_{2p}, and M_{3p} vanish. The constitutive equations for M_{1s}, M_2, and M_3 yield then a system of three coupled nonlinear differential equations for $\{u_2 \;\; u_3 \;\; \varphi_1\}$. The axial constitutive equation may be solved for u_1 in terms of u_2, u_3, and φ_1. The quantities at end b, $\mathbf{F}^b$, $\mathbf{G}_s^b$, $\mathbf{u}^b$, and $\boldsymbol{\varphi}^b$, are to be determined together with the integration constants by means of the boundary conditions at both member ends. Boundary conditions are discussed in more detail in the next section. If there is a distributed span load, $(\mathbf{p}, \mathbf{m}_s)$, the particular equilibrium solution is expressed in integral form. The solution at point x is obtained by integration over the interval (x, l) of the effect of elementary loads $\mathbf{p}d\xi$, $\mathbf{m}_s d\xi$. The integrands are thus obtained by applying Eqs. (11–2) to a member of length ξ subjected at its end b to concentrated actions $\mathbf{p}d\xi$, $\mathbf{m}_s d\xi$.

Exact solutions of the governing differential equations are not generally possible. Numerical methods may be appropriate for solving individual member problems, but for general structural analysis including member problems, there is need for a discrete formulation such as the stiffness method. The differential equations are useful, however, theoretically and for member linearized stability analysis which is described next.

12. DIFFERENTIAL EQUATIONS OF LINEARIZED STABILITY ANALYSIS

Consider a member in an initial equilibrium state $(\mathbf{u}, \varphi_1)$ caused by a certain loading condition. To investigate stability, a load level is sought at which an alternate equilibrium state $(\mathbf{u} + \delta\mathbf{u}, \varphi_1 + \delta\varphi_1)$ becomes possible to within second-order terms in $\delta\mathbf{u}$ and $\delta\varphi_1$. The governing differential equations for $(\delta\mathbf{u}, \delta\varphi_1)$ may be obtained by variation of the equations for a general equilibrium state, keeping only linear terms in $\delta\mathbf{u}$ and $\delta\varphi_1$ and setting in the result $\mathbf{u}$ and φ_1 to their initial values.

Derivation of governing equations for stability may start with the differential equations of equilibrium. It is advantageous, however, to start with Eqs. (11–2). A simplified formulation of stability equations is made in what follows in which the effect of initial displacements is neglected. The problem appears then as one of bifurcation or buckling from an initially undeformed but stressed state. To do this, Eqs. (11–2) are linearized with respect to displacements and are then varied with respect to displacements and forces. A similar procedure is used in Eqs. (11–3) to express $\delta\mathbf{G}_s^b$ in terms of $\delta\mathbf{G}^b$. Only problems with constant axial loads will be considered so that $\delta F_1^b = \delta N_{1p} = 0$. We thus obtain $\delta N = 0$, and

$$\begin{aligned}\delta M_{1s} = {} & G_2^b\delta\varphi_3 - G_3^b\delta\varphi_2 - F_1^b(z_s\delta\varphi_3 + y_s\delta\varphi_2) \\ & - F_2^b[\delta u_3^b - \delta u_3 + (l-x)\delta\varphi_2 - y_s\delta\varphi_1^b] \\ & + F_3^b[\delta u_2^b - \delta u_2 - (l-x)\delta\varphi_3 + z_s\delta\varphi_1^b] \\ & + \delta G_1^b + z_s\delta F_2^b - y_s\delta F_3^b + \delta M_{1sp}\end{aligned} \tag{12–1a}$$

$$\begin{aligned}\delta M_2 = {} & -G_1^b\delta\varphi_3 + G_3^b\delta\varphi_1 + F_1^b[\delta u_3^b - \delta u_3 + y_s(\delta\varphi_1 - \delta\varphi_1^b)] \\ & + F_2^b[(l-x)\delta\varphi_1 - z_s\delta\varphi_3] + F_3^b[z_s\delta\varphi_2^b + y_s(\delta\varphi_3 - \delta\varphi_3^b)] \\ & + \delta G_2^b - (l-x)\delta F_3^b + \delta M_{2p}\end{aligned} \tag{12–1b}$$

$$\begin{aligned}\delta M_3 = {} & G_1^b\delta\varphi_2 - G_2^b\delta\varphi_1 - F_1^b[\delta u_2^b - \delta u_2 - z_s(\delta\varphi_1 - \delta\varphi_1^b)] \\ & + F_2^b[y_s\delta\varphi_3^b + z_s(\delta\varphi_2 - \delta\varphi_2^b)] + F_3^b[(l-x)\delta\varphi_1 - y_s\delta\varphi_2] \\ & + \delta G_3^b + (l-x)\delta F_2^b + \delta M_{3p}\end{aligned} \tag{12–1c}$$

In the foregoing equations, δM_{1s}, δu_2, and δu_3 are defined at the shear center, and δM_2 and δM_3 at the centroid. δM_{1sp}, δM_{2p}, and δM_{3p} depend on the span load and are considered in an example. Moments $\delta\mathbf{G}^b$ are components in the fixed axes and are defined at the centroid. Elements of $\mathbf{F}^b$ and $\mathbf{G}^b$ could be prescribed or could be reactions at supports. Elements of $\delta\mathbf{F}^b$ and $\delta\mathbf{G}^b$ are zero if corresponding elements of $\mathbf{F}^b$ and $\mathbf{G}^b$ are prescribed with fixed orientations. In general, elements of $\delta\mathbf{F}^b$ and $\delta\mathbf{G}^b$ may be nonzero because of the change in geometry.

For expressing δM_{1s}, δM_2, and δM_3 in terms of displacement variations,

Eqs. (10–8b, c, d, f) are linearized with respect to strains, and these are then varied. Variations of strains and rotations are those of the linear theory. Using a prime to denote derivatives, we obtain

$$(\delta\varphi_2, \delta\varphi_3, \delta\chi_1, \delta\chi_2, \delta\chi_3) = (-\delta u_3', \delta u_2', \delta\varphi_1', -\delta u_3'', \delta u_2'') \tag{12–2}$$

$$(GJ + \rho^2 N_1 + \beta_2 M_2 - \beta_3 M_3)\delta\varphi_1' - EK\delta\varphi_1''' = \delta M_{1s} \tag{12–3a}$$

$$(EI_2 \delta u_3'', EI_3 \delta u_2'') = (-\delta M_2, \delta M_3) \tag{12–3b}$$

Substituting from Eqs. (12–1) into Eqs. (12–3) and deleting the superscript from $\mathbf{F}^b$ and $\mathbf{G}^b$, we obtain

$$\begin{aligned} &EK\delta\varphi_1''' - (GJ + \rho^2 N_1 + \beta_2 M_2 - \beta_3 M_3)\delta\varphi_1' + G_2 \delta u_2' + G_3 \delta u_3' \\ &\quad - F_1(z_s \delta u_2' - y_s \delta u_3') + F_2[(l-x)\delta u_3' + \delta u_3 - \delta u_3^b + y_s \delta\varphi_1^b] \\ &\quad - F_3[(l-x)\delta u_2' + \delta u_2 - \delta u_2^b - z_s \delta\varphi_1^b] \\ &\quad + \delta G_1 + z_s \delta F_2 - y_s \delta F_3 + \delta M_{1sp} = 0 \end{aligned} \tag{12–4a}$$

$$\begin{aligned} &EI_2 \delta u_3'' - G_1 \delta u_2' + G_3 \delta\varphi_1 - F_1[\delta u_3 - \delta u_3^b - y_s(\delta\varphi_1 - \delta\varphi_1^b)] \\ &\quad + F_2[(l-x)\delta\varphi_1 - z_s \delta u_2'] + F_3[y_s(\delta u_2' - \delta u_2'^b) - z_s \delta u_3'^b] \\ &\quad + \delta G_2 - (l-x)\delta F_3 + \delta M_{2p} = 0 \end{aligned} \tag{12–4b}$$

$$\begin{aligned} &EI_3 \delta u_2'' + G_1 \delta u_3' + G_2 \delta\varphi_1 - F_1[\delta u_2 - \delta u_2^b + z_s(\delta\varphi_1 - \delta\varphi_1^b)] \\ &\quad + F_2[z_s(\delta u_3' - \delta u_3'^b) - y_s \delta u_2'^b] - F_3[(l-x)\delta\varphi_1 + y_s \delta u_3'] \\ &\quad - \delta G_3 - (l-x)\delta F_2 - \delta M_{3p} = 0 \end{aligned} \tag{12–4c}$$

Equations (12–4) form a linear system of seventh order whose general solution depends on seven integration constants. It also depends on three undetermined end displacements $\delta u_2^b, \delta u_3^b, \delta\varphi_1^b$ and five undetermined end forces $\delta F_2, \delta F_3, \delta G_1$, δG_2, δG_3. However, the integration constants are related to the undetermined displacements by the conditions $(\delta u_2, \delta u_3, \delta\varphi_1) = (\delta u_2^b, \delta u_3^b, \delta\varphi_1^b)$ at end b. There remains thus 12 independent constants. In a specific problem there are as many boundary conditions, consisting of six static or kinematic conditions at each end. Further force variations in a stability problem are either zero or linear in the displacements, and kinematic boundary conditions are homogeneous. The result is a linear characteristic-value problem.

The following examples are limited to making Eqs. (12–4) specific to particular cases. Solutions of classical problems may be seen in references [6]. Solutions by energy methods are obtained in Sections 15 and 16. The direct derivation of governing differential equations and static boundary conditions is instructive even if a solution by the strain energy method is to be obtained. It allows a comparison with the Euler differential equations and the natural boundary conditions of the variational formulation.

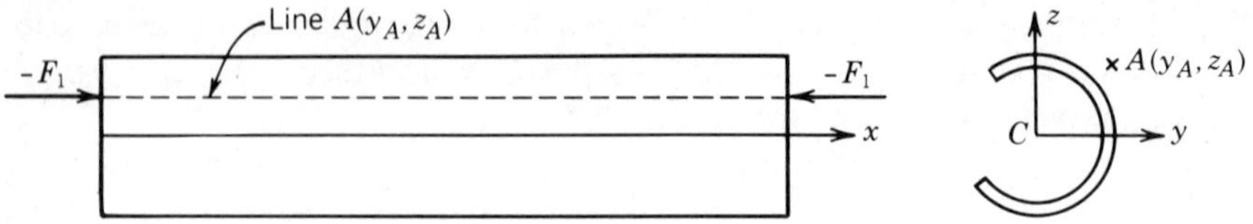

FIG. 12–1. Stability in eccentric compression, Section 12, Example 1; Section 15, Example 4.

Example 1. Stability Equations in Eccentric Compression

Consider a uniform member subjected at end b to an axial force F_1 applied at point $A(y_A, z_A)$ and supported in a manner to be described later, Fig. 12–1. We have then $N_1 = F_1$, $F_2 = F_3 = G_1 = 0$, $M_2 = G_2 = F_1 z_A$, and $M_3 = G_3 = -F_1 y_A$. In order to allow for various cases of support conditions, force, moment, and displacement variations at end b are all kept in Eqs. (12–4) which become

$$EK\delta\varphi_1''' - [GJ + F_1(\rho^2 + \beta_2 z_A + \beta_3 y_A)]\delta\varphi_1' + F_1(z_A - z_s)\delta u_2'$$
$$- F_1(y_A - y_s)\delta u_3' + \delta G_1 + z_s\delta F_2 - y_s\delta F_3 = 0 \qquad (12\text{–}5a)$$

$$EI_2\delta u_3'' - F_1(y_A - y_s)\delta\varphi_1 - F_1\delta u_3$$
$$- (l - x)\delta F_3 + \delta G_2 + F_1(\delta u_3^b - y_s\delta\varphi_1^b) = 0 \qquad (12\text{–}5b)$$

$$EI_3\delta u_2'' + F_1(z_A - z_s)\delta\varphi_1 - F_1\delta u_2$$
$$- (l - x)\delta F_2 - \delta G_3 + F_1(\delta u_2^b + z_s\delta\varphi_1^b) = 0 \qquad (12\text{–}5c)$$

Case of Simple Supports

A simple support is one that restrains φ_1, u_2, and u_3 but allows free warping and flexural rotations. The kinematic end conditions are then $\delta\varphi_1 = \delta u_2 = \delta u_3 = 0$, and the static conditions are $\delta\varphi_1'' = \delta u_2'' = \delta u_3'' = 0$, at $x = 0$ and $x = l$. These conditions applied to Eqs. (12–5b, c) require that $\delta F_2 = \delta F_3 = \delta G_2 = \delta G_3 = 0$. Equations (12–5) reduce then to

$$EK\delta\varphi_1''' - [GJ + F_1(\rho^2 + \beta_2 z_A + \beta_3 y_A)]\delta\varphi_1'$$
$$+ F_1(z_A - z_s)\delta u_2' - F_1(y_A - y_s)\delta u_3' + \delta G_1 = 0 \qquad (12\text{–}6a)$$
$$EI_2\delta u_3'' - F_1(y_A - y_s)\delta\varphi_1 - F_1\delta u_3 = 0 \qquad (12\text{–}6b)$$
$$EI_3\delta u_2'' + F_1(z_A - z_s)\delta\varphi_1 - F_1\delta u_2 = 0 \qquad (12\text{–}6c)$$

The conditions $\delta u_2'' = \delta u_3'' = 0$ at $x = 0$ and $x = l$ now follow from the kinematic end conditions. For satisfying the remaining eight boundary conditions, there are seven integration constants and δG_1.

If $y_A \neq y_s$ and $z_A \neq z_s$, Eqs. (12–6) form a coupled system of differential equations, and buckling involves a general state of combined bending and torsion. Equations (12–6) have constant coefficients and can be solved by exact methods. The critical load $F_{1,cr}$ is the lowest in magnitude of the characteristic values of the system.

If the force F_1 is applied at the shear center, the three equations become uncoupled. The first represents then the case of purely torsional buckling seen in Section 10, and the two others represent Euler buckling in the two principal planes.

Partial uncoupling occurs if, say, $z_A = z_s$ but $y_A \neq y_s$.

Case of Restrained Ends

The boundary conditions for restrained ends with restrained warping are $\delta\varphi_1 = \delta u_2 = \delta u_3 = 0$ and $\delta\varphi_1' = \delta u_2' = \delta u_3' = 0$ at both ends. The five force and moment variations in Eqs. (12–5) may be considered as statical redundants. They form with the seven integration constants the required number of 12 constants by which to satisfy the 12 boundary conditions.

Case of a Cantilever

For a cantilever fixed at end a, the boundary conditions are $\delta\varphi_1 = \delta u_2 = \delta u_3 = 0$ and $\delta\varphi_1' = \delta u_2' = \delta u_3' = 0$ at $x = 0$. At end b warping is assumed to be free so that $\delta\varphi_1'' = 0$. The remaining conditions consist of $\delta F_2 = \delta F_3 = 0$ and of three equations expressing δG_1, δG_2, and δG_3. Moments at the centroid are related to moments at point A by equations similar to Eqs. (11–3). Since $\delta G_{1A} = \delta G_{2A} = \delta G_{3A} = 0$, we obtain

$$(\delta G_1, \delta G_2, \delta G_3) = (0, F_1 y_A \delta\varphi_1^b, F_1 z_A \delta\varphi_1^b) \qquad (12\text{–}7)$$

The undetermined constants that remain in Eqs. (12–5) are the displacements at end b that, as explained earlier, are dependent on the seven integration constants of the solution. The seven boundary conditions to be satisfied are the six kinematic ones at end a and the free warping condition at end b.

Example 2. Stability Equations in Pure Bending

Consider a simply supported beam in a state of pure bending in the (x, y) plane, Fig. 12–2. We let then in Eqs. (12–4), $(F_1, F_2, F_3, G_1, G_2) = 0$, $(N_1, M_2) = 0$, $M_3 = G_3$, and $(\delta\varphi_1^b, \delta u_2^b, \delta u_3^b) = 0$. In a buckled configuration $(\delta M_2, \delta M_3) = 0$ at $x = 0$ and $x = l$. It follows from Eqs. (12–1b, c) that $(\delta F_2, \delta F_3, \delta G_2, \delta G_3) = 0$. This holds whether the applied moments rotate or not with the cross section. Equations (12–4) reduce to

$$EK\delta\varphi_1''' - (GJ - \beta_3 G_3)\delta\varphi_1' + G_3 \delta u_3' + \delta G_1 = 0 \qquad (12\text{–}8a)$$

$$EI_2 \delta u_3'' + G_3 \delta\varphi_1 = 0 \qquad (12\text{–}8b)$$

$$\delta u_2'' = 0 \qquad (12\text{–}8c)$$

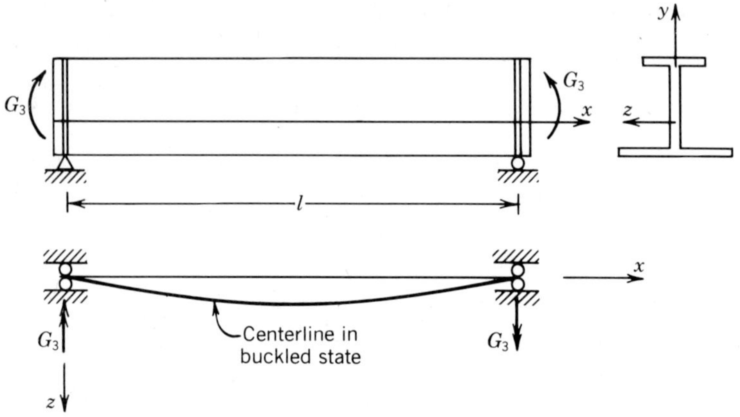

FIG. 12–2. Stability in pure bending, Section 12, Example 2; Section 15, Example 1.

and there remains eight boundary conditions. Two of these are $\delta u_2 = 0$ at both ends. The solution of Eq. (12–8c) for δu_2 thus vanishes. The problem reduces then to solving Eqs. (12–8a, b) with the six boundary conditions $(\delta\varphi_1, \delta\varphi_1'', \delta u_3) = 0$ at both ends. The five integration constants of Eqs. (12–8a, b) and δG_1 form the required number of six undetermined constants. For a uniform member the equations have constant coefficients and may be solved by exact methods. Solution by the energy method may be seen in Section 15.

Example 3. Stability Equations of a Cantilever Subjected to a Concentrated Force

Consider a uniform cantilever fixed at end a and subjected at end b to a transverse force F_2 applied at a point $A(y_A, z_A)$ fixed in the cross section, Fig. 12–3. F_2 is assumed to pass through the shear center, so that $z_A = z_s$, and the cantilever is in a state of planar bending in the principal plane (x, y). We have then $F_1 = F_3 = G_2 = G_3 = 0$, $\delta F_2 = \delta F_3 = 0$, and $G_1 = -F_2 z_s$. In a buckled configuration moments of F_2 about the centroid vary by $\delta\mathbf{G}$. These moments are obtained by varying equations similar to Eqs. (11–3) obtained by replacing point S with

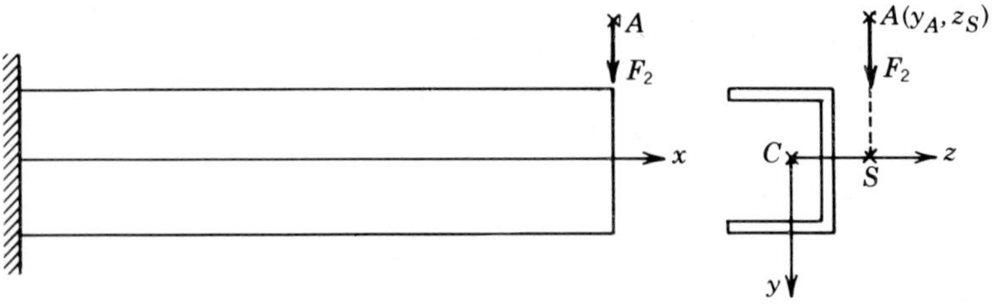

FIG. 12–3. Stability of cantilever, Section 12, Example 3; Section 15, Example 2; Section 16, Example 1.

point A. Since $\delta\mathbf{G}_A = 0$, we obtain

$$(\delta G_1, \delta G_2, \delta G_3) = (-F_2 y_A \delta\varphi_1^b, 0, -F_2(z_s \delta u_3'^b + y_A \delta u_2'^b)) \qquad (12\text{–}9)$$

Equations (12–4) become

$$EK\delta\varphi_1''' - [GJ - \beta_3 F_2(l-x)]\delta\varphi_1' + F_2(l-x)\delta u_3'$$
$$+ F_2[\delta u_3 - \delta u_3^b + (y_s - y_A)\delta\varphi_1^b] = 0 \qquad (12\text{–}10a)$$

$$EI_2 \delta u_3'' + F_2(l-x)\delta\varphi_1 = 0 \qquad (12\text{–}10b)$$

$$EI_3 \delta u_2'' + F_2(y_A - y_s)\delta u_2'^b = 0 \qquad (12\text{–}10c)$$

The first two equations form a separate system in $\delta\varphi_1$ and δu_3. The boundary conditions are the kinematic conditions at the fixed end, as described in an earlier example. With $F_2 = F_{2,\mathrm{cr}}$ determined from the solution of these two equations, Eq. (12–10c) has only the trivial solution $\delta u_2 = 0$. Solutions by energy methods may be seen in Sections 15 and 16.

Example 4. Stability Equations of a Cantilever Subjected to a Distributed Transverse Load

Consider a cantilever subjected to a distributed load p_y and bending in the (x, y) plane, Fig. 12–4. The shear force and bending moment are

$$N_2 = N_{2p} = \int_x^l p_y^\xi \, d\xi \qquad (12\text{–}11a)$$

$$M_3 = M_{3p} = \int_x^l (\xi - x) p_y^\xi \, d\xi \qquad (12\text{–}11b)$$

where p_y^ξ is function of the integration variable ξ. In the present problem the forces and their variations at end b vanish. The effect of the distributed load is represented by the terms δM_{1sp}, δM_{2p}, and δM_{3p} in Eqs. (12–4). To obtain these terms, an elementary load $p_y^\xi d\xi$ is treated similarly to a load F_2 applied at the tip of a cantilever of span ξ. Assuming the load is applied along a line ($y = y_A, z = z_A = z_s$), the stability equations may be obtained from Eqs. (12–10) by replacing l

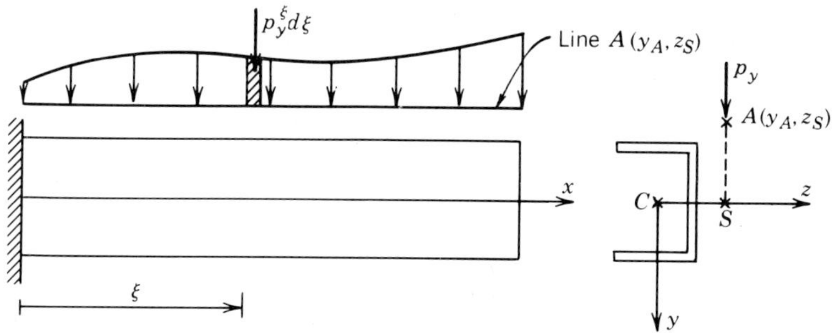

FIG. 12–4. Stability of cantilever, Example 4.

by ξ, F_2 by $p_y^\xi d\xi$, and integrating the load terms with respect to ξ in the interval (x, l). Noting that displacements at end b in Eqs. (12–10) are now defined at the variable location ξ, we obtain

$$EK\delta\varphi_1''' - (GJ - \beta_3 M_3)\delta\varphi_1' + M_3\delta u_3' + N_2\delta u_3 + \int_x^l [(y_s - y_A)\delta\varphi_1^\xi - \delta u_3^\xi] p_y^\xi d\xi = 0 \qquad (12\text{–}12a)$$

$$EI_2\delta u_3'' + M_3\delta\varphi_1 = 0 \qquad (12\text{–}12b)$$

$$EI_3\delta u_2'' + \int_x^l (y_A - y_s)\delta u_2'^\xi p_y^\xi d\xi = 0 \qquad (12\text{–}12c)$$

The first and last equations become differential equations upon differentiation with respect to x. Noting that $M_3' = -N_2$ and $N_2' = -p_y$, there comes for a uniform member,

$$EK\delta\varphi_1'''' - (GJ - \beta_3 M_3)\delta\varphi_1'' - \beta_3 N_2\delta\varphi_1' - p_y(y_s - y_A)\delta\varphi_1 + M_3\delta u_3'' = 0 \qquad (12\text{–}13a)$$

$$EI_3\delta u_2''' - (y_A - y_s)p_y\delta u_2' = 0 \qquad (12\text{–}13b)$$

The equations obtained here apply to an arbitrary load distribution and include, in particular, the case of a concentrated force.

Example 5. Stability Equations of a Simply Supported Beam Subjected to a Transverse Load

Consider a simply supported member bending in the (x, y) plane under an arbitrary distributed load p_y as considered in the preceding example. The

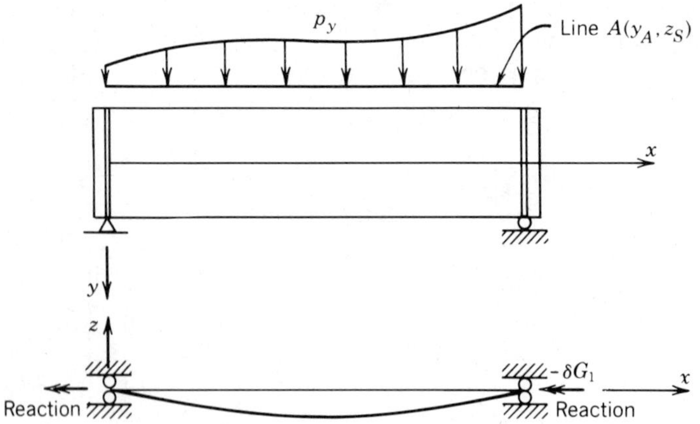

FIG. 12–5. Stability of simply supported member, Section 12, Example 5; Section 15, Example 3.

reactions at supports before buckling are statically determinate. At end b the reaction is F_2. For simple supports, as considered in Example 1, the only change in reactions due to buckling is the occurrence of torsional moments caused by the shift in the z direction of the line of action of the applied load, Fig. 12–5. The torsional reaction δG_1 at end b adds a constant term to the torsional moment δM_{1s}, as expressed in Eq. (12–1a), and does not affect δM_2 and δM_3. Thus $\delta M'_{1s}$, δM_2, and δM_3 have the same expressions in terms of p_y and F_2 as in the two preceding examples. Further the effect of F_2 is included in Eqs. (12–12b) and (12–13). These equations govern then also the present problem, if the shear force N_2 and the bending moment M_3 are evaluated for the simply supported member. The case of a concentrated load at midspan is solved in Sections 15 and 16.

13. ENERGY METHOD

The energy method for establishing generalized stiffness equations and for formulating stability problems is presented in this section. An important observation is made distinguishing between a single member and an assemblage of members.

Expressions needed for applications are derived in Section 14. Member stability is treated with applications in Section 15 and by the method of virtual forces in Section 16.

A member stiffness matrix suitable for structural analysis by the direct stiffness method is established in Chapter 16.

Generalized Stiffness Equations

Consider a general system having a strain energy $\bar{U}$ and an external loading consisting of a conservative part having a potential $\bar{V}$ and of a nonconservative part. $\bar{U}$ and $\bar{V}$ are functionals of independent displacement functions $\mathbf{u}$. $\mathbf{u}$ will be understood here to include the rotation component φ_1. By the principle of virtual displacements an equilibrium state $\mathbf{u}$ is characterized by the variational equation

$$\delta\bar{U} + \delta\bar{V} - \delta W_n = 0 \tag{13–1}$$

for arbitrary but admissible virtual displacements $\delta\mathbf{u}$. δW_n is the virtual work of external nonconservative forces.

A discretization $\mathbf{u} = \boldsymbol{\psi}\mathbf{q}$, where $\boldsymbol{\psi}$ are shape functions and $\mathbf{q}$ are independent generalized displacements, turns $\bar{U}$ and $\bar{V}$ into ordinary functions of $\mathbf{q}$ and δW_n into a linear form in $\delta\mathbf{q}$,

$$\delta W_n = \mathbf{Q}_n^T \delta\mathbf{q} \tag{13–2}$$

A nonconservative load is characterized by the property that $\mathbf{Q}_n^T \delta\mathbf{q}$ is not an exact differential. The discrete form of Eq. (13–1) is the generalized stiffness equation

$$\frac{\partial \bar{U}}{\partial \mathbf{q}} + \frac{\partial \bar{V}}{\partial \mathbf{q}} - \mathbf{Q}_n = 0 \tag{13–3}$$

Consider now an infinitesimal increment in the load that causes displacement increments $d\mathbf{q}$. Incremental stiffness equations are obtained by incrementing Eq. (13–3) with respect to $\mathbf{q}$ and including external force increments $d\mathbf{Q}$ that are displacement independent. There comes

$$\mathbf{K} d\mathbf{q} = d\mathbf{Q} \tag{13–4a}$$

where

$$\mathbf{K} = \frac{\partial^2(\bar{U} + \bar{V})}{\partial \mathbf{q} \partial \mathbf{q}^T} + \mathbf{K}_n \tag{13–4b}$$

$$\mathbf{K}_n = -\frac{\partial \mathbf{Q}_n}{\partial \mathbf{q}^T} \tag{13–4c}$$

$\mathbf{K}$ is a generalized tangent stiffness matrix.

Equation (13–4b) shows that the part of $\mathbf{K}$ due to $(\bar{U} + \bar{V})$ is symmetric. By contrast, $\mathbf{K}_n$ is not symmetric because $\mathbf{Q}_n^T \delta\mathbf{q}$ is not an exact differential.

Choice of Generalized Displacements

For analysis of a single member the only essential consideration for choosing generalized displacements $\mathbf{q}$ is to represent the displacement functions u_1, u_2, u_3, and φ_1 over the interval $(0, l)$. Usual representations in the form of trigonometric series and polynomials are often appropriate for that purpose. Polynomials may in turn be expressed in terms of nodal values of displacements and displacement derivatives as done in previous theories of member discretization.

For analysis of an assemblage of members, however, there is an additional essential requirement in the choice of $\mathbf{q}$ which is the geometric continuity at the joints. It is also practically important to be able to assemble the structure tangent stiffness from element tangent stiffnesses by the direct stiffness method. Both of these requirements are satisfied by the formulation of Section 8 in which the generalized displacements are components of the translation vector $\vec{u}$ and rotation vector $\vec{\varphi}$.

By contrast, if displacement derivatives u_2', u_3' at member ends are chosen together with φ_1 as member generalized displacements, a transformation to joint rotations remains often necessary in order to ensure continuity at the joints. This transformation is nonlinear, as is apparent in Eqs. (10–1). Consequently the transformation law for the tangent stiffness is not totally congruent but involves also a geometric stiffness. A case where the transformation is not necessary is that of a one-member structure subdivided into elements. In that case displacement derivatives are continuous at the joints and may then be chosen as generalized displacements for the member as a whole.

The preceding considerations are applied in Chapter 16. The examples to be treated in Section 15 of the present chapter pertain all to one-member problems.

Variational Formulations of Stability

1. Work Criterion

Consider first a conservative system, and let $\Pi = \bar{U} + \bar{V}$ be its potential energy. An equilibrium state is characterized by the variational equation $\delta\Pi = 0$, as seen earlier. If the equilibrium state is also stable, then by the work criterion (Chapter 6, Section 5.3), $\delta^2\Pi$ is positive for any admissible $\delta\mathbf{u}$. Assuming a loading defined in terms of a parameter γ, a typical problem in stability analysis is to determine the lowest γ at which there is some $\delta\mathbf{u}$ for which $\delta^2\Pi = 0$.

From Eq. (13–4b) in which the nonconservative term is here assumed to be zero, the discretized form of $\delta^2\Pi$ is

$$\delta^2\Pi = \delta\mathbf{q}^T\mathbf{K}\delta\mathbf{q} \tag{13–5}$$

For a stable system at low γ, $\mathbf{K}$ is positive definite; that is, $\delta^2\Pi > 0$ for any $\delta\mathbf{q} \neq 0$. Assuming $\mathbf{K}$ is continuous in γ, $\delta^2\Pi$ cannot become negative except by passing through zero at which point $\mathbf{K}$ is singular. The homogeneous incremental stiffness equations $\mathbf{K}\delta\mathbf{q} = 0$ have then a nontrivial solution $\delta\mathbf{q}$, and state $(\mathbf{q} + \delta\mathbf{q})$ is an alternate equilibrium state to within second-order terms in $\delta\mathbf{q}$. $\delta\mathbf{q}$ represents a neutral equilibrium path and is also called the buckling displacement. In order to avoid confusion with virtual displacements, buckling displacements are renamed $\dot{\mathbf{q}}$, and the stability criterion becomes

$$\mathbf{K}\dot{\mathbf{q}} = 0 \tag{13–6}$$

2. Alternate Equilibrium State Method

The result of the preceding approach to stability may also be arrived at by applying the principle of virtual displacements, Eq. (13–1), as a condition for a neighboring state $(\mathbf{u} + \delta\mathbf{u})$ to be an alternate equilibrium state to within second-order terms in $\delta\mathbf{u}$. To avoid confusion of $\delta\mathbf{u}$ with virtual displacements and to allow a notation for virtual displacements taking place from the alternate equilibrium state, $\delta\mathbf{u}$ is renamed $\mathbf{v}$. In applying Eq. (13–1) to state $(\mathbf{u} + \mathbf{v})$, $\mathbf{u}$ is fixed and $\delta\bar{U}$ and $\delta\bar{V}$ are variations with respect to $\mathbf{v}$. To express these variations analytically, $\Pi = \bar{U} + \bar{V}$ is developed in a Taylor expansion in orders of $\mathbf{v}$ of the form

$$\Pi(\mathbf{u} + \mathbf{v}) = \Pi_0 + \Pi_1 + \tfrac{1}{2}\Pi_2 + \ldots \tag{13–7}$$

If $\mathbf{v}$ is considered as a variation of $\mathbf{u}$ then $\Pi_0 = \Pi(\mathbf{u})$, $\Pi_1 = \delta\Pi$, and $\Pi_2 = \delta^2\Pi$.

Since $\mathbf{u}$ is an equilibrium state, Π_1 vanishes identically, that is, for any $\mathbf{v}$. The variational condition for $(\mathbf{u}+\mathbf{v})$ to be an alternate equilibrium state to within second-order terms in $\mathbf{v}$ is then

$$\delta\Pi_2 = 0 \tag{13–8}$$

for arbitrary admissible $\delta\mathbf{v}$.

In discrete form Π_2 is the same as $\delta^2\Pi$ in Eq. (13–5) with $\delta\mathbf{q}$ renamed $\dot{\mathbf{q}}$ and representing generalized displacements for $\mathbf{v}$. Thus

$$\Pi_2 = \dot{\mathbf{q}}^T\mathbf{K}\dot{\mathbf{q}} \tag{13–9}$$

With $\delta\dot{\mathbf{q}}$ arbitrary, Eq. (13–8) yields $\mathbf{K}\dot{\mathbf{q}} = 0$ which is the same equation as obtained through the work criterion.

3. Nonconservative Forces

Consider now the case where some external forces are nonconservative. Their virtual work δW_n evaluated in state $(\mathbf{u}+\mathbf{v})$ is a linear functional in $\delta\mathbf{v}$ whose coefficients are functionals of $(\mathbf{u}+\mathbf{v})$. The Taylor expansion of δW_n in orders of $\mathbf{v}$ has the form

$$\delta W_n = \delta W_{n0} + \delta W_{n1} + \dots \tag{13–10}$$

The equation of virtual displacements in state $\mathbf{u}$ is $\delta\Pi_1 - \delta W_{n0} = 0$. In state $(\mathbf{u}+\mathbf{v})$ it is $\delta\Pi_1 + \frac{1}{2}\delta\Pi_2 = \delta W_{n0} + \delta W_{n1}$, and it reduces to

$$\tfrac{1}{2}\delta\Pi_2 - \delta W_{n1} = 0 \tag{13–11}$$

In a formal variational method Eq. (13–11) yields as Euler equations the differential equations of equilibrium and the static boundary conditions, expressed in terms of $\mathbf{v}$. The admissibility conditions on $\delta\mathbf{v}$ are the kinematic boundary conditions (see Section 15, Example 5).

In discrete form $\delta W_n = \mathbf{Q}_n^T\delta\dot{\mathbf{q}}$, and $\mathbf{Q}_n$ is function of $(\mathbf{q}+\dot{\mathbf{q}})$. The linear term in the expansion of $\mathbf{Q}_n$ in powers of $\dot{\mathbf{q}}$ is

$$\mathbf{Q}_{n1} = \left(\frac{\partial\mathbf{Q}_n}{\partial\dot{\mathbf{q}}^T}\right)_0 \dot{\mathbf{q}} = -\mathbf{K}_n\dot{\mathbf{q}} \tag{13–12a}$$

and ∂W_{n1} in Eq. (13–11) is

$$\delta W_{n1} = \mathbf{Q}_{n1}^T\delta\dot{\mathbf{q}} = -(\mathbf{K}_n\dot{\mathbf{q}})^T\delta\dot{\mathbf{q}} \tag{13–12b}$$

Since $\mathbf{Q}_n$ is function of $(\mathbf{q}+\dot{\mathbf{q}})$, partial derivatives of $\mathbf{Q}_n$ with respect to $\dot{\mathbf{q}}$ at $\dot{\mathbf{q}} = 0$ coincide with partial derivatives with respect to $\mathbf{q}$ in state $\mathbf{q}$. $\mathbf{K}_n$ as obtained in Eq. (13–12a) coincides with that found earlier in Eq. (13–4c). The discrete stability equations obtained from Eq. (13–11) are then $\mathbf{K}\dot{\mathbf{q}}=0$, where $\mathbf{K}$ is the tangent stiffness of the incremental equations (13–4).

To conclude this outline, needed for deriving incremental stiffness equations and for stability analysis by the energy method are expressions of $\bar{U}_2=\delta^2\bar{U}$, $\bar{V}_2=\delta^2\bar{V}$ and of δW_{n1} or $\mathbf{K}_n$. This is done in the next section.

14. EXPRESSIONS NEEDED IN ENERGY METHOD

The expression of $\delta^2\bar{U}$ and expressions of $\delta^2\bar{V}$ and of the tangent stiffness for basic loadings are obtained in this section.

Expression of $\delta^2\bar{U}$

$\delta^2\bar{U}$ is obtained as $\delta(\delta\bar{U})$, that is, as the first variation of $\delta\bar{U}$. Variation of $\delta\bar{U}$ in Eq. (10–12) yields an expression of the form

$$\delta^2\bar{U}=\int_0^l(\delta N_1\delta\varepsilon_{1c}+N_1\delta^2\varepsilon_{1c}+\delta\mathbf{M}^T\delta\boldsymbol{\chi}+\mathbf{M}^T\delta^2\boldsymbol{\chi})\,dx \tag{14–1}$$

where $\mathbf{M}=\{M_{1s}\ M_2\ M_3\}$ and $\boldsymbol{\chi}=\{\chi_1\ \chi_2\ \chi_3\}$. δN_1 and $\delta\mathbf{M}$ are obtained by variation of Eqs. (10–8a, b, c, d). The term $EK\delta\chi_1\delta\chi_1''$ which occurs in $\delta M_{1s}\delta\chi_1$ is integrated by parts, thus recovering the initial form $EK(\delta\chi_1')^2$ which appears in the strain energy density, Eq. (10–9a). We thus obtain

$$\begin{aligned}\delta^2\bar{U}=&\int_0^l[EA(\delta\varepsilon_{1c})^2+GJ(\delta\chi_1)^2+EK(\delta\chi_1')^2+EI_2(\delta\chi_2)^2+EI_3(\delta\chi_3)^2]\,dx\\&+\int_0^l[N_1\delta^2\varepsilon_{1c}+\mathbf{M}^T\delta^2\boldsymbol{\chi}+(\rho^2N_1+\beta_2M_2-\beta_3M_3)(\delta\chi_1)^2]\,dx\\&+2\int_0^l(EA\rho^2\delta\varepsilon_{1c}+EI_2\beta_2\delta\chi_2-EI_3\beta_3\delta\chi_3)\chi_1\delta\chi_1\,dx\\&+\int_0^l E(3H-\tfrac{1}{2}A\rho^4-\tfrac{1}{2}I_2\beta_2^2-\tfrac{1}{2}I_3\beta_3^2)(\chi_1\delta\chi_1)^2\,dx\end{aligned} \tag{14–2}$$

$\delta^2\bar{V}$ and Tangent Stiffness for Conservative Forces and Moments

A fundamental loading for expressing the external potential consists of prescribed force components $\mathbf{P}=\{P_x,\ P_y,\ P_z\}$ fixed in direction and applied at some cross section at a material point $A(y_A,z_A)$. Then $\bar{V}=-\mathbf{P}^T\mathbf{u}^A$. The displacement distribution over a cross section is given in Eq. (5.6–1) which should be applied here with $\mathbf{u}$ at the shear center, and thus

$$\mathbf{r}=\{0\quad y_A-y_s\quad z_A-z_s\} \tag{14–3}$$

Equation (5.6–1) is simplified according to the second-order theory through use of Eq. (9–3) and reduces to

$$\mathbf{u}^A = \mathbf{u} - \underset{\sim}{\boldsymbol{\varphi}}(\mathbf{I} - \tfrac{1}{2}\underset{\sim}{\boldsymbol{\varphi}})\mathbf{r} = \mathbf{u} + \underset{\sim}{\mathbf{r}}\boldsymbol{\varphi} + \tfrac{1}{2}[\boldsymbol{\varphi}\boldsymbol{\varphi}^T - (\boldsymbol{\varphi}^T\boldsymbol{\varphi})\mathbf{I}]\mathbf{r} \qquad (14\text{–}4)$$

With rearrangement of terms, $\bar{V}$ takes the form

$$\bar{V} = -\mathbf{P}^T\mathbf{u}^A = -\mathbf{P}^T\mathbf{u} - \mathbf{P}^T\underset{\sim}{\mathbf{r}}\boldsymbol{\varphi} - \tfrac{1}{2}\boldsymbol{\varphi}^T[\mathbf{P}\mathbf{r}^T - (\mathbf{P}^T\mathbf{r})\mathbf{I}]\boldsymbol{\varphi} \qquad (14\text{–}5)$$

Because elements of $\mathbf{u}$ are independent $\delta^2\mathbf{u} = 0$. $\delta^2\bar{V}$ depends then only on the terms in $\boldsymbol{\varphi}$. Forming $\delta\bar{V} = (\partial\bar{V}/\partial\boldsymbol{\varphi}^T)\delta\boldsymbol{\varphi}$ and then $\delta^2\bar{V} = \delta(\delta\bar{V})$, we obtain

$$\frac{\partial\bar{V}}{\partial\boldsymbol{\varphi}} = \underset{\sim}{\mathbf{r}}\mathbf{P} - \tfrac{1}{2}[\mathbf{P}\mathbf{r}^T + \mathbf{r}\mathbf{P}^T - 2(\mathbf{P}^T\mathbf{r})\mathbf{I}]\boldsymbol{\varphi} \qquad (14\text{–}6a)$$

$$\delta^2\bar{V} = -\frac{1}{2}\delta\boldsymbol{\varphi}^T[\mathbf{P}\mathbf{r}^T + \mathbf{r}\mathbf{P}^T - 2(\mathbf{P}^T\mathbf{r})\mathbf{I}]\delta\boldsymbol{\varphi} + \frac{\partial\bar{V}}{\partial\boldsymbol{\varphi}^T}\delta^2\boldsymbol{\varphi} \qquad (14\text{–}6b)$$

The tangent stiffness due to the conservative load is the matrix of the quadratic form expressing $\delta^2\bar{V}$ in terms of independent generalized displacements $\delta\mathbf{q}$. Letting

$$\mathbf{G}_\varphi = \{G_{1\varphi} \;\; G_{2\varphi} \;\; G_{3\varphi}\} = -\frac{\partial\bar{V}}{\partial\boldsymbol{\varphi}}$$

the tangent stiffness is obtained from Eq. (14–6b) in the form

$$\mathbf{k}_{eq} = -\frac{1}{2}\frac{\partial\boldsymbol{\varphi}^T}{\partial\mathbf{q}}[\mathbf{P}\mathbf{r}^T + \mathbf{r}\mathbf{P}^T - 2(\mathbf{P}^T\mathbf{r})\mathbf{I}]\frac{\partial\boldsymbol{\varphi}}{\partial\mathbf{q}^T} - \sum_i G_{i\varphi}\frac{\partial^2\varphi_i}{\partial\mathbf{q}\partial\mathbf{q}^T} \qquad (14\text{–}7a)$$

A possible choice of $\mathbf{q}$ is $\mathbf{q} = \{\varphi_1 \;\; -u_3' \;\; u_2'\}$. $\boldsymbol{\varphi}$ is then defined in terms of $\mathbf{q}$ by Eqs. (10–1). If $\boldsymbol{\varphi}$ is chosen as $\mathbf{q}$, however, then the tangent stiffness is

$$\mathbf{k}_{e\varphi} = -\tfrac{1}{2}(\mathbf{P}\mathbf{r}^T + \mathbf{r}\mathbf{P}^T) + (\mathbf{P}^T\mathbf{r})\mathbf{I} \qquad (14\text{–}7b)$$

Equations (14–4) to (14–7) may be applied in a wider context. Point A may be any material point of a rigid arm connected to the member at some cross section B where the displacements are $\mathbf{u}$ and $\boldsymbol{\varphi}$. $\mathbf{r}$ consists then of the components in the undeformed state of the lever arm $\overrightarrow{BA}$. For a couple consisting of $\mathbf{P}$ applied at A and $-\mathbf{P}$ applied at B, the potential is $\bar{V} = -\mathbf{P}^T(\mathbf{u}^A - \mathbf{u}^B)$. Equations (14–5) to (14–7) may then be applied by deleting $\mathbf{u}$ from Eq. (14–5) and letting $\mathbf{r}$ be the components in the undeformed state of the lever arm $\overrightarrow{BA}$.

Generalized Forces and Tangent Stiffness of Nonconservative Moments

Two types of nonconservative moments are considered in what follows, and their tangent stiffnesses are determined.

External moment components applied at an arbitrary cross section and defined on the fixed triad $\vec{\mathbf{e}}_0$ are denoted $\mathbf{G} = \{G_1 \ \ G_2 \ \ G_3\}$. As seen in Section 7, the virtual work of $\mathbf{G}$ is $\delta W = \mathbf{G}^T \boldsymbol{\delta\omega} = \mathbf{G}_\varphi^T \delta\boldsymbol{\varphi}$. In the second-order theory

$$\delta W = \mathbf{G}_\varphi^T \delta\boldsymbol{\varphi} = (\mathbf{G} - \tfrac{1}{2}\underset{\sim}{\mathbf{G}}\boldsymbol{\varphi})^T \delta\boldsymbol{\varphi} \tag{14–8a}$$

For a given discretization in terms of generalized displacements $\mathbf{q}$, δW takes the form $\mathbf{Q}^T \delta\mathbf{q}$, where

$$\mathbf{Q} = \frac{\partial \boldsymbol{\varphi}^T}{\partial \mathbf{q}} \mathbf{G}_\varphi = \frac{\partial \boldsymbol{\varphi}^T}{\partial \mathbf{q}} \left(\mathbf{G} - \frac{1}{2} \underset{\sim}{\mathbf{G}} \boldsymbol{\varphi} \right) \tag{14–8b}$$

A moment vector fixed in direction and magnitude has constant components $\mathbf{G}$. In that case δW contains the term $\boldsymbol{\varphi}^T \underset{\sim}{\mathbf{G}} \delta\boldsymbol{\varphi}$ which is not an exact differential because $\underset{\sim}{\mathbf{G}}$ is antisymmetric. The moment is thus nonconservative, δW and $\mathbf{Q}$ are identified with δW_n and $\mathbf{Q}_n$, and the tangent stiffness due to $\mathbf{G}$ is, using Eq. (13–4c),

$$\mathbf{k}_n = -\frac{\partial \mathbf{Q}_n}{\partial \mathbf{q}^T} = \frac{1}{2} \frac{\partial \boldsymbol{\varphi}^T}{\partial \mathbf{q}} \underset{\sim}{\mathbf{G}} \frac{\partial \boldsymbol{\varphi}}{\partial \mathbf{q}^T} - \sum_i G_{i\varphi} \frac{\partial^2 \varphi_i}{\partial \mathbf{q} \partial \mathbf{q}^T} \tag{14–9}$$

Let now $\mathbf{G}_\varphi$ and $\mathbf{Q}$ be expressed in terms of the components $\mathbf{G}'$ of the moment vector on the cross-section-bound triad $\vec{\mathbf{e}}$. The virtual work expression is $\mathbf{G}'^T \boldsymbol{\delta\omega}'$, where, from Eq. (3–7b), $\boldsymbol{\delta\omega}' = \boldsymbol{\mu} \delta\boldsymbol{\varphi}$. Thus

$$\delta W = \mathbf{G}_\varphi^T \delta\boldsymbol{\varphi} = (\mathbf{G}' + \tfrac{1}{2}\underset{\sim}{\mathbf{G}}'\boldsymbol{\varphi})^T \delta\boldsymbol{\varphi} \tag{14–10a}$$

$$\mathbf{Q} = \frac{\partial \boldsymbol{\varphi}^T}{\partial \mathbf{q}} \mathbf{G}_\varphi = \frac{\partial \boldsymbol{\varphi}^T}{\partial \mathbf{q}} \left(\mathbf{G}' + \frac{1}{2} \underset{\sim}{\mathbf{G}}' \boldsymbol{\varphi} \right) \tag{14–10b}$$

Equations (14–10) and (14–8) are equivalent in that they express the same quantities but in terms of moment vector components defined on different axes. Incrementally, however, they are not necessarily equivalent. They are only if $\mathbf{G}_\varphi^T \delta\boldsymbol{\varphi}$ is an exact differential, in which case the moment is conservative.

Another example of a nonconservative moment is a moment vector bound to the cross section or a follower moment. In that case $\mathbf{G}'$ is constant, and the tangent stiffness is

$$\mathbf{k}_n = -\frac{\partial \mathbf{Q}_n}{\partial \mathbf{q}^T} = -\frac{1}{2} \frac{\partial \boldsymbol{\varphi}^T}{\partial \mathbf{q}} \underset{\sim}{\mathbf{G}}' \frac{\partial \boldsymbol{\varphi}}{\partial \mathbf{q}^T} - \sum_i G'_{i\varphi} \frac{\partial^2 \boldsymbol{\varphi}_i}{\partial \mathbf{q} \partial \mathbf{q}^T} \tag{14–11}$$

Note that if $\mathbf{q}$ is chosen as $\boldsymbol{\varphi}$, then $\mathbf{k}_n = \frac{1}{2}\underset{\sim}{\mathbf{G}}$ for the fixed moment and $\mathbf{k}_n = -\frac{1}{2}\underset{\sim}{\mathbf{G}}'$ for the follower moment.

15. LINEARIZED STABILITY ANALYSIS BY ENERGY METHOD

Linearized stability analysis was formulated by means of differential equations in Section 12. It is based on assuming that the initial equilibrium state is undeformed and thus reduces the problem to one of bifurcation of equilibrium. The same assumption is implemented in the variational formulation, Eq. (13–11), by neglecting the initial displacements $\mathbf{u}$ in the expressions of Π_2 and δW_{n1} but keeping the initial internal forces. It is recalled that $\Pi_2 = \delta^2\Pi = \delta^2\bar{U} + \delta^2\bar{V}$ and that $\delta\mathbf{u}$ is replaced by $\mathbf{v}$ so that Π_2 is a functional of the buckling displacements $\mathbf{v}$.

Expression of $\bar{U}_2$

The simplified expression of $\delta^2\bar{U}$ consists of the first two integrals in Eq. (14–2). To express these integrals, we form the first and second variations of strains from the undeformed state. First variations are obtained from the linear terms. From Eqs. (10–1, 2, 3) and (10–7a) we obtain

$$\delta\varphi_2, \delta\varphi_3 = -\delta u_3', \delta u_2' \tag{15–1}$$

$$\delta\chi_1, \delta\chi_2, \delta\chi_3 = \delta\varphi_1', -\delta u_3'', \delta u_2'' \tag{15–2a}$$

$$\delta\varepsilon_{1c} = \delta u_1' + z_s\delta u_3'' + y_s\delta u_2'' \tag{15–2b}$$

Second variations are twice the second-order terms in the displacements. From the expressions of Section 10 we obtain

$$\delta^2\varphi_2, \delta^2\varphi_3 = \delta\varphi_1\delta u_2', \delta\varphi_1\delta u_3' \tag{15–3}$$

$$\delta^2\chi_1 = \delta u_3'\delta u_2'' - \delta u_2'\delta u_3'' \tag{15–4a}$$

$$\delta^2\chi_2 = 2\delta\varphi_1\delta u_2'' \tag{15–4b}$$

$$\delta^2\chi_3 = 2\delta\varphi_1\delta u_3'' \tag{15–4c}$$

$$\delta^2\varepsilon_{1c} = (\delta u_2')^2 + (\delta u_3')^2 - 2z_s\delta\varphi_1\delta u_2'' + 2y_s\delta\varphi_1\delta u_3'' \tag{15–4d}$$

Renaming $\delta^2\bar{U}$ as $\bar{U}_2$ and

$$\{\delta u_1 \;\; \delta u_2 \;\; \delta u_3 \;\; \delta\varphi_1\} = \{v_1 \;\; v_2 \;\; v_3 \;\; \theta_1\} \tag{15–5}$$

the first two integrals in Eq. (14–2) yield

$$\bar{U}_2 = \int_0^l [EA(v_1' + z_s v_3'' + y_s v_2'')^2 + GJ(\theta_1')^2 + EK(\theta_1'')^2 + EI_2(v_3'')^2 + EI_3(v_2'')^2]\,dx$$

$$+\int_0^l N_1[(v_2')^2+(v_3')^2-2z_s\theta_1 v_2''+2y_s\theta_1 v_3''+\rho^2(\theta_1')^2]\,dx$$

$$+\int_0^l M_{1s}(v_3' v_2''-v_2' v_3'')\,dx+\int_0^l M_2[2\theta_1 v_2''+\beta_2(\theta_1')^2]\,dx$$

$$+\int_0^l M_3[2\theta_1 v_3''-\beta_3(\theta_1')^2]\,dx \tag{15–6}$$

The last four integrals represent a change in strain energy due to initial internal forces. They appear in certain derivations as part of the external potential.

Expression of $\bar{V}_2$

For the conservative forces and moments considered in Section 14, $\bar{V}_2$ is obtained from Eq. (14–6b) in which $\delta\boldsymbol{\varphi}=\{\theta_1 \;\; -v_3' \;\; v_2'\}$, $\partial\bar{V}/\partial\boldsymbol{\varphi}=\underset{\sim}{\mathbf{r}}\mathbf{P}$, and $\delta^2\varphi_1=0$. $\delta^2\varphi_2$ and $\delta^2\varphi_3$ are obtained from Eq. (15–3) in treating one-member problems using displacement derivatives as generalized displacements. For the case of a force $\mathbf{P}=\{P_x \;\; P_y \;\; P_z\}$ fixed in direction and applied at point $A(y_A, z_A)$ of a cross section, we obtain

$$\begin{aligned}\bar{V}_2 = 2P_x[(y_A-y_s)\theta_1 v_3'-(z_A-z_s)\theta_1 v_2']+P_y[(y_A-y_s)(\theta_1^2+v_2'^2)\\ +(z_A-z_s)v_2' v_3']+P_z[(z_A-z_s)(\theta_1^2+v_3'^2)+(y_A-y_s)v_2' v_3']\end{aligned} \tag{15–7}$$

A similar procedure is used for the other cases of conservative forces and moments considered in Section 14.

Expression of δW_{n1} and k_n for Nonconservative Moments

The tangent stiffness matrix for nonconservative moments was determined in Eqs. (14–9) and (14–11). These formulas are applied with

$$\mathbf{q}=\{\varphi_1 \;\; -u_3' \;\; u_2'\} \tag{15–8a}$$

and with partial derivatives at $\mathbf{q}=0$. The buckling displacements at the point of application of the moment are

$$\dot{\mathbf{q}}=\{\theta_1 \;\; -v_3' \;\; v_2'\} \tag{15–8b}$$

For a moment fixed in orientation Eq. (14–9) yields

$$\mathbf{k}_n=\frac{1}{2}\underset{\sim}{\mathbf{G}}-\sum_i G_i\frac{\partial^2\varphi_i}{\partial\mathbf{q}\partial\mathbf{q}^T}=\begin{bmatrix}0 & G_3 & -G_2\\ 0 & 0 & \frac{1}{2}G_1\\ 0 & -\frac{1}{2}G_1 & 0\end{bmatrix} \tag{15–9a}$$

and from Eq. (13–12b),

$$\delta W_{n1} = -(\mathbf{k}_n \dot{\mathbf{q}})^T \delta\dot{\mathbf{q}} = (G_3 v_3' + G_2 v_2')\delta\theta_1 + \tfrac{1}{2}G_1 v_2' \delta v_3' - \tfrac{1}{2}G_1 v_3' \delta v_2' \quad (15\text{–}9\text{b})$$

For a follower moment Eq. (14–11) applies with $\mathbf{G}' = \mathbf{G}$, since $\mathbf{G}'$ is also defined in the initial state $\boldsymbol{\varphi} = 0$. There comes

$$\mathbf{k}_n = -\frac{1}{2}\underset{\sim}{\mathbf{G}} - \sum_i G_i \frac{\partial^2 \varphi_i}{\partial \mathbf{q} \partial \mathbf{q}^T} = \begin{bmatrix} 0 & 0 & 0 \\ G_3 & 0 & -\tfrac{1}{2}G_1 \\ -G_2 & \tfrac{1}{2}G_1 & 0 \end{bmatrix} \quad (15\text{–}10\text{a})$$

and

$$\delta W_{n1} = -(\mathbf{k}_n \dot{\mathbf{q}})^T \delta\dot{\mathbf{q}} = (G_3\theta_1 - \tfrac{1}{2}G_1 v_2')\delta v_3' + (G_2\theta_1 + \tfrac{1}{2}G_1 v_3')\delta v_2' \quad (15\text{–}10\text{b})$$

Example 1. Stability of Simply Supported Beam in Pure Bending

Consider a simply supported beam in a state of pure bending in the principal plane (x, y), Fig. 12–2. We have then $(N_1, M_{1s}, M_2) = 0$ and $M_3 = G_3$. Assume first that the applied end moments are fixed in orientation. The virtual work term δW_{n1} for one end moment is obtained from Eq. (15–9b) as $G_3 v_3' \delta\theta_1$ and is zero because of the support condition $\delta\theta_1 = 0$. The same result is obtained from Eq. (15–10b) for follower moments because $\theta_1 = 0$ at the member ends. If the end moments are assumed caused each by a couple of conservative forces applied in the (x, y) plane perpendicularly to the cross section, the potential $\bar{V}_2$ is obtained by application of Eq. (15–7) and is zero because $\theta_1 = 0$. Thus for any of these cases the stability criterion reduces to $\delta\bar{U}_2 = 0$. An outcome of this variational equation is that both the EA and EI_3 terms in $\bar{U}_2$, Eq. (15–6), vanish—that is, the axial force and bending moment M_3 remain unchanged through buckling. $\bar{U}_2$ reduces to

$$\bar{U}_2 = \int_0^l [(GJ - \beta_3 G_3)(\theta_1')^2 + EK(\theta_1'')^2 + EI_2(v_3'')^2]\, dx + 2G_3 \int_0^l \theta_1 v_3''\, dx \quad (15\text{–}11)$$

Admissible displacements need only satisfy the kinematic conditions $\theta_1 = v_3 = 0$ at $x = 0$ and $x = l$. The representation

$$\theta_1, v_3 = a \sin\frac{\pi x}{l}, b \sin\frac{\pi x}{l}$$

satisfies also the static conditions $\theta_1'' = v_3'' = 0$ at $x = 0$ and $x = l$. Evaluating $\bar{U}_2$ for a uniform member, and setting its partial derivatives with respect to a and b

equal to zero, we obtain the characteristic-value problem

$$\begin{bmatrix} GJ - \beta_3 G_3 + EK\dfrac{\pi^2}{l^2} & -G_3 \\ -G_3 & \dfrac{\pi^2 EI_2}{l^2} \end{bmatrix} \begin{Bmatrix} a \\ b \end{Bmatrix} = 0$$

whose characteristic equation is

$$G_3^2 + \beta_3 \frac{\pi^2 EI_2}{l^2} G_3 - \frac{\pi^2 EI_2}{l^2}\left(GJ + \frac{\pi^2 EK}{l^2}\right) = 0 \qquad (15\text{–}12)$$

The representation used for θ_1 and v_3 in this example satisfies exactly the governing differential equations and boundary conditions seen in Section 12, Example 2. Equation (15–12) yields therefore the exact solution for the critical moment G_3. If the z axis is an axis of symmetry, or if the centroid is a center of symmetry, then β_3 as given in Eq. (10–8i) is zero, and Eq. (15–12) yields a single value for G_3. In general, Eq. (15–12) has two roots of opposite signs, each of which is a critical moment depending on the sense of the applied G_3.

Example 2. Lateral Buckling of a Cantilever Subjected to a Concentrated Force

The cantilever under consideration is the same one as in Section 12, Example 3, Fig. 12–3. The initial internal forces are $M_3 = F_2(l-x)$ and $N_1 = M_{1s} = M_2 = 0$. As in Example 1 the EA and EI_3 terms may be deleted from $\bar{U}_2$. From Eq. (15–6),

$$\bar{U}_2 = \int_0^l \{[GJ - \beta_3 F_2(l-x)](\theta_1')^2 + EK(\theta_1'')^2 + EI_2(v_3'')^2\}\, dx$$

$$+ \int_0^l 2F_2(l-x)\theta_1 v_3''\, dx \qquad (15\text{–}13)$$

The force F_2 is applied at a point $A(Y_A, z_s)$ at end b, has a fixed direction, and passes through the shear center in the initial state. From Eq. (15–7) with $P_y = F_2$, $\theta_1 = \theta_1^b$, and $v_2' = 0$,

$$\bar{V}_2 = F_2(y_A - y_s)(\theta_1^b)^2 \qquad (15\text{–}14)$$

The admissibility conditions are the kinematic end conditions $\theta_1 = v_3 = 0$ and $\theta_1' = v_3' = 0$ at $x = 0$. If the formal variational method is applied, the Euler equations consist of the differential equations of equilibrium and the static conditions at end b. These equations may be integrated into a set of lower order, and it may be verified that they reduce to the differential equations established in Section 12, Example 3.

For a simple application of the direct variational method, consider a cross section symmetric about the z axis and having a negligible warping constant. Then $\beta_3 = 0$, and $EK = 0$. Also let F_2 be applied at the shear center which must be on the axis of symmetry. Thus $y_A = y_s = 0$, and $\bar{V}_2$ vanishes.

In choosing a representation for θ_1 and v_3, it is advantageous for accuracy and for lessening the calculations involved to enforce the differential equation relating them. This equation may be found by the formal variational method, and it reduces to Eq. (12–10b), or

$$EI_2 v_3'' + F_2(l - x)\theta_1 = 0$$

Expressing $\bar{U}_2$ in terms of θ_1, we obtain

$$\bar{U}_2 = \int_0^l [GJ(\theta_1')^2 - \frac{F_2^2}{EI_2}(l - x)^2\theta_1^2]\,dx$$

A representation of θ_1 must satisfy the kinematic condition $\theta_1 = 0$ at $x = 0$. Using the two-term polynomial

$$\theta_1 = a\frac{x}{l} + b\left(\frac{x}{l}\right)^2$$

we obtain

$$\bar{U}_2 = \frac{GJ}{l}\left(a^2 + \frac{4}{3}b^2 + 2ab\right) - \frac{F_2^2 l^3}{EI_2}\left(\frac{a^2}{30} + \frac{b^2}{105} + \frac{ab}{30}\right)$$

The conditions $\partial\bar{U}_2/\partial a = \partial\bar{U}_2/\partial b = 0$ result in a characteristic-value problem whose lowest characteristic value is $\lambda = 4.08$, where

$$\lambda = \frac{F_2 l^2}{\sqrt{EI_2 GJ}} \tag{15–15}$$

Use of more terms would make λ tend by decreasing values to the exact value. It will be seen that, with the same two-term polynomial shape, the complementary energy method yields $\lambda_{cr} = 4.013$ with four accurate digits.

Consider now the case where F_2 is applied at point $A(y_A, z_s)$. $\bar{V}_2$ is evaluated through Eq. (15–14) in which $\theta_1^b = a + b$.
Letting

$$\alpha = \frac{y_s - y_A}{l}\sqrt{\frac{EI_2}{GJ}} \tag{15–16}$$

the condition $\delta(\bar{U}_2 + \bar{V}_2) = 0$ yields

$$\left(1 - \frac{\lambda^2}{30} - \alpha\lambda\right)a + \left(1 - \frac{\lambda^2}{60} - \alpha\lambda\right)b = 0$$

$$\left(1 - \frac{\lambda^2}{60} - \alpha\lambda\right)a + \left(\frac{4}{3} - \frac{\lambda^2}{105} - \alpha\lambda\right)b = 0$$

The determinantal equation may be solved by successive approximations. A first approximation obtained by the Newton-Raphson method, starting at the solution λ_0 for $\alpha = 0$, is found to be $\lambda = \lambda_0(1 - 1.11\alpha)$. A more accurate solution would require a better representation of θ_1. The solution given by Timoshenko [6] is also approximate and is $\lambda_0(1 - \alpha)$ with $\lambda_0 = 4.013$.

The y axis is oriented in the sense of F_2 so that F_2 and λ are positive. If F_2 points from point A toward the shear center, then $y_s - y_A > 0, \alpha > 0$, and λ_{cr} is decreased. In the other case $\alpha < 0$ and λ_{cr} is decreased.

Example 3. Lateral Buckling of a Simply Supported Beam Subjected to a Transverse Load

The beam under consideration is simply supported and is in equilibrium under a distributed transverse load p_y fixed in orientation and applied on the line $(y = y_A, z = z_s)$, Fig. 12–5. Then $N_1 = M_{1s} = M_2 = 0$, and as in the preceding example, the EA and v_2 terms are deleted. $\bar{U}_2$ and $\bar{V}_2$ reduce to

$$\bar{U}_2 = \int_0^l [(GJ - \beta_3 M_3)(\theta_1')^2 + EK(\theta_1'')^2 + EI_2(v_3'')^2]\,dx + \int_0^l 2M_3\theta_1 v_3''\,dx \tag{15–17a}$$

$$\bar{V}_2 = (y_A - y_s)\int_0^l p_y\theta_1^2\,dx \tag{15–17b}$$

The stability criterion is $\delta\bar{U}_2 + \delta\bar{V}_2 = 0$ subject to the conditions $\delta\theta_1 = \delta v_3 = 0$ at $x = 0$ and $x = l$.

The equilibrium equation relating v_3 and θ_1 may be found by the formal variational method to be the same as the one found in Section 12, Example 5; that is

$$EI_2 v_3'' + M_3\theta_1 = 0$$

As an application, consider a cross section for which β_3 and K are zero. A concentrated force $P_y = P$ is applied at midspan at the shear center. Then $y_A = y_s$ and $\bar{V}_2$ vanishes. Letting the origin of x be at midspan, we have

$$M_3 = \frac{P}{2}\left(\frac{l}{2} - x\right), \qquad 0 \leqslant x \leqslant \frac{l}{2}$$

Expressing v_3'' in terms of θ_1 and noting symmetry with respect to midspan, we obtain

$$\frac{1}{2}\bar{U}_2 = \int_0^{l/2} \left[GJ(\theta_1')^2 - \frac{P^2(l/2-x)^2}{4EI_2}\theta_1^2 \right] dx \tag{15–18}$$

The representation $\theta_1 = a\cos(\pi x/l)$ satisfies symmetry and the admissibility conditions $\theta_1 = 0$ at $x = \pm l/2$ and yields

$$\frac{1}{2}\bar{U}_2 = \left[GJ\frac{\pi^2}{4l} - \frac{P^2}{4EI_2}\frac{l^3}{8}\left(\frac{1}{6}+\frac{1}{\pi^2}\right)\right]a^2$$

whence

$$\lambda = \frac{Pl^2}{\sqrt{EI_2 GJ}} = 4\pi^2\sqrt{\frac{3}{\pi^2+6}} = 17.16$$

The exact value to four digits is given by Timoshenko [6] as $\lambda_{cr} = 16.94$.

Example 4. Stability in Eccentric Compression

The problem considered in Section 12, Example 1, Fig. 12–1 is formulated here by the energy method. The eccentric axial force applied at point $A(y_A, z_A)$ causes the bending moments $M_2 = F_1 z_A$ and $M_3 = -F_1 y_A$. The EA term is deleted from $\bar{U}_2$ because N_1 remains unchanged through buckling. From Eqs. (15–6) and (15–7), with $P_x = F_1$,

$$\begin{aligned}\bar{U}_2 = &\int_0^l [(GJ + \beta_2 z_A F_1 + \beta_3 y_A F_1)(\theta_1')^2 + EK(\theta_1'')^2 + EI_2(v_3'')^2 + EI_3(v_2'')^2]\,dx \\ &+ F_1 \int_0^l [2(z_A - z_s)\theta_1 v_2'' - 2(y_A - y_s)\theta_1 v_3'' + (v_2')^2 + (v_3')^2 + \rho^2(\theta_1')^2]\,dx\end{aligned} \tag{15–19a}$$

$$\bar{V}_2 = 2F_1[(y_A - y_s)\theta_1 v_3' - (z_A - z_s)\theta_1 v_2'] \tag{15–19b}$$

The expression of $\bar{U}_2 + \bar{V}_2$ may be simplified by integrating by parts the terms in $\theta_1 v_2''$ and $\theta_1 v_3''$ in Eq. (15–19a).

It is noted that if $y_A = y_s$ and $z_A = z_s$, there are no coupling terms between θ_1, v_2, and v_3. The stability problem may then be separated into three independent problems as seen in Section 12 in describing the governing differential equations. Various boundary conditions have also been discussed in Section 12.

Example 5. Stability in Torsion by Formal Variational Method

Consider a member simply supported at its ends against transverse displacements and supported at end a against axial rotation and displacement. A torque

G_1 is applied at end b so that $N_1 = M_2 = M_3 = 0$ and $M_{1s} = G_1$. The θ_1 terms in $\bar{U}_2$ are then uncoupled from the remaining displacements. Further the various types of applied torque to be considered do not contribute θ_1 terms to the boundary conditions. Terms in θ_1 and EA are thus deleted from $\bar{U}_2$, Eq. (15–6), which reduces to

$$\bar{U}_2 = \int_0^l [EI_2(v_3'')^2 + EI_3(v_2'')^2 + G_1(v_3' v_2'' - v_2' v_3'')]\, dx \qquad (15\text{–}20a)$$

Forming $\delta\bar{U}_2$, integrating by parts twice terms in $\delta v_3''$ and $\delta v_2''$, and applying the kinematic conditions $\delta v_2 = \delta v_3 = 0$ at the ends, there comes

$$\tfrac{1}{2}\delta\bar{U}_2 = -\int_0^l [\delta v_3(EI_2 v_3' - G_1 v_2)''' + \delta v_2(EI_3 v_2' + G_1 v_3)''']\, dx$$
$$+ [(EI_2 v_3'' - \tfrac{1}{2}G_1 v_2')\delta v_3' + (EI_3 v_2'' + \tfrac{1}{2}G_1 v_3')\delta v_2']_0^l \qquad (15\text{–}20b)$$

In the present problem the terms $\delta\bar{V}_2$ and δW_{n1} in the stability criterion, Eq. (13–11), are boundary terms. The Euler differential equations are thus obtained by equating to zero the coefficients of δv_3 and δv_2 in the preceding expression. There comes

$$(EI_2 v_3' - G_1 v_2)''' = 0 \qquad (15\text{–}21a)$$

$$(EI_3 v_2' + G_1 v_3)''' = 0 \qquad (15\text{–}21b)$$

Various types of external torques are now considered.

Fixed External Torques

External torques fixed in orientation are nonconservative. The stability criterion is then

$$\tfrac{1}{2}\delta\bar{U}_2 - \delta W_{n1} = 0 \qquad (15\text{–}22)$$

where δW_{n1} is obtained from Eq. (15–9b) as

$$\delta W_{n1} = [\tfrac{1}{2}G_1(v_2'\delta v_3' - v_3'\delta v_2')]_0^l \qquad (15\text{–}23)$$

The natural boundary conditions are obtained by equating to zero the coefficients of $\delta v_3'$ and $\delta v_2'$ at $x = 0$ and $x = l$ in Eq. (15–22). There comes

$$EI_2 v_3'' - G_1 v_2' = 0 \qquad (15\text{–}24a)$$

$$EI_3 v_2'' + G_1 v_3' = 0 \qquad (15\text{–}24b)$$

Integrating each of Eqs. (15–21) twice, the integration constants are found to be zero by the preceding boundary conditions which also become the differential equations. The remaining boundary conditions are the kinematic ones, $v_3 = v_2 = 0$ at $x = 0$ and $x = l$. Equations (15–24) are conveniently solved using complex functions by setting

$$z = \frac{v_3}{\sqrt{EI_3}} + i\frac{v_2}{\sqrt{EI_2}} \tag{15–25}$$

$$\lambda = \frac{G_1}{\sqrt{EI_2 EI_3}} \tag{15–26}$$

and combining Eqs. (15–24) into

$$z'' + i\lambda z' = 0 \tag{15–27}$$

The general solution for z has the form

$$z = Ae^{-i\lambda x} + B$$

the boundary conditions $z = 0$ at $x = 0$ and $x = l$ yield $B = -A$ and $A(e^{-i\lambda l} - 1) = 0$. A nontrivial solution exists if $e^{-i\lambda l} = 1$, whence $\lambda_{cr} = 2\pi/l$ and

$$G_{1,cr} = \frac{2\pi}{l}\sqrt{EI_2 EI_3} \tag{15–28}$$

Follower External Torques

For follower external torques δW_{n1} is obtained from Eq. (15–10b) as

$$\delta W_{n1} = \tfrac{1}{2}[G_1(v_3'\delta v_2' - v_2'\delta v_3')]_0^l \tag{15–29}$$

The natural boundary conditions of Eq. (15–22) are obtained as $v_3'' = v_2'' = 0$ at $x = 0$ and $x = l$. In terms of the complex function z the problem consists of the differential equation

$$z'''' + i\lambda z''' = 0 \tag{15–30}$$

with the boundary conditions $z'' = z = 0$ at $x = 0$ and $x = l$. The general solution has the form

$$z = Ae^{-i\lambda x} + \frac{Bx^2}{2} + Cx + D$$

and the boundary conditions yield

$$A + D = 0 \tag{15–31a}$$

$$Ae^{-i\lambda l} + \frac{Bl^2}{2} + Cl + D = 0 \tag{15–31b}$$

$$-\lambda^2 A + B = 0 \tag{15–31c}$$

$$-\lambda^2 Ae^{-i\lambda l} + B = 0 \tag{15–31d}$$

For $A \neq 0$ the last two equations yield $e^{-i\lambda l} = 1$. Then $B = \lambda^2 A, D = -A$, and $C = -Bl/2$. The critical λ is the same as in the previous case.

Quasi-Tangential Torque

Consider an external torque at $x = l$ produced by a couple consisting of a force P_z applied at point $A(y_A, z_A = 0)$ and an opposite force applied at point $B(y_B, z_B = 0)$, Fig. 15–1*a*. The torque is

$$G_1 = P_z(y_A - y_B) \tag{15–32}$$

Assuming the forces of the couple to be fixed in orientation, the couple is conservative and is similar in linearized stability analysis to what is called by Ziegler [18] a quasi-tangential torque. The reaction torque at end *a* is assumed to be of the same type as at end *b*. The potential of the end torques is obtained from Eq. (15–7) by replacing y_s by y_B. Thus

$$\bar{V}_2 = G_1 v_2' v_3' |_0^l \tag{15–33}$$

The stability criterion $\delta \bar{U}_2 + \delta \bar{V}_2 = 0$ yields the differential equations (15–21) and

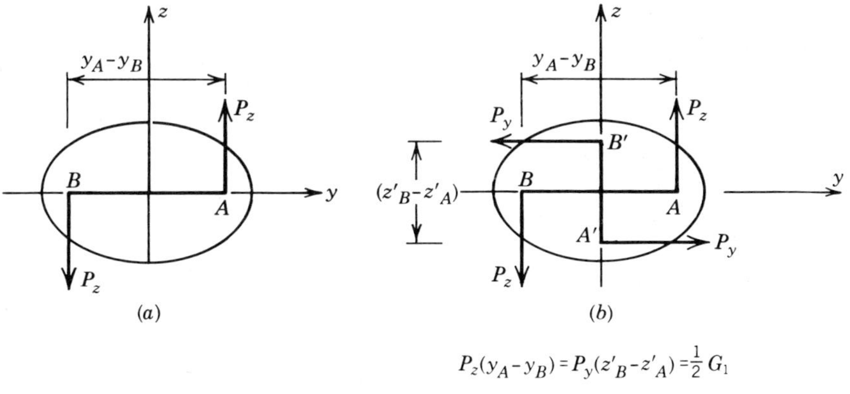

FIG. 15–1. Conservative torsional moments. (*a*), Quasitangential (*b*), Semitangential.

the natural boundary conditions

$$EI_2 v_3'' = 0 \tag{15–34a}$$

$$EI_3 v_2'' + G_1 v_3' = 0 \tag{15–34b}$$

at $x = 0$ and $x = l$. These and the conditions $v_3 = v_2 = 0$ at $x = 0$ and $x = l$ result in a characteristic-value problem whose solution is found similarly to previous cases to be

$$G_{1,\text{cr}} = \frac{\pi}{l}\sqrt{EI_2 EI_3} \tag{15–35}$$

Semitangential Torque

Consider an external torque at $x = l$, half of which is produced as in the preceding quasi-tangential case and the other half by a force P_y applied at $A'(y_A = 0, z_A')$ and an opposite force applied at $B'(y_A = 0, z_B')$, Fig. 15–1b. A similar torque is assumed at end a. The half torque is

$$\tfrac{1}{2}G_1 = P_z(y_A - y_B) = -P_y(z_A' - z_B') \tag{15–36}$$

Equation (15–7) applied as in the previous case yields $\bar{V}_2 = 0$. The stability criterion is then $\delta \bar{U}_2 = 0$, and the natural boundary conditions are obtained from Eq. (15–20b) as

$$EI_2 v_3'' - \tfrac{1}{2}G_1 v_2' = 0 \tag{15–37a}$$

$$EI_3 v_2'' + \tfrac{1}{2}G_1 v_3' = 0 \tag{15–37b}$$

at $x = 0$ and $x = l$. These equations combine into

$$z'' + \tfrac{1}{2}i\lambda z' = 0 \tag{15–37c}$$

The characteristic equation for λ is found to be

$$e^{-i\lambda l} = \frac{6 + i\lambda l}{6 - i\lambda l} \tag{15–38a}$$

Letting $t = \lambda l/6$, the equation reduces to

$$\left(t - \frac{1}{t}\right)\tan 6t = 2 \tag{15–38b}$$

A solution by trial and error yields $t_{\text{cr}} = 0.818548$, or

$$G_{\text{cr}} = \frac{4.9113}{l}\sqrt{EI_2 EI_3} = 1.563\frac{\pi}{l}\sqrt{EI_2 EI_3} \tag{15–39}$$

16. METHOD OF VIRTUAL FORCES IN LINEARIZED STABILITY ANALYSIS

The method of virtual forces in lateral-torsional stability analysis is similar to the one presented for planar stability in Chapter 6, Section 6. It is based on the equation of virtual work in which the force system is virtual and the kinematic system is the incremental one taking place from the initial equilibrium state. Before outlining the method, a notation is adopted that avoids confusion between incremental quantities due to buckling and virtual quantities. The incremental displacements are designated $\mathbf{v} = \{\theta_1 \ v_2 \ v_3\}$, as done earlier, and the incremental strains $\delta\chi_1, \delta\chi_2, \delta\chi_3$ are now designated $\boldsymbol{\kappa} = \{\kappa_1 \ \kappa_2 \ \kappa_3\}$. The incremental strain-displacement relations are then

$$(\kappa_1 \ \kappa_2 \ \kappa_3) = (\theta_1', -v_3'', v_2'') \tag{16–1}$$

Similarly moment increments occurring through buckling δM_{1s}, δM_2, and δM_3 are now denoted μ_{1s}, μ_2, and μ_3, respectively. The bending constitutive equations are then

$$\kappa_2, \kappa_3 = (EI_2)^{-1}\mu_2, (EI_3)^{-1}\mu_3 \tag{16–2a}$$

For torsion, Eq. (12–3a) is replaced by the equivalent system of equations

$$\kappa_1, \kappa_1' = (GJ_N)^{-1}\mu_{1\chi}, (EK)^{-1}\eta \tag{16–2b}$$

$$\mu_{1s} = \mu_{1\chi} - \eta' \tag{16–3}$$

where

$$GJ_N = GJ + \rho^2 N_1 + \beta_2 M_2 - \beta_3 M_3 \tag{16–4}$$

With this new notation the equilibrium solution, Eq. (12–1), takes the form

$$\begin{aligned}\mu_{1s} = {} & G_2 v_2' + G_3 v_3' - F_1(z_s v_2' - y_s v_3') - F_2[v_3^b - v_3 - (l-x)v_3' - y_s\theta_1^b] \\ & - F_3[-v_2^b + v_2 + (l-x)v_2' - z_s\theta_1^b] + \delta G_1 + z_s\delta F_2 - y_s\delta F_3 + \mu_{1sp}\end{aligned} \tag{16–5a}$$

$$\begin{aligned}\mu_2 = {} & -G_1 v_2' + G_3\theta_1 - F_1[-v_3^b + v_3 - y_s(\theta_1 - \theta_1^b)] + F_2[(l-x)\theta_1 - z_s v_2'] \\ & + F_3[y_s(v_2' - v_2'^b) - z_s v_3'^b] + \delta G_2 - (l-x)\delta F_3 + \mu_{2p}\end{aligned} \tag{16–5b}$$

$$\begin{aligned}\mu_3 = {} & -G_1 v_3' - G_2\theta_1 - F_1[v_2^b - v_2 - z_s(\theta_1 - \theta_1^b) + F_2[z_s(v_3'^b - v_3') + y_s v_2'^b] \\ & + F_3[(l-x)\theta_1 + y_s v_3'] + \delta G_3 + (l-x)\delta F_2 + \mu_{3p}\end{aligned} \tag{16–5c}$$

The equation of (complementary) virtual work is a necessary and sufficient condition for geometric compatibility, that is, for $\boldsymbol{\kappa}$ to be related to $\mathbf{v}$ through Eq. (16–1) and for $\mathbf{v}$ and $\mathbf{v}'$ to take prescribed values $\bar{\mathbf{v}}$ and $\bar{\mathbf{v}}'$ at $x=0$ and $x=l$. It has the form

$$\int_0^l (\kappa_2 M_2^* + \kappa_3 M_3^* + \kappa_1 M_{1\chi}^* + \kappa_1' B^*)\,dx = \int_0^l (v_2 p_y^* + v_3 p_z^* + \theta_1 m_{xs}^*)\,dx$$

$$+ (\bar{v}_2 N_2^* + \bar{v}_3 N_3^* + \bar{\theta}_1 M_{1s}^* - \bar{v}_3' M_2^* + \bar{v}_2' M_3^* + \bar{\theta}_1' B^*)_0^l \tag{16–6}$$

where the virtual force system is arbitrary but statically admissible in the undeformed geometry. The statical admissibility conditions are

$$p_y^*, p_z^*, m_{xs}^* = M_3^{*\prime\prime}, -M_2^{*\prime\prime}, (-M_{1\chi}^* + B^{*\prime})' \tag{16–7a}$$

$$(N_2^*, N_3^*, M_{1s}^*)_0^l = (-M_3^{*\prime}, M_2^{*\prime}, M_{1\chi}^* - B^{*\prime})_0^l \tag{16–7b}$$

The internal virtual forces, which appear on the right-hand side of Eq. (16–7a), may be chosen arbitrarily, and Eqs. (16–7) are then definition equations of the external virtual forces. Integration by parts transforms Eq. (16–6) into

$$\int_0^l [M_2^*(\kappa_2 + v_3'') + M_3^*(\kappa_3 - v_2'') + M_{1\chi}^*(\kappa_1 - \theta_1') + B^*(\kappa_1' - \theta_1'')]\,dx$$

$$+ [N_2^*(v_2 - \bar{v}_2) + N_3^*(v_3 - \bar{v}_3) + M_{1s}^*(\theta_1 - \bar{\theta}_1) - M_2^*(v_3' - \bar{v}_3')$$

$$+ M_3^*(v_2' - \bar{v}_2') + B^*(\theta_1' - \bar{\theta}_1')]_0^l = 0 \tag{16–8}$$

Since the internal virtual forces are arbitrary in the interval $(0, l)$ including the end points, Eq. (16–8) is seen to be a necessary and sufficient condition for compatibility, as stated earlier.

The boundary terms in Eqs. (16–6) and (16–8) are modified, depending on which elements of $\mathbf{v}$ and $\mathbf{v}'$ are prescribed at $x=0$ and $x=l$. Nonprescribed elements should be designated by nonbarred symbols in Eq. (16–6), and they vanish from Eq. (16–8).

Eq. (16–8) is transformed into a governing equation for θ_1, v_2, and v_3 by enforcing the constitutive equations (16–2) and the equilibrium solution (16–5). This yields

$$\int_0^l \left[M_2^*\left(\frac{\mu_2}{EI_2} + v_3''\right) + M_3^*\left(\frac{\mu_3}{EI_3} - v_2''\right) + M_{1\chi}^*\left(\frac{\mu_{1\chi}}{GJ_N} - \theta_1'\right)\right.$$

$$\left. + B^*\left(\frac{\eta}{EK} - \theta_1''\right)\right] dx + [N_2^*(v_2 - \bar{v}_2) + \cdots + B^*(\theta_1' - \bar{\theta}_1')]_0^l = 0 \tag{16–9}$$

where μ_2, μ_3, and $\mu_{1s} = \mu_{1\chi} - \eta'$ are as given in Eqs. (16–5). η is not subject to any equilibrium requirement in the interval $(0, l)$ but is subject to the end condition

$\eta = 0$ where warping is free. η is thus an independent statical unknown, and $\mu_{1\chi}$ is obtained from Eq. (16–3).

In a stability problem, prescribed incremental displacements have zero values. The left hand side of Eq. (16–9) is thus a linear homogeneous functional in $\theta_1, v_2, v_3, \eta, \delta\mathbf{F}$, and $\delta\mathbf{G}$.

To solve a given problem, a representation of θ_1, v_2, v_3, and η is assumed in terms of shape functions and undetermined parameters. η must satisfy the end conditions of free warping where applicable. If support conditions are statically determinate, then elements of $\delta\mathbf{G}$ and $\delta\mathbf{F}$ are either zero or have known linear expressions in θ_1, v_2 and v_3 as seen in examples in Section 12. If support conditions are statically indeterminate, some elements of $\delta\mathbf{F}$ and $\delta\mathbf{G}$ are statical redundants. They form with the undetermined parameters a set of n unknowns defining the buckled state. An internal virtual force system depending on as many arbitrary parameters is chosen. To each virtual force parameter corresponds a discrete compatibility equation obtained from Eq. (16–9). The n equations thus obtained define a linear characteristic-value problem.

Note that in the general procedure outlined here, the assumed displacement representation needs not satisfy the kinematic end conditions as these are an outcome of Eq. (16–9). However, it is usually simple enough to satisfy such conditions and thus to use fewer undetermined parameters and to delete the corresponding boundary terms from Eq. (16–9). The freedom with which virtual force systems may be chosen produces many possibilities, some of which have been explored in the case of planar behavior in Chapter 6, Section 6. We will consider in what follows a choice of virtual forces that has the advantage of leading to a minimum problem for the critical load parameter and to an interpretation of the method as a variational complementary energy method. This is

$$(M_2^*, M_3^*, M_{1\chi}^*, B^*) = (\delta\mu_2, \delta\mu_3, \delta\mu_{1\chi}, \delta\eta) \tag{16–10}$$

where $\delta\mu_2, \delta\mu_3$, and $\delta\mu_{1s}$ are variations of the expressions in Eqs. (16–5) with respect to the displacements and to the statical redundants. δ_η is arbitrary but vanishes at a member end free to warp, and $\delta\mu_{1\chi}$ is defined as $\delta\mu_{1\chi} = \delta\mu_{1s} + \delta\eta'$. In Eq. (16–9) the term $B^*\theta_1'' = \delta\eta\theta_1''$ is integrated by parts, and it is noted that $\delta\eta = 0$ if warping is free and $\theta_1' = 0$ if warping is restrained. Equation (16–9) takes the form

$$\delta\bar{U}_c + \delta\bar{V}_c = 0 \tag{16–11a}$$

where $\delta\bar{U}_c$ is the variation of

$$\bar{U}_c = \int_0^l \frac{1}{2}\left(\frac{\mu_2^2}{EI_2} + \frac{\mu_3^2}{EI_3} + \frac{(\mu_{1s} + \eta')^2}{GJ_N} + \frac{\eta^2}{EK}\right) dx \tag{16–11b}$$

and

$$\delta \bar{V}_c = -\int_0^l (v_2'' \delta\mu_3 - v_3'' \delta\mu_2 + \theta_1' \delta\mu_{1s})\, dx$$
$$+ [-(v_2 - \bar{v}_2)\delta\mu_3' + (v_3 - \bar{v}_3)\delta\mu_2' + (\theta_1 - \bar{\theta}_1)\delta\mu_{1s}$$
$$- (v_3' - \bar{v}_3')\delta\mu_2 + (v_2' - \bar{v}_2')\delta\mu_3]_0^l \qquad (16\text{–}11c)$$

Equation (16–11a) is a variational equation with respect to displacements, to statical redundants in a statically indeterminate case, and to η. The only admissibility condition apart from continuity requirements is that $\delta\eta$ vanishes at a member end free to warp.

If the displacement representation is chosen such that it satisfies the kinematic boundary conditions, the boundary terms may be deleted from Eq. (16–11c).

The existence of a functional $\bar{V}_c$ of which Eq. (16–11c) represents the variation may be established for conservative forces and related to similar terms that occur in the potential energy method. This will be shown as part of the following example.

Example 1. Cantilever Subjected to a Concentrated Force

The cantilever under consideration is the same one as in Section 12, Example 3, Fig. 12–3, and for which it was found, using the present notation,

$$\mu_2 = F_2(l - x)\theta_1 \qquad (16\text{–}12a)$$
$$\mu_3 = F_2(y_s - y_A)v_2'^b \qquad (16\text{–}12b)$$
$$\mu_{1s} = F_2(l - x)v_3' + F_2(v_3 - v_3^b) + F_2(y_s - y_A)\theta_1^b \qquad (16\text{–}12c)$$

For a simple application the cross section is assumed to have a negligible warping constant and to be symmetric with respect to the z axis. Terms in η are then deleted from Eqs. (16–11), and $GJ_N = GJ$. The assumed displacement representation will be chosen so as to satisfy the kinematic conditions at the fixed end. The boundary terms are thus also deleted from Eq. (16–11c). The terms involving μ_3 and v_2 are uncoupled from the remaining terms and result in a separate equation for v_2 that does not affect the critical load. Deleting these terms, Eq. (16–11a) takes the form

$$\int_0^l \left[\delta\mu_2 \left(\frac{\mu_2}{EI_2} + v_3'' \right) + \delta\mu_{1s} \left(\frac{\mu_{1s}}{GJ} - \theta_1' \right) \right] dx = 0 \qquad (16\text{–}13)$$

Starting with an assumed shape for θ_1, the shape for v_3 is chosen so that the coefficient of $\delta\mu_2$ vanishes, that is,

$$v_3'' = -\frac{F_2(l - x)}{EI_2}\theta_1 \qquad (16\text{–}14)$$

If the term $\delta\mu_{1s}\theta_1'$ is integrated by parts and Eqs. (16–12c) and (16–14) are used to form $\delta\mu_{1s}'$ and $\delta v_3''$, it is found that

$$\int_0^l \delta\mu_{1s}\theta_1'\,dx = F_2(y_s - y_A)\theta_1^b\,\delta\theta_1^b + \int_0^l \frac{F_2^2(l-x)^2}{EI_2}\theta_1\,\delta\theta_1\,dx \qquad (16\text{–}15)$$

Equation (16–13) may now be put in the variational form $\delta\Pi_c = 0$, where

$$\Pi_c = \frac{1}{2}\int_0^l \frac{\mu_{1s}^2}{GJ}dx - \frac{1}{2}F_2(y_s - y_A)(\theta_1^b)^2 - \frac{1}{2}\int_0^l \frac{F_2^2(l-x)^2}{EI_2}\theta_1^2\,dx \qquad (16\text{–}16)$$

The solution that makes $\delta\Pi_c = 0$ also makes $\Pi_c = 0$.

Referring to Section 15, Example 2, which deals with the present problem by the potential energy method, it is noted that $\frac{1}{2}(\bar{U}_2 + \bar{V}_2)$ transforms into Π_c after deleting the v_2 terms, enforcing Eq. (16–14), and transforming the torsional strain energy term into its expression in terms of μ_{1s} using the torsional constitutive equations.

Let now the assumed shape be

$$\theta_1 = at + bt^2$$

where $t = x/l$. v_3 is determined by integration of Eq. (16–14) with the boundary conditions $v_3 = 0$, $v_3' = 0$ at $x = 0$. The result is

$$v_3 = -\frac{F_2 l^3}{2EI_2}t^3\left(\frac{a}{3} + \frac{b-a}{6}t - \frac{b}{10}t^2\right)$$

Evaluation of Π_c involves integration of polynomial terms. Factoring out a constant A, we obtain

$$\begin{aligned}\frac{\Pi_c}{A} &= \frac{F_2^2 l^4}{GJEI_2}\left[\frac{a^2}{504} + \frac{181}{415{,}800}b^2 + \frac{139}{75{,}600}ab\right] + \frac{a^2}{30} + \frac{b^2}{105} + \frac{ab}{30} \\ &\quad + \left[\frac{EI_2}{F_2 l^3} - \frac{EI_2}{GJl^2}(y_s - y_A)\right](y_s - y_A)(a+b)^2 \\ &\quad - \frac{F_2 l}{GJ}(y_s - y_A)(a+b)\left(\frac{a}{15} + \frac{b}{30}\right)\end{aligned}$$

The case where F_2 is applied at the shear center is treated first. Then $y_A = y_s$, and the smallest characteristic value of the equations $\partial\Pi_c/\partial a = \partial\Pi_c/\partial b = 0$ is found to be

$$\lambda_{\text{cr}} = \frac{F_{2,\text{cr}} l^2}{\sqrt{EI_2 GJ}} = 4.013 \qquad (16\text{–}17)$$

and is accurate to four digits. The same shape used in the potential energy method yields a less accurate value but also involves fewer calculations.

For $y_A \neq y_s$, and α as defined in Eq. (15–16), the terms in $(y_s - y_A)$ may be put in the form

$$\alpha\left(\alpha - \frac{1}{\lambda}\right)(a+b)^2 + \frac{\lambda\alpha}{30}(a+b)(2a+b)$$

A closed-form solution for λ_{cr} ceases to be possible. If the characteristic equation is denoted $f(\lambda, \alpha) = 0$, and $\lambda_0 = 4.013$ is considered as a starting value, a one-cycle Newton-Raphson procedure yields the correction

$$\delta\lambda = -\frac{f(\lambda_0, \alpha)}{\dfrac{\partial f}{\partial \lambda}(\lambda_0, \alpha)}$$

Linearization of this result in α yields the approximation $\lambda = \lambda_0(1 - 0.97\alpha)$ which is close enough to $\lambda = \lambda_0(1 - \alpha)$.

Example 2. Simply Supported Beam Subjected to a Point Load

A cross section with nonzero warping constant K is assumed in the present example. Consider a simply supported beam of span $2l$ subjected to a point load $P_y = -P$ at midspan at the centroid. Because of symmetry the stability analysis may be formulated for half the beam. The left half is treated as a member of span l supported at end a and subjected at end b to $F_2 = -P/2$ and $G_3 = Pl/2$, Fig. 16–1. The remaining forces at end b are zero; that is, $F_1 = F_3 = 0$ and $G_1 = G_2 = 0$. Equilibrium in the buckled state requires torsional moment reactions at supports

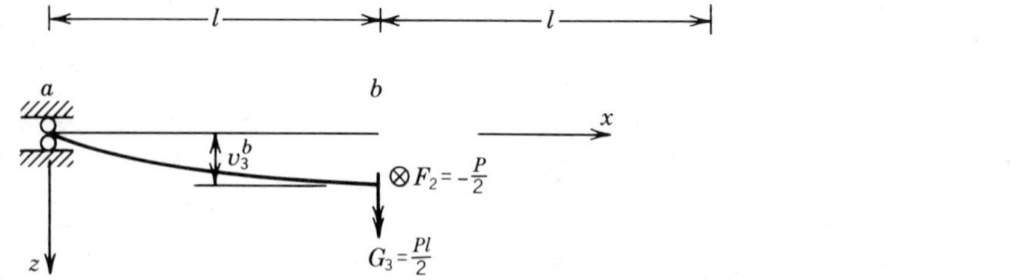

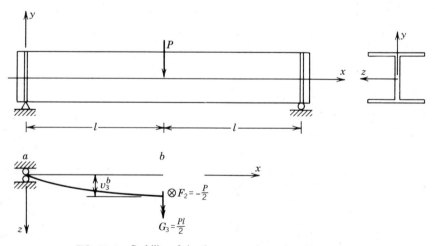

FIG. 16–1. Stability of simply supported member, Example 2.

due to the shift of the line of action of the applied load. At midspan, however, the torsional moment is zero because of symmetry, and the other force and moment components in the fixed axes remain unchanged. Thus $\delta F_1 = \delta F_2 = \delta F_3 = 0$, and $\delta G_1 = \delta G_2 = \delta G_3 = 0$.

Assume a cross section whose shear center coincides with the centroid, so that $y_s = z_s = 0$. Substituting all the preceding conditions into Eqs. (16–5), we obtain $\mu_3 = 0$, and

$$\mu_{1s} = \frac{P}{2}(v_3^b - v_3 + xv_3')$$

$$\mu_2 = \frac{P}{2}x\theta_1$$

Following the same procedure as in the preceding example, the representation of v_3 is made in terms of θ_1 by satisfying the constitutive equation

$$v_3'' = -\frac{\mu_2}{EI_2} = -\frac{Px\theta_1}{2EI_2}$$

The boundary terms are dropped from Eq. (16–11c), and the term $\theta_1' \delta\mu_{1s}$ is integrated by parts, noting that $\theta_1 \delta\mu_{1s}$ vanishes at $x = 0$ and $x = l$. The resulting form of Eq. (16–11) is $\delta\Pi_c = 0$, where

$$\Pi_c = \frac{1}{2}\int_0^l \left\{ \frac{\eta^2}{EK} + \frac{1}{GJ}\left[\frac{P}{2}(v_3^b - v_3 + xv_3') + \eta'\right]^2 - \frac{P^2}{4EI_2}x^2\theta_1^2 \right\} dx$$

The simplest polynomial representation of θ_1 is

$$\theta_1 = at(2 - t)$$

where $t = x/l$. This satisfies the kinematic conditions $\theta_1 = 0$ at $x = 0$ and $\theta_1' = 0$ at $x = l$. v_3 is obtained by integrating the preceding expression of v_3'' and satisfying the kinematic end conditions $v_3 = 0$ at $x = 0$ and $v_3' = 0$ at $x = l$. This yields

$$v_3 = -\frac{Pl^3}{2EI_2}a\left(\frac{t^4}{6} - \frac{t^5}{20} - \frac{5}{12}t\right)$$

The representation of η must satisfy the condition of free warping, $\eta = 0$ at $x = 0$, and the symmetry condition $\eta' = 0$ at $x = l$. A choice similar to that of θ_1 is

$$\eta = bt(2 - t)$$

Evaluation of Π_c involves only integration of polynomial terms and results in a

$\frac{GJL^2}{EK}$	0	0.4	4.0	8	16	24	32	48	64	80	96	160	240	320	400	∞
λ, Eq. (16–19)	64.8	52.5	29.3	24.4	21.2	19.9	19.2	18.5	18.1	17.9	17.8	17.5	17.3	17.2	17.2	16.96
λ, Ref. [6]	—	86.4	31.9	25.6	21.8	20.3	19.6	18.8	18.3	18.1	17.9	17.5	17.4	17.2	17.2	—

FIG. 16–2. Values of critical stability parameter λ, Example 2.

quadratic form in a and b. The solution of the characteristic value problem is found to be

$$P_{cr} = \lambda \frac{\sqrt{EI_2 GJ}}{L^2} \tag{16–18}$$

where $L = 2l$ is the length of the original member and

$$\lambda = 8\sqrt{\frac{145[1 + (1/10)(GJL^2/EK)}{2039/924 + (532/165)(GJL^2/EK)}} \tag{16–19}$$

The solution for λ is compared in Fig. 16–2 to that given by Timoshenko [6] and obtained by the strain energy method using two trigonometric terms. It is noted that values of λ by the present method are smaller and are thus closer to the exact solution. The two sets of values are close for large values of the ratio L^2GJ/EK but differ substantially at the lower end of the range. For $L^2GJ/EK \to \infty$ warping restraint is negligible, and the exact λ is 16.94. The value obtained here is 16.965.

EXERCISES

Sections 2 to 5

1. Determine the coordinate transformation matrix $\boldsymbol{\lambda}$ in the relation $\vec{\mathbf{e}} = \boldsymbol{\lambda}\vec{\mathbf{e}}_0$ if $\vec{\mathbf{e}}$ is obtained by rotating $\vec{\mathbf{e}}_0$ thirty degrees about an axis oriented by the vector having the components $\{1 \ \ 1 \ \ 1\}$.
2. For the rotation of Exercise 1, verify Eqs. (2–27) and (2–28).
3. Determine $\delta\vec{\varphi}$ if the triad $\vec{\mathbf{e}}$ in Exercise 1 is given an infinitesimal rotation $\delta\omega_1$ about $\vec{e}_{1,0}$. Do the same if the rotation is $\delta\omega_1'$ about $\vec{e}_1$.
4. a. Determine the deformation vector $\vec{\varphi}_r$ for a member whose end a is rotated by the angle ψ_2 about $\vec{e}_{2,0}$ and whose end b is rotated by the angle ψ_3 about $\vec{e}_{3,0}$.
   b. Determine the coordinate transformation matrix $\boldsymbol{\lambda}_r$ in the relation $\vec{\mathbf{e}}^b = \boldsymbol{\lambda}_r \vec{\mathbf{e}}^a$.
   c. Determine the angle of rotation ψ_r of $\vec{\mathbf{e}}^b$ relative to $\vec{\mathbf{e}}^a$.

5. Vefify that the deformation vectors $\vec{u}_r$ and $\vec{\varphi}_r$ in Eq. (5.2–1) vanish for any rigid body displacement of the member as a whole. Do the same for the strain vectors in Eq. (5.3–1).

Section 6

6. Write the components on $\vec{\mathbf{e}}_0$ of the vector differential equations of equilibrium (6–7) using stress resultant components on $\vec{\mathbf{e}}_0$. Write in detail Eqs. (6–9).

Sections 7, 8

7. Express in detail the generalised moments $\mathbf{G}_\varphi$ in terms of the moments $\mathbf{G}$, Eq. (7–8).

8. Use the incremental strain-displacement relations to establish an integration by parts formula for the left-hand side of Eq. (7–5a); then deduce the principles of virtual displacements and virtual forces.

9. Obtain the geometric stiffness $\mathbf{k}_G$ in the relation $d\mathbf{BV} = \mathbf{k}_G d\mathbf{s}$ by incrementing Eq. (6–4). Equations (3–10) and (3–7a) may be found helpful. Find that $\mathbf{k}_G$ is not symmetric.

Sections 9, 10

10. Derive the generalized static-kinematic matrix $\mathbf{B}_\varphi$ in the second-order theory using Eqs. (7–11a) and (9–14, 15).

11. Derive scalar differential equations of equilibrium consistent with the second-order theory of Section 9.

12. Determine the geometric constants in the constitutive equations (10–8) for a channel and I cross sections.

Section 12

13. Specialize Eqs. (12–5) to the case of a simply supported member in pure compression with the shear center coinciding with the centroid. Determine $K = 0$, the critical load for torsional buckling is obtained as in Eq. (10–14).

14. Specialize Eqs. (12–5) as in the preceding exercise but with the shear center distinct from the centroid. Seek a solution in the form

$$\{\delta\varphi_1 \;\; \delta u_2 \;\; \delta u_3\} = \{a_1 \;\; a_2 \;\; a_3\} \sin\frac{\pi x}{l}$$

Reduce the characteristic equation to a cubic equation in the axial force.

Show that the lowest characteristic value is smaller than the three characteristic values of the preceding exercise.

15. For the cantilever of Section 12, Example 3, derive a fourth-order differential equation for $\delta\varphi_1$ by eliminating δu_3 from Eqs. (12–10a, b). Show that in the case $K=0$ and $\beta_3=0$ the equation reduces to the Bessel equation:

$$\delta\varphi_1'' + \frac{\lambda^2}{l^2}\left(1-\frac{x}{l}\right)^2 \delta\varphi_1 = 0$$

where λ is as defined in Eq. (15–15).

16. For the case of pure bending of Section 12, Example 2, derive a differential equation for $\delta\varphi_1$ by eliminating δu_3 from Eqs. (12–8a, b). Solve the buckling problem with $\delta\varphi_1 = a\sin \pi x/l$, and check with the solution of Section 15, Example 1.

17. Specialize Eqs. (12–4) to the case of a cantilever subjected to an end moment G_3 of fixed orientation. Do the same assuming the moment vector follows the motion of the cross section. Show that in both cases a nontrivial solution does not exist.

Sections 13 to 15

18. Modify Example 1 of Section 15 so that end rotations v_3' are restrained, and assume $EK=0$, $\beta_3=0$. Using the assumed shape $v_3 = a(1-\cos 2\pi x/l)$ obtain $G_{3,\text{cr}} = (2\pi/l)\sqrt{GJEI_2}$. Show that this is an exact solution.

19. In Section 15, Example 2, let a portion of length a at the right end of the cantilever be infinitely rigid. Establish the expression of $\bar{V}_2$, and obtain a solution by the energy method.

20. Solve any of the problems of Section 15, Example 5, by a direct energy method.

21. Refer to the four types of external torque considered in Section 15, Example 5. In each case interpret statically the static (natural) boundary conditions.

22. For any of the Examples of Section 12, derive the differential equations and statical boundary conditions by a formal variational method.

23. Derive the Euler differential equations and natural boundary conditions for buckling of a cantilever of narrow rectangular cross section subjected at its free end to a conservative bending moment G_3. Assume G_3 is caused by a couple of conservative forces applied in the (x, y) plane perpendicularly to the cross section. Show that

$$G_{3,\mathrm{cr}} = \frac{\pi}{2l}\sqrt{GJEI_2}$$

24. Compare the cantilever of Exercise 23 to a quarter span of the beam of Exercise 18.

Section 16

25. Apply the method of virtual forces to the stability problem of pure bending of Section 15, Example 1.

16

ELEMENT MATRIX FOR NONLINEAR SPATIAL ANALYSIS

1. INTRODUCTION

In this chapter an element tangent stiffness matrix suitable for assembly by the direct stiffness method is developed for geometrically nonlinear analysis and for linearized stability analysis of spatial structures. The underlying theory was developed in Chapter 15, Section 8, and needs the element stiffness in the bound reference, $\mathbf{k}_\varphi$, in order to be applicable numerically. $\mathbf{k}_\varphi$ will be also referred to as the cantilever or reduced element matrix. The subscript φ is used to emphasize that the rotational degrees of freedom are represented by the rotations $\boldsymbol{\varphi} = \{\varphi_1\ \varphi_2\ \varphi_3\}$ and not by displacement derivatives. The significance of this point is explained in Chapter 15, Section 13, under the heading "Choice of Generalised Displacements." It is considered again in Section 7 of the present chapter.

In deriving $\mathbf{k}_\varphi$, a simplifying assumption is made by which incremental deformations are assumed to take place from an undeformed, but stressed, state. The assumption is made only for displacements relative to the bound reference and becomes more accurate as the element becomes smaller. $\mathbf{k}_\varphi$ will thus be applicable to general nonlinear analysis including the particular problem of linearized stability analysis.

Transformation of $\mathbf{k}_\varphi$ into the complete stiffness $\mathbf{k}_{c\varphi}$ in fixed axes is described by Eq. (15/8–6b) and may be carried out as appropriate with exact static-kinematic matrices or within the second-order theory.

2. METHOD FOR DERIVING REDUCED ELEMENT STIFFNESS $\mathbf{k}_\varphi$

The desired reduced stiffness relations are defined in Eqs. (15/8–1, 2), or

$$d\mathbf{V}_\varphi = \mathbf{k}_\varphi d\mathbf{v} \tag{2–1}$$

$$\mathbf{k}_\varphi = \frac{\partial^2 \bar{U}}{\partial \mathbf{v} \partial \mathbf{v}^T} \tag{2–2}$$

where $\mathbf{v} = \{\mathbf{u}_r \ \boldsymbol{\varphi}_r\}$ and $\mathbf{V}_\varphi = \{\mathbf{F}_r \ \mathbf{G}_{\varphi r}\}$. $\mathbf{u}_r$ and $\boldsymbol{\varphi}_r$ are the components on the bound axes of the translation and rotation vectors of end b relative to the bound axes (Section 15/5–2). $\mathbf{F}_r$ and $\mathbf{G}_{\varphi r}$ are components on these same axes of the force and generalized moment vectors at end b. $\mathbf{G}_{\varphi r}$ is defined in terms of the moments $\mathbf{G}_r$ in Eq. (15/7–8c). A direct choice of $\boldsymbol{\varphi}_r$ as generalized coordinates for discretization of the strain energy is not convenient, however, because $\bar{U}$ is expressed in terms of displacement derivatives

$$\theta_2, \theta_3 = -u_3', u_2' \tag{2–3}$$

We thus make a preliminary choice of generalized displacements

$$\mathbf{v}_\theta = \{\mathbf{u}_r \ \theta_r\} = \{u_{1r} \ u_{2r} \ u_{3r} \ \varphi_{1r} \ \theta_{2r} \ \theta_{3r}\} \tag{2–4}$$

and then transform to

$$\mathbf{v} = \{\mathbf{u}_r \ \boldsymbol{\varphi}_r\} = \{u_{1r} \ u_{2r} \ u_{3r} \ \varphi_{1r} \ \varphi_{2r} \ \varphi_{3r}\} \tag{2–5}$$

$\mathbf{v}_\theta$ and $\mathbf{v}$ differ only by θ_{2r} and θ_{3r} which are defined at end b by applying Eq. (2–3) relative to the bound reference.

To $\mathbf{v}_\theta$ correspond generalized force $\mathbf{V}_\theta$ and incremental stiffness relations similar to Eq. (2–1), or

$$d\mathbf{V}_\theta = \mathbf{k}_\theta d\mathbf{v}_\theta \tag{2–6}$$

where

$$\mathbf{k}_\theta = \frac{\partial^2 \bar{U}}{\partial \mathbf{v}_\theta \partial \mathbf{v}_\theta^T} \tag{2–7}$$

$\mathbf{k}_\theta$ is determined in subsequent sections. Its transformation to $\mathbf{k}_\varphi$ is obtained in what follows.

The transformation from $\mathbf{v}_\theta$ to $\mathbf{v}$ is nonlinear because θ_{2r} and θ_{3r} are related nonlinearly to φ_{2r} and φ_{3r}. A general treatment of such a transformation is given now because it will be useful in a similar instance later on. From the relations defining $\mathbf{v}_\theta$ as functions of $\mathbf{v}$ we obtain

$$\delta \mathbf{v}_\theta = \mathbf{T} \delta \mathbf{v} \tag{2–8a}$$

where

$$\mathbf{T} = \frac{\partial \mathbf{v}_\theta}{\partial \mathbf{v}^T} \tag{2–8b}$$

$\mathbf{V}_\theta$ and $\mathbf{V}_\varphi$ satisfy the virtual work relation

$$\delta \bar{U} = \mathbf{V}_\theta^T \delta \mathbf{v}_\theta = \mathbf{V}_\varphi^T \delta \mathbf{v} \tag{2–9a}$$

from which we obtain the contragradient transformation

$$\mathbf{V}_\varphi = \mathbf{T}^T \mathbf{V}_\theta \tag{2–9b}$$

Differentiation of both sides of Eq. (2–9b) and use of Eqs. (2–6) and (2–8a) yields $\mathbf{k}_\varphi$ in the form

$$\mathbf{k}_\varphi = \mathbf{T}^t \mathbf{k}_\theta \mathbf{T} + \mathbf{k}_g \tag{2–10}$$

where $\mathbf{k}_g$ is a geometric stiffness obtained from the relation

$$d\mathbf{T}^t \mathbf{V}_\theta = \mathbf{k}_g d\mathbf{v} \tag{2–11a}$$

By considering individual columns of $d\mathbf{T}^T$ and the corresponding elements of $\mathbf{V}_\theta$, we obtain

$$\mathbf{k}_g = \sum_i V_{i\theta} \frac{\partial^2 v_{i\theta}}{\partial \mathbf{v} \partial \mathbf{v}^T} \tag{2–11b}$$

A shorter derivation of this result that bypasses, however, the static part of the transformation is obtained by forming the second variation of $\bar{U}$ considering $\mathbf{v}$ as the independent variables. Then from Eq. (2–9a)

$$\delta^2 \bar{U} = \delta \mathbf{V}_\theta^T \delta \mathbf{v}_\theta + \mathbf{V}_\theta^T \delta^2 \mathbf{v}_\theta = \delta \mathbf{V}_\varphi^T \delta \mathbf{v} \tag{2–12a}$$

Substituting from Eqs. (2–6) and (2–8a) and noting the definition equation of $\mathbf{k}_g$

(2–11b), we obtain

$$\delta^2 \bar{U} = \delta \mathbf{v}^T (\mathbf{T}^T \mathbf{k}_\theta \mathbf{T} + \mathbf{k}_g) \delta \mathbf{v} = \delta \mathbf{v}^T \mathbf{k}_\varphi \delta \mathbf{v} \qquad (2\text{–}12b)$$

In applying Eq. (2–11b), the only elements of $\mathbf{v}_\theta$ that are nonlinear in $\mathbf{v}$ are θ_{2r} and θ_{3r}. They are obtained by solving Eqs. (15/10–1) for the displacement derivatives and neglecting φ_{1r}^2 with respect to unity. There comes

$$\theta_{2r} = \varphi_{2r} - \tfrac{1}{2}\varphi_{1r}\varphi_{3r} \qquad (2\text{–}13a)$$

$$\theta_{3r} = \varphi_{3r} + \tfrac{1}{2}\varphi_{1r}\varphi_{2r} \qquad (2\text{–}13b)$$

The assumption that incremental deformations may be considered to take place from the undeformed state allows us to neglect the displacements in $\mathbf{T}$. The result is $\mathbf{T} = \mathbf{I}$, $d\mathbf{v}_\theta = d\mathbf{v}$, and $\mathbf{V}_\theta = \mathbf{V}$. Equation (2–10) reduces to

$$\mathbf{k}_\varphi = \mathbf{k}_\theta + \left[\begin{array}{c|c} 0 & 0 \\ \hline 0 & \mathbf{k}_{\theta\varphi} \end{array}\right] \qquad (2\text{–}14a)$$

where

$$\mathbf{k}_{\theta\varphi} = \begin{bmatrix} 0 & \tfrac{1}{2}G_{3r} & -\tfrac{1}{2}G_{2r} \\ \tfrac{1}{2}G_{3r} & 0 & 0 \\ -\tfrac{1}{2}G_{2r} & 0 & 0 \end{bmatrix} \qquad (2\text{–}14b)$$

3. DERIVATION OF $\mathbf{k}_\theta$

A representation of incremental displacements in terms of $\delta\mathbf{v}_\theta$ as generalized displacements turns $\delta^2 \bar{U}$ into the quadratic form

$$\delta^2 \bar{U} = \delta \mathbf{v}_\theta^T \mathbf{k}_\theta \delta \mathbf{v}_\theta \qquad (3\text{–}1)$$

For a treatment that includes the effect of initial deformations, both initial and incremental displacements are discretized, and $\delta^2 \bar{U}$ is as obtained in Eq. (15/14–2). The assumption neglecting initial deformations will be adopted, however, and $\delta^2 \bar{U}$ is obtained as $\bar{U}_2$ in Eq. (15/15–6).

The displacements $(v_1, v_2, v_3, \theta_1)$ in $\bar{U}_2$ are increments $(du_1, du_2, du_3, d\varphi_1)$, and the translations are defined at the shear center. These displacements are considered here to be relative to the bound reference. In accordance with this notation $d\mathbf{v}_\theta$ is renamed $\mathbf{q}_\theta$, the subscript r is not used, and a superscript b is used temporarily to distinguish between displacement functions and values at end b. The desired set of generalized displacements is then

$$d\mathbf{v}_\theta = \mathbf{q}_\theta = \{\mathbf{v}_c^b \;\; \boldsymbol{\theta}^b\} = \{v_{1c}^b \;\; v_{2c}^b \;\; v_{3c}^b \;\; \theta_1^b \;\; \theta_2^b \;\; \theta_3^b\} \qquad (3\text{–}2)$$

By comparison with Eq. (2–4), $\mathbf{v}_c^b$ replaces $d\mathbf{u}_r$, θ_1^b replaces $d\varphi_{1r}$, and θ_2^b and θ_3^b replace $d\theta_{2r}$ and $d\theta_{3r}$, respectively. $\mathbf{v}_c^b$ is defined at the centroid.

A direct use of $\mathbf{q}_\theta$ as generalized displacements for discretization of $\bar{U}_2$ is not practical because the translational displacements in $\bar{U}_2$ are defined at the shear center and because a linear representation of the twisting rotation θ_1 in terms of θ_1^b is not good enough. We thus start with the generalized displacements

$$d\mathbf{v}'_\theta = \mathbf{q}'_\theta = \{v_{1c}^b \ \ v_2^b \ \ v_3^b \ \ \theta_1^b \ \ \theta_2^b \ \ \theta_3^b \ \ \kappa_1^a \ \ \kappa_1^b\} \tag{3–3}$$

where v_2^b and v_3^b are defined at the shear center, and $\kappa_1 = \theta_{1,x}$. It is convenient to keep v_{1c}^b in $\mathbf{q}'$ because the EA term in $\bar{U}_2$ is $EA(v'_1 + z_s v''_3 + y_s v''_2)^2 = EA(v'_{1c})^2$ and is uncoupled from the remaining terms.

The discretized $\bar{U}_2$ in terms of $\mathbf{q}'_\theta$ has the form

$$\bar{U}_2 = \mathbf{q}'^T \mathbf{k}'_\theta \mathbf{q}' \tag{3–4}$$

The shape functions are the polynomials used in the planar theory. Letting $t = x/l$, the displacement representation is

$$v_{1c} = v_{1c}^b t \tag{3–5a}$$

$$v_2 = v_2^b(3t^2 - 2t^3) + \theta_3^b l(t^3 - t^2) \tag{3–5b}$$

$$v_3 = v_3^b(3t^2 - 2t^3) - \theta_2^b l(t^3 - t^2) \tag{3–5c}$$

$$\theta_1 = \theta_1^b(3t^2 - 2t^3) + \kappa_1^b l(t^3 - t^2) + \kappa_1^a l[(1-t)^2 - (1-t)] \tag{3–5d}$$

The internal forces in $\bar{U}_2$ are expressed in terms of $\mathbf{V} = \{\mathbf{F}_r \ \ \mathbf{G}_r\}$, Fig. 3–1, by

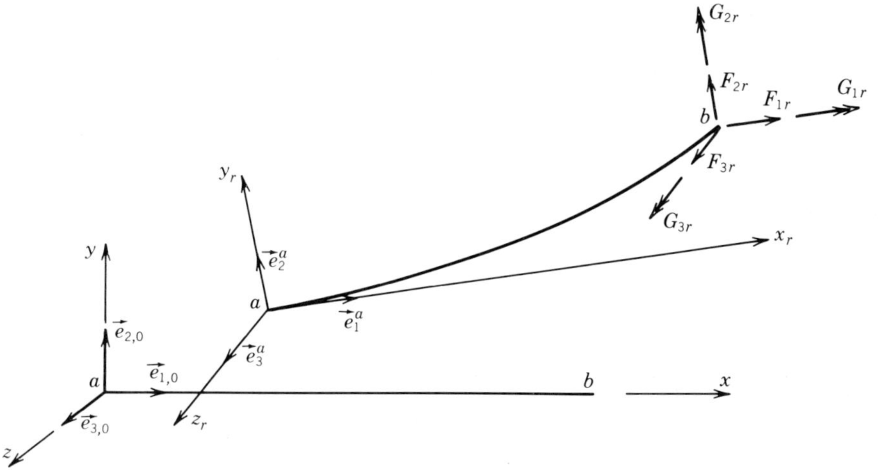

FIG. 3–1. Element reduced forces, $\mathbf{V} = \{\mathbf{F}_r \ \ \mathbf{G}_r\}$.

statics in the undeformed state, or

$$N = F_{1r} \tag{3–6a}$$

$$M_{1s} = G_{1r} + z_s F_{2r} - y_s F_{3r} \tag{3–6b}$$

$$M_2 = G_{2r} - F_{3r} l(1-t) \tag{3–6c}$$

$$M_3 = G_{3r} + F_{2r} l(1-t) \tag{3–6d}$$

Evaluation of $\bar{U}_2$ yields $\mathbf{k}'_\theta$ as shown in Eq. (3–9). The geometric part of $\mathbf{k}'_\theta$ is expressed in terms of moments defined at the shear center,

$$G^s_{1r} = G_{1r} + z_s F_{2r} - y_s F_{3r} \tag{3–7a}$$

$$G^s_{2r} = G_{2r} - z_s F_{1r} \tag{3–7b}$$

$$G^s_{3r} = G_{3r} + y_s F_{1r} \tag{3–7c}$$

It is also convenient to define the modified torsional rigidity

$$GJ_N = GJ + \rho^2 F_{1r} + \beta_2 G_{2r} - \beta_3 G_{3r} \tag{3–8}$$

$\mathbf{k}'_\theta$ is found as shown next:

$\frac{EA}{l}$	0	0	0	0	0	0	0
	$\frac{12EI_3}{l^3} + \frac{6}{5}\frac{F_{1r}}{l}$	0	$\frac{F_{3r}}{10} - \frac{6}{5}\frac{G^s_{2r}}{l}$	$\frac{G^s_{1r}}{l}$	$-\frac{6EI_3}{l^2} - \frac{F_{1r}}{10}$	$\frac{G^s_{2r}}{10}$	$\frac{G^s_{2r}}{10} - \frac{lF_{3r}}{10}$
		$\frac{12EI_2}{l^3} + \frac{6}{5}\frac{F_{1r}}{l}$	$-\frac{F_{2r}}{10} - \frac{6}{5}\frac{G^s_{3r}}{l}$	$\frac{6EI_2}{l^2} + \frac{F_{1r}}{10}$	$\frac{G^s_{1r}}{l}$	$\frac{G^s_{3r}}{10}$	$\frac{lF_{2r}}{10} + \frac{G^s_{3r}}{10}$
			$\frac{6GJ_N}{5l} + \frac{12EK}{l^3} - \frac{3}{5}\beta_2 F_{3r} - \frac{3}{5}\beta_3 F_{2r}$	$-\frac{lF_{2r}}{5} - \frac{11}{10}G^s_{3r}$	$-\frac{lF_{3r}}{5} + \frac{11}{10}G^s_{2r}$	$-\frac{GJ_N}{10} - \frac{6EK}{l^2} + \frac{l}{10}(\beta_2 F_{3r} + \beta_3 F_{2r})$	$-\frac{GJ_N}{10} - \frac{6EK}{l^2}$

$$\mathbf{k}'_\theta = \begin{bmatrix} \text{symmetric} & \dfrac{4EI_2}{l} + \dfrac{2lF_{1r}}{15} & 0 & \dfrac{l^2F_{2r}}{30} + \dfrac{2lG^s_{3r}}{15} & -\dfrac{lG^s_{3r}}{30} \\ & & \dfrac{4EI_3}{l} + \frac{2}{15}lF_{1r} & \dfrac{l^2F_{3r}}{30} - \frac{2}{15}lG^s_{2r} & \dfrac{lG^s_{2r}}{30} \\ & & & \frac{2}{15}GJ_Nl + \dfrac{4EK}{l} - \dfrac{l^2}{30}(\beta_2F_{3r} + \beta_3F_{2r}) & -\frac{1}{30}GJ_Nl + \dfrac{2EK}{l} + \dfrac{l^2}{60}(\beta_2F_{3r} + \beta_3F_{2r}) \\ & & & & \frac{2}{15}GJ_Nl + \dfrac{4EK}{l} - \dfrac{l^2}{10}(\beta_2F_{3r} + \beta_3F_{2r}) \end{bmatrix} \tag{3–9}$$

The first row and column of $\mathbf{k}'_\theta$ contain only the EA/l term because the initial state is assumed undeformed. (An incremental axial stiffness taking into account initial deformations was derived for planar behavior in Chapter 6.)

There remains to transform $\mathbf{k}'_\theta$ to $\mathbf{k}_\theta$, that is, to transform from $d\mathbf{v}'_\theta$, Eq. (3–3), to $d\mathbf{v}_\theta$, Eq. (3–2). Two problems are involved here: the first is the elimination of κ_1^a and κ_1^b, and the second a transformation from displacements at the shear center to displacements at the centroid. For a general solution of the first problem one must consider the case of a member subdivided into several elements, as was done in Chapter 14, Section 3.3, under the title "Effective Torsional Stiffness." Except for the case where the warping stiffness EK is zero, the problem is more complex in the nonlinear theory and will be considered in Section 8. For now we consider the case of a member represented by a single element with end conditions idealized into either restrained or free warping. If warping is restrained at a member end, $\kappa_1 = 0$, and the corresponding row and column of $\mathbf{k}'_\theta$ are deleted. If warping is free, κ_1 is eliminated by static condensation. This latter case also applies if the warping stiffness EK is zero.

In what follows $\mathbf{k}'_\theta$ refers to the 6×6 matrix resulting from elimination of κ_1^a and κ_1^b. The transformation from displacements at the shear center to displacements at the centroid is obtained by applying Eqs. (15/5.6–1) and using Eq. (15/9–3). These equations are applied here relative to the bound reference, and rotations are expressed in terms of displacement derivatives. Only the transverse displacements need be considered. Letting a superscript s refer to the shear center, we obtain

$$u_{2r}^s = u_{2r} - z_s\varphi_{1r} - \frac{y_s}{2}(\varphi_{1r}^2 + \theta_{3r}^2) + \frac{z_s}{2}\theta_{2r}\theta_{3r} \tag{3–10a}$$

$$u_{3r}^s = u_{3r} + y_s\varphi_{1r} + \frac{y_s}{2}\theta_{2r}\theta_{3r} - \frac{z_s}{2}(\varphi_{1r}^2 + \theta_{2r}^2) \tag{3–10b}$$

The transformation from $\mathbf{v}_\theta^s = \{u_{1r}\ u_{2r}^s\ u_{3r}^s\ \varphi_{1r}\ \theta_{2r}\ \theta_{3r}\}$ to $\mathbf{v}_\theta = \{u_{1r}\ u_{2r}\ u_{3r}\ \varphi_{1r}\ \theta_{2r}\ \theta_{3r}\}$ follows the method presented in Section 2 for transforming from $\mathbf{v}_\theta$ to $\mathbf{v}$. We thus obtain similarly to Eq. (2–10)

$$\mathbf{k}_\theta = \mathbf{T}^T\mathbf{k}'_\theta\mathbf{T} + \mathbf{k}_g \tag{3–11a}$$

where

$$\mathbf{T} = \frac{\partial \mathbf{v}_\theta^s}{\partial \mathbf{v}_\theta^T} = \begin{bmatrix} \mathbf{I} & \mathbf{T}_{u\theta} \\ \mathbf{0} & \mathbf{I} \end{bmatrix} \tag{3–11b}$$

$$\mathbf{T}_{u\theta} = \begin{bmatrix} 0 & 0 & 0 \\ -z_s - y_s\varphi_{1r} & \frac{z_s}{2}\theta_{3r} & \frac{z_s}{2}\theta_{2r} - y_s\theta_{3r} \\ y_s - z_s\varphi_{1r} & \frac{y_s}{2}\theta_{3r} - z_s\theta_{2r} & \frac{y_s}{2}\theta_{2r} \end{bmatrix} \tag{3–11c}$$

and

$$\mathbf{k}_g = F_{2r}\frac{\partial^2 u_{2r}^s}{\partial \mathbf{v}_\theta \partial \mathbf{v}_\theta^T} + F_{3r}\frac{\partial^2 u_{3r}^s}{\partial \mathbf{v}_\theta \partial \mathbf{v}_\theta^T} = \begin{bmatrix} 0 & 0 \\ 0 & \mathbf{k}_{cs} \end{bmatrix} \tag{3–11d}$$

where

$$\mathbf{k}_{cs} = \begin{bmatrix} -y_sF_{2r} - z_sF_{3r} & 0 & 0 \\ 0 & -z_sF_{3r} & \frac{z_s}{2}F_{2r} + \frac{y_s}{2}F_{3r} \\ 0 & \frac{z_s}{2}F_{2r} + \frac{y_s}{2}F_{3r} & -y_sF_{2r} \end{bmatrix} \tag{3–11e}$$

The assumption by which initial deformations are neglected makes $\varphi_{1r} = \theta_{2r} = \theta_{3r} = 0$ in Eq. (3–11c).

The problem of obtaining $\mathbf{k}_\theta$ for a one-element member is now solved. The desired element matrix $\mathbf{k}_\varphi$ is obtained from $\mathbf{k}_\theta$ through Eqs. (2–14).

4. ELEMENT REDUCED STIFFNESS $\mathbf{k}_\varphi$ WITHOUT WARPING STIFFNESS

The member stiffness is specialized here to the case of a cross section having a negligible warping stiffness EK. Also, to make formal derivations manageable, the cross section is assumed to have two axes of symmetry. The shear center coincides then with the centroid, and the geometric constants β_2 and β_3 in Eq. (3–8) are zero. $\mathbf{k}_\theta$ is obtained by static condensation of κ_1^a and κ_1^b from $\mathbf{k}'_\theta$, Eq. (3–9). $\mathbf{k}_\varphi$ is then obtained through Eqs. (2–14). Elements of $\mathbf{k}_\varphi$ are listed next by rows, starting at the main diagonal. The ordering is that of $\mathbf{v}$ in Eq. (2–5). The relationship to elements of $\mathbf{k}_\theta$ is indicated for k_{45} and k_{46}.

$$k_{11} = \frac{EA}{l} \tag{4–1a}$$

$$k_{1i} = 0, i = 2, \ldots, 6 \tag{4–1b}$$

$$k_{22} = \frac{12EI_3}{l^3} + \frac{6}{5}\frac{F_{1r}}{l} - \frac{2}{GJ_N l}\left[\frac{G_{2r}^2}{10} - \frac{l}{10}F_{3r}G_{2r} + \frac{l^2}{25}F_{3r}^2\right] \tag{4–1c}$$

$$k_{23} = -\frac{2}{GJ_N l}\left[\frac{G_{2r}G_{3r}}{10} + \frac{l}{20}F_{2r}G_{2r} - \frac{l}{20}F_{3r}G_{3r} - \frac{l^2}{25}F_{2r}F_{3r}\right] \tag{4–1d}$$

$$k_{24} = -\frac{G_{2r}}{l} \tag{4–1e}$$

$$k_{25} = \frac{G_{1r}}{l} - \frac{2}{GJ_N}\left[\frac{G_{2r}G_{3r}}{20} + \frac{l}{60}F_{2r}G_{2r} - \frac{l^2}{300}F_{2r}F_{3r}\right] \tag{4–1f}$$

$$k_{26} = -\frac{6EI_3}{l^2} - \frac{F_{1r}}{10} + \frac{2}{GJ_N}\left[\frac{G_{2r}^2}{20} - \frac{l}{60}F_{3r}G_{2r} + \frac{l^2}{300}F_{3r}^2\right] \tag{4–1g}$$

$$k_{33} = \frac{12EI_2}{l^3} + \frac{6}{5}\frac{F_{1r}}{l} - \frac{2}{GJ_N l}\left[\frac{G_{3r}^2}{10} + \frac{l}{10}F_{2r}G_{3r} + \frac{l^2}{25}F_{2r}^2\right] \tag{4–1h}$$

$$k_{34} = -\frac{G_{3r}}{l} \tag{4–1i}$$

$$k_{35} = \frac{6EI_2}{l^2} + \frac{F_{1r}}{10} - \frac{2}{GJ_N}\left[\frac{G_{3r}^2}{20} + \frac{l}{60}F_{2r}G_{3r} + \frac{l^2}{300}F_{2r}^2\right] \tag{4–1j}$$

$$k_{36} = \frac{G_{1r}}{l} + \frac{2}{GJ_N}\left[\frac{G_{2r}G_{3r}}{20} - \frac{l}{60}F_{3r}G_{3r} - \frac{l^2}{300}F_{2r}F_{3r}\right] \tag{4–1k}$$

$$k_{44} = \frac{GJ_N}{l} \tag{4–1l}$$

$$k_{45} = (k_{45})_\theta + \frac{1}{2}G_{3r} = -\frac{1}{2}G_{3r} - \frac{lF_{2r}}{6} \tag{4–1m}$$

$$k_{46} = (k_{46})_\theta - \frac{1}{2}G_{2r} = \frac{1}{2}G_{2r} - \frac{lF_{3r}}{6} \tag{4–1n}$$

$$k_{55} = \frac{4EI_2}{l} + \frac{2l}{15}F_{1r} - \frac{2l}{GJ_N}\left[\frac{G_{3r}^2}{15} + \frac{l}{30}F_{2r}G_{3r} + \frac{l^2}{225}F_{2r}^2\right] \tag{4–1o}$$

$$k_{56} = \frac{2l}{GJ_N}\left[\frac{G_{2r}G_{3r}}{15} - \frac{l}{60}F_{3r}G_{3r} + \frac{l}{60}F_{2r}G_{2r} - \frac{l^2}{225}F_{2r}F_{3r}\right] \tag{4–1p}$$

$$k_{66} = \frac{4EI_3}{l} + \frac{2l}{15}F_{1r} - \frac{2l}{GJ_N}\left[\frac{G_{2r}^2}{15} - \frac{l}{30}F_{3r}G_{2r} + \frac{l^2}{225}F_{3r}^2\right] \tag{4–1q}$$

Before describing the use of the element matrix $\mathbf{k}_\phi$ for an assemblage of members, its use for linearized stability analysis of a one-element cantilever is illustrated in two examples. In a third example use of $\mathbf{k}_\theta$ is illustrated.

Example 1

For the cantilever of Fig. 4–1 lateral buckling involves degrees of freedom 3, 4, and 5 of the 6 × 6 matrix $\mathbf{k}_\varphi$. Extracting the corresponding rows and columns from Eq. (4–1), we obtain, with $F_{2r} = -P$ and zero values for the other initial forces,

$$\mathbf{k}_\varphi = \begin{bmatrix} \frac{12EI_2}{l^3} - \frac{2}{25}\frac{lP^2}{GJ} & 0 & \frac{6EI_2}{l^2} - \frac{l^2P^2}{150GJ} \\ & \frac{GJ}{l} & \frac{lP}{6} \\ \text{symmetric} & & \frac{4EI_2}{l} - \frac{2l^3P^2}{225GJ} \end{bmatrix} \tag{4–2}$$

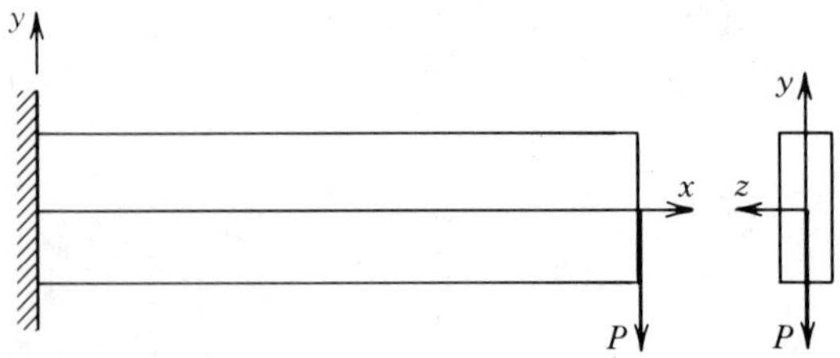

FIG. 4–1. Lateral stability of cantilever, Section 4, Example 1.

Assuming P to be fixed in orientation, $d\mathbf{V}_\varphi = 0$, and the stiffness equations reduce to $\mathbf{k}_\varphi \mathbf{q} = 0$. The determinantal equation may be put in the form

$$13\lambda^4 - 3060\lambda^2 + 54{,}000 = 0 \qquad (4\text{–}3)$$

where

$$\lambda^2 = \frac{P^2 l^4}{EI_2 GJ} \qquad (4\text{–}4)$$

The critical root is found to be $\lambda = 4.38$. Use of more elements would approach the correct value $\lambda_{cr} = 4.013$.

Example 2

The example treats the stability problem of a cantilever connected to a rigid arm which is subjected to a load as shown in Figs. 4–2*a* and 4–2*d*. The problem of Fig. 4–2*a* is considered first.

The incremental stiffness equations for the cantilever are

$$\mathbf{k}_\varphi d\mathbf{v} - d\mathbf{V}_\phi = 0 \qquad (4\text{–}5)$$

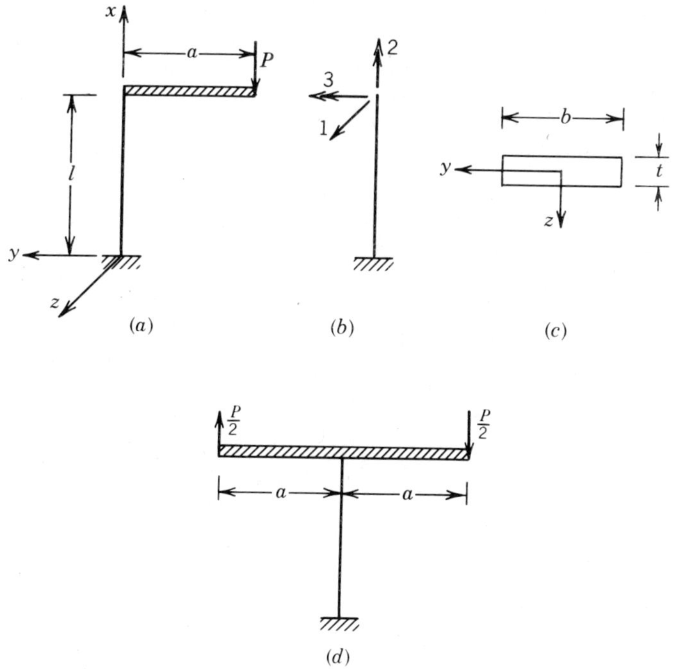

FIG. 4–2. Lateral stability, Section 4, Examples 2, 3

The effective degrees of freedom for out-of-plane buckling are identified and ordered in Fig. 4–2*b*. They coincide with degrees of freedom 3, 4, 5 of the 6×6 matrix $\mathbf{k}_\varphi$ in which $F_{1r} = -P, G_{3r} = -Pa$ and $F_{2r} = F_{3r} = G_{1r} = G_{2r} = 0$. We thus obtain

$$\mathbf{k}_\varphi = \left[\begin{array}{c|c|c} \dfrac{12EI_2}{l^3} - \dfrac{6}{5}\dfrac{P}{l} - \dfrac{2P^2a^2}{10GJ_N l} & \dfrac{Pa}{l} & \dfrac{6EI_2}{l^2} - \dfrac{P}{10} - \dfrac{P^2a^2}{10GJ_N} \\ \hline & \dfrac{GJ_N}{l} & \dfrac{Pa}{2} \\ \hline \text{symmetric} & & \dfrac{4EI_2}{l} - \dfrac{2lP}{15} - \dfrac{2P^2a^2 l}{15GJ_N} \end{array}\right] \tag{4–6}$$

The notation $d\mathbf{V}_\varphi = \{d\mathbf{F}_r \;\; d\mathbf{G}_{\varphi r}\}$ is changed to $\{d\mathbf{F} \;\; d\mathbf{G}_\varphi\}$ because the subscript r serves no purpose for the cantilever whose bound reference is fixed. Assuming that the applied load has a fixed direction, we have $d\mathbf{F} = 0$, and $d\mathbf{G}_\varphi$ is obtained from the potential $\bar{V}$ of the conservative force through

$$d\mathbf{G}_\varphi = -\frac{\partial^2 \bar{V}}{\partial \boldsymbol{\varphi} \partial \boldsymbol{\varphi}^T} d\boldsymbol{\varphi} \tag{4–7}$$

The problem is solved in Eq. (15/14–7b) in which $\mathbf{P} = \{-P \;\; 0 \;\; 0\}$ and $\mathbf{r} = \{0 \; -a \;\; 0\}$. There comes

$$\frac{\partial^2 \bar{V}}{\partial \boldsymbol{\varphi} \partial \boldsymbol{\varphi}^T} = -\frac{1}{2}(\mathbf{P}\mathbf{r}^T + \mathbf{r}\mathbf{P}^T) + (\mathbf{P}^T\mathbf{r})\mathbf{I} = \begin{bmatrix} 0 & -\frac{1}{2}Pa & 0 \\ -\frac{1}{2}Pa & 0 & 0 \\ 0 & 0 & 0 \end{bmatrix} \tag{4–8}$$

The effective $d\mathbf{G}_\varphi$ is $\{dG_{1\varphi} \;\; dG_{2\varphi}\}$. Its tangent stiffness is the first 2×2 submatrix in the preceding result. With $d\mathbf{V}_\varphi = \{0 \;\; dG_{1\varphi} \;\; dG_{2\varphi}\}$ and the ordering of Fig. 4–2*b*, we obtain $d\mathbf{V}_\varphi = -\mathbf{k}_{e\varphi} d\mathbf{v}$, where

$$\mathbf{k}_{e\varphi} = \begin{bmatrix} 0 & 0 & 0 \\ 0 & 0 & -\frac{1}{2}Pa \\ 0 & -\frac{1}{2}Pa & 0 \end{bmatrix} \tag{4–9}$$

The structure tangent stiffness is

$$\mathbf{K} = \mathbf{k}_\varphi + \mathbf{k}_{e\varphi} \tag{4–10}$$

It is noted that elements (2, 3) in K are zero. Letting

$$\lambda^2 = \frac{P^2a^2l^2}{GJ_N EI_2} \tag{4–11a}$$

$$\mu = \frac{Pl^2}{EI_2} = \frac{l}{a}\sqrt{\frac{GJ_N}{EI_2}}\lambda \tag{4–11b}$$

$$x = \frac{1}{10}(\lambda^2 + \mu) \tag{4–11c}$$

the determinantal equation reduces to

$$15x^2 - 52x + 12 = 0 \tag{4–12}$$

To a root x correspond two roots of Eq. (4–11c) for λ which are of opposite sign and which correspond to the two possible senses of application of P. The smaller root for x yields the lower critical load. We find

$$x = 0.248596 \tag{4–13}$$

$$\lambda = -\frac{1}{2}\frac{l}{a}\sqrt{\frac{GJ_N}{EI_2}} \pm \sqrt{\frac{1}{4}\frac{l^2}{a^2}\frac{GJ_N}{EI_2} + 2.4860} \tag{4–14}$$

GJ_N depends on P. From Eq. (3–8), with $\beta_2 = \beta_3 = 0$,

$$GJ_N = GJ - \rho^2 P = GJ - \frac{I}{A}P \tag{4–15}$$

where I is the polar moment of inertia about the centroid. For the narrow rectangular cross section of Fig. 4–2c, we have $J = \frac{1}{3}bt^3$, $I \simeq I_z = bt^3/12$, and $A = bt$. Thus

$$\frac{GJ_N}{EI_2} = \frac{GJ}{EI_2} - \frac{b^2}{12l^2}\mu = \frac{GJ}{EI_2} - \frac{b^2}{12l^2}(10x - \lambda^2) \tag{4–16}$$

For a given ratio l/a, λ may be determined by iteration, whereby Eq. (4–16) determines GJ_N/EI_2 for a current λ and Eq. (4–14) determines λ for the next iteration cycle. Starting with $GJ_N = GJ$ yields the first approximation to λ which could be accurate enough if the term in b^2/l^2 is negligible.

In the limiting case $a \to 0$, the negative root in λ tends to infinity, which corresponds to the impossibility of buckling with a tensile axial force as $a \to 0$. The other root tends to zero, and from Eq. (4–11c), $\mu = 10x = 2.486$. This solution is that of planar buckling of a compressed cantilever, and $\mu = 2.486$ is a good approximation to the exact value $\pi^2/4$.

The case of Fig. 4–2d is dealt with similarly. The only difference with the preceding is that the axial force in the cantilever vanishes. The characteristic equation remains Eq. (4–12), but $\mu = 0$ and $\lambda^2 = 10x = 2.4860$. In Eq. (4–11a)

$GJ_N = GJ$, and Pa is the applied moment. The critical value is thus found to be

$$Pa = \frac{\lambda}{l}\sqrt{GJEI_2} = \frac{1.577}{l}\sqrt{GJEI_2}$$

The exact numerical factor is $\pi/2 = 1.571$.

Example 3

This example is to show that the preceding one could also be solved using the element stiffness $\mathbf{k}_\theta$ instead of $\mathbf{k}_\varphi$. Equation (4–5) is then replaced with

$$\mathbf{k}_\theta d\mathbf{v}_\theta - d\mathbf{V}_\theta = 0 \tag{4–17}$$

$d\mathbf{V}_\theta$ is obtained from the same potential function $\bar{V}$ as for $d\mathbf{V}_\varphi$ but with the displacement derivatives $\boldsymbol{\theta}$ as generalized coordinates instead of $\boldsymbol{\varphi}$. The problem is solved in Eq. (15/14–7a) in which $\mathbf{q}$ is $\boldsymbol{\theta}$ and which yields in addition to the matrix of Eq. (4–8) the term

$$\sum -G_i \frac{\partial^2 \varphi_i}{\partial\boldsymbol{\theta}\partial\boldsymbol{\theta}^T} = \begin{bmatrix} 0 & +\frac{1}{2}G_3 & -\frac{1}{2}G_2 \\ +\frac{1}{2}G_3 & 0 & 0 \\ -\frac{1}{2}G_2 & 0 & 0 \end{bmatrix} \tag{4–18}$$

In the present example $G_2 = 0$, and $G_3 = -Pa$. It may be verified that the additional stiffness obtained here exactly cancels the difference between $\mathbf{k}_\theta$ and $\mathbf{k}_\varphi$.

5. ELEMENT COMPLETE STIFFNESS MATRIX $\mathbf{k}_{c\varphi}$ TANGENT STIFFNESS OF EXTERNAL LOADS

The complete element tangent stiffness $\mathbf{k}_{c\varphi}$ in the local axes is obtained through Eq. (15/8–6b), or

$$\mathbf{k}_{c\varphi} = \mathbf{B}_\varphi \mathbf{k}_\varphi \mathbf{B}_\varphi^T + \mathbf{k}_{G\varphi} \tag{5–1}$$

$\mathbf{B}_\varphi$ and $\mathbf{k}_{G\varphi}$ are obtained without approximations in Eqs. (15/7–11b) and (15/8–11, 12).

For a second-order theory in the fixed reference, $\mathbf{B}_\varphi$ is readily obtained from the deformation-displacement relations, Eqs. (15/9–14, 15), through the property $\mathbf{B}_\varphi = \partial\mathbf{v}^T/\partial\mathbf{s}$, and $\mathbf{k}_{G\varphi}$ is obtained in Eq. (15/9–25).

For linearized stability analysis the intitial displacements are neglected in $\mathbf{B}_\varphi$ which reduces then to the static-kinematic matrix $\mathbf{B}$ of the linear theory, Eqs. (14/3.1–5, 10).

The structure tangent stiffness matrix is formed of two parts. The first is assembled from $\mathbf{k}_{c\varphi}$ by the direct stiffness method and may be referred to as intrinsic or internal. The second depends on the external loading and may be called external. Fundamental loading cases and their tangent stiffnesses have been considered in Chapter 15, Sections 14 and 15. It should be noted, however, that contrary to the one-member problems of Chapter 15, Section 15, the tangent stiffness of external loads to be used with $\mathbf{k}_{c\varphi}$ should be obtained with respect to $\boldsymbol{\varphi}$ as generalized displacement. A summary review of basic cases follows. For simplicity, tangent stiffnesses of external moments are limited to the second-order theory.

Fixed External Force

An external force of fixed orientation whose point of application is a reference or nodal point at which the independent displacements are defined has no tangent stiffness. If the force is applied at a non-nodal point, its moment at the nodal point has a tangent stiffness, and this is treated later on in the case of a rigid moment arm.

Fixed External Moment

From Chapter 15, Section 14, external moment components $\mathbf{G}$ having a fixed orientation and applied at a cross section or at a joint where the rotation components on the same axes are $\boldsymbol{\varphi}$ have the tangent stiffness

$$\mathbf{k}_{n\varphi} = \tfrac{1}{2}\underset{\sim}{\mathbf{G}} = \begin{bmatrix} 0 & \tfrac{1}{2}G_3 & -\tfrac{1}{2}G_2 \\ -\tfrac{1}{2}G_3 & 0 & \tfrac{1}{2}G_1 \\ \tfrac{1}{2}G_2 & -\tfrac{1}{2}G_1 & 0 \end{bmatrix} \tag{5-2}$$

The degrees of freedom of $\mathbf{k}_{n\varphi}$ are $\varphi_1, \varphi_2, \varphi_3$.

Follower External Moments

For follower moment components

$$\mathbf{k}_{n\varphi} = -\tfrac{1}{2}\underset{\sim}{\mathbf{G}}' \tag{5-3}$$

here $\mathbf{G}'$ are the components on axes that have been rotated from the initial state through the rotation $\boldsymbol{\varphi}$. Note that the components $\boldsymbol{\varphi}$ of the rotation vector are the same on the initial and on the rotated axes. Note also that $\mathbf{G}'$ are equal to the components on the fixed axes of the vector obtained by rotating the follower moment back to the initial state.

Moment of Conservative Forces

Consider a couple of fixed external forces $\mathbf{P}$ and $-\mathbf{P}$ applied, respectively, at two points A and B of a rigid body. Letting $\mathbf{r}$ be the components of the lever arm BA, and $\boldsymbol{\varphi}$ be the rotations of the rigid body, the tangent stiffness of the couple is obtained from Eq. (15/14–7b) as

$$\mathbf{k}_{e\varphi} = \begin{bmatrix} r_y P_y + r_z P_z & -\tfrac{1}{2} r_y P_x - \tfrac{1}{2} r_x P_y & -\tfrac{1}{2} r_z P_x - \tfrac{1}{2} r_x P_z \\ & r_z P_z + r_x P_x & -\tfrac{1}{2} r_z P_y - \tfrac{1}{2} r_y P_z \\ \text{symmetric} & & r_x P_x + r_y P_y \end{bmatrix} \tag{5–4}$$

The rigid body in question may be a rigid arm attached to a structure or simply a cross section of a member. Equation (5–4) may also be applied to the moment of a single force taken at the reference point where the translational displacements are defined. Particular cases are now considered.

Quasi-Tangential Torques G_1

The two cases $\mathbf{r} = \{0 \; r_y \; 0\}$, $\mathbf{P} = \{0 \; 0 \; P_z\}$ and $\mathbf{r} = \{0 \; 0 \; r_z\}$, $\mathbf{P} = \{0 \; P_y \; 0\}$ correspond, respectively, to torsional moments $G_1^{(1)} = r_y P_z$ and $G_1^{(2)} = -r_z P_y$ for which

$$\mathbf{k}_{e\varphi}^{(1)} = \begin{bmatrix} 0 & 0 & 0 \\ 0 & 0 & -\tfrac{1}{2} G_1^{(1)} \\ 0 & -\tfrac{1}{2} G_1^{(1)} & 0 \end{bmatrix} \tag{5–5a}$$

$$\mathbf{k}_{e\varphi}^{(2)} = \begin{bmatrix} 0 & 0 & 0 \\ 0 & 0 & +\tfrac{1}{2} G_1^{(2)} \\ 0 & +\tfrac{1}{2} G_1^{(2)} & 0 \end{bmatrix} \tag{5–5b}$$

Quasi-Tangential Bending Moments G_2

The two cases $\mathbf{r} = \{0 \; 0 \; r_z\}$, $\mathbf{P} = \{P_x \; 0 \; 0\}$ and $\mathbf{r} = \{r_x \; 0 \; 0\}$, $\mathbf{P} = \{0 \; 0 \; P_z\}$ correspond, respectively, to bending moments $G_2^{(1)} = r_z P_x$ and $G_2^{(2)} = -r_x P_z$ for which

$$\mathbf{k}_{e\varphi}^{(1)} = \begin{bmatrix} 0 & 0 & -\tfrac{1}{2} G_2^{(1)} \\ 0 & 0 & 0 \\ -\tfrac{1}{2} G_2^{(1)} & 0 & 0 \end{bmatrix} \tag{5–6a}$$

$$\mathbf{k}_{e\varphi}^{(2)} = \begin{bmatrix} 0 & 0 & \tfrac{1}{2} G_2^{(2)} \\ 0 & 0 & 0 \\ \tfrac{1}{2} G_2^{(2)} & 0 & 0 \end{bmatrix} \tag{5–6b}$$

Quasi-Tangential Bending Moments G_3

The two cases $\mathbf{r} = \{r_x \; 0 \; 0\}$, $\mathbf{P} = \{0 \; P_y \; 0\}$ and $\mathbf{r} = \{0 \; r_y \; 0\}$, $\mathbf{P} = \{P_x \; 0 \; 0\}$ correspond, respectively, to bending moments $G_3^{(1)} = r_x P_y$ and $G_3^{(2)} = -r_y P_x$ for which

$$\mathbf{k}_{e\varphi}^{(1)} = \begin{bmatrix} 0 & -\frac{1}{2}G_3^{(1)} & 0 \\ -\frac{1}{2}G_3^{(1)} & 0 & 0 \\ 0 & 0 & 0 \end{bmatrix} \tag{5–7a}$$

$$\mathbf{k}_{e\varphi}^{(2)} = \begin{bmatrix} 0 & \frac{1}{2}G_3^{(2)} & 0 \\ \frac{1}{2}G_3^{(2)} & 0 & 0 \\ 0 & 0 & 0 \end{bmatrix} \tag{5–7b}$$

Semitangential Moments

A semitangential moment component is formed of a pair of equal quasi-tangential moments of types (1) and (2), respectively. More generally, it may be formed of any number of such pairs defined each on any set of orthogonal axes. For such moments Eqs. (5–5, 6, 7) yield

$$\mathbf{k}_{e\varphi} = \mathbf{k}_{e\varphi}^{(1)} + \mathbf{k}_{e\varphi}^{(2)} = 0 \tag{5–8}$$

Thus semitangential moments have no tangent stiffness with respect to $\boldsymbol{\varphi}$.

6. LINEARIZED LATERAL STABILITY OF A PLANAR STRUCTURE

Consider a planar structure in planar behavior in the global (X, Y) plane. The member local axes (x, y) are all in that plane. For an out-of-plane linearized stability analysis the effective member end displacements are

$$d\mathbf{s} = \{d\mathbf{s}^a \;\; d\mathbf{s}^b\} = \{v_3^a \;\; \theta_1^a \;\; \theta_2^a \;\; v_3^b \;\; \theta_1^b \;\; \theta_2^b\} \tag{6–1}$$

These are incremental displacements defined in the member local reference. For denoting deformations, we revert to using the subscript r. The effective incremental deformations are

$$d\mathbf{v} = \{v_{3r} \;\; \theta_{1r} \;\; \theta_{2r}\} \tag{6–2}$$

The effective element stiffness in local coordinates is formed through Eq. (5–1) in which $\mathbf{B}_\varphi$ is replaced by $\mathbf{B}$, or

$$\mathbf{k}_{c\varphi} = \mathbf{B}\mathbf{k}_\varphi\mathbf{B}^T + \mathbf{k}_{G\varphi} \tag{6–3}$$

The effective $\mathbf{k}_\varphi$ is obtained by extracting rows and columns 3, 4, 5 of the 6×6 matrix established in Eqs. (4–1). $\mathbf{k}_{G\varphi}$ is obtained by extracting rows and columns

3, 4, 5, 9, 10, 11 from Eq. (15/9–25a) and is found to be

$$\mathbf{k}_{G\varphi} = \begin{bmatrix} 0 & -F_2^b & F_1^b & 0 & 0 & 0 \\ & 0 & \dfrac{lF_2^b}{2} & F_2^b & 0 & -\dfrac{G_3^b}{2} \\ & & -lF_1^b & -F_1^b & \dfrac{G_3^b}{2} & 0 \\ & \text{symmetric} & & 0 & 0 & 0 \\ & & & & 0 & 0 \\ & & & & & 0 \end{bmatrix} \qquad (6\text{–}4)$$

The effective forces in the relation $\mathbf{S} = \mathbf{BV}$ are

$$\mathbf{S} = \{\mathbf{S}^a \; \mathbf{S}^b\} = \{F_3^a \; G_1^a \; G_2^a \; F_3^b \; G_1^b \; G_2^b\} \qquad (6\text{–}5)$$

and

$$\mathbf{V} = \mathbf{S}^b \qquad (6\text{–}6)$$

By statics in the undeformed geometry,

$$\begin{Bmatrix} \mathbf{S}^a \\ \mathbf{S}^b \end{Bmatrix} = \begin{bmatrix} -1 & 0 & 0 \\ 0 & -1 & 0 \\ l & 0 & -1 \\ 1 & 0 & 0 \\ 0 & 1 & 0 \\ 0 & 0 & 1 \end{bmatrix} \begin{Bmatrix} F_3^b \\ G_1^b \\ G_2^b \end{Bmatrix} = \begin{bmatrix} \mathbf{B}^a \\ \hline \mathbf{I} \end{bmatrix} \mathbf{V} \qquad (6\text{–}7)$$

With the partitioning of $\mathbf{B}$ as shown, Eq. (6–3) yields

$$\mathbf{k}_{c\varphi} = \begin{bmatrix} \mathbf{B}^a \mathbf{k}_\varphi \mathbf{B}^{aT} & \mathbf{B}^a \mathbf{k}_\varphi \\ \mathbf{k}_\varphi \mathbf{B}^{aT} & \mathbf{k}_\varphi \end{bmatrix} + \mathbf{k}_{G\varphi} \qquad (6\text{–}8)$$

Elements of $\mathbf{k}_{c\varphi}$ are listed next by rows starting at the main digonal. The ordering of the degrees of freedom is shown in Fig. 6–1*a*. The superscript *b* is deleted from the forces at end *b*, Fig. 6–1*b*.

$$k_{11} = \frac{12EI_2}{l^3} + \frac{6}{5}\frac{F_1}{l} - \frac{2}{GJ_N l}\left(\frac{G_3^2}{10} + \frac{lF_2 G_3}{10} + \frac{l^2 F_2^2}{25}\right) \qquad (6\text{–}9a)$$

$$k_{12} = -F_2 - \frac{G_3}{l} \qquad (6\text{–}9b)$$

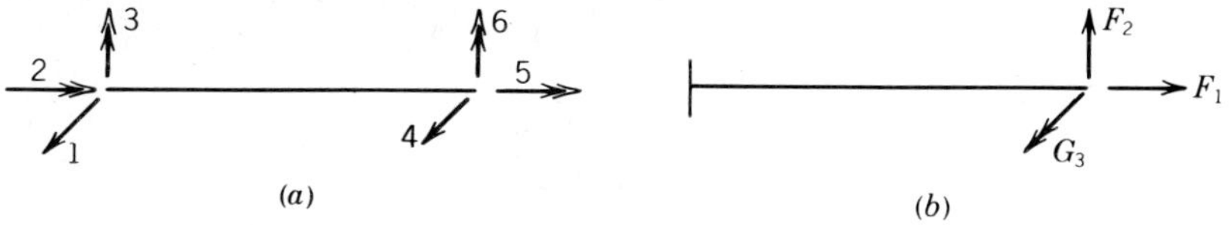

FIG. 6–1. (*a*) Effective degrees of freedom in Eq. (6–9); (*b*) initial forces in Eq. (6–9).

$$k_{13} = -\frac{6EI_2}{l^2} - \frac{F_1}{10} + \frac{2}{GJ_N}\left(\frac{G_3^2}{20} + \frac{lF_2G_3}{12} + \tfrac{11}{300}l^2F_2^2\right) \tag{6–9c}$$

$$k_{14} = -\frac{12EI_2}{l^3} - \frac{6}{5}\frac{F_1}{l} + \frac{2}{GJ_Nl}\left(\frac{G_3^2}{10} + \frac{lF_2G_3}{10} + \frac{l^2}{25}F_2^2\right) \tag{6–9d}$$

$$k_{15} = \frac{G_3}{l} \tag{6–9e}$$

$$k_{16} = -\frac{6EI_2}{l^2} - \frac{F_1}{10} + \frac{2}{GJ_N}\left(\frac{G_3^2}{20} + \frac{lF_2G_3}{60} + \frac{l^2F_2^2}{300}\right) \tag{6–9f}$$

$$k_{22} = \frac{GJ_N}{l} \tag{6–9g}$$

$$k_{23} = \frac{G_3}{2} + \tfrac{1}{3}lF_2 \tag{6–9h}$$

$$k_{24} = F_2 + \frac{G_3}{l} \tag{6–9i}$$

$$k_{25} = -\frac{GJ_N}{l} \tag{6–9j}$$

$$k_{26} = \frac{lF_2}{6} \tag{6–9k}$$

$$k_{33} = \frac{4EI_2}{l} + \frac{2}{15}lF_1 - \frac{2l}{GJ_N}\left(\frac{G_3^2}{15} + \frac{lF_2G_3}{10} + \tfrac{17}{450}l^2F_2^2\right) \tag{6–9l}$$

$$k_{34} = \frac{6EI_2}{l^2} + \frac{F_1}{10} - \frac{2}{GJ_N}\left(\frac{G_3^2}{20} + \frac{lF_2G_3}{12} + \tfrac{11}{300}l^2F_2^2\right) \tag{6–9m}$$

$$k_{35} = \frac{lF_2}{6} \tag{6–9n}$$

$$k_{36} = \frac{2EI_2}{l} - \frac{lF_1}{30} + \frac{2l}{GJ_N}\left(\frac{G_3^2}{60} + \frac{lF_2G_3}{60} + \frac{l^2F_2^2}{900}\right) \tag{6–9o}$$

$$k_{44} = \frac{12EI_2}{l^3} + \frac{6}{5}\frac{F_1}{l} - \frac{2}{GJ_N l}\left(\frac{G_3^2}{10} + \frac{lF_2G_3}{10} + \frac{l^2F_2^2}{25}\right) \tag{6–9p}$$

$$k_{45} = -\frac{G_3}{l} \tag{6–9q}$$

$$k_{46} = \frac{6EI_2}{l^2} + \frac{F_1}{10} - \frac{2}{GJ_N}\left(\frac{G_3^2}{20} + \frac{lF_2G_3}{60} + \frac{l^2F_2^2}{300}\right) \tag{6–9r}$$

$$k_{55} = \frac{GJ_N}{l} \tag{6–9s}$$

$$k_{56} = -\tfrac{1}{2}G_3 - \frac{lF_2}{6} \tag{6–9t}$$

$$k_{66} = \frac{4EI_2}{l} + \tfrac{2}{15}lF_1 - \frac{2l}{GJ_N}\left(\frac{G_3^2}{15} + \frac{lF_2G_3}{30} + \frac{l^2F_2^2}{225}\right) \tag{6–9u}$$

Example 1

The critical load of the frame shown in Fig. 6–2*a* is obtained in what follows using one element to represent each member. The result will be compared to that reported in [19].

For using Eq. (6–9), the members are assumed to have a narrow rectangular cross section shown in local axes in Fig. 6–2*b*. Local z axis and global Z axis coincide.

The effective degrees of freedom for out-of-plane buckling are identified and ordered in Fig. 6–2*c*. The effective stiffness of member 1 was determined with a different ordering in Eq. (4–6) in which a is now equal to l. Its merging sequence for the ordering of Fig. 6–2*c* is $(1, 3, -2)$. For member 2 the local and global orderings are the same, and its stiffness is the 6×6 matrix of Eq. (6–9). The only nonzero initial force in Eq. (6–9) is $F_2 = -P$.

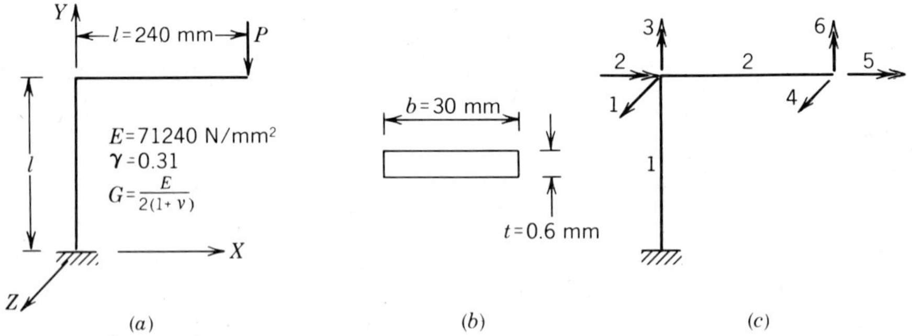

FIG. 6–2. Lateral stability of frame, Section 6, Example 1.

Assuming that the load is applied at the centroid and has a fixed orientation, it has no tangent stiffness. The structure tangent stiffness is thus obtained by assembling the member matrices as described earlier.

To write the assembled matrix in concise form and in a form suitable for further calculations, the incremental variables are made physically homogeneous by dividing translations by l and multiplying forces by l. Letting

$$\lambda^2 = \frac{P^2 l^4}{GJEI_2} \qquad \text{(6–10a)}$$

$$\lambda_n^2 = \frac{GJ}{GJ_N}\lambda^2 = \frac{GJ}{GJ - \rho^2 P}\lambda^2 \qquad \text{(6–10b)}$$

$$\mu = \frac{Pl^2}{EI_2} \qquad \text{(6–10c)}$$

Factoring out EI_2/l, the nondimensionalized assembled tangent stiffness is found to be

$$\mathbf{K} = \begin{bmatrix}
24 - \frac{2\lambda^2}{25} - \frac{\lambda_n^2}{5} - \frac{6\mu}{5} & -6 + \frac{\lambda_n^2}{10} + \frac{11\mu}{10} & -6 + \frac{11\lambda^2}{150} + \mu & -12 + \frac{2\lambda^2}{25} & 0 & -6 + \frac{\lambda^2}{150} \\
 & 4 - \frac{2\lambda_n^2}{15} - \frac{2\mu}{15} + \frac{GJ}{EI_2} & -\frac{5\mu}{6} & -\mu & -\frac{GJ}{EI} & -\frac{\mu}{6} \\
 & & 4 - \frac{17}{225}\lambda^2 + \frac{GJ_N}{EI_2} & 6 - \frac{11\lambda^2}{150} & -\frac{\mu}{6} & 2 + \frac{\lambda^2}{450} \\
 & \text{symmetric} & & 12 - \frac{2\lambda^2}{25} & 0 & 6 - \frac{\lambda^2}{150} \\
 & & & & \frac{GJ}{EI_2} & \frac{\mu}{6} \\
 & & & & & 4 - \frac{2\lambda^2}{225}
\end{bmatrix} \qquad \text{(6–10d)}$$

A formal reduction of **K** to simplify the evaluation of the determinant is manageable by hand. The determinantal equation $|\mathbf{K}| = 0$ may be put in the form of Eq. (6–11c), where x is defined in terms of λ in Eqs. (6–11a, b):

$$\frac{GJ_N}{GJ} = 1 - \frac{\rho^2}{l^2}\sqrt{\frac{EI_2}{GJ}}\lambda \tag{6–11a}$$

$$x = \frac{1}{10}\left(\frac{\lambda^2}{GJ_N/GJ} + \sqrt{\frac{GJ}{EI_2}}\lambda\right) \tag{6–11b}$$

$$\frac{5400 - 306\lambda^2 + 1.3\lambda^4}{18\lambda^2(1200 - 11\lambda^2)} - \frac{GJ}{EI_2}\frac{1-x}{12 - 52x + 15x^2} = 0 \tag{6–11c}$$

Equation (6–11c) is solved for λ by trial and error. There should be at least one positive root corresponding to the sense of application of P in Fig. 6–2*a* and one negative root corresponding to the opposite sense. Both critical loads should be smaller in absolute value than those of Example 2, Section 4, which correspond to member 2 being infinitely rigid.

For a numerical application and a comparison with a solution of the same problem obtained by different means, the data shown in Fig. 6–2 are used. The critical values are found to be

$$\lambda_1 = 0.8262 \quad \text{and} \quad \lambda_2 = -1.3181$$

for which

$$P_{1,\mathrm{cr}} = 0.6818N$$
$$P_{2,\mathrm{cr}} = -1.0877N$$

These results agree with those reported in [19] where more elements had to be used to achieve the same accuracy. Values higher in the last digit than those obtained here are reported using, respectively, 4 and 10 elements per member. The element matrix in [19] is based on a linear shape function for the twisting angle.

Example 2

The frame of Fig. 6–3*a* is simply supported in its own plane and is fully restrained at the supports with respect to twisting and out-of-plane bending. The frame is in a state of pure bending due to the applied moments M. The structure stiffness for lateral stability is formed in what follows using one element per member.

Because of the support conditions the effective member stiffness in local coordinates is that of a cantilever, and the effective displacements are those

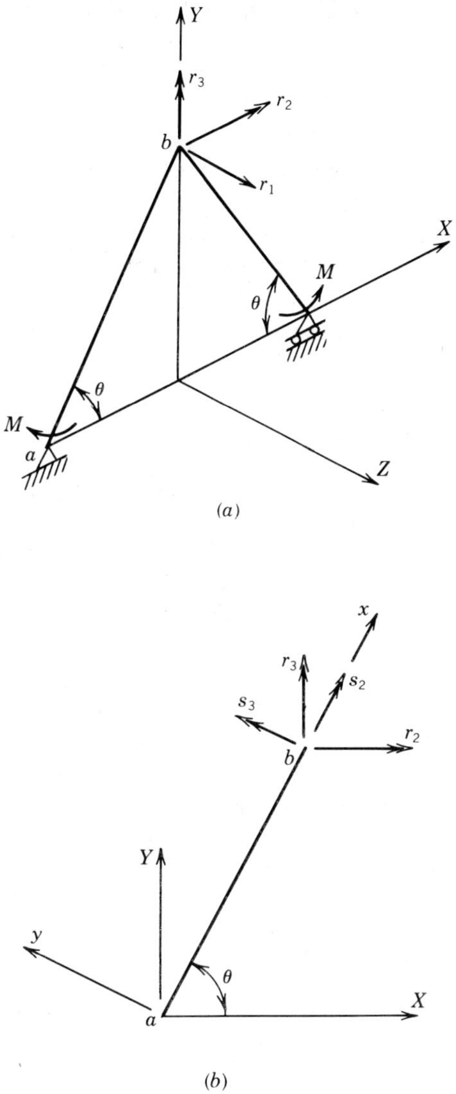

FIG. 6–3. Lateral stability of frame, Section 6, Example 2.

numbered (4, 5, 6) in Fig. 6–1*a*. They are denoted (s_1, s_2, s_3) and are shown together with the joint displacement **r** in Fig. 6–3*b*. The connectivity relation $\mathbf{s} = \mathbf{ar}$ is

$$\begin{Bmatrix} s_1 \\ s_2 \\ s_3 \end{Bmatrix} = \begin{bmatrix} 1 & 0 & 0 \\ 0 & \cos\theta & \sin\theta \\ 0 & -\sin\theta & \cos\theta \end{bmatrix} \begin{Bmatrix} r_1 \\ r_2 \\ r_3 \end{Bmatrix}$$

In global coordinates the member stiffness transforms into

$$\bar{\mathbf{k}}_\varphi = \mathbf{a}^T \mathbf{k}_\varphi \mathbf{a} \tag{6–12}$$

In applying the general formulas for $\mathbf{k}_\varphi$, $G_3 = M$, and the remaining initial forces are zero. For the second member $G_3 = -M$, and θ is changed into $\pi - \theta$.

The applied moments have no tangent stiffness in the present problem irrespective of how they are generated. This property follows from the condition of restrained twisting and would remain in effect if the lateral bending rotation were free.

Superposition of member matrices yields the structure stiffness

$$\mathbf{K} = 2\begin{bmatrix} 12\dfrac{EI_2}{l^3} - \dfrac{2}{10}\dfrac{M^2}{GJl} & -\dfrac{Mc}{l} - \left(\dfrac{6EI_2}{l^2} - \dfrac{1}{10}\dfrac{M^2}{GJ}\right)s & 0 \\ \text{symmetric} & \dfrac{GJ}{l}c^2 + Msc + \left(\dfrac{4EI_2}{l} - \dfrac{2}{15}\dfrac{M^2 l}{GJ}\right)s^2 & 0 \\ & & \dfrac{GJ}{l}s^2 - Msc + \left(\dfrac{4EI_2}{l} - \dfrac{2}{15}\dfrac{M^2 l}{GJ}\right)c^2 \end{bmatrix} \tag{6–13}$$

in which $c = \cos\theta$ and $s = \sin\theta$. The determinantal equation separates into two parts corresponding to antisymmetric and symmetric buckling modes. The latter is the critical one. Its characteristic equation is obtained by setting the determinant of the first 2×2 submatrix equal to zero and may be put in the form

$$(60 - \lambda^2)(12 - \lambda^2)\sin^2\theta + 72\frac{GJ}{EI_2}\cos^2\theta(10 - \lambda^2) = 0 \tag{6–14}$$

where

$$\lambda = \frac{Ml}{\sqrt{EI_2 GJ}} \tag{6–15}$$

A property to be discussed subsequently is that the exact solution for λ_{cr} is independent of θ and of GJ/EI_2 and is $\lambda_{cr} = \pi$. The lowest root of Eq. (6–14) is found to be in the range

$$\frac{\sqrt{10}}{\pi} \leqslant \frac{\lambda}{\lambda_{cr}} \leqslant \frac{\sqrt{12}}{\pi} \tag{6–16}$$

The lower limit is obtained in the case $\theta = 0$ for which the frame reduces to a single member of span $2l$. The upper limit is obtained in the case $\theta = \pi/2$ in which the frame is formed of two adjoining members. For $\theta = \pi/4$ and a narrow rectangular cross section with a Poisson's ratio $v = 0.3$, the error is found to be 3.7%. These results are obtained using only one element per member.

The fact that the exact solution for λ is independent of θ and of GJ/EI may be shown by reducing the problem to a cantilever subjected at its tip to a bending moment M, and restrained against the rotation r_3. The moment M must be assumed to be semitangential in order for it not to have an external tangent stiffness. The governing differential equations of the cantilever have constant coefficients and may be solved exactly.

7. COMMENT ON THE ENERGY AND DIRECT STIFFNESS METHODS

An element tangent stiffness for linearized stability analysis is derived in [20] by applying the energy method in a fixed reference and using displacement derivatives to represent rotational degrees of freedom. This matrix will be denoted here $\mathbf{k}_{c\theta}$ and its displacements $\mathbf{s}_\theta$. To concentrate on the basic difference between $\mathbf{k}_{c\theta}$ and the particular form of $\mathbf{k}_{c\varphi}$ in linearized stability, the comparison is limited to a cross section having zero warping stiffness, and it is assumed that warping degrees of freedom are eliminated by static condensation. $\mathbf{k}_{c\varphi}$ expresses the strain energy functional $\bar{U}_2$ in the form $\bar{U}_2 = d\mathbf{s}_\theta^T \mathbf{k}_{c\theta} d\mathbf{s}_\theta$. The associated incremental stiffness relations are $d\mathbf{S}_\theta = \mathbf{k}_{c\theta} d\mathbf{s}_\theta$.

It would be incorrect, in general, to assemble a structure stiffness matrix from element matrices $\mathbf{k}_{c\theta}$. A kinematic reason for this was given in Chapter 15, Section 13. In the context of linearized stability, we have $d\mathbf{s}_\theta = d\mathbf{s}$ but $d\mathbf{S}_\theta \neq d\mathbf{S}_\varphi$. The direct stiffness method may be used with $\mathbf{k}_{c\varphi}$ because all elements of $d\mathbf{S}_\varphi$ behave as vector components and because moment equilibrium equations are formulated in terms of generalized moments $\mathbf{G}_\varphi$ in the same way as with moments $\mathbf{G}$. This property does not hold, in general, for the generalized moments in $\mathbf{S}_\theta$. The transformation from $\mathbf{k}_{c\theta}$ to $\mathbf{k}_{c\varphi}$ is similar to the one from $\mathbf{k}_\theta$ to $\mathbf{k}_\varphi$, Eq. (2–14), and takes the form

$$\mathbf{k}_{c\varphi} = \mathbf{k}_{c\theta} + \begin{bmatrix} & \mathbf{k}^a_{\varphi\theta} & & \\ & & & \\ & & & \mathbf{k}^b_{\varphi\theta} \end{bmatrix} \tag{7–1}$$

where $\mathbf{k}^a_{\varphi\theta}$ and $\mathbf{k}^b_{\varphi\theta}$ are obtained from Eq. (2–14b) by replacing (G_{2r}, G_{3r}) by (G_2, G_3) at ends a and b, respectively.

For a string of elements connected in a straight line, the energy method may be applied both with $\mathbf{k}_{c\varphi}$ and $\mathbf{k}_{c\theta}$ and results in either case in an assembly of element stiffnesses by the direct stiffness method. However, the external tangent stiffness must be formed consistently with the choice of generalized displacements. It is left as an exercise to show as in Section 4, Example 3, that the difference between $\mathbf{k}_{c\varphi}$

and $\mathbf{k}_{c\theta}$ is exactly compensated by the difference in external tangent stiffness matrices.

In [19] a geometric stiffness matrix for linearized stability is derived starting from three-dimensional theory. Because rotations are linearized, however, the matrix obtained is similar to $\mathbf{k}_{c\theta}$. It is subsequently "corrected" on finding inconsistencies in its application to the stability analysis of the angle frame of Example 1, Section 6. The correction is justified by a heuristic argument concerning the incremental behavior of the member end moments treated as external actions. This question does not arise here, and this is as it should be for internal generalized forces. Only for external actions is there a need to prescribe incremental behavior.

The key then to a correct formulation, as was outlined in Chapter 15, Section 13, is a choice of generalized displacements that satisfy geometric continuity, and use of associated Lagrangian generalized forces.

8. DIRECT STIFFNESS METHOD WITH WARPING STIFFNESS

The problem to be considered in this section is the formulation of the direct stiffness method in general nonlinear analysis in the case where a member of nonzero warping stiffness is subdivided into several elements. The case of a one-element member with idealized end conditions was treated in Section 3. In the present problem end conditions are also idealized, and the question of interest is how to form the member stiffness from element stiffnesses in the presence of the warping degrees of freedom.

Formulation of stiffness equations for the member in question requires inclusion of warping degrees of freedom in element stiffness equations in fixed axes. What needs to be done then is first to transform the reduced stiffness $\mathbf{k}'_\theta$ derived in Section 3 to a reduced stiffness $\mathbf{k}'_\varphi$ similar to $\mathbf{k}_\varphi$ but including warping degrees of freedom. $\mathbf{k}'_\varphi$ must then be transformed into a complete stiffness $\mathbf{k}'_{c\varphi}$ in fixed axes.

The generalized displacements of $\mathbf{k}'_\varphi$ are designated $d\mathbf{v}'$, and $\mathbf{v}'$ is an extended set of deformations that includes the twisting strains χ_1^a and χ_1^b at ends a and b of the element, respectively. We thus let

$$\mathbf{v}' = \{\mathbf{v} \;\; \chi_1\} = \{u_{1r} \;\; u_{2r} \;\; u_{3r} \;\; \varphi_{1r} \;\; \varphi_{2r} \;\; \varphi_{3r} \;\; \chi_1^a \;\; \chi_1^b\} \tag{8–1}$$

The generalized displacements for $\mathbf{k}'_\theta$ are $d\mathbf{v}'_\theta$. Elements of $\mathbf{v}'_\theta$ are

$$\mathbf{v}'_\theta = \{\mathbf{v}^s_\theta \;\; \chi_1\} = \{u_{1r} \;\; u^s_{2r} \;\; u^s_{3r} \;\; \varphi_{1r} \;\; \theta_{2r} \;\; \theta_{3r} \;\; \chi_1^a \;\; \chi_1^b\} \tag{8–2}$$

Since χ_1 is common to $\mathbf{v}'$ and $\mathbf{v}'_\theta$, transformation of $\mathbf{k}'_\theta$ into $\mathbf{k}'_\varphi$ involves the transformation of Section 2 from $\{\theta_{2r} \;\; \theta_{3r}\}$ to $\{\varphi_{2r} \;\; \varphi_{3r}\}$ and the transformation of Section 3 from $\{u^s_{2r} \;\; u^s_{3r}\}$ to $\{u_{2r} \;\; u_{3r}\}$.

The transformation of $\mathbf{k}'_\varphi$ into a complete stiffness matrix $\mathbf{k}'_{c\varphi}$ is similar to the transformation of $\mathbf{k}_\varphi$. The underlying theory is an extension of that of Chapter 15, Sections 7 and 8, and is represented by the relations

$$\mathbf{k}'_{c\varphi} = \mathbf{B}'_\varphi \mathbf{k}'_\varphi \mathbf{B}'^T_\varphi + \mathbf{k}'_{G\varphi} \tag{8–3}$$

$$\mathbf{k}'_{G\varphi} = \sum V'_{i\varphi} \frac{\partial^2 v'_i}{\partial \mathbf{s}' \partial \mathbf{s}'^T} \tag{8–4}$$

where

$$\mathbf{B}'^T_\varphi = \frac{\partial \mathbf{v}'}{\partial \mathbf{s}'^T} \tag{8–5}$$

$$\mathbf{V}'_\varphi = \{\mathbf{V}_\varphi \ \mathbf{V}_\chi\} = \{\mathbf{V}_\varphi \ V^a_\chi \ V^b_\chi\} \tag{8–6}$$

$$\mathbf{s}' = \{\mathbf{s} \ \boldsymbol{\varphi}'_1\} = \{\mathbf{s} \ \varphi'^a_1 \ \varphi'^b_1\} \tag{8–7}$$

In these equations $\mathbf{s}'$ is an extended set of element displacements obtained by appending to $\mathbf{s}$ the two warping displacements φ'^a_1 and φ'^b_1. $\mathbf{V}'_\varphi$ is similarly obtained by appending to $\mathbf{V}_\varphi$ the bimoments V^a_χ and V^b_χ. The matrices needed to form $\mathbf{k}'_{c\varphi}$, that is, $\mathbf{B}'_\varphi$ and $\mathbf{k}'_{G\varphi}$, are obtained by appending to $\mathbf{B}_\varphi$ and adding to $\mathbf{k}_{G\varphi}$ the contributions of the two additional deformations χ^a_1 and χ^b_1:

$$\mathbf{B}'^T_\varphi = \begin{bmatrix} \mathbf{B}^T_\varphi & \mathbf{0} \\ \dfrac{\partial \chi_1}{\partial \mathbf{s}^T} & \dfrac{\partial \chi_1}{\partial \boldsymbol{\varphi}'^T_1} \end{bmatrix} \tag{8–8}$$

$$\mathbf{k}'_{G\varphi} = \begin{bmatrix} \mathbf{k}_{G\varphi} & \mathbf{0} \\ \mathbf{0} & \mathbf{0} \end{bmatrix} + V^a_\chi \frac{\partial^2 \chi^a_1}{\partial \mathbf{s}' \partial \mathbf{s}'^T} + V^b_\chi \frac{\partial^2 \chi^b_1}{\partial \mathbf{s}' \partial \mathbf{s}'^T} \tag{8–9}$$

what is needed then are the expressions of χ^a_1 and χ^b_1 in terms of $\mathbf{s}' \cdot \chi_1$ is obtained from the strain-displacement relations (15/5.3–3b) as

$$\chi_1 = \frac{1}{1 + \frac{1}{4}\boldsymbol{\varphi}^T\boldsymbol{\varphi}} \left(\varphi'_1 + \frac{1}{2}\varphi_3\varphi'_2 - \frac{1}{2}\varphi_2\varphi'_3 \right) \tag{8–10}$$

To obtain the desired expressions of χ^a_1 and χ^b_1, φ'^a_2, φ'^a_3, φ'^b_2, and φ'^b_3 need to be expressed in terms of $\mathbf{s}'$. The basis for this is the shape function representation of the deformed element. φ'_2 and φ'_3 are thus determined in terms of $\boldsymbol{\chi}$ by solving Eqs. (15/5.3–3b) for $\boldsymbol{\varphi}'$. Noting that $\boldsymbol{\mu}^{-1}$ is obtained in Eq. (15/2–17), there comes

$$\boldsymbol{\varphi}' = \boldsymbol{\mu}^{-1}\boldsymbol{\chi} = (\mathbf{I} - \tfrac{1}{2}\underset{\sim}{\boldsymbol{\varphi}} + \tfrac{1}{4}\boldsymbol{\varphi}\boldsymbol{\varphi}^T)\boldsymbol{\chi} \tag{8–11}$$

Extracting φ'_2 and φ'_3 and substituting into Eq. (8–10) yield

$$\chi_1 = \frac{1}{1 + \frac{1}{4}\varphi_1^2} \left[\varphi'_1 + \frac{1}{2}\left(\varphi_3 - \frac{1}{2}\varphi_1\varphi_2\right)\chi_2 - \frac{1}{2}\left(\varphi_2 + \frac{1}{2}\varphi_1\varphi_3\right)\chi_3 \right] \tag{8–12}$$

For expressing χ_2 and χ_3 at ends a and b, we have the strain-displacement relations (15/10–3b, c) and the shape function representations (3–5b, c). The displacements in these equations are relative to the bound reference and are defined at the shear center. Transformation to centroidal displacements is made using Eqs. (3–10), and transformation from displacement derivatives to rotations is made using Eqs. (2–13). All the equations in the bound reference are limited to the second-order theory, but the deformation-displacement relations remain those of the general theory. We thus obtain

$$\chi_2^a = -\frac{6}{l^2}u_{3r} - \frac{2}{l}\left(\varphi_{2r} - \frac{1}{2}\varphi_{1r}\varphi_{3r}\right) - \frac{6}{l^2}\left[y_s\varphi_{1r} + \frac{y_s}{2}\varphi_{2r}\varphi_{3r} - \frac{z_s}{2}(\varphi_{1r}^2 + \varphi_{2r}^2)\right] \tag{8–13a}$$

$$\chi_3^a = \frac{6}{l^2}u_{2r} - \frac{2}{l}\left(\varphi_{3r} + \frac{1}{2}\varphi_{1r}\varphi_{2r}\right) - \frac{6}{l^2}\left[z_s\varphi_{1r} - \frac{z_s}{2}\varphi_{2r}\varphi_{3r} + \frac{y_s}{2}(\varphi_{1r}^2 + \varphi_{3r}^2)\right] \tag{8–13b}$$

χ_1^a is now obtained by applying Eq. (8–12) at end a and substituting for χ_2^a and χ_3^a from these equations. The contribution of χ_1^a to $\mathbf{B}'_\varphi$ and $\mathbf{k}'_{G\varphi}$, Eqs. (8–8) and (8–9), is obtained by taking first and second derivatives with respect to $\mathbf{s}'$. The algebra is straightforward but tedious and will not be carried out here. It is noted, however, that derivatives of deformations $\mathbf{u}_r$ and $\boldsymbol{\varphi}_r$ have already seen evaluated as elements of $\mathbf{B}_\varphi$.

The expression of χ_1^b may be obtained similarly to that of χ_1^a. However, the contribution of χ_1^b to $\mathbf{B}'_\varphi$ and $\mathbf{k}'_{G\varphi}$ should be obtainable by interchanging the roles of ends a and b.

In the second-order theory only the linear terms in Eqs. (8–13) need be kept because they contribute second-order terms to χ_1^a. The linearized expressions of Eqs. (8–13) and the similar expressions at end b are given as follows:

$$\chi_2^a = -\frac{6}{l^2}(u_3^b - u_3^a) - \frac{6y_s}{l^2}(\varphi_1^b - \varphi_1^a) - \frac{2}{l}(\varphi_2^b + 2\varphi_2^a) \tag{8–14a}$$

$$\chi_3^a = \frac{6}{l^2}(u_2^b - u_2^a) - \frac{6z_s}{l^2}(\varphi_1^b - \varphi_1^a) - \frac{2}{l}(\varphi_3^b + 2\varphi_3^a) \tag{8–14b}$$

$$\chi_2^b = \frac{6}{l^2}(u_3^b - u_3^a) + \frac{6y_s}{l^2}(\varphi_1^b - \varphi_1^a) + \frac{2}{l}(\varphi_2^a + 2\varphi_2^b) \tag{8–14c}$$

$$\chi_3^b = -\frac{6}{l^2}(u_2^b - u_2^a) + \frac{6z_s}{l^2}(\varphi_1^b - \varphi_1^a) + \frac{2}{l}(\varphi_3^a + 2\varphi_3^b) \tag{8–14d}$$

Returning now to the original problem of assembling the member stiffness from element stiffnesses, $\mathbf{k}'_{c\varphi}$ is assembled by the direct stiffness method. The resulting matrix may be reduced to degrees of freedom defined at the member ends by static condensation of the displacements at the internal joints. The warping degrees of freedom at the member ends are dealt with in the same manner as in the case of the one-member element.

In the case of linearized stability analysis the initial displacements are neglected in $\mathbf{B}'_\varphi$, which reduces to

$$\mathbf{B}' = \begin{bmatrix} \mathbf{B} & \mathbf{0} \\ \mathbf{0} & \mathbf{I} \end{bmatrix} \tag{8-15}$$

$\mathbf{k}'_{G\varphi}$ is the same as in the second-order theory.

EXERCISES

Exercises 1 to 5

Use the element reduced stiffness matrix $\mathbf{k}_\varphi$ for stability analysis of the cantilevers in Exercises 1 to 5.

1. Cantilever subjected to a transverse load applied at the top of the cross section at the free end.
2. Cantilever formed of a deformable part and of an infinitely rigid part subjected at its tip to a transverse force.
3. Cantilever subjected at its tip to a quasi-tangential torque.
4. Cantilever subjected at is tip to a quasi-tangential bending moment G_3 of type (1) and (2), respectively (see Section 5).
5. The two preceding cantilevers subjected to semitangential moments.
6. Do Exercise 2 using the element matrix $\mathbf{k}_\theta$ instead of $\mathbf{k}_\varphi$.

Exercises 7 to 17

Form the tangent stiffness matrix for lateral stability of the structures in Exercises 7 to 17. Use symmetry where possible to reduce problem size. Assume forces have fixed directions.

7. Simply supported beam in pure bending. Assume that the supports restrain twisting but allow bending rotations.
8. The preceding beam with a bending moment applied only at one end.
9. Simply supported beam subjected to a transverse load at midspan.

10. Simply supported beam with an overhang subjected at its tip to a transverse force. Let the span between supports be l and the length of the overhang be a.

11. The beam of Exercise 10 modified to have a second overhang symmetric of the first and loaded symmetrically.

12. The angle frame of Fig. 6–2*a* fixed at both ends and subjected to a compressive force in the column.

13. The angle frame of Fig. 6–2*a* supported at both ends and subjected at the corner to a force bisecting the right angle. Assume pin supports in the plane of the frame and full fixity for out-of-plane behavior.

14. The structure of the preceding problem subjected at the supports to equal and symmetric bending moments.

15. The structure and loadings of the two preceding problems with supports modified to allow out-of-plane bending rotations, but keeping twist restrained.

16. The structures and loadings of Exercises 13, 14, and 15 with one support modified to allow a translation along the line joining it to the other support.

17. A portal frame of three identical members fixed at the supports and subjected to column compressive forces $P/2$ and P, respectively.

REFERENCES

1. Crandall, S. H., Dahl, N. C., and Lardner, T. J., *An Introduction to the Mechanics of Solids*, 2nd ed., McGraw-Hill, New York (1978).
2. Timoshenko, S., *Strength of Materials*, Part II, 3rd ed., Kreiger Publishing, Huntington, N.Y. (1976).
3. DeVeubeke, B. M. F., *A Course in Elasticity*, Springer-Verlag, New York (1979).
4. Chen, W. F., and Atsuta, T., *Theory of Beam-Columns*, Vol. 1, McGraw-Hill, New York (1976), sec. 5.4, p. 142.
5. Timoshenko, S. P., *History of Strength of Materials*, McGraw-Hill, New York (1953).
6. Timoshenko, S. P., and Gere, J. M., *Theory of Elastic Stability*, 2nd ed., McGraw-Hill, New York (1961).
7. Shanley, F. R., Inelastic Column Theory, *J. Aero. Sci.*, **14**, 261 (1947).
8. Bergan, P. G., and Soreide, T. H., Solution of Large Displacement and Stability Problems Using the Current Stiffness Parameter, in *Finite Elements in Nonlinear Mechanics*, Vol. 2, TAPIR Publishers, Trondheim (1978), p. 647.
9. Smith, E. A., and Smith, G. D., Collapse Analysis of Space Trusses, in "Long Span Roof Structures," Proc. Symp., St. Louis, Missouri, October 26–30, 1981, ASCE, New York, N.Y. 10017.
10. Bathe, K. J., and Wilson, E. L., *Numerical Methods in Finite-Element Analysis*, Prentice-Hall, Englewood, N.J. (1976).
11. *STRUDL User Manual*, McDonnell Douglas Automation Company, Box 516, St. Louis, Mo. 63166.
12. Vlassov, V. Z., *Thin Walled Elastic Beams*, 2nd ed., National Science Foundation, Washington, D.C. (1961).
13. Norris, C. H., Wilbur, J. B., and Utku, S., *Elementary Structural Analysis*, 3rd. ed., McGraw-Hill, New York (1976).
14. Cuoco, D. A., State-of-the-Art of Space Frame Roof Structures, in "Long Span Roof Structures," Proc. Symp., St. Louis, Missouri, October 26-30, 1981, ASCE, New York, N.Y. 10017.
15. *Domebook 2*, Pacific Domes, Box 219, Bolinas, Calif. 94924 (1971).
16. Reissner, E., On One-Dimensional Large-Displacement Finite-Strain Beam Theory, Studies in Applied Mathematics, Vol. 52, No. 2, June 1973.
17. Cosserat, F., and Cosserat, F., *Theorie des corps deformables*, Librairie Scientifique, Hermann, Paris (1909).
18. Ziegler, H., *Principles of Structural Stability*, Blaisdell, Waltham, Mass. (1968).

19. Argyris, J. H., O. Hilpert, G. A. Malejannakis, and D. W. Scharpf, on the Geometrical Stiffness of a Beam in Space—A Consistent V. W. Approach, *Comp. Meths. Appl. Mech. Eng.*, **20**, 105–131 (1979).

20. Barsoum, R. S., and Gallagher, R. H., Finite Element Analysis of Torsional and Torsional-Flexural Stability Problems, *Int. J. Num. Meths. Eng.*, **2**, 335–352 (1970).

INDEX